Deltas of the World

Series Editor
Abdelazim Negm, Zagazig, Egypt

Worldwide, over 600 million people live in deltas. Extensive human activities in deltas, their supplying rivers, and related maritime currents increase the pressure on deltas' natural resources and habitat equilibrium, leading to degradation of the available resources and jeopardizing inhabitants' livelihoods – and in some cases even their lives.

This book series is intended to promote a sustainable future for deltas around the globe and the communities that live in them. Accordingly, we publish studies focusing on various aspects of deltas, including hydrogeology, soil pollution, water pollution, environmental hydrology, environmental hydraulics, groundwater engineering and management, morphology, anthropology, socioeconomics, wastewater management, water resources engineering and management, agriculture, greenhouse agriculture, irrigation and drainage engineering, hydraulic structures, geophysics and geology, sedimentology, mineralogy, remote sensing and GIS studies, climate change variabilities and impacts, marine-delta interactions and sustainability, and solid waste management. The findings presented here support the SDGs and Agenda 2030 with regard to helping countries with deltas and related environments to preserve their natural resources for future generations.

The editors invite scientists, researchers and scientific communities to contribute to the series by submitting delta atlases, edited books, monographs, and high-quality conference proceedings focused on deltas around the world.

Abd-Alla Gad

Digital Land Resources to Support Sustainable Development in Egypt

Nile Delta, Nile Valley and Promising Desert Areas

Springer

Abd-Alla Gad
Environmental Studies and Land Use Division
National Authority for Remote Sensing and Space Sciences (NARSS)
Cairo, Egypt

ISSN 2731-832X ISSN 2731-8338 (electronic)
Deltas of the World
ISBN 978-3-031-37169-1 ISBN 978-3-031-37167-7 (eBook)
https://doi.org/10.1007/978-3-031-37167-7

Cover image: NASA image created by Jesse Allen, Earth Observatory, using data obtained from the University of Maryland's Global Land Cover Facility. https://commons.wikimedia.org/w/index.php?curid=3471590.

This Springer imprint is published by the registered company Springer Nature Switzerland AG
The registered company address is: Gewerbestrasse 11, 6330 Cham, Switzerland

Paper in this product is recyclable.

Preface

Reclamation and farming of new lands is an activity that must continue because of its vital necessity for the life. There is no room for the horizontal agricultural expansion except in arid areas. The accomplishment of this target faces technical and economic difficulties considering the condition of soils and irrigation water. Nevertheless, some of these difficulties may be overcome through scientific approach in research and development, implementation, extension, and management. Thus, providing precise and linked information related to natural resources is necessary, particularly with advancement of information related techniques. The collected information would be the plan in basement and decision tool and research needs support tools.

The progress of information technology has resulted in introducing various surveying techniques (i.e., Global Positioning System—GPS, Geographic Information Systems—GIS, and Remote Sensing). The synoptic view and multi-concept of remote sensing (i.e., multi-temporality, multi-sensors, and multi-spectrality) allow operational acquisition and monitoring of land resources. GIS has proven to be a powerful tool to store, manipulate, and analyze thematic data.

It could be confirmed that obtaining such detailed digital natural basic land resources data for local administration is a considerable step leading toward the realization of sustainable development and program management. The achieved digital database is regarded as an exhaustive, geographically accurate, and updatable. Currently, with the distinctive progress in information technology, such data can be exchanged, improved, and shared by different users and authorities. The most striking finding noticed was the urban expansion on the account of most cultivable lands; hence, diminution of fertile agriculture areas. Conversely, urban sprawl doesn't extend largely to low capable land. It was also observed a common unbalance between both irrigation and drainage network systems, which often leads to progress of salinization and land degradation.

Cairo, Egypt

Abd-Alla Gad

Acknowledgments

Sincere appreciation is recorded to late Prof. Dr. El-Desouky, Ex-chairman of NARSS, and Prof. Dr. Atef Shref, the former chairman of NARSS, for their keen and unyielding efforts in supporting all research activities, needless to say that without their keen and continuous encouragement this publication couldn't have been easily produced.

Special thanks goes to booth eminent professors Prof. Dr. Farida Rabi, Faculty of Agriculture, Ain Shams University, and late Prof. Dr. Faiez Hanna, Soil and Water Department, National Research Center, for their accurate and diligent review of manuscript drafts and for their dedication, hard work, long hours, and high professional standards during the intiation period of land resources database development. Special appreciation goes to:

Associate Scientific Research Staff:

Dr. Raffat Ali

Dr. Adel Shalaby

Dr. Ahmed Afifi

Technical Staff:

Dr. Mohamed Abdel Rahman El-Sayed

Mr. Ahmed Hanafy

Dr. Ihab Yosef

Gratitude is gifted to their efforts and contribution along with their kind support in different disciplines during execution of this work. Also, the author thanks Prof. Abdelazim Negm for his review and constructive advice throughout all stages of the book preparation. Thanks are also due to the Academy of Scientific Research and Technology for awarding special priority supporting the preparation of land research database for agriculture use.

Special gratitude is credited to all staff members of the National Authority of Remote Sensing and Space Sciences for their institutional and personal support.

Author

Prof. Dr. Abd-Alla Gad

Summary

The National Development Plan essentially desires to maintain the arable lands in the Nile Valley and Delta, in addition to the western desert oases and Nile basin zones, interfering desert fringes. Moreover, increasing land productivity is an essential national goal for narrowing the gap between both food production and consumption. Therefore, providing accurate and integrated land resources information is necessary, especially with the sustainable information technology progress that would be the base for planning, decision support, and research needs.

The project *"Preparation of Egyptian land resources database for agricultural purposes"* is an important scientific documentation mechanism that has long been expected to make the best use of great national investments in thematic mapping.

The progress of information technology has resulted in introducing various surveying techniques (i.e., Global Positioning Systems (GPS), Geographic Information Systems (GIS), and remote sensing). The synoptic view and multi-concept of remote sensing images (i.e., multi-temporality, multi-sensors, and multi-spectrally) allow the operational acquisition and monitoring of land resources data. GIS has proven to be a powerful tool to store, manipulate, and analyze thematic data.

Variable information and database systems were utilized within the project (e.g., ArcGIS, ERDAS IMAGINE, and ENVI). The outcomes of previously accomplished soil survey projects (e.g., High Dam Soil Survey, Soil Map of Egypt, and Land Master Plane) were the basis for the created digital database. The analogue data of the project (Soil Map of Egypt) were digitized, geo-referenced, re-projected, and attached with additional attribute tables, including landscape properties and soil characteristics. A number of 42 soil map sheets, produced through the soil map of Egypt project at scale of 1:100,000, were adjusted to be GIS-ready maps. Moreover, additional thematic information layers were added to the GIS model representing the networks of irrigation and drainage, roads, railways, and utilities, each layer includes different categories. Residential settlements were updated using recent ETM+ and SPOT satellite images. The Digital Elevation Model (DEM) was elaborated using SRTM space images, in addition to spot heights and contour lines derived from the topographic maps (scale 1: 50,000) produced by the Egyptian Survey Authority

(ESA). The resulted thematic layers were incorporated in the created GIS land resources database.

A mosaicking was performed on the thematic map sheets of the Nile Delta and Valley including their desert fringes (71182.3 km^2). Another mosaic was prepared for the North-western coastal zone (14117.7 km^2). Separate mosaics were created for each of the four western desert oases (i.e., Siwa, Bahariya, Kharga, and Dakhla), which cover a total area of 23510.4 km^2.

An intensive field investigation was performed in all regions, with the purpose of representing the soil units and collecting ground control points and soil samples for laboratory analyses. The Global Positioning System (GPS) was used for precise localization of the investigated sites. A total number of 300 soil samples, collected from 106 soil profiles were laboratory analyzed for representing the soil units. The American soil taxonomy system was followed for classifying the studied soils, based on the version of USDA, 2008. It was found that the *Entisols* soil order characterize most of the alluvial soils of Nile Delta and Valley, while *Aridisols* cover a limited area. The northwestern coastal zone is exhibited by *Entisols* and *Aridisols*, having a frequency of 57.9 and 42.1% respectively. The soil orders *Entisols* and *Aridisols* were found dominating the western desert oases; however, Kharga unlike other oases attains more frequency of alluvial soils. The *Aridsols* were found dominating Siwa oases.

The elaborated DEM, derived from ESA contour maps and spot heights revealed a landscape homogeny within the Nile Delta and Valley. This one elaborated from the 62 SRTM images, covering Egypt, showed the landscape heterogeneity within the territories. The elevation ranges from 0 to 40 meters ASL within the Nile Delta and Valley, however reaching 1200 meters in the Red mountainous region and 2000 meters in southern Sinai.

The created digital land resources database was used to evaluate and map land capabilities on the bases of FAO, 1985 methodology. The results showed that 45% of the Delta soils and 15.5% of Nile Valley are classified as highly capable, corresponding to *Vertic Torrifluvents* soil sub-great groups. Moderately capable soils, which correspond to *Typic Haplogypsids, Typic Torriorthents, Typic Haplocalsids,* and *Typic Quartzipssaments* sub-great groups, exhibit 9.0 and 55.8% of the Nile Delta and Valley respectively. The low capable soils, corresponding to sub-great groups *Typic Petrogypsids, Typic Haplosalids,* and *Typic Petrogypsids*, represent 1.7 of the Nile Valley and 46.0% of the Delta located mainly around its northern lakes. Soil capability of the northwestern coast shows the occurrence of class II (12.9%) and class III (27.1%) corresponding to sub-great groups *Typic Torriorthents* and *Typic Torripsamments,* respectively. Class IV of soil capability characterizes 1.6% of the northwestern coast, corresponding to soil sub-great groups *Aquic Haplosalides* and *Typic Petrogypsids.* Soil Capability Classification of the western desert oasis showed that 20.2% of the area is classified as class II, 11.0% as class III, and 27.4% as class IV. Non-capable soils reach 41.4% of total area.

The study activities included the establishment of digital land resources databases of 21 Governorates, where different thematic layers were included. Layouts were produced for each Governorate, integrating a number of thematic layers (i.e., soil, DEM, urban areas, networks of irrigation, drainage, roads, railways, and utility). Such database area prepared to be the base of detailed studies within each Governorate. Two examples for elaborating the detailed digital land resources database were fulfilled on Al-Fayoum and Al-Qalyubia. Al-Fayoum Governorate was chosen as an example representing the alluvial soils of the Nile Valley to be followed in other Middle and Upper Egypt Governorates. It is located in a depression, at the southwest of Cairo, where the distance to Al-Fayoum town is 90 km. It is connected to the Nile Valley by Hawara canal, through which Bahr Yousef is transporting the Nile water. Administratively, Al-Fayoum Governorate includes a number of five districts, (i.e., Atsa, Al-Fayoum, Ibsheway, Senouris, and Tamyah). However, the regions of North Qarun lakes and Wadi El-Rayan depression were studied and added to the created land resources database. The total area of Al-Fayoum Governorate is 6068.7 km^2; however, the used land includes only 1849.6 km^2 (i.e., 30.5% of the total territories). The agricultural land exhibits 1609.3 km^2 (i.e., 26.5% of the total area) classified into different grades of productivity. Land resources database of the whole Al-Fayoum Governorate, as well as its five administrative districts was elaborated. Physiographic study was applied on the three regions (i.e., Fayoum depression, north Qarun Lake and Wadi El-Rayan depression). Soil mapping was based on the findings of physiographic analysis. Field validation was performed for describing the studied landscape and collecting representative soil samples, which were analyzed in laboratories for characterizing the soil units.

Al-Qalyubia Governorate is located on the eastern side of the River Nile, near the Delta head. It is bounded from the south by both Cairo and Al-Giza Governorates, from the north by Al-Dakahlia and Al-Gharbia Governorates, from the east by Al-Sharqia Governorate and from the west by Al-Monufia Governorate. It is characterized by a specific location since it is considered the meeting point of the main transport lines between northern Governorates. The total area of the Governorate is estimated to be 1001.1 km^2, where the cultivated area is 810.0 km^2, representing 81.0% of its total coverage. Al-Qalyubia Governorate includes a number of eight districts, (i.e., Al-Khankah, Al-Qanatir Al-Khayriyyah, Awal Shubra Al-Khaymah, Banha, Kafr Shokr, Qelyoub, Shibin Al-Qanatir, and Tokh). The common activity in the Governorate is mostly agriculture, in addition to existence of some industrial parks. The digital land resources database of Al-Qalyubia Governorate, as well as its eight administrative districts was elaborated. Physiographic study for the whole Governorate was implemented, upon which soil mapping was based. Field validation was performed for describing the studied landscape and collecting representative soil samples, which were analyzed in laboratories for characterizing the soil units.

It is possible to confirm that generating such details of natural land resources for local administrations is a great step toward the elaboration of sustainable development and management scheme. It is categorized by its comprehensiveness, geographical accuracy, and updatability. In the current era of distinctive progress in information technology, such data can be handled, enhanced, and exchanged by different users and authorities. The most striking finding noticed was the urban encroachment on the account of most fertile soils; hence, shrinkage in areas of high agricultural capabilities. Conversely, urbanization doesn't extend largely to low capable land. It was also noticed that a common unbalance between irrigation and drainage system networks is often and may lead to formation of soil salinity and initiation of land degradation process.

Contents

1 Induction

1.1 General

The rapid increase in population rate in Egypt, with the decrease in natural land resources, caused serious shortage in food requirements. Thus, raising the productivity of the land and its protection against degradation, besides reclamation of desert areas are the foremost purpose of sustainable agriculture development plans. However, realizing these objectives must be based on available accurate data for soil resources. Remote sensing and Geographical Information Systems (GIS) are the most suitable means for inventory monitoring and documentation of natural resources, as they are characterized by high resolution and multi-spectrally. Moreover, distribution and changes in natural resources are accessible by multi-temporal space data nature.

The Project of *"Preparation of land resources data base for agricultural usage"* is a great step taken by the Academy of Scientific Research and Technology towards digitizing and documenting one of the most important Egyptian scientific wealth (i.e. thematic maps). The project was undertaken following a suggestion to the Soils and Water Resources Research Council of the Academy of Scientific Research and Technology in 2003. It is the primary step in Egypt to prepare a digital land resources information geographic database for the Nile alluvial soils and the interfering desert fringes.

The Egyptian Government, represented by the academy of scientific research and technology, in cooperation with some International organization, has invested many funds to produce the soil maps of Egypt. Three major projects, namely High Dam Soil Survey (FAO 1965), Soil Map of Egypt (ASRT 1982) and Land Master Plan of Egypt (ARE and Kingdom of the Netherlands 1986) were the main projects covering agricultural potential areas. Moreover, many of M.Sc. and Ph.D. theses and research papers dealing with mapping soil resources are available in the academic libraries. The link between these

A.-A. Gad, *Digital Land Resources to Support Sustainable Development in Egypt*, Deltas of the World,
https://doi.org/10.1007/978-3-031-37167-7_1

different activities is missed, sometimes difficult to obtain particular soil information due to their wide distribution. The nature of the available data in analogue format makes it difficult to manipulate and upgrade their information. Following different land classification and mapping schemes, lead to contradiction in estimating the cultivated areas and crop acreages. Thus, the National Authority for Remote Sensing and Space Sciences (NARSS), in cooperation with the Academy of Scientific Research and Technology realized the importance of creating a digital land resources database, on basis of remote sensing and GIS capabilities possessed by NARSS.

1.2 Work Objectives

The current scientific work aims to employ the geographic information systems in designing and establishing a comprehensive geographical land resources database for the Nile Valley and Delta, in addition to cultivable lands in the northwestern coastal area and the western desert oases and promising eastern desert interference zones, at scale of 1: 100,000. Data available for soil resources are geometrically corrected and updated, by using space images. Integrated thematic data are added from the terrestrial maps produced by the Egyptian General Survey Authority (EGSA), Geologic maps produced by CONCO 1989, climatic data collected by Central Lab of Agriculture Climate (CLAC) and performed field validation. Also, recent satellite images were used in producing recent soil maps (Scale 1: 50,000) for Al-Fayoum and Al-Qalubia Governorates. These two cases are selected to represent the soils in the Nile Valley and Delta, and are considered as examples to be followed in the future to produce the soil maps for all Egyptian territories.

The detailed outlines of the objectives can be referred as follows:

1. Establishing a basic reference database including the previous studies and research works related to the Egyptian agricultural lands at scale of 1: 100,000.
2. Using the previous studies as a base for establishing the GIS land resources numerical database for agricultural use of sustained arable land (i.e. Nile Valley and Delta, in addition to the northwestern coast and western desert oases).
3. Usage of remote sensing data for updating the existing data and detecting multi-temporal environmental changes.
4. Usage of GIS technology in storing, documenting, retrieving and analyzing land resources data, including the tabulated geographic attributes (e.g. Physical and chemical soil properties).
5. Producing semi detailed soil maps at scale of 1: 50,000, in context of developing an updated land resource database for the Al-Fayoum and Al-Qalubia governorates.
6. Employing the created database in mathematical modelling for the purpose of assessing land capability, land degradation hazard and environmental sensitivity.

The progress in the introduced scientific work passed through dpfferent phases. The task packages, within the wok phases include transformation of 42 analogue soil map sheets, covering the whole Egyptian Nile Delta, Valley, Northwestern coast and the western desert oases, produced through the *"Soil map of Egypt"* project Also, variety of thematic data driven from different sources were combined.

An extensive fieldwork was carried out to represent the mapped soil units with well-localized soil profiles. Laboratory analyses for soil samples, representing the soil units of the studied regions were achieved. The outcomes of purposive field missions to represent and update mapping of different land and soil units were fulfilled The outcomes include also the developed Digital Elevation Model (DEM) of whole Egyptian territories, driven from the SRTM satellite data.

The comprehensive database of the studied regions was elaborated, including a number of 13 thematic layers. Soil classification was reviewed and adopted to the soil taxonomy classification (USDA 2006). Each Governorate was shown in separate layouts. The northwestern coastal area, belonging to Matrouh Governorate, is also produced in one layout. Moreover each of the western desert oases, belonging to El-Wadi El-Gedid Governorate, is shown in separate layouts. Distribution of land resources in each Governorate was discussed on basis of the developed land resources database. The final chapter of current book was dedicated for detailed land resources database of Al-Fayoum and Al-Qalubia Governorates respectively. These two Governorates were chosen to represent the alluvial soils of the Nile Valley and Delta and to be considered as examples for developing updated land resources numerical database for the entire territory of Egypt.

The book includes demonstration and analyses of detailed land resources database thematic layers and related attribute tables. Concerning Al-Fayoum Governorate, the studies dealt with each of the governorate's five administrative districts and the regions of North Qarun Lake and wadi Al-Rayan depression. Concerning Al-Qalubia Governorate, the studies dealt with each of its eight districts as a separate study unit. Physiographic study was applied on the study regions and soil mapping was based on the findings of physiographic analysis. Field work was performed for describing the studied landscape and collecting representative soil samples, which were analyzed in laboratories for characterizing the soil units.

The current book targets to demonstrate the outcomes of the whole research activities along the land reoures dcatabase development elabotrared tasks. Inclusion of different layouts, discussions and annexes are achieved. An atlas, including variety of thematic maps is combined, in addition to an operational access database, on a CD media, for retrieving the details of the project.

References

ASRT (1982) Soil Map of Egypt-Final Report. Academy of Scientific Research and Technology, Cairo, Egypt

Continental Oil Company CONOCO (1989) WebCONOCO, Geologic Map of Egypt, Scale 1.500.000, cited by the following article: TITLE: Relationships between Hydraulic Parameters
FAO (1965) High Dam Soil Survey, Egypt. UNDP – FAO/SF Project 16/UAR
Hammad MA (1986) Land master plan, Northwest Coast. Regional Report. General Authority for Rehabilitation Projects and Agricultural Development, Cairo, Egypt
USDA (2006) "Keys to Soil Taxonomy" United State Department of Agriculture, Natural Resources Conservation Service (NRCS) tenth edition, 2006

2 Materials Used

2.1 Bibliography

A survey of bibliography related to the studied areas and topics was performed through available National and International publishers. It resulted in collecting satisfactory information concerning soil, geology, land use/cover, human resources development data and hydrogeology of the studied. The collected review materials include theses, reports and scientific papers available in libraries of Universities and Nile delta soils.

For the importance of these materials, as integral portion of the current work, annex (I) was attached referring to soil studies.

2.2 Soil Maps

The soil maps of Egypt (ASRT 1982) are the main documented soil information collected and converted to the digital format. The collected soil maps cover the arable land, Nile Valley-desert fringes as well as the western desert oases and northwestern coastal region. The maps are attached to the final report in addition to a set of eight progress reports representing the different areas of Egypt as follows;

1. The Northern part of the Delta.
2. The middle delta region between the two branches of the Nile.
3. The western side of the Delta.
4. The eastern side of the Delta.
5. The north-western Mediterranean Coastal littoral.
6. The Valley from Cairo till Aswan including the Fayoum Depression.

A.-A. Gad, *Digital Land Resources to Support Sustainable Development in Egypt*, Deltas of the World,
https://doi.org/10.1007/978-3-031-37167-7_2

7. The Oases in the Western Desert.
8. Selected two areas representing the Western and Eastern desert.

The collected maps consist of a number of 42 map sheets at scale of 1: 100,000 in analogue format, Fig. 2.1 shows the index of the collected soil maps.

2.3 Topographic Maps

Topographic maps at scale 1: 50,000 covering the whole basine of the Nile Valley and Delta, produced by the EGSMA in 1992 were integrated. These maps were converted to the GIS digital format by NARSS in 2001. The maps are available in mosaics covering different Governorates of the Nile Valley and Delta and amount 202 map sheets (Fig. 2.2).

2.4 Geologic Maps

The geological map at scale 1: 500,000 produced by CONOCO (1989) is available in digital format. This map is added to the created GIS database of Egyptian land resources.

2.5 Satellite Images

ETM images of 2004, covering the whole territory of Egypt (Fig. 2.3) amount 57 scenes. These scenes, covering the Nile valley and Delta, were collected and processed to be included in the GIS land resources database and thematic mapping processes. The recent SPOT images of 2006 were obtained from NARSS receiving station, and processed for updating different thematic maps and detecting changes in land use/land cover.

2.6 Meteorological Data

FAO (1970a, b, c) determind a number of four main parameters, that characterize the climate along the Mediterranean Egyotian Coastal t: (1) the regional situation regarding the general circulation of the atmosphere; (2) the adjacency of the Mediterranean Sea; (3) the placement of the coast and (4) the landscape. The collected data include the geographical coordinates and elevations of metrological stations. Also, different climatic parameters summarizing the characteristics of rainfall, temperature, humidity, wind velocity and solar radiation parameters were collected.

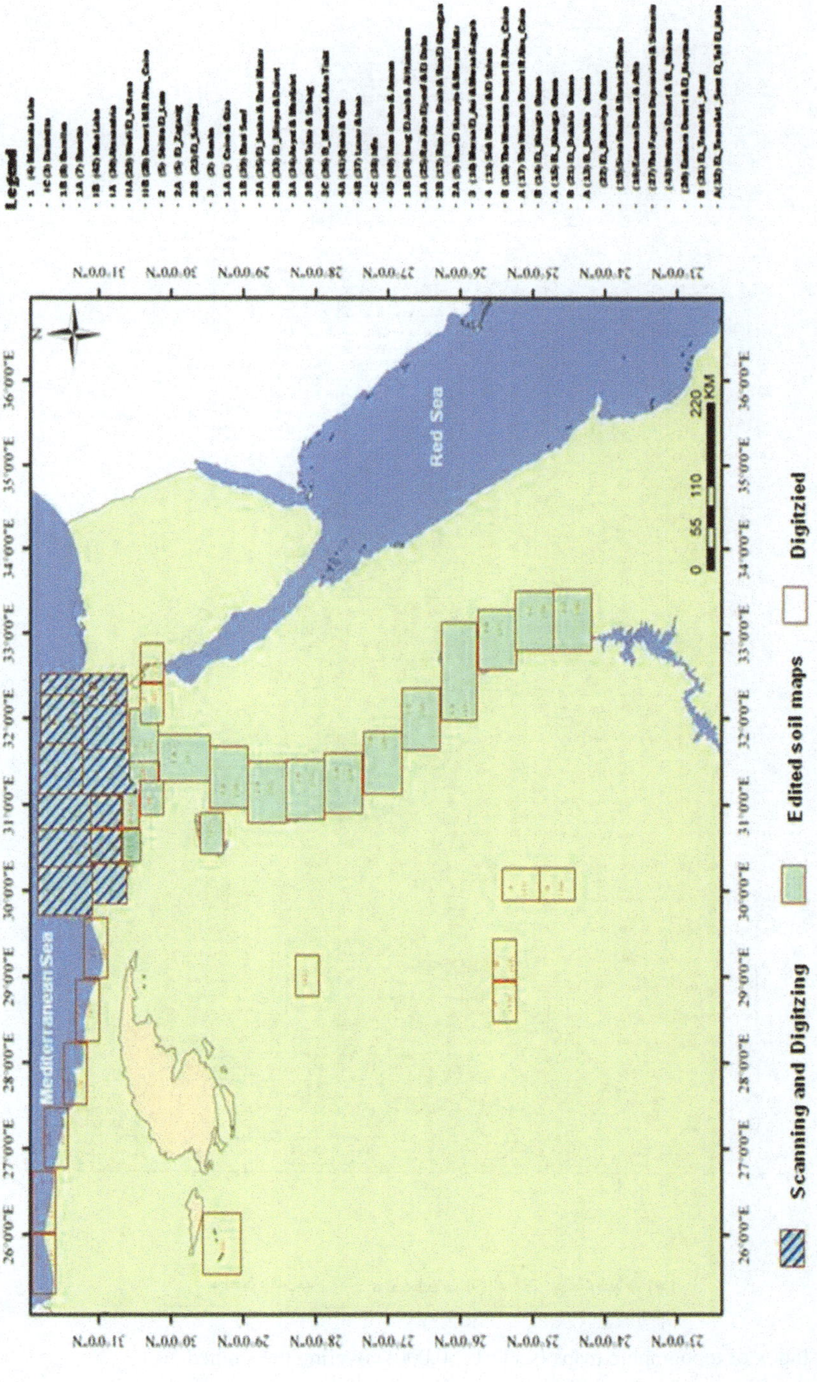

Fig. 2.1 Index of the collected soil maps (Soil Map of Egypt, 1982)

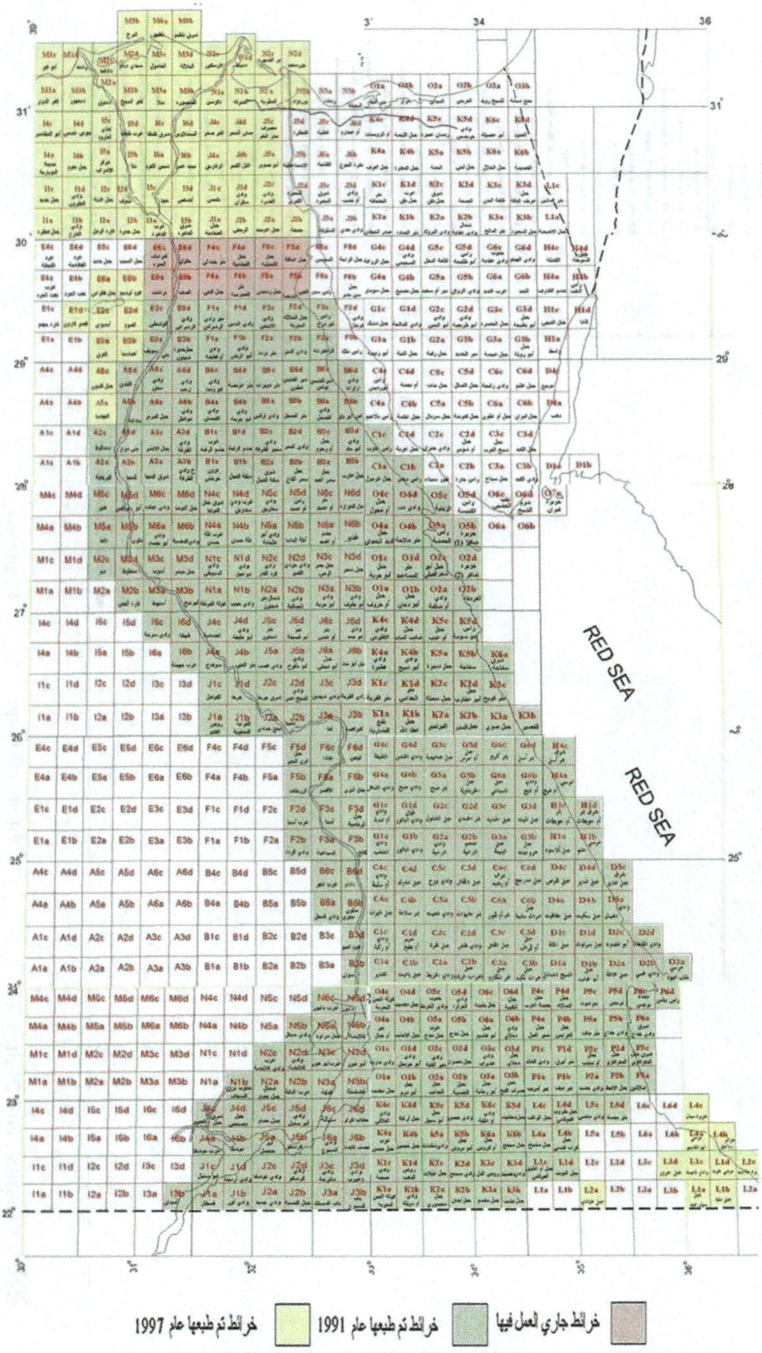

Fig. 2.2 Index of topographic maps (scale 1: 50,000) covering the studied area

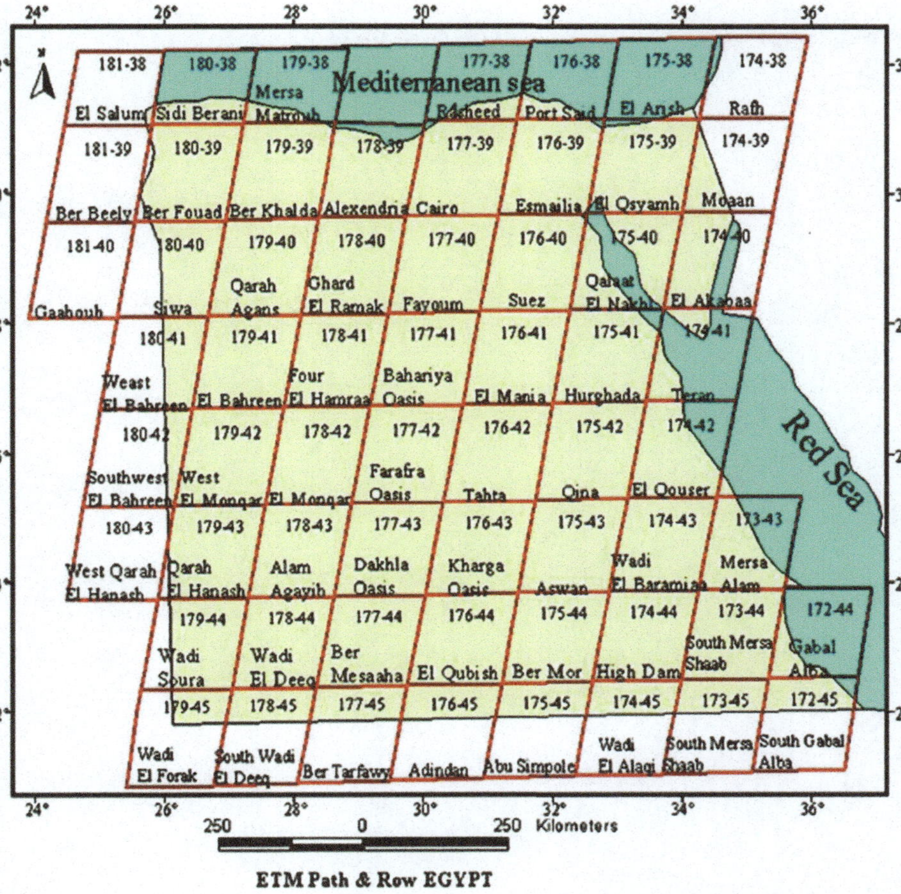

Fig. 2.3 Index of LANDSAT ETM images covering the whole territory of Egypt

References

ASRT (1982) Soil Map of Egypt-Final Report. Academy of Scientific Research and Technology, Cairo, Egypt

FAO (1970a) Key to soil units for the soil map of the world: Soil resources, development and conservation services. Land and Water Development Division, FAO, U.N., Rome

FAO (1970b) Ground water resources of North Coastal Zone, Egypt: University of Alabama, USA

FAO (1970c) Pre-investment survey of the North Western Coastal region: comprehensive account of the project. Technical Report No. 1, ESE: SF/UAR 49, 109p + 5 Maps

Methodology 3

3.1 Coding the Soil Units

In order to input the soil maps in the geographic information database, it was necessary to codify the mapping units. It was found that the number of soil sub-great groups identified in the soil map of Egypt (ASRT 1982) amount to 20. These units belong to three soil orders (i.e. *Entisoil, Vertisols* and *Aridisols*). Table 3.1 demonstrates the different categories of soils classification on basis of USDA (1975) including soil orders, sub-orders, great groups and ssub-great group. Coding the soil units ought to be indicative to the following;

1. Soil properties (surface and sub-surface texture, soil salinity, calcium carbonate content).
2. Landscape properties (slope).

The considered coding system in this phase was based on the soil formula. The formula adopted in the soil map represents the following items:

Soil sub great group (Surface soil texture/sub-surface soil texture), slope, salinity & calcium carbonate content.

Where;

The surface and sub-surface soil texture rage from course to fine. The slope was considered as an indication of landscape. Codes were assigned according to Table 3.2.

Salinity and calcium carbonate content classes were input in the GIS database by their ranges. Example (Eftv3/4a3) refers to *Vertic Torrifluvents*, medium-fine surface soil texture, fine sub-surface soil texture, level landscape and highly saline. Codes of different parameters are explained in the following paragraphs. Areas of "non-soils" are shown on the sheets as miscellaneous land units.

A.-A. Gad, *Digital Land Resources to Support Sustainable Development in Egypt*, Deltas of the World,
https://doi.org/10.1007/978-3-031-37167-7_3

Table 3.1 Orders, sub-orders, great groups and sub-great groups of Egyptian Soils, (ASRT 1982), based on, soil taxonomy (USDA 1975)

Soil order	Sub order	Great group	Sub great group
Entisols (E)	Fluvents (f)	Torrifluvents (t)	Typic Torrifluvents (Eftt)
			Vertic Torrifluvents (Efvt)
		Ustifluvents (u)	Typic Ustifluvents (Efut)
			Vertic Ustifluvents (Efuv)
	Orthents (o)	Torrioethents (t)	Typic Torriorthents (Eott)
			Calcic Torrioethents (Eotca)
	Aquents (a)	Haplaquents (h)	Typic Haplaquents (Eaht)
	Psamments (p)	Quartzipsammets (q)	Typic Quartzipsammets (Epqt)
		Torripsamments (t)	Typic Torripsamments (Eptt)
			Lithic Torripsamments (Eptl)
		Calcipsamments (ca)	Oolitic Calcipsamments (Epcao)
Vertisols (V)	Torrerts (t)	Torrerts (t)	Typic Torrerts (Vtt)
			Paleustollic Torrerts (Vtp)
		Chromusterts (ucr)	Paleustollic Chromusterts (Vucrpl)
Aridisols (A)	Orthids (o)	Calciorthids (ca)	Typic Calciorthids (Aocat)
		Gypsyorthids (g)	Typic Gypsyorthids (Aogt)
			Calcic Gypsyorthids (Aogca)
			Petrogypsic Gypsyorthids (Aogp)
		Salorthids (s)	Typic Salorthids (Aost)
			Aqualic Salorthids (Aosa)

Table 3.2 Assigned codes for the soil texture and slope classes

Soil texture class	Soil texture code	Slope class	Slope code	salinity class	Salinity code
Coarse	1	Level	a	Slightly saline	1
Medium	2	Sloping	b	Moderately saline	2
Medium-Fine	3	Moderately steep	c	Highly saline	3
Fine	4	Steep	d		

Table 3.3 shows the developed data dictionary of soil maps on the basis of the above-mentioned criteria.

Table 3.3 Developed soil data dictionary of the mapped soil units

Soil code	Soil order	Sub order	Great group	Sub great group	Surface texture	Subsurface texture	Slope	EC dS/m	% CaCO3
Eftt 3/3a	*Entisols*	*Fluvents*	*Torrifluvents*	*Typic Torrifluvents*	Medium fine	Medium fine	Level	< 8	< 4
Eftt 3/3a1	*Entisols*	*Fluvents*	*Torrifluvents*	*Typic Torrifluvents*	Medium fine	Medium fine	Level	8–16	< 4
Eftt 3/3a7	*Entisols*	*Fluvents*	*Torrifluvents*	*Typic Torrifluvents*	Medium fine	Medium fine	Level	< 8	4–8
Eftt 4/3a	*Entisols*	*F.0 luvents*	*Torrifluvents*	*Typic Torrifluvents*	Fine	Medium fine	Level	< 8	< 4
Eftt 4/4a	*Entisols*	*Fluvents*	*Torrifluvents*	*Typic Torrifluvents*	Fine	Fine	Level	< 8	< 4
Eftt 2/3a	*Entisols*	*Fluvents*	*Torrifluvents*	*Typic Torrifluvents*	Medium	Medium fine	Level	< 8	< 4
Eftt 2/2a	*Entisols*	*Fluvents*	*Torrifluvents*	*Typic Torrifluvents*	Medium	Medium	Level	< 8	< 4
Eftt 3/2a	*Entisols*	*Fluvents*	*Torrifluvents*	*Typic Torrifluvents*	Medium fine	Medium	Level	< 8	< 4
Eftt 1/1b	*Entisols*	*Fluvents*	*Torrifluvents*	*Typic Torrifluvents*	Coarse	Coarse	Sloping	< 8	< 4
Eftt 1/1c	*Entisols*	*Fluvents*	*Torrifluvents*	*Typic Torrifluvents*	Coarse	Coarse	Moderately Steep	< 8	< 4
Eftt 4/2a	*Entisols*	*Fluvents*	*Torrifluvents*	*Typic Torrifluvents*	Fine	Medium	Level	< 8	< 4
Eftt 2–3/3a	*Entisols*	*Fluvents*	*Torrifluvents*	*Typic Torrifluvents*	Medium to Medium fine	Medium fine	Level	< 8	< 4
Eftt 3/4a	*Entisols*	*Fluvents*	*Torrifluvents*	*Typic Torrifluvents*	Medium fine	Fine	Level	< 8	< 4
Eftv 4/4a	*Entisols*	*Fluvents*	*Torrifluvents*	*Vertic Torrifluvents*	Fine	Fine	Level	< 8	< 4
Eftv 4/4a1	*Entisols*	*Fluvents*	*Torrifluvents*	*Vertic Torrifluvents*	Fine	Fine	Level	8–16	< 4
Eftv 4/3a	*Entisols*	*Fluvents*	*Torrifluvents*	*Vertic Torrifluvents*	Fine	Medium fine	Level	< 8	< 4
Eftt 3/3a3	*Entisols*	*Fluvents*	*Torrifluvents*	*Typic Torrifluvents*	Medium fine	Medium fine	Level	8–16	4–8

(continued)

Table 3.3 (continued)

Soil code	Soil order	Sub order	Great group	Sub great group	Surface texture	Subsurface texture	Slope	EC dS/m	% CaCO3
Eftl 3/2a	*Entisols*	*Fluvents*	*Torrifluvents*	*Lithic Torrifluvents*	Medium fine	Medium	Level	< 8	< 4
Eptl 1/1a	*Entisols*	*Psamments*	*Torripsamments*	*Lithic Torripsamments*	Coarse	Coarse	Level	< 8	< 4
Eptl 1/1b	*Entisols*	*Psamments*	*Torripsamments*	*Lithic Torripsamments*	Coarse	Coarse	Sloping	< 8	< 4
Eptl 2/1b	*Entisols*	*Psamments*	*Torripsamments*	*Lithic Torripsamments*	Medium	Coarse	Sloping	< 8	< 4
Eptl 1/1c	*Entisols*	*Psamments*	*Torripsamments*	*Lithic Torripsamments*	Coarse	Coarse	Moderately Steep	< 8	< 4
Eptl 2/1c	*Entisols*	*Psamments*	*Torripsamments*	*Lithic Torripsamments*	Medium	Coarse	Moderately Steep	< 8	< 4
Eptl 1/1d	*Entisols*	*Psamments*	*Torripsamments*	*Lithic Torripsamments*	Coarse	Coarse	Steep	< 8	< 4
Eptl 2/1d	*Entisols*	*Psamments*	*Torripsamments*	*Lithic Torripsamments*	Medium	Coarse	Steep	< 8	< 4
Eptl 2/1a	*Entisols*	*Psamments*	*Torripsamments*	*Lithic Torripsamments*	Medium	Coarse	Level	< 8	< 4
Eptt 1/1a	*Entisols*	*Psamments*	*Torripsamments*	*Typic Torripsamments*	Coarse	Coarse	Level	< 8	< 4
Eptt 1/1a3	*Entisols*	*Psamments*	*Torripsamments*	*Typic Torripsamments*	Coarse	Coarse	Level	8–16	4–8
Eptt 2/1a	*Entisols*	*Psamments*	*Torripsamments*	*ipsamments*	Medium	Coarse	Level	< 8	< 4
Eptt 2/1a3	*Entisols*	*Psamments*	*Torripsamments*	*Typic Torripsamments*	Medium	Coarse	Level	8–16	4–8
Eptt 1/1a8	*Entisols*	*Psamments*	*Torripsamments*	*Typic Torripsamments*	Coarse	Coarse	Level	< 8	8–15
Eptt 1/1b	*Entisols*	*Psamments*	*Torripsamments*	*Typic Torripsamments*	Coarse	Coarse	Sloping	< 8	< 4
Eptt 2/1b	*Entisols*	*Psamments*	*Torripsamments*	*Typic Torripsamments*	Medium	Coarse	Sloping	< 8	< 4
Eptt 1/1c	*Entisols*	*Psamments*	*Torripsamments*	*Typic Torripsamments*	Coarse	Coarse	M. steep	< 8	< 4
Eptt 2/1c	*Entisols*	*Psamments*	*Torripsamments*	*Typic Torripsamments*	Medium	Coarse	M. steep	< 8	< 4
Eptt	*Entisols*	*Psamments*	*Torripsamments*	*Typic Torripsamments*				< 8	< 4
Efut 1/3a	*Entisols*	*Fluvents*	*Ustifluvents*	*Typic Ustifluvents*	Coarse	Medium fine	Level	< 8	< 4
Efut 4/3a	*Entisols*	*Fluvents*	*Ustifluvents*	*Typic Ustifluvents*	Fine	Medium fine	Level	< 8	< 4
Efut 4/3a1	*Entisols*	*Fluvents*	*Ustifluvents*	*Typic Ustifluvents*	Fine	Medium fine	Level	8–16	< 4
Efut 4/4a	*Entisols*	*Fluvents*	*Ustifluvents*	*Typic Ustifluvents*	Fine	Fine	Level	< 8	< 4

Efut 3/2a	*Entisols*	*Fluvents*	*Ustifluvents*	*Typic Ustifluvents*	Medium fine	Medium	Level	< 8	< 4
Efuv 3/2a	*Entisols*	*Fluvents*	*Ustifluvents*	*Vertic Ustifluvents*	Medium fine	Medium	Level	< 8	< 4
Efuv 4/3a	*Entisols*	*Fluvents*	*Ustifluvents*	*Vertic Ustifluvents*	Fine	Medium fine	Level	< 8	< 4
Efuv 4/4a	*Entisols*	*Fluvents*	*Ustifluvents*	*Vertic Ustifluvents*	Fine	Fine	Level	< 8	< 4
Efuv 4/4a1	*Entisols*	*Fluvents*	*Ustifluvents*	*Vertic Ustifluvents*	Fine	Fine	Level	8–16	< 4
Epqt 1/3a	*Entisols*	*Psamments*	*Quartizipsamments*	*Typic Quartizipsamments*	Coarse	Medium fine	Level	< 8	< 4
Epqt 1/1b3	*Entisols*	*Psamments*	*Quartizipsamments*	*Typic Quartizipsamments*	Coarse	Coarse	Sloping	8–16	4–8
Epqt 1/1a	*Entisols*	*Psamments*	*Quartizipsamments*	*Typic Quartizipsamments*	Coarse	Coarse	Level	< 8	< 4
Epqt 1/1a1	*Entisols*	*Psamments*	*Quartizipsamments*	*Typic Quartizipsamments*	Coarse	Coarse	Level	8–16	< 4
Epqt 1/1a7	*Entisols*	*Psamments*	*Quartizipsamments*	*Typic Quartizipsamments*	Coarse	Coarse	Level	< 8	4–8
Epqt 1/1a2	*Entisols*	*Psamments*	*Quartizipsamments*	*Typic Quartizipsamments*	Coarse	Coarse	Level	> 16	< 4
Epqt 2/1a	*Entisols*	*Psamments*	*Quartizipsamments*	*Typic Quartizipsamments*	Medium	Coarse	Level	< 8	< 4
Epqt 3/1a	*Entisols*	*Psamments*	*Quartizipsamments*	*Typic Quartizipsamments*	Medium fine	Coarse	Level	< 8	< 4
Epqt 4/1a	*Entisols*	*Psamments*	*Quartizipsamments*	*Typic Quartizipsamments*	Fine	Coarse	Level	< 8	< 4
Epqt 1/1b	*Entisols*	*Psamments*	*Quartizipsamments*	*Typic Quartizipsamments*	Coarse	Coarse	Sloping	< 8	< 4
Epqt 2/1b	*Entisols*	*Psamments*	*Quartizipsamments*	*Typic Quartizipsamments*	Medium	Coarse	Sloping	< 8	< 4
Epqt 1/1c	*Entisols*	*Psamments*	*Quartizipsamments*	*Typic Quartizipsamments*	Coarse	Coarse	Moderately Steep	< 8	< 4
Epqt 4/3a	*Entisols*	*Psamments*	*Quartizipsamments*	*Typic Quartizipsamments*	Fine	Medium fine	Level	< 8	< 4
Epqt 1/1b-c	*Entisols*	*Psamments*	*Quartizipsamments*	*Typic Quartizipsamments*	Coarse	Coarse	Sloping - Moderately Steep	< 8	< 4
Epqt 1/1b1	*Entisols*	*Psamments*	*Quartizipsamments*	*Typic Quartizipsamments*	Coarse	Coarse	Sloping	8–16	< 4
Vtt 1/1 a	*Vertisols*	*Torrerts*	*Torrerts*	*Typic Torrerts*	Coarse	Coarse	Level	< 8	< 4
Vtt 4/3 a	*Vertisols*	*Torrerts*	*Torrerts*	*Typic Torrerts*	Fine	Medium fine	Level	< 8	< 4

(continued)

Table 3.3 (continued)

Soil code	Soil order	Sub order	Great group	Sub great group	Surface texture	Subsurface texture	Slope	EC dS/m	% CaCO3
Vtt 3/4 a	*Vertisols*	*Torrerts*	*Torrerts*	*Typic Torrerts*	Medium fine	Fine	Level	< 8	< 4
Vtt 4/4 a	*Vertisols*	*Torrerts*	*Torrerts*	*Typic Torrerts*	Fine	Fine	Level	< 8	< 4
Vtt 4/4 a1	*Vertisols*	*Torrerts*	*Torrerts*	*Typic Torrerts*	Fine	Fine	Level	8–16	< 4
Vtt 4/4 a3	*Vertisols*	*Torrerts*	*Torrerts*	*Typic Torrerts*	Fine	Fine	Level	8–16	4–8
Vtt 4/4 a8	*Vertisols*	*Torrerts*	*Torrerts*	*Typic Torrerts*	Fine	Fine	Level	< 8	8–15
Vtpl 4/4a	*Vertisols*	*Torrerts*	*Torrerts*	*Paleustollic Torrerts*	Fine	Fine	Level	< 8	< 4
Eott 1/2a3	*Entisols*	*Orthents*	*Torriorthents*	*Typic Torriorthents*	Coarse	Medium	Level	8–16	4–8
Eott 1/1a	*Entisols*	*Orthents*	*Torriorthents*	*Typic Torriorthents*	Coarse	Coarse	Level	< 8	< 4
Eott 1/1a5	*Entisols*	*Orthents*	*Torriorthents*	*Typic Torriorthents*	Coarse	Coarse	Level	> 16	4–8
Eott 1/1a7	*Entisols*	*Orthents*	*Torriorthents*	*Typic Torriorthents*	Coarse	Coarse	Level	< 8	4–8
Eott 2/1a	*Entisols*	*Orthents*	*Torriorthents*	*Typic Torriorthents*	Medium	Coarse	Level	< 8	< 4
Eott 2/1a3	*Entisols*	*Orthents*	*Torriorthents*	*Typic Torriorthents*	Medium	Coarse	Level	8–16	4–8
Eott 2/1a5	*Entisols*	*Orthents*	*Torriorthents*	*Typic Torriorthents*	Medium	Coarse	Level	> 16	4–8
Eott 1/2a	*Entisols*	*Orthents*	*Torriorthents*	*Typic Torriorthents*	Coarse	Medium	Level	< 8	< 4
Eott 1/2a2	*Entisols*	*Orthents*	*Torriorthents*	*Typic Torriorthents*	Coarse	Medium	Level	> 16	< 4
Eott 1/1b	*Entisols*	*Orthents*	*Torriorthents*	*Typic Torriorthents*	Coarse	Coarse	Sloping	< 8	< 4
Eott 1/1b3	*Entisols*	*Orthents*	*Torriorthents*	*Typic Torriorthents*	Coarse	Coarse	Sloping	8–16	4–8
Eott 1/1b6	*Entisols*	*Orthents*	*Torriorthents*	*Typic Torriorthents*	Coarse	Coarse	Sloping	> 16	8–15
Eott 1/1b7	*Entisols*	*Orthents*	*Torriorthents*	*Typic Torriorthents*	Coarse	Coarse	Sloping	< 8	4–8
Eott 2/1b	*Entisols*	*Orthents*	*Torriorthents*	*Typic Torriorthents*	Medium	Coarse	Sloping	< 8	< 4
Eott 2/2b	*Entisols*	*Orthents*	*Torriorthents*	*Typic Torriorthents*	Medium	Medium	Sloping	< 8	< 4
Eott 3/3b3	*Entisols*	*Orthents*	*Torriorthents*	*Typic Torriorthents*	Medium fine	Medium fine	Sloping	8–16	4–8
Aocat 1/2 b	*Aridisols*	*Orthids*	*Gypsiorthids*	*Typic Calciorthids*	Coarse	Medium	Sloping	< 8	> 15
Aocat	*Aridisols*	*Orthids*	*Gypsiorthids*	*Typic Calciorthids*				< 8	> 15
Eott	*Entisols*	*Orthents*	*Torriorthents*	*Typic Torriorthents*				< 8	> 15
Aosat 4/4a8	*Aridisols*	*Orthids*	*Salorthids*	*Typic Salorthids*	Fine	Fine	Level	< 8	8–15

Aogca 1/1a1	*Aridisols*	*Orthids*	*Gypsiorthids*	*Typic Gypsiorthids*	Coarse	Coarse	Level	8–16	< 4
Aogca 1/1a2	*Aridisols*	*Orthids*	*Gypsiorthids*	*Typic Gypsiorthids*	Coarse	Coarse	Level	> 16	< 4
Aogca 3/3a	*Aridisols*	*Orthids*	*Gypsiorthids*	*Typic Gypsiorthids*	Medium fine	Medium fine	Level	< 8	< 4
Aogca 3/3a2	*Aridisols*	*Orthids*	*Gypsiorthids*	*Typic Gypsiorthids*	Medium fine	Medium fine	Level	> 16	< 4
Aogca	*Aridisols*	*Orthids*	*Gypsiorthids*	*Typic Gypsiorthids*				< 8	< 4
Aogca 3/3a8	*Aridisols*	*Orthids*	*Gypsiorthids*	*Typic Gypsiorthids*	Medium fine	Medium fine	Level	< 8	8–15
Vtt 4/4a7	*Vertisols*	*Torrerts*	*Torrerts*	*Typic Torrerts*	Fine	Fine	Level	< 8	4–8
Aogp 1/1a	*Aridisols*	*Orthids*	*Gypsiorthids*	*Petrogypsic Gypsiorthids*	Coarse	Coarse	Level	< 8	< 4
Aogp 1/1a1	*Aridisols*	*Orthids*	*Gypsiorthids*	*Petrogypsic Gypsiorthids*	Coarse	Coarse	Level	8–16	< 4
Aogp 1/4a	*Aridisols*	*Orthids*	*Gypsiorthids*	*Petrogypsic Gypsiorthids*	Coarse	Fine	Level	< 8	< 4
Aogp 1/1b	*Aridisols*	*Orthids*	*Gypsiorthids*	*Petrogypsic Gypsiorthids*	Coarse	Coarse	Sloping	< 8	< 4
Aogt 1/1a1	*Aridisols*	*Orthids*	*Gypsiorthids*	*Typic Gypsiorthids*	Coarse	Coarse	Level	8–16	< 4
Aogp 1/3a1	*Aridisols*	*Orthids*	*Gypsiorthids*	*Petrogypsic Gypsiorthids*	Coarse	Medium fine	Level	8–16	< 4
Aogt 2/1a	*Aridisols*	*Orthids*	*Gypsiorthids*	*Typic Gypsiorthids*	Medium	Coarse	Level	< 8	< 4
Aogt 1/3a2	*Aridisols*	*Orthids*	*Gypsiorthids*	*Typic Gypsiorthids*	Coarse	Medium fine	Level	> 16	< 4
Aogt 1/3a5	*Aridisols*	*Orthids*	*Gypsiorthids*	*Typic Gypsiorthids*	Coarse	Medium fine	Level	> 16	4–8
Aogt 1/1b	*Aridisols*	*Orthids*	*Gypsiorthids*	*Typic Gypsiorthids*	Coarse	Coarse	Sloping	< 8	< 4
Aogt 1/3a3	*Aridisols*	*Orthids*	*Gypsiorthids*	*Typic Gypsiorthids*	Coarse	Medium fine	Level	8–16	4–8
Aogt 1/1b1	*Aridisols*	*Orthids*	*Gypsiorthids*	*Typic Gypsiorthids*	Coarse	Coarse	Sloping	8–16	< 4
Aosca	*Aridisols*	*Orthids*	*Saliorthids*	*Calcic Saliorthids*				< 8	> 15
Aoct 2/3a	*Aridisols*	*Orthids*	*Gypsiorthids*	*Typic Calciorthids*	Medium	Medium fine	Level	< 8	> 15
Aogc 2/3a	*Aridisols*	*Orthids*	*Gypsiorthids*	*Calcic Gypsiorthids*	Medium	Medium fine	Level	< 8	> 15
Aosah	*Aridisols*	*Orthids*	*Hemists*	*Salorthic Hemists*				< 8	< 4
Aost 2/2a	*Aridisols*	*Orthids*	*Salorthids*	*Typic Salorthids*	Medium	Medium	Level	< 8	< 4

(continued)

Table 3.3 (continued)

Soil code	Soil order	Sub order	Great group	Sub great group	Surface texture	Subsurface texture	Slope	EC dS/m	% CaCO3
Vu cr pl 4/4a	*Vertisols*	*Torrerts*	*Chromusterts*	*Paleustollic Chromusterts*	Fine	Fine	Level	< 8	< 4
Eott 1/1b1	*Entisols*	*Orthents*	*Torriorthents*	*Typic Torriorthents*	Coarse	Coarse	Sloping	8–16	4–8
Eptt & Eott	*Entisols*	*Psamments & Orthents*	*Torripsamments & Torriorthents*	*Typic Torripsamments & Typic Torriorthents*				< 8	< 4
Eott 1/2 b3	*Entisols*	*Orthents*	*Torriorthents*	*Typic Torriorthents*	Coarse	Medium	Sloping	8–16	< 4
Eott 2/2b3	*Entisols*	*Orthents*	*Torriorthents*	*Typic Torriorthents*	Medium	Medium	Sloping	8–16	4–8
Eptt 1/1a & Eott 1/1a	*Entisols*	*Psamments & Orthents*	*Torripsamments & Torriorthents*	*Typic Torripsamments & Typic Torriorthents*	Coarse	Coarse	Level	< 8	4–8
Vtt	*Vertisols*	*Torrerts*	*Torrerts*	*Typic Torrerts*				< 8	< 4
Epqt 1/1a & Eott 1/1a	*Entisols*	*Psamments & Orthents*	*Quartizipsamments & Torriorthents*	*Typic Quartizipsamments & Typic Torriorthents*	Coarse	Coarse	Level	< 8	< 4
Sand dune				Sand dune					
High dune				High dune					
Active sand dune				Active sand dune					
Shifting sand dune				Shifting sand dune					
Rock escarpment				Rock escarpment					
Rock land				Rock land					
Stone land				Stone land					
Hilly gravel and cobble				Hilly gravel and cobble					
Consolidated rocky ridge				Consolidated rocky ridge					
Plateau surface				Plateau surface					
Mountains and hills				Mountains and hills					

Swamps				Swamps					
Dry swamps				Dry swamps					
Salt marshes				Salt marshes					
Water				Water					
Urban				Urban					
Aoct 1/1a	*Aridisols*	*Orthids*	*Calciorthids*	Typic Calciorthids	Coarse	Coarse	Level	< 8	> 15
Aoct 1/2b	*Aridisols*	*Orthids*	*Calciorthids*	Typic Calciorthids	Coarse	Medium	Sloping	< 8	> 15
River Nile				River Nile		*Aoct* Coarse			
Aoct & Eott	*Aridisols & Entisols*	*Orthids & Orthents*	*calciorthids & Torriorthents*	Typic calciorthids & typic Torriorthents				< 8	> 15
Epc	*Entisols*	*Psamments*	*Torripsamments*	Calcic Torripsamments				< 8	> 15
Aosa 4/3a1	*Aridisols*	*Orthids*	*Salorthids*	Aquollic Salorthids	Fine	Medium fine	Level	8–16	< 4
Aosa 3/3a1	*Aridisols*	*Orthids*	*Salorthids*	Aquollic Salorthids	Medium fine	Medium fine	Level	8–16	< 4
Aosa 3/4a1	*Aridisols*	*Orthids*	*Salorthids*	Aquollic Salorthids	Medium fine	Fine	Level	8–16	< 4
Aosa4/4a1	*Aridisols*	*Orthids*	*Salorthids*	Aquollic Salorthids	Fine	Fine	Level	8–16	< 4
Aosa1/4a1	*Aridisols*	*Orthids*	*Salorthids*	Aquollic Salorthids	Coarse	Fine	Level	8–16	< 4

Table 3.4 Average particle size distribution of soil textural classes

Code	Textural class	0–2 μ	2–50 μ	50–2000 μ
5	Very fine	> 60%	<40%	<40%
4	Fine	35–60%	<65%	<15%
3	Medium-fine	<35%	<50%	>15%
2	Medium	<35%	<85%	>15%
1	Coarse	<18%	<35%	>65%

3.1.1 Textural Classes

Soil texture is permanent features and is considered one of the most important character directly related to other soil characteristics; namely, constancy, soil-moisture constants, structure, permeability, and cation exchange capacity (CEC). The texture of surfae layer is of a chief importance for tillage and water holding. Sub-surface texture also governs the drainage conditions and root scattering. Therefore, both surface and sub-surface soil texture are included in soil coding units. The different textural classes are coded as 1, 2, 3, 4, and 5 for very fine, fine, medium-fine, medium, and coarse respectively. Table 3.4 shows, the average grain size distribution of different textural classes.

3.1.2 Slopes

The topographic parameters, particularly slope influences overspill and erosion. Slope classes are coded in the created diital land resources database as a, b, c & d. Table 3.5 points out the coding system of different slope classes.

3.1.3 The Phases

Phases are reflected as divisions of the soil units based on characteristics significant to the use and/or management of the land. They are however, not diagnostic to separate the soil units themselves. The phases that are recognized on the soil Map of Egypt are salinity and calcium carbonate content.

The saline phase on the map defines soils that contain, in some horizons within 100 cm of the surface, electrical conductivity values in the saturation extract, higher than16 mmohs/cm. It should be noted that the saline is not shown for the *salorthids* great soil group as their definition implies a high salt content. Subdivision of the degree of salinity is introduced on the Soil Map of Egypt as follows:

- Slightly saline <8 dS/m
- Moderately saline 8–16 dS/m
- Highly saline >16 dS/m

Table 3.5 Cods of different slope classes

Code	Slope classes	%
a	Level	0–3
b	Sloping	3–10
c	Moderately steep	10–25
d	Steep	>25

Calcium carbonate content has been introduced on the Soil Map of Egypt, 1982 and subdivided into the following:

- *Crypto* contains <4% $CaCO_3$
- *Hypo* contains 4–8 $CaCO_3$
- *Hyper* contains 8–12% $CaCO_3$

Soils containing higher calcium carbonate are classified as calciorthids by the definition. Coding soil phase was done by assigning a number indicating both salinity and calcium carbonate content (Table 3.6).

3.2 Pre-processing of Satellite Images

Satellite images are principally a set of pixels that are often less than an exact image representation. Pre-processing appications maycause some undesired variations/noises which can be reduced, while desired features get enhanced. The image variation/noise sources f can be attributed to certain conditions, as shown in Table 3.7.

Pre-processing of space images prior to theirclassification and change detection is essential. It includsa grou of sequential operations, including radiometric correction or normalization, image registration, geometric correction, masking and image enhancement (e.g., for clouds, water, irrelevant features) (Coppin and Bauer 1996). In the current project different functions of ERDAS IMAGINE (Ver. 8.6) are used for all processing steps.

3.2.1 Radiometric Correction or Normalization

The normalization of satellite imagery considers the integrated, measurable reflectance of the atmosphere, aerosol scattering and absorption, and the earth's surface (Kim and Elman 1990). It is the volatility of the atmosphere, which can introduce variation between the reflectance values or digital numbers (DN's) of satellite images acquired at different times. Although the effects of the atmosphere upon remotely sensed data are not considered errors, since they are part of the signal received by the sensing device (Bernstein et al. 1983), consideration of these effects is important. The objectev suitably stated by Hall et al. (1991), should be that following image preprocessing, all imageries should showas if they

Table 3.6 Codes of different soil phases

	Soil salinity		Calcium carbonates	
Code	Class	EC value (dS/m)	Class	$CaCO_3$ (%)
1	Slightly saline	<8	*Crypto*	0–4
2	Moderately saline	8–16	*Crypto*	0–4
3	Slightly saline	<8	*Hypo*	4–8
4	Slightly saline	<8	*Hyper*	8–12
5	Moderately saline	8–16	*Hypo*	4–8
6	Moderately saline	8–16	*Hyper*	8–12
7	Slightly saline	<8	*Hypo*	4–8
8	Slightly saline	<8	*Hyper*	8–12

Table 3.7 Electronic noise sources and ways of manipulations

Source of variation/ noise	Resulted image fault
Electronic noise	Interfering with image spectral signals, resulting in a low Signal to Noise Ratio (SNR).
Lighting variations	Inconsistent lighting across an image
Equipment defects	Artefact, such as stripes, or pixels stuck off or on.

were acquired from the same sensor. Figure 3.1 shows the steps of satellite image radiometric correction.

3.2.2 Geometric Correction

Geometric rectification of the imagery converts the pixel grid to map projection system or another reference image. This becomes rather important when scene to scene comparisons of individual pixels in applications such as deticting changes (ERDAS 1999).

To confirm the pixel grids and remove any geometric distortions in the satellite imagery, the individual images are registered to the Egyptian Transverse Mercator (ETM) system. Ground Control Points (GCP) collection, first order transformation, and nearest neighborhood geographic coordination of the uncorrected imageryis performed. First order conversion is also known as a linear transformation that applies the standard linear equation ($y = ax + b$) to the X and Y coordinates of the GCP's. The nearest neighbor resembling method uses the value of the closest pixel to assign to the output pixel value and thus transfers original data values without averaging them as other methods do, therefore, the extremes and subtleties of the data values are not lost (ERDAS 1999).

Recorded vectors (road, canals and drains network) were overlapped upon the 2001 un-registered imagery. Image recognizable points were selected and matched to vectors until a semi-regular grid of GCP's covered the entire scene. Those GCP's were then utilised

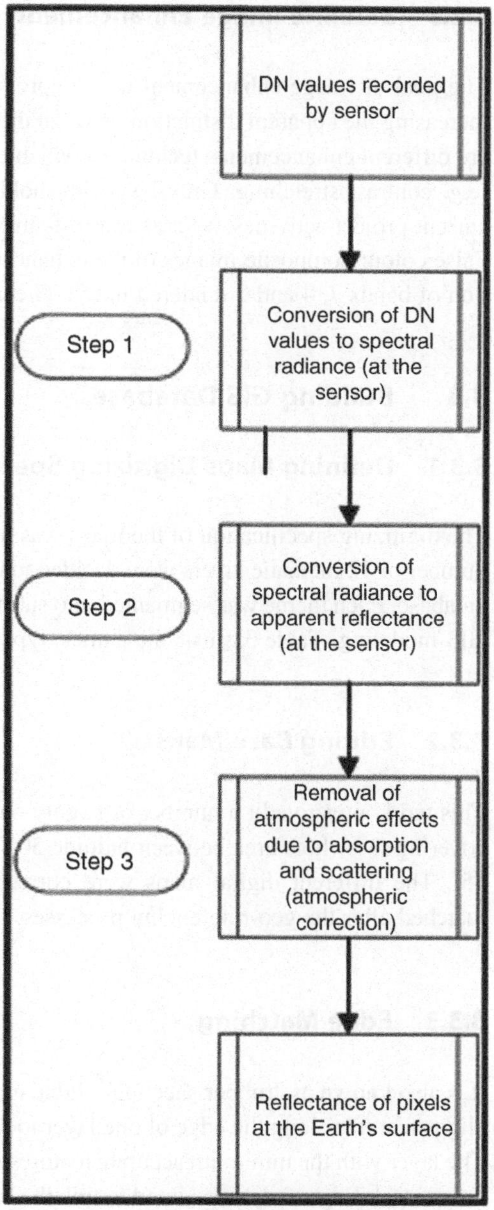

Fig. 3.1 Processes of radiometric correction

to project the uncorrected imagery to an ETM coordinate system. Each GCP was ordered by the residual error it subsidized to the polynomial fit. Points with high error were discarded before registration. Image fit was considered acceptable if the RMS error was <15 m or one-half pixel wide (RMS = 0.5). Overall, RMS errors of less than 0.5 pixels were obtained for each transformation. RMS error is the distance between the input (source) location of a GCP, and the resample location of the same GCP (ERDAS 1999).'

3.2.3 Satellite Image Enhancement

The goal of image enhancement is to improve the visual interpretability of an image by increasing the apparent distinction between the features (Lillesand and Kiefer 1994). There are different enhancements techniques which can be applied on the image for that purpose (e.g. contrast stretching, Gray-level threshold, Level slicing, and spatial stretch). In the current project activities, several contrast-stretching processes were applied to the scenes. False colour composite images of the enhanced scenes were produced using the combination of bands 7, 4 and 2 rendered in red, Green and Blue, respectively.

3.3 Building GIS Database

3.3.1 Defining Maps Digitizing Specifications

The digitizing specification of the maps was determined based on the available themes. A number of 12 thematic layers were decided to be included in the geographic land resources database. Each theme was segmented into sub-themes, which were coded for the purpose of GIS modeling. Table 3.8 lists the names, types, code and description of the sub-themes.

3.3.2 Editing Base Maps

This work aims to edit a number of twenty one topographic digital maps (scale 1:50,000) covering the whole area between latitude 30° 30′ to 31° 15′ and longitude 26° 45′ to 27° 45′. The different digital maps were corrected for different error resourses and edge-matched after the geo-referencing processes.

3.3.3 Edge Matching

It is also known as "rubber sheeting". Edge matching is the spatial adjustment process that aligns features along the edge of one layer to features of an adjoining layer (Tomlin 1990). The layer with the minimum accurate features is adjusted, and the adjoining layer is used as the control. Edge matching aimed to link the base maps (scale 1:50,000) of the Nile Valley and Delta Governorates, in addition to the northwestern coast which were produced by the Egyptian General Survey Authority. The same edge matching technique was performed on all thematic layers included in the 42 soil map heets covering the study regions (scale 1: 100,000). This process resulted in acheving mosaic maps covering the whole study area.

Table 3.8 List and specifications of sub-themes included in different thematic layers

No.	Layer	Layer type	Sub-theme	Code
1	Roads	Lines	Double direction road	101
			Main Paved road	102
			Secondary paved road	103
			Compacted un-paved road	104
			Un-paved road	105
			Passage	106
			Tunnel	107
2	Railways	Lines	Single railway	201
			Double railway	202
3	Irrigation	Lines	Irrigation canals >25 m wide	301
			Irrigation canals 10–25 m wide	302
			Irrigation canals 5–10 m wide	303
			Drain >25 m wide	304
			Drain 10–25 m wide	305
			Drain 5–10 m wide	306
4	Water bodies	Polygons	Nile stream	401
			Main channel (Rayah)	402
			Nile islands	403
			Marsh	404
			Ponds and sabkha	405
			Sea	406
			Fish ponds	407
			Swamp	408
5	Urban	Polygons	Habitation	601
			Architectural sites	606
			Others	999
6	Cultivated	Polygons	Cultivation boundaries	602
			Others	999
7	Cemetery	Polygons	Islamic cemetery	603
			Christian cemetery	604
			Judaism cemetery	605
8	Spot hights	Points	Levels	Altitude ASL
9	Utilities	Lines	Electrical power lines	701
			Gas and petroleum pip lines	702
			Water pip lines	703
			Telephone lines	704
10	Bridges	Lines	Bridge 70 ton	801
			Bridge <70 ton	802
			Wooden Bridge	803
11	Contour	Lines	Contain the contour elevation value	Altitude OSL
12	Others	Polygons	Any barriers cut the contour lines	999

3.3.4 Editing and Refining Mosaic Maps

In order to sustain a effective database, another editing assembly has been taken place after edge- matching of the thematic maps covering the Nile Delta and valley. The influential integrated functionalities (i.e. Topology rules, networks and relations) in ArcGIS system enabled truthful refinement. Editing during this phase was to modify geometry of the features and/or their attributes without disturbing the spatial context of the map neither the previously stored attribute data Bonham—(Carter and Miller 1994).

3.3.5 Input of the Digital Geological Map

The geologic map of the whole study areas and their surroundings was obtainrd from CONOCO, 1989 (Fig. 3.2). The surface geology includes sedimentary rocks belonging to the Tertiary and Quaternary. The cultivable soils in the study area are concluded from the sedimentary rocks, developed mainly by fluvial activity (i.e. Sediments of alluvial fans and flood plains). Aeolian processes have contributed to the soil formation processes in some locations (coastal and inland dunes). The subsurface soil layers of the fans are related to their calcareous origin. Soil properties differ according to their texture, depth, calcium carbonate contents, salt contents and secondary formations

3.3.6 Spatial Adjustment on Basis of LANDSAT Images and Topographic Maps

It was noticed, after edge matching, that there is a kind of deviation (constant in many places in its direction and magnitude) between the produced maps and the well registered topographic maps as well as the satellite images of the study area. It was possible to limk the deviation to two parameters; lack of coordinate system in some original map sheets and the rubber-sheeting accompanied the edge-matching task (Boullie 1978). In order to avoid such falt, another spatial adjustment (transformation) has been carried out. Well registered topographic maps and accurately georeferenced satellite image have been used to elaborate the transformation process. The transformation tools of ArcGIS systems were choosen to effectively performi the spatial adjustment of the soil maps (Burrough 1986 and Tomlin 1990).

3.3.7 Compilation of Laboratory Analysis Results

The field studied and gsthered soil samples, representing the variable soil units of the studied regions have been analyzed in the lab for soil chemical and physical properties. The results of these analyses have been input in database tables and then incorporated into the

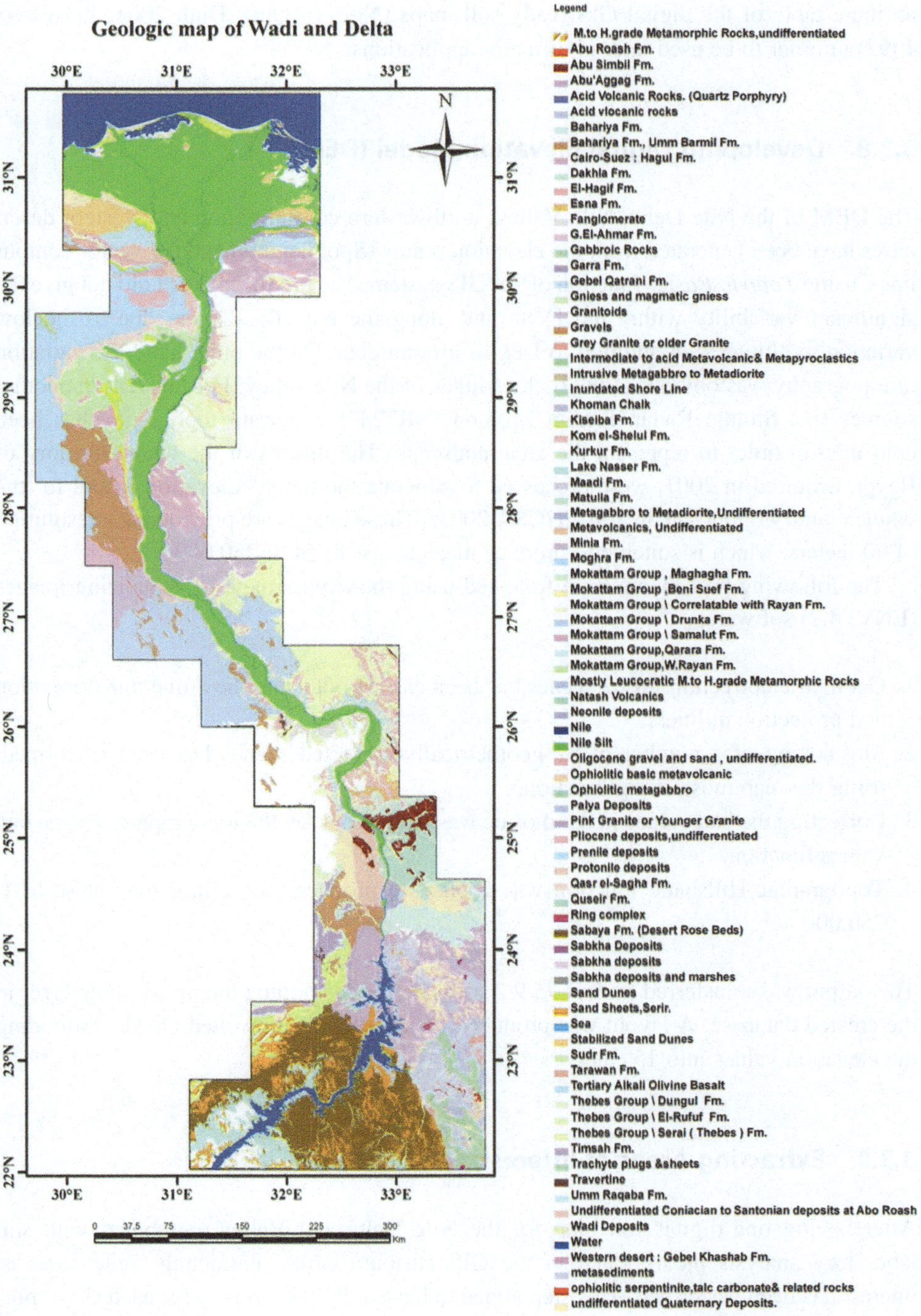

Fig. 3.2 The geologic map of the whole study regions CONOCO, 1989

attribute table of the digital GIS ready soil maps (Nguyen Quec Dinh 2001; Pavasovic 1993) in order to be used later in different applications.

3.3.8 Developing Digital Elevation Model (DEM)

The DEM of the Nile Delta, Nile Valley, northwestern coastal region and western desert oases have been generated from the elevation points (Spot heights) and the vector contour lines, using *Topo to Raster* function of ArcGIS system. The produced DEM did not give the significant variability within the Delta and along the Nile flood plain due to the low variation in altitudes, in addition to lack of ground data. On the other hand the variation in topography was obvious at the rocky fringes of the Nile valley. For this reason another sources (i.e. Shuttle Radar Terrain Mission "SRTM") of terrain topography has been consulted in order to represent the area landscape. The images of the whole territory of Egypt, acquired in 2001, were processed to generate the digital Elevation model for the whole country (Dobos et al. 2002; NCSA 2005). These images are provided at a resolution of 90 meters, which is suitable to produce maps at a scale of 1: 250,000.

The following procedures were followed using the environment of visualizing images (ENVI 4.2) software;

1. Geometric correction for 62 scenes has been carried out using the affine transformation and projection utilities.
2. Mosaicking of a number of 62 geometrically corrected scenes has been performed, using the map mosaicking function.
3. Correcting the bad values in the mosaic was carried out, by the topographic replace bad values function.
4. Topographic-Hillshade function was applied to produce the surface map at scale 1: 250,000.

The output was transferred to *ArcGIS 9.2* software for developing the topographic layer in the created database. A layout was produced for the classified resulted DEM subdividing the elevation values into 15 categories.

3.3.9 Extracting Areas of Interest

After having one digital soil map for the Nile Delta and Valley, combined with soil laboratory analysis incorporated in the GIS attribute tables, extracting some areas of interest has been performed. This step aimed to have a digital soil map for each Governorate in the Nile Delta and Valley. The *clip* function in ArcGIS system has been used to extract the soil map from the soil mosaic using the administrative boundaries of each Governorate.

3.3.10 Layouting

Layouts of the produced maps were done using a different template sizes according to the dimensions of map sheets and included information. An important task that has been considered before the layouting process is the development of a unified and standard map legend for the soil maps. The comprehensive color pallets and renders of ArcGIS systems enabled the rendering of different mapping layers in one map sheet without creating any kind of interference between them.

3.4 Field Observation and Laboratory Soil Analyses

Field missions were carried out to study the different soil-mapping units. A number of 166 soil profiles were studied according to FAO guidelines (1990) and ISRIC (1991). Moreover, a number of 25 soil profiles, included in the "Soil Map of Egypt" project reports were additionally studied. Figure 3.3 shows the geographical distribution of the studied soil profiles, and Annex II demonstrates the field description sheets for each soil profile. The soil laboratory analyses (i.e. soil texture, CaCO3 content, CEC, EC, ESP, pH, soluble cations and anions and organic matter content) were carried out for 422 disturbed soil samples representing the different soil profile layers using the soil survey laboratory methods manual (USDA 2004). Soil color was defined by the Munssel Color Charts (Soil Survey Staff 1975). Annex III reveals the results of soil laboratory analysis.

3.5 Land Capability Classification

Land Capability classes were demarcated according to the rating of soil properties adapted from Sideruis (1989). The impact of each land quality is determined by a set of interacting single or compound land characteristics. Each land quality is assessed qualitatively according to its liability to the concerned constraints as shown in Table 3.9.

The class distinctions are expressed in common numerical terms, thus permitting objective comparison between different land use kinds The assessment of land quality was based on several factors (i.e. drainage conditions, soil texture and structure, % coarse fragments, soil depth, % $CaCO_3$, CEC, soil salinity-EC and exchangeable sodium percentage-ESP). The rating values, assigned to each of the land quality, are demonstrated in Tables 3.10, 3.11, 3.12, 3.13, 3.14.

The land capability maps of different studied regions were produced using Arc—GIS 9.0 software, depending on the capability classes calculated in the established database.

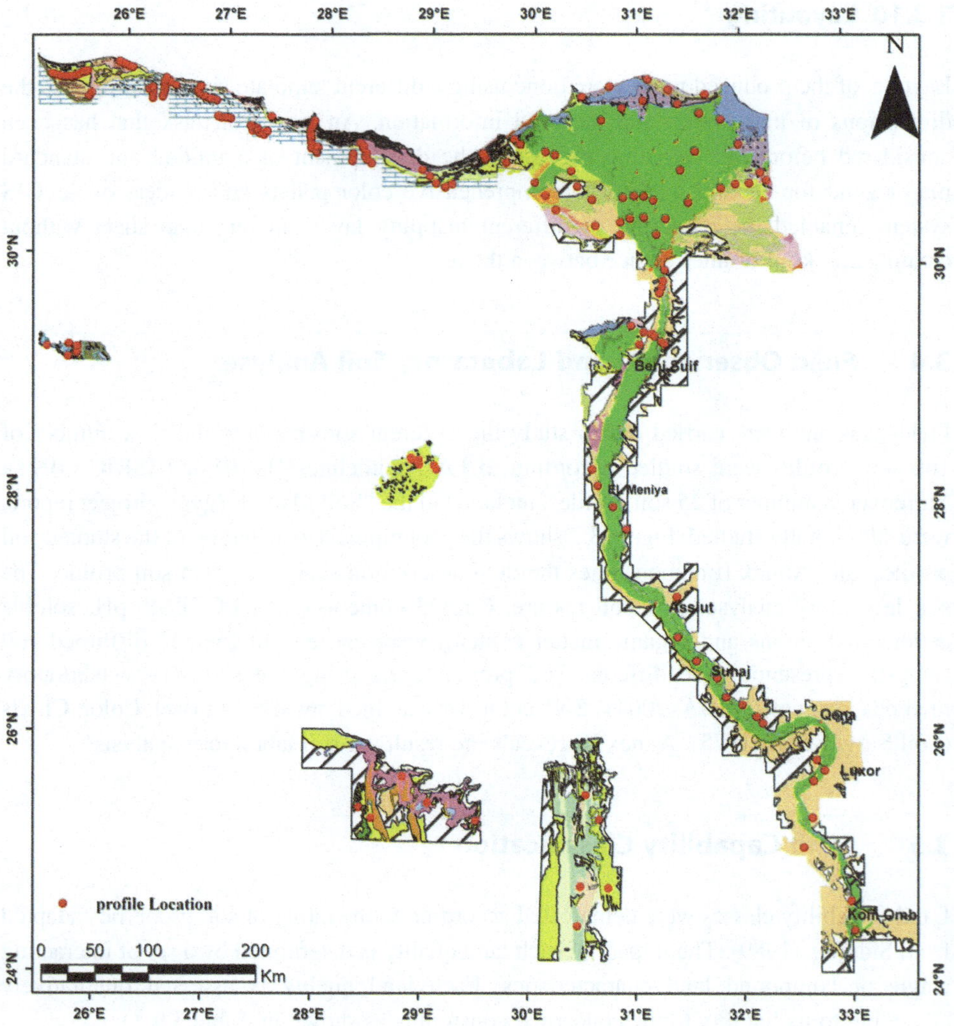

Fig. 3.3 Geographical distribution of the studied soil profiles

Table 3.9 Description of land capability classes

Rating	Capability classes	Description[a]
I	Very highly capable	Very low grade of liability / absence of risk
II	Highly capable	Low grade of liability / low risk
III	Moderately capable	Medium grade of liability / medium risk
IV	Low capable	High grade of liability / high risk
V	Very low capable	Very high grade of liability / very high risk

[a]Adapted to arid condition

Adapted from Sideruis (1989) and FAO (1985)

Table 3.10 Rating drainage conditions land quality

Drainage class	Class	Rating
Excessively, somewhat excessively and well drained	Very low liability	I
Moderately well drained	Low liability	II
Imperfectly drained	Moderate liability	III
Poorly drained	High liability	IV
Very poorly drained	Very high liability	V

Table 3.11 Rating soil consistency and structure for land qualities

	Consistence			
Structure	Dry	Moist	Wet	Rating
Granular	Loose, slightly Hard	Loose, friable	Non-sticky, non plastic	I
Medium blocky	Hard	Friable, firm	Slight sticky, slight plastic	II
Coarse blocky	Very hard	Very firm	Sticky & plastic	III
Platy, massive	Extremely hard	Extremely firm	Very sticky, very plastic	IV
–	Rock	Rock	–	V

Table 3.12 Rating soil depth for land quality

Soil depth in (cm)	Class	Rating
Very deep (>120)	Very low liability	I
Deep (80–120)	Low liability	II
Moderately deep (50–80)	Moderate liability	III
Shallow (25–50)	High liability	IV
Very shallow (<25)	Very high liability	V

Table 3.13 Rating %$CaCO_3$ for land quality

$CaCO_3$%	Class	Rating
(<8)	Very low liability	I
(8–12)	Low liability	II
(12–16)	Moderate liability	III
(16–35)	High liability	IV
(>35)	Very high liability	V

Table 3.14 Rating CEC, EC and ESP land qualities

CEC Meq/100 g. Soil	EC dS/m	ESP	Class	Rating
>24	0–2	0–10	Very low liability	I
16–24	2–4	10–20	Low liability	II
12–16	4–8	20–35	Moderate liability	III
6–12	8–16	35–50	High liability	IV
<6	>16	>50	Very high liability	V

References

ASRT (1982) Soil Map of Egypt-Final Report. Academy of Scientific Research and Technology, Cairo, Egypt

Bernstein RS, Thornton JC, Yang MU, Wang J, Redmond AM, Pierson RN Jr et al (1983) Prediction of the resting metabolic rate in obese patients. Am J Clin Nutr 37(4):595–602

Boullie F (1978) Structuring cartographic data and spatial processes with the hypergraph-based data structure. In: Dutton G (ed) First international advanced study symposium on topological data structures for geographic information systems, vol 5. Laboratory for Computer Graphics and Spatial Analysis, Harvard

Burrough PA (1986) Principals of geographic Information Systems for land resources assessment. Clarendon, Oxford

Carter GA, Miller RL (1994) Early detection of plant stress by digital imaging within narrow stress-sensitive wavebands. Remote Sens Environ 50(3):295–302

Continental Oil Company CONOCO (1989) WebCONOCO, Geologic Map of Egypt, Scale 1.500.000, cited by the following article: TITLE: Relationships between Hydraulic Parameters

Coppin PR, Bauer ME (1996) Digital change detection in forest ecosystems with remote sensing imagery. Remote Sens Rev 13(3–4):207–234

Dobos E, Norman B, Bruee W, Luca M, Chris J, Erika M (2002) "The use of DEM and Satellite images for regional scale soil database" 17th World Congress of Soil Science (WCSS), 14–21 August 2002, Bangkok, Thailand

ERDAS (1999) ERDAS field guide revised and expanded, 5th edn. ERDAS Inc., Atlanta

FAO (1985). "Land evaluation for irrigated agriculture" soils bulletin 55, FAO, Rome

Hall P, Sheather SJ, Jones MC, Marron JS (1991) On optimal data-based bandwidth selection in kernel density estimation. Biometrika 78(2):263–269

ISRIC (1991) "FAO – soil – database" version 2.0, FAO publication, Rome

Kim HH, Elman GC (1990) Normalization of satellite imagery. Int J Remote Sens 11(8):1331–1347

Lillesand TM, Kiefer RW (1994) Remote Sensing and Image Interpretation, xvi+ 750 pp. New York, Chichester, Brisbane, Toronto, Singapore: John Wiley & Sons. Price £ 67.00 (hard covers), £ 19.95 (paperback). ISBN 0 471 30575 8 (pb). Geol Mag 132(2):248–249

NCSA (2005) "Extracting topographic features from shuttle radar topography mission (SRTM) Images" Technical Report, alg. 05–002, July 18, 2005, National Center for Supercomputing Applications (NCSA), 605 East Springfield Avenue, Champaign, IL 61820

Nguyen Quec Dinh (2001) Cave database development, spatial analysis and 3D visualization with GIS – Case study in Son La (Vietnam). VUBrussels, Master dissertation (unpubl)

Pavasovic K (1993) Land suitability models. Workshop on Geographical Information Systems in Integrated Coastal Management, Alexandria, Egypt

Siderius W (1989) "Selected Readings in Land Evaluation Part I Rating of Soils Derived Land Qualities" lecture notes, ITC, Enschade, and the Netherlands

Soil survey staff (1975) "Soil munssel color charts" USDA soil conserve Washington, DC

Tomlin CD (1990) Geographic information systems and cartographic modeling. Prentice Hall, Englewood Cliffs, NJ

USDA (1975) Soil Survey Staff, Soil Taxonomy, A basic system of soil classification for making and interpreting soil surveys, Agricultural Handbook, 436 (1975), Washington DC

USDA (2004) "Soil Survey Laboratory Methods Manual" Soil Survey Investigation Report No. 42 Version 4.0 November 2004

Results and Discussion

4

4.1 Transformation of Soil Maps into the Digital Format

Transformation of the analogue soil maps, driven from the report of soil map of Egypt, 1982, to a digital format was the main source of information for the current Egyptian land resources database (Egy.LRDB). The main thematic layer is the soil map at scale 1: 100,000. The soil database include a detailed description of the mapping units regarding sub-great group, surface and sub-surface soil texture, slope and soil phase (salinity and calcium carbonate content). Moreover, layers of irrigation and drainage channels, road network, geology, settlements, climate and utilities were added to the created digital Egy. LRDB. The attached attribute tables include areas of polygons (e.g. soil units and urban areas), length of line features (e.g. irrigation and drainage channels, roads, railways and utility networks). Both irrigation/drainage channels and roads are subdivided into different classes. Roads are subdivided to main, secondary, un-paved and tracks. Irrigation and drainage channels are subdivided on basis of their width (i.e. >15, 10–15 and <10 meters). Railways are also subdivided into double and single ways.

For the purpose of analyzing the included information, the following paragraphs will deal with the studied areas on basis of map sheets.

Digital GIS database of 42 map sheets was created. Refrence is drawn to the preceded figure (Fig. 4.1) showing the map index. The accomplished Egy.LRDB includes the following 42 map sheets;

1. Manzala lake area (sheet no. 1)
2. Damitta area (sheet no. 1C)
3. Burullus area (sheet no.1B)
4. Rosetta area (sheet no. 1A)

A.-A. Gad, *Digital Land Resources to Support Sustainable Development in Egypt*, Deltas of the World,
https://doi.org/10.1007/978-3-031-37167-7_4

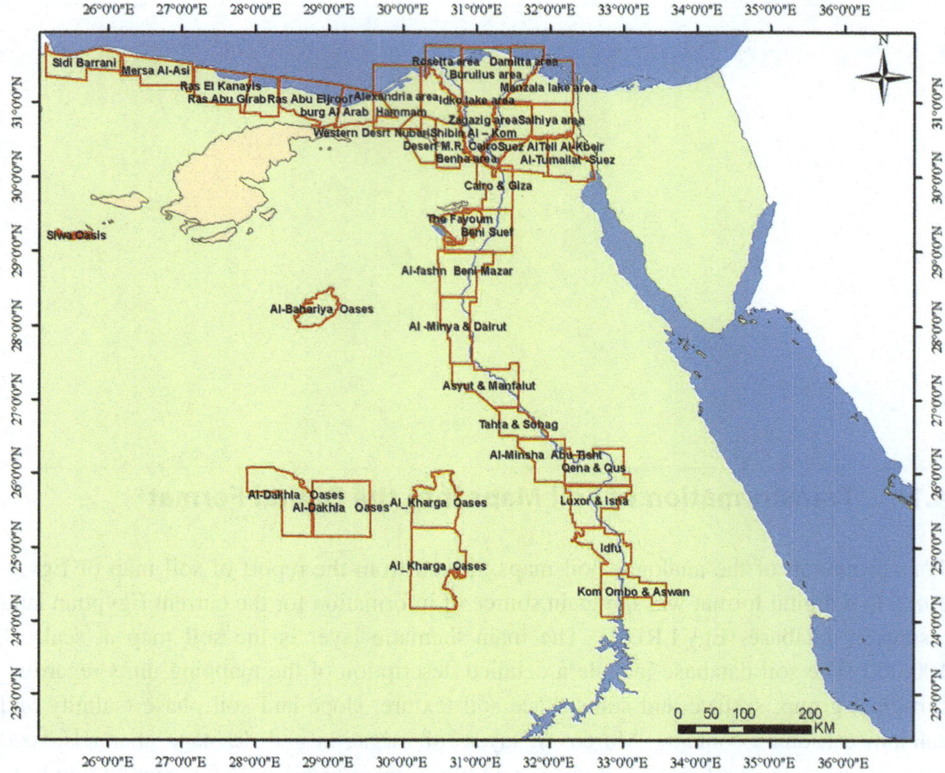

Fig. 4.1 Index of soil map after mosaiking

5. Idko lake area (sheet no. IB)
6. Alexandria area (sheet no. IA)
7. Wadi Al-Natrun area (sheet no. IIA)
8. Desert M.R. Cairo–Alex area (sheet no. IIB)
9. Shibin Al-Kom area (sheet no. 2)
10. Zagazig area (sheet no. 2A)
11. Salhiya area (sheet no. 2B)
12. Benha area (sheet no. 3)
13. 1A Cairo & Giza
14. 1B Beni Suef
15. 2A Al-fashn & Beni Mazar
16. 2B Al-Minya & Dairut
17. 3A Asyut & Manfalut
18. 3B Tahta & Sohag

19. 3C Al-Minsha & Abu Tisht
20. 4A Qena & Qus
21. 4B Luxor & Isna
22. 4C Idfu
23. 4D Kom Ombo & Aswan
24. 1B burg Al-Arab & Al-Hammam
25. 1A Ras Abu Eljroof & Al-Daba
26. 2B Ras Abu Girab & Ras Al-Ghergan
27. 2A Ras Al-Kanayis & Merse Matrouh
28. 3 Mersa Al-Asi & Mersa Gargub
29. 4 Sidi Barrani & Al-Salum
30. B The Western Desert R Alex_Cairo
31. A The Western Desert R Alex_Cairo
32. B Al_Kharga Oases
33. A Al_Kharga Oases
34. B Al-Dakhla Oases
35. A Al-Dakhla Oases
36. Al-Bahariya Oases
37. Siwa Oasis & Berket Zeitun
38. Eastern Desert & Atfih
39. Fayoum Depression & Sinnuris
40. Western Desert & Al-Nubaria
41. B Al-Tumailat-Suez
42. A Al-Tumailat-Suez Al-Tell Al-Kabeir

4.1.1 Nile Delta

4.1.1.1 Manzala Lake

Figure 4.2 shows soil map of Manzala lake area, which extends from latitudes 31° 00′ to 31° 40′ N and longitudes 31° 00′ to 32° 20′ E. Figure 4.3 shows the ETM+ image. The main taxonomic units of this area include the following sub-great groups:

- *Typic Torrerts*
- *Aquollic Saliorthids*
- *Vertic Torrifluvents*
- *Typic Torrifluvents*

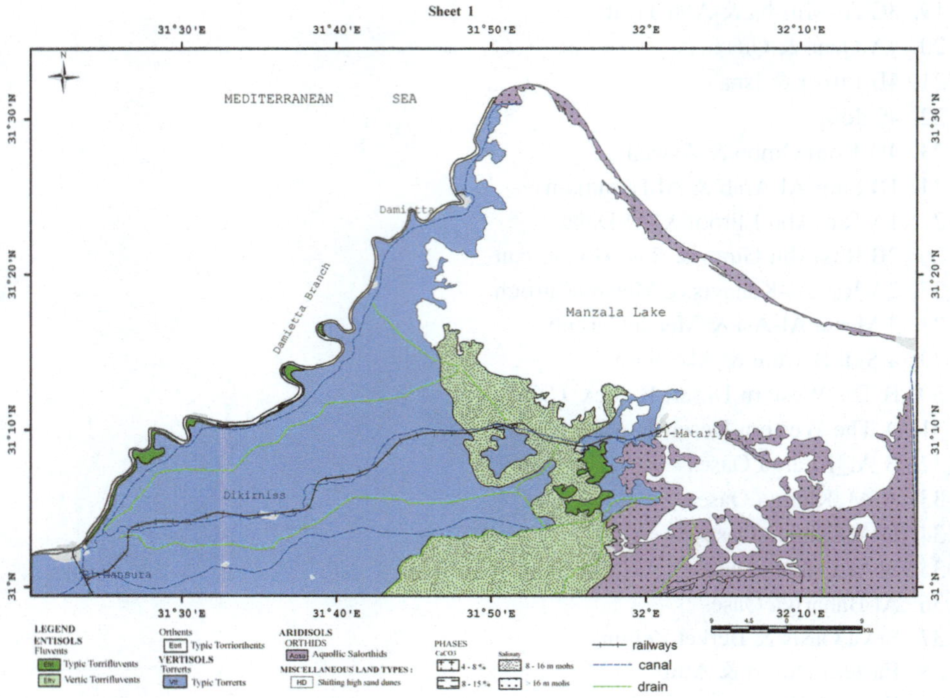

Fig. 4.2 The soil map of Al-Manzala lake

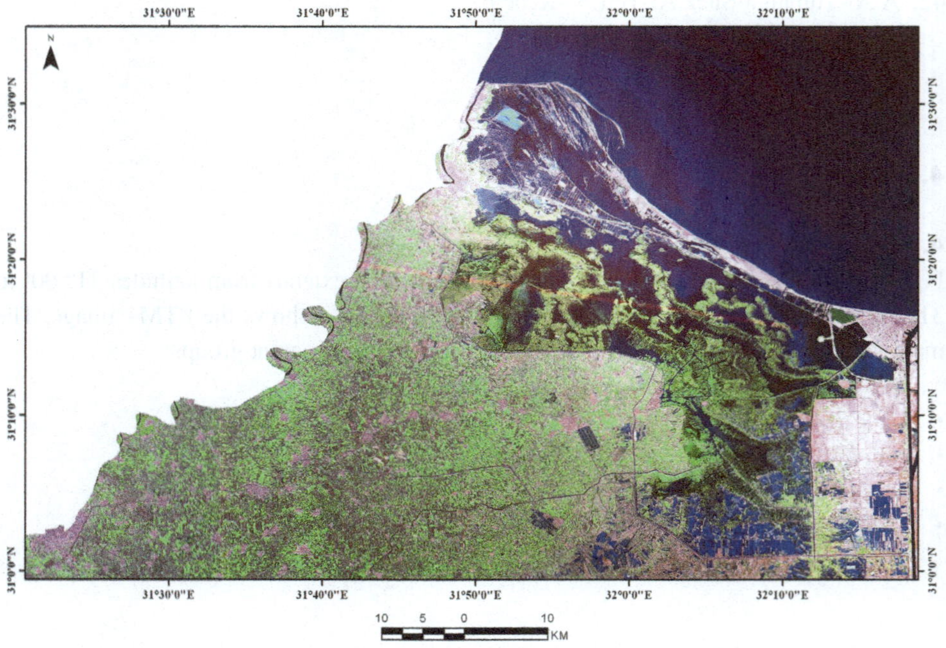

Fig. 4.3 ETM Landsat images of Al-Manzala lake

Table 4.1 Distribution of the soil units in Al-Manzala area

Soil Code	Area (km^2)	Area %
Aosa 1/1 a	254.5	5.9
Aosa 2/3 a	3.7	0.1
Aosa 3/3 a	14.3	0.3
Aosa 3/4 a	5.8	0.1
Aosa 4/4 a	153.7	3.6
Eftt 3/3a	18.1	0.4
Eftt 4/3a	31.9	0.3
Eftv 4/4a	326.8	7.6
Vtt 4/4 a	1112.0	25.9
River Nile	30.0	0.7
Urban	16.0	0.4
Water	2336.2	54.5
total	4284.9	100.0

Table 4.1 shows the areas of different soil units and their percentage. The soils of Typic Torrerts sub-great group (Vtt) represent about 25.9% of the map sheet amounting to 1112.0 km^2. It is characterized by fine soil texture in both surface and sub-surface layers and low salinity level. These soils dominate the area between Damitta branch and Manzala Lake. The Vertic Torrifluvents (Eftv), soils sub-great group exists in southern The Manzala lake covering an area of 326.8 km^2 representing 7.6% of the map sheet area. The texture of these soils is fine in both surface and sub-surface layers, while salinity is medium (8–16 dS/m) and $CaCO_3$ is low (<4%). The *Aquollic Saliorthids* (Aosa) sub-great group is mainly existent in the south east of Lake Manzala covering a total area of 431.9 km^2 representing about 10% of the map sheet. These soils cover a wide range of textural classes, while salinity is high (> 16 dS/m). *Typic Torrifluvents* (Eftt) sub–great groups is found in small scattered areas around the lake and the Damietta branch covering an area of 31.9 km^2, representing about 0.7% of map sheet.

The additional infra structure Figs. 4.4, 4.5, 4.6, 4.7, 4.8 and (Table 4.2) include urban areas covering 84.4 km^2, irrigation and drainage channels with a total length of 3325.9 km, a road network of 3608.2 km, length of railway of 81.7 km and utilities (electric power, water pipes and telephone lines) with a total length of 427.6 km.

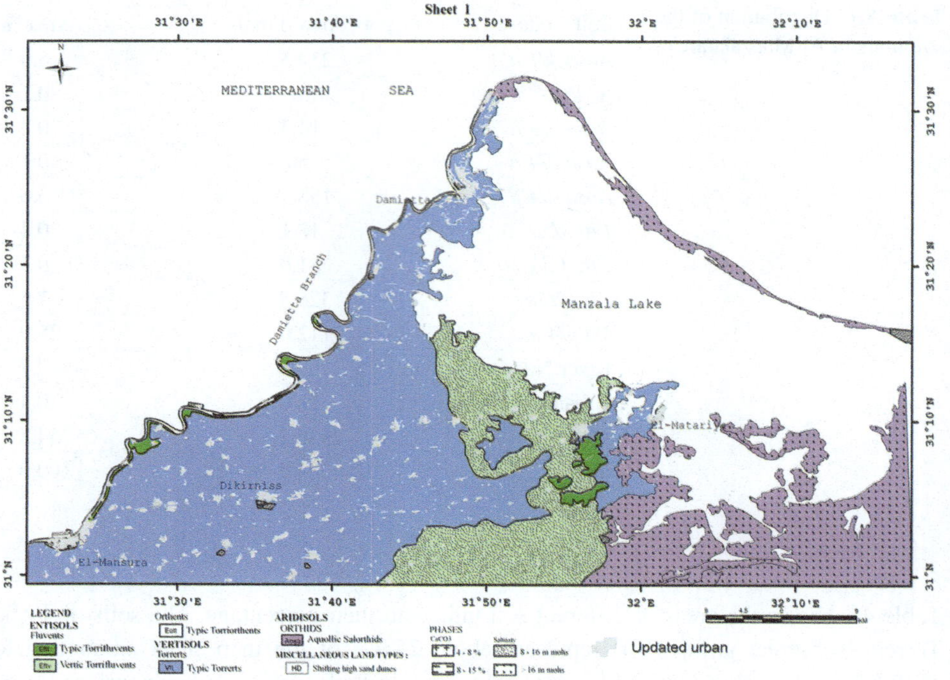

Fig. 4.4 Urban areas superimposed on the soil map, Al-Manzala

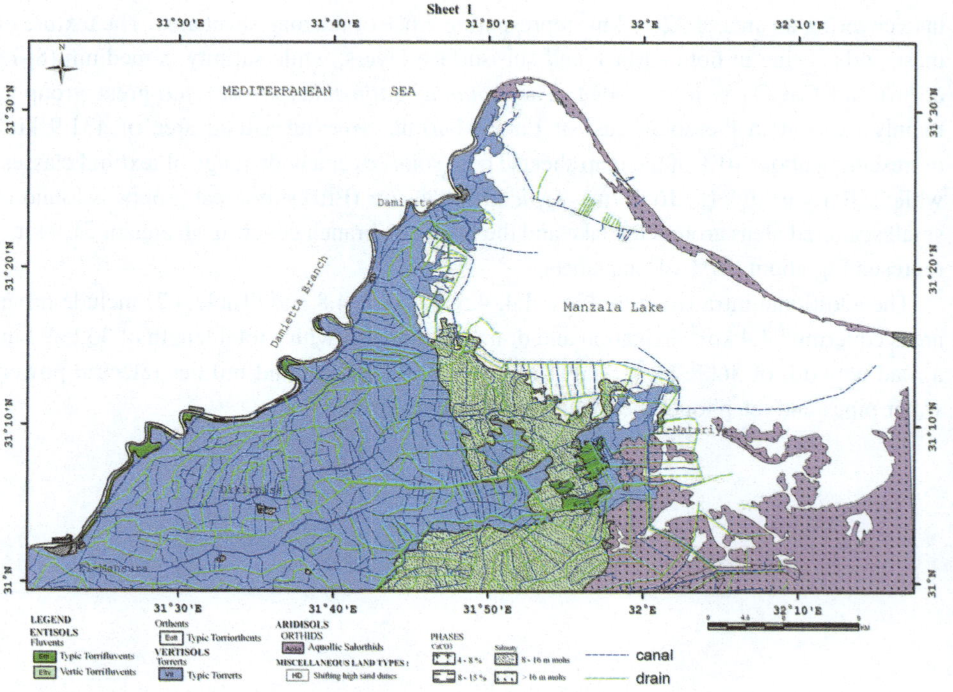

Fig. 4.5 Irrigation and drainage canals superimposed on the soil map, Al-Manzala

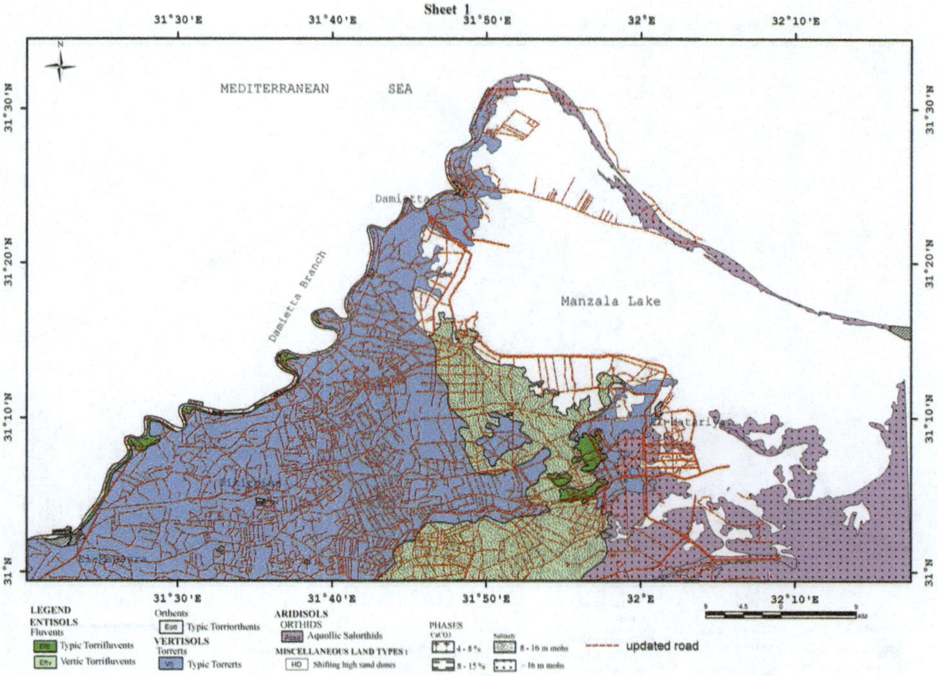

Fig. 4.6 Roads network superimposed on the soil map, Al-Manzala

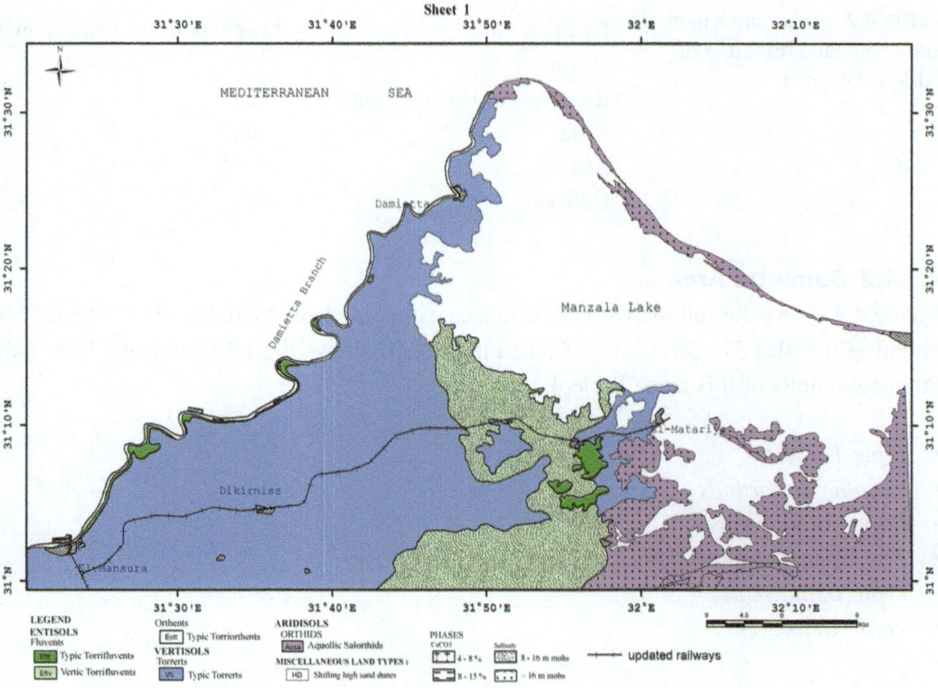

Fig. 4.7 Railways superimposed on the soil map, Al-Manzala

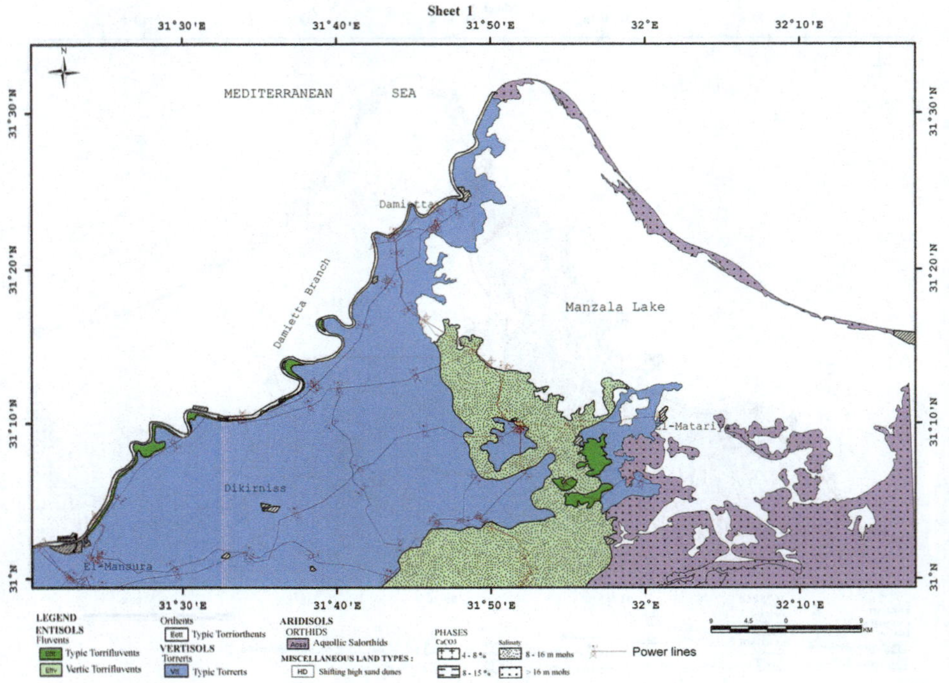

Fig. 4.8 Utilities superimposed on the soil map, Al-Manzala

Table 4.2 Areas and length of additional land resources of Al-Manzala area

Layer	Length (km)	Area (km^2)
Urban		84.4
Irrigation & Drainage canals	3325.9	
Roads	3608.2	
Railway	81.7	
Utilities	427.6	

4.1.1.2 Damietta Area

Figure 4.9 shows the soil map of Damietta area extending from latitudes 31° 10′ to 31° 50′ N and longitudes 31° 30′ to 32° 00′ E. Figure 4.10 shows the ETM+ image. The main taxonomic units of this area (Table 4.3) are:

- *Typic Torrerts*
- *Aquollic Saliorthids*
- *Typic Torrifluvents*
- *Typic Quartizpsamments*
- *Typic Ustifluvents*
- *Vertic Ustifluvents*

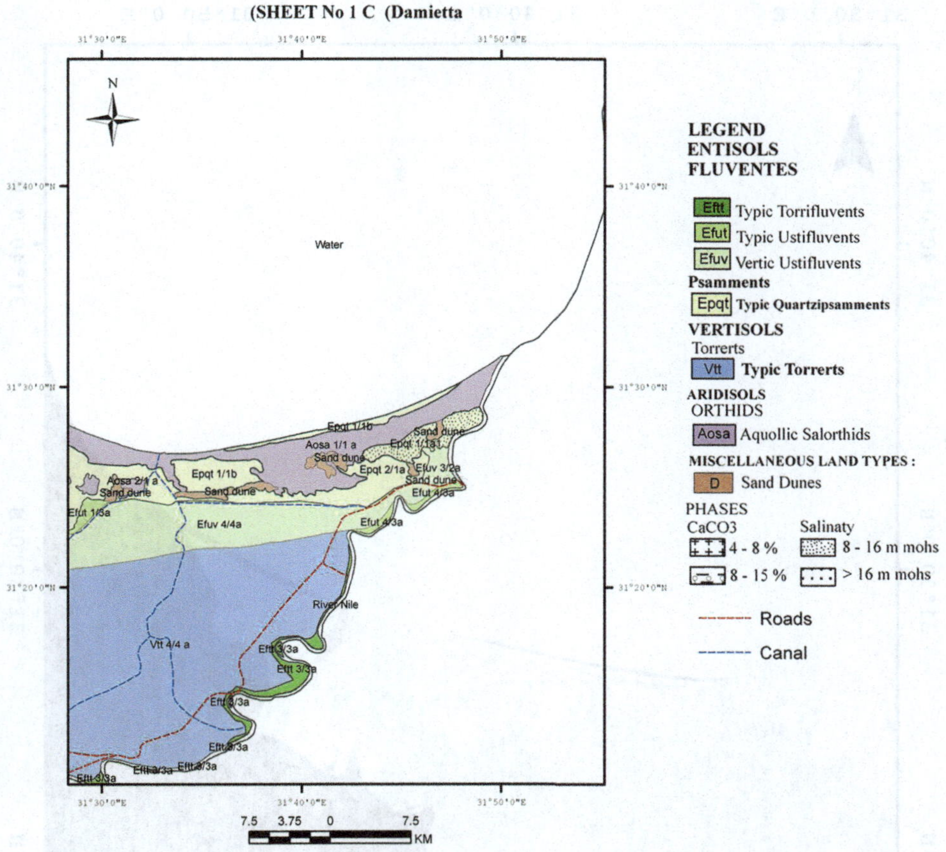

Fig. 4.9 The soil map of Damietta

Also sand dunes are found in limited locations (13.21 km^2) near the Mediterranean coastal zone, these sand dunes are considered as miscellaneous land type. The soils of *Typic Torrerts (Vtt)* dominate the soutern part of this map sheet covering an area of 347.2 km^2, representing 16.6% of the map sheet area. The soils located in the coastal zone area are mainly *Aquollic Saliorthids* (Aosa) having an area of 83.5 km^2, where the *Typic Torrifluvents (Eftt), Typic Ustifluvents (Efut)* and *Vertic Ustifluvents (Efuv)* are located southwards covering areas 14.2, 8 and 113.5 km^2 respectively. The sub-great group *Typic Quartizpsamments (Epqt)* soil is located near the shoreline covering an area of 100.3 km^2. The soil texture of the *Typic Torrerts (Vtt)*, which cover an area of 347.2 km^2, ranges between fine to very fine, while the *Typic Quartizpsamments* (Epqt) is characterized by a coarse texture. The slope gradient in Damietta area is generally less than 2%. The soil salinity is less than 8 dS/m in most of the area except being high in the northern corner part of the map sheet area. Calcium carbonates content is less than 4% of the area.

Figures 4.11, 4.12, 4.13, 4.14, 4.15 and Table 4.4 show the thematic database of Demietta area. The urban coverage attains a total area of 27.3 km^2. The irrigation and

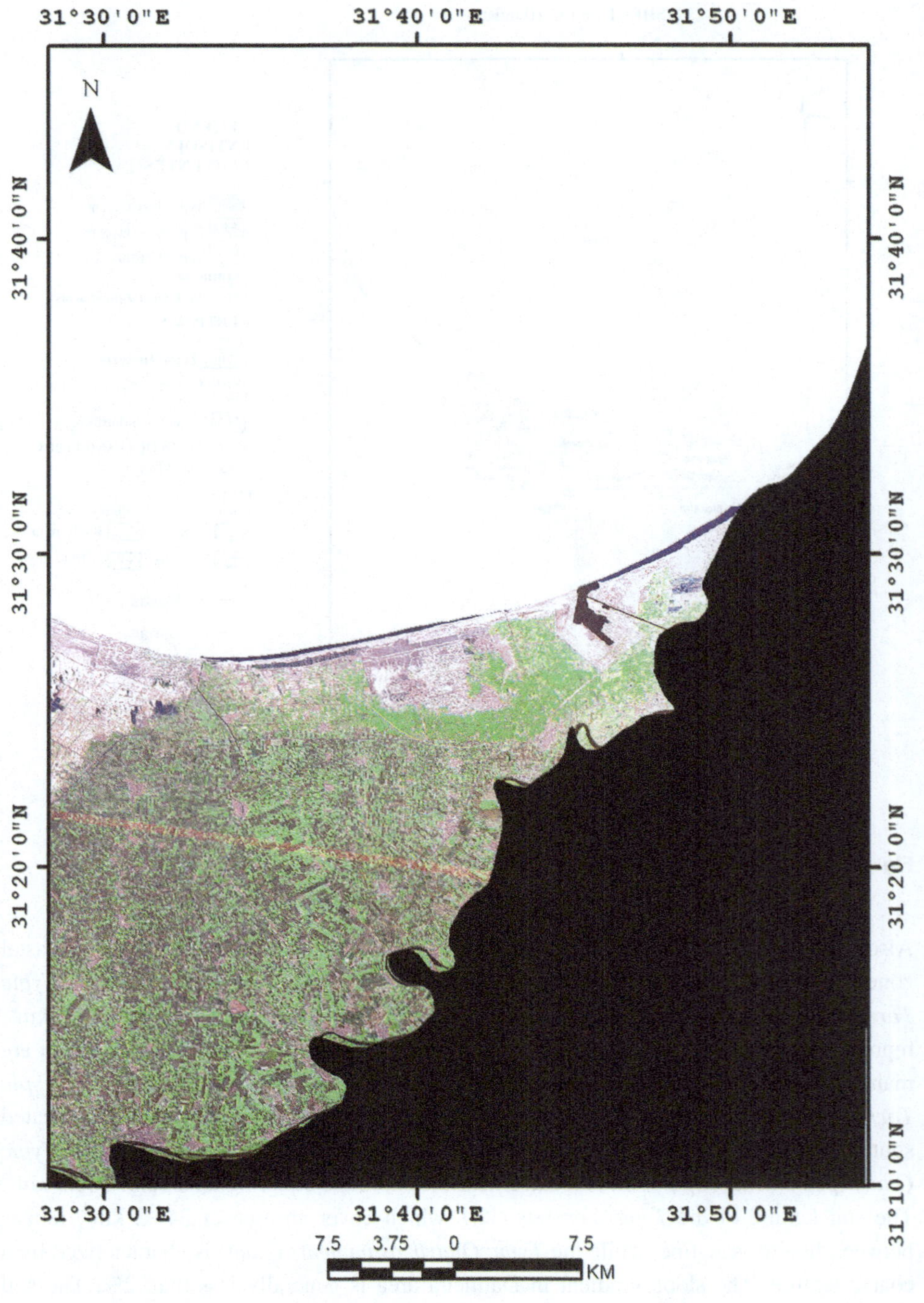

Fig. 4.10 ETM Landsat images of Damietta

Table 4.3 Distribution of the soil units in Damietta area

Soil Code	Area (km^2)	Area %
Aosa 1/1 a	81.7	3.9
Aosa 2/1 a	1.8	0.1
Eftt 3/3a	14.2	0.7
Efut 1/3a	3.9	0.2
Efut 4/3a	4.1	0.2
Efuv 3/2a	6.6	0.3
Efuv 4/4a	106.9	5.1
Epqt 1/1a1	16.5	0.8
Epqt 1/1b	26.1	1.3
Epqt 2/1a	57.7	2.8
Vtt 4/4 a	347.2	16.7
River Nile	23.1	1.1
Sand dune	13.2	0.6
Water	1382.7	66.3
total	2085.7	100.0

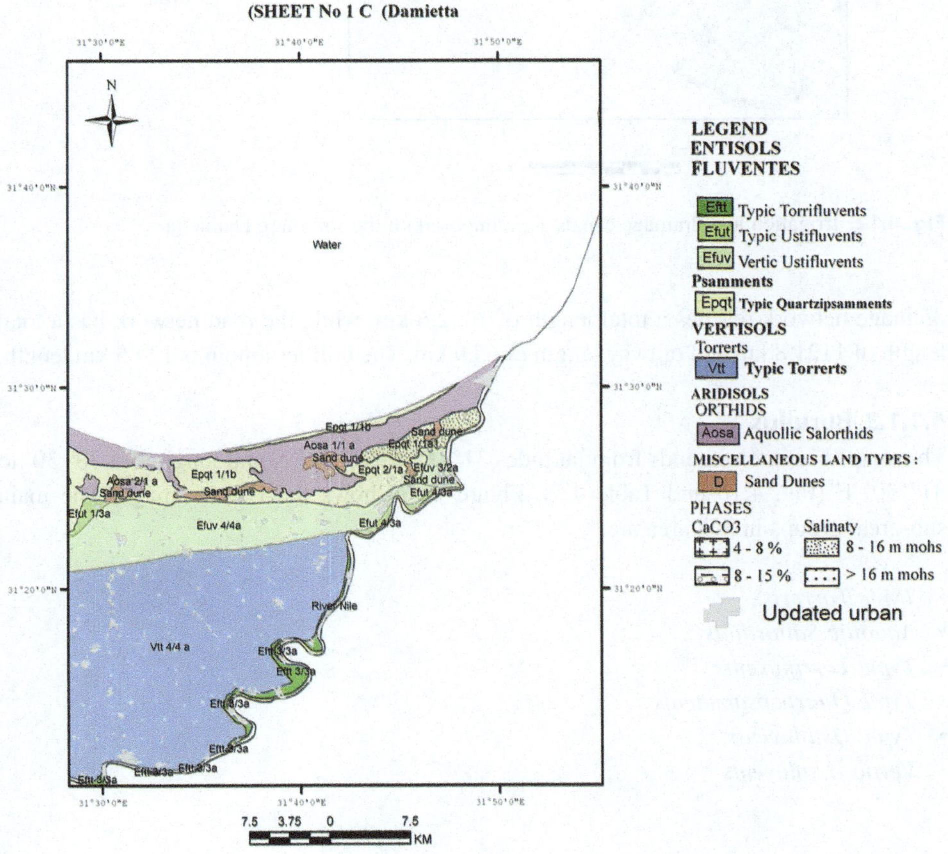

Fig. 4.11 Urban areas superimposed on the soil map, Damietta

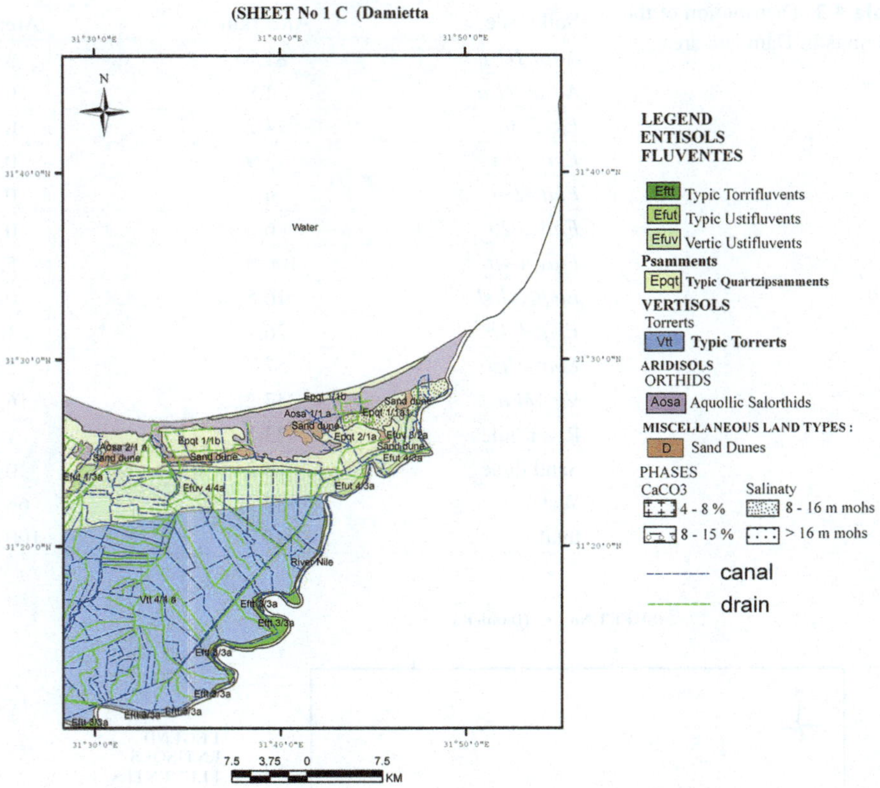

Fig. 4.12 Irrigation and drainage canals superimposed on the soil map, Damietta

drainage network reaches a total length of 1022.6 km, while the road network has a total length of 1121.8 km and railway length of 65.9 km. The utilities amounts 124.5 km length.

4.1.1.3 Burullus

This map sheet area extends from latitudes 31° 00′ to 31° 50′ N and longitudes 30° 30′ to 31° 30′ E (Fig. 4.16 and Table 4.5). Figure 4.17 shows the ETM+ image. The main sub-great groups in this area are:

- *Typic Torrerts*
- *Aquollic Saliorthids*
- *Typic Torrifluvents*
- *Typic Quartizpsamments*
- *Typic Ustifluvents*
- *Vertic Ustifluvents*

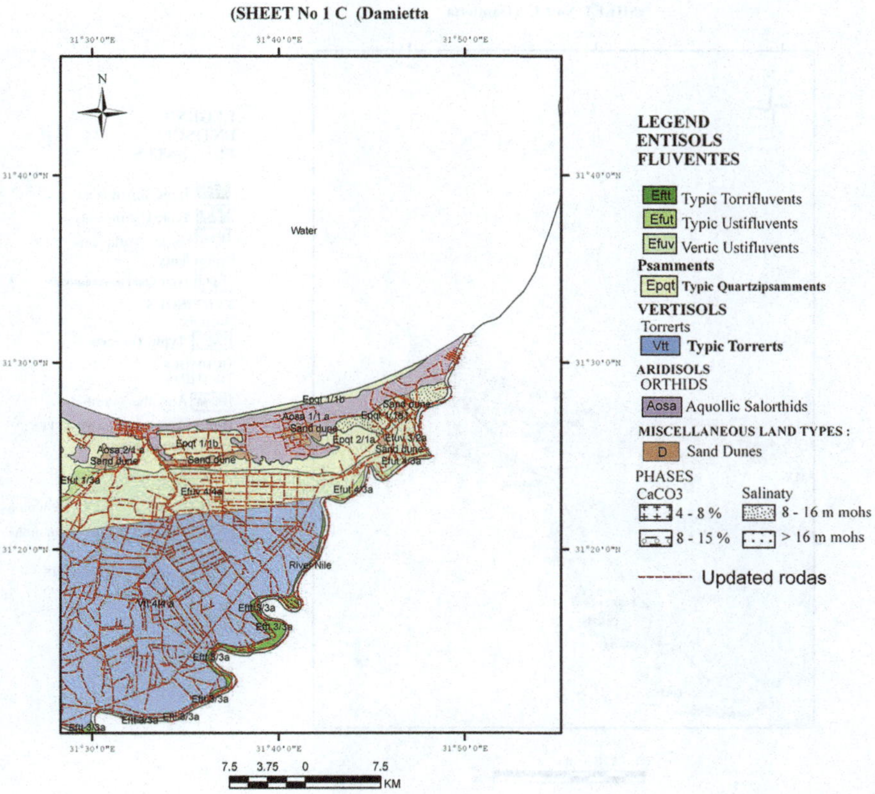

Fig. 4.13 Roads network superimposed on the soil map, Damietta

Also sand dunes are found in a large area, covering 100.9 km^2, near the Mediterranean coastal zone and are considered as miscellaneous land type. The soils of *Typic Torrerts* (Vtt) sub-great group dominate the southern part of this map sheet covering an area of 2100.5 km^2, with the inclusion of *Typic Torrifluvents (Eftt)* sub-great group (43.3 km^2). The soils located near to the Coast of El-Burullus Lake mainly belong to sub-great groups *Aquollic Saliorthids* (Aosa) sub-great group (745.1 km^2), *Typic Ustifluvents (Efut)* sub-great group (70.3 km^2) and *Vertic Ustifluvents (Efuv)* sub-great group (190.1 km^2). These soil units are characterized by a medium to fine soil texture and high salinity level (>16 dS/m), specially in the soils adjacent to the Lake. The *Typic Quartizpsamments* (*Epqt*) sub-great group (194.3 km^2) is found mainly in the north of the map near to the Mediterranean Sea, distributed in small patches. The soil texture ranges between very fine to fine in the soils of *Typic Torrerts (Vtt)* and *Aquollic Saliorthids (Aosa)* sub-great groups, while becoming coarse in the *Typic Quartizpsamments (Epqt)* in both surface and

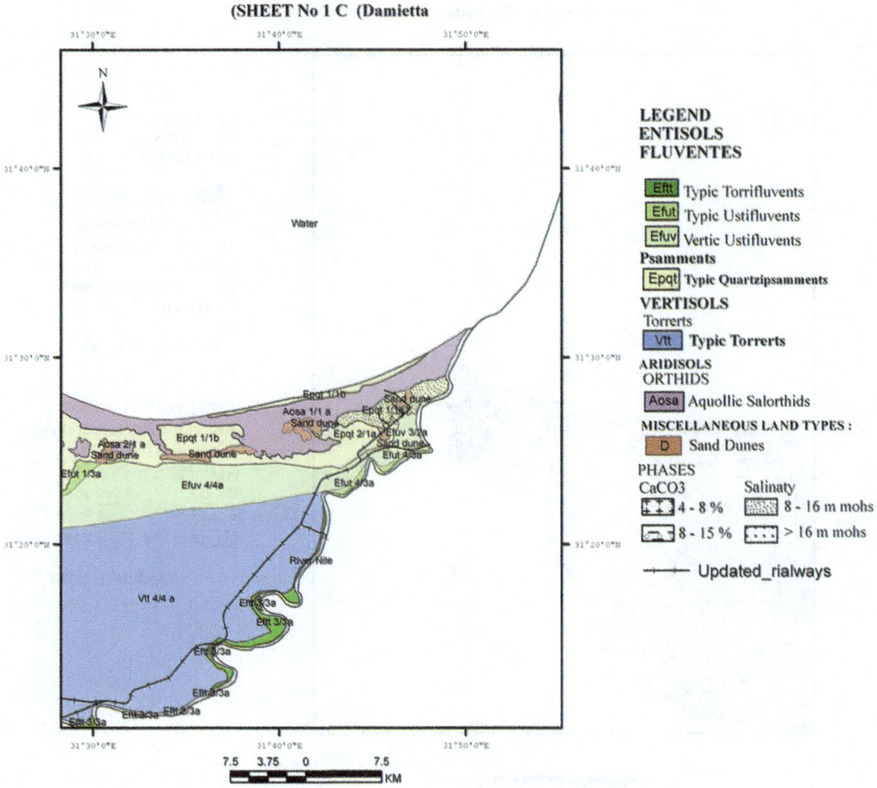

Fig. 4.14 Railways superimposed on the soil map, Damietta

sub-surface layers. The slope gradient is less than 2% and the Calcium carbonates content is less than 4% in different parts of this area. The soil salinity is often less than 8 dS/m, but exhibits the range 8 to 16 dS/m in the *Ustifluvents (Efut)* great group, and partly *Typic Torrerts (Vtt) sub-great group*.

Figures 4.18, 4.19, 4.20, 4.21, 4.22 and Table 4.6 show the area coverages of urban settelements, irrigation and drainage network, roads, railways and utilities. The urban settlements cover an area of 112.6 km^2, irrigation and drainage network attains total length of (5414.7 km), roads networks (5879.7 km), railways (121.6 km) and utilities is (598.5 km).

4.1.1.4 Rosetta

Figure 4.23 and Table 4.7 show soil map and units coverage in the Rosetta area which extends from latitudes 31° 00′ to 31° 45′ N and longitudes 30° 00′ to 30° 50′ E. Figure 4.24 shows the ETM+ image. The main taxonomic units of this area are:

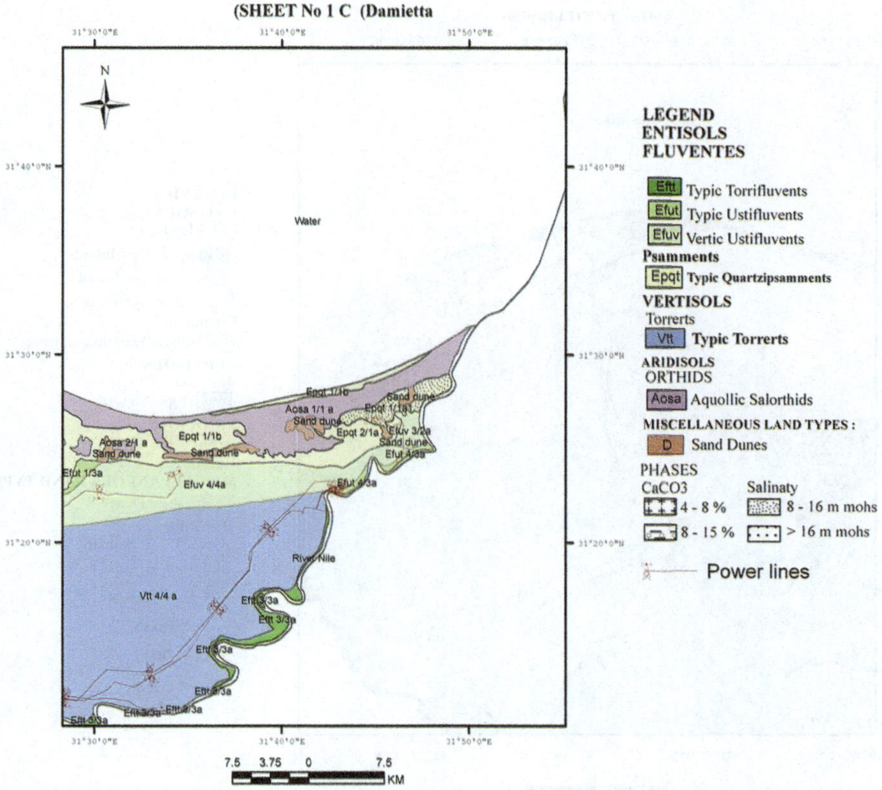

Fig. 4.15 Utilities superimposed on the soil map, Damietta

Table 4.4 Areas and length of additional land resources of Damietta area

Layer	Length (km)	Area (km^2)
Urban		27.3
Irrigation&Drainage	1022.6	
Roads	1121.8	
Railway	65.9	
Utilities	124.5	

- *Typic Torrerts*
- *Aquollic Saliorthids*
- *Typic Torrifluvents*
- *Typic Quartizpsamments*
- *Typic Ustifluvents*
- *Vertic Ustifluvents*

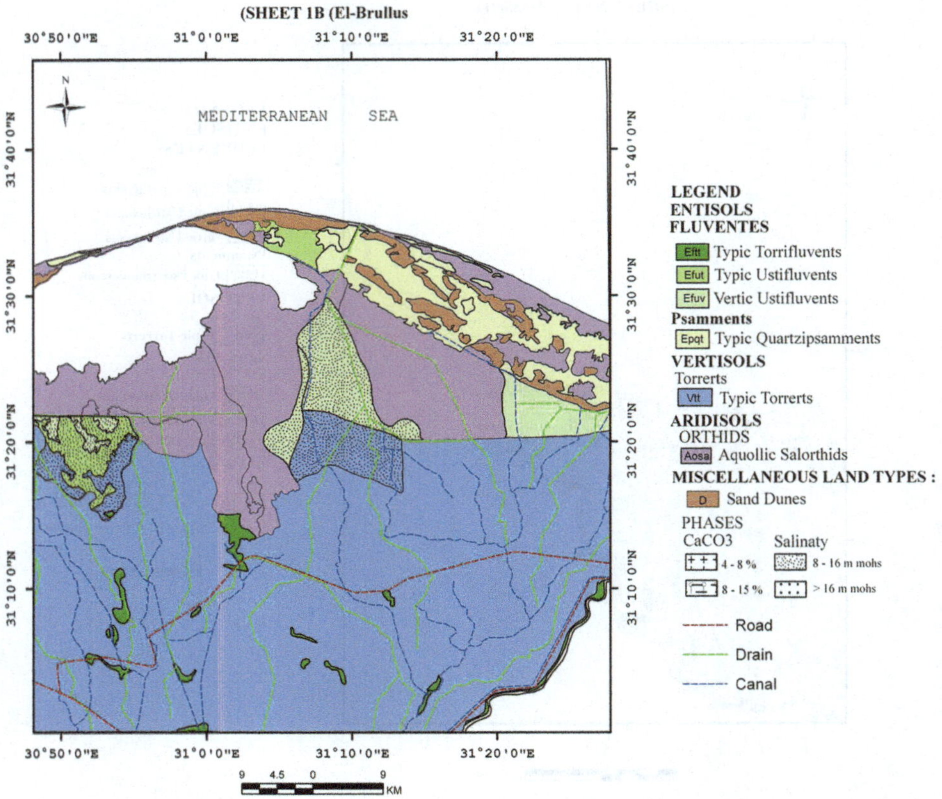

Fig. 4.16 The soil map of Al-Burullus

The soils of *Typic Torrerts (Vtt)* and *Typic Torrifluvents (Eftt)* sub-great groups dominate the middle and southern part of this area, covering areas of 783.3 km^2 and 26.1 km^2 respectively. The soils around Al-Burllus Lake are mainly *Aquollic Saliorthids (Aosa)* covering an area of 239.9 km^2 and is characterized by a coarse to fine soil textural classes in the surface and the sub-surface layers. The sub-great groups of *Typic Torrifluvents (Eftt)*, *Typic Quartizpsamments (Epqt)*, and *Typic Ustifluvents (Efut)* cover scattered areas of 26.1, 9.1 and 3.6 km^2 respectively around the lake and Rosetta branch. The soil texture is fine in both the surface and subsurface layers except for the soils of *Typic Quartizpsamments (Epqt)* which is characterized by medium texture. The slope gradient is less than 2% in different parts of the area, and soil salinity is mostly in the range of 4–8 dS/m, except in a limited coverage of *Vertic Ustifluvents (Efut)*, at the west of the lake, ranges from 8–16 dS/m. Calcium carbonates content ranges between 4–8% in the whole sheet area.

Table 4.5 Distribution of the soil units in Al-Burllus area

Soil Code	Area (km^2)	Area %
Aosa 1/1 a	83.9	1.6
Aosa 2/1 a	21.9	0.4
Aosa 4/3 a	240.3	4.5
Aosa 4/4 a	398.9	7.5
Eftt 3/3a	33.9	0.6
Eftt 4/3a	9.4	0.2
Efut 3/2a	21.5	0.4
Efut 4/3a1	48.8	0.9
Efuv 4/3a1	22.2	0.4
Efuv 4/4a	59.2	1.1
Efuv 4/4a1	108.6	2.1
Epqt 1/1a	14.7	0.3
Epqt 1/1b3	0.8	0.0
Epqt 1/1c	9.1	0.2
Epqt 2/1a	4.6	0.1
Epqt 2/1b	165.1	3.1
River Nile	8.1	0.2
Sand dune	100.9	1.9
Vtt 4/4 a	1982.9	37.5
Vtt 4/4 a1	117.6	2.2
Water	1842.9	34.8
Total	5295.5	100.0

Figures 4.25, 4.26, 4.27, 4.28, 4.29 the created database of this area, Table 4.8 indicates that the total urban settlement area is 30.1 km^2, while the total length of irrigation and drainage canals is 1714.4 km, roads network length is 1869.5 km, railway network is 62.6 km and utilities is 195.9 km length.

4.1.1.5 Idko lake

Figure 4.30 and Table 4.9 show the soil map and the soil units coverage of Idko lake area, which extends from latitudes 30° 40′ to 31° 30′ N and longitudes 30° 10′ to 30° 50′ E. Figure 4.31 shows the ETM+ image. The main taxonomic units of this area include the following sub-great groups:

- *Typic Torrerts (Vtt4/4a)*
- *Aquollic Saliorthids (Aosa1/1a)*
- *Typic Torrifluvents (Eftt)*
- *Typic Calciorthids (Aoct)*

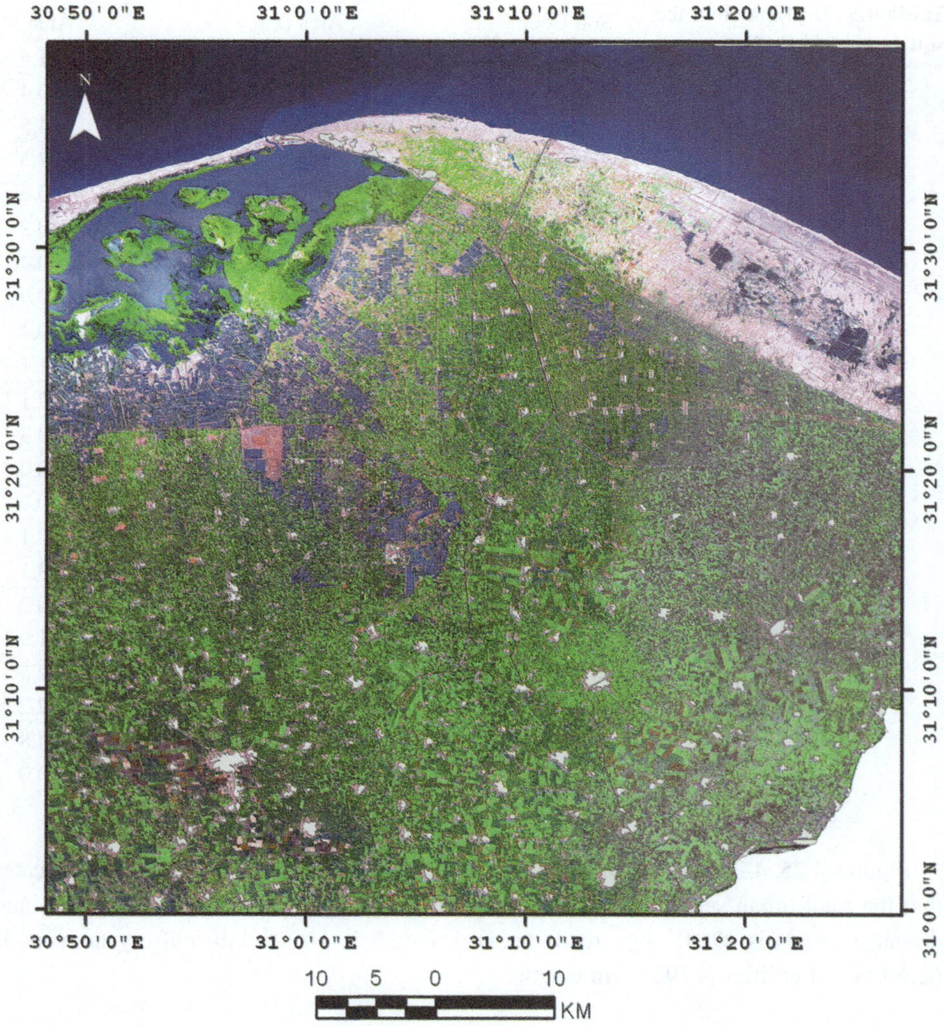

Fig. 4.17 ETM Landsat images of Al-Burullus

- *Calcic Gypsiorthids (Aogc)*
- *Typic torripsamments (Eptt)*

The soils of *Typic Torrerts (Vtt)* sub-great group dominate the map sheet area (1780.5 km^2) as it represents 58.7% of the map area. *Typic Torrifluvents (Eftt)* sub-great group (100.2 km^2) exhibits areas adjacent to Rosetta branch stream. *Vertic Torrifluvents (Efvt)* is occupying the northeastern corner of Idko Lake having an area of 240.1 km^2. The soils of

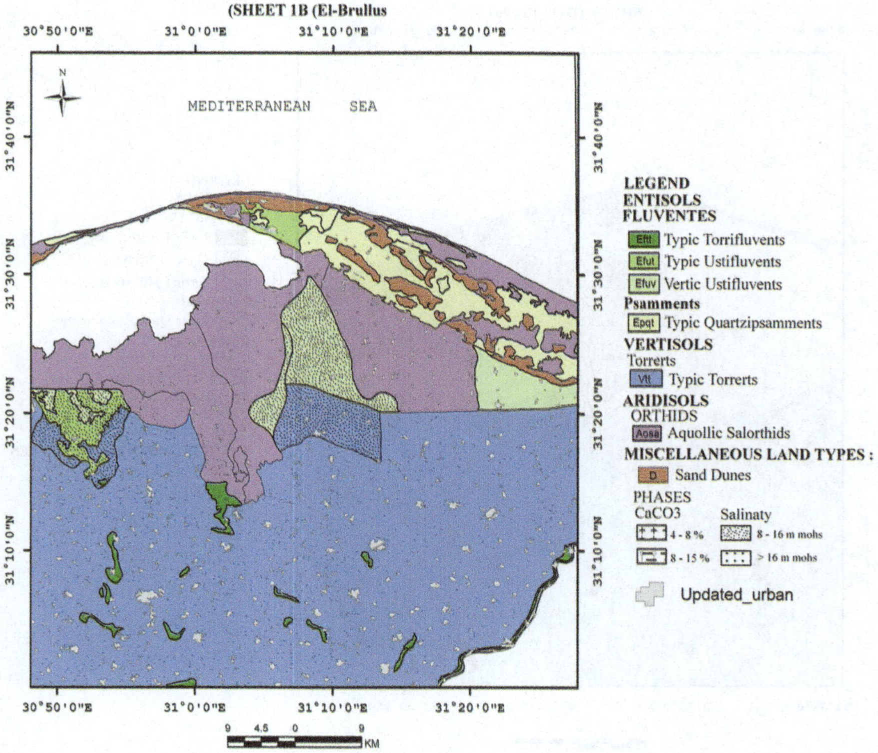

Fig. 4.18 Urban areas superimposed on the soil map, Al-Burullus

Aquollic Saliorthids (Aosa) and *Typic Torripsamments (Eptt)* sub-great groups exist in small area (85.7 and 433.6 km^2 respectively) adjacent to the Lake and Abu-Qir bay. The soil texture, in general, ranges between fine to very fine in both surface and subsurface layers. However the *Typic Torripsamments (Eptt)* are characterized by a coarse texture. The slope gradient is less than 2% in different parts of the area. Soil salinity is less than 8 dS/m in most of the area but reaches 8–16 dS/m in some scattered patches south of the lake. Calcium carbonates content ranges generally between 4–8% in this area.

The created database of this area (Figs. 4.32, 4.33, 4.34, 4.35, 4.36 and Table 4.10) indicates that the total urban settlement area is 85.8 km^2, the total length of irrigation and drainage canals is 4885.6 km, roads network length is 5240.4 km, railways network is 162.7 km and utilities network is 469.4 km length.

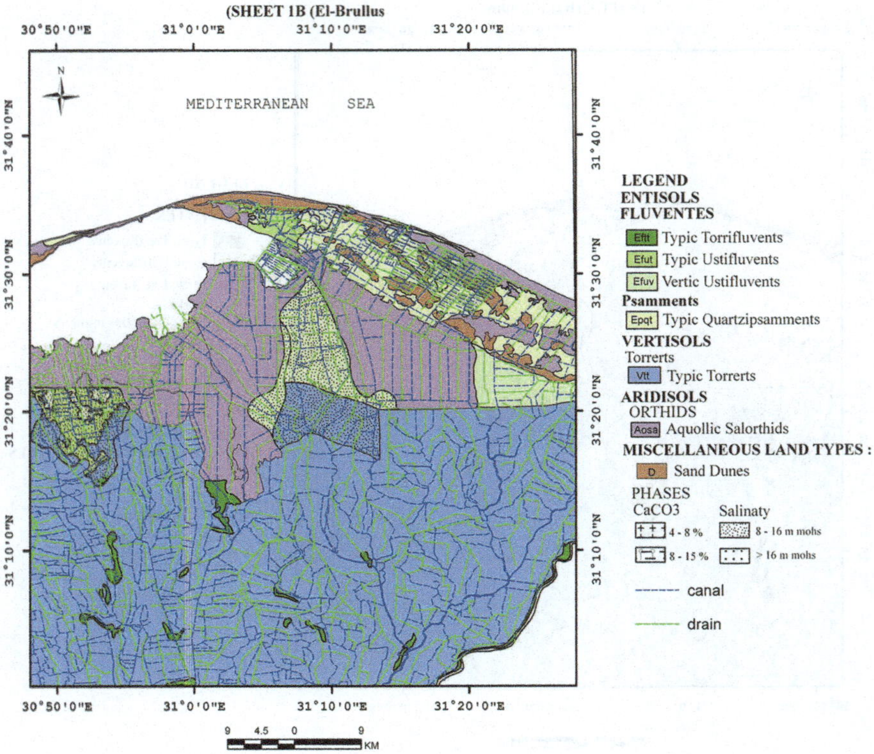

Fig. 4.19 Irrigation and drainage canals superimposed on the soil map, Al-Burullus

4.1.1.6 Alexandria

Figure 4.37 and Table 4.11 show soil map and units coverage of Alexandria area which extends from latitudes of 30° 40′ to 31° 30′ N and longitudes 29° 35′ to 31° 15′ E. Figure 4.38 shows the ETM+ image. The main soil sub-great groups in this area are:

- *Typic Torrerts*
- *Aquollic Saliorthids*
- *Typic Torrifluvents*
- *Vertic Torrifluvents*
- *Typic Quartizpsamments*
- *Typic Torripsamments*
- *Typic Calciorthids*
- *Typic Gypsiorthids*
- *Aquollic Salorthids*

The soils of *Typic Calciorthids (Aoct)*, with inclusion of *Typic Gypsiorthids (Aogc)*, *Typic Torripsamments (Aptt)* and *Typic Quartizpsamments (Epqt)* sub-great groups dominate the

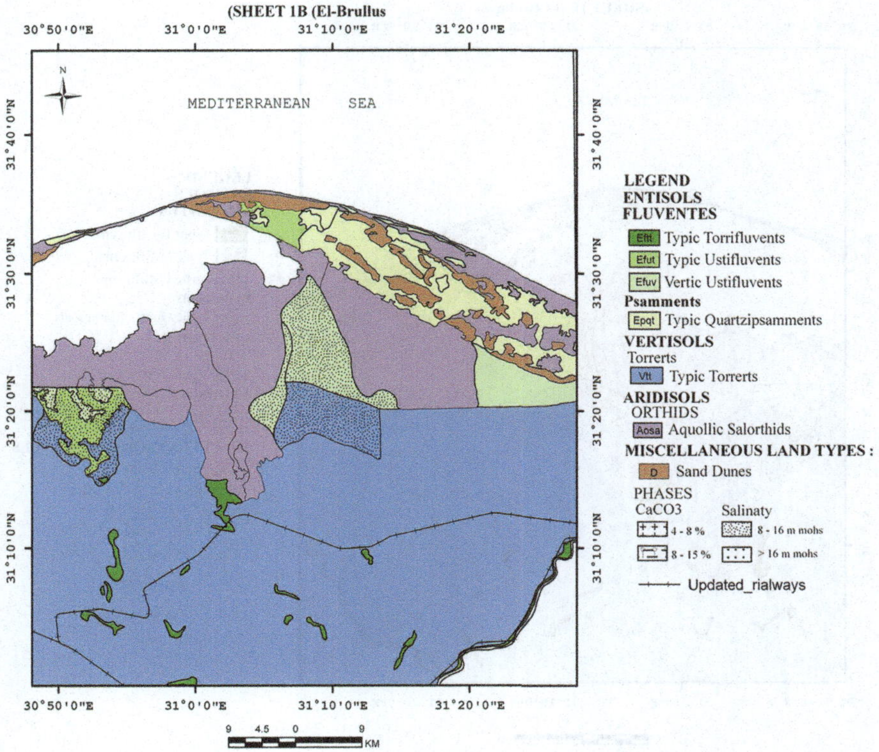

Fig. 4.20 Roads network superimposed on the soil map, Al-Burullus

southern part of this area covering an area of 756.5, 4.1, 25.4 and 92.9 km^2 respectively. The soils close to Idko lake coast are mainly *Aquollic Saliorthids (Aosa)*, covering an area 8.5 km^2. The soils of the area located between Idko and Maryiut Lakes belong to the great sub-groups *Typic Torrivluvents (Eftt)*, *Vertic Torrifluvents (Eftv)* and *Typic Torrerts (Vtt)* covering areas 436.8, 383.8 and 204.9 km^2 respectively. The soil texture ranges between fine to very fine in both the surface and subsurface layers at the middle part of the sheet area. The soil texture is coarse at southern *Petrogypsic Gypsiorthids (Aogp), Typic Calciorthids (Aoct)* and *Typic Gypsiorthids (Aogt)* soils while in the soils of the north and south parts the texture is coarse to very coarse. The slope gradient is nearly level to slopping in different parts of the area. Soil salinity is generally less than 8 dS/m.

The created database of this area Figs. 4.39, 4.40, 4.41, 4.42, 4.43 and Table 4.12 indicates that the total urban settlement area is 118.7 km^2, the total length of irrigation and drainage canals is 4335.2 km, roads network length is 4444.7 km, railway network 197.9 km and utilities of 631.0 km length.

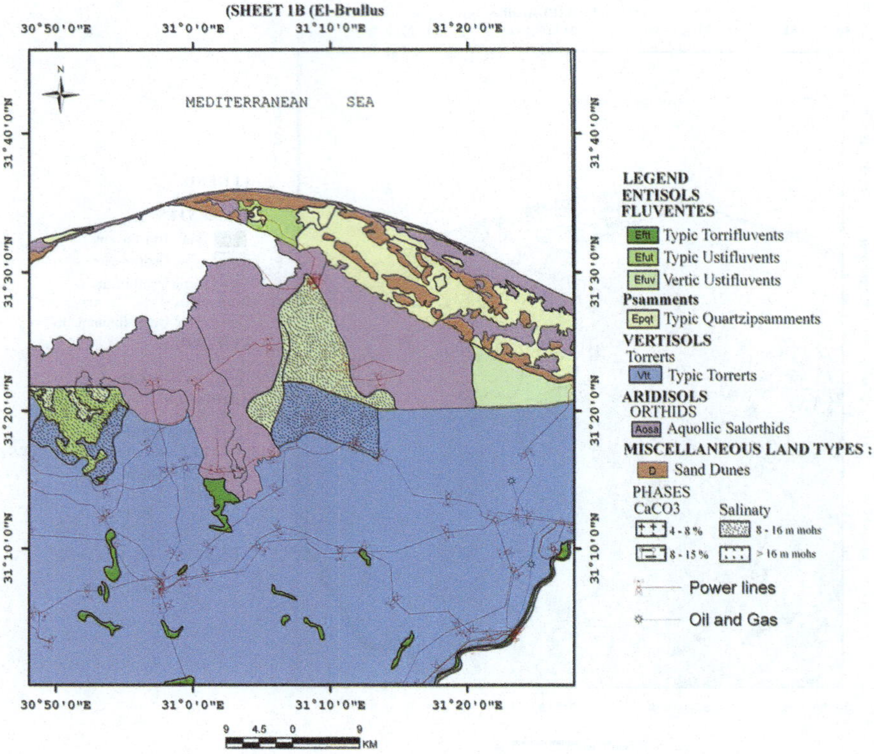

Fig. 4.21 Railways superimposed on the soil map, Al-Burullus

4.1.1.7 Wadi Al-Natrun

Figure 4.44 and Table 4.13 show soil map and units coverage in Wadi Al-Natrun area which extends from latitudes 30° 10′ to 30° 50′ N and longitudes 29° 50′ to 30° 25′ E. Figure 4.45 shows the ETM+ image.The main taxonomic units of this area are the following sub-great groups:

- *Typic Torrerts*
- *Typic Torriorthents*
- *Typic Torrifluvents*
- *Typic Gypsiorthids*
- *Typic Quartizpsamments*
- *Typic Torripsamments*
- *Typic Calciorthids*

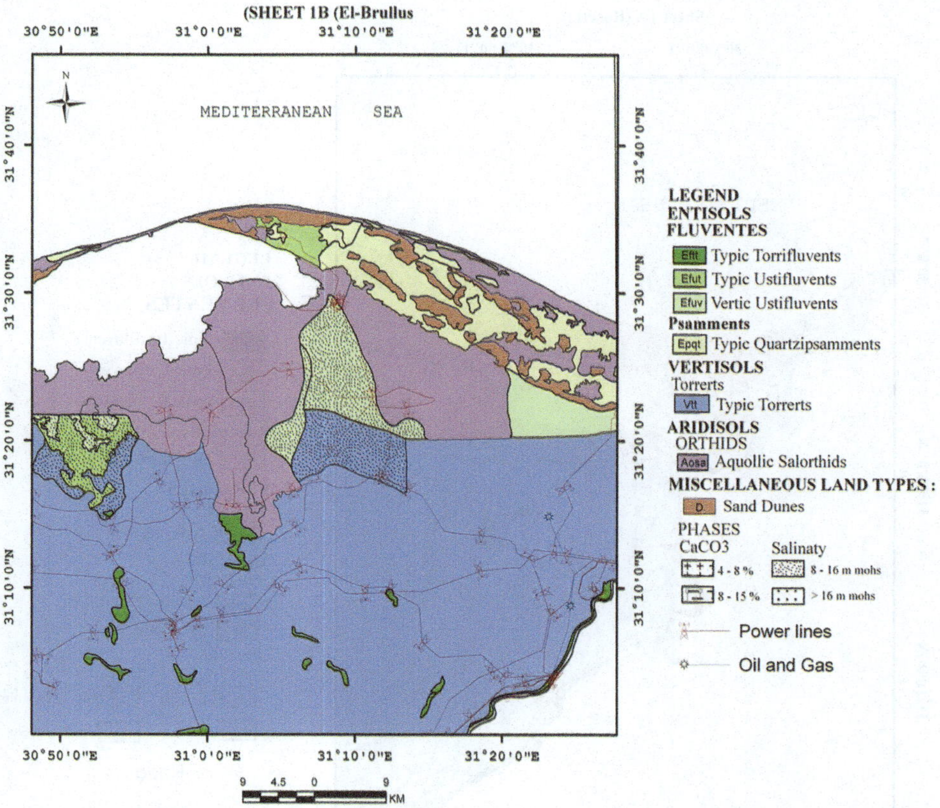

Fig. 4.22 Utilities superimposed on the soil map, Al-Burullus

Table 4.6 Areas and length of additional land resources of Al-Burllus area

Layer	Length (km)	Area (km^2)
Urban		112.6
Irrigation&Drainage	5414.7	
Roads	5879.7	
Railway	121.6	
Utilities	598.5	

- *Typic Torripsamments & Typic Torriorthents*
- *Calcic Gypsiorthids*

The soils of *Typic Quartzipsamments (Epqt)*, dominate the area in the middle and northern parts covering an area of 880.5 km^2, the Typic Torripsamments (Eptt) and Typic Torriorthents (Eott) sub-great groups are found in the northern part of the area, exhibit an area of 754.7 km^2. The sub–great groups of *Typic Calciorthids (Aocat 1/1b)* and *Calci*

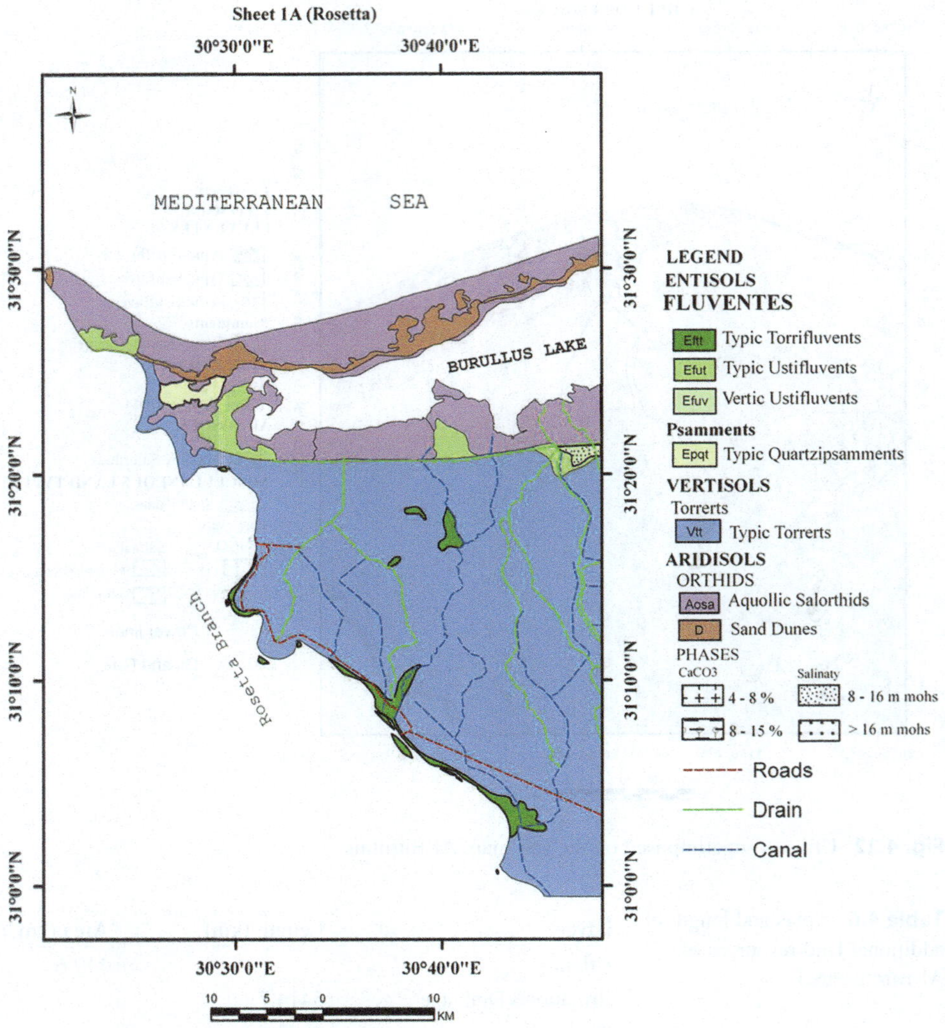

Fig. 4.23 The soil map of Rosetta

Gypsiorthids (Aogt1/1b) represents the soils of Wadi Al-Natrun depression covering an area of 64.5, and 386.3 km^2 respectively. Also the miscellaneous land type of rock land and shifting sand dunes exhibit the southern and North Western parts of the area.

The soil texture in general ranges between medium to a coarse in both the surface and subsurface soil layers, and the slope gradient is classified as nearly level to slopping. Soil salinity is less than 8 dS/m, while calcium carbonates content ranges between 4–16%.

Table 4.7 Distribution of soil units in Rosetta

Soil Code	Area (km^2)	Area %
Aosa 1/1 a	94.3	3.3
Aosa 4/3 a	65.7	2.3
Aosa 4/4 a	79.9	2.8
Eftt 3/3a	26.1	0.9
Efut 4/3a	7.5	0.3
Efut 4/4a	26.3	0.9
Efuv 4/3a	3.6	0.1
Epqt 2/1b	9.1	0.3
Vtt 4/4 a	738.3	25.8
River Nile	40.1	1.4
Sand dune	39.1	1.4
Water	1736.4	60.6
Total	2866.3	100.0

Regarding the thematic layers of urban areas, irrigation and drainage networks, roads, railways and utilities, Figs. 4.46, 4.47, 4.48, 4.49, 4.50 and Table 4.14 shows their area coverage. The urban settlements cover an area of 6.8 km^2, irrigation and drainage network attains total length of 1002.1 km, roads networks 1591.5 km, railways 32.5 km and utilities is 272.9 km.

4.1.1.8 Desert Main Road Cairo–Alex

Figure 4.51 and Table 4.15 show the soil map and its units areas of the Desert M.R. Cairo–Alex., which is located between latitudes 30° 10′ to 30° 50′ N and longitudes 30° 15′ to 31° 00′ E. Figure 4.52 shows the ETM+ image.The main taxonomic units of this area are the following sub-great groups:

- *Typic Torrerts*
- *Typic Torriorthents*
- *Typic Torrifluvents*
- *Typic Gypsiorthids*
- *Typic Quartizpsamments*
- *Typic Torripsamments*
- *Typic Calciorthids*

The northern part of this area is dominated by the *Typic Torripsamments* (*Eptt 1/1a*) covering an area of 51.1 km^2, while the middle part is dominated by the *Typic Torriorthents* soil sub-great groups (*Eott 1/1b*) covering an area of 726.7 km^2. These soils are characterized by a medium to coarse texture classes in the surface and

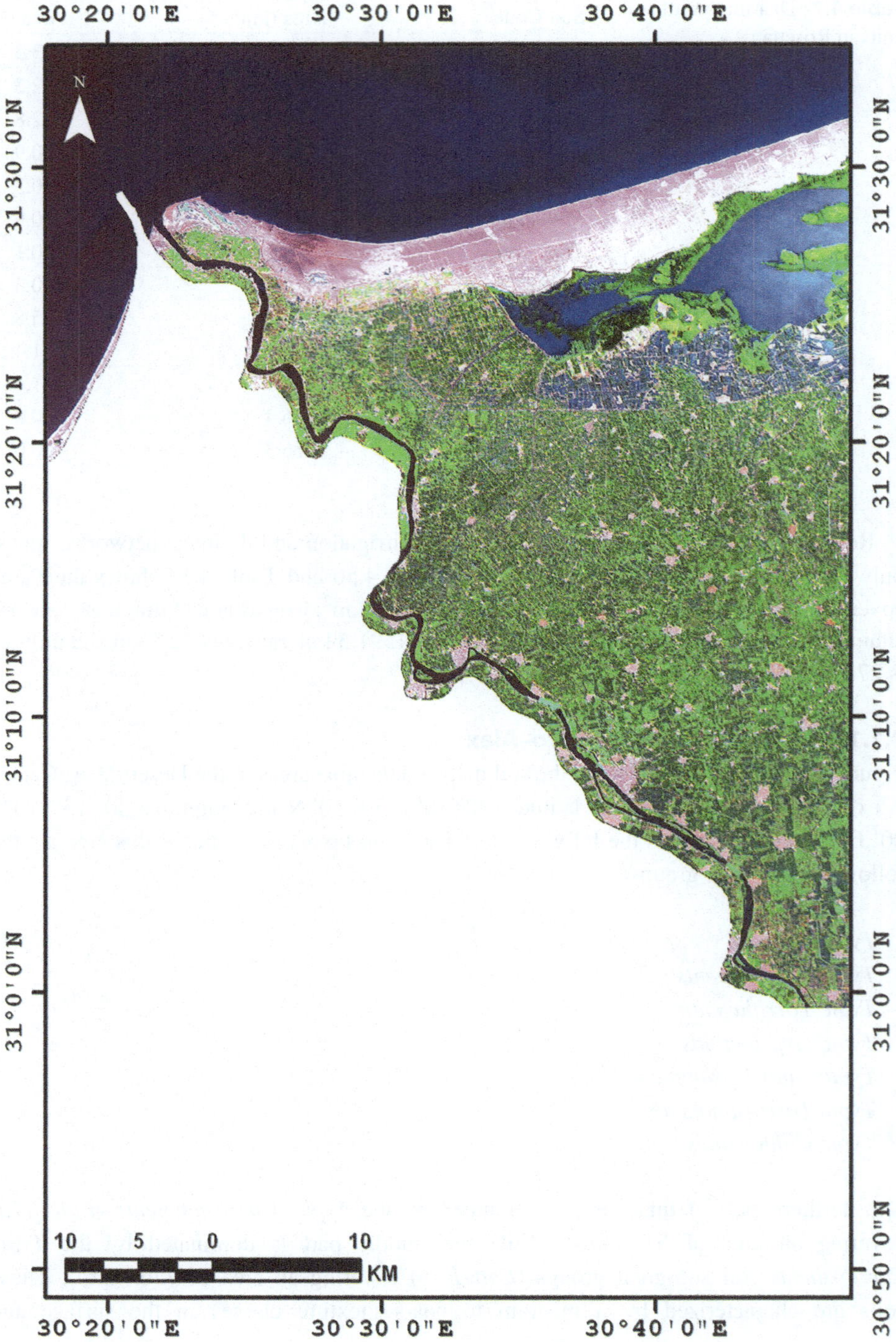

Fig. 4.24 ETM Landsat images of Rosetta

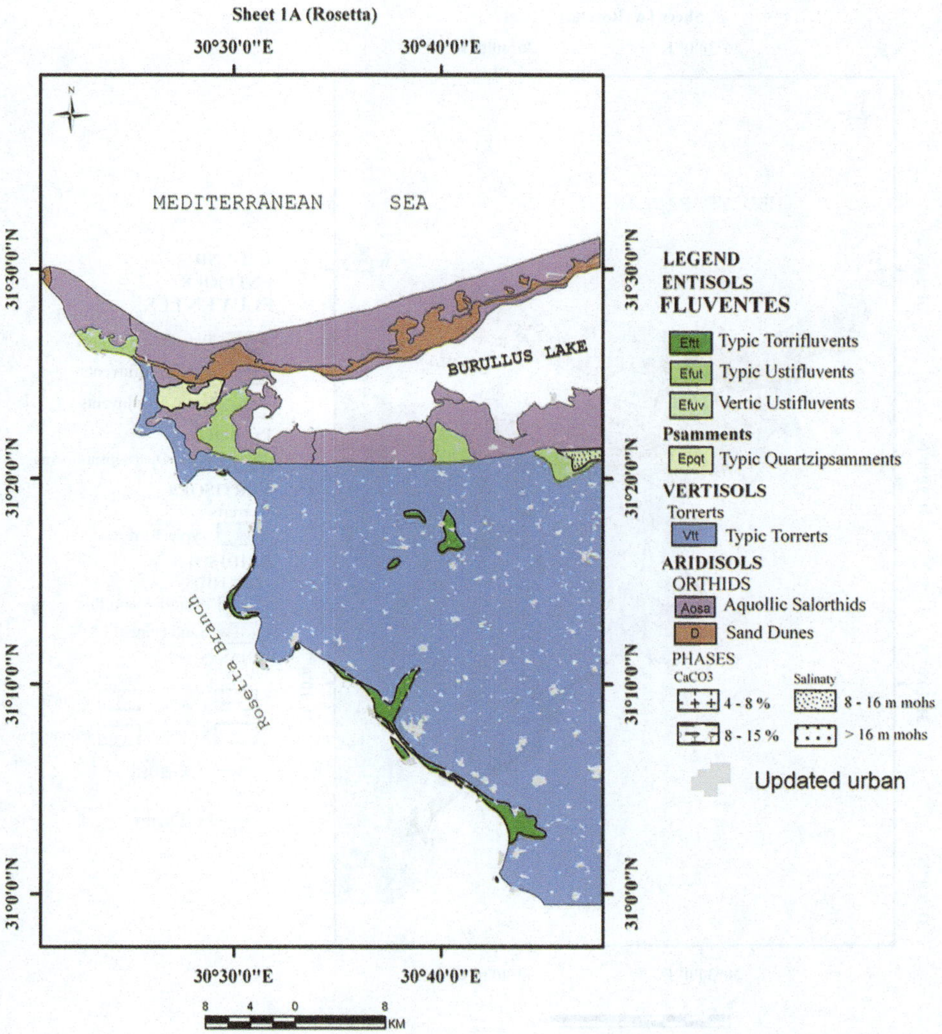

Fig. 4.25 Urban areas superimposed on the soil map, Rosetta

sub-surface layers. The *Typic Quartizpsamments* (soil sub-great group is found mainly in the northern of this area occupying 884.7 km^2, it is also found in small patches at the southern part. The shifting sand dunes are found in the north western part, while miscellaneous rock land exhibits the southern parts. The *Typic Calciorthids* (197.1 km^2) and *Typic Gypsiorthids* (*Aogt 1*/1b), covering an area of 147.9 km^2. They are found in the south west of the area, north of the miscellaneous rock land. The *Typic Torrifluvents* and *Typic*

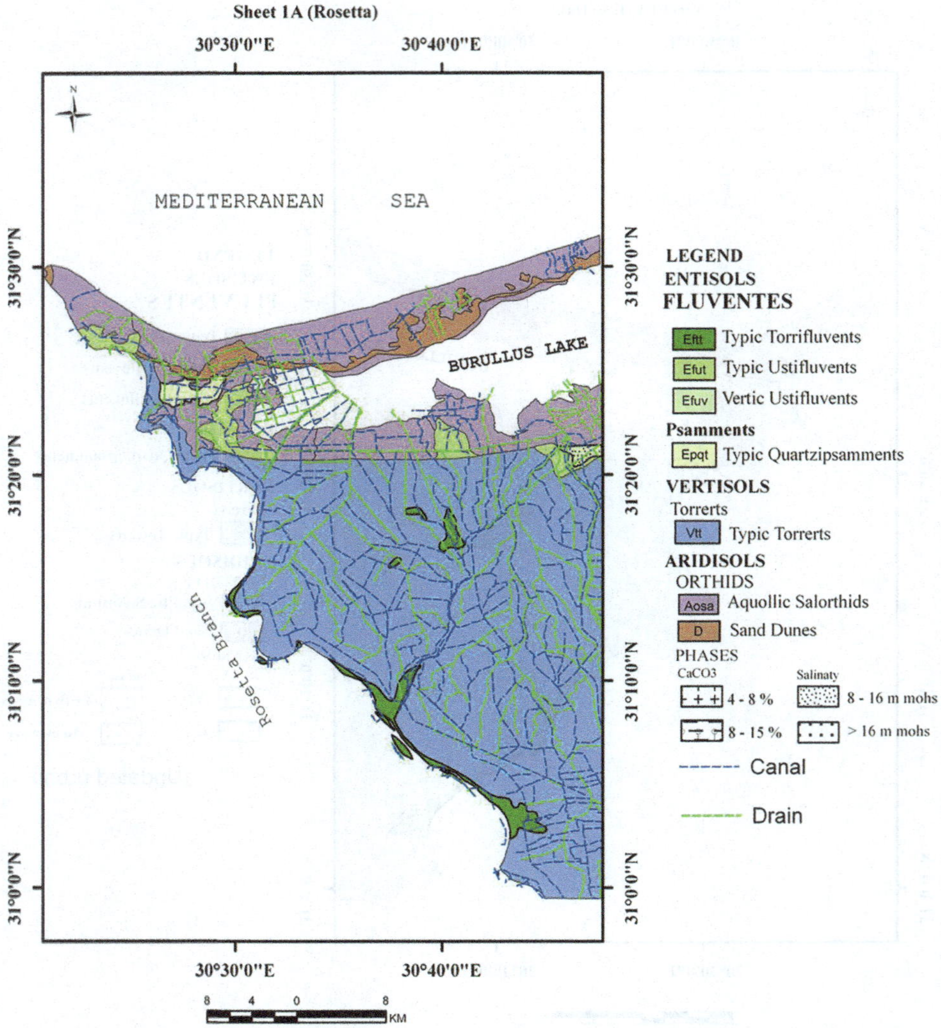

Fig. 4.26 Irrigation and drainage canals superimposed on the soil map, Rosetta

Torrerts (Vtt 4/4a) soils are found along the Rosetta branch with a total area of 109.4 and 36.7 km^2 respectively. The slope gradient is classified as nearly level to slopping in different parts of the area. The soil salinity in general is less than 8 dS/m but it is more than 16 dS/m in the some areas of *Typic Torrerts* soil. The Calcium carbonates content ranges between medium to high.

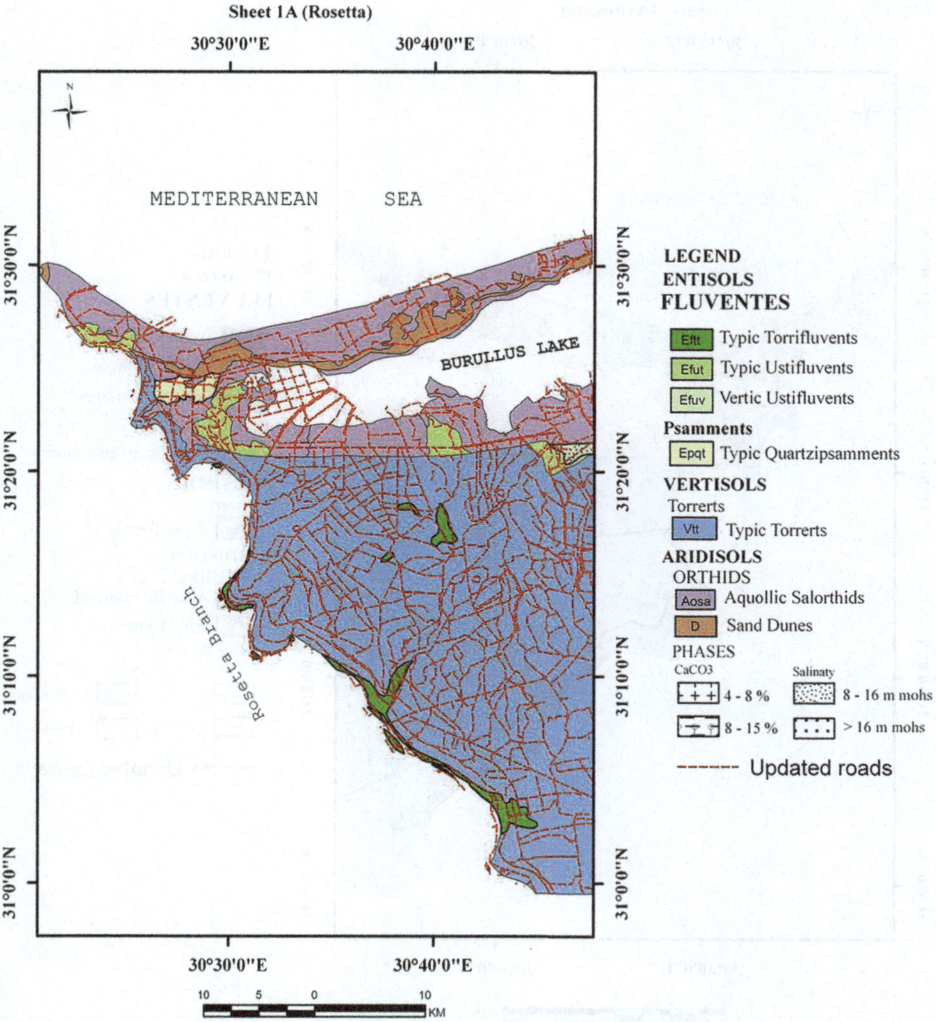

Fig. 4.27 Roads network superimposed on the soil map, Rosetta

Regarding the thematic layers of urban areas, irrigation and drainage networks, roads, railways and utilities, Figs. 4.53, 4.54, 4.55, 4.56, 4.57 and Table 4.16 shows their area coverage. The urban settlements cover an area of 17.9 km^2, irrigation and drainage network attains total length of 667.5 km, roads networks 2746.1 km, railways 105.7 km and utilities is 298.4 km.

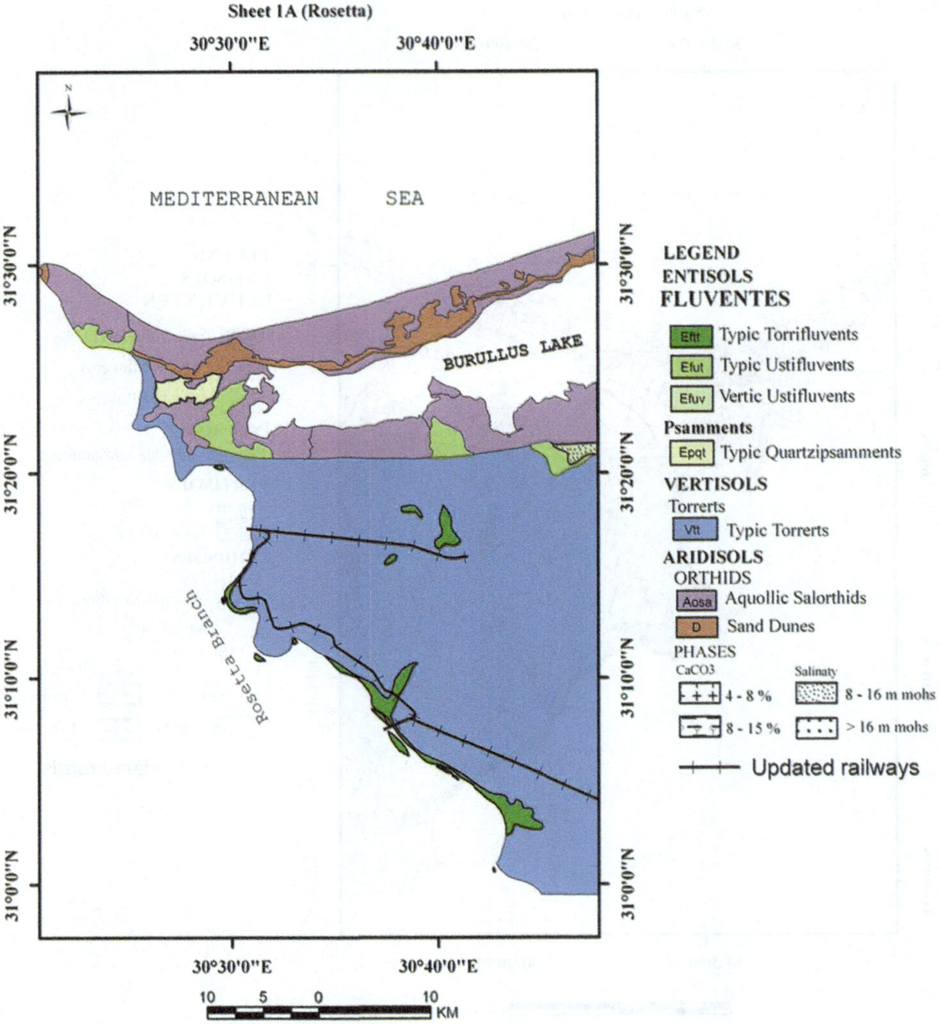

Fig. 4.28 Railways superimposed on the soil map, Rosetta

4.1.1.9 Shibin El–Kom

Figure 4.58 shows soil map of Shibin El-Kom area, which is located between latitudes 30° 10′ to 31° 00′ N and longitudes 30° 40′ to 31° 15′ E. Figure 4.59 shows the ETM+ image. The main taxonomic units of this soils are as the following sub-great groups:

- *Typic Torrerts*
- *Typic Torrifluvents*
- *Typic Quartizipsamment*

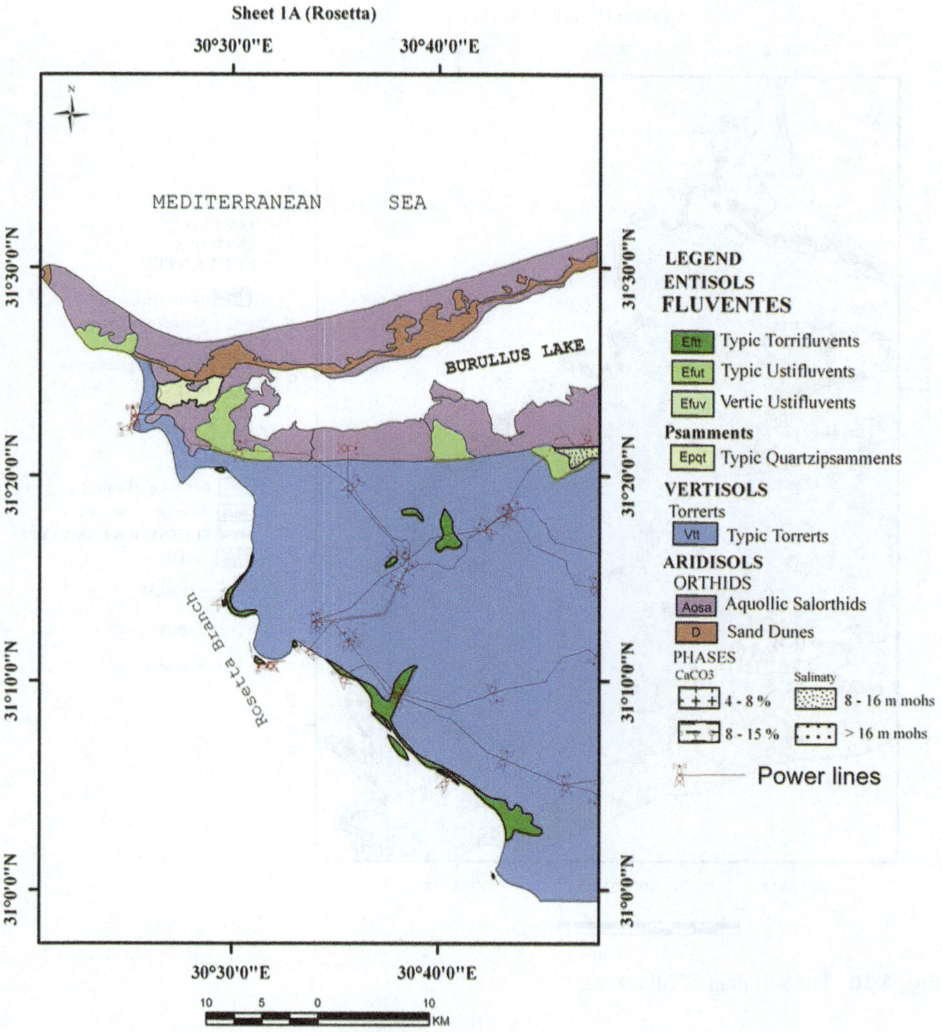

Fig. 4.29 Utilities superimposed on the soil map, Rosetta

Table 4.8 Areas and length of land resources of Rosetta area

Layer	Length (km)	Area (km^2)
Urban		30.1
Irrigation & Drainage	1714.4	
Roads	1869.5	
Railway	62.6	
Utilities	195.9	

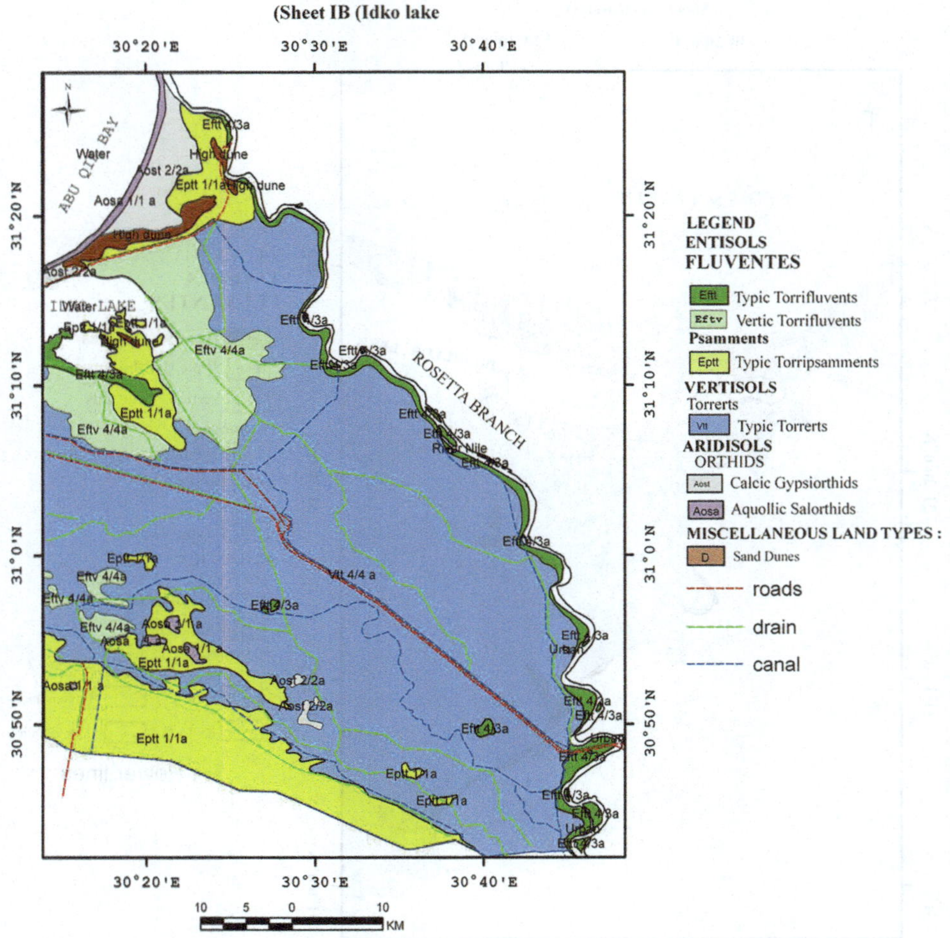

Fig. 4.30 The soil map of Idko Lake

The area is dominanted by the *Typic Torrerts (Vtt 4/4 a)* soil sub-great group (2867.9 km^2) which represents 86% of the sheet area. The soils are generally characterized by fine to very fine texture class. Soil salinity and calcium carbonate content are very low, as they don not exceed 4 dS/m and 4% respectively. The soils of *Typic Torrifluvents* are found included within the *Typic Torrerts* sub-great group, covering an area of 457.8 km^2, where the soil texture is very fine in the surface layers and fine to very fine in the sub-surface. The soils of *Typic Quartizpsamments (Apqt 1/1b)* sub-great group are found in small patches around

Table 4.9 Distribution of soil units in Idko lake

Soil Code	Area (km^2)	Area %
Aosa 1/1 a	22.4	0.7
Aosa 2/2a	63.3	2.1
Eftt 4/3a	100.2	3.3
Eftv 4/4a	240.1	7.9
Eptt 1/1a	433.6	14.3
Vtt 4/4 a	1780.5	58.7
High dune	25.8	0.9
River Nile	76.2	2.5
Urban	1.5	0.1
Water	288.7	9.5
Total	3032.2	100.0

Queesna area (9.3 km^2), these soils have a coarse to medium soil texture classes in both surface and sub-surface layers. Table 4.17 demonstrates the areas and existent percentages of each soil unit.

Regarding the thematic layers of urban areas, irrigation and drainage networks, roads, railways and utilities, Figs. 4.60, 4.61, 4.62, 4.63, 4.64 and Table 4.18 show their area coverage. The urban settlements cover an area of 173.3 km^2, irrigation and drainage network attains a total length of 4654.2 km, roads networks 4612.0 km, railways 355.7 km and utilities is 587.8 km.

4.1.1.10 Zagazig Area

Figure 4.65 and Table 4.19 show the soil map of Zagazig area, which is located between latitudes 30° 30′ to 31° 00′ N and longitudes 31° 10′ to 31° 45′ E. Figure 4.66 shows the ETM+ image. The main taxonomic units of this area include the following sub-great groups:

- *Typic Torrerts*
- *Typic Torrifluvents*
- *Vertic Torrifluvents*
- *Typic Torripsamments*
- *Typic Quartizpsamments*
- *Typic Torriorthents*

The soils of *Typic Torrerts* (*Vtt 4/4 a*) dominate the map with an area (2410.8 km^2) as it represents 91.1% of map area. The soil texture of this soil sub-great group is fine to very fine; soil salinity and % $CaCO_3$ are low as they do not exceed 4 dS/m and 4% respectively. *Typic Torrifluvents* soil (31.4 km^2) exhibits areas adjacent to Damietta branch stream in

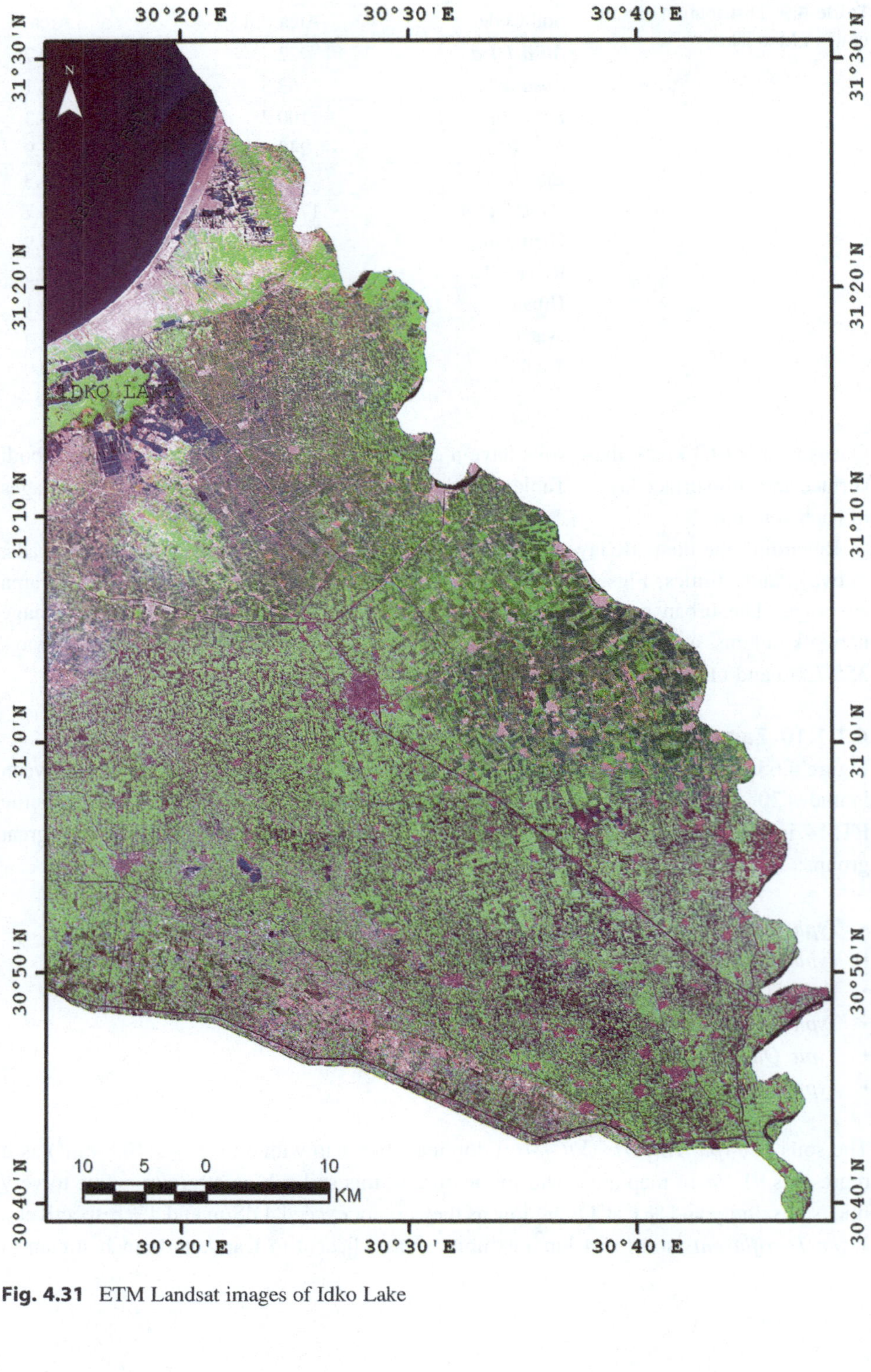

Fig. 4.31 ETM Landsat images of Idko Lake

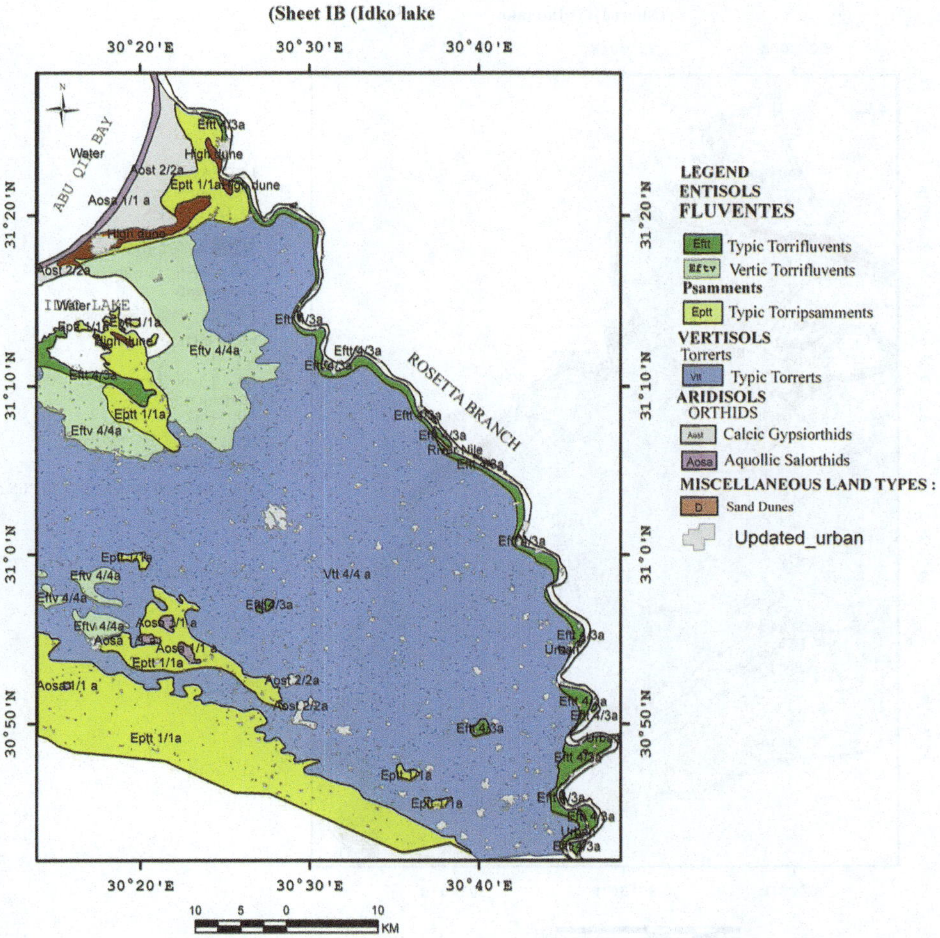

Fig. 4.32 Urban areas superimposed on the soil map, Idko

addition to some small patches among the *Typic Torrerts sub-great group*. *Vertic Torrifluvents* (*Eftv 4/4a)* is occupying the northeastern corner of the area (49.4 km^2). The soils of *Typic Quartzipsamments* and *Typic Torripsamments* (Eptt 1/1 a3) sub-great groups exist in small scattered areas (61.2 and 29.6 km^2 respectively). The Typic *Torriorthents (Aott 1/15)* exhibit a small area (30.39 km^2) in the south east of the map sheet area. The soil texture, in general, ranges between fine to very fine in both the surface and subsurface layers. However the *Typic Quartizpsamments* and *Typic Torripsamments* are characterized by a coarse texture. The slope gradient is less than 2% in different parts of the area. Soil salinity is less than 8 dS/m in most of the area but reaches 8–16 dS/m in some scattered patches in the soils of *Typic Torripsamments and Typic Torriorthents*. Calcium carbonates

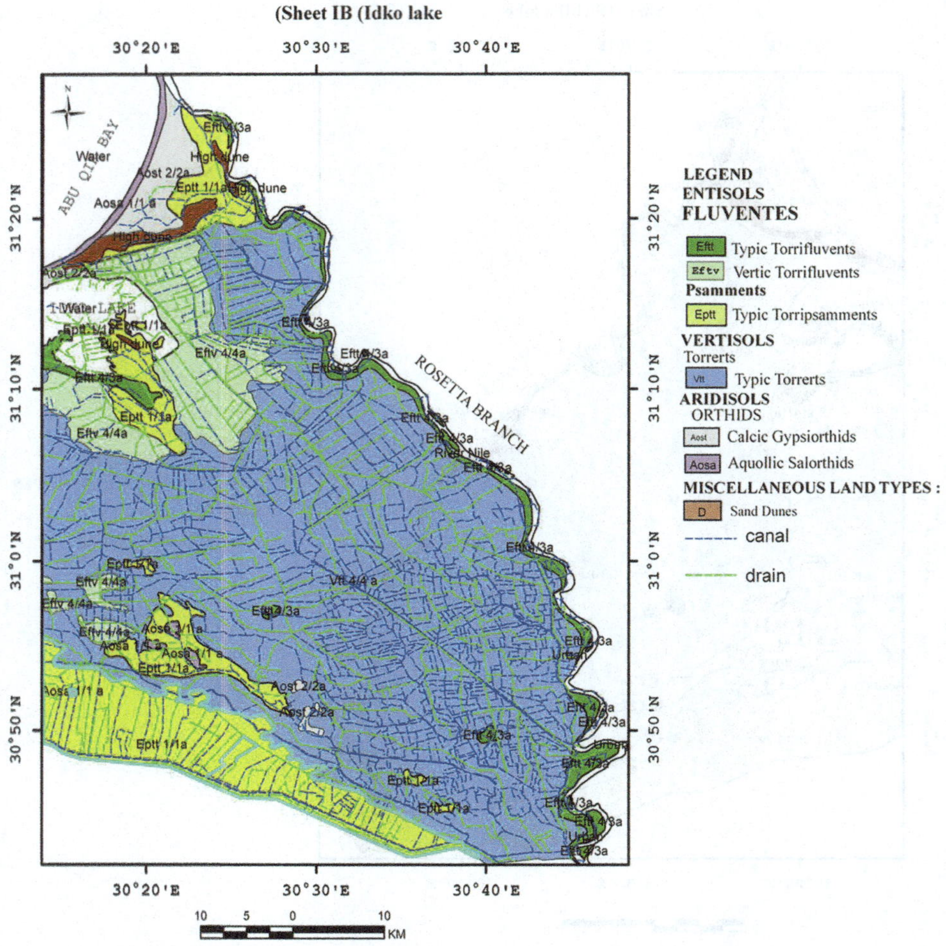

Fig. 4.33 Irrigation and drainage canals superimposed on the soil map, Idko

content ranges generally between 4–8% in this area, the high values are found in the *Typic Torriorthents* soils.

The created database of this area, Figs. 4.67, 4.68, 4.69, 4.70, 4.71 and Table 4.20 indicates that the total urban settlement area is 163.1km^2 and the total length of irrigation and drainage canals is 4138.9 km. The roads network length, railway network and the utilities network attains lengths of 4071.8, 156.1 and 348.1 km respectively.

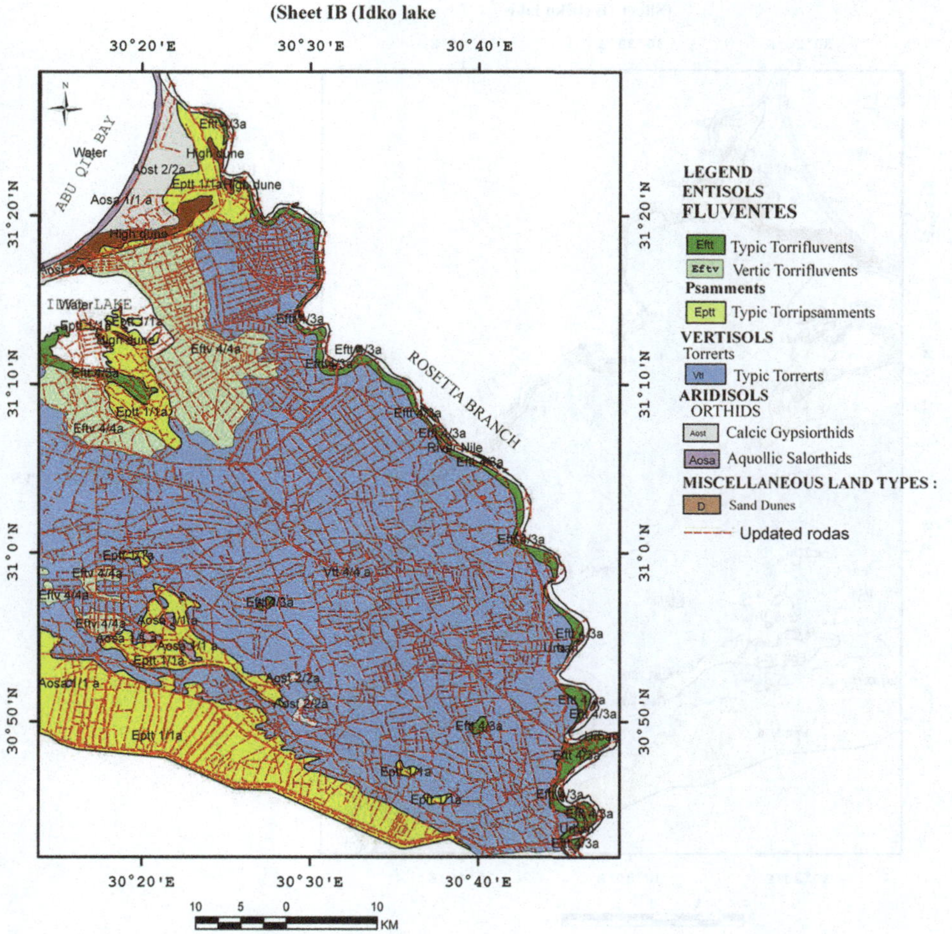

Fig. 4.34 Roads network superimposed on the soil map, Idko

4.1.1.11 Salhiya Area

Figure 4.72 shows soil map of Salhiya area, which is located between latitudes 30° 30′ to 31° 00′ N and longitudes 31° 30′ to 32° 00′ E. Figure 4.73 shows the ETM+ image of the area.The main taxonomic units Table 4.21 of the soils in this area are the following sub-great groups:

• *Typic Torrerts*	• *Typic Torriorthents*
• *Typic Torrifluvents*	• *Aquallic Salorthids*
• *Vertic Torrifluv*	• *Typic Quartizpsamments*
• *Typic Torripsamments*	

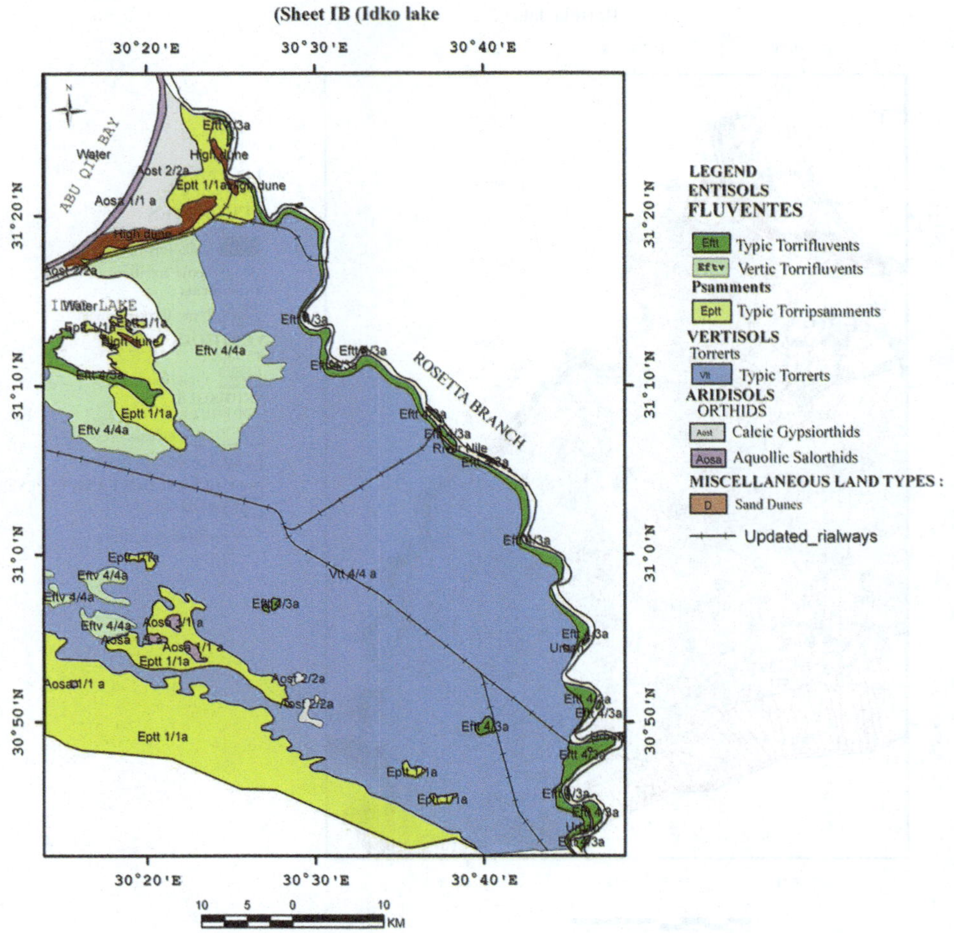

Fig. 4.35 Railways superimposed on the soil map, Idko

The northeastern part of this area is dominated by the soils of *Aquallic Salorthids* soils covering an area of 423.7 km^2. This sub-great group is characterized by a high soil salinity (more than 16 dS/m). The soil texture is coarse to very fine in both surface and sub-surface layers. In the north western parts of this area the soils of *Vertic Torrifluventst (Eftv4/4a1)* are found with an area of 382.5 km^2, while the soils of *Typic Torrifluvents(Eftt)* are found in a scattered small patches in this map with an area of 37.5 km^2. The soils of *Typic Torrerts (Vtt4/4a)* dominate the western part of the area covering a total of 424.7 km^2. The southern and south eastern parts are dominated by the soil sub-great groups *Typic Quartzipsamments (Epqt)*, *Typic Torripsamments*, *Typic Torriorthents* and *Petrogypsic Ggypsiorthids(Aogp1/*

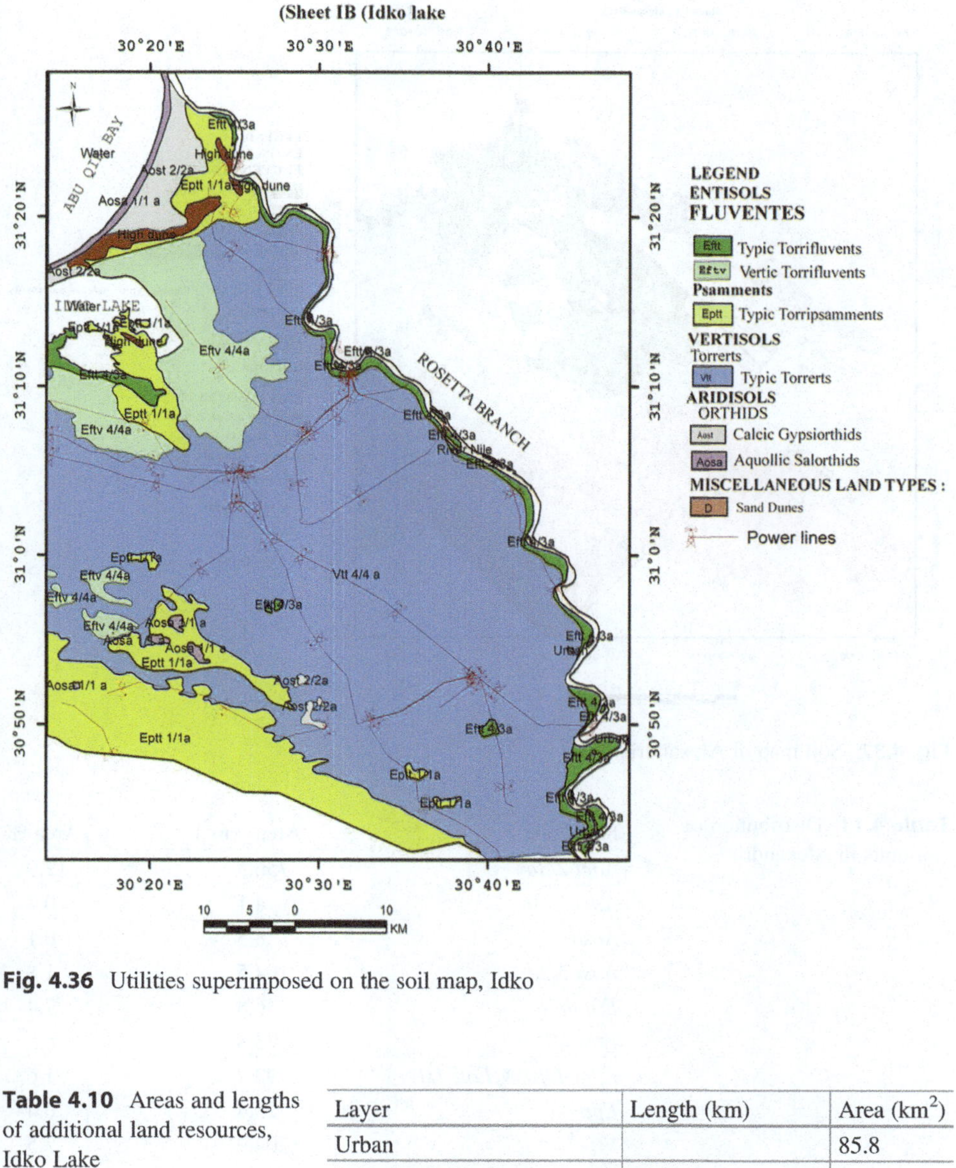

Fig. 4.36 Utilities superimposed on the soil map, Idko

Table 4.10 Areas and lengths of additional land resources, Idko Lake

Layer	Length (km)	Area (km^2)
Urban		85.8
Irrigation & Drainage	4885.6	
Roads	5240.4	
Railway	162.7	
Utilities	469.4	

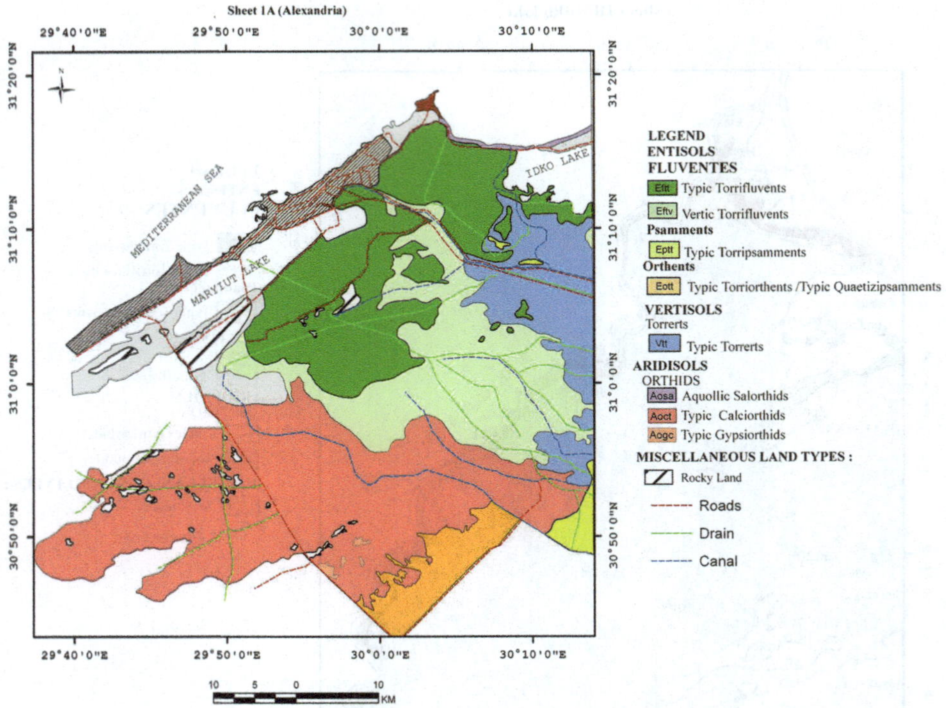

Fig. 4.37 Soil map of Alexandria

Table 4.11 Distribution of soil units in Alexandria

Soil Code	Area (km^2)	Area %
Aoct 2/3a	756.5	12.9
Aogc 2/3a	4.1	0.1
Aosa	8.5	0.1
Aost 2/2a	104.5	1.8
Eftt 4/3a	436.8	7.4
Eftv 4/4a	383.8	6.5
Epqt 1/1a & Eott 1/1	92.9	1.6
Eptt	25.4	0.4
Vtt	204.9	3.5
High dune	3.6	0.1
Rock land	57.3	0.9
urban	93.5	1.6
Water	3713.2	63.1
total	5884.9	100.0

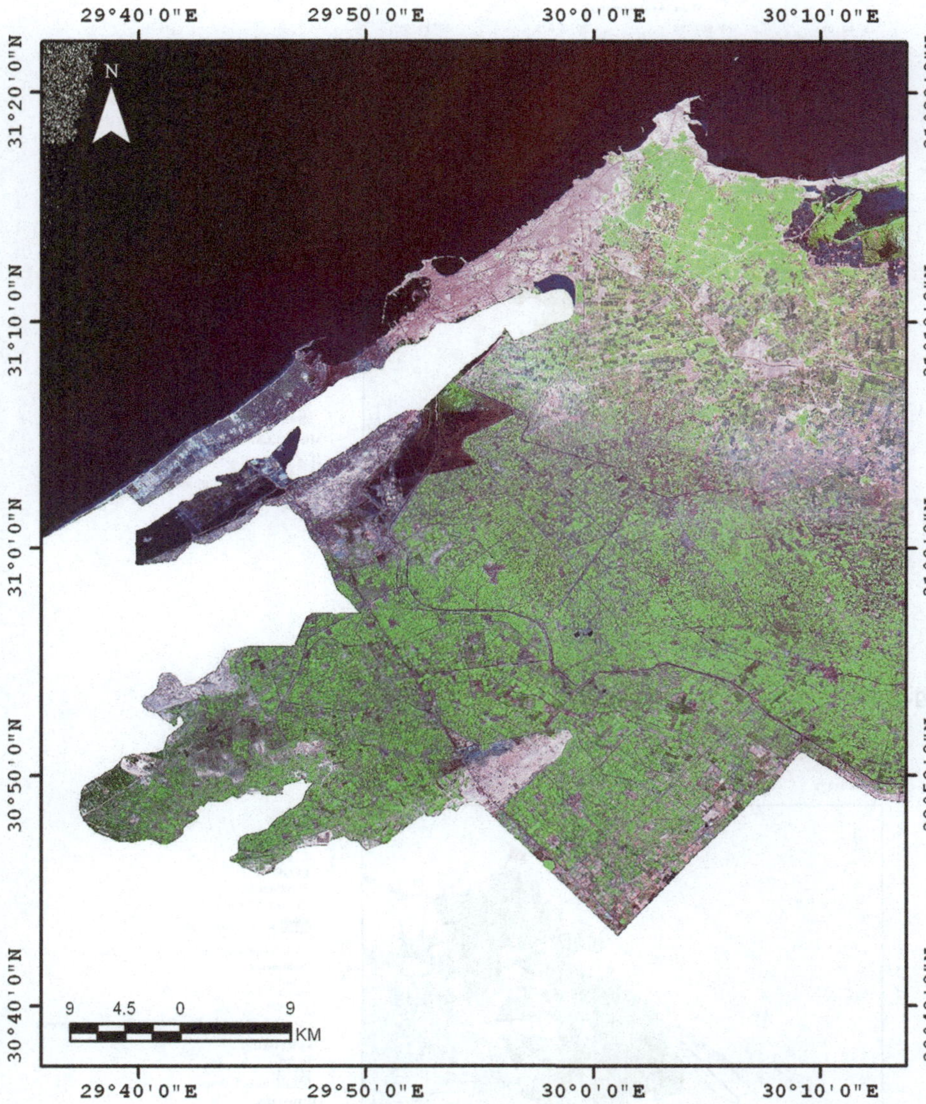

Fig. 4.38 ETM Landsat images of Alexandria

1a1) having areas of 703.5, 235.8, 312.8 and 100.3 km^2 respectively. The soil texture is generally coarser in the rest of the area. The soil texture is generally coarser in the southern parts than the northern one, also the CaCO3 percentage ranges between 4–8% in the southern parts and doesn't exceed 4% in the rest of the area.

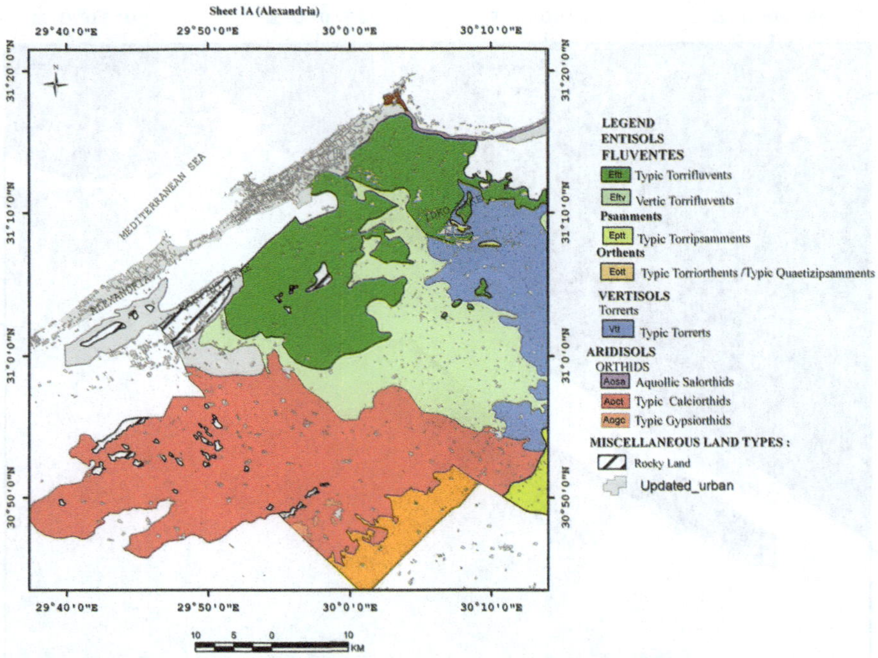

Fig. 4.39 Urban areas superimposed on the soil map, Alexandria

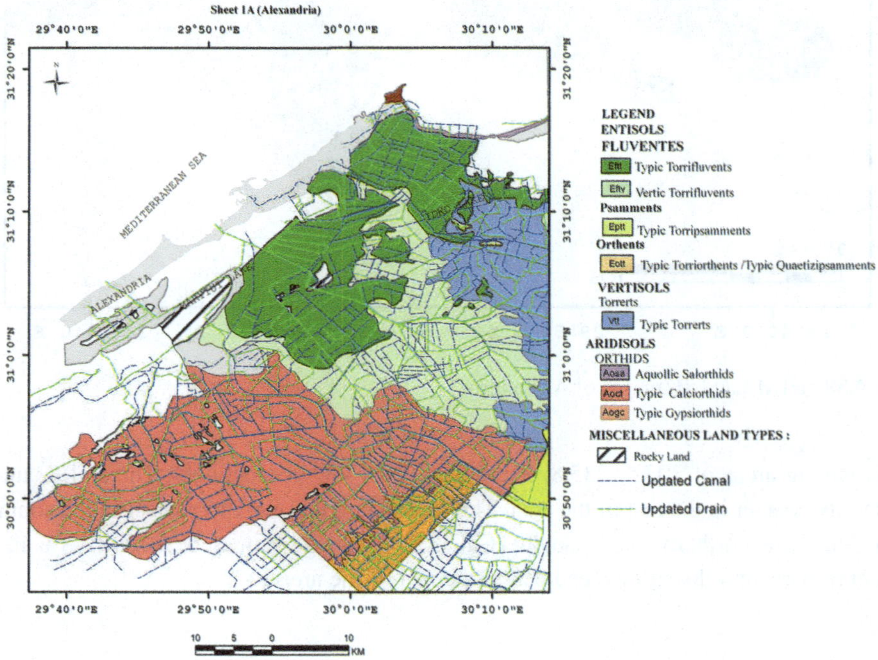

Fig. 4.40 Irrigation and drainage canals superimposed on the soil map, Alexandria

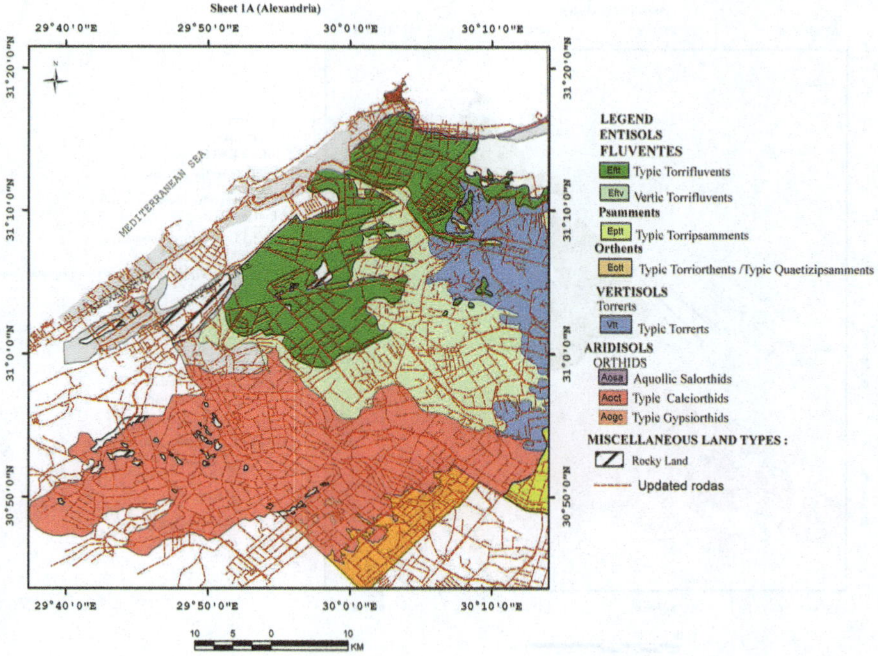

Fig. 4.41 Roads network superimposed on the soil map, Alexandria

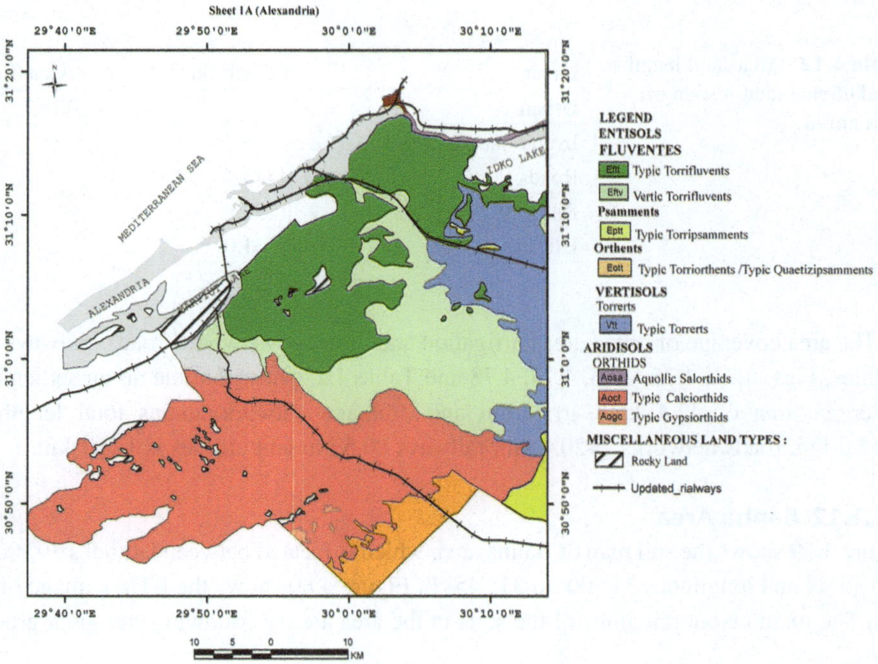

Fig. 4.42 Railways superimposed on the soil map, Alexandria

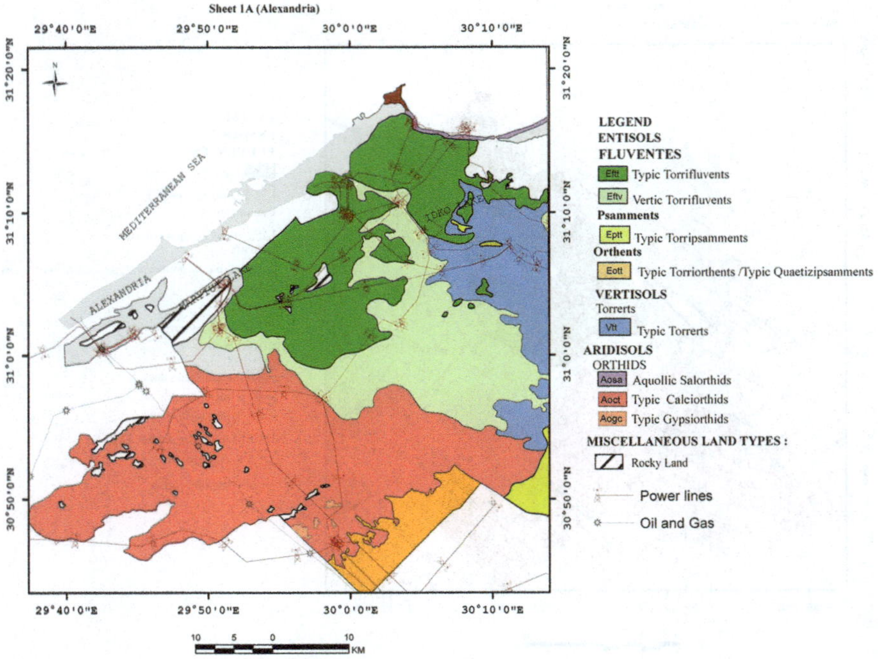

Fig. 4.43 Utilities superimposed on the soil map, Alexandria.

Table 4.12 Areas and lengths of additional land resources, Alexandria

Layer	Length (km)	Area (km^2)
Urban		118.7
Irrigation&Drainage	4335.2	
Roads	4444.7	
Railway	197.9	
Utilities	631.0	

The area coverage of urban areas, irrigation and drainage networks, roads, railways and utilities, Figs. 4.74, 4.75, 4.76, 4.77, 4.78 and Table 4.22 show that the urban settlements cover an area of 42.4 km^2, irrigation and drainage network attains total length of 1755.2 km, roads networks 2420.4 km, railways 86.5 km and utilities is 124.1 km.

4.1.1.12 Banha Area

Figure 4.79 shows the soil map of Banha area, which is located between latitudes 30° 00′ to 30° 30′ N and longitudes 31° 00′ to 31° 45′ E. Figure 4.80 shows the ETM+ image of the area.The main taxonomic units of the soils in the area are the following sub-great groups:

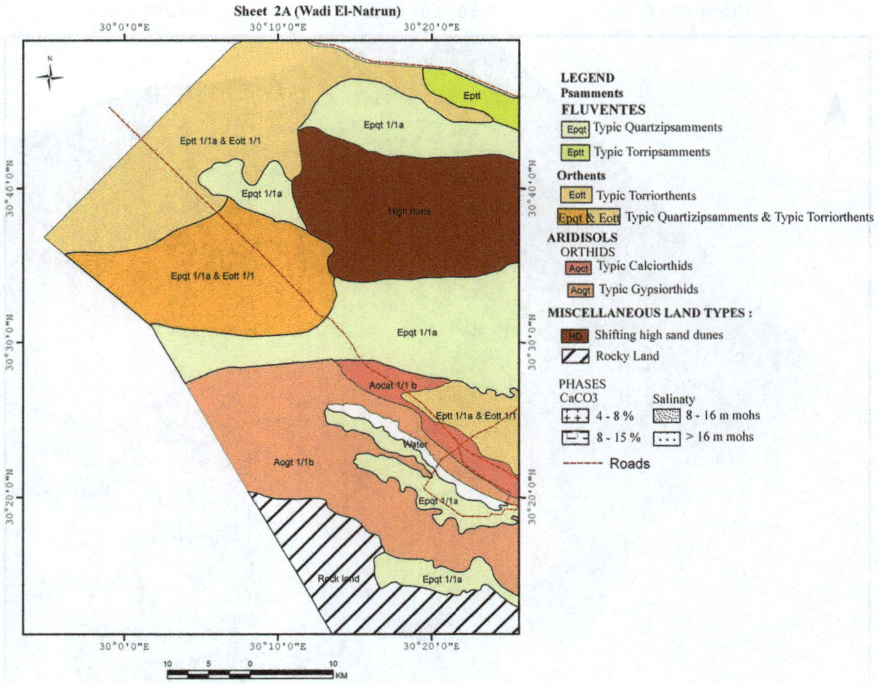

Fig. 4.44 The soil map of Wadi Al-Natrun

Table 4.13 Distribution of soil units in Wadi Al-Natrun

Soil Code	Area (km^2)	Area %
Aocat 1/1 b	64.5	2.7
Aogt 1/1b	386.3	16.1
Epqt 1/1a	593.9	24.7
Epqt 1/1a & Eott 1/1	286.7	11.9
Eptt	33.9	1.4
Eptt 1/1a & Eott 1/1	467.9	19.5
High dune	332.6	13.8
Rock land	214.6	8.9
Water	25.1	1.0
total	2405.5	100.0

- *Typic Torrifluvents*
- *Typic Quartizpsamments*
- *Typic Torrerts*

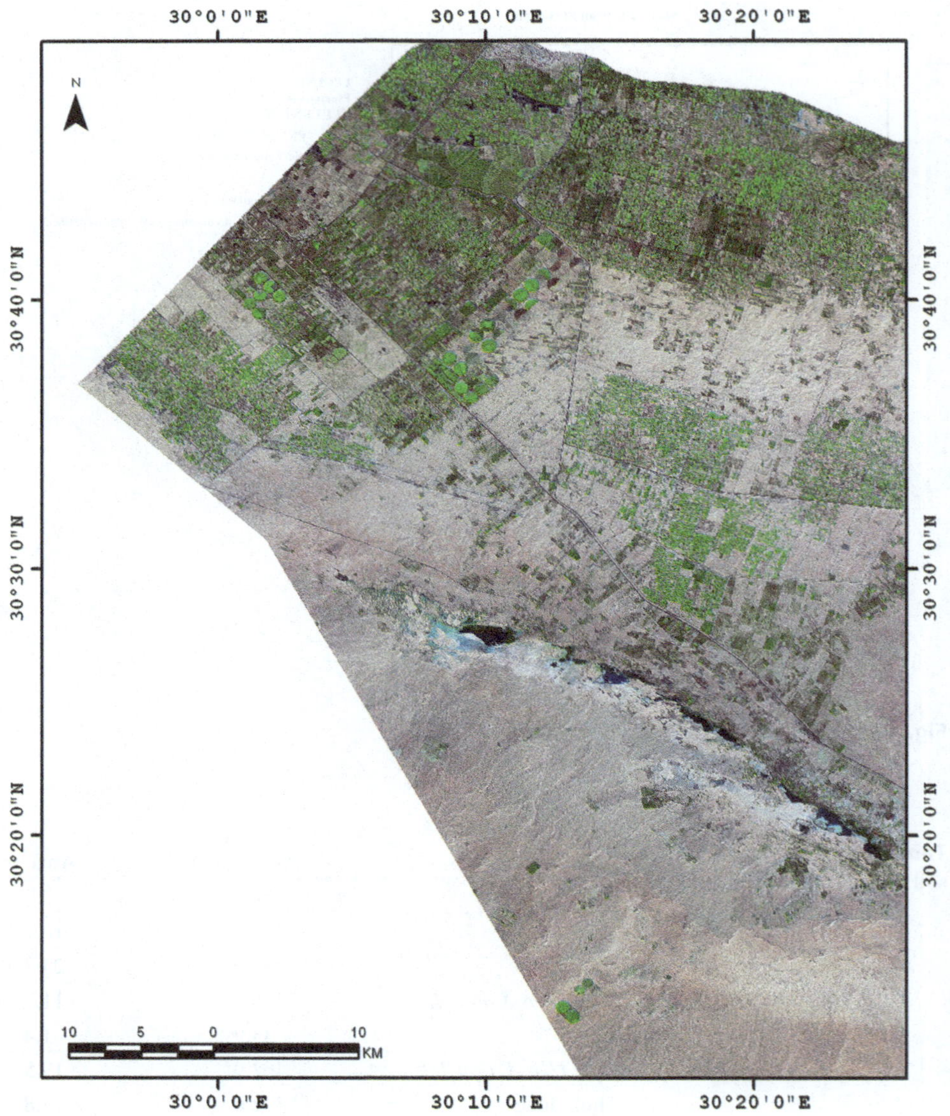

Fig. 4.45 ETM Landsat images of Al-Natrun

Table 4.23 shows that the area is generally dominated by *Typic Torrerts (Vtt)* sub-great group which cover an area of 1090.0 km^2, representing 78.7% of sheet area. The *Typic Quartzipsamments (Epqt)* sub-great group exhibits patches along the northeastern corner of the map sheet, covering an area of 69.2 km^2, representing 4.9% of the sheet area. The *Typic Torrifluvents (Eftt)* sub-great group is distributed along Damiatta branch covering an area

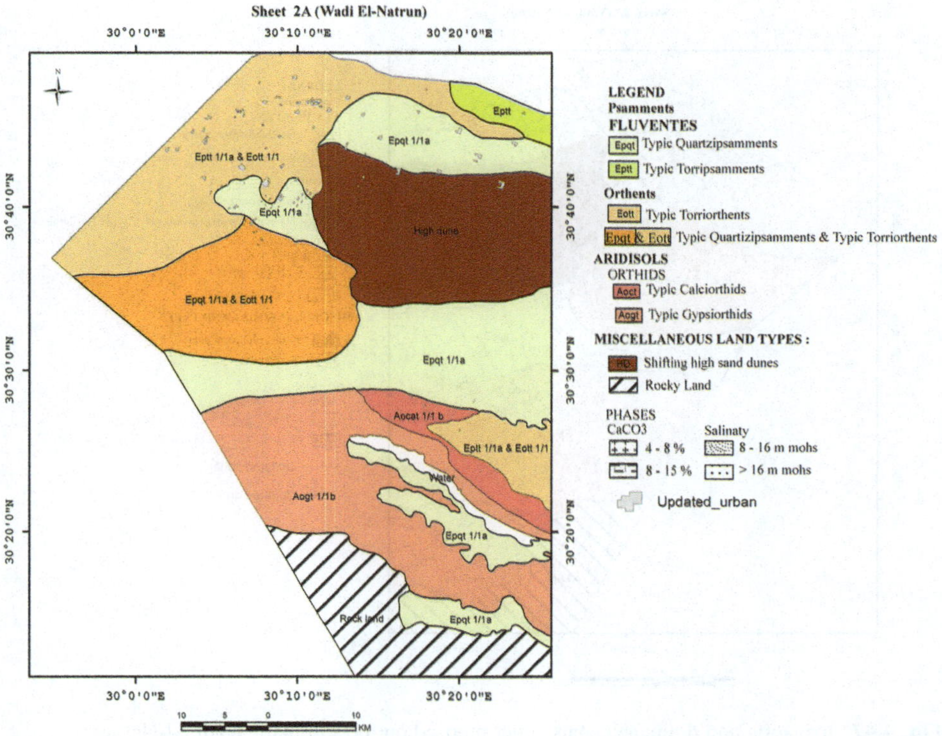

Fig. 4.46 Urban areas superimposed on the soil map, Wadi Al-Natrun

of 130.4 km^2, representing 8.1%. The soil texture is fine to very fine in both surface and sub-surface layers, except the *Typic Quartzipsamments* sub-great group which characterize by coarse to medium texture. The soils in this area are characterized by low salinity and $CaCO_3$ contents.

Regarding the thematic layers of urban areas, irrigation and drainage networks, roads, railways and utilities, Figs. 4.81, 4.82, 4.83, 4.84, 4.85 and Table 4.24 show the area coverage. The urban settlements cover an area of 113.7 km^2, irrigation and drainage network attains total length of 2182.7 km, roads networks 1834.4 km, railways 164.4 km and utilities is 256.8 km.

4.1.1.13 Cairo & Giza Area

Figure 4.86 show the soil map of Cairo and Giza. The area extends from latitudes 29° 30′ to 30° 15′ N and longitudes 30° 45′ to 31° 30′ E. Figure 4.87 shows the ETM+ image.The main taxonomic units of this area include the following sub-great groups:

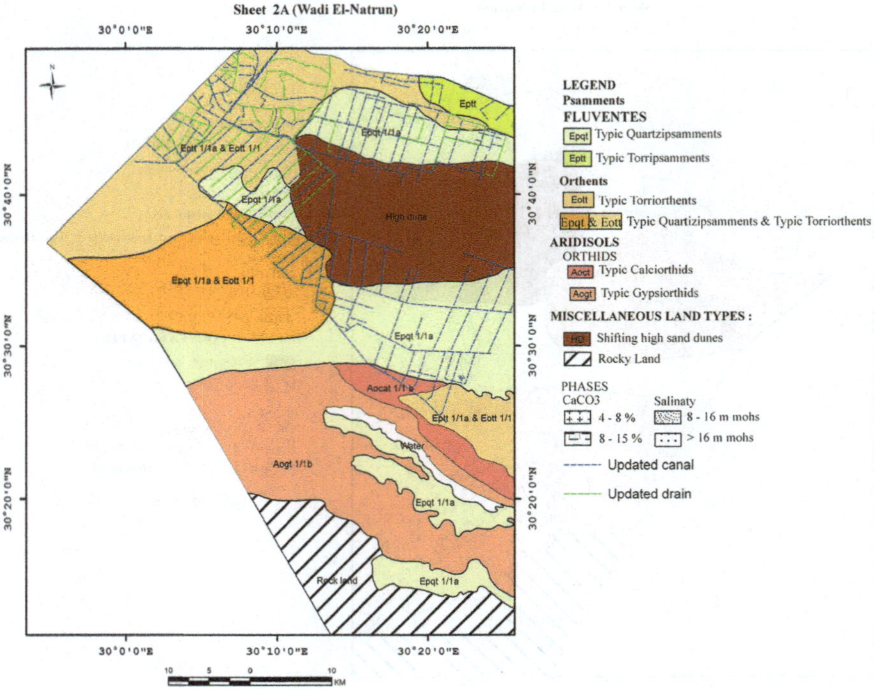

Fig. 4.47 Irrigation and drainage canals superimposed on the soil map, Wadi Al-Natrun

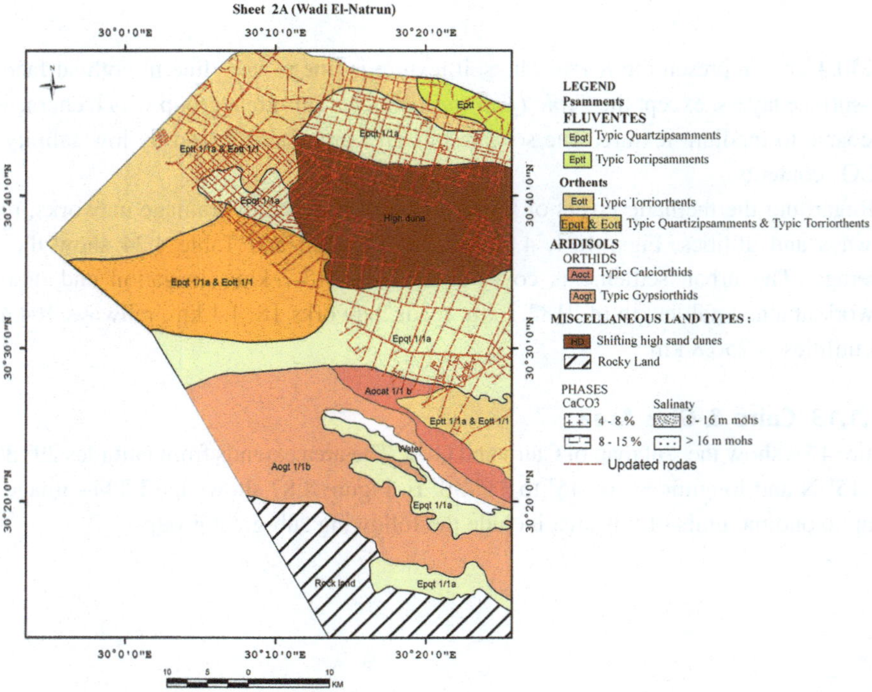

Fig. 4.48 Roads network superimposed on the soil map, Wadi Al-Natrun

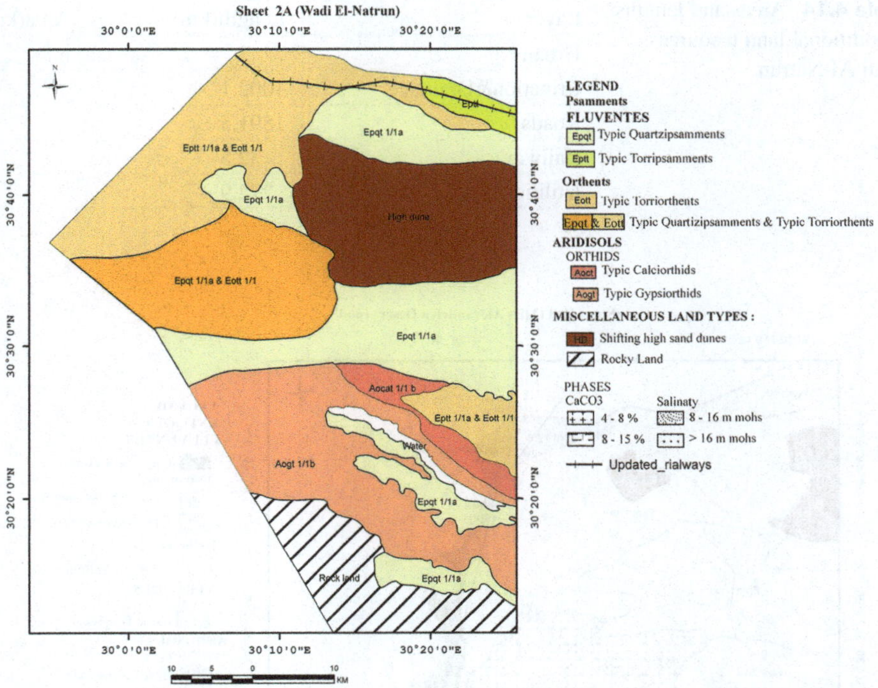

Fig. 4.49 Railways superimposed on the soil map, Wadi Al-Natrun

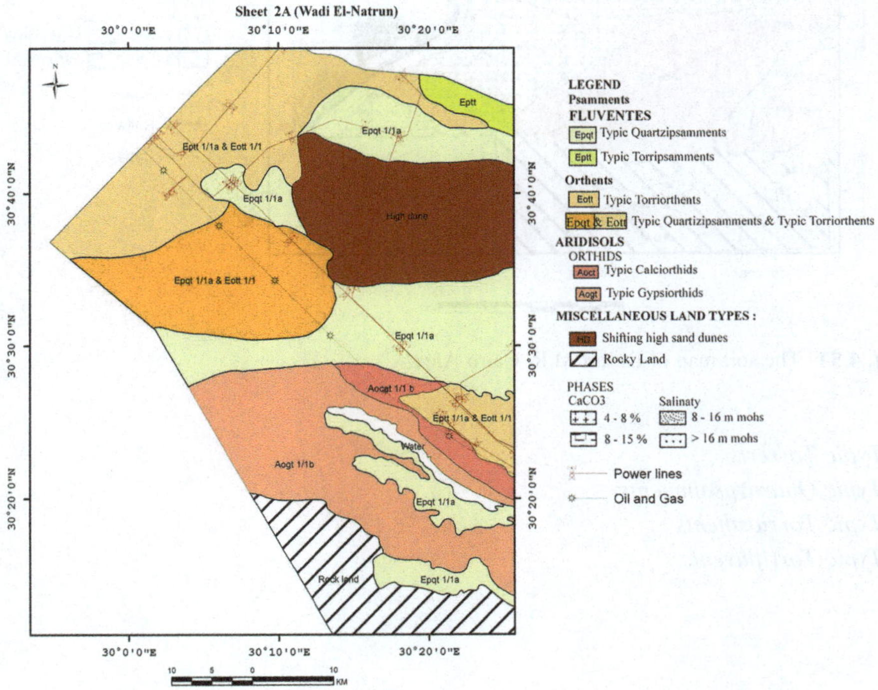

Fig. 4.50 Utilities, superimposed on the soil map, Wadi Al-Natrun

Table 4.14 Areas and lengths of additional land resources, Wadi Al-Natrun

Layer	Length(km)	Area(km^2)
Urban		6.8
Irrigation&Drainage	1002.1	
Roads	1591.5	
Railway	32.5	
Utilities	272.9	

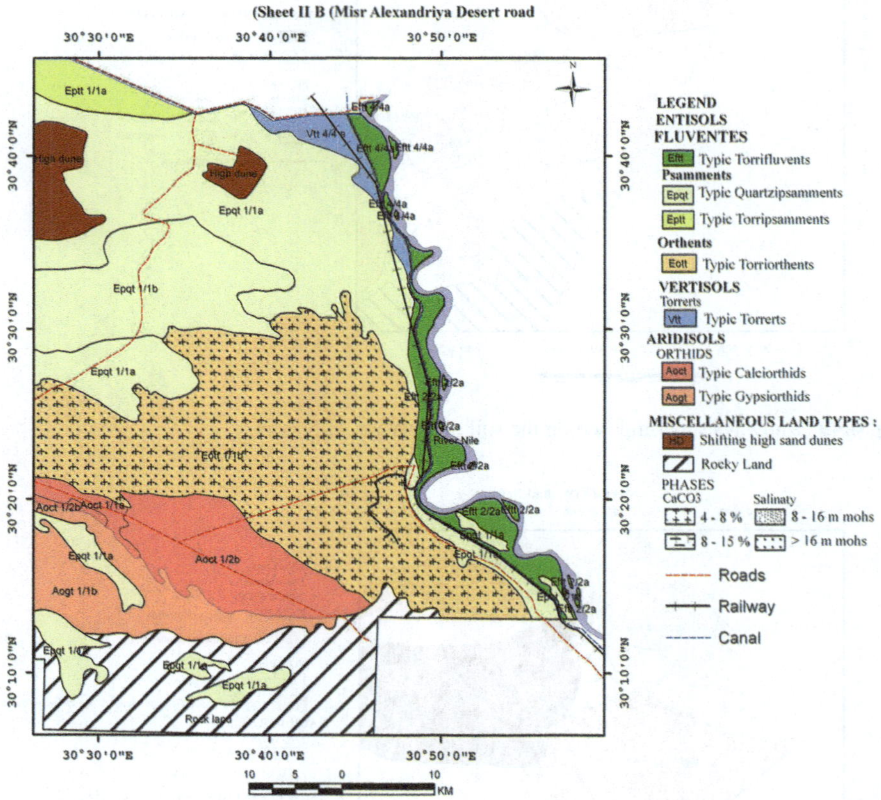

Fig. 4.51 The soil map of Desert M.R. Cairo-Alex

- *Typic Torrerts*
- *Typic Quartizpsamments*
- *Typic Torriorthents*
- *Typic Torrifluvents*

Table 4.15 Distribution of soil units in Desert Main road (M.R.) Cairo–Alex

Soil Code	Area (km^2)	Area %
Aoct 1/1a	20.4	0.8
Aoct 1/2b	176.7	6.8
Aogt 1/1b	147.9	5.7
Eftt 2/2a	90.8	3.5
Eftt 4/4a	18.6	0.7
Eott 1/1b	726.7	28.2
Epqt 1/1a	663.8	25.7
Epqt 1/1b	220.9	8.6
Eptt 1/1a	51.1	1.9
High dune	71.9	2.8
River Nile	53.7	2.1
Rock land	302.2	11.7
Vtt 4/4 a	36.7	1.4
Total	2581.4	100.0

Table 4.25 shows the areas of the different soil units and their percentage. The soils of *Typic Torrerts* sub-great group *(Vtt)* represent about 15.6% of the map sheet amounting to 435.8 km^2.

This area is characterized by fine soil texture in both surface and sub-surface layers and low salinity level. These soils dominate the western bank of the River Nile. The *Typic Quartizpsamments (Epqt)* sub-great group exhibits small patches covering an area of 102.9 km^2 representing 3.6% of the map. The texture of these soils is medium to coarse in both surface and sub-surface layers, while salinity and $CaCO_3$ is low (0–4 dS/m) and (< 4%) respectively. The *Typic Torriorthents (Eott) sub-great group is* mainly existent in both eastern and western sides of the Nile Valley covering an area of 552.0 km^2 representing about 19.7% of the map sheet. These soils characterized by a coarse texture, while salinity is high (>16 dS/m). *Typic Torrifluvents (Eftt),* sub-great groups is found in small scattered areas in the valley, with an area of 88.70 km^2, representing about 3.1% of the map.

The additional infra structure, Figs. 4.88, 4.89, 4.90, 4.91, 4.92 and Table 4.26 include urban areas covering 154.6 km^2, irrigation and drainage channels with a total length of 1546.8 km, a road network of 1340.7 km, a railway of 202.4 km and utilities (electric power, water pipes and telephone lines) with a total length of 255.8 km.

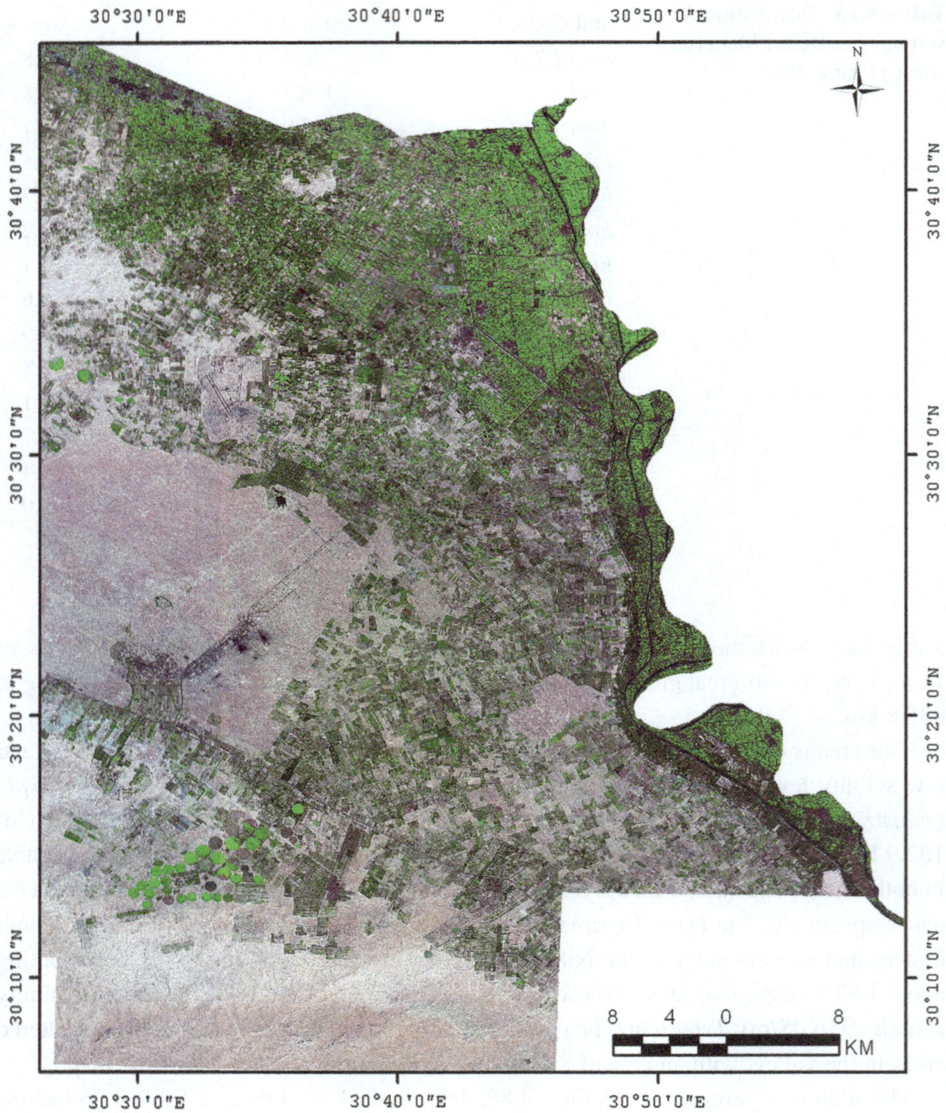

Fig. 4.52 ETM Landsat images of Desert M.R. Cairo-Alex

4.1.2 NIle Valley

4.1.2.1 Beni Suef Area

Figure 4.93 shows the soil map of Beni Suef area, which is located between latitudes 29° 00′ to 29° 30′ N and longitudes 30° 30′ to 31° 30′ E. Figure 4.94 shows the ETM+ image. The main sub-great groups in this area are:

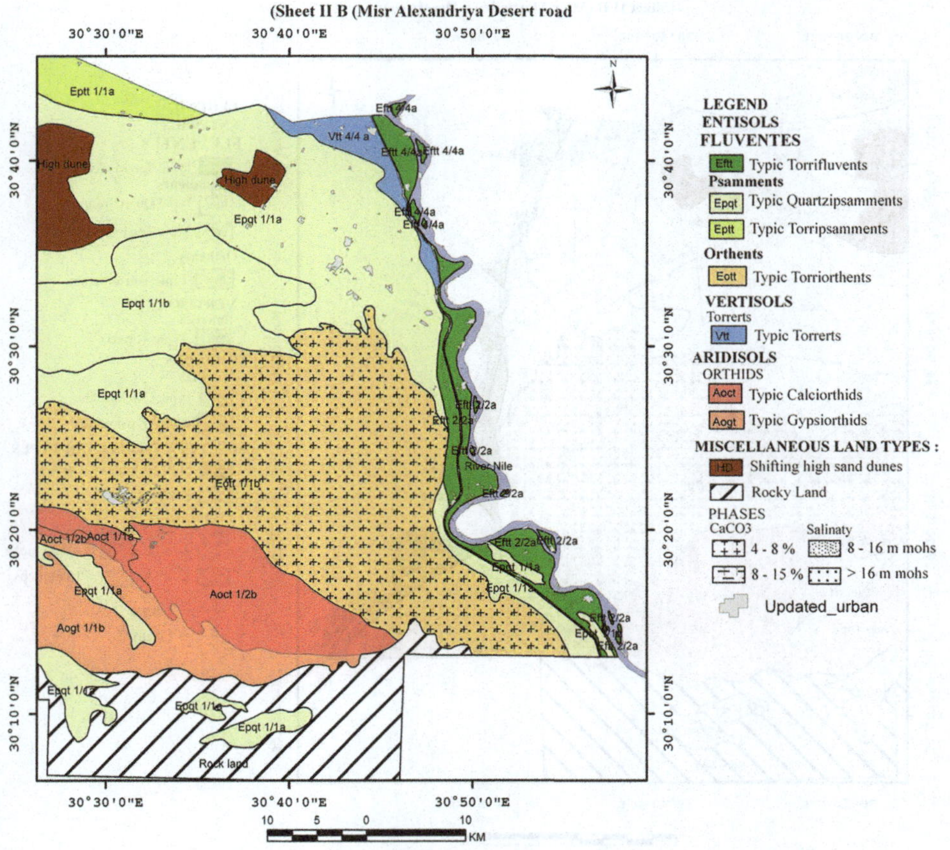

Fig. 4.53 Urban areas superimposed on the soil map, Desert M.R. Cairo-Alex

- *Typic Torrerts*
- *Typic Torrifluvents*
- *Typic Quartizpsamments*
- *Typic Torripsamments*
- *Typic Torriorthents*
- *Typic Calciorthids*
- *Calcic Gypsiorthids*
- *Petrogypsic Gypsiorthids*

The obtained data Table 4.27 reveal that the soils of *Typic Torrerts (Vtt)* covers an area of 1084.20 km^2 representing about 24.6% of the total area, the *Typic Torrerts* in this area is found mainly in the western bank of the river Nile. The soils of *Typic Torrifluvents (Eftt)*

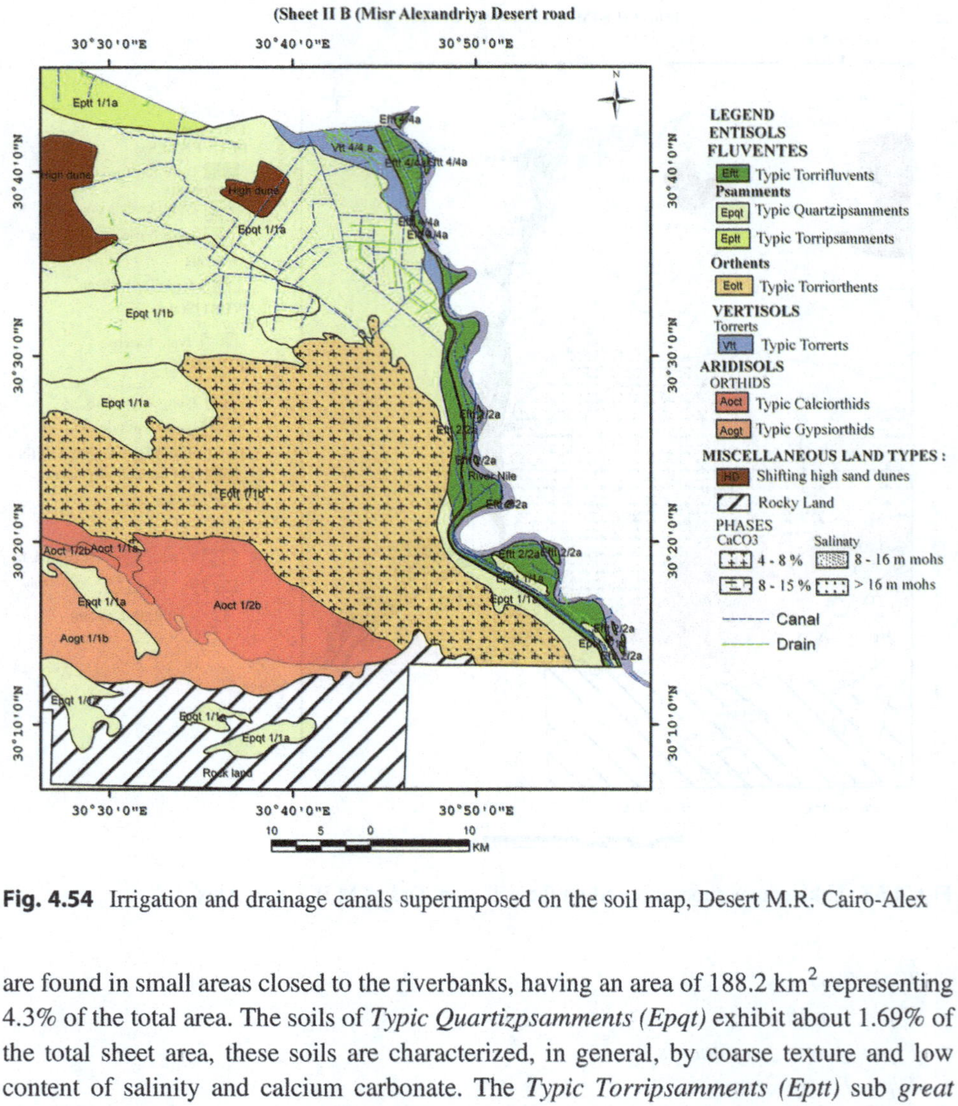

Fig. 4.54 Irrigation and drainage canals superimposed on the soil map, Desert M.R. Cairo-Alex

are found in small areas closed to the riverbanks, having an area of 188.2 km^2 representing 4.3% of the total area. The soils of *Typic Quartizpsamments (Epqt)* exhibit about 1.69% of the total sheet area, these soils are characterized, in general, by coarse texture and low content of salinity and calcium carbonate. The *Typic Torripsamments (Eptt)* sub *great* group is found also in small patches and narrow strip around the River stream covering an area of 0.3% of the total sheet coverage. The soils of Typic *Calciorthids (Aocat), Calcic Gypsiorthids (Aogca)* and *Petrogypsic Gypsiorthids (Aogp)* are found in different locations outside the Nile valley, more distributed in the western side of the Nile. The total area of these subgroups amount 465.2 km^2.

Regarding the thematic layers, Figs. 4.95, 4.96, 4.97, 4.98, 4.99 and Table 4.28 show the area and lengths coverage of different resources, the urban settlements cover an area of

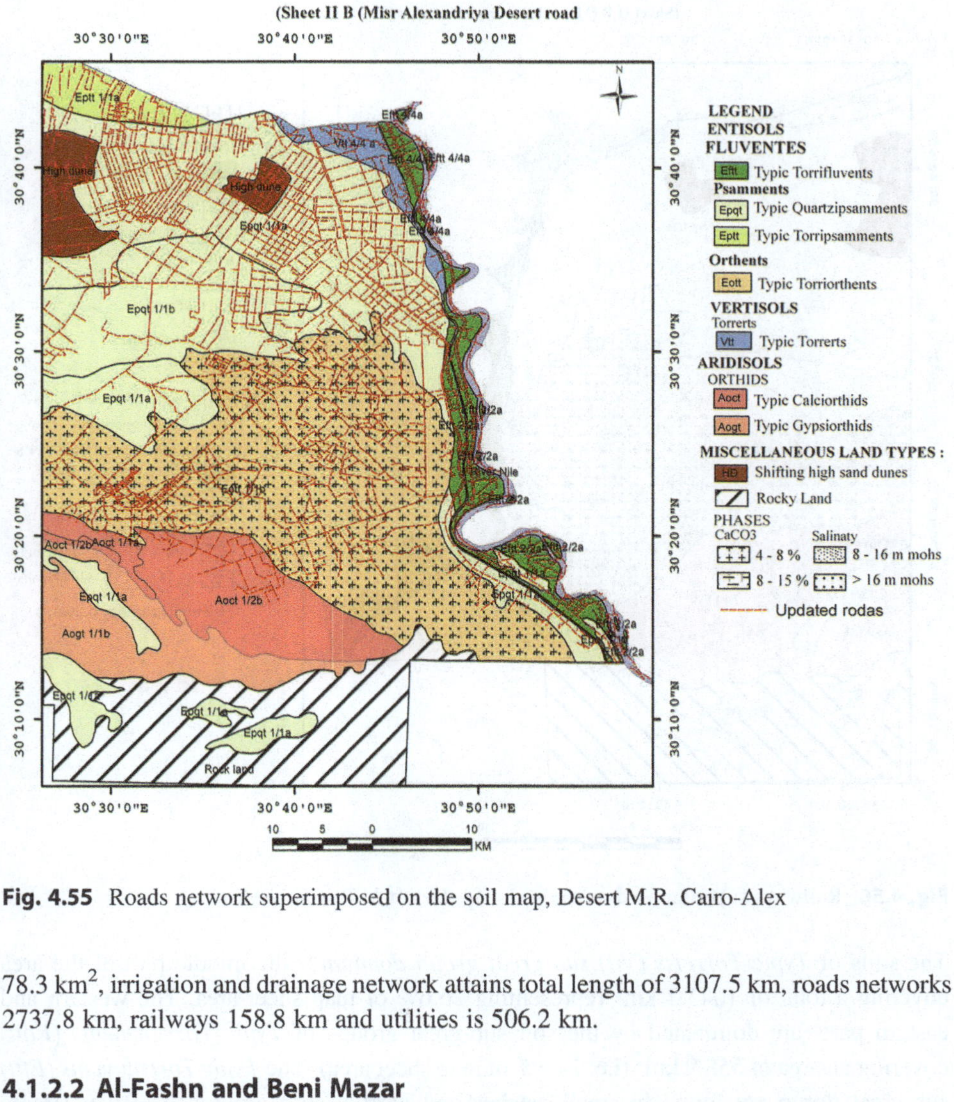

Fig. 4.55 Roads network superimposed on the soil map, Desert M.R. Cairo-Alex

78.3 km^2, irrigation and drainage network attains total length of 3107.5 km, roads networks 2737.8 km, railways 158.8 km and utilities is 506.2 km.

4.1.2.2 Al-Fashn and Beni Mazar

Figure 4.100 shows the soil map of Al-fashn & Beni Mazar area, which is located between latitudes 28° 20′ to 29° 00′ N and longitudes 30° 29′ to 31° 10′ E. Figure 4.101 shows the ETM+ image. The main soil taxonomic units Table 4.29 of the soils in this area are the following sub-great groups:

- *Typic Torrerts*
- *Typic Torriorthents*
- *Typic Torrifluvents*
- *Typic Quartizpsamments*

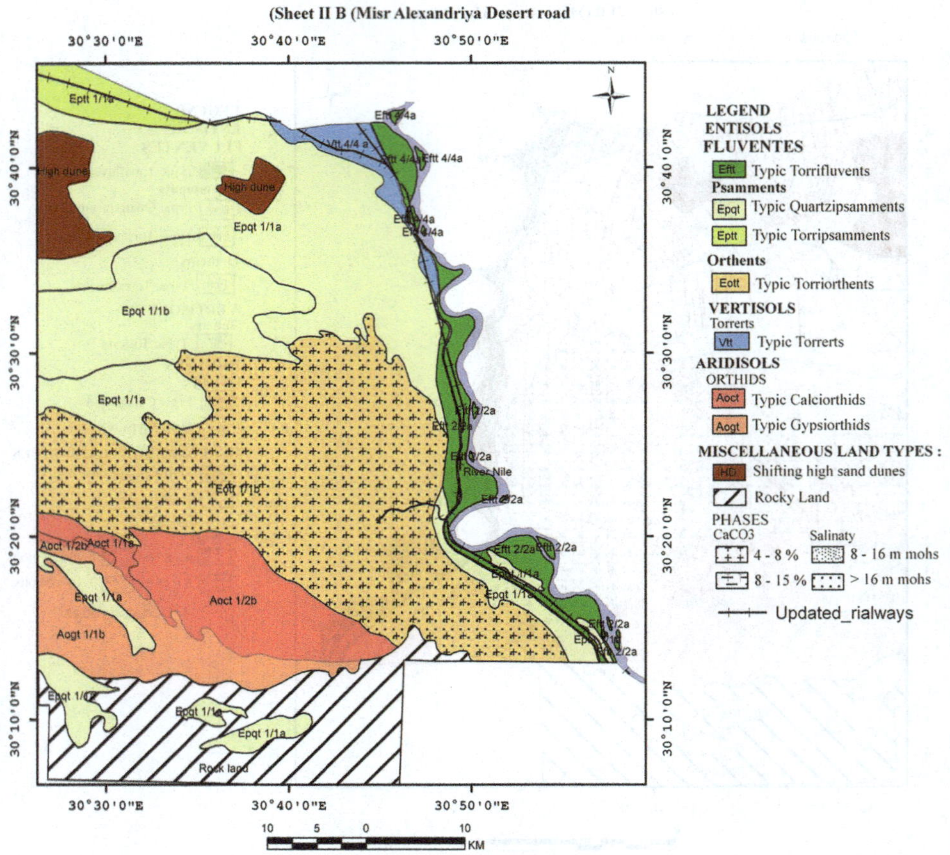

Fig. 4.56 Railways superimposed on the soil map, Desert M.R. Cairo-Alex

The soils of *Typic Torrerts (Vtt) sub-great group dominates* the middle part of the area covering a total of 1043.2 km^2 representing 26.6% of map sheet area. The western and eastern parts are dominated by the soil sub-great groups of *Typic Torriorthents (Eott)* covering an area of 558.9 km^2 (i.e. 14.3% of map sheet area). The *Typic Torrifluvents (Eftt)* sub-great group are found in small patches and narrow strip around the River stream covering an area of 138.3 km^2 representing 3.5% of the area. The soils of Typic *Quartizpsamments (Epqt)* are found in an elongated strip along the interference zone with western desert fringes, having a total area of 320.4 km^2. These soils are characterized by soil texture ranging from fine sand to coarse sand. While, the $CaCO_3$ content ranging between 4% and 8%.

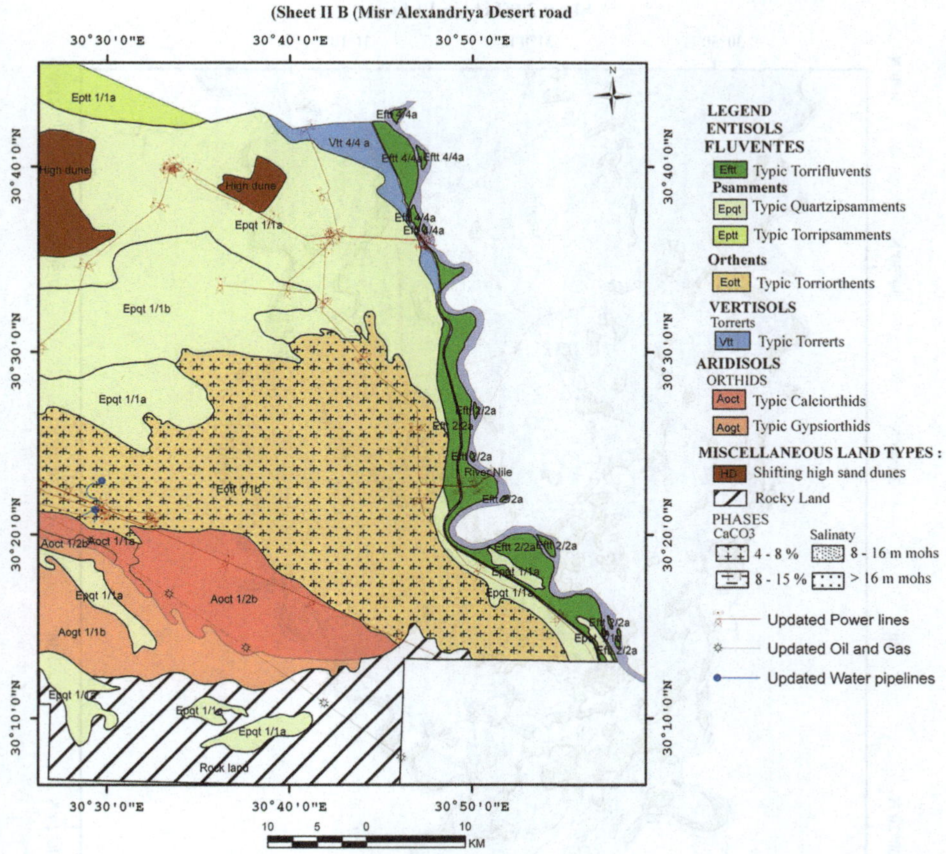

Fig. 4.57 Utilities, superimposed on the soil map, Desert M.R. Cairo-Alex

Table 4.16 Areas and lengths of additional land resources, Desert Main Road (M.R.) Cairo–Alex

Layer	Length (km)	Area (km^2)
Urban		17.9
Irrigation&Drainage	667.5	
Roads	2746.1	
Railway	105.7	
Utilities	298.4	

Regarding the thematic layers of urban areas, irrigation and drainage networks, roads, railways and utilities, Figs. 4.102, 4.103, 4.104, 4.105, 4.106 and Table 4.30 show the area coverage. The urban settlements cover an area of 52.2 km^2, irrigation and drainage network

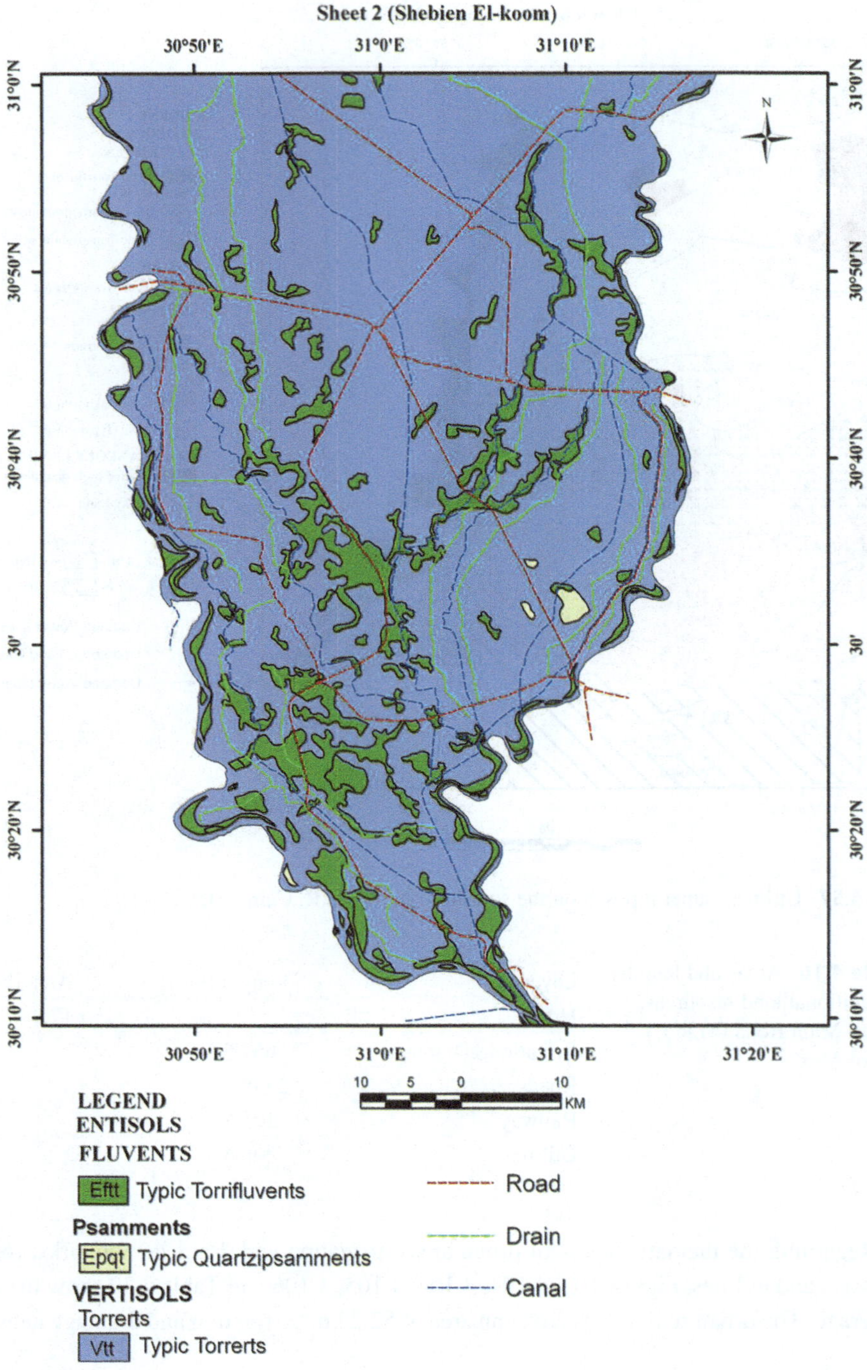

Fig. 4.58 The soil map of Shibin Al-Kom

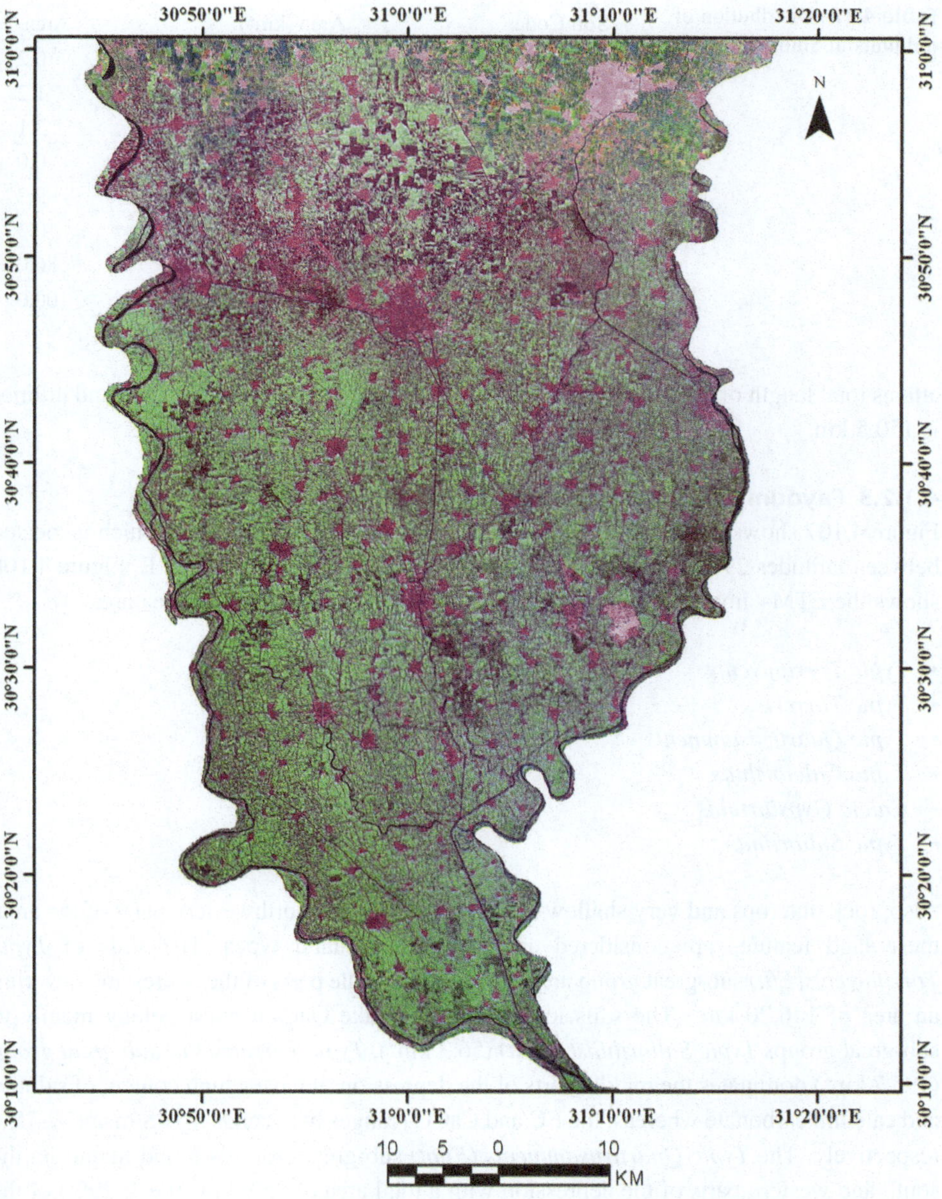

Fig. 4.59 ETM Landsat images of Shibin Al-Kom

Table 4.17 Distribution of soil units in Shibin Al-Kom

Soil Code	Area (km^2)	Area %
Eftt 2/2a	5.9	0.2
Eftt 3/2a	39.4	1.2
Eftt 3/3a	104.6	3.1
Eftt 3/3al	1.1	0.0
Eftt 4/3a	306.8	9.2
Epqt 1/1b	9.3	0.3
Vtt 4/4 a	2867.9	86.0
Total	3334.9	100.0

attains total length of 1949.1 km, roads networks 1643.8 km, railways 74.2 km and utilities is 150.5 km.

4.1.2.3 Fayoum Depression & Sinnuris

Figure 4.107 shows the soil map of Fayoum Depression & Sinnuris area, which is located between latitudes 29° 00′ to 29° 30′ N and longitudes 30° 15′ to 31° 10′ E. Figure 4.108 shows the ETM+ image. The main sub-great groups Table 4.31 in this area are:

- *Typic Torrifluvents*
- *Typic Torrerts*
- *Typic Quartizpsamments*
- *Typic Calciorthids*
- *Calcic Gypsiorthids*
- *Typic Saliorthids*

Also, rock outcrops and very shallow soils are found in the northwestern parts of the area, these land features are considered as miscellaneous land types. The soils of *Typic Torrifluvents(Eftt)* sub-great group are found in the middle parts of the depression covering an area of 146.20 km^2. The soils located near to Lake Qaroun coast belong mainly to sub-great groups *Typic Saliorthids(Aosat)* (56.3 km^2), *Typic Torrerts(Vtt) sub great group* (795.7 km^2) dominates the middle parts of the depression, it have a high content of salinity and calcium carbonate whereas, the EC and $CaCO_3$ ranges between 8–16 dS/m and 8–15% respectively. The *Typic Quartizpsamments (Epqt)* sub-great group is found mainly in the south and western parts of the depression with a total area of 75.9 km^2 (i.e. 3.79%) of the map sheet area. The soil texture ranges from very fine-to-fine in the soils of *Typic Torrerts* and *Typic Saliorthids(Aosat)*sub-great groups, while becoming much more coarser in both surface and sub-surface layers of the *Typic Quartizpsamments*. The *Calcic Gypsiorthids (Aogca) and Typic Calciorthids (Aocat)* sub great groups are found mainly around the depression and exhibit the areas of high elevation with an area of 567.6 km^2.

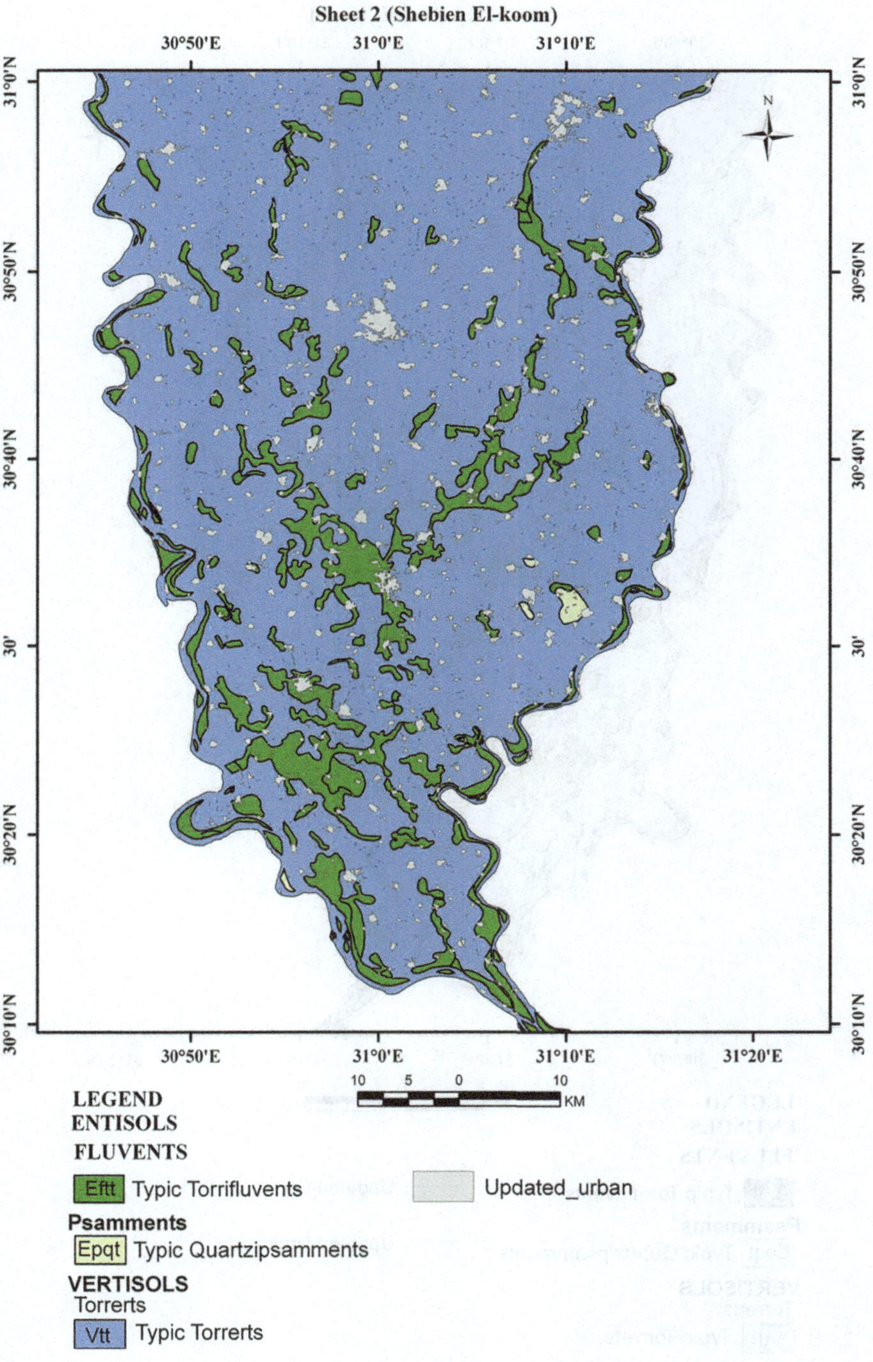

Fig. 4.60 Urban areas superimposed on the soil map, Shibin Al-Kom

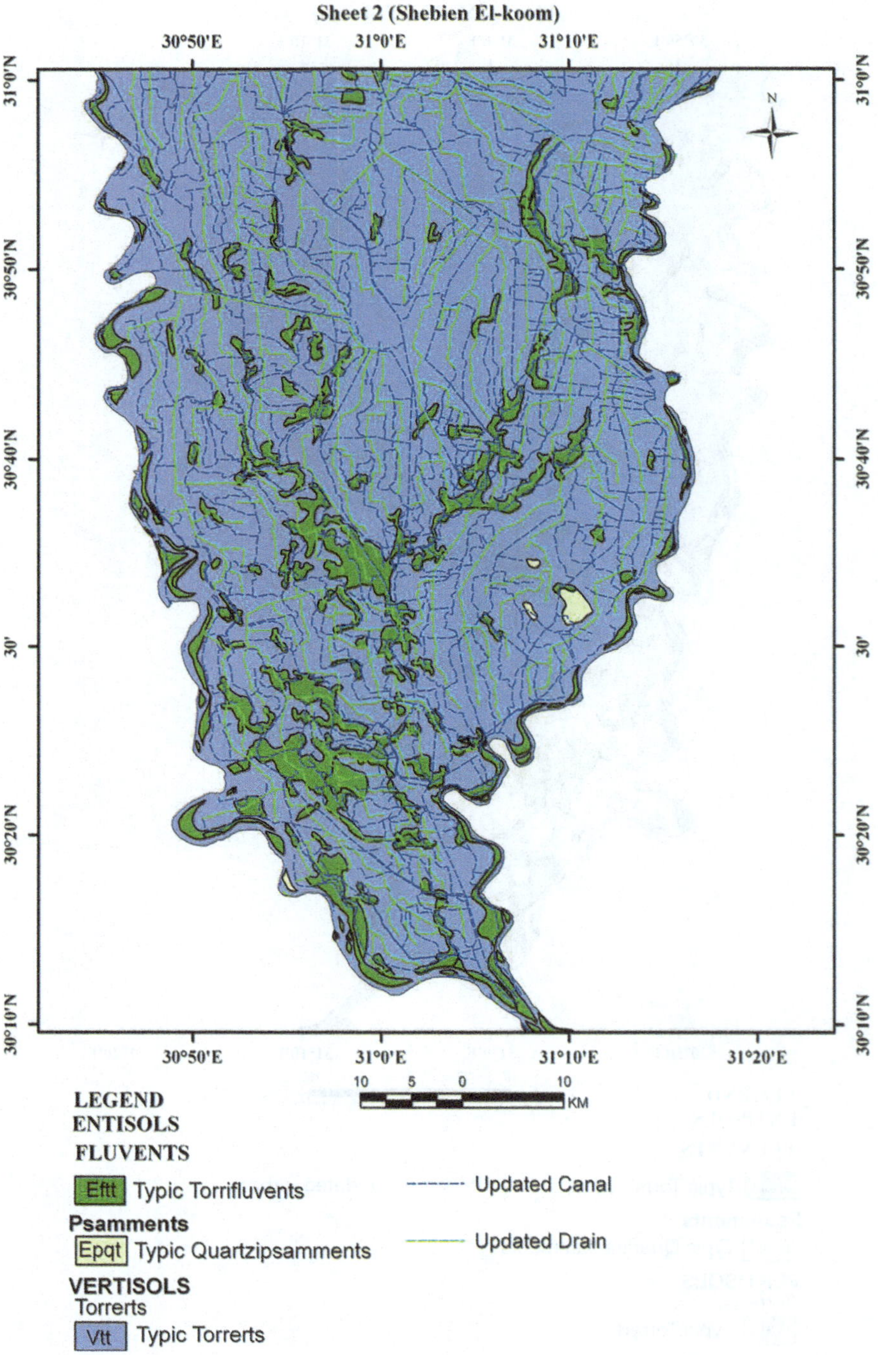

Fig. 4.61 Irrigation and drainage canals superimposed on the soil map, Shibin Al-Kom

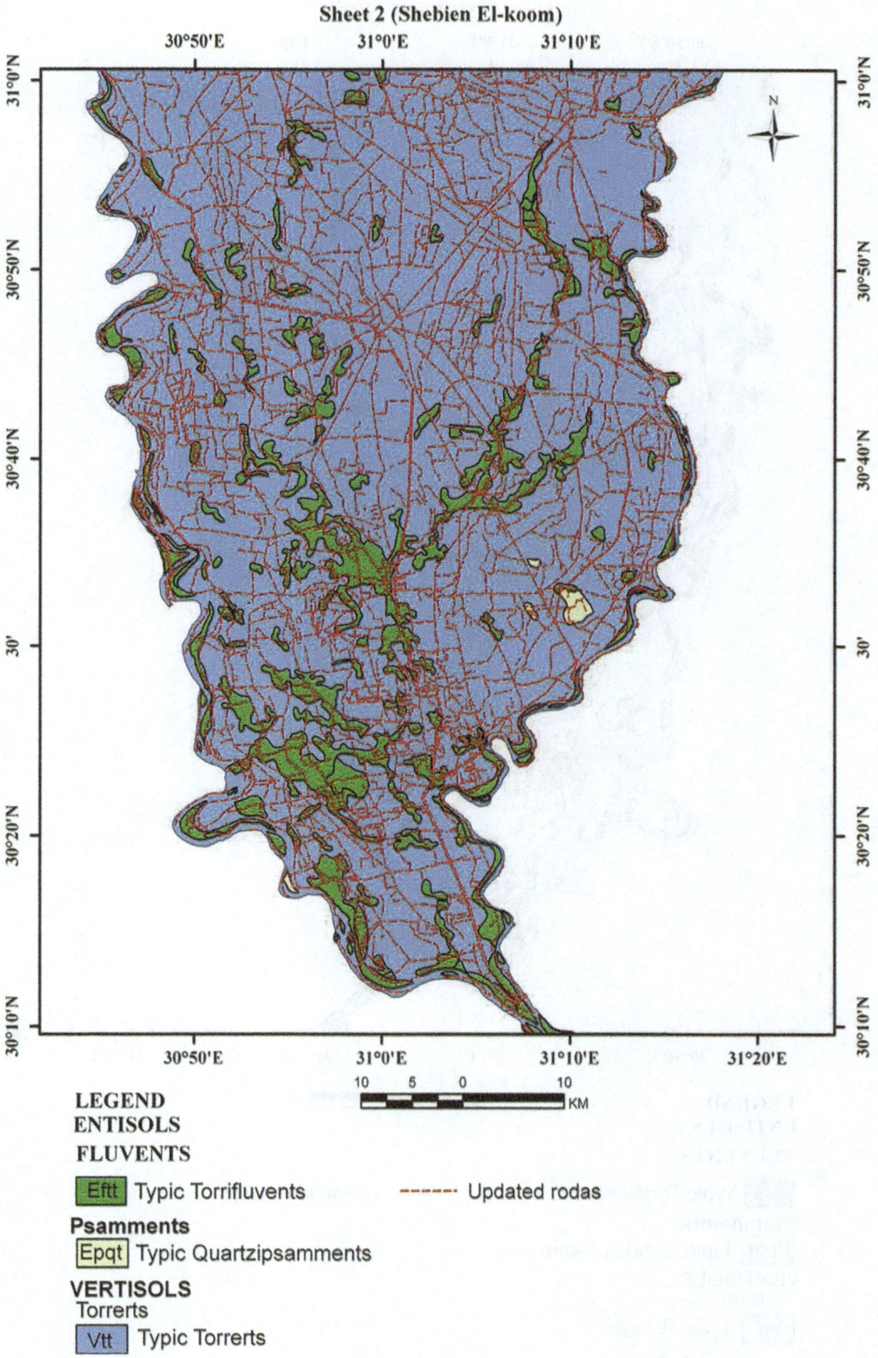

Fig. 4.62 Roads network superimposed on the soil map, Shibin Al-Kom

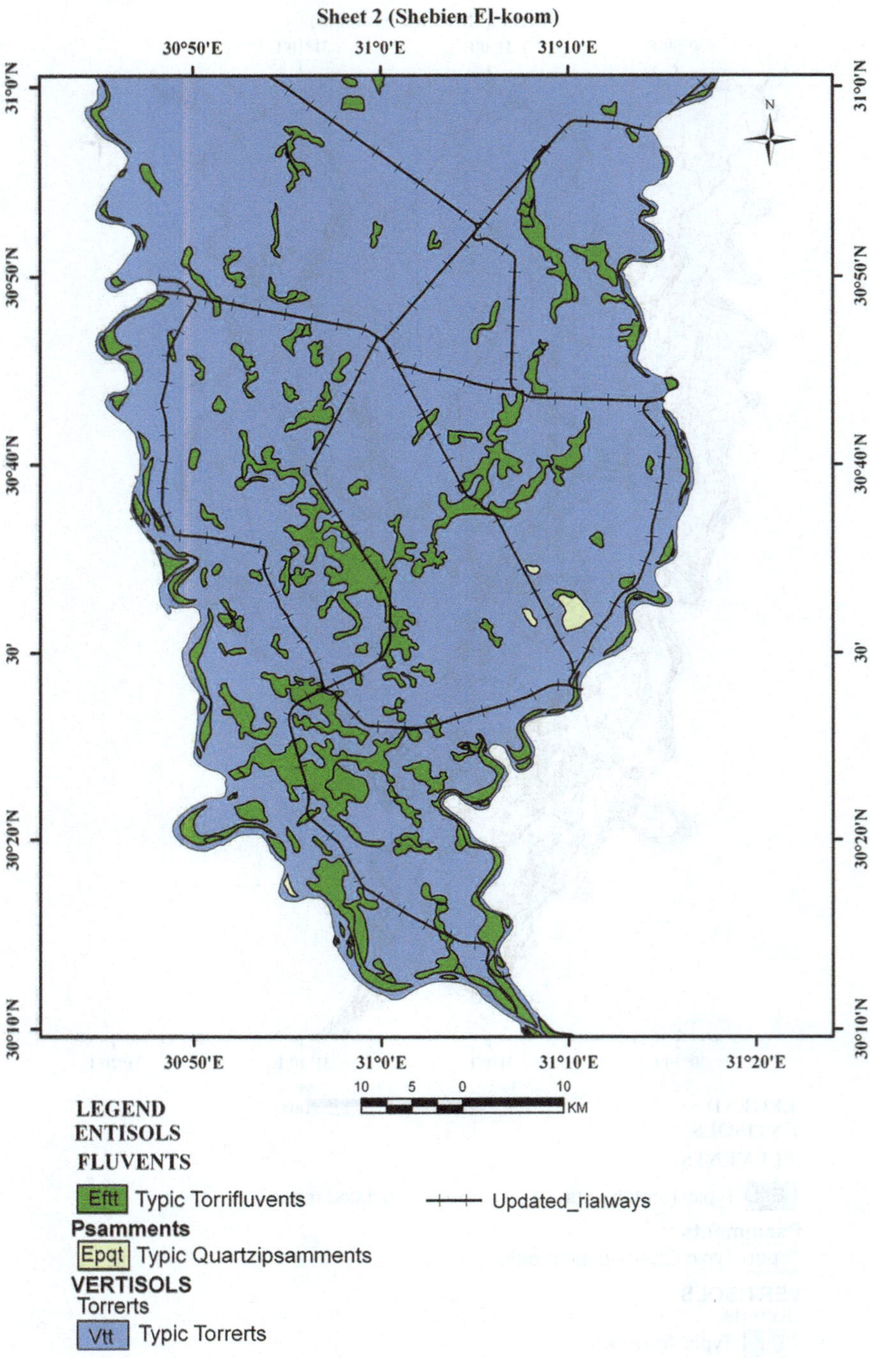

Fig. 4.63 Railways superimposed on the soil map, Shibin Al-Kom

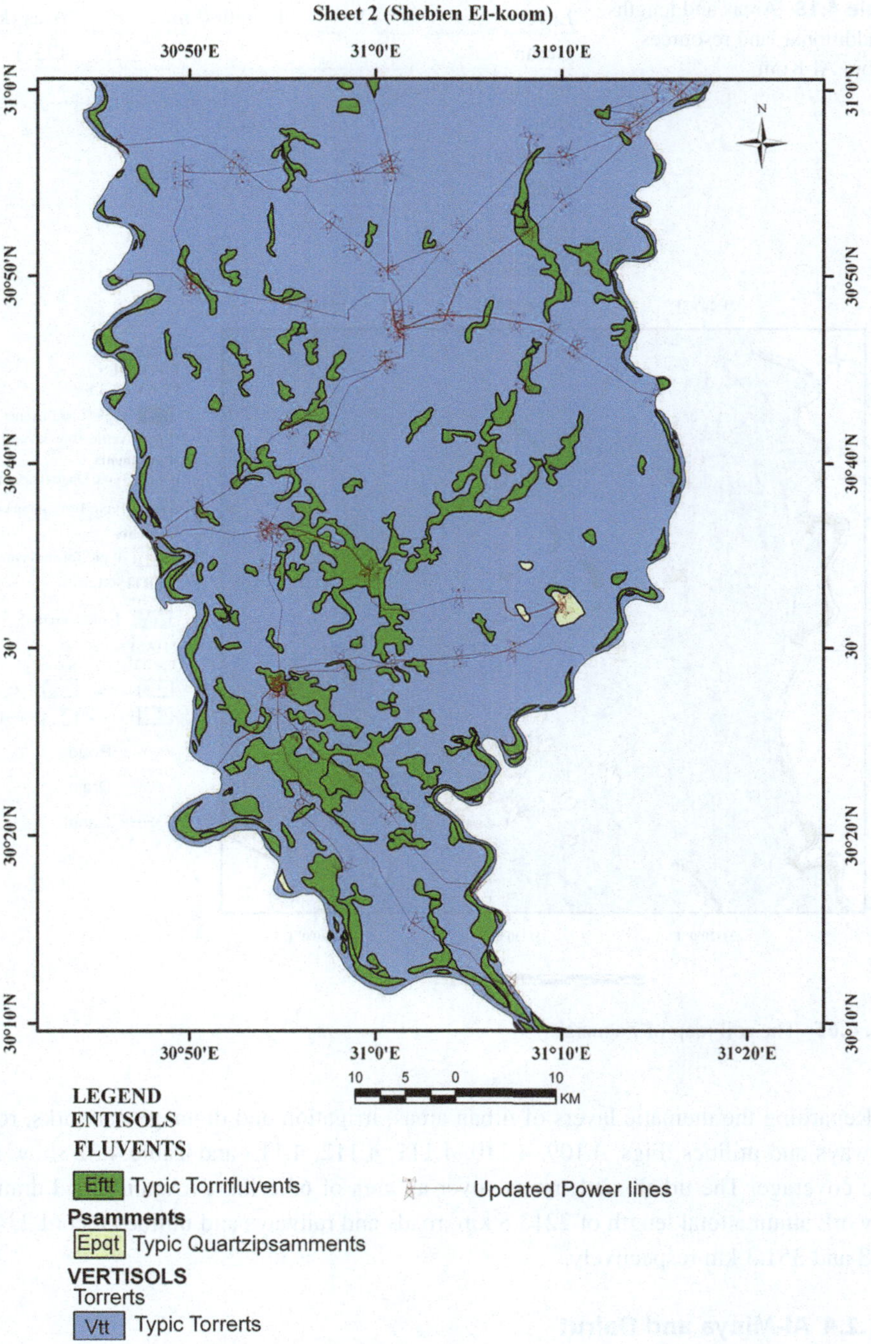

Fig. 4.64 Utilities, superimposed on the soil map, Shibin Al-Kom

Table 4.18 Areas and lengths of additional land resources, Shibin Al-Kom

Layer	Length (km)	Area (km^2)
Urban		173.3
Irrigation&Drainage	4654.2	
Roads	4612.0	
Railway	355.7	
Utilities	587.8	

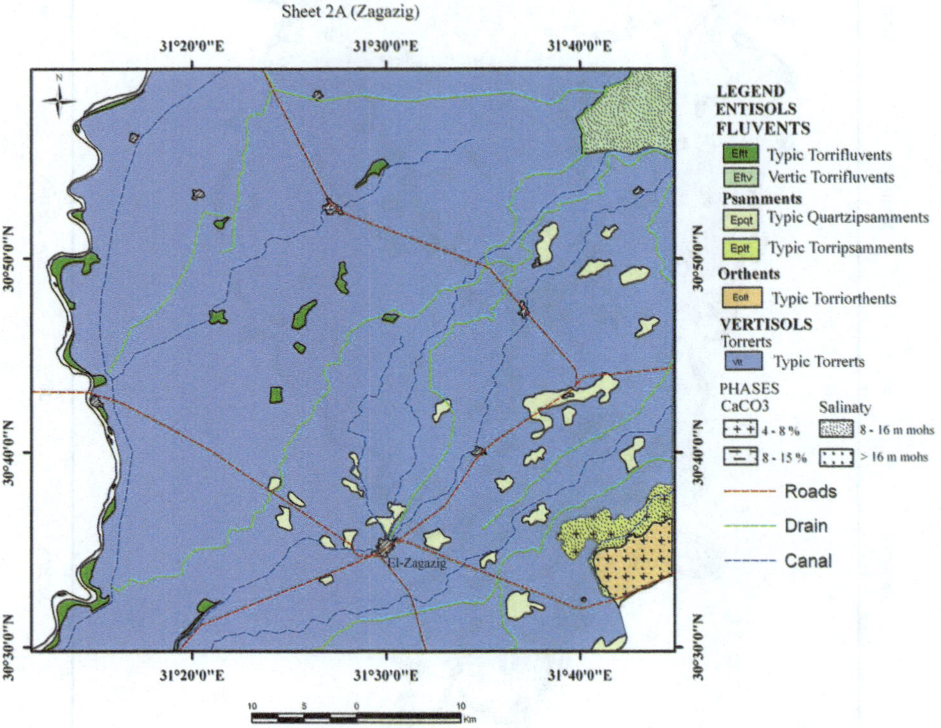

Fig. 4.65 The soil map of Zagazig

Regarding the thematic layers of urban areas, irrigation and drainage networks, roads, railways and utilities, Figs. 4.109, 4.110, 4.111, 4.112, 4.113 and Table 4.32 show their area coverage. The urban settlements cover an area of 63.8 km^2, irrigation and drainage network attain a total length of 2213.8 km, roads and railways and utilities record 2344.4, 58.8 and 351.3 km respectively.

4.1.2.4 Al-Minya and Dairut

Figure 4.114 shows the soil map of Al-Minya and Dairut area, which is located between latitudes 27° 30′ to 28° 25′ N and longitudes 30° 15′ to 31° 15′ E. Figure 4.115 shows the ETM+ image. The main sub-great groups Table 4.33 in this area are:

Table 4.19 Distribution of soil units in Zagazig

Soil Code	Area (km^2)	Area %
Eftt 2/2a	12.1	0.5
Eftt 3/3a	9.2	0.3
Eftt 4/2a	1.4	0.1
Eftt 4/3a	8.8	0.3
Eftv 4/4a	49.4	1.9
Eott 1/1a5	30.4	1.1
Epqt 1/1a	36.0	1.4
Epqt 4/1a	25.2	0.9
Eptt 1/1a3	29.6	1.1
River Nile	25.5	0.9
Urban	7.6	0.3
Vtt 4/4 a	2410.8	91.1
Total	2645.9	100.0

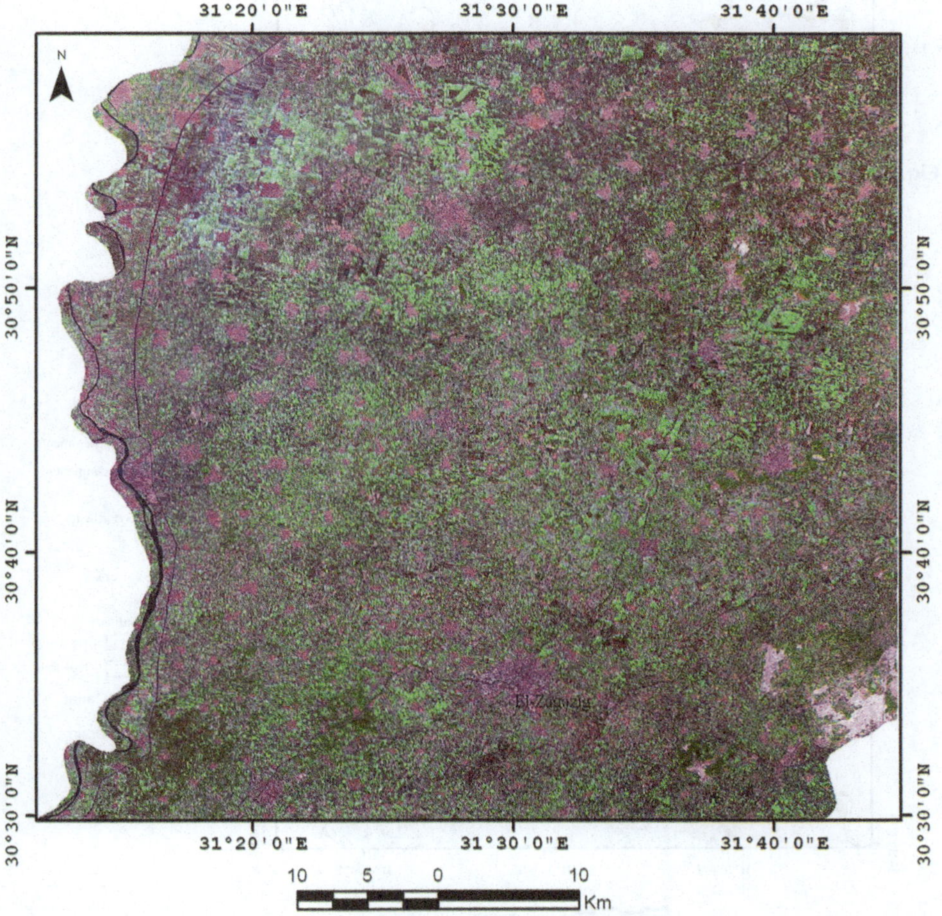

Fig. 4.66 ETM Landsat images of Zagazig

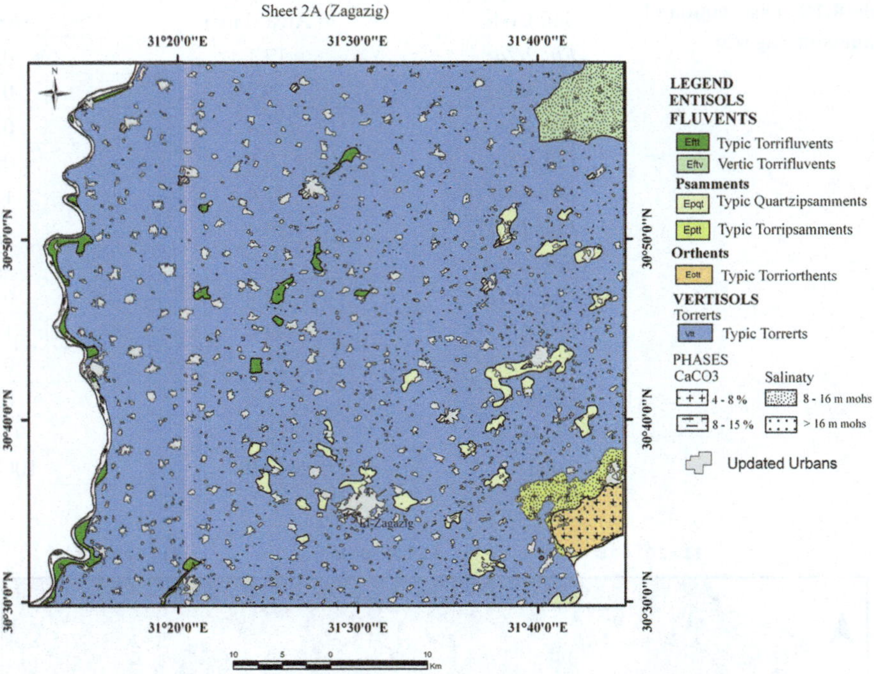

Fig. 4.67 Urban areas superimposed on the soil map, Zagazig

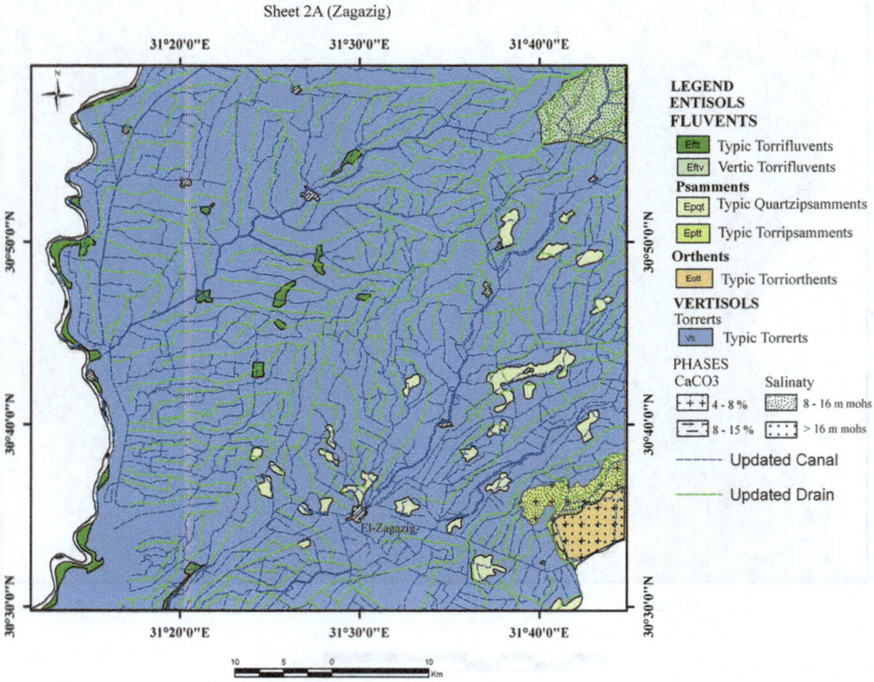

Fig. 4.68 Irrigation and drainage canals superimposed on the soil map, Zagazig

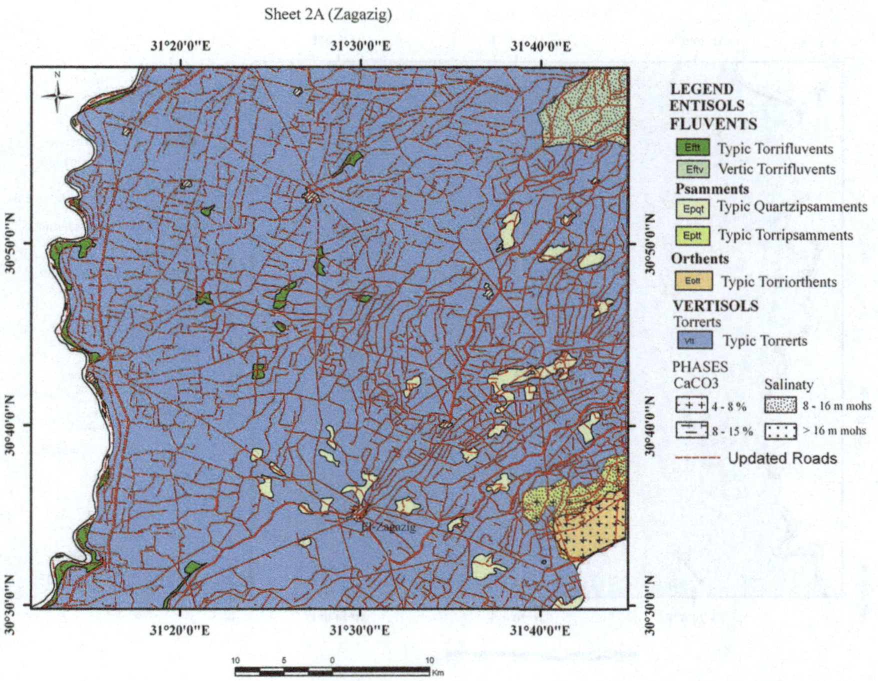

Fig. 4.69 Roads network superimposed on the soil map, Zagazig

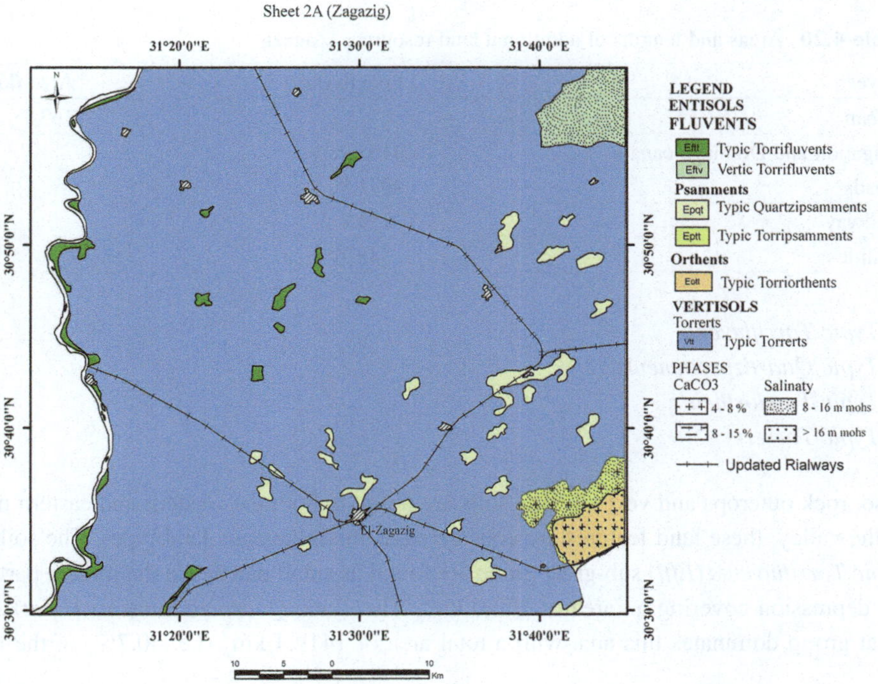

Fig. 4.70 Railways superimposed on the soil map, Zagazig

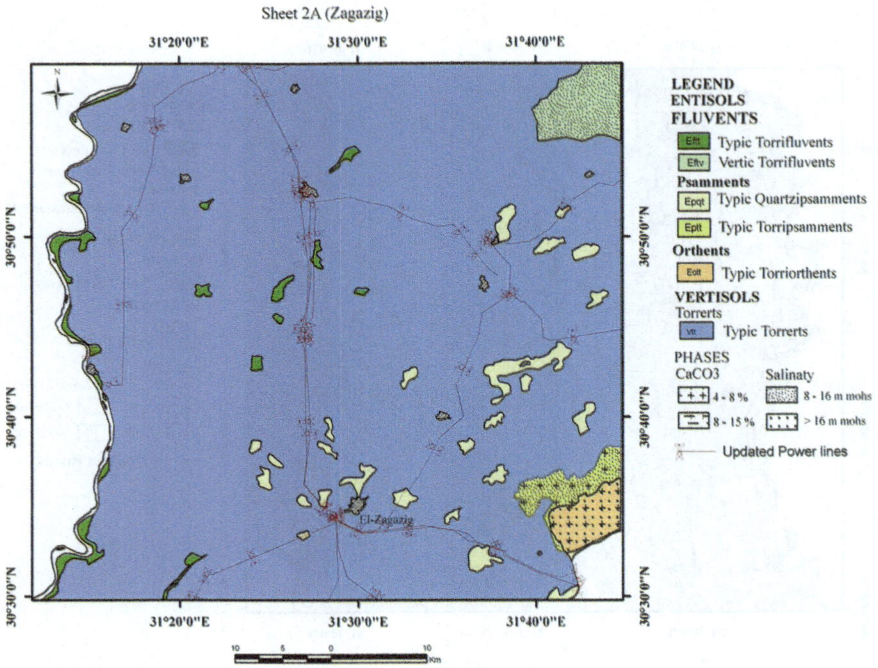

Fig. 4.71 Utilities, superimposed on the soil map, Zagazig

Table 4.20 Areas and lengths of additional land resources, Zagazig

Layer	Length (km)	Area (km^2)
Urban		163.1
Irrigation and Drainage canals	4138.9	
Roads	4071.8	
Railway	156.1	
Utilities	348.1	

- *Typic Torrifluvents*
- *Typic Quartizpsamments*
- *Typic Torriorthents*
- *Typic Torrerts*

Also, rock outcrops and very shallow soils are found in both the western and eastern parts of the valley, these land features are considered as miscellaneous land types. The soils of *Typic Torrifluvents (Eftt)* sub-great group are found as small patches in the middle parts of the depression covering an area of 193.0 km^2. The *Typic Quartizpsamments (Epqt) sub-*great group dominates this area with a total area of 1419.4 km^2 (i.e. 40.7%) of the map

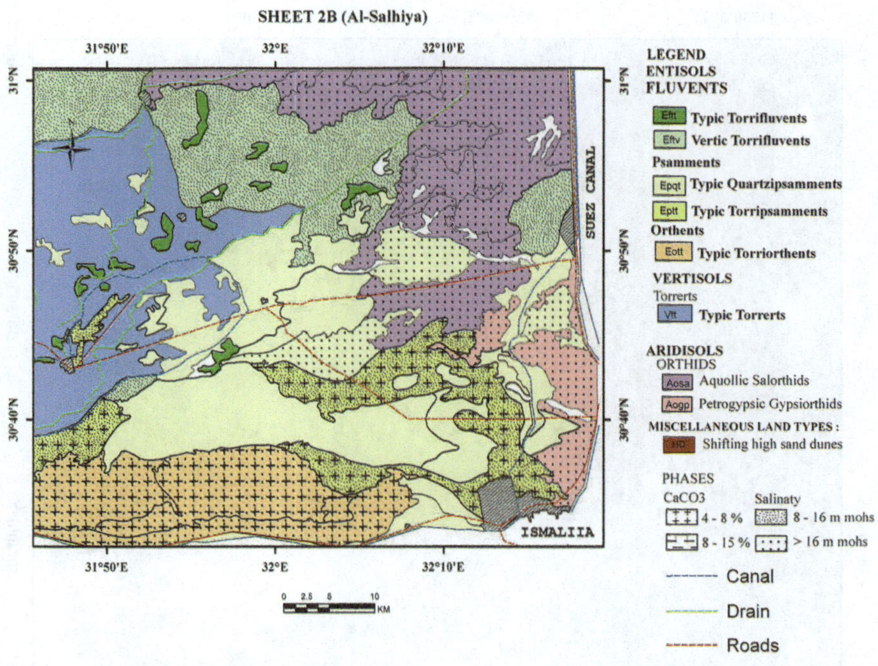

Fig. 4.72 The soil map of Salhia

sheet area. The *Typic Torriorthents (Eott) sub-great groups are found mainly in the western and eastern parts of the area covering an area of* 357.6 km^2. *Typic Torrerts (Vtt)* sub-great-group are found in small parts near to the Nile stream with a total area of 31.9 km^2.

The created database of this area Figs. 4.116, 4.117, 4.118, 4.119, 4.120 and Table 4.34 indicates that the total urban settlement area is 73.4 km^2; the total length of canals is 2205.3 km. The roads, railway and utilities network lengths are 2052.9, 114.3 and 215.6 km respectively.

4.1.2.5 Asyut and Manfalut

Figure 4.121 shows the soil map of Asyut & Manfalut area, which is located between latitudes 26° 05′ to 27° 30′ N and longitudes 30° 45′ to 31° 30′ E. Figure 4.122 shows the ETM+ image.The main taxonomic units of this area include the following sub-great groups, Table 4.35 shows the coverage distribution of different soil units in the area are as follow:

- *Typic Torrerts*
- *Typic Torrifluvents*
- *Typic Torriorthents*
- *Typic Quartizpsamments*

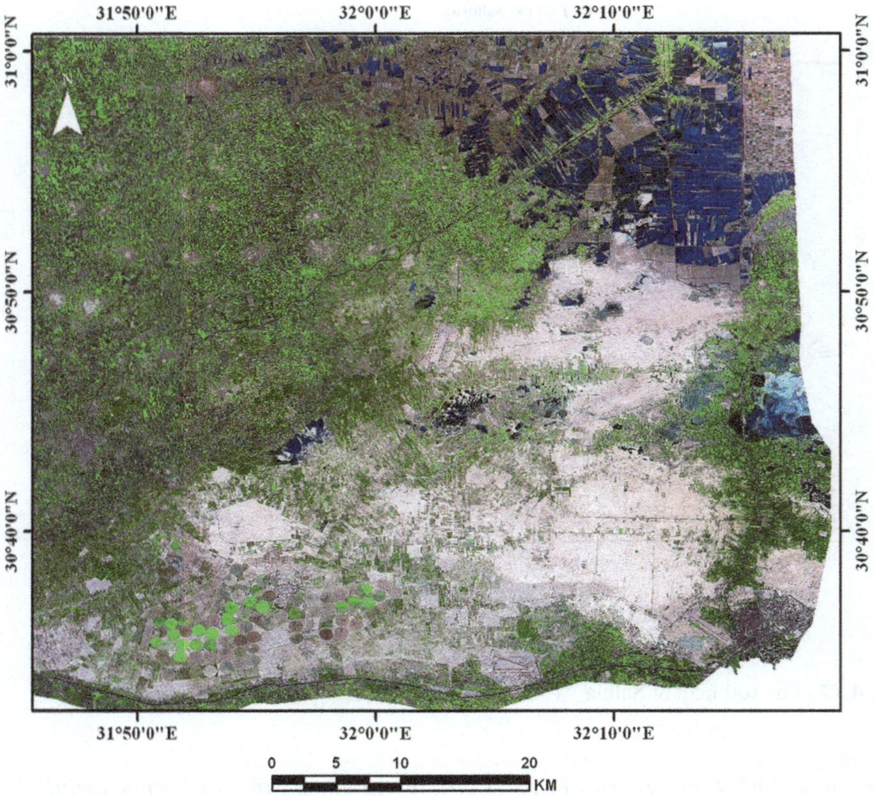

Fig. 4.73 ETM Landsat images of Salhia

The soils of *Typic Torrerts (Vtt)* sub-great group dominate the map sheet area (929.8 km^2) as they represent 21.8% of the coverage. The soil texture of this soil sub-great group is fine to very fine; soil salinity and %$CaCO_3$ are low as they don't exceed 4 dS/m and 4% respectively. *Typic Torrifluvents (Eftt)* sub-great group (202.2 km^2) exhibits areas adjacent to the Nile stream, in addition to some small patches mixed with the *Typic Torrerts* sub-great group. The soils of *Typic Quartzipsamments (Epqt)* exist in small scattered areas (150.6 km^2) around the Nile banks. These soils texture range from fine to coarse sand, low salinity and $CaCO_3$ contents. The *Typic Torriorthents (Eott) sub* great group represents 19.4% of the total coverage, characterized by high content of $CaCO_3$ and a coarse texture class.

The created database of this area Figs. 4.123, 4.124, 4.125, 4.126, 4.127 and Table 4.36 indicate that the total urban settlement area is 63.4 km^2; the total length of irrigation and drainage canals is 1709.5 km. The roads, railways and utilities network lengths are 1553.7, 111.0 and 263.5 km respectively.

Table 4.21 Distribution of soil units in Salhiya

Soil Code	Area (km^2)	Area %
Aogp 1/1a1	100.3	3.7
Aosa 3/3a1	105.2	3.9
Aosa 3/4a1	100.5	3.7
Aosa 4/1 a 1	16.1	0.5
Aosa 4/3a1	25.8	0.9
Aosa 4/4 a	12.5	0.5
Aosa1/4a1	3.1	0.1
Aosa4/4a1	160.5	5.9
Eftt 4/2a	4.4	0.2
Eftt 4/3a	28.2	1.0
Eftt 4/4a	4.9	0.2
Eftv 4/4a1	382.5	14.3
Eott 2/1a3	128.3	4.8
Eott1/1c3	87.2	3.2
Eott2/1a3	90.2	3.4
Eott2/1b3	7.1	0.3
Epqt 1/1a	216.4	8.1
Epqt 1/1a1	93.1	3.5
Epqt 1/1b	318.8	11.9
Epqt 1/1b1	19.6	0.7
Epqt 1/1c	39.7	1.5
Epqt 4/1a	15.9	0.6
Eptt 1/1a3	10.3	0.4
Eptt 2/1a3	225.5	8.4
HD	17.4	0.6
Urban	33.5	1.2
Vtt 4/4 a	424.7	15.8
W	12.3	0.5
Total	2683.8	100.0

4.1.2.6 Tahta and Sohag

Figure 4.128 shows the soil map of Tahta & Sohag area, which is located between latitudes 26° 30′ and 27° 00′ N and longitudes 31° 15′ and 32° 00′ E. Figure 4.129 shows the ETM+ image of the area. The main soil taxonomic units Table 4.37 include the following sub-great group:

- *Typic Torrerts*
- *Typic Torriorthents*
- *Typic Torrifluvents*
- *Typic Quartizpsamments*

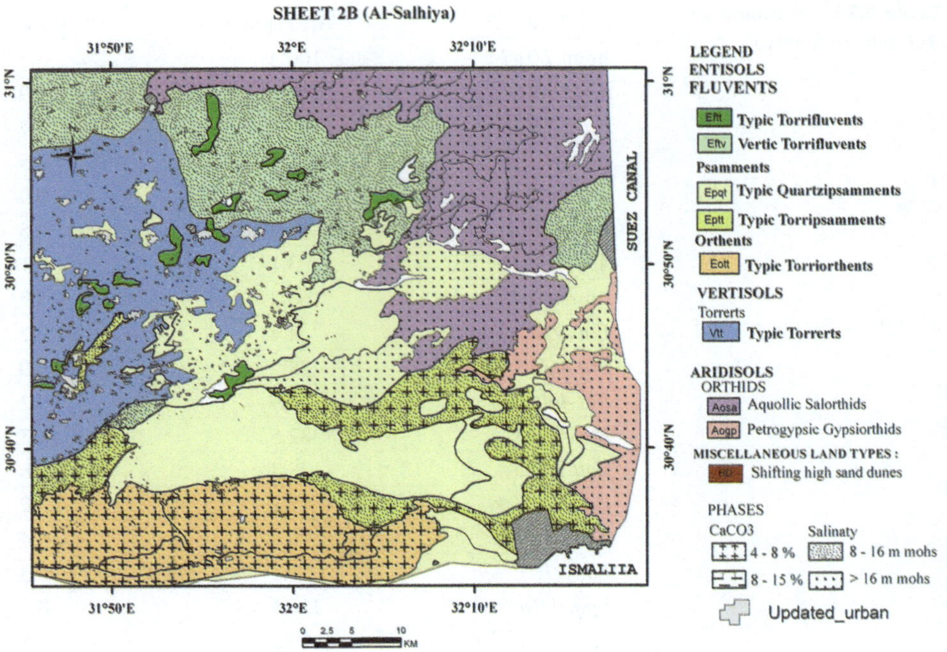

Fig. 4.74 Urban areas superimposed on the soil map, Salhia

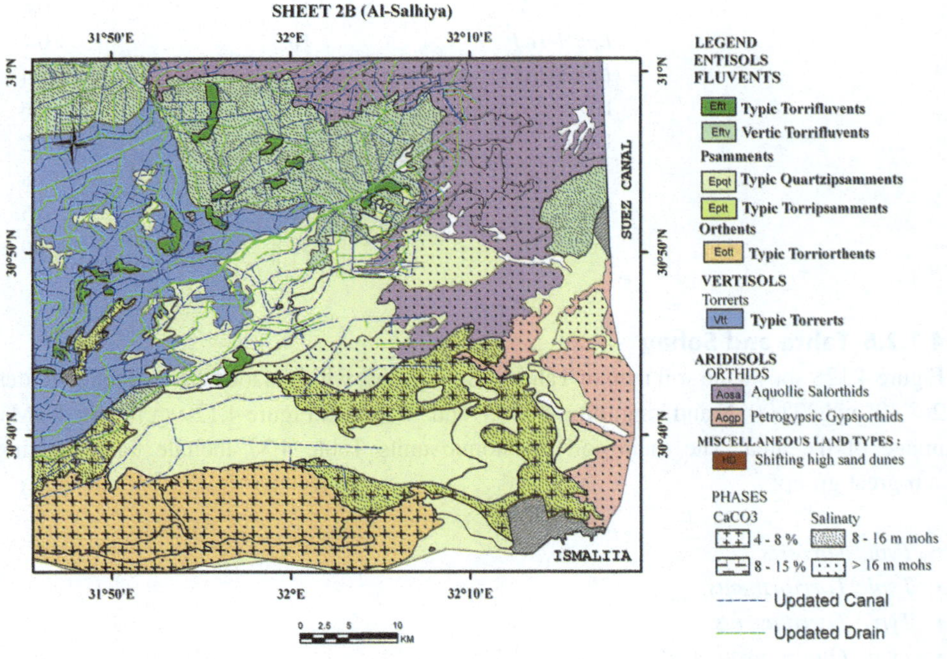

Fig. 4.75 Irrigation and drainage canals superimposed on the soil map, Salhia

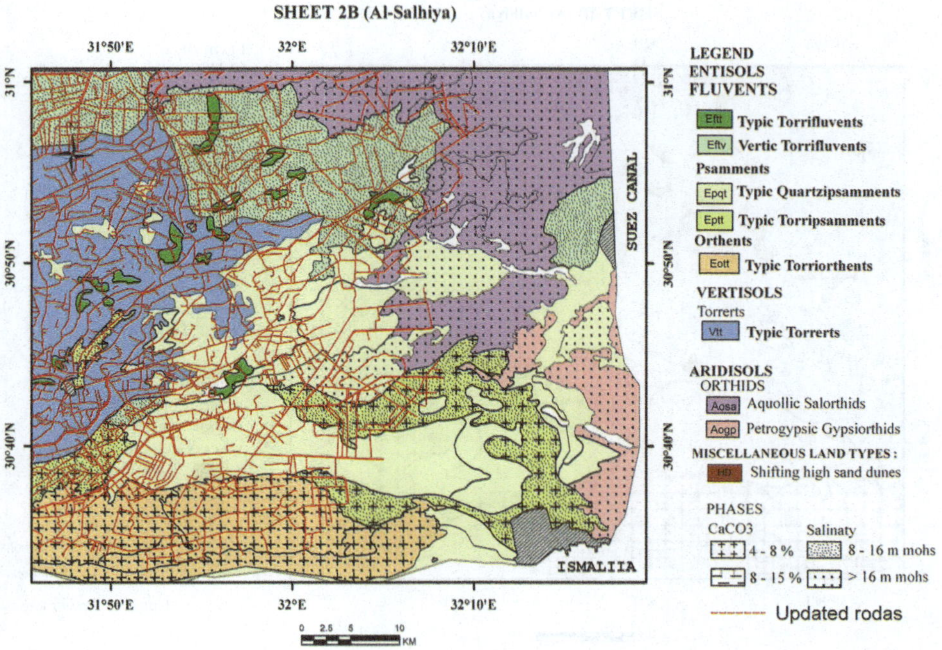

Fig. 4.76 Roads network superimposed on the soil map, Salhia

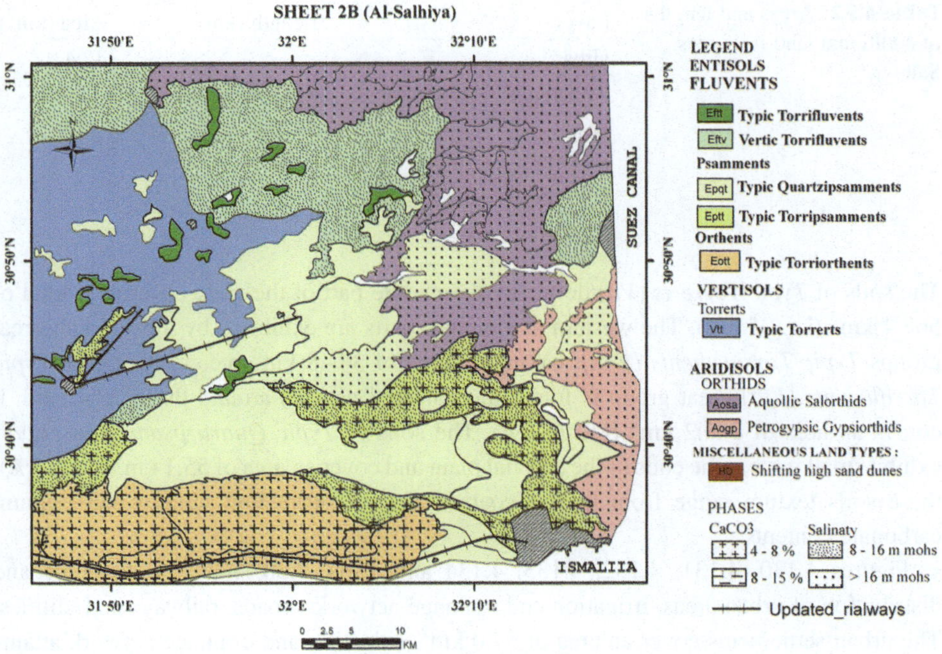

Fig. 4.77 Railways superimposed on the soil map, Salhia

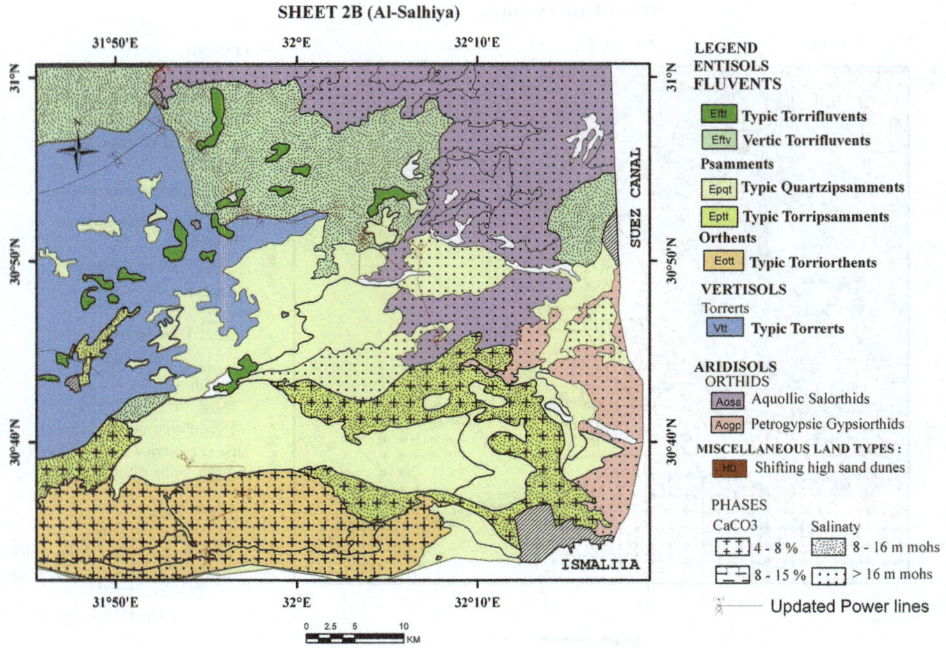

Fig. 4.78 Utilities, superimposed on the soil map, Salhia

Table 4.22 Areas and lengths of additional land resources, Salhiya

Layer	Length (km)	Area (km^2)
Urban		42.4
Irrigation& Drainage	1755.2	
Roads	2420.4	
Railway	86.5	
Utilities	124.1	

The soils of *Typic Torrerts (Vtt)* dominate the middle part of the area, covering a total of 568.4 km^2 (i.e. 25.3%). The western and eastern parts are exhibited by the soil sub-great groups *Typic Torriorthents (Eott),* having an area of 347.8 km^2 (i.e. 15.5%). The *Typic Torrifluvents (Eftt)* great group is found in elongated patches around the Nile stream, it covers an area of 266.2 km^2 (i.e. 11.9%). The soils of *Typic Quartizpsamments (Epqt)* exhibit small parts at the ends of the alluvial plain and cover an area of 55.1 km^2 (i.e. 2.5%); these soils texture range from fine to coarse sand and have low salinity and calcium carbonate contents.

Figures 4.130, 4.131, 4.132, 4.133, 4.134 and Table 4.38 show the coverage and distribution of urban areas, irrigation and drainage networks, roads, railways and utilities. The urban settlements cover an area of 57.0 km^2, irrigation and drainage network attains

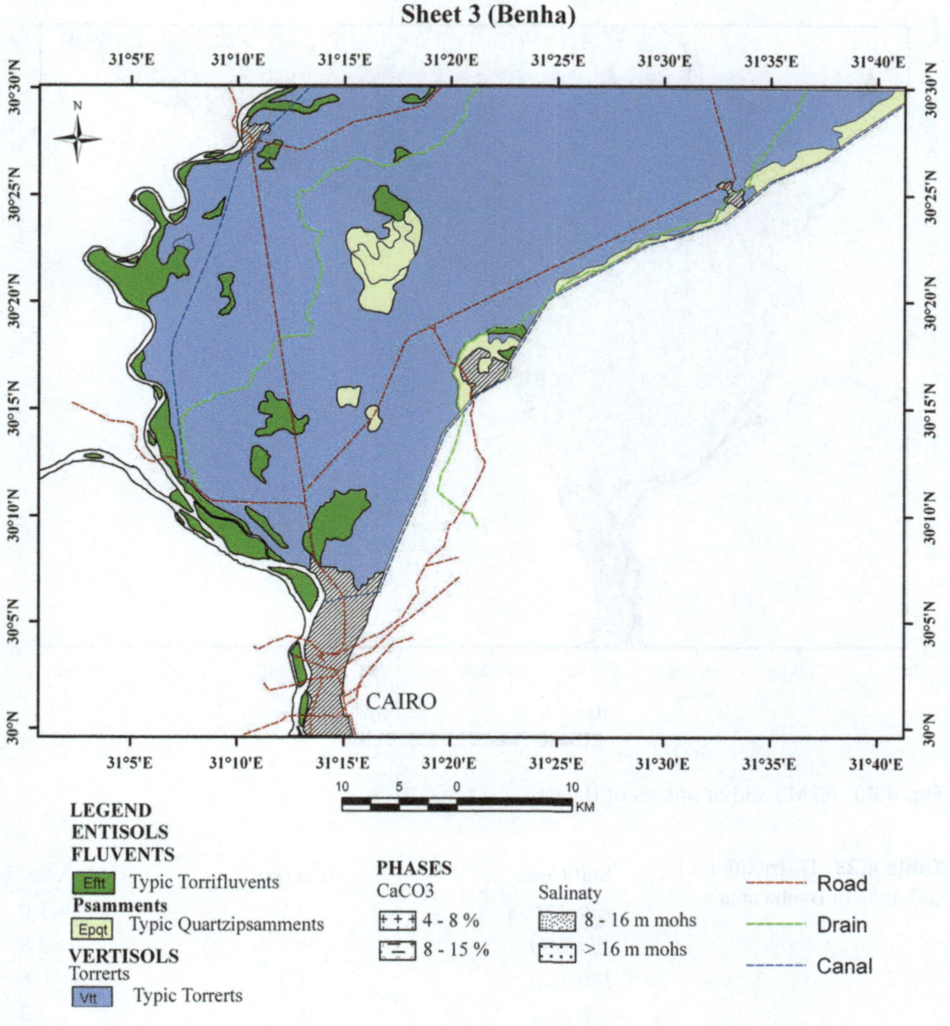

Fig. 4.79 Soil map of Banha

total length of 1378.2 km, roads networks 1409.3 km, railways 64.0 km and utilities is 143.0 km.

4.1.2.7 Al-Minsha & Abu Tisht

Figure 4.135 shows soil map of Al-Minsha & Abu Tisht area, which is located between latitudes 28° 20′ to 29° 00′ N and longitudes 30° 30′ to 31° 15′ E. Figure 4.136 shows the ETM+ image. The main taxonomic units Table 4.39 of the soils in this area include the following sub-great groups:

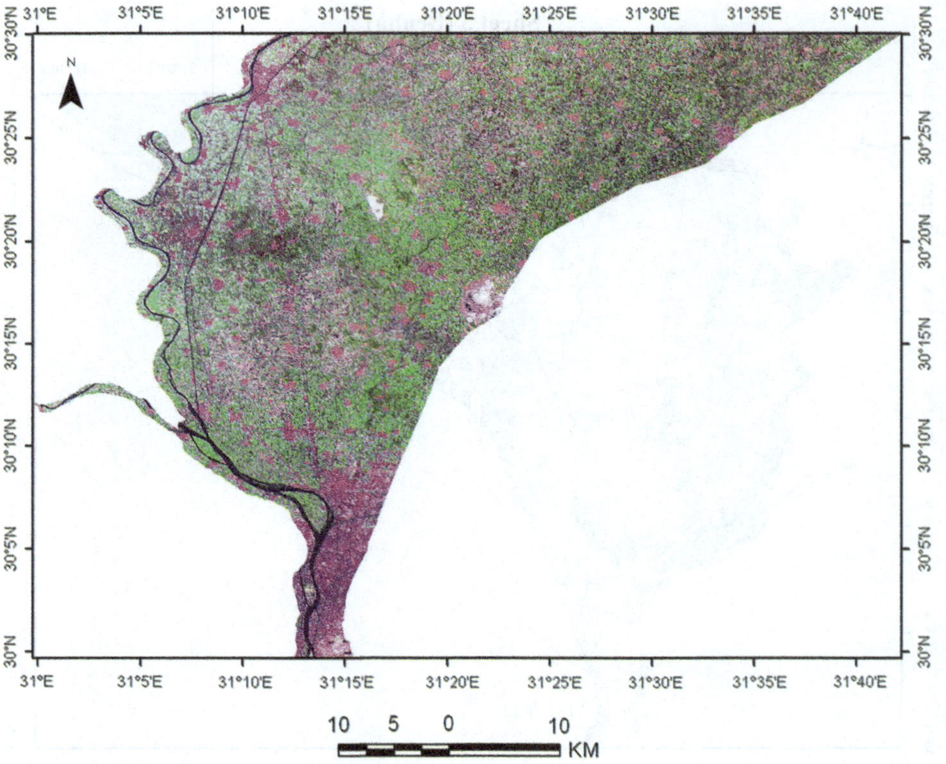

Fig. 4.80 ETM Landsat images of Banha

Table 4.23 Distribution of soil units in Banha area

Soil Code	Area (km^2)	Area %
Eftt 2–3/3a	13.9	1.0
Eftt 2/3a	38.9	2.8
Eftt 3/2a	8.7	0.6
Eftt 3/3a	46.4	3.3
Eftt 4/3a	5.4	0.4
Epqt 1/1a	46.4	3.3
Epqt 2/1a	22.8	1.6
River Nile	51.4	3.7
Urban	61.8	4.5
Vtt 3/4 a	1.3	0.1
Vtt 4/4 a	1088.8	78.6
Total	1385.8	100.0

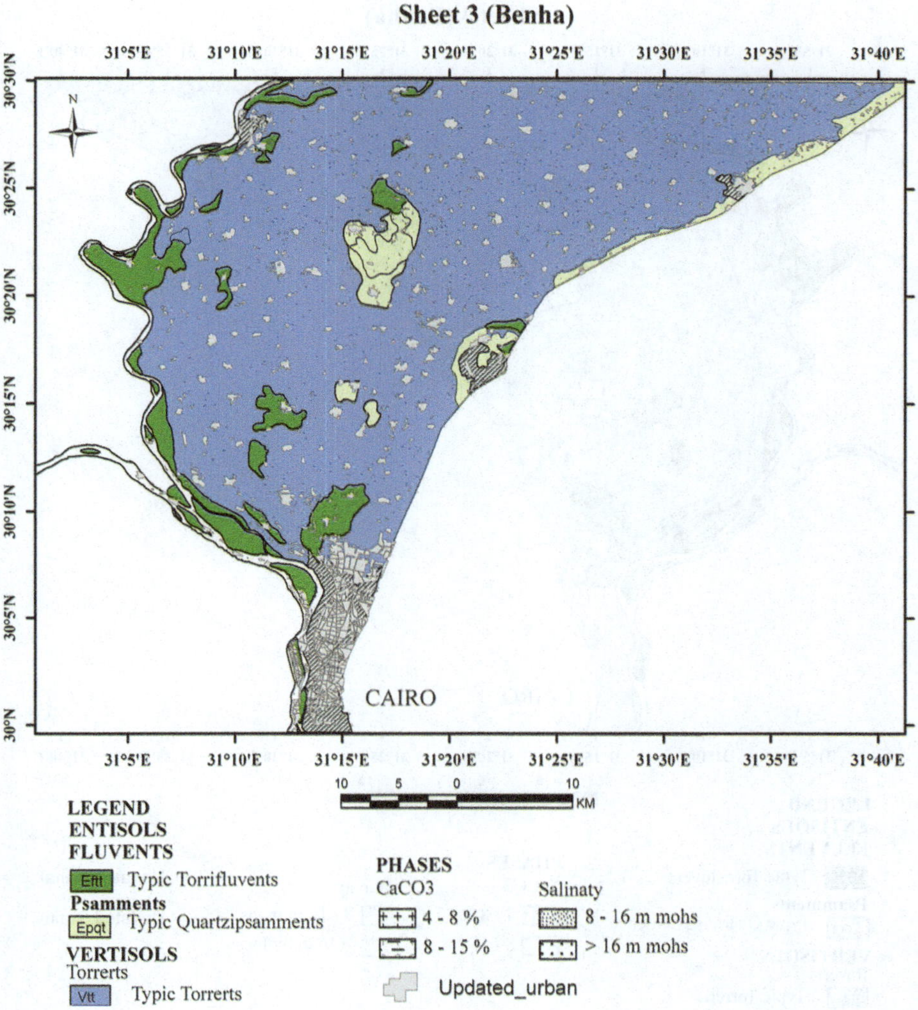

Fig. 4.81 Urban areas superimposed on the soil map, Banha

- *Typic Torrerts*
- *Typic Torriorthents*
- *Typic Torrifluvents*
- *Typic Quartizpsamments*

The soils of *typic Torrerts (Vtt)* dominate the middle part of the area covering a total of 572.80 km^2 (18.2% of the total area). The western and eastern parts are dominated by the soil sub-great groups *Typic Torriorthents (Eott),* having an area of 992.2 km^2 (i.e.31.5%). The *Typic Torrifluvents (Eftt)* great group is found in small patches and narrow strip around the River stream, covering a total area of 275.7 km^2 (8.8% of the total area).

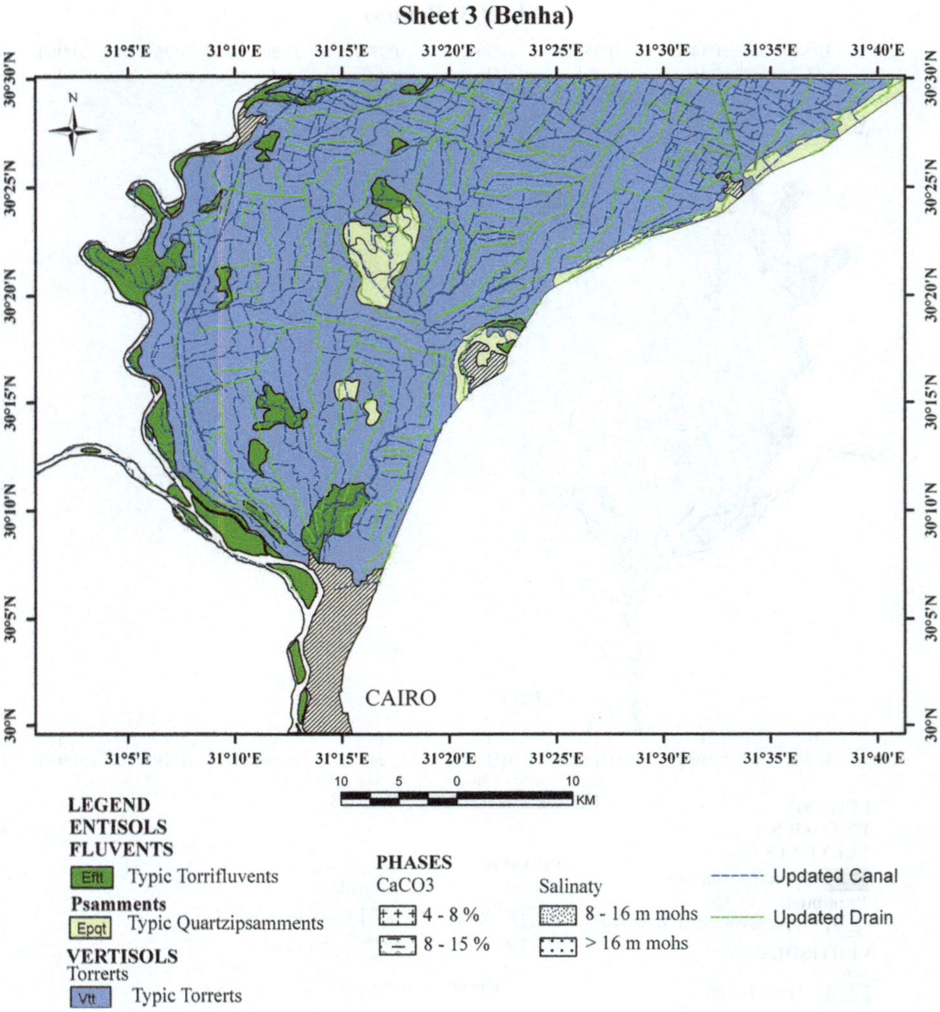

Fig. 4.82 Irrigation and drainage canals superimposed on soil map, Banha

The soils of Typic *Quartizpsamments (Epqt)* exhibit small patches (33.7 km^2), thier texture ranges from fine to coarse sand and low content of salinity and calcium carbonate. It should be referred that the soil texture is generally coarser in the eastern and western fringes of the valley than the situation at the middle parts. The $CaCO_3$ content ranges between 4 to 8% in the western and eastern fringes, while it doesn't exceed 4% in the rest of the area.

Urban areas, irrigation and drainage networks, roads, railways and utilities, as shown in Figs. 4.137, 4.138, 4.139, 4.140, 4.141 and Table 4.40 show their area coverage. The urban

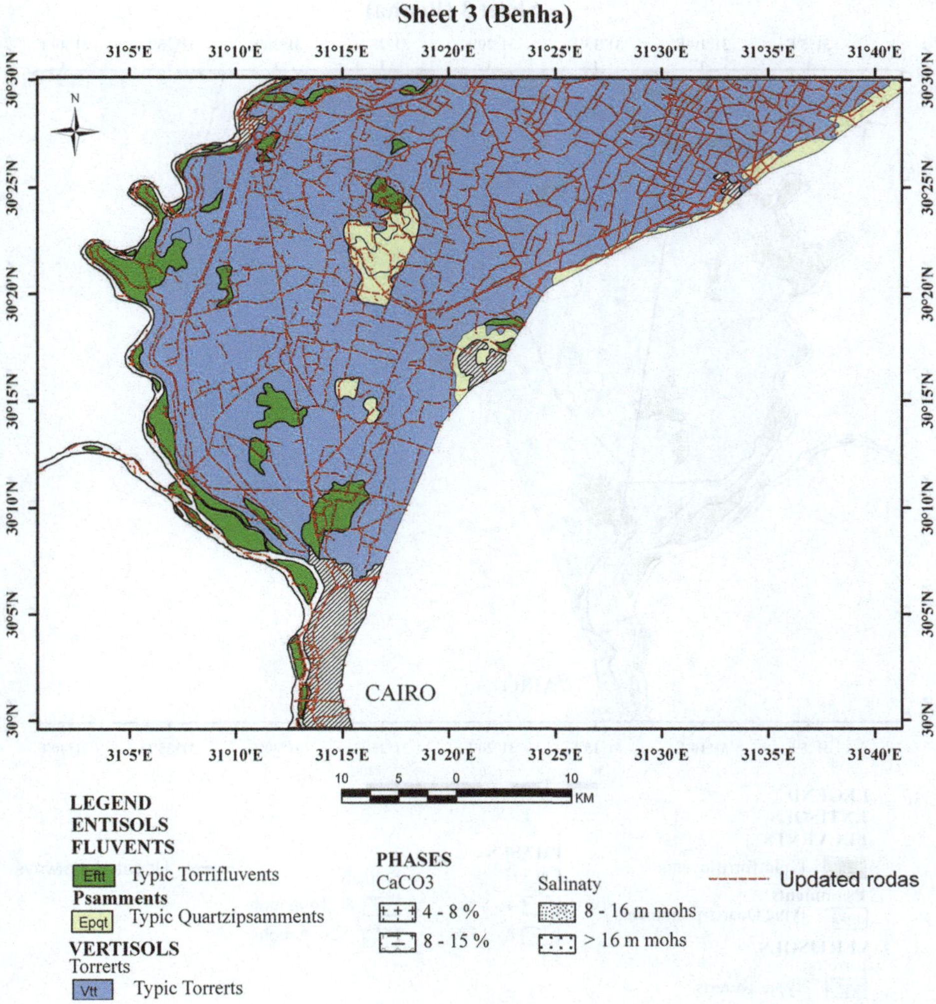

Fig. 4.83 Road network superimposed on soil map, Banha

settlements cover an area of 50.1 km^2, irrigation and drainage network attains total length of 1452.3 km, roads networks 1273.9 km, railways 257.0 km and utilities is 142.9 km.

4.1.2.8 Qena & Qus

Figure 4.142 shows the soil map of Qena & Qus area, which is located between latitudes 25° 50′ to 26° 20′ N and longitudes 32° 10′ to 33° 00′ E. Figure 4.143 shows the ETM+

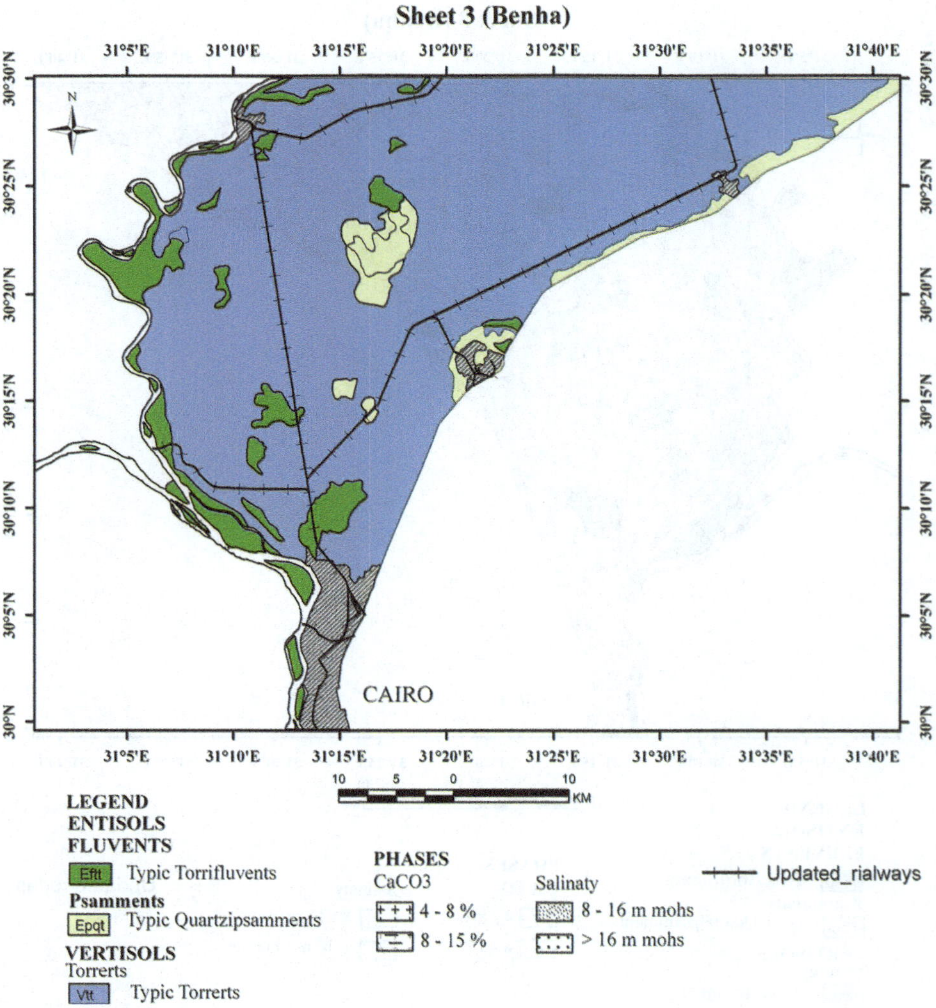

Fig. 4.84 Railway superimposed on the soil map, Banha

image of the area. The main taxonomic units (Table 4.41) of the soils in this area include the following sub-great groups:

- *Typic Torrifluvents*
- *Typic Torriorthents*
- *Typic Quartizpsamments*
- *Typic Torrerts*

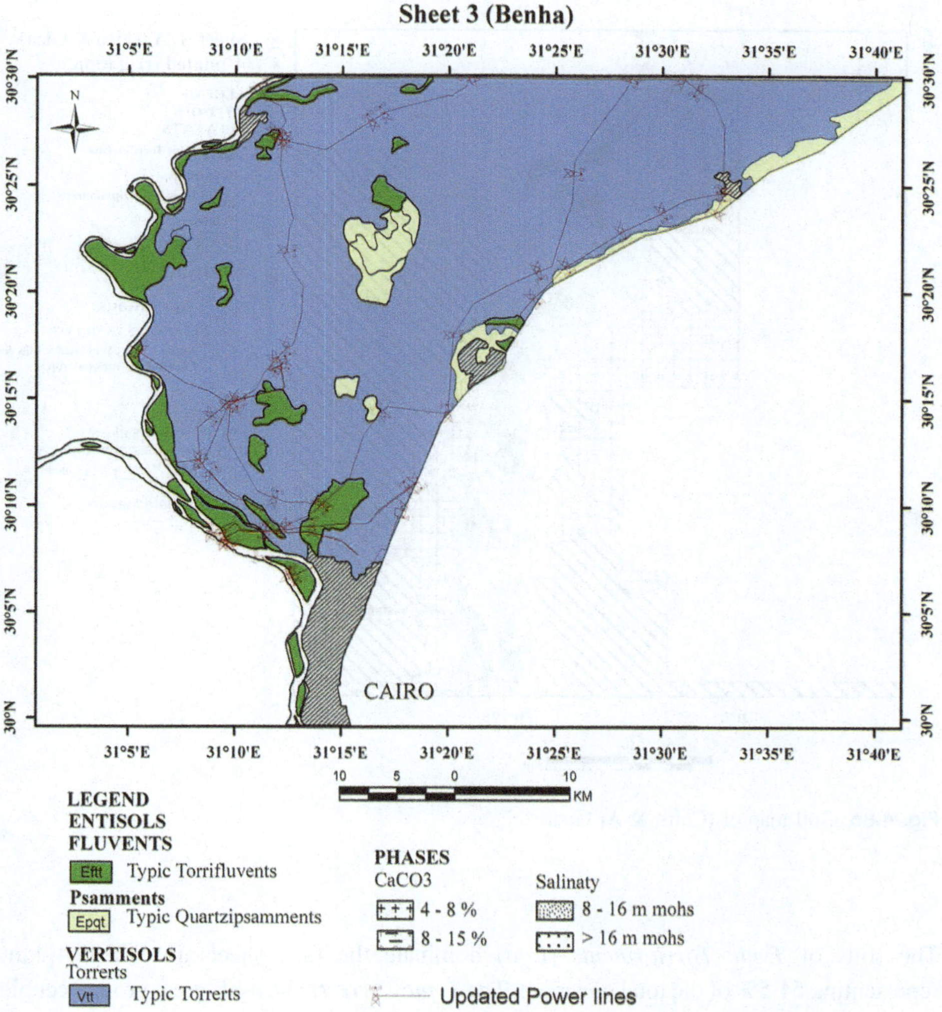

Fig. 4.85 Utilities superimposed on the soil map, Banha

Table 4.24 Areas and lengths of additional land resources, Banha

Layer	Length (km)	Area (km^2)
Urban		113.7
Irrigation& Drainage	2182.7	
Roads	1834.4	
Railway	164.4	
Utilities	256.8	

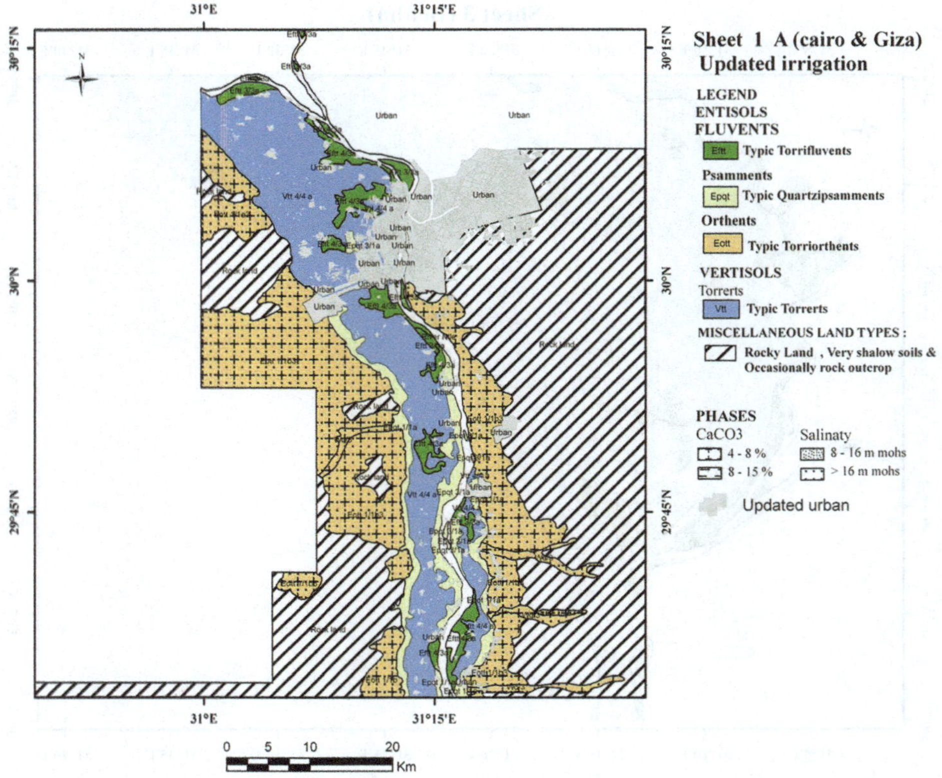

Fig. 4.86 Soil map of (Cairo & Al-Giza)

The soils of *Typic Torriorthents (Eott)* dominate the map sheet area (2295.4 km^2) representing 51.5% of the total coverage. The *Typic Torrerts (Vtt)* sub-great group occupies most of the Nile plain, with an area of 598.8 km^2 that represents 13.4% of area coverage. The *Typic Torrifluvents (Eftt)* sub-great group (207.5 km^2) *is* distributed on both sides of the Nile stream representing 4.7% of the area, interfering with the *Typic Torrerts* sub-great group.

The *Typic Quartizpsamments (Epqt)* exists on the interference zone with both western and eastern deserts, covering an area of 60.5 km^2 (i.e. 1.3%). The CaCO3 content in the desert fringes is likely to be in the range 4–8%. The soil salinity is significant in the Typic torriorthents as it ranges between 8 to 16 dS/m.

Figures 4.144, 4.145, 4.146, 4.147, 4.148 and Table 4.42 demonstrate the area of lands occupied by urban settlements and the lengths of railways, roads and utilities. The urban settlements cover an area of 50.9 km^2, irrigation and drainage network attains total length of 1299.2 km, roads networks 1512.7 km, railways 128.6 km and utilities is 159.7 km.

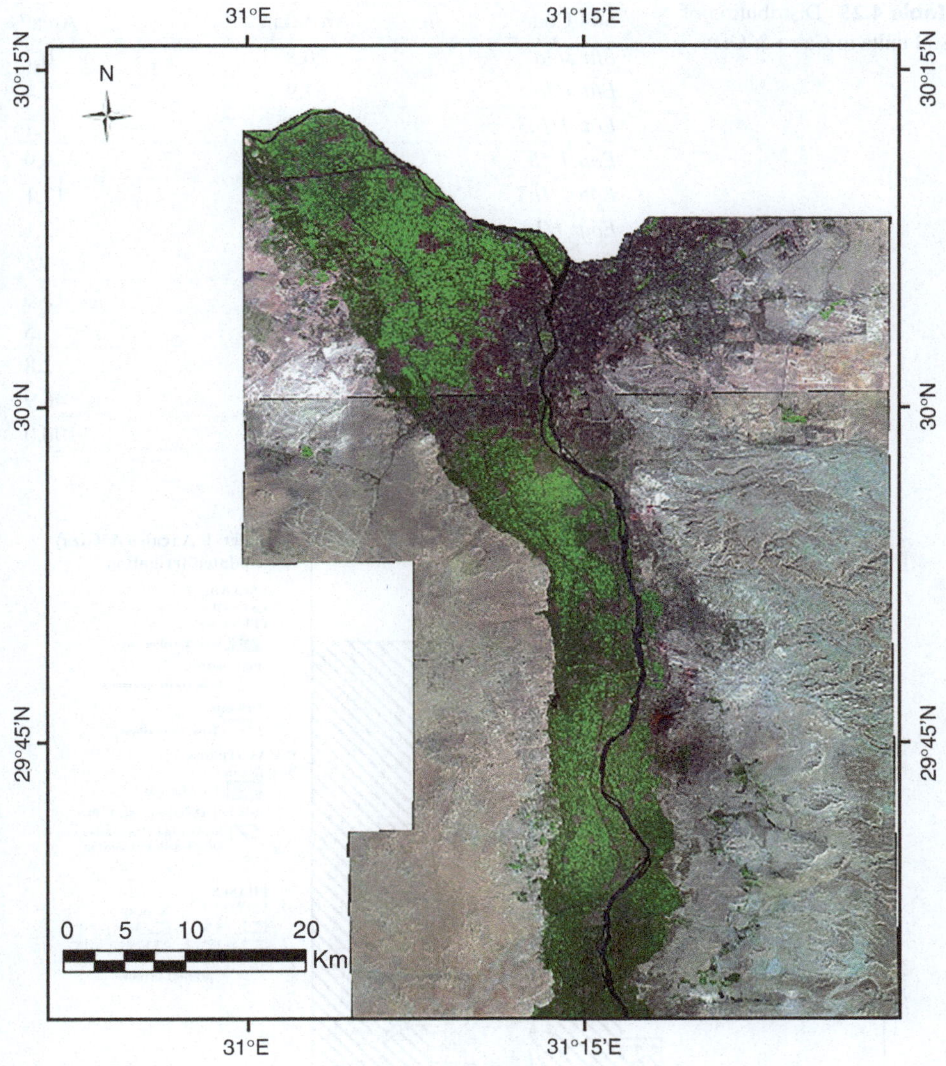

Fig. 4.87 ETM Landsat images of (Cairo & Al-Giza)

4.1.2.9 Luxor & Isna

Figure 4.149 shows soil map of Luxor & Isna area, which is located between latitudes 25° 15′ and 25° 50′ N and longitudes 32° 15′ and 33° 00′ E. Figure 4.150 shows the ETM+ image of the area. The main soil taxonomic units Table 4.43 of the soils in this area represent the following sub-great groups:

- *Typic Torrerts*
- *Typic Torriorthents*

Table 4.25 Distribution of soil units in Cairo & Giza

Soil Code	Area (km^2)	Area %
Eftt 3/3a	20.8	0.7
Eftt 4/3a	67.9	2.4
Eott 1/1a3	47.8	1.7
Eott 1/1b	24.7	0.9
Eott 1/1b3	479.5	17.1
Epqt 1/1a	60.0	2.1
Epqt 3/1a	42.9	1.5
Vtt 4/4 a	435.8	15.6
Wd/3	42.1	1.5
River Nile	78.6	2.8
Rock land	1313.1	46.9
Total	2801.3	100.0

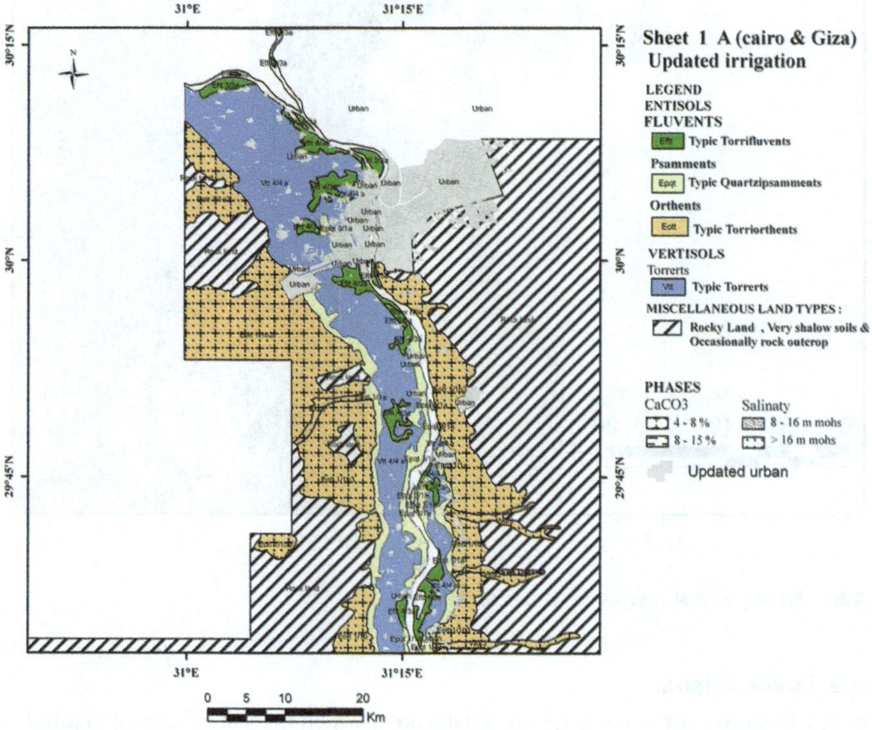

Fig. 4.88 Urban areas superimposed on the soil map, (Cairo & Al-Giza)

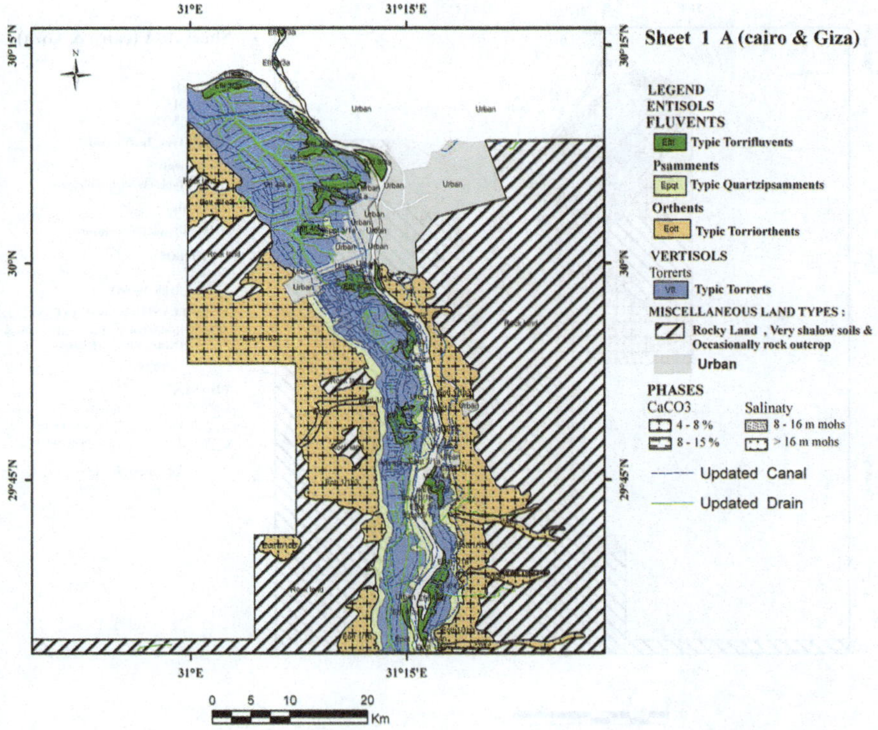

Fig. 4.89 Irrigation and drainage canals superimposed on soil map, (Cairo & Al-Giza)

- *Typic Torrifluvents*
- *Typic Quartizpsamments*

The soils of *Typic Torrerts (Vtt) sub-great group* dominate the Nile flood plain covering a total of 423.5 km^2, representing 11.1% of sheet area. The western and eastern Nile plain fringes are dominated by the soil sub-great groups, Typic *Torriorthents (Eott)* having an areas of 1655.4 km^2 representing 43.4% of total area. The *Typic Torrifluvents (Eftt) sub*-great group is found in small patches and narrow strip around the River stream covering an area of 96.2 km^2 representing 2.50%. The soils of *Typic Quartizpsamments (Epqt)* exhibit only 2.7% of the sheet area (i.e. 102.6 km^2), these soils texture range from fine to coarse and low contents of salinity and calcium carbonate.

The soil texture is generally coarser in both *Typic Torriorthents (Eott)* and Typic *Quartizpsamments (Epqt)* sub-great groups and the $CaCO_3$ content ranges between 4 to 8% in these soils.

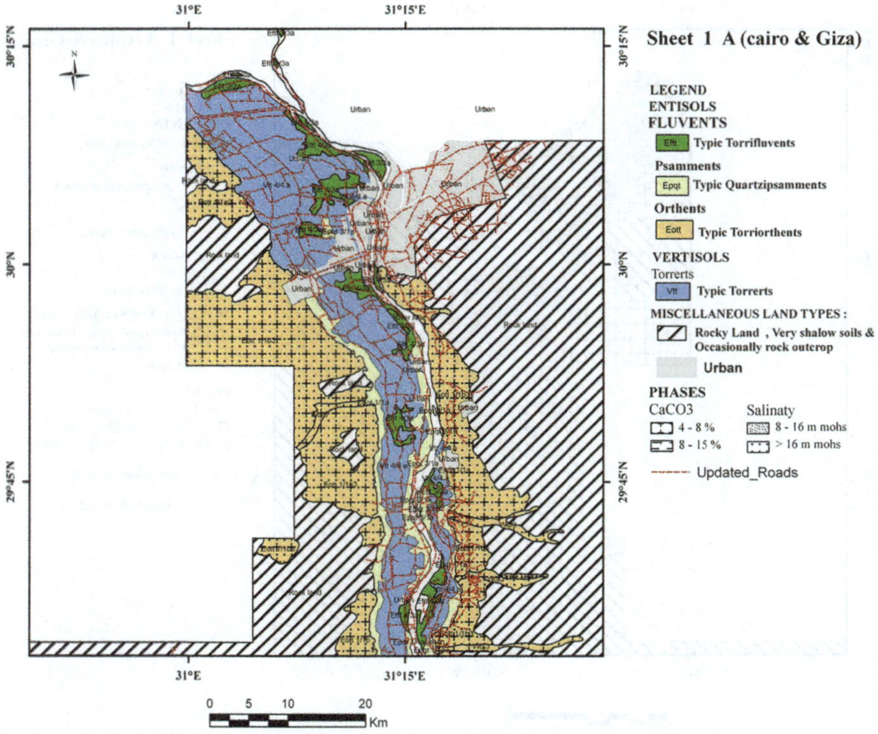

Fig. 4.90 Road network superimposed on soil map, (Cairo & Giza)

Regarding the thematic layers of urban areas, irrigation and drainage networks, roads, railways and utilities, Figs. 4.151, 4.152, 4.153, 4.154, 4.155 and Table 4.44 show their area coverage. The urban settlements cover an area of 33.9 km^2, irrigation and drainage network attains a total length of 1010.1 km, roads networks 1110.1 km, railways 83.2 km and utilities is 272.9 km.

4.1.2.10 Idfu

Figure 4.156 shows the soil map of Idfu area, which is located between latitudes 24° 40′ and 25° 15′ N and longitudes 32° 00′ and 33° 05′ E. Figure 4.157 shows the ETM+ image. The main taxonomic units of this area Table 4.45 include the following sub-great groups:

- *Typic Torrerts*
- *Typic Torrifluvents*
- *Typic Torriorthents*
- *Typic Quartizpsamments*

The soils of *Typic Torrerts (Vtt)* sub-great group exhibit small parts of this map sheet area (170.7 km^2) as they represents 5.4% of map area. The soil texture of this soil sub-great

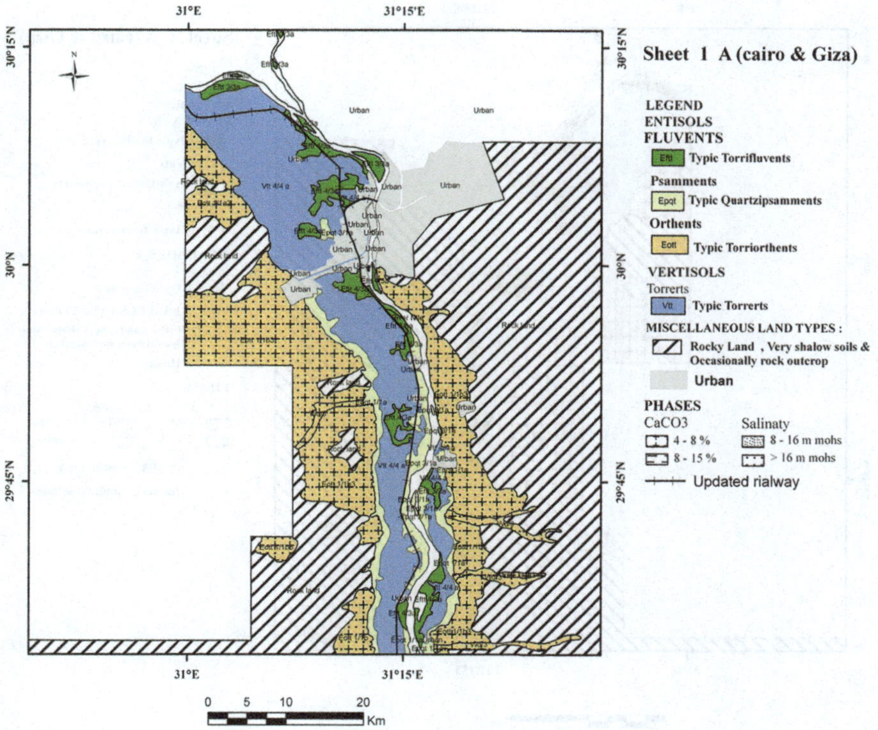

Fig. 4.91 Railway superimposed on the soil map, (Cairo & Giza)

group is fine to very fine; soil salinity and $\%CaCO_3$ content are low, as they not exceed 4 dS/m and 4% respectively. *Typic Torrifluvents (Eftt)* sub-great group (158.1 km^2) exhibits areas adjacent to Nile stream, in addition to some small patches interfering with the *Typic Torrerts sub-great group*. The soils of *Typic Quartzipsamments* (*Epqt 1/1a*) exist in very small scattered areas having a total of 21.3 km^2 distributed at the margins of the Nile banks, characterized by a coarse texture, low salinity and low $CaCO_3\%$. The *Typic Torriorthents (Eott)* sub great group is the main sub great groups in this area as it represents 53.7% of the total coverage (i.e. 1699.2 km^2).

They are characterized by high content of $CaCO_3$ and coarse texture class. The rock outcrops and very shallow soils also are found especially in the south of the area, covering a total area of 1090.3 km^2.

The created database of this area Figs. 4.158, 4.159, 4.160, 4.161, 4.162 and Table 4.46 indicate that the total urban settlement area is 18.1 km^2 while the total length of irrigation and drainage canals is 666.8 km. The length of roads network, railways and the utilities attain lengths of 520.1, 87.9 and 222.3 km respectively.

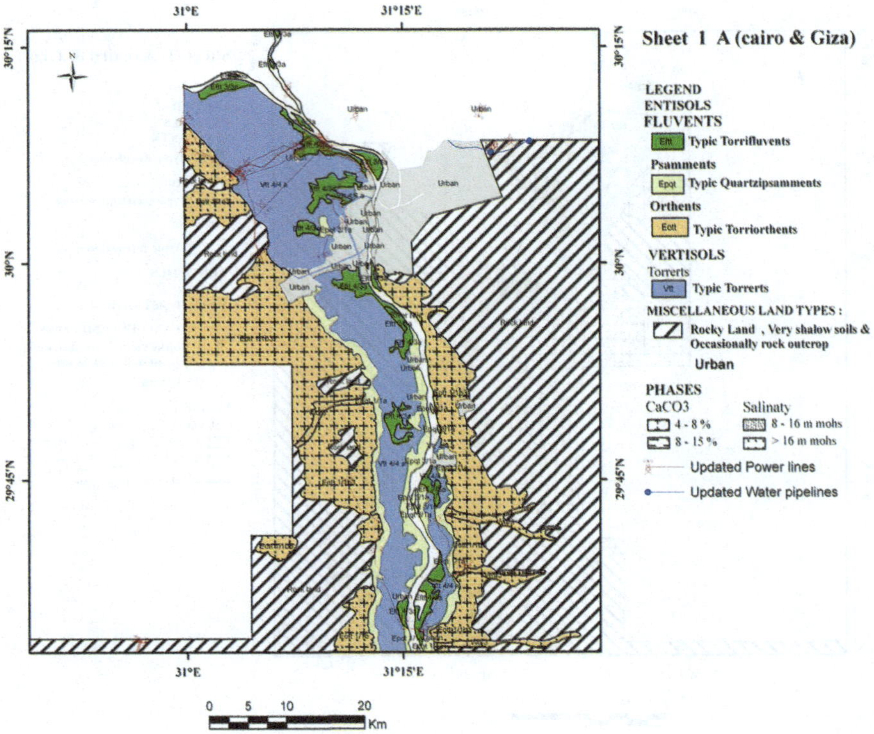

Fig. 4.92 Utilities, superimposed on the soil map, (Cairo & Giza)

Table 4.26 Areas and lengths of additional land resources, Cairo and Giza areas

Layer	Length (km)	Area(km^2)
Urban		154.6
Irrigation&Drainage	1546.8	
Road	1340.7	
Railway	202.4	
Utilities	255.8	

4.1.2.11 Kom Ombo & Aswan

Figure 4.163 shows the soil map of Kom Ombo & Aswan area, which is located between latitudes 24° 00′ to 24° 40′ N and longitudes 32° 40′ to 33° 30′ E. Figure 4.164 shows the ETM+ image of the area.The main taxonomic units of this area Table 4.47 includes the area coverage of the following sub-great groups;

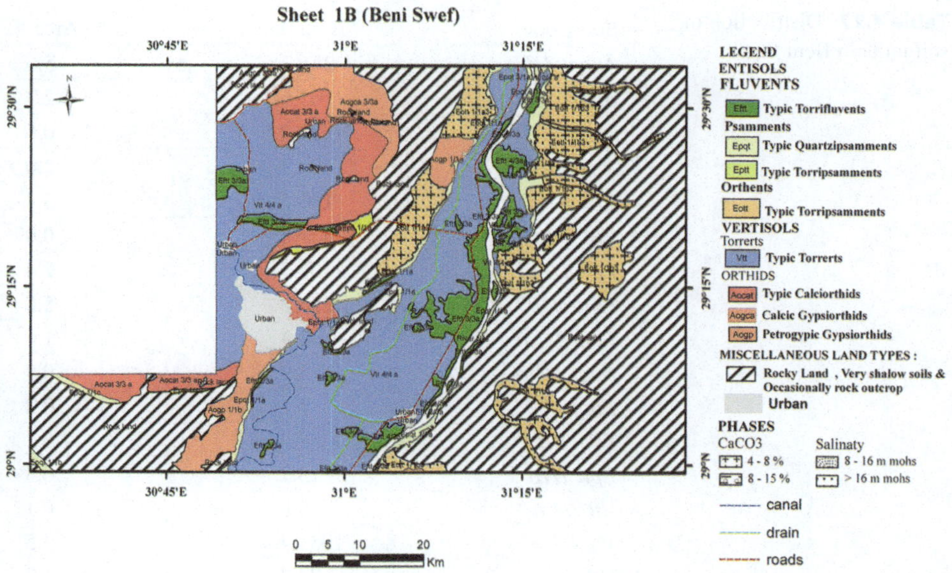

Fig. 4.93 Soil map of (Beni Swief)

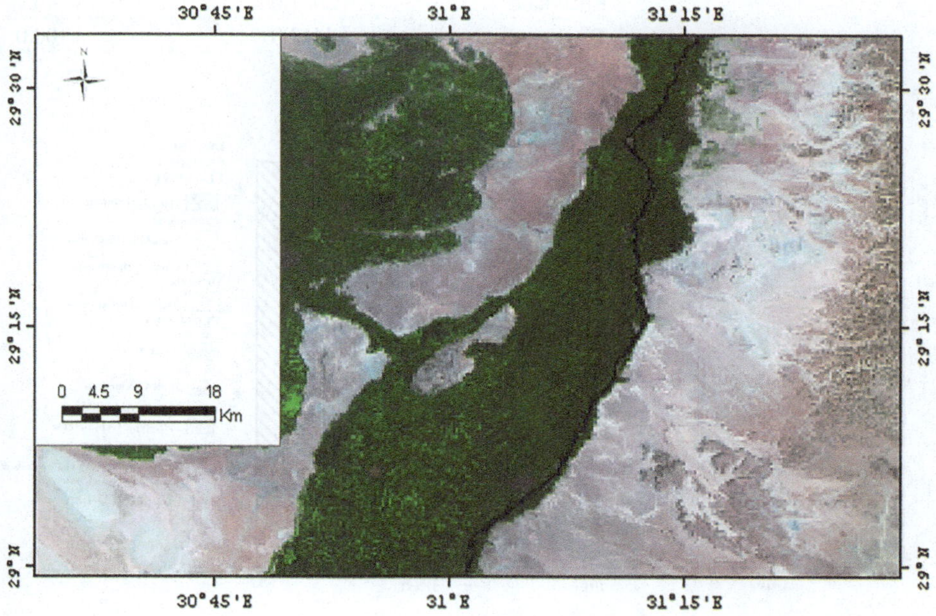

Fig. 4.94 ETM Landsat images of (Beni Swief)

Table 4.27 Distribution of soil units in Beni Suef

Soil Code	Area (km^2)	Area %
Aocat 3/3 a	246.7	5.6
Aogca 3/3a	103.6	2.3
Aogp 1/1a	28.0	0.6
Aogp 1/1b	86.9	2.0
Eftt 3/3a	150.4	3.4
Eftt 4/3a	37.8	0.09
Eott 1/1a3	107.7	2.4
Eott 1/1b3	258.2	5.8
Epqt 1/1a	72.4	1.6
Epqt 1/1b	3.9	0.9
Epqt 1/1c	7.1	0.2
Epqt 3/1a	4.9	0.1
Eptt 1/1a	13.2	0.3
Vtt 3/4 a	2.8	0.1
Vtt 4/4 a	1081.4	24.5
Urban	73.3	1.7
Wd/3	29.2	0.7
River Nile	64.5	1.5
Rock land	2041.0	46.2
Total	4413.0	100.0

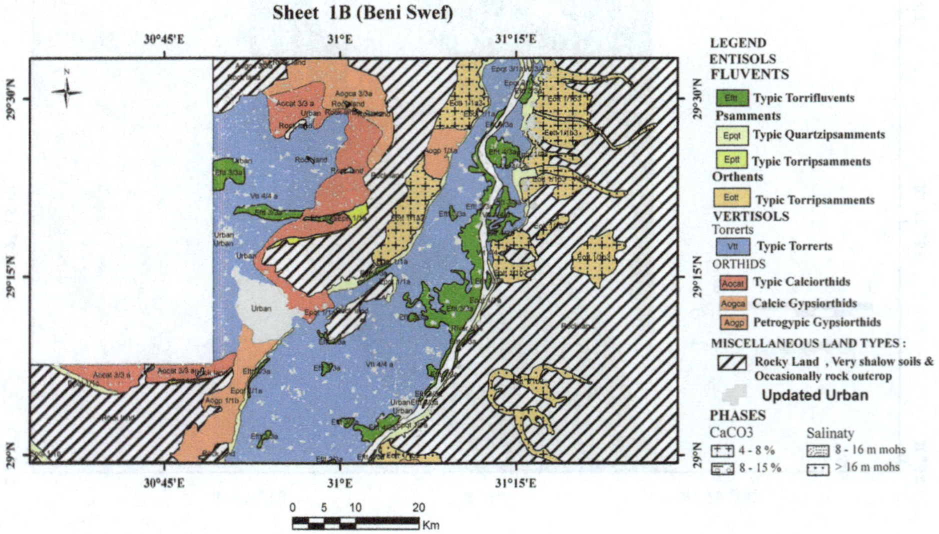

Fig. 4.95 Urban areas superimposed on the soil map, (Beni Swief)

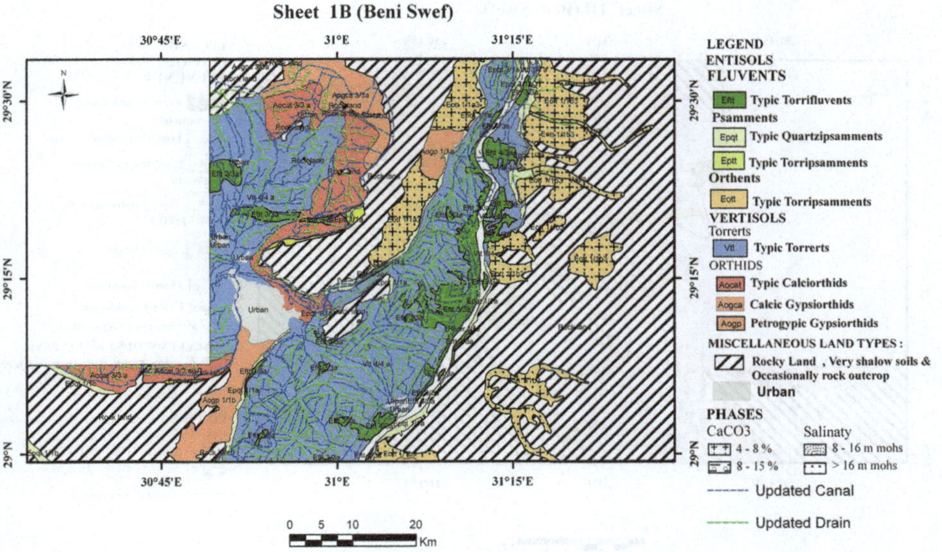

Fig. 4.96 Irrigation and drainage canals superimposed on soil map, (Beni Swief)

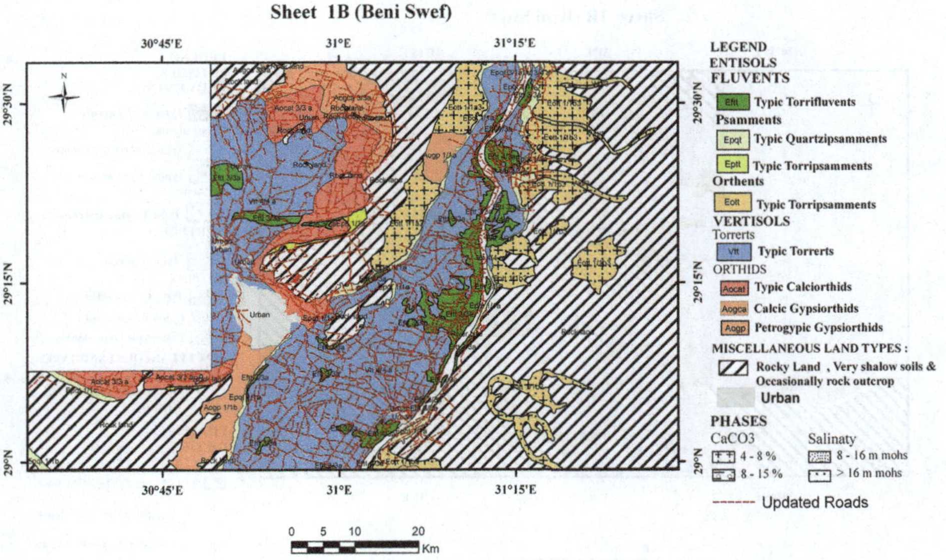

Fig. 4.97 Road network superimposed on soil map, (Beni Swief)

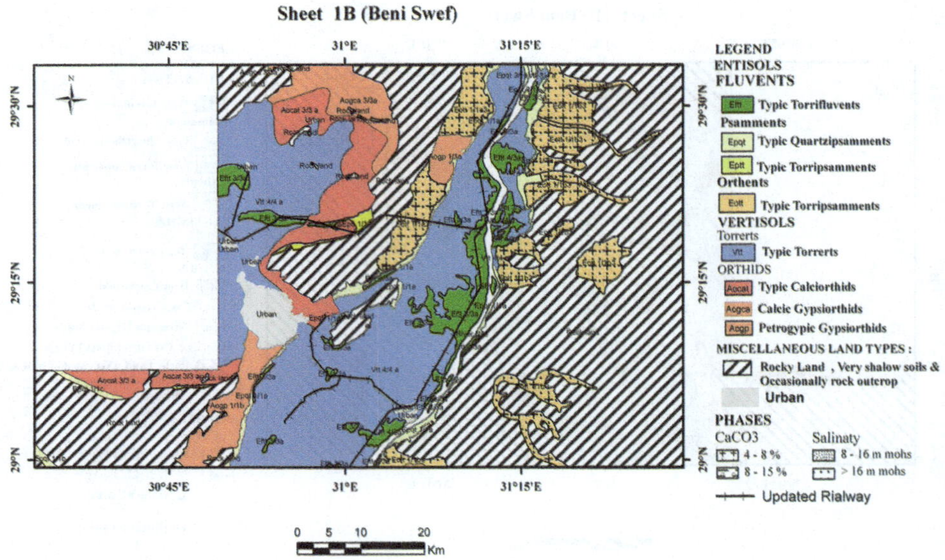

Fig. 4.98 Railway superimposed on the soil map, (Beni Swief)

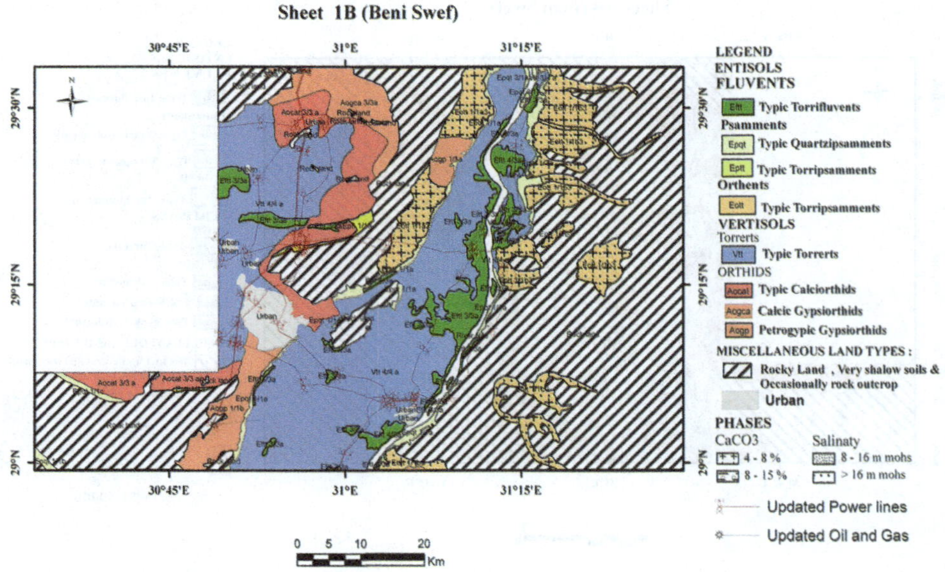

Fig. 4.99 Utilities, superimposed on the soil map, (Beni Swief)

Table 4.28 Areas and lengths of additional land resources, Beni Suef area

Layer	Length (km)	Area (km^2)
Urban		78.3
Irrigation&Drainage	3107.5	
Roads	2737.8	
Railways	158.8	
Utilities	506.2	

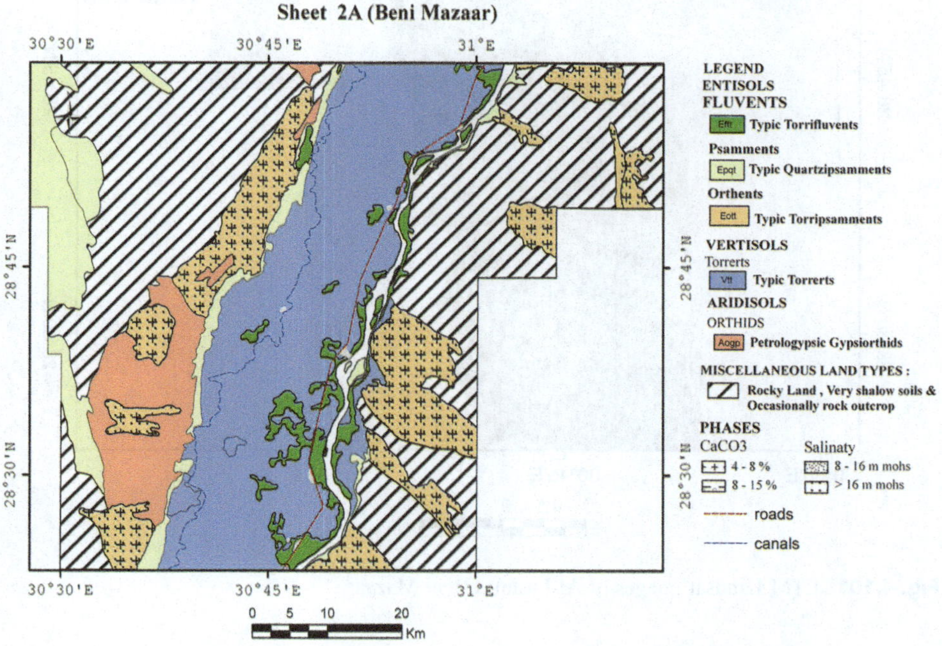

Fig. 4.100 Soil map of Al-Fashn &Beni Mazaar

- *Typic Torrerts*
- *Typic Torrifluvents*
- *Typic Torriorthents*
- *Typic Quartizpsamments*

The soils of *Typic Torriorthents (Eott)* sub great group dominate the map sheet area, representing 49.3% of the total area (i.e. 1804.3 km^2). These soils are characterized by high content of $CaCO_3$ and coarse textural class. *Typic Torrerts* sub-great group (*Vtt*) covers an area of 287.8 km^2 represents 7.8% of map area. The soil texture of this sub-great group is fine to very fine; soil salinity and %$CaCO_3$ are low as they do not exceed 4 dS/m and 4% respectively. *Typic Torrifluvents* (*Eftt*) sub-great group (140.2 km^2) exhibits areas adjacent to Nile stream in addition to some small patches among the *Typic Torrerts sub-great group*. The soils of *Typic Quartzipsamments* (*Epqt*) exist in very small-scattered areas covering a

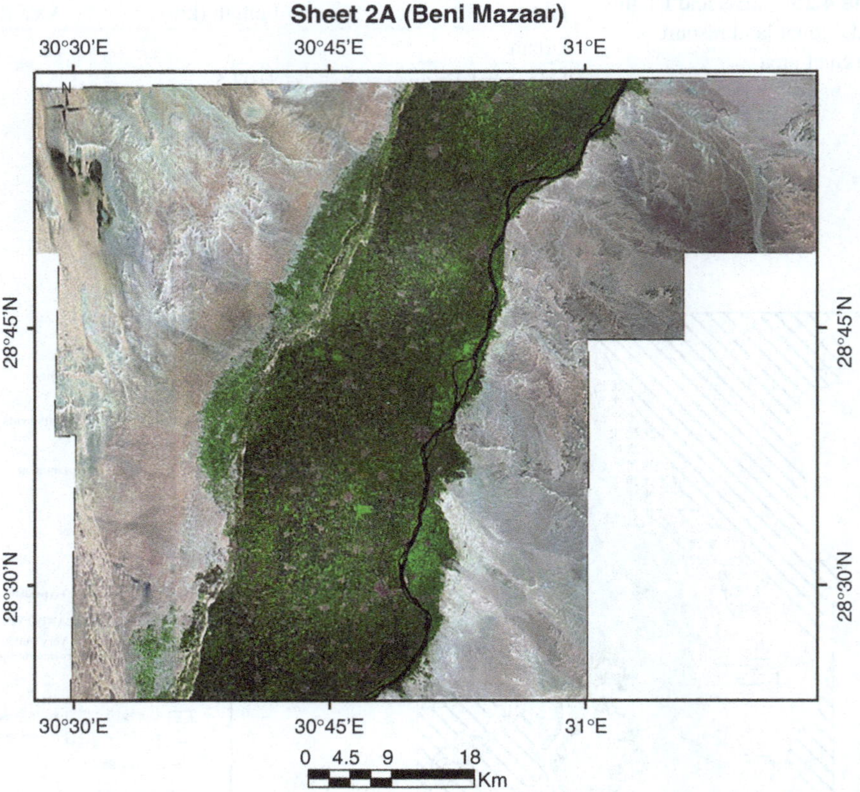

Fig. 4.101 ETM Landsat images of Al-Fashn &Beni Mazaar

Table 4.29 Distribution of soil units in Al-fashn & Beni Mazar

Soil Code	Area (km^2)	Area %
Aogp 1/1b	252.3	6.4
Eftt 3/3a	102.1	2.6
Eftt 4/3a	36.2	0.9
Eott 1/1b3	403.9	10.3
Eott 3/3b3	155.0	4.0
Epqt 1/1a	66.3	1.7
Epqt 1/1al	2.9	0.1
Epqt 1/1c	239.6	6.1
Epqt 3/1a	3.8	0.1
Epqt 4/1a	7.8	0.2
Vtt 4/4 a	1027.8	26.2
Vtt 4/4 al	15.4	0.4
Wd/3	10.0	0.3
Rock land	1523.9	38.8
River Nile	70.5	1.8
Total	3921.4	100.0

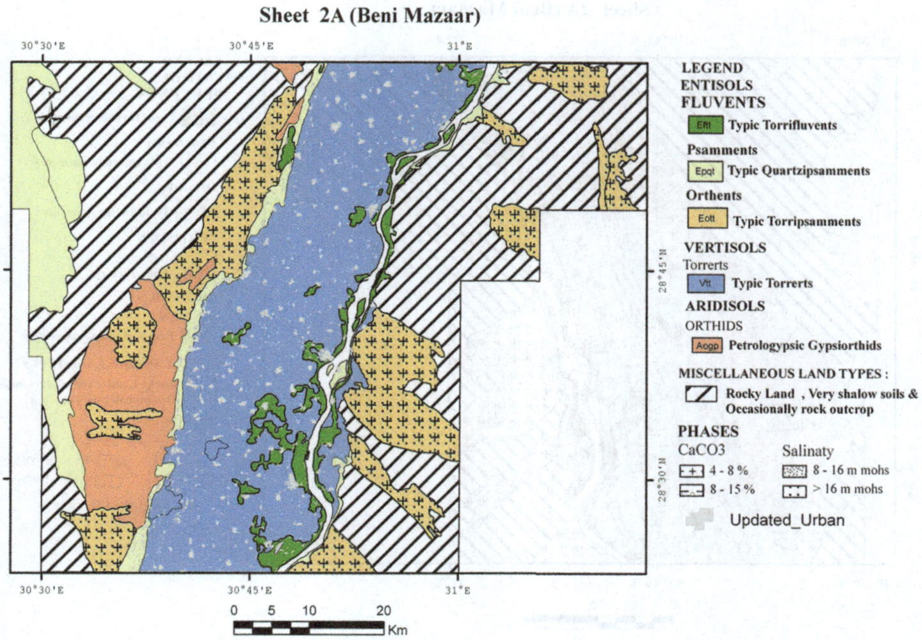

Fig. 4.102 Urban areas superimposed on the soil map, Al-Fashn &Beni Mazaar

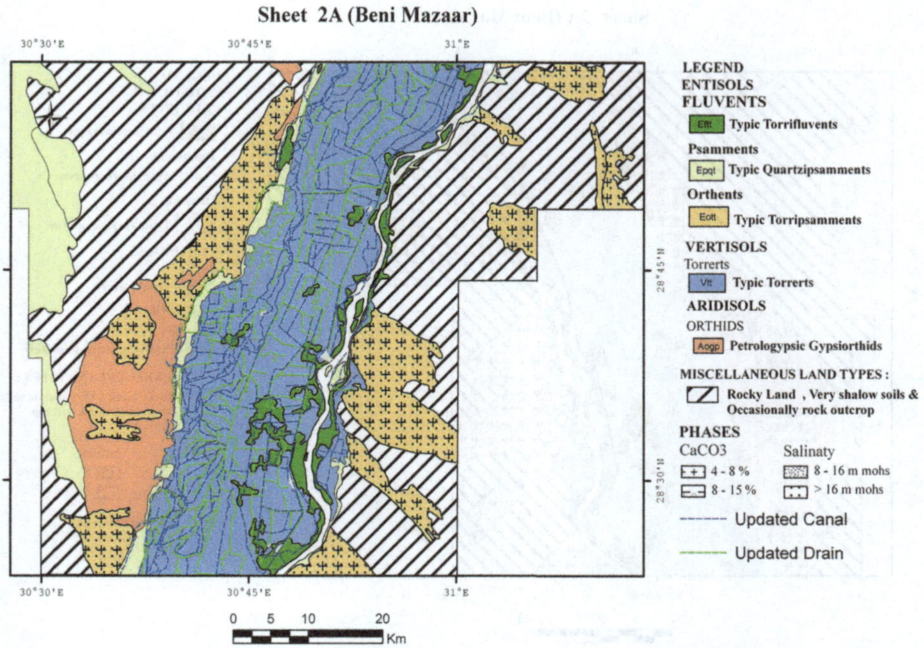

Fig. 4.103 Irrigation and drainage canals superimposed on soil map, Al-Fashn & Beni Mazaar

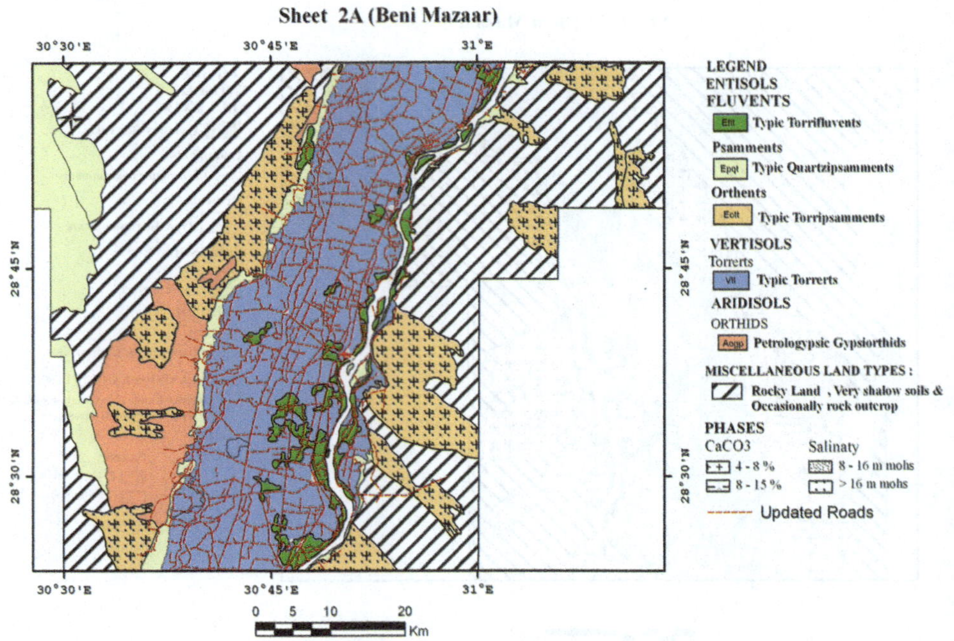

Fig. 4.104 Road network superimposed on soil map, Al-Fashn & Beni Mazaar

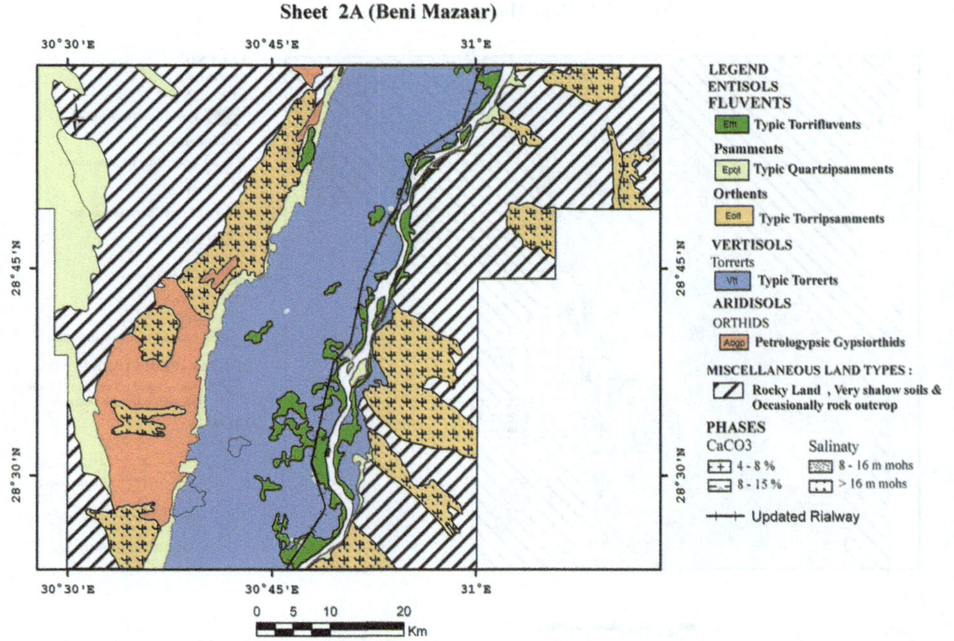

Fig. 4.105 Railway superimposed on the soil map, Al-Fashn & Beni Mazaar

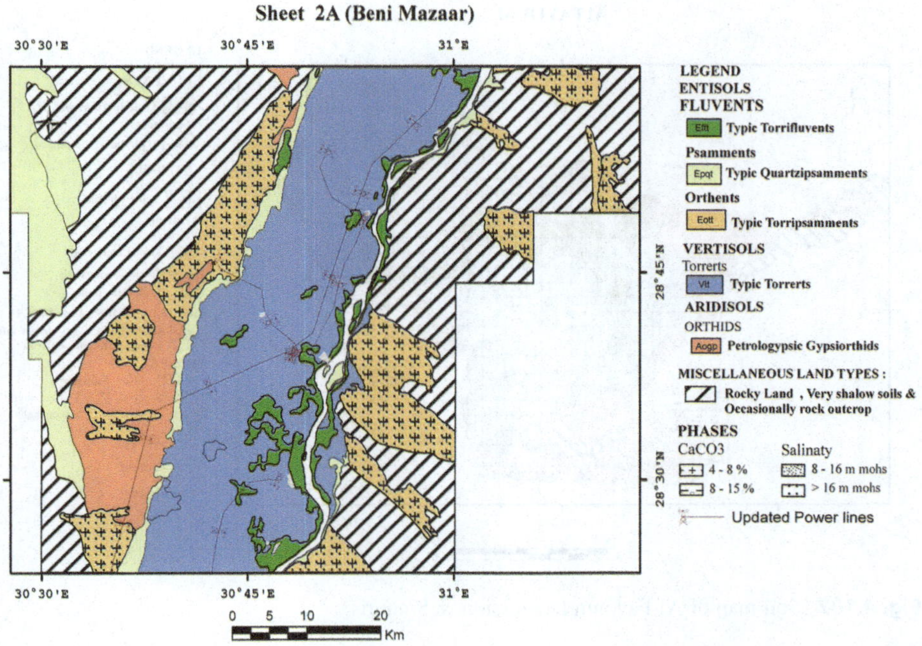

Fig. 4.106 Utilities, superimposed on the soil map, Al-Fashn & Beni Mazaar

Table 4.30 Areas and lengths of additional land resources, Al-fashn & Beni Mazar

Layer	Length (km)	Area (km^2)
Urban		52.2
Irrigation&Drainage	1949.1	
Roads	1643.8	
Railways	74.2	
Utilities	150.5	

total of 21.2 km^2 around the Nile banks. These soils are characterized by a coarse texture, low salinity and low $CaCO_3$%. The rock land and very shallow soils are found mainly in the southern parts of this area covering a total of 1325.6 km^2 (i.e. 36.2%) of the total area.

The created database of this area Figs. 4.165, 4.166, 4.167, 4.168, 4.169 and Table 4.48 indicates that the total urban settlement area is 28.5 km^2, while the total length of the irrigation and drainage canals is 512.1 km. The lengths of roads, railway and the utilities networks attain 644.7, 71.0 and 89.4 km respectively.

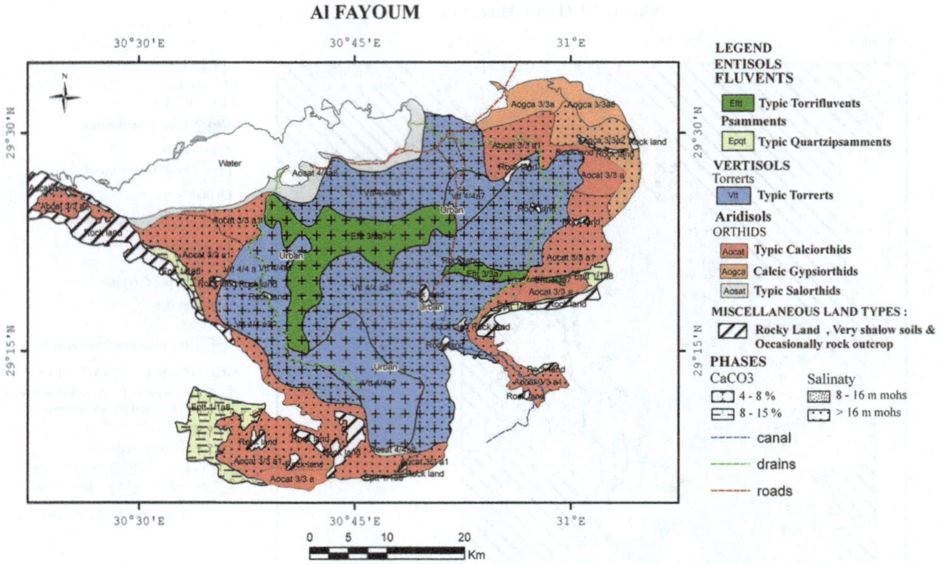

Fig. 4.107 Soil map of Al-Fayoum Depression & Sinnuris

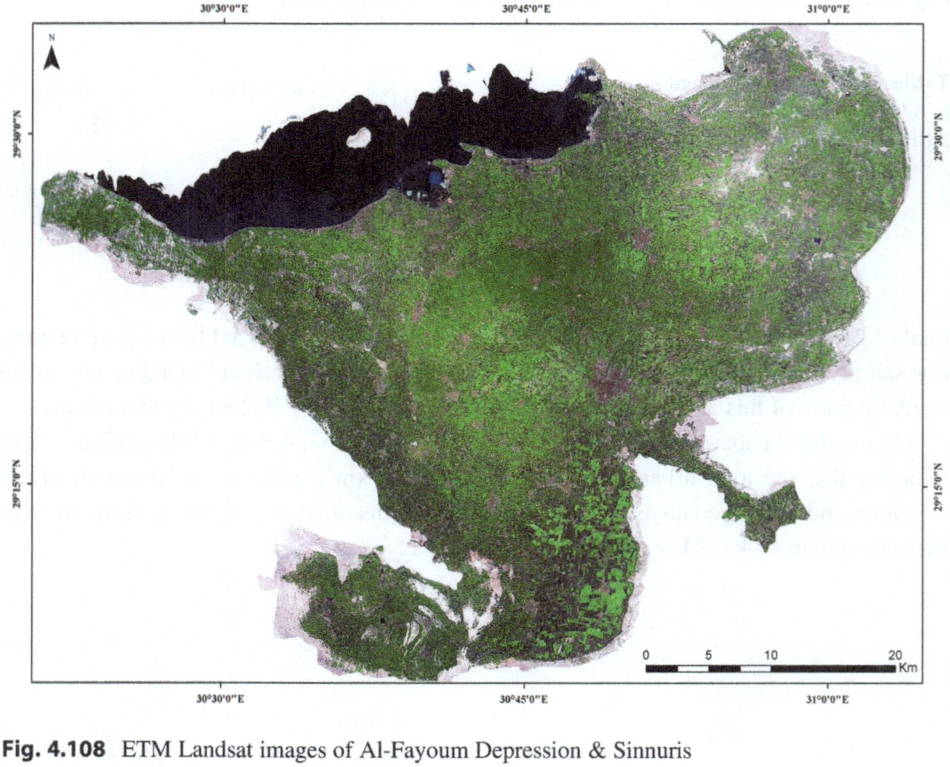

Fig. 4.108 ETM Landsat images of Al-Fayoum Depression & Sinnuris

Table 4.31 Distribution of soil units in Fayoum Depression

Soil Code	Area (km^2)	Area %
Aocat 3/3 a	57.4	2.9
Aocat 3/3 a1	386.4	19.3
Aocat 3/3 a2	17.8	0.9
Aogca 3/3a	31.2	1.6
Aogca 3/3a2	51.5	2.6
Aogca 3/3a8	23.3	1.2
Aosat 4/4a8	56.3	2.8
Eftt 3/3a7	146.2	7.3
Epqt 1/1a8	75.9	3.8
Vtt 4/4 a	13.6	0.7
Vtt 4/4 a3	605.2	30.2
Vtt 4/4a7	176.9	8.7
Water	239.3	11.9
Rock land	119.0	5.9
Urban	4.2	0.2
Total	2004.2	100.0

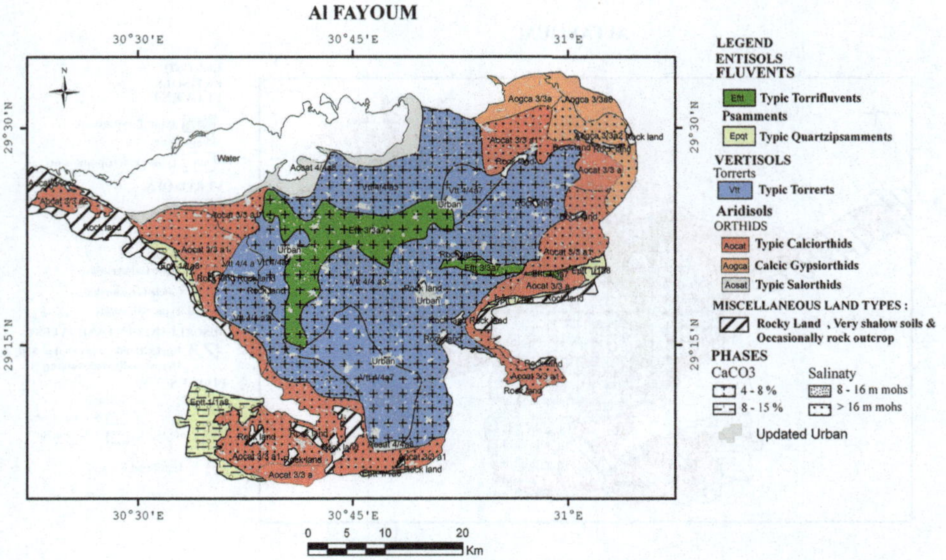

Fig. 4.109 Urban areas superimposed on the soil map, Al-Fayoum Depression & Sinnuris

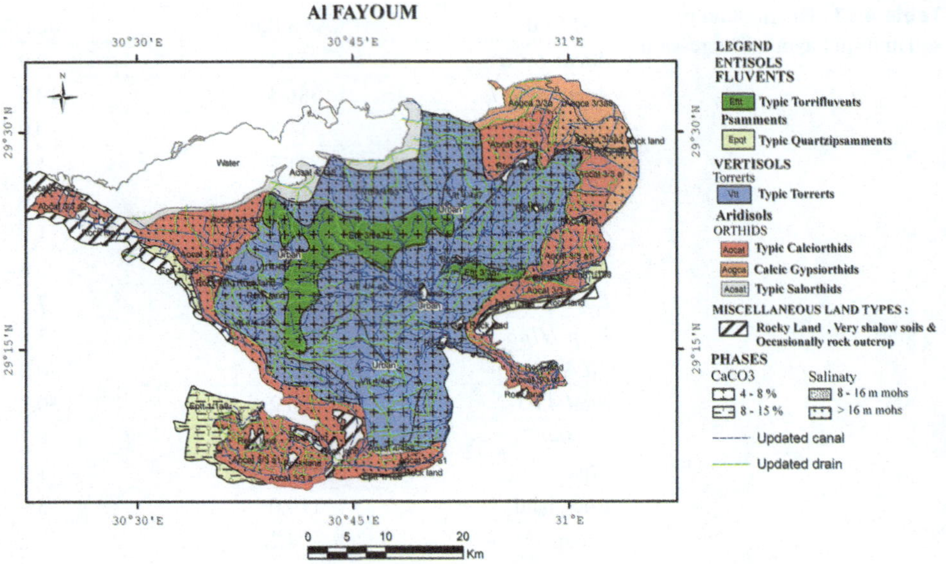

Fig. 4.110 Irrigation and drainage canals superimposed on soil map, Al-Fayoum Depression & Sinnuris

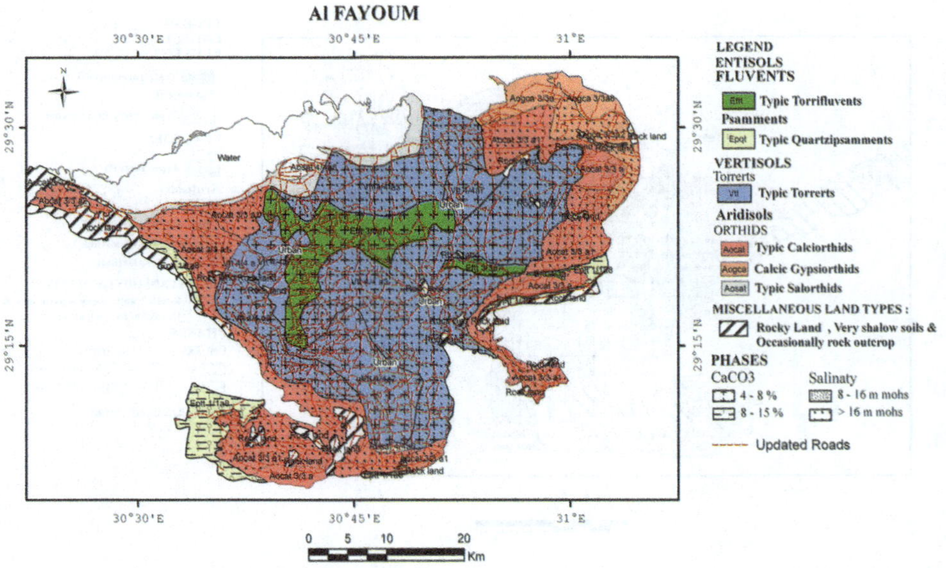

Fig. 4.111 Road network superimposed on soil map, Al-Fayoum Depression & Sinnuris

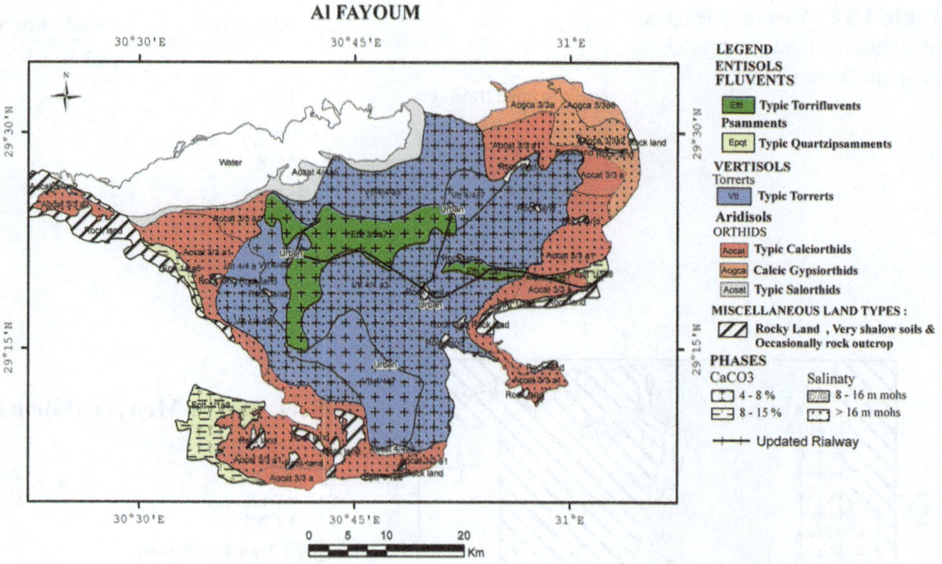

Fig. 4.112 Railway superimposed on the soil map, Al-Fayoum Depression & Sinnuris

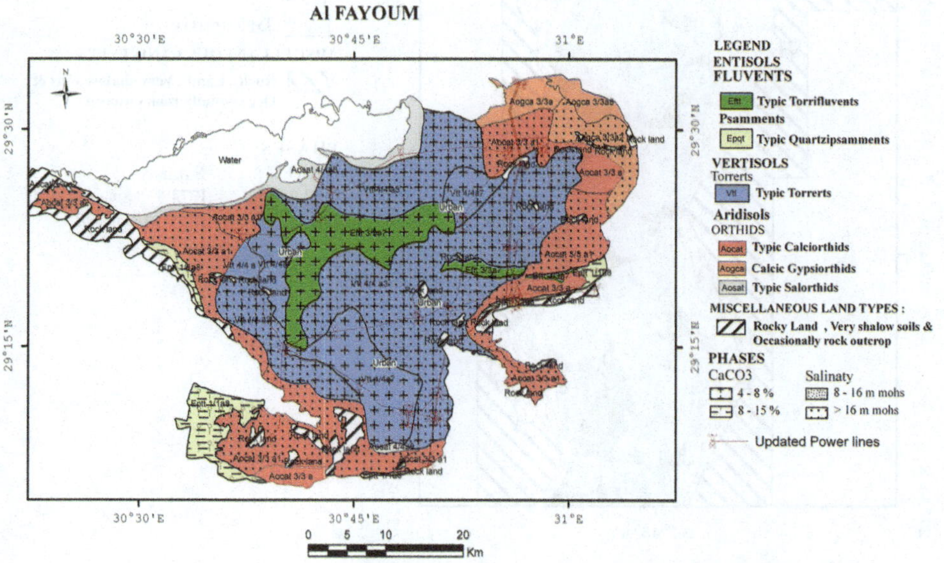

Fig. 4.113 Utilities, superimposed on the soil map, Al-Fayoum Depression & Sinnuris

Table 4.32 Areas and lengths of additional land resources, Fayoum Depression

Layer	Length (km)	Area (km^2)
Urban		63.8
Irrigation&Drainage	2213.8	
Roads	2344.4	
Railways	58.8	
Utilities	351.3	

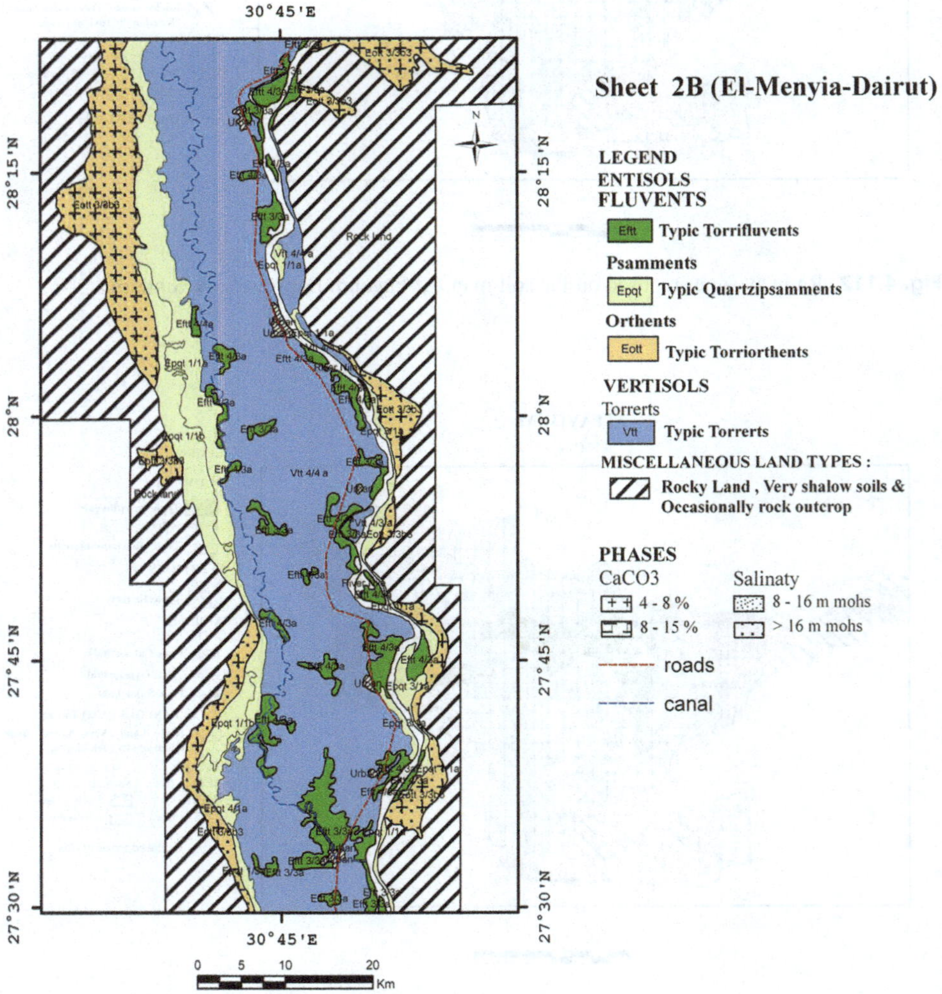

Fig. 4.114 Soil map of Al-Minya and Dairut

Fig. 4.115 ETM Landsat images of Al-Minya and Dairut

Table 4.33 Distribution of soil units in Al-Minya and Dairu

Soil Code	Area (km^2)	Area %
Eftt 3/3a	96.9	2.8
Eftt 4/3a	93.8	2.7
Eftt 4/4a	2.3	0.1
Eott 3/3b3	357.6	10.2
Epqt 1/1a	186.3	5.3
Epqt 1/1b	95.1	2.7
Epqt 1/3a	6.5	0.2
Epqt 3/1a	1125.9	32.3
Epqt 4/1a	5.6	0.2
River Nile	90.4	2.6
Rock land	1386.7	39.76
Urban	11.7	0.36
Vtt 4/3 a	4.1	0.2
Vtt 4/4 a	27.8	0.8
Total	3490.7	100.0

4.2 Soil Taxonomy of Egyptian Nile Basine

4.2.1 Nile Delta Soils

Three soil orders can only be discoursed here, as the Nile Delta soils can only be classified under one of these orders (i.e. *Vertisols, Aridisols and Entisols*).

4.2.1.1 Soil Order "*Vertisols*"

The soils classified under the *Vertisols* order are those soils, which have the fikkiwing description sccording to (USDA, 2008).

1. A layer of 25 cm or thicker, with an upper boundary within 100 cm of the mineral soil surface, that has either slickensides close enough to intersect or wedge-shaped aggregates which have their long axes tiled 10 to 60 degrees from the horizontal; and
2. The cjay fractioncontents represent 30 percent or more of the fine-earth fraction either between the mineral soil surface and a depth of 18 cm or in Ap horizon, whichever is thicker, and 30 percent or more clay in the fine-earth fraction of all horizons between a depth of 18 cm and either depth of 50 cm, or a densic, lithic, or paralithic contact, duripan, or petrocalcic horizon if shallower; and
3. Cracks, a very dustinket characteristic that open and close periodically.

All previous characteristics cannot be generalized on the Nile Delta alluvial soils mainly due to irrigation. *Vertisols* however could be found in small unmappable patches under the

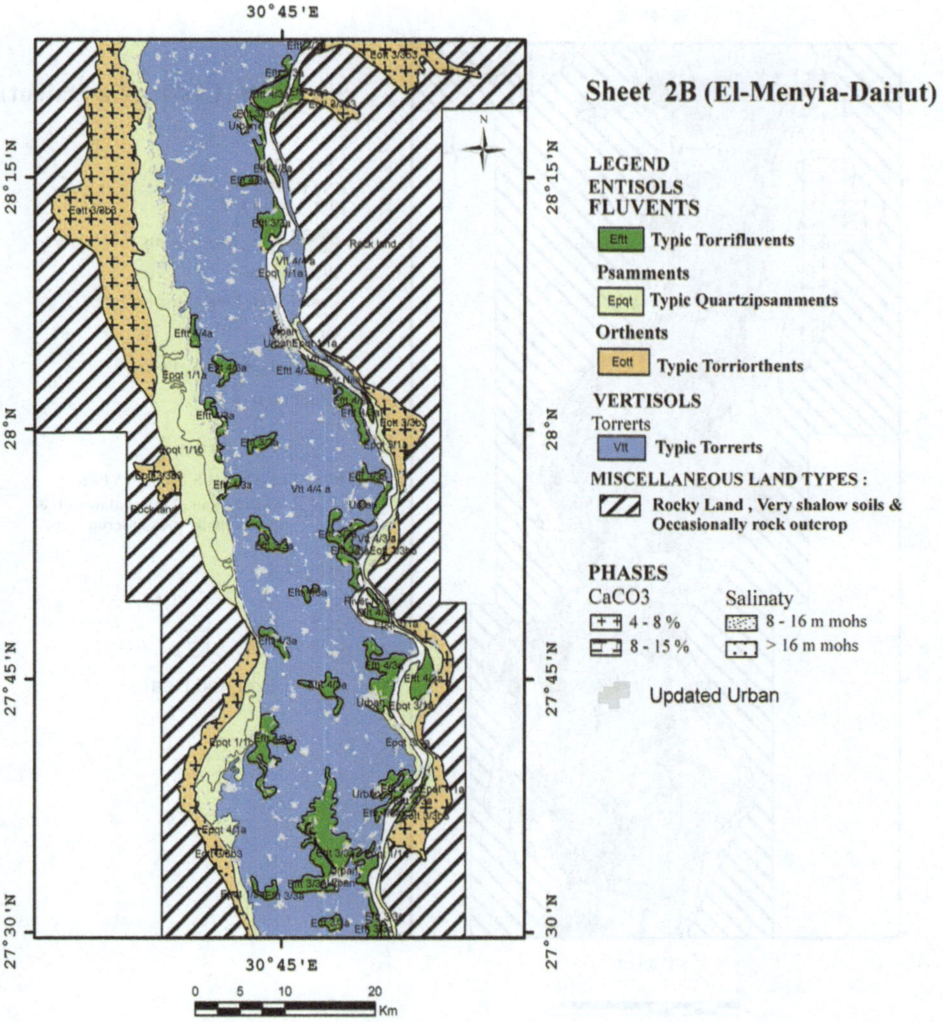

Fig. 4.116 Urban areas superimposed on the soil map, Al-Minya and Dairut

used scale (1: 100,000). The soil map sheets of the Nile Delta produced through the project of *"Soil Map of Egypt"* (ASRT, 1982) showed that the majority of the soils are classified under the order *"Vertisols"*. However, revising the required characteristics of this soil order, on basis of USDA, 2006 revealed that these soils can generally be classified under the sub-great group *Vertic Torrifluvents*, which will be explained in detail under the *Entisols* soil order.

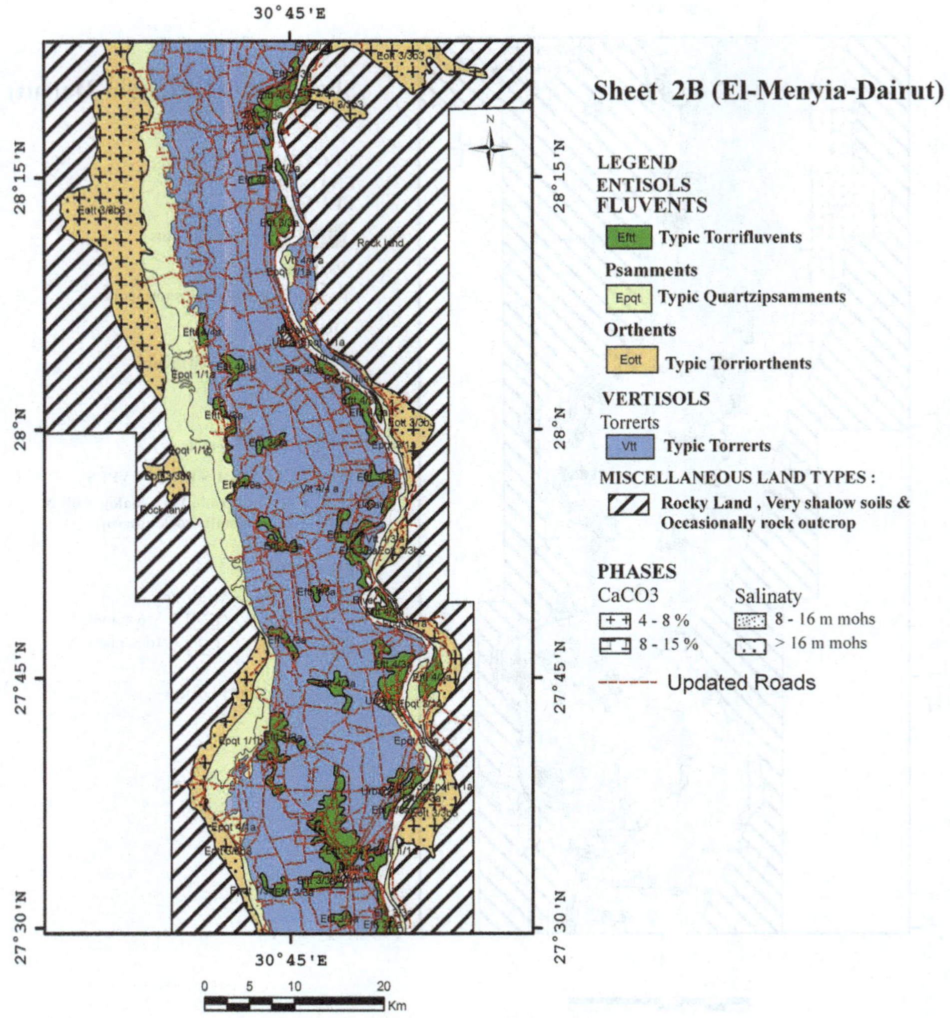

Fig. 4.117 Irrigation and drainage canals superimposed on soil map, Al-Minya and Dairut

4.2.1.2 Soil Order *"Aridisols"*

The characteristics of this soil order refer that these soils are characterized as follow;

(1) Having the following conditions:

a. Aridic soil moisture regime and
b. An ochric or anthropic epipidon and
c. One or more of the following with the upper boundary within 100 cm of the soil surface.

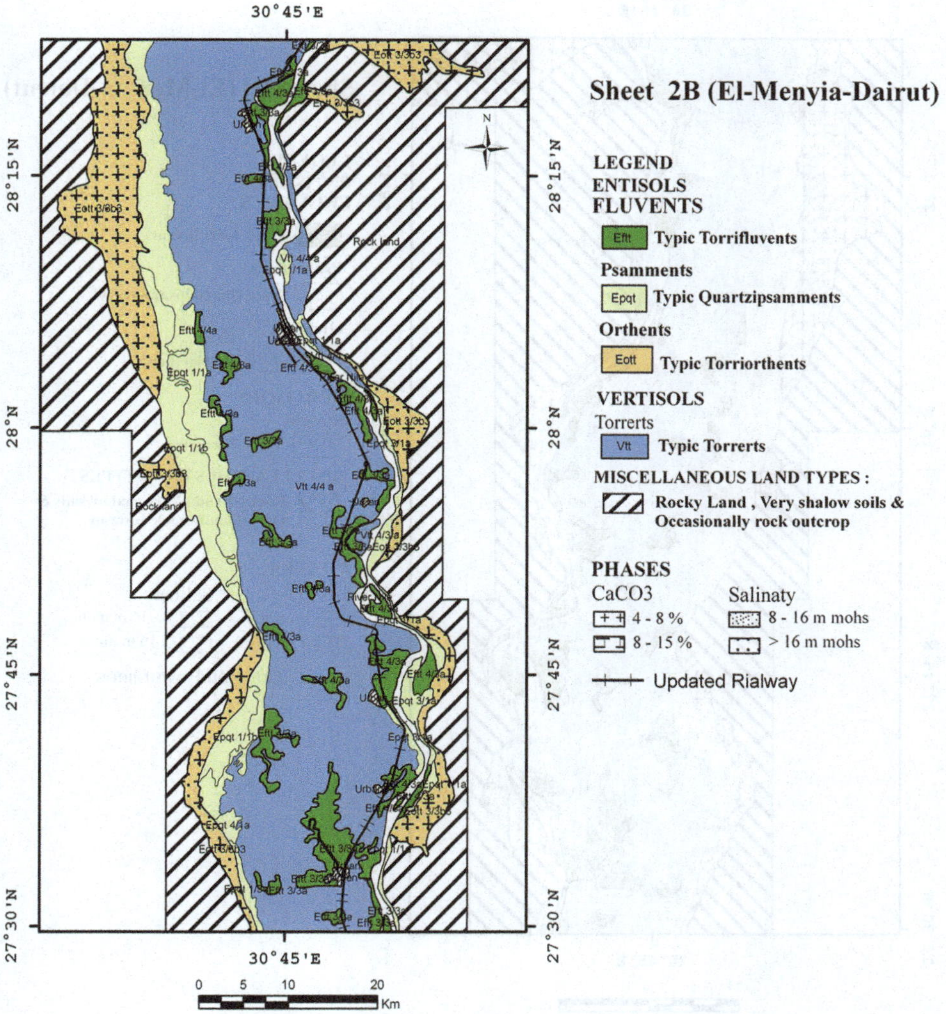

Fig. 4.118 Road network superimposed on soil map, Al-Minya and Dairut

A cambic horizon with a lower depth of 25 cm or more; A cryic temperature regime and a cambic horizon; A calcic, Gypsic, Petrocalcic, Petrogypsic or Salic horizon; or a duripan; or

d. An argilic or natric horizon; or

(2) Have a salic horizon and

a. Saturation with water in one or more layers within 100 cm of the soil surface for one month or more during a normal year; and

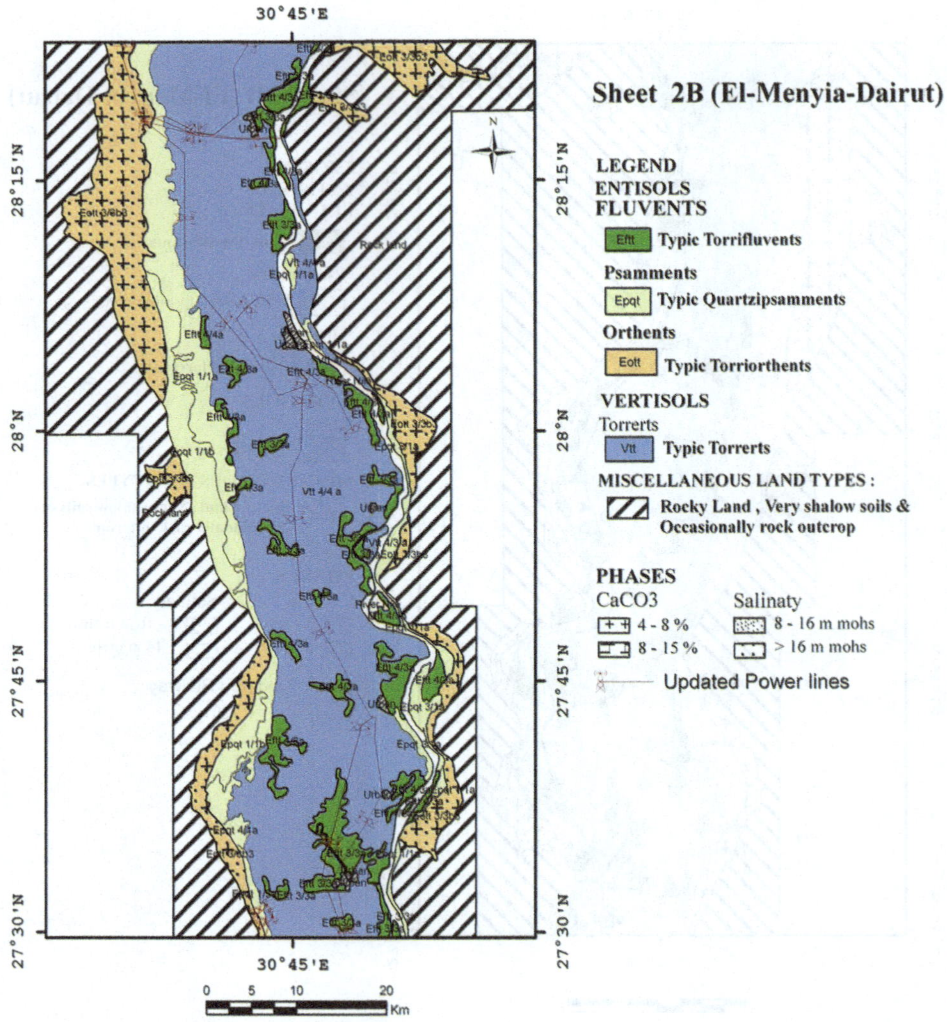

Fig. 4.119 Railway superimposed on the soil map, Al-Minya and Dairut

b. A moisture control section that is dry in some or all parts at some time during normal years; and
c. No sulfuric horizon that has its upper boundary within 150 cm of the mineral soil surface

Three sub-orders (Fig. 4.170) were found under this soil order (i.e. *Calcids, Gypsids* and *Salids*). Table 4.49 summarizes the areas of different great groups included in those suborders.

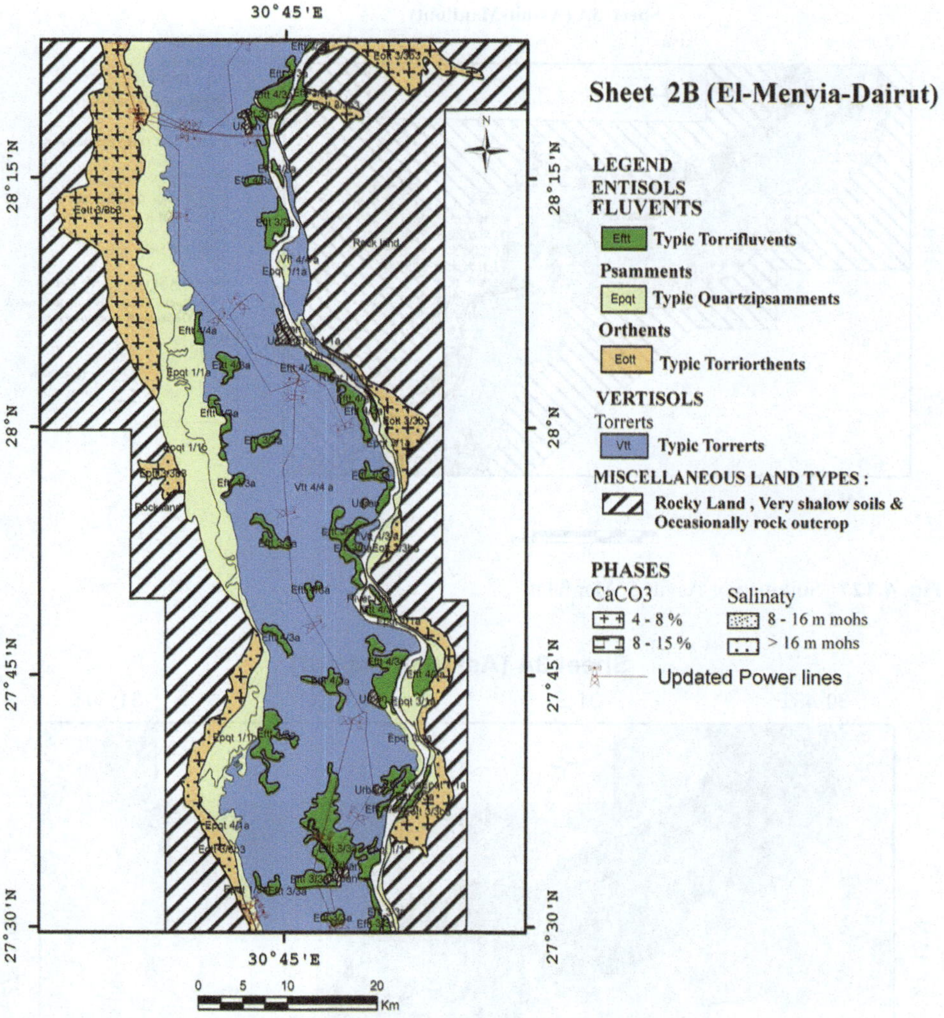

Fig. 4.120 Utilities, superimposed on the soil map, Al-Minya and Dairut

Table 4.34 Areas and lengths of additional land resources, Al-Minya-Dairut

Layer	Length (km)	Area (km^2)
Urban		73.4
Irrigation& Drainage	2205.3	
Roads	2052.9	
Railways	114.3	
Utilities	215.6	

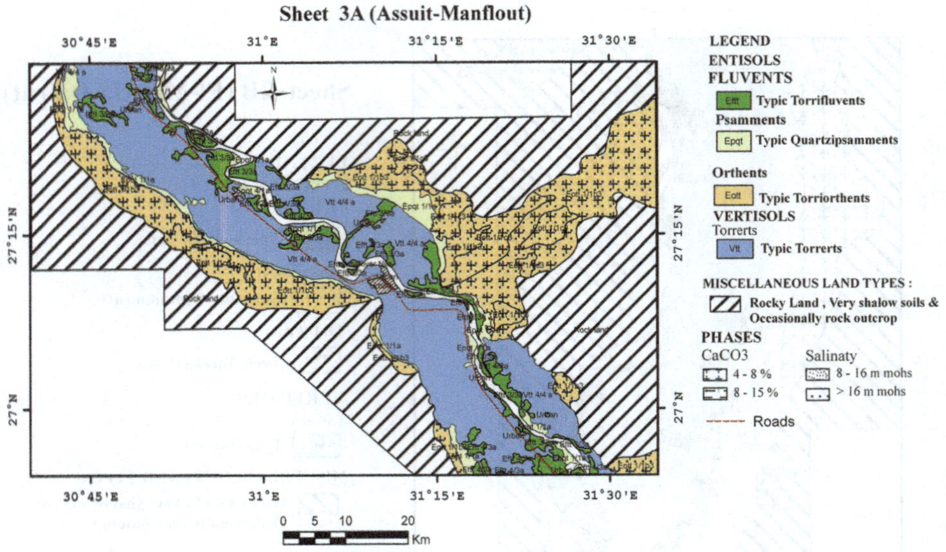

Fig. 4.121 Soil map of Asyut & Manfalut

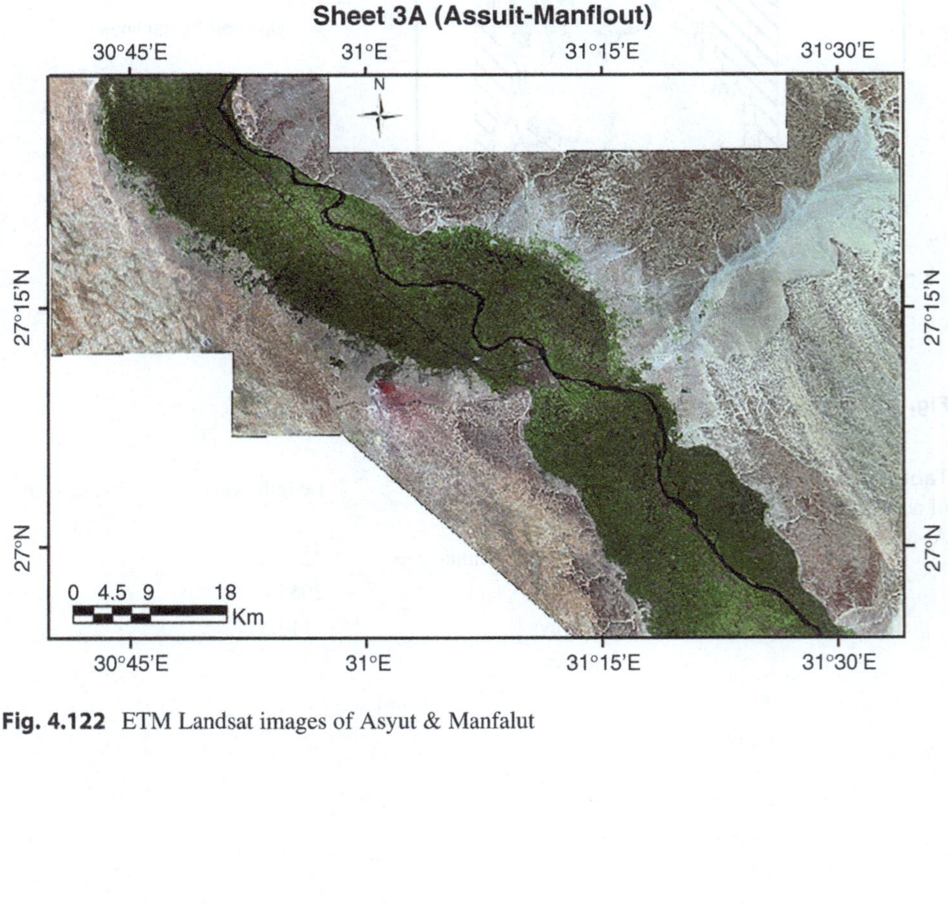

Fig. 4.122 ETM Landsat images of Asyut & Manfalut

Table 4.35 Distribution of soil units in Asyout

Soil Code	Area (km^2)	Area %
Eftt 3/3a	161.6	03.8
Eftt 4/3a	40.6	01.0
Eott 1/1b3	632.2	14.8
Eott 1/1c3	194.3	04.6
Epqt 1/1a	140.0	03.3
Epqt 1/1b	4.6	00.1
Epqt 4/1a	6.0	00.1
Vtt 4/4 a	929.8	21.8
Rock land	2048.0	48.0
Urban	14.24	00.3
River Nile	93.5	02.2
Total	4264.8	100.0

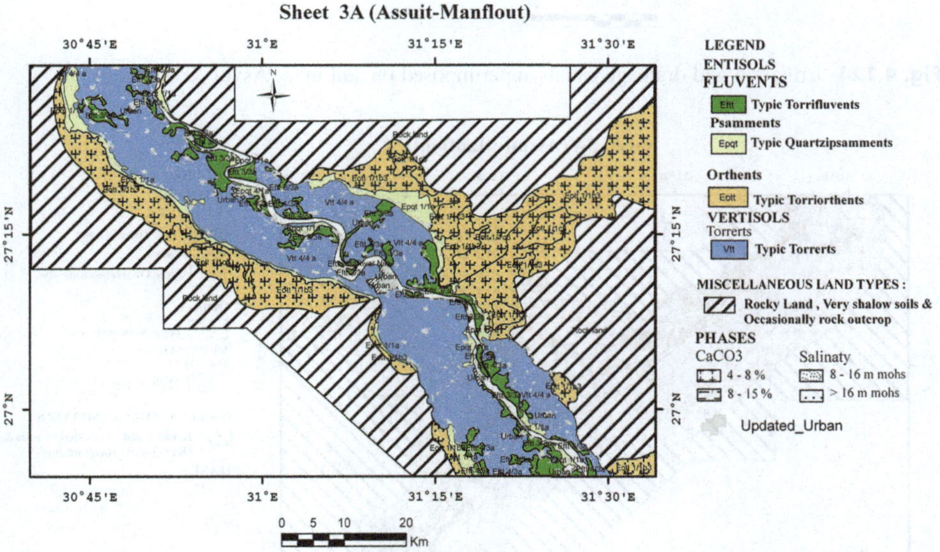

Fig. 4.123 Urban areas superimposed on the soil map, Asyut & Manfalut

(1) The *Calcids* are *Aridisols* that have a calcic or petrocalcic horizon having its upper boundary within 100 cm of the soil surface. Only *Haplocalcids* soil great group, under which *Typic Haplocalcids* sub-great group is found, covering an area of 1012.6 km^2 representing 23.7% of the Nile Delta's *Aridisols*. The *Haplocalcids* are *Calsids* soils that have no petrocaclic horizon, where the *Typic Haplocalcids* includes soils having no lithic, ustic, vertic, aquic, duric, sodic properties.

(2) The *Gypsids* are *Aridisols* that have a gypsic or petrogypsic horizon having its upper boundary within 100 cm of the soil surface and do not have a petrocalcic horizon overlying

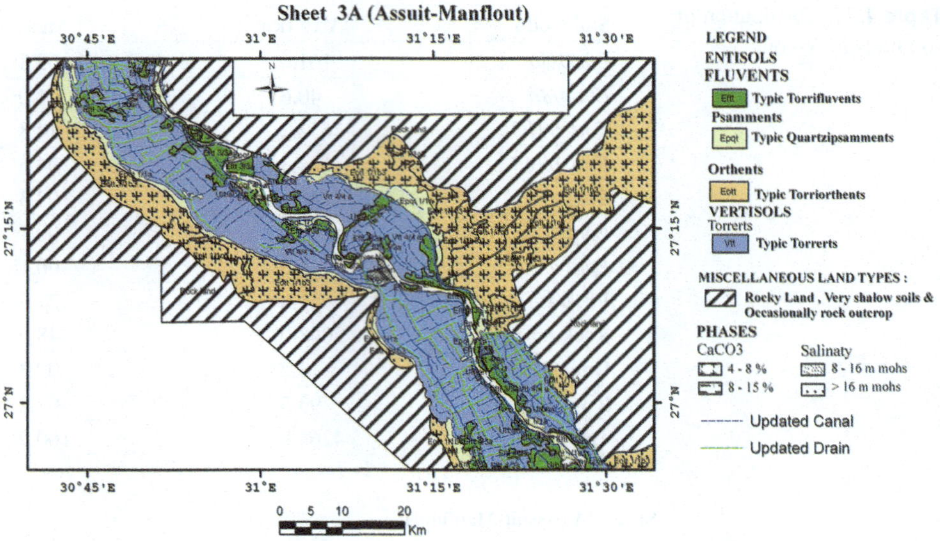

Fig. 4.124 Irrigation and drainage canals superimposed on soil map, Asyut & Manfalut

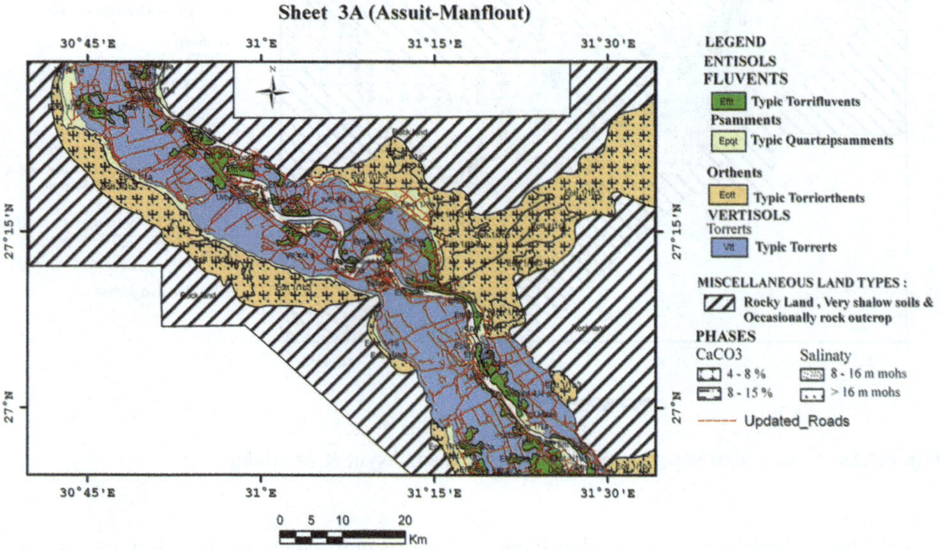

Fig. 4.125 Road network superimposed on soil map, Asyut & Manfalut

these horizons nor having cryic, salic, duripan properties. Three great groups are found belonging to this sub-order *(i.e. Calcigypsids, Petrogypsids and Haplogypsids)*. The *Calcigypsids* great group soils are *Gypsids* having a calcic horizon with its upper boundary within 100 cm of the soil surface. These soils exhibits an area of 4.1 km^2 representing 0.1%

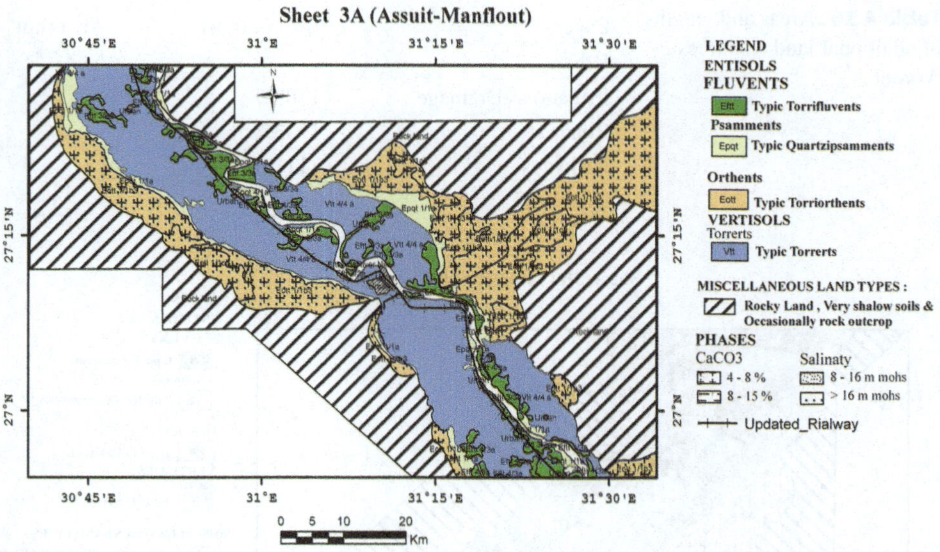

Fig. 4.126 Railway superimposed on the soil map, Asyut & Manfalut

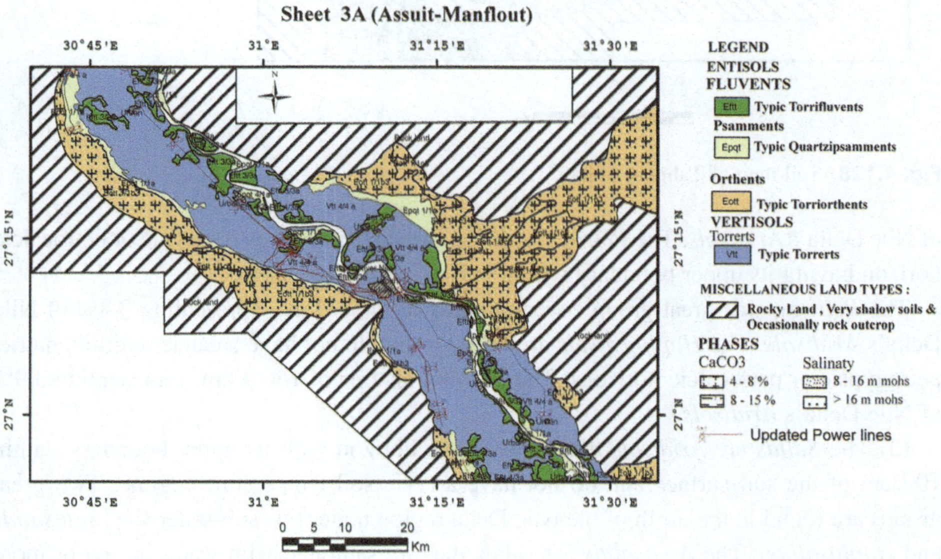

Fig. 4.127 Utilities, superimposed on the soil map, Asyut & Manfalut

Table 4.36 Areas and lengths of additional land resources, Asyout

Layer	Length (km)	Area (km^2)
Urban		63.4
Irrigation&Drainage	1709.5	
Roads	1553.7	
Railways	111.0	
Utilities	263.5	

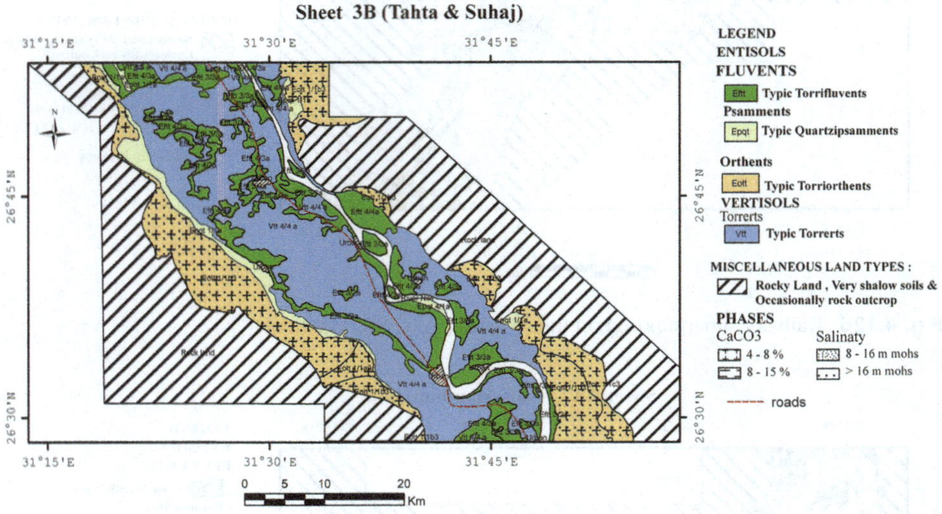

Fig. 4.128 Soil map of Tahta & Sohag

of Nile Delta's *Aridisols*. The *Petrogypsids* are *Gypsids* having a petrogypsic or petrocalcic horizon having its upper boundary within 100 cm of the soil surface.

The *Petrogypsids* great group covers an area of 145.9 km^2 representing 3.4% of Nile Delta's *Aridisols*. The *Haplogypsids* are *Gypsids* that do not have a calcic, argillic, natric, petrogypsic or petrocalcic horizons. They exhibit an area of 166.9 km^2 representing 3.9% of Nile Delta's *Aridisols*.

(3) The *Salids* are *Aridisols* that have a salic horizon with its upper boundary within 100 cm of the soil surface and do not have a cryic soil temperature regime. Two great groups are found in the north of the Nile Delta region under this sub-order (i.e. *Aquisalids* and *Haplosalids*). The *Aquisalids* are salids that are saturated with water in one or more layers within 100 cm of the mineral soil surface for one month or more in normal years. They cover an area of 1928.8 km^2 representing 45.2% of *Aridisols* of the Nile Delta region. The *Haplosalids* are other *salids*, which aren't qualified for *Aquisalids* great group. These cover an area of 166.9 km^2 representing 3.9% of Nile Delta *Aridisols*. Only *Typic* sub-great groups of the previous great groups are found within the Nile Delta soils.

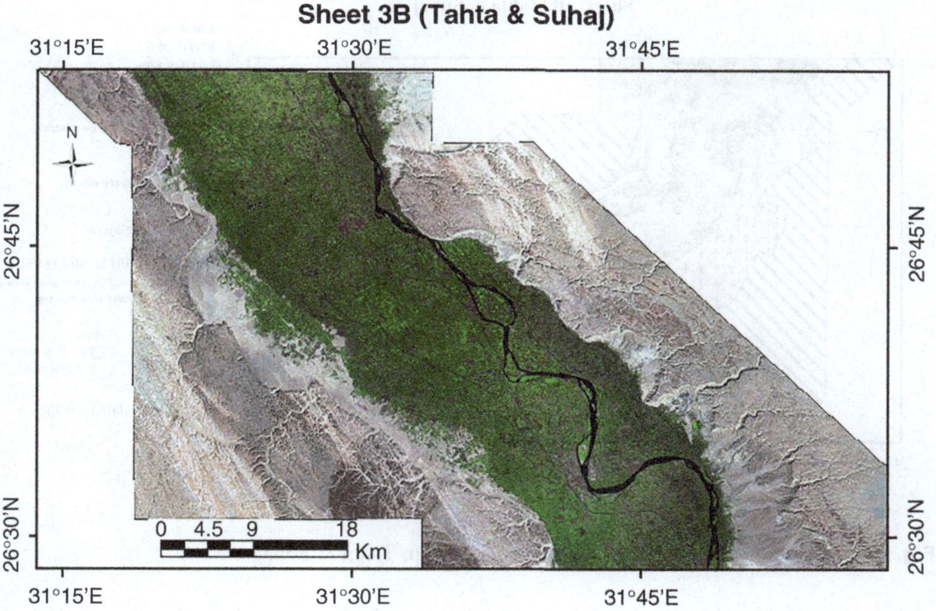

Fig. 4.129 ETM Landsat images Tahta & Sohag

Table 4.37 Distribution of soil units in Tahta & Sohag

Soil Code	Area (km^2)	Area %
Eftt 3/2a	5.9	0.3
Eftt 3/3a	155.7	6.9
Eftt 4/3a	85.2	3.8
Eftt 4/4a	19.4	0.9
Eott 1/1a3	15.4	0.7
Eott 1/1b3	298.0	13.3
Eott 1/1c3	34.4	1.5
Epqt 1/1a	52.0	2.3
Epqt 4/1a	3.1	0.1
Vtt 4/4 a	568.4	25.3
River Nile	57.2	2.6
Rock land	941.1	41.9
Urban	8.5	0.4
Total	2244.3	100.0

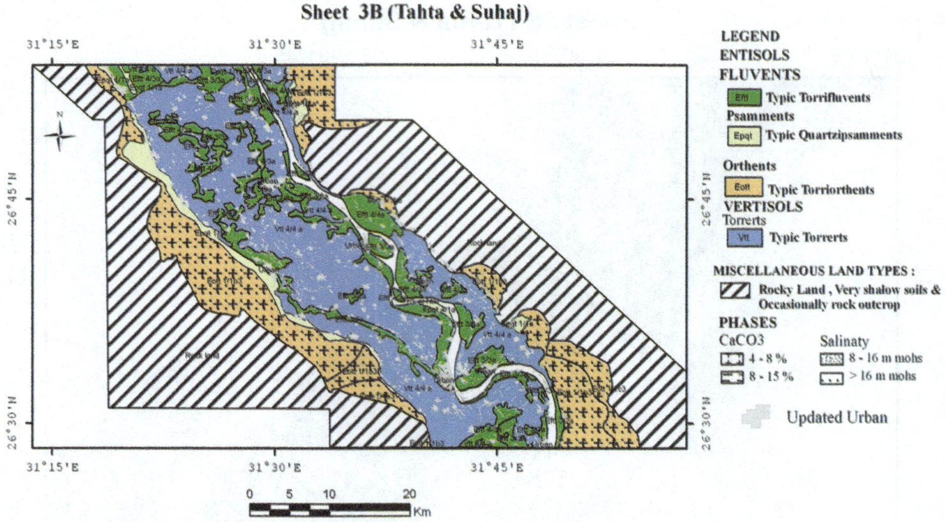

Fig. 4.130 Urban areas superimposed on the soil map, Tahta & Sohag

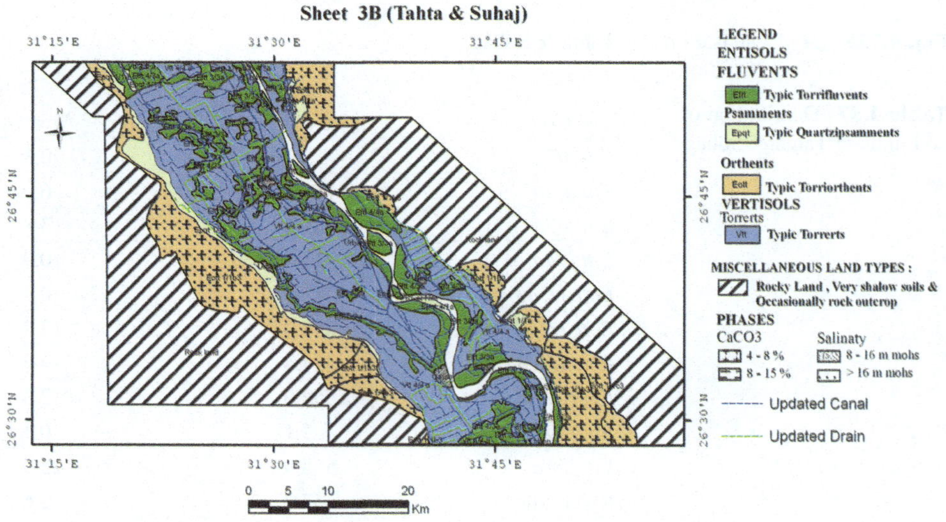

Fig. 4.131 Irrigation and drainage canals superimposed on the soil map, Tahta & Sohag

4.2.1.3 Soil Order *"Entisols"*

This soil order includes recently developed soils, which do not have the requirements of the other soil orders. A number of three sub-orders are found in the Nile Delta soils (i.e. *Psamments, Fluvents and Orthents)*. Figure 4.171 and Table 4.50 demonstrate the geographical distribution and areas of different great groups, belonging to the *Entisols* of the Nile Delta region.

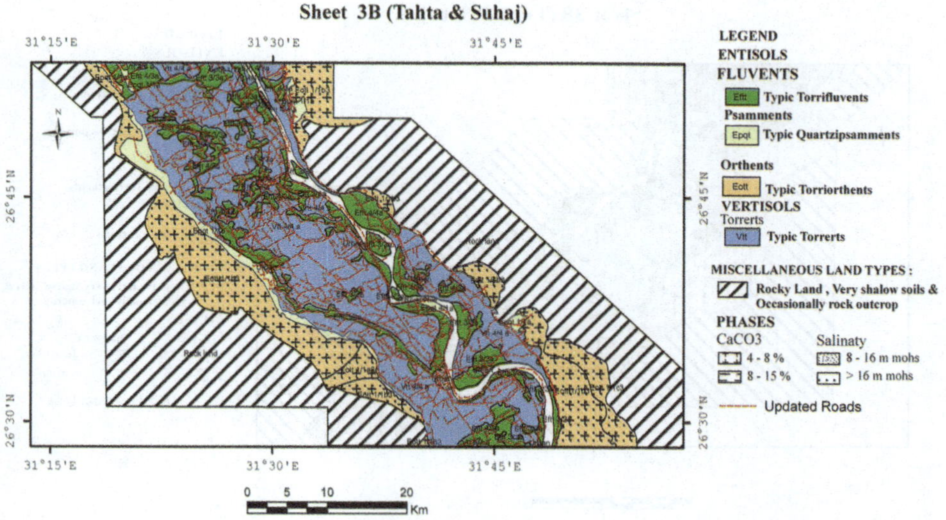

Fig. 4.132 Road network superimposed on soil map, Tahta & Sohag

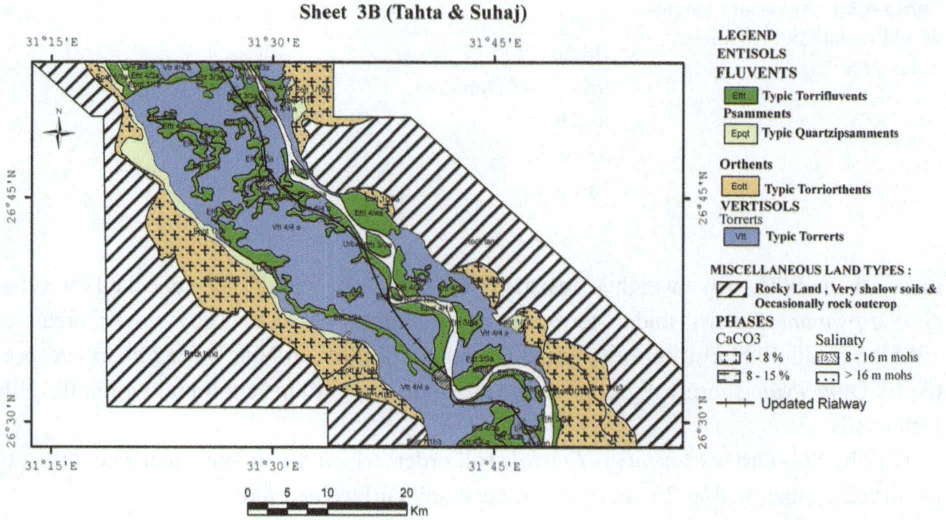

Fig. 4.133 Railway superimposed on the soil map, Tahta & Sohag

(1) The Sub-Order *Psamments* is *Entisols* that have less than 35% by volume rock fragments, and a texture of loamy fine sand or coarser in all layers within the profile control section. Two great groups are found in the Nile Delta under this sub-order (i.e. *Torripsamments* and *Quartzipsammets*). The great group *Torripsamments* soils are *Psamments* that have an aridic (or torric) moisture regime, while the *Quartzipsammets* are *Psamments* that have a sand texture within the particle size control texture consisting of

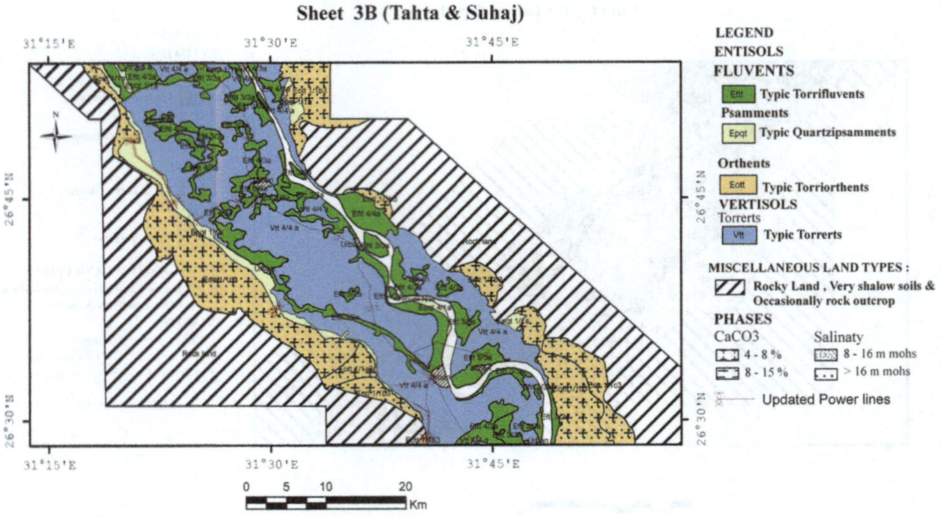

Fig. 4.134 Utilities, superimposed on the soil map, Tahta & Sohag

Table 4.38 Areas and lengths of additional land resources, Tahta & Sohag

Layer	Length (km)	Area (km^2)
Urban		57.0
Irrigation&Drainage	1378.2	
Roads	1409.3	
Railways	64.0	
Utilities	143.0	

more than 90% (by weight) resistant minerals. It should be checked weather *Quartzipsammets* exist under torric moisture regime or not. They cover areas of 1385 km^2 and 4082 km^2 representing 5.8% and 17.2% of the Nile Delta *Entisols* respectively. Only *Typic* sub-great groups of the previous great groups are found within the Nile Delta soils.

(2) The Sub-Order *Fluvents* is *Entisols* soil order, which do not have a densic, lithic, or paralithic contact within 25 cm of the mineral soil surface and have:

1. Slope of less than 25%
2. Either 0.2% or more organic carbon at a depth of 125 cm below the mineral soil surface or an irregular decrease in organic carbon content from a depth of 25 cm to a depth of 125 cm, or to a densic, lithic, or paralithic contact if shallower; and
3. A mean annual soil temperature above 0°

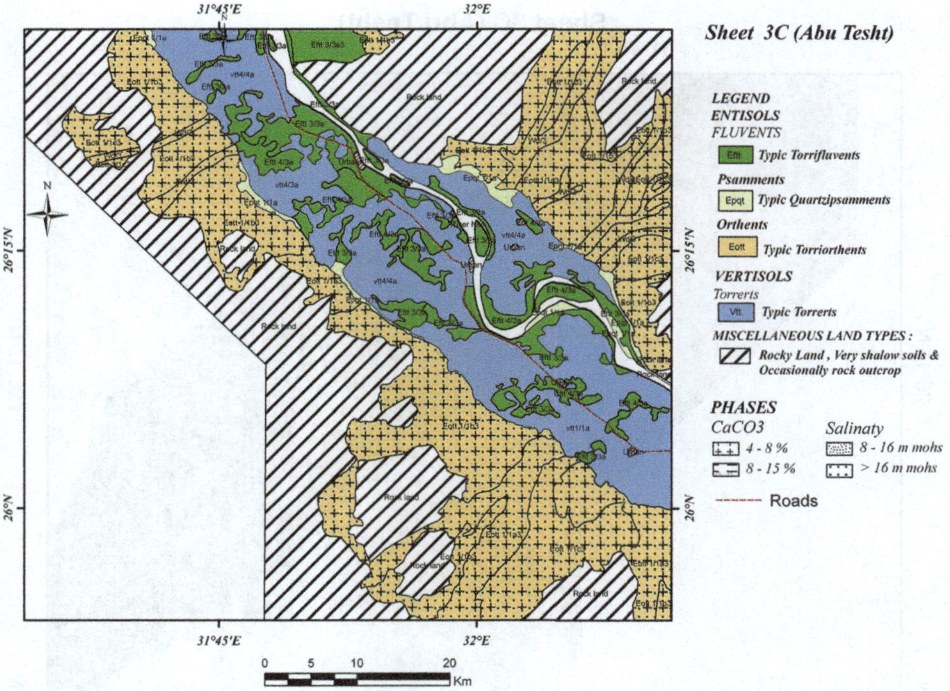

Fig. 4.135 Soil map of Al-Minsha & Abu Tisht

Only one great group was found in the Nile Delta region (i.e. *Torrifluvents*), which are *Fluvents* in areas that have an aridic (or torric) moisture regime. The *Vertic Torrifluvents* sub-great group is found dominating the alluvial Nile Delta soils. They cover an area of 15893.3 km^2 representing 66.9% of the Nile Delta *Entisols*. These soils are characterized by the following;

1. Torrifluvents that have cracks within 125 cm of the soil surface that are 5 mm or more wide through a thickness of 30 cm or more for some time in most normal years and slickensides or wedge shaped aggregates in a layer 15 cm or more thick that has its upper boundary within 125 cm of the soil surface; or
2. A linear extensibility or 6.0 cm or more between the soil surface and either a depth of 100 cm or a densic, lithic, or paralithic contact, whichever is shallower.

The *Aquic torrifluvents* sub-great group includes other torrifluvents that have in one or more horizons, within 100 cm of the soil surface, both redox depletions with chroma of 2 or less and aquic conditions for some time in normal years (or artificial drainage).

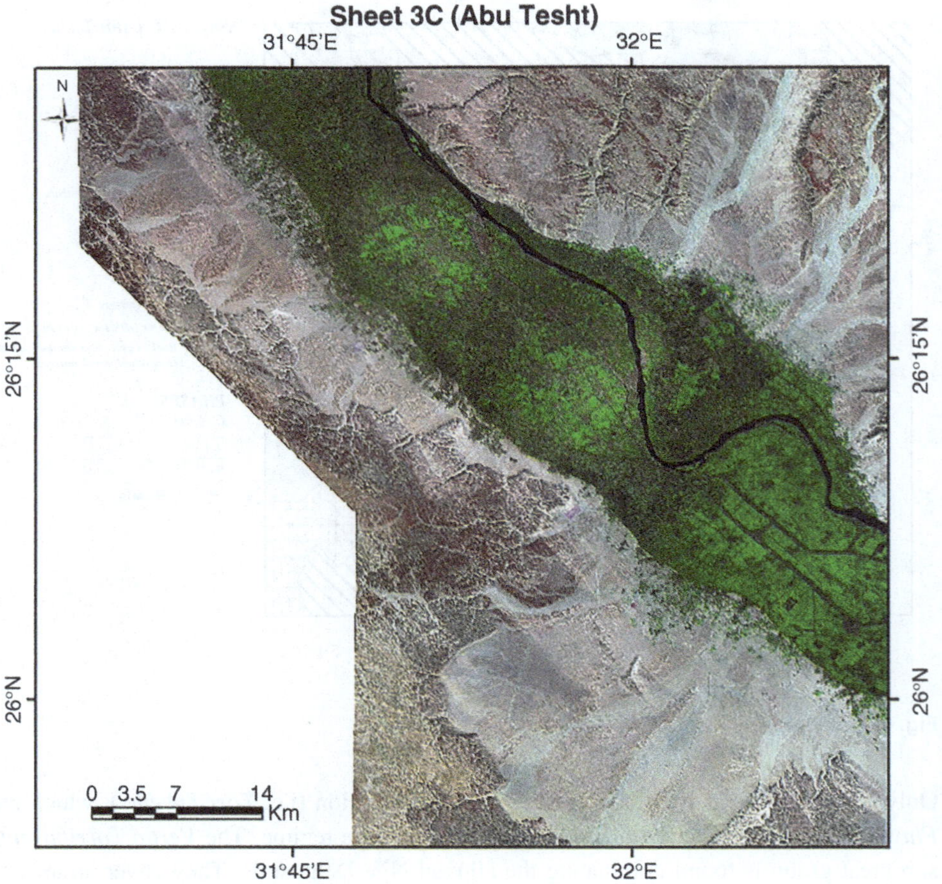

Fig. 4.136 ETM Landsat images of Al-Minsha & Abu Tisht

Also, *Typic Torrifluvents* sub-great group is found interfering in the previous one. These soils are characterized by gentle slope and texture of sandy loam or finer, they are of arid climate and have a torric moisture regime. Nevertheless, the *Typic Torrifluvents* is soils that do not have the requirements of the other *Torrifluvents* sub-great group.

(3) The sub-order *Orthents* is other *Entisols* that do not have the requirements of *Aquents, Arents, Psamments and Fluvents*. Only the sub-great group *Torriorthents* is found exhibiting an area of 2390.9 km^2 representing 10.1% of the Nile Delta *Entisols*. These soils are *orthents* that have an aridic or torric moisture regime. The existent sub great groups *Typic Torriorthents* is identified as other *torriorthents*, which do not have the requirements of other *Torriorthents*, sub great groups.

Table 4.39 Distribution of soil units in Al-Minsha & Abu Tisht

Soil Code	Area (km^2)	Area %
Eftt 3/3a	144.2	4.6
Eftt 3/3a3	30.8	1.0
Eftt 4/3a	100.7	3.2
Eott 1/1a3	198.3	6.3
Eott 1/1b3	509.5	16.2
Eott 1/1c3	284.4	9.0
Epqt 1/1a	33.7	1.1
vtt1/1a	213.0	6.8
vtt4/3a	44.9	1.4
vtt4/4a	314.9	10.0
Wd/3	92.7	2.9
Urban	3.9	0.1
River Nile	62.8	2.0
Rock land	1116.4	35.4
Total	3150.2	100.0

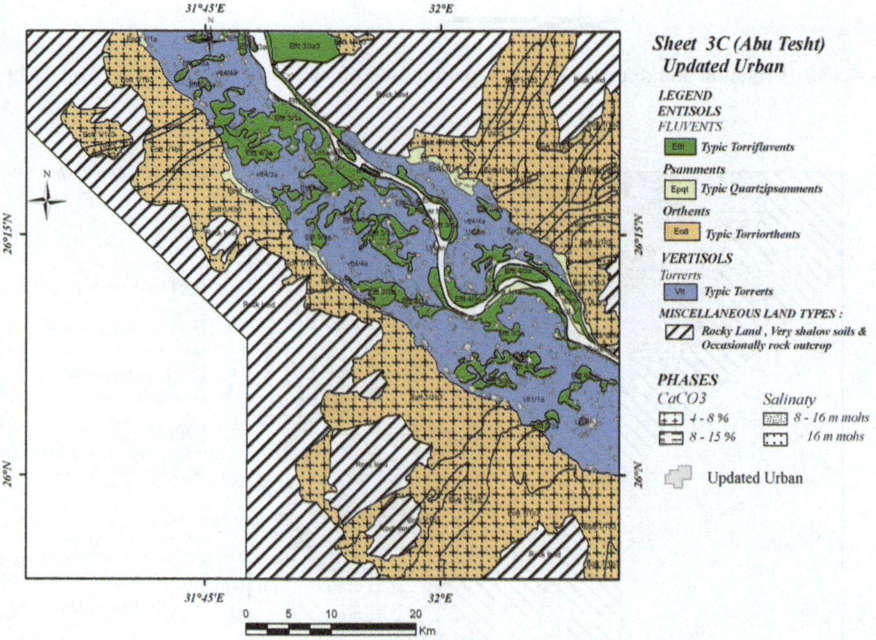

Fig. 4.137 Urban areas superimposed on the soil map, Al-Minsha & Abu Tisht

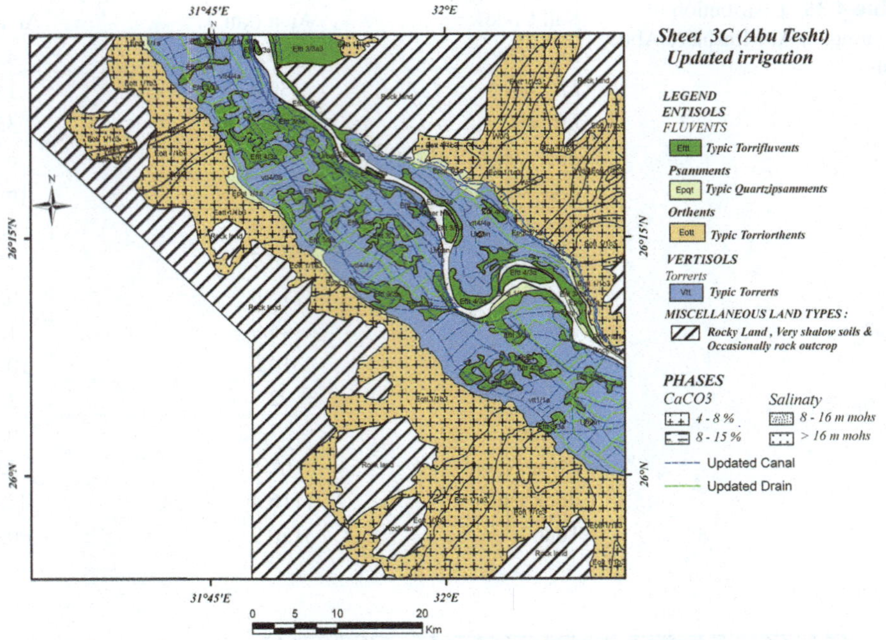

Fig. 4.138 Irrigation and drainage canals superimposed on soil map, Al-Minsha & Abu Tisht

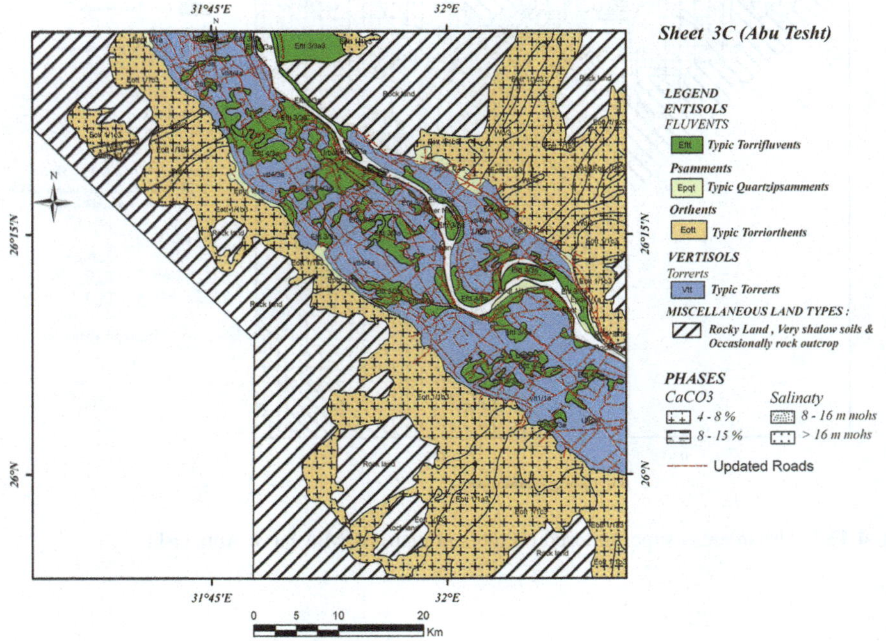

Fig. 4.139 Road network superimposed on soil map, Al-Minsha & Abu Tisht

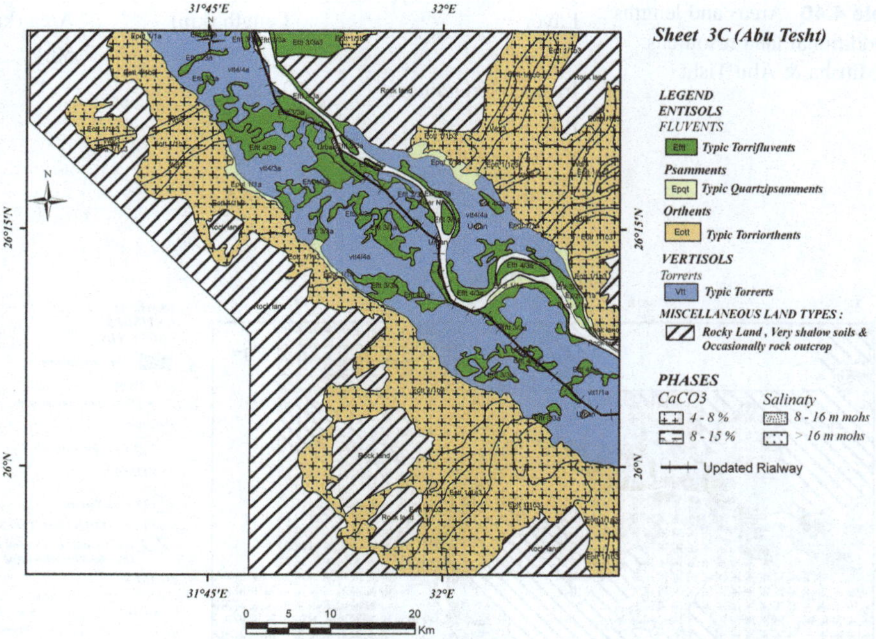

Fig. 4.140 Railway superimposed on the soil map, Al-Minsha & Abu Tisht

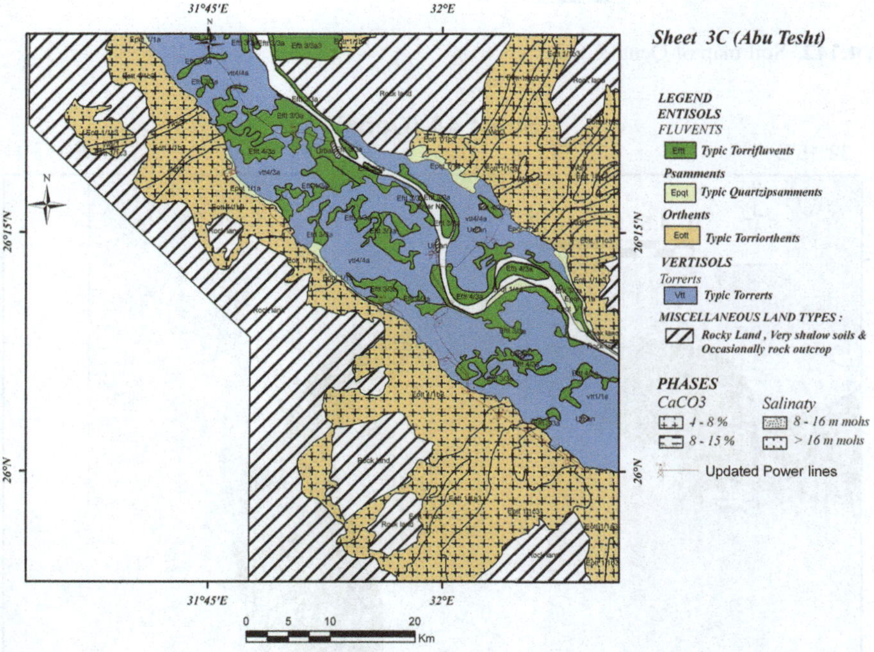

Fig. 4.141 Utilities, superimposed on the soil map, Al-Minsha & Abu Tisht

Table 4.40 Areas and lengths of additional land resources, Al-Minsha & Abu Tisht

Layer	Length (km)	Area (km^2)
Urban		50.1
Irrigation&Drainage	1452.3	
Roads	1273.9	
Railways	257.0	
Utilities	142.9	

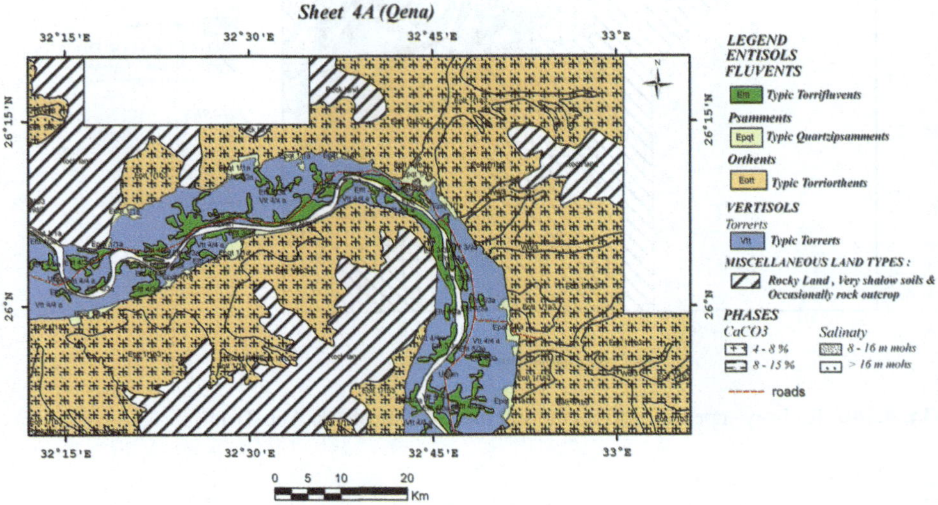

Fig. 4.142 Soil map of Qena & Qus

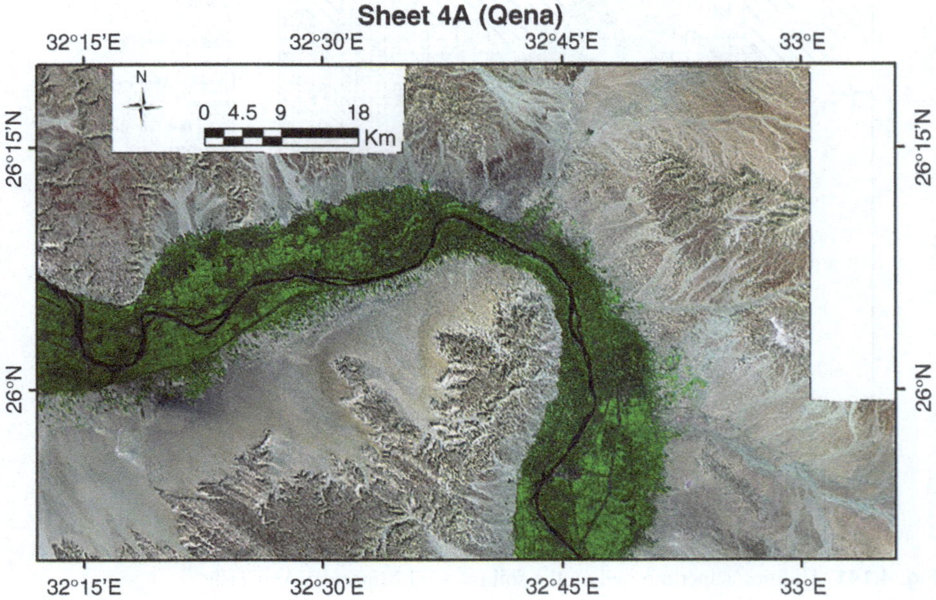

Fig. 4.143 ETM Landsat images of Qena & Qus

Table 4.41 Distribution of soil units in Qena & Qus

Soil Code	Area (km^2)	Area %
Eftt 3/3a	106.2	2.4
Eftt 4/3a	101.3	2.3
Eott 1/1a3	478.0	10.7
Eott 1/1b	11.3	0.3
Eott 1/1b3	1778.0	39.9
Eott 1/1c3	28.1	0.6
Epqt 1/1a	56.9	1.3
Epqt 2/1a	1.8	0.04
Epqt 4/3a	1.8	0.04
River Nile	85.1	1.9
Rock land	1053.5	23.6
Urban	7.1	0.2
Vtt 4/4 a	598.8	13.4
Wd/3	149.7	3.4
Total	4457.6	100.0

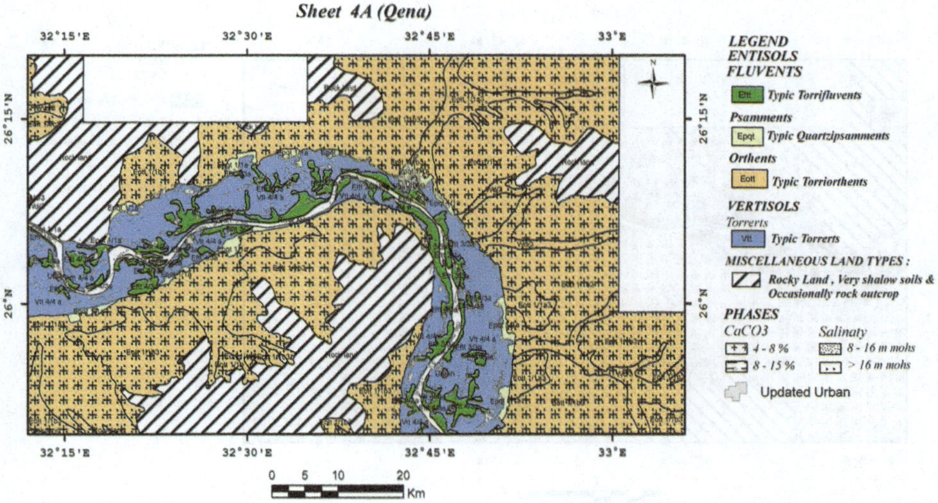

Fig. 4.144 Urban areas superimposed on the soil map, Qena & Qus

4.2.2 Nile Valley

Table 4.51 and Fig. 4.172 show the geographical distribution of soil units in the Nile Valley region. Both *Arididsols* and Entisols soil orders are found. The *Arididsols* are found in the

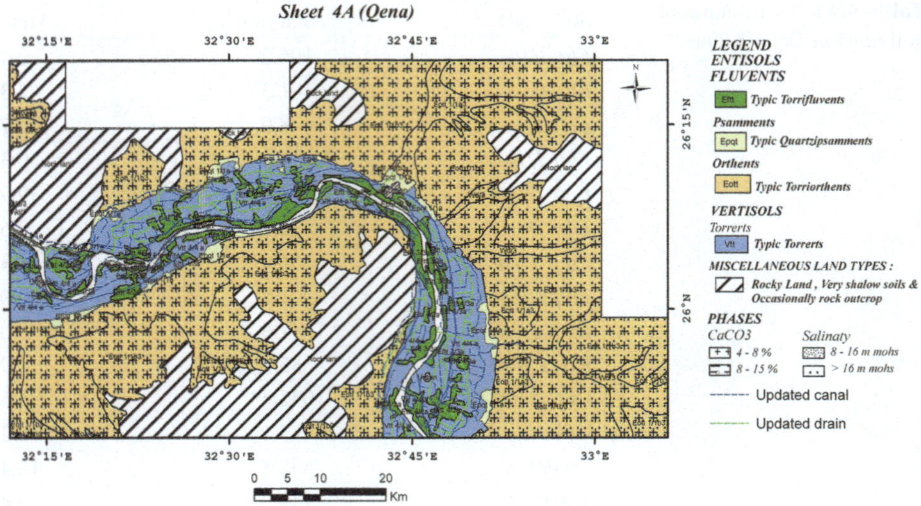

Fig. 4.145 Irrigation and drainage canals superimposed on soil map, Qena & Qus

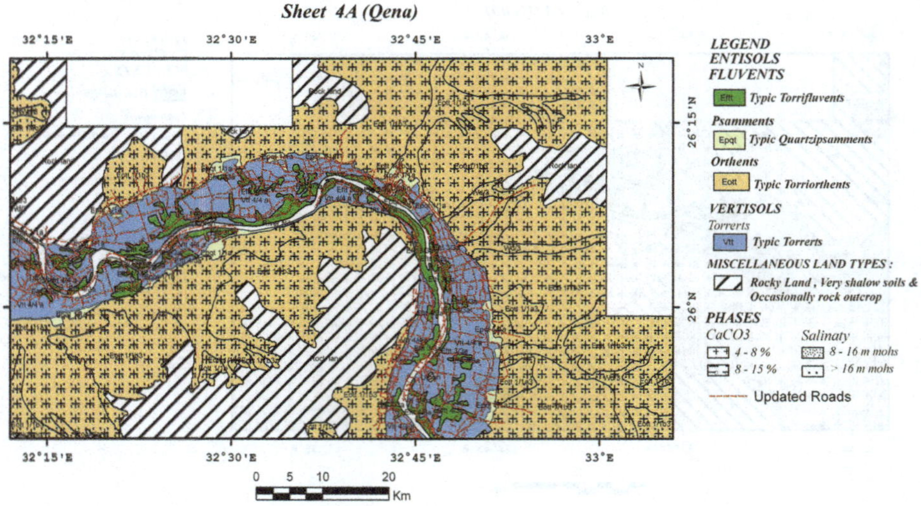

Fig. 4.146 Road network superimposed on soil map, Qena & Qus

western desert fringes of Beni Sweif Governorate and around Al-Fayoum depression, exhibiting 2.6% of the mapped soils. The soils are classified under the sub-orders *Calcids, Gypsids and Salids*. The most pronounced great group is the *Petrogypsids*, which covers an area of 24.3 km^2 representing 1.4% of the soils.

The *Entisols* soil order dominates the study area representing 97.4% of studied soils, including the three sub-orders *Psamments, Fluvents* and *Orthents,* Fig. 4.173. The most

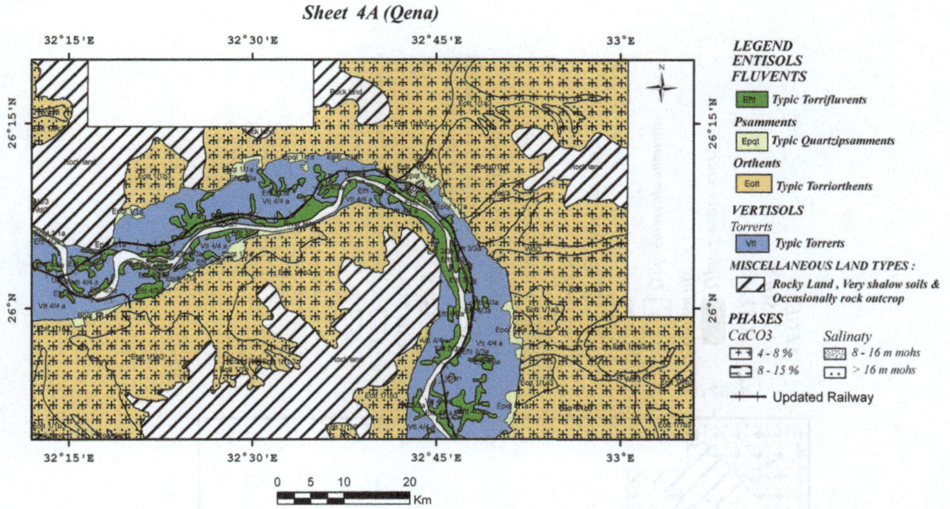

Fig. 4.147 Railway superimposed on the soil map, Qena & Qus

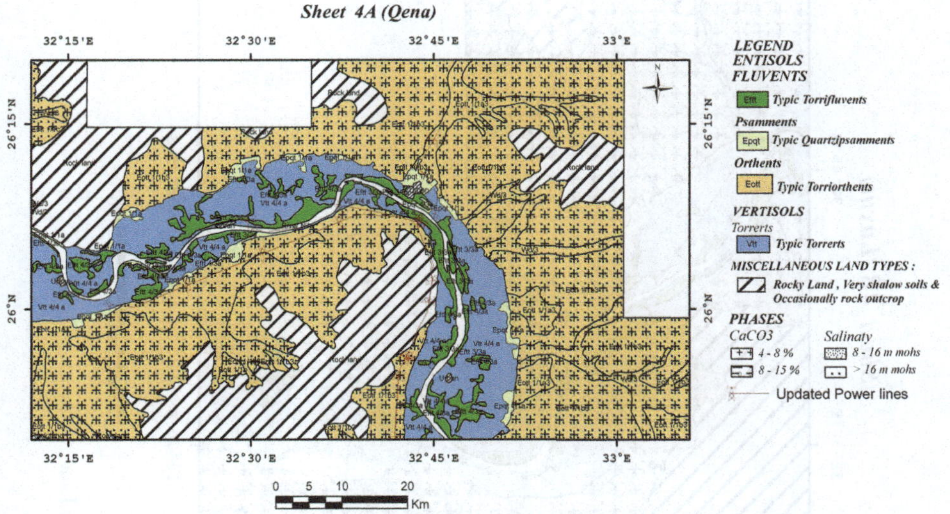

Fig. 4.148 Utilities, superimposed on the soil map, Qena & Qus

Table 4.42 Areas and lengths of additional land resources, Qena & Qus

Layer	Length (km)	Area (km^2)
Urban		50.9
Irrigation&Drainage	1299.2	
Roads	1512.7	
Railways	128.6	
Utilities	159.7	

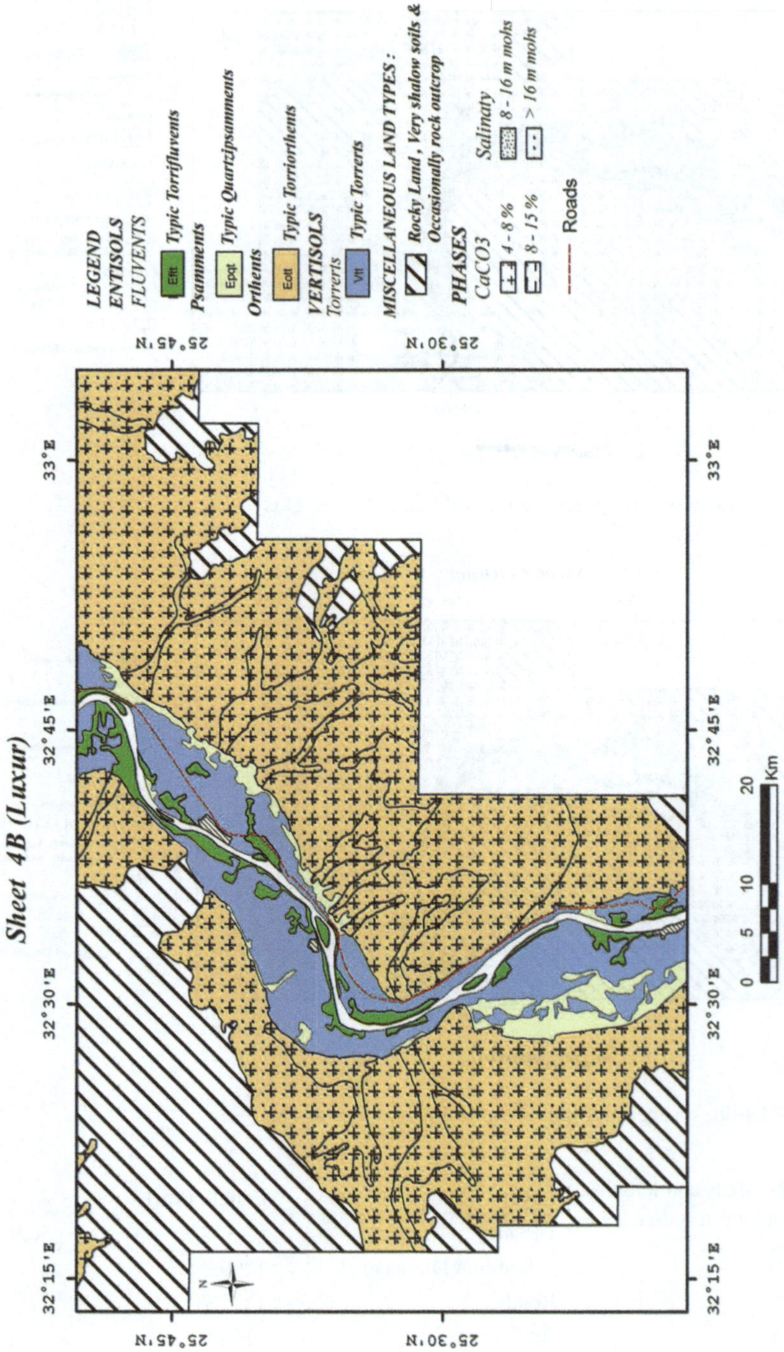

Fig. 4.149 Soil map of Luxor & Isna

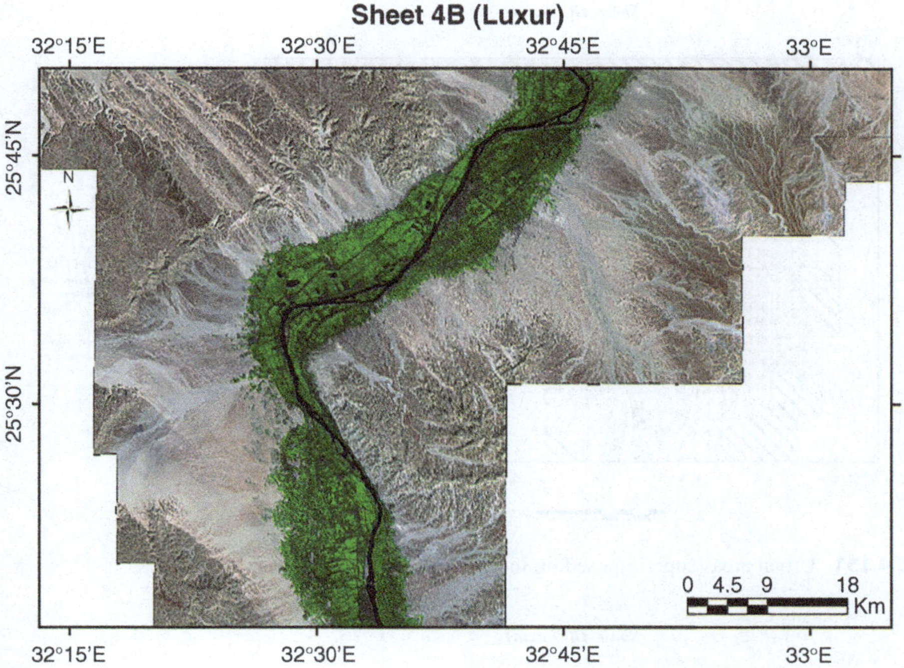

Fig. 4.150 ETM Landsat images of Luxor & Isna

Table 4.43 Distribution of soil units in Luxor & Isna

Soil Code	Area (km^2)	Area %
Eftt 3/3a	8.6	0.2
Eftt 4/3a	87.6	2.3
Eott 1/1b3	1665.4	43.4
Epqt 1/1a	89.8	2.4
Epqt 4/1a	12.8	0.3
Vtt 4/4 a	24.5	11.1
Wd/3	495.6	13.0
Urban	6.4	0.2
Rock land	971.4	25.5
River Nile	63.3	1.7
Total	3814.4	100.0

pronounced great group is the *Torrifluvents* covering an area of 910.2 km^2 representing 52.1% of all soils.

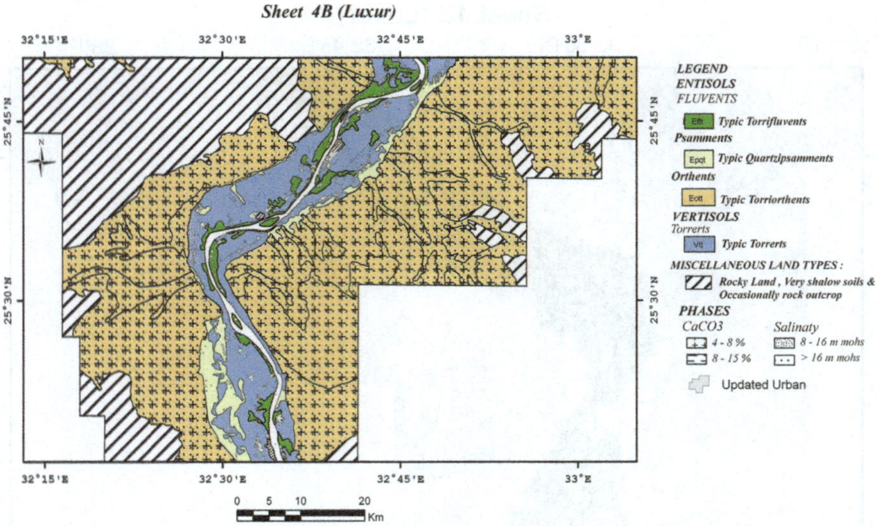

Fig. 4.151 Urban areas superimposed on the soil map, Luxor & Isna

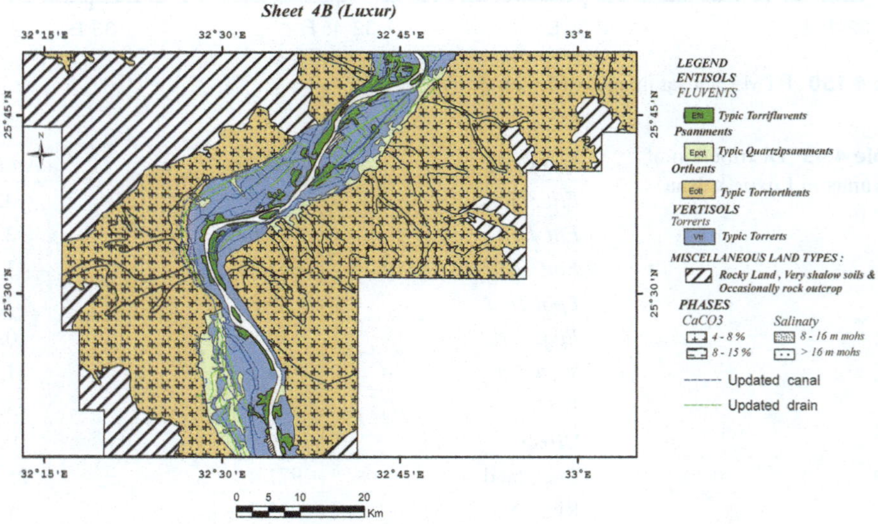

Fig. 4.152 Irrigation and drainage canals superimposed on soil map, Luxor & Isna

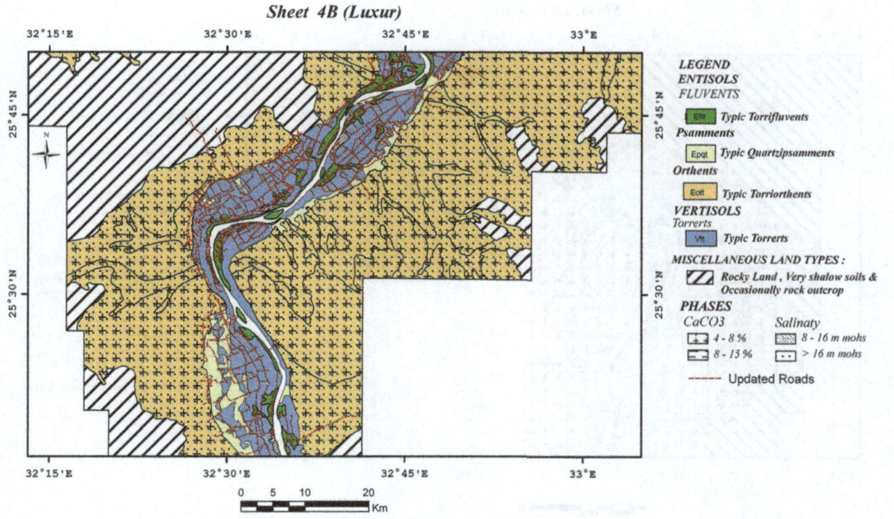

Fig. 4.153 Road network superimposed on soil map, Luxor & Isna

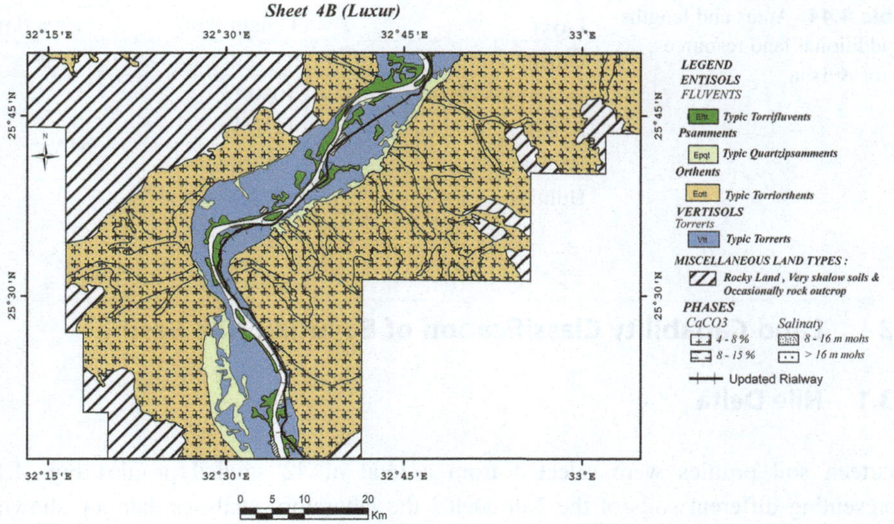

Fig. 4.154 Railway superimposed on the soil map, Luxor & Isna

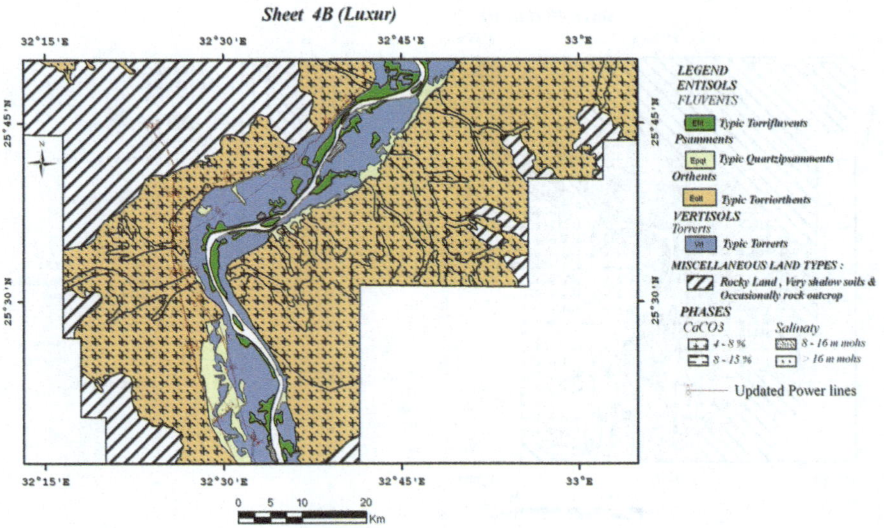

Fig. 4.155 Utilities, superimposed on the soil map, Luxor & Isna

Table 4.44 Areas and lengths of additional land resources, Luxor & Isna

Layer	Length (km)	Area (km^2)
Urban		33.9
Irrigation&Drainage	1010.1	
Roads	1110.1	
Railways	83.2	
Utilities	272.9	

4.3 Land Capability Classification of Egyptian Nile Basine

4.3.1 Nile Delta

Fourteen soil profiles were selected from a total of 42 studied profiles Fig. 4.174 representing different soils of the Nile Delta, the laboratory analyses data are shown in Tables 4.52 and 4.53. These data were used to assess the land capability for each soil unit (i.e. sub-great groups). The assessment of land quality was based on several factors (i.e. drainage condition, texture and structure, % coarse fragments, soil depth, % $CaCO_3$, CEC, soil salinity-EC and exchangeable sodium percentage-ESP as shown in Table 4.54.

The results indicate that the soil salinity, soil depth, soil texture, CEC and coarse fragments content are the main limiting factors for land capability in the Nile Delta and its desert fringes. Figure 4.175 shows the land capability classification map and the main land qualities of the Nile Delta, the different classes and their areas are shown in Table 4.90.

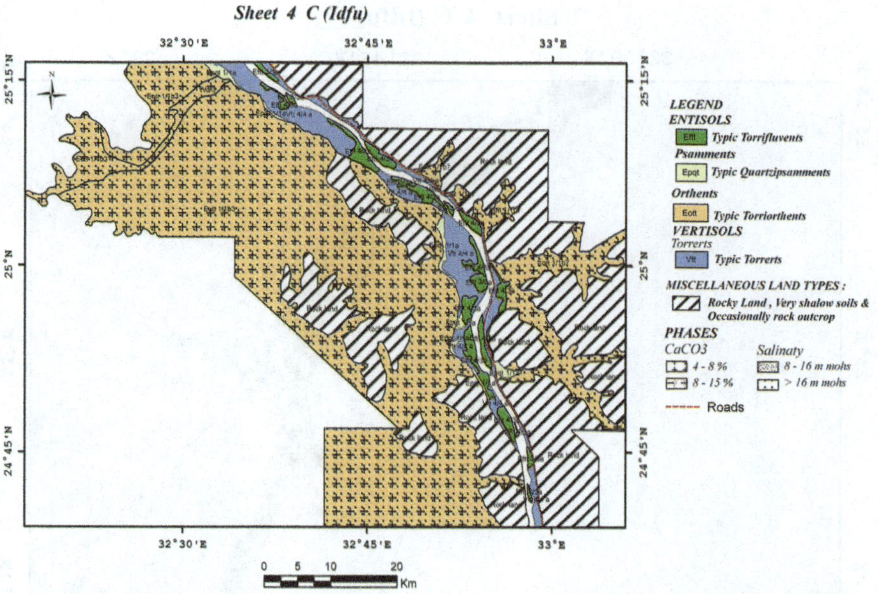

Fig. 4.156 Soil map of Idfu

It is clear that the very highly capable soils (class I) represents about 45.1% of the total area, it is associated with the *Vertic Torrifluvents* sub-great group, and dominates the middle parts of the Nile Delta. The moderately capable soils represent 8.9% of the total area, it is found in scattered parts in the middle, northwest and northeast of the Nile Delta, as they are associated with the soils of *Typic Torrifluvents*. Low and very low capability classes represent 15.0% and 20.7% respectively, they are dominating the rest of the area as they are found around the Northern Lakes and limited parts of the desert frings. They are mostly associated with *Calcids, Gypsids, Salids* soils, in addition to some of sub-great groups under the suborders *Psamments, Fluvents* and *Orthents,* The rocky areas and water bodies represent about 10.3% of the total area (Tables 4.55, 4.56).

4.3.2 Nile Valley

Nine soil profiles (i.e. profiles number 46, 47, 48, 52, 53, 62, 65, 71, 72 and 73) were selected from a total of 40 studied profiles representing different sub-great groups of the Nile Valley Fig. 4.287, the morphological description of the selected profiles can be checked in Annex II. The laboratory analyses data are shown in Annex III. These data were used to assess the land capability for each soil unit (i.e. sub-great groups). The assessment of land quality is based on several factors (i.e. drainage condition, texture

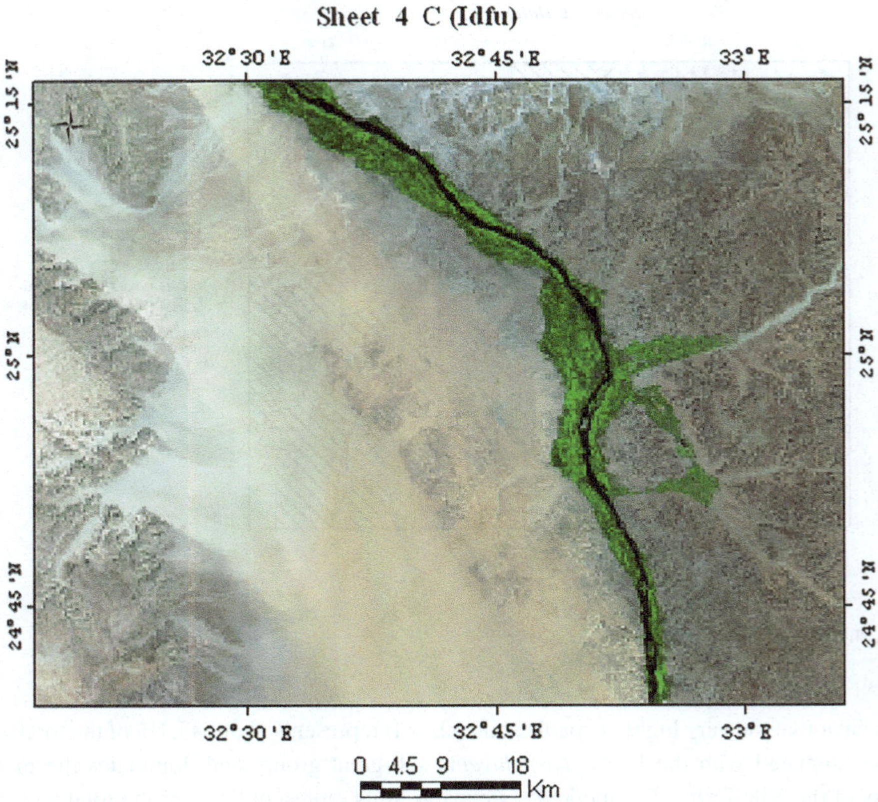

Fig. 4.157 ETM Landsat images of Idfu

Table 4.45 Distribution of soil units in Idfu

Soil Code	Area (km^2)	Area %
Eftt 3/3a	88.2	2.8
Eftt 4/3a	69.9	2.2
Eott 1/1b3	1511.8	47.8
Eott 1/1b7	187.4	5.9
Epqt 1/1a	21.3	0.7
Rock land	1090.3	34.5
Vtt 4/4 a	170.7	5.4
Wd/3	20.7	0.7
Total	3160.3	100.0

and structure, % coarse fragments, soil depth, % $CaCO_3$, CEC, soil salinity-EC and exchangeable sodium percentage-ESP) in Table 4.91.

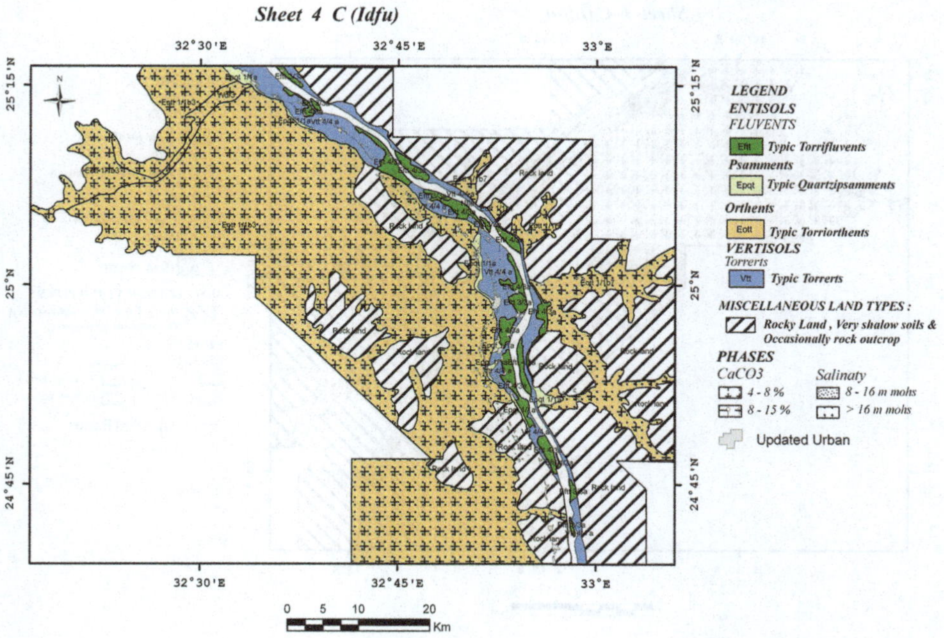

Fig. 4.158 Urban areas superimposed on the soil map, Idfu

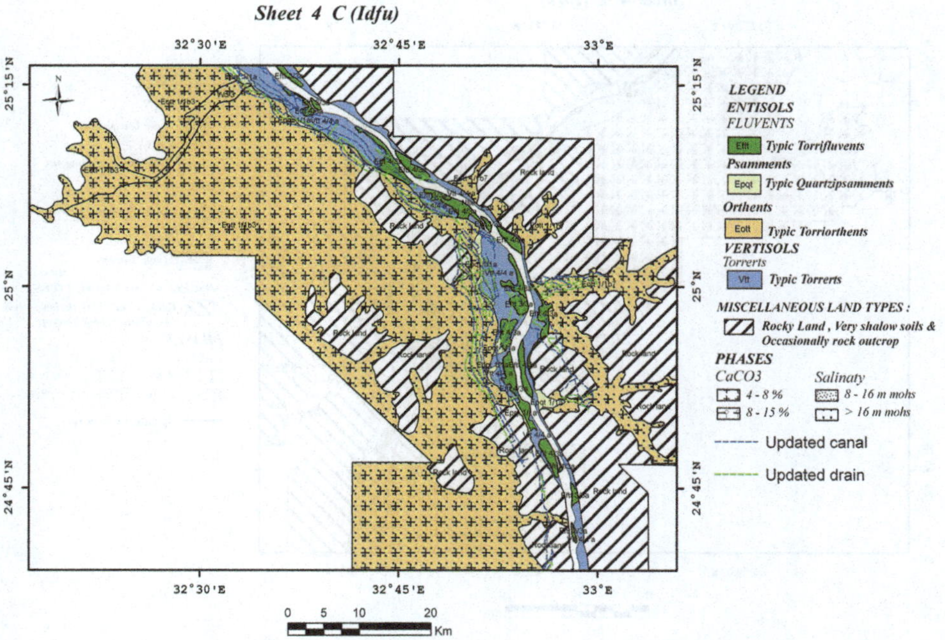

Fig. 4.159 Irrigation and drainage canals superimposed on soil map, Idfu

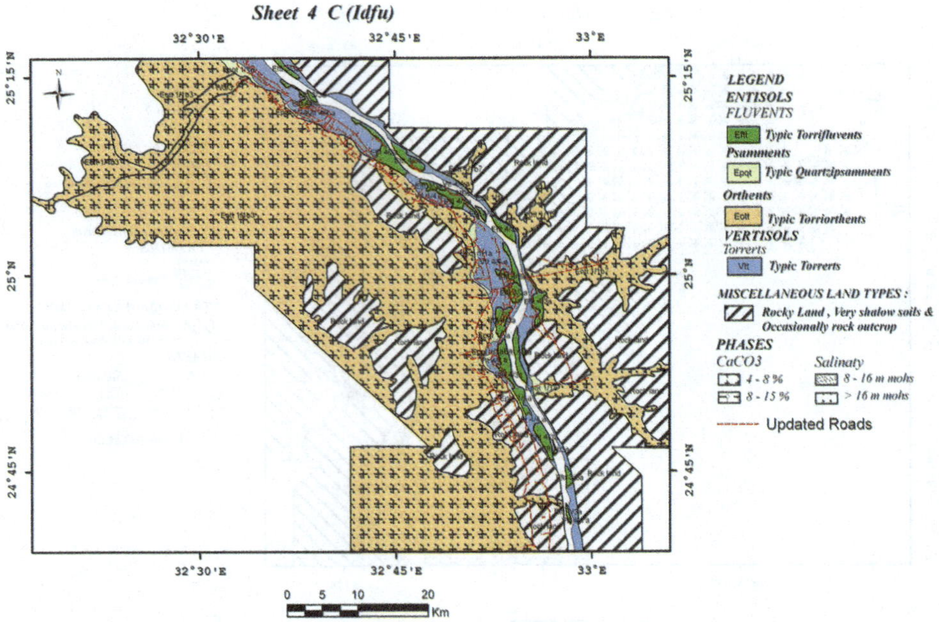

Fig. 4.160 Road network superimposed on soil map, Idfu

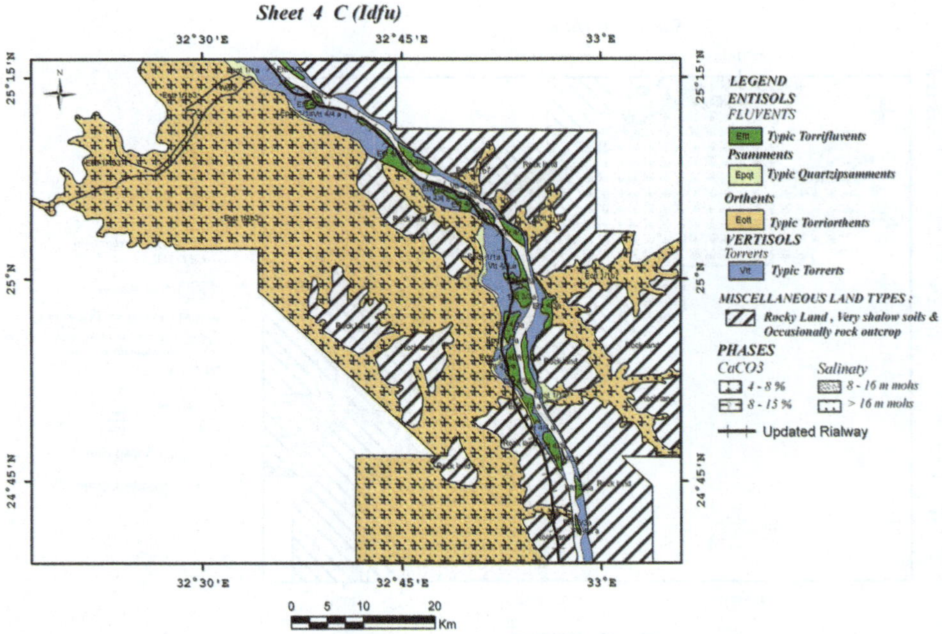

Fig. 4.161 Railway superimposed on the soil map, Idf

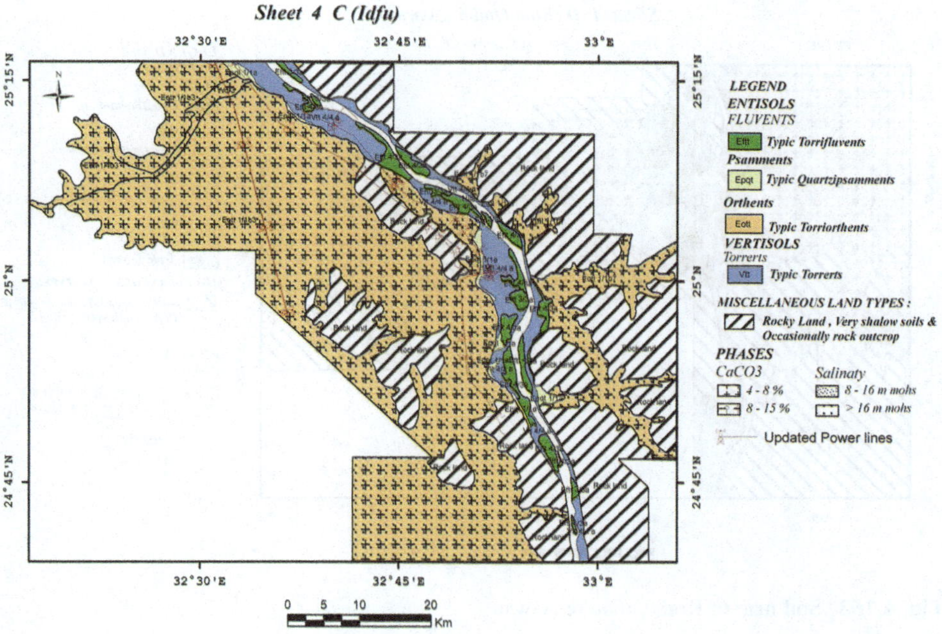

Fig. 4.162 Utilities, superimposed on the soil map, Idfu

Table 4.46 Areas and lengths of additional land resources, Idfu

Layer	Length (km)	Area (km^2)
Urban		18.1
Irrigation&Drainage	666.8	
Roads	520.1	
Railways	87.9	
Utilities	222.3	

The results indicate that the soil salinity (E), soil depth (D), CEC (F) and $CaCO_3$% (K) are the main limiting factors in soils of the Nile Valley and its desert fringes. Figure 4.176 shows the land capability classification map and the main limiting factors of the Nile Valley soils, the different classes and their areas are shown in Table 4.57. It is clear that the highly capable soils (class, I) represents about 10.38% of the total area, it is associated with the *Vertic Torrifluvents* sub-great group, and dominates the middle parts of the Nile valley. The moderately capable soils (Class II) exhibit 5.13% of the total area, it is found in scattered parts along the Nile Valley, as they are associated with the soils of *Typic Torrifluvents*. Low soil capability class (Class III) covers 55.8% of the soils existing in the interference desert zones and wadis (Fig. 4.177).

These soils crossponds the sub-great groups *Typic Quartizpsamments*, *Typic Haplocalsids*, *Typic Torriorthents* and *Typic haplogypsids*. The very low soil capability

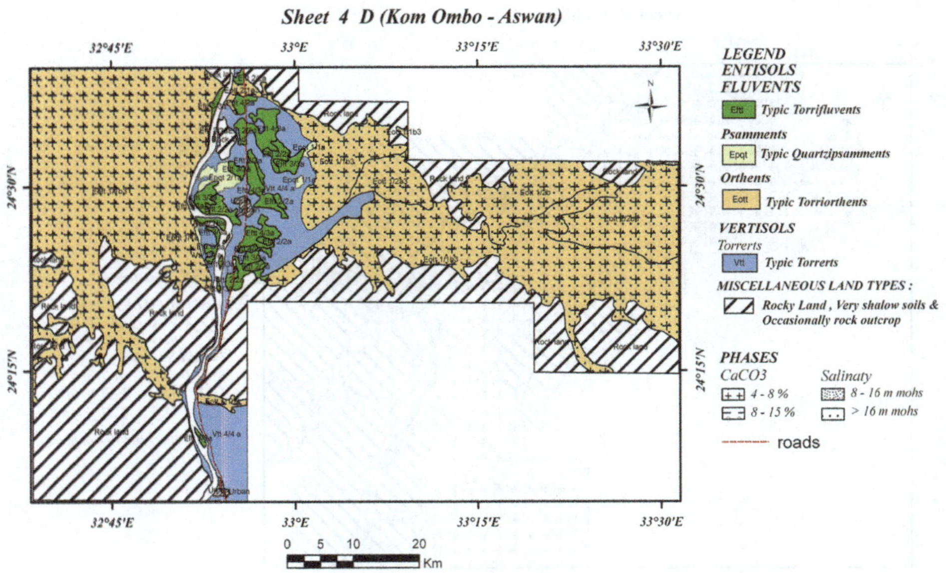

Fig. 4.163 Soil map of Kom Ombo & Aswan

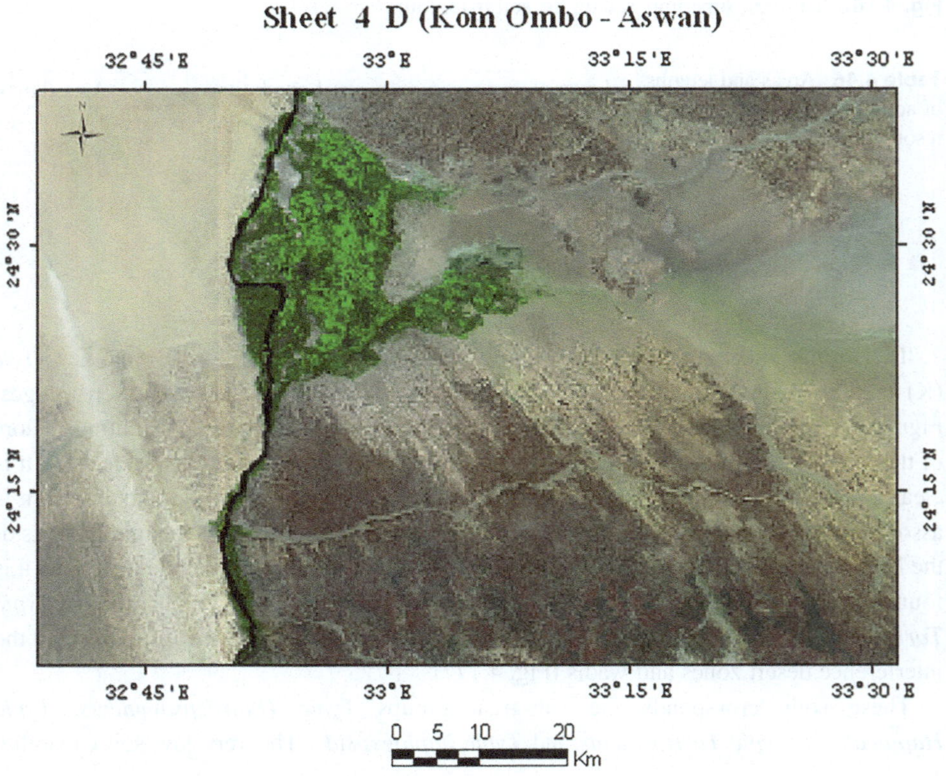

Fig. 4.164 ETM Landsat images of Kom Ombo & Aswan

Table 4.47 Distribution of soil units in Kom Ombo & Aswan

Soil Code	Area (km^2)	Area %
Eftt 2/2a	36.3	1.0
Eftt 3/3a	49.6	1.4
Eftt 4/2a	2.1	0.5
Eftt 4/3a	52.2	1.4
Eott 1/1b3	1451.5	39.7
Eott 1/2a3	58.2	1.5
Eott 1/2b	128.3	3.5
Eott 2/1a3	9.7	0.3
Eott 2/2b3	156.6	4.3
Epqt 1/1a	5.4	0.1
Epqt 2/1a	15.8	0.4
River Nile	61.0	1.7
Rock land	1325.6	36.2
Urban	7.0	0.2
Vtt 4/4 a	287.8	7.8
Total	3647.1	100.0

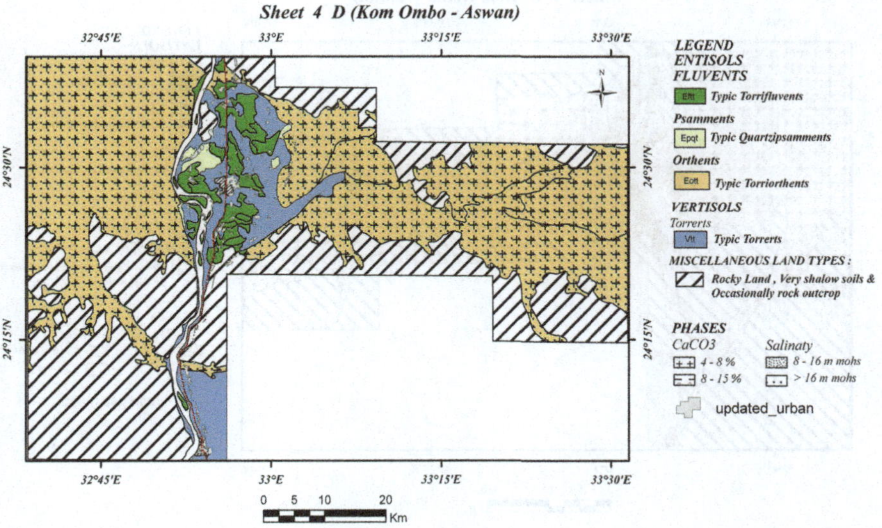

Fig. 4.165 Urban areas superimposed on the soil map, Kom Ombo & Aswan

class (Class IV) covers 1.7% of total soil areas spread in the latitudes between Al-Fayoum and Beni Sweif Governorates. These soils coincide the sub-great group *Typic Petrogypsids, Typic Haplosalids* and *Typic Torripsamments.* It can be noticed that the low and very low capability classes are mostly associated with the soil great groups *Calcids, Gypsids* and *Salids* soils.

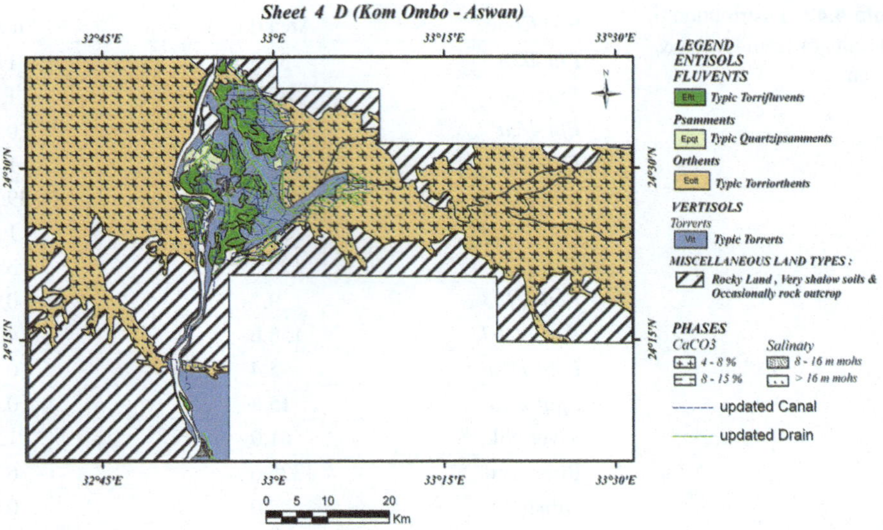

Fig. 4.166 Irrigation and drainage canals superimposed on soil map, Kom Ombo & Aswan

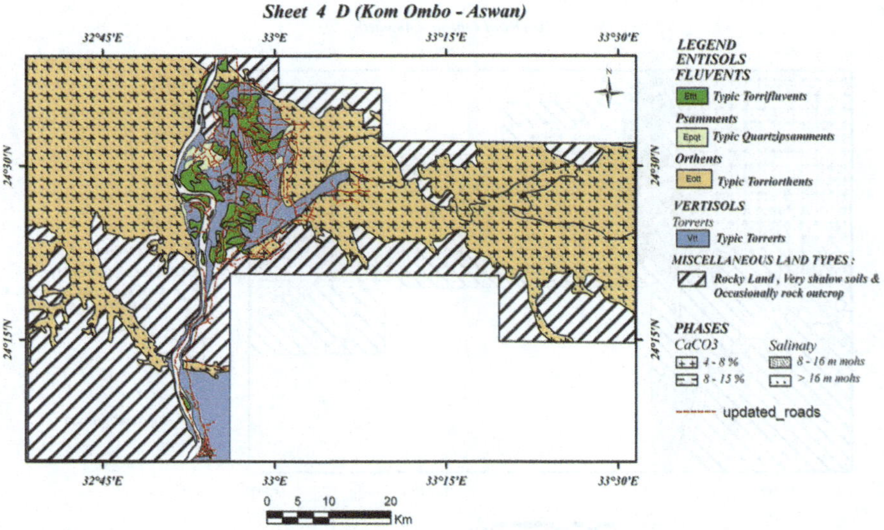

Fig. 4.167 Road network superimposed on soil map, Kom Ombo & Aswan

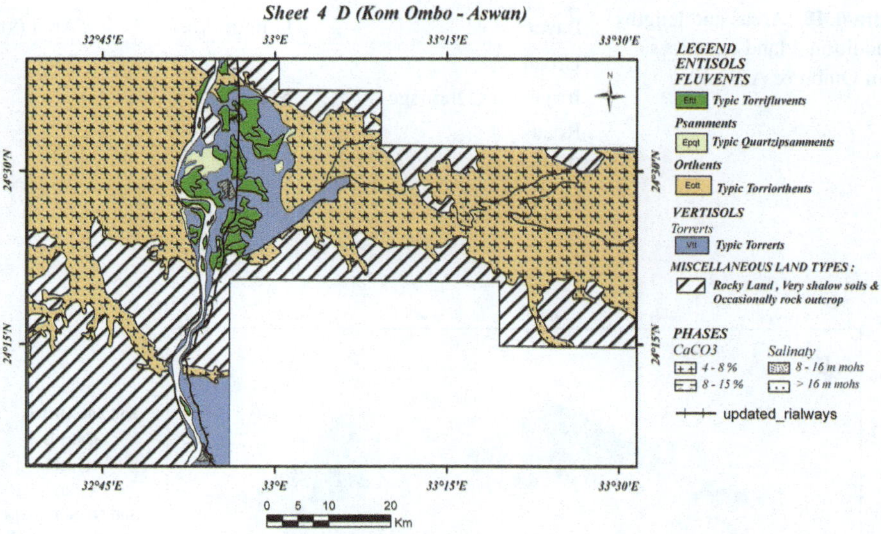

Fig. 4.168 Railway superimposed on the soil map, Kom Ombo & Aswan

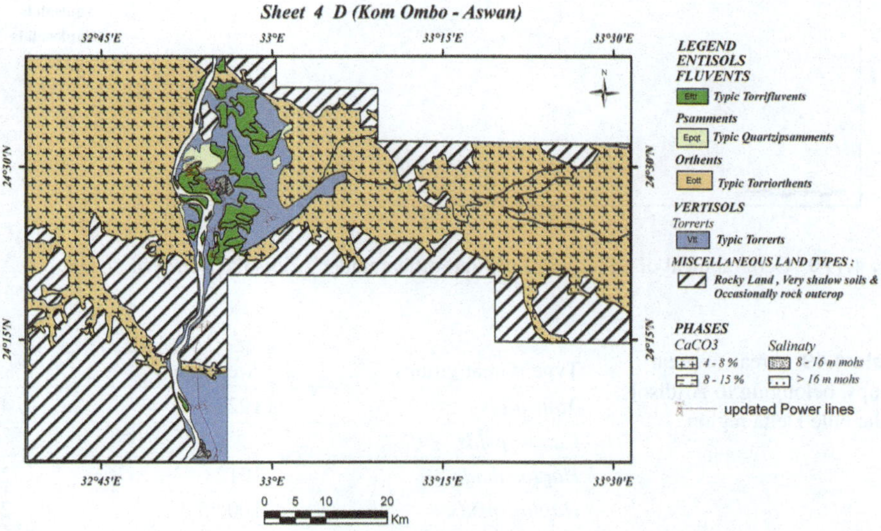

Fig. 4.169 Utilities, superimposed on the soil map, Kom Ombo & Aswan

Table 4.48 Areas and lengths of additional land resources, Kom Ombo & Aswan

Layer	Length (km)	Area (km^2)
Urban		28.5
Irrigation&Drainage	512.1	
Roads	644.7	
Railways	71.0	
Utilities	89.4	

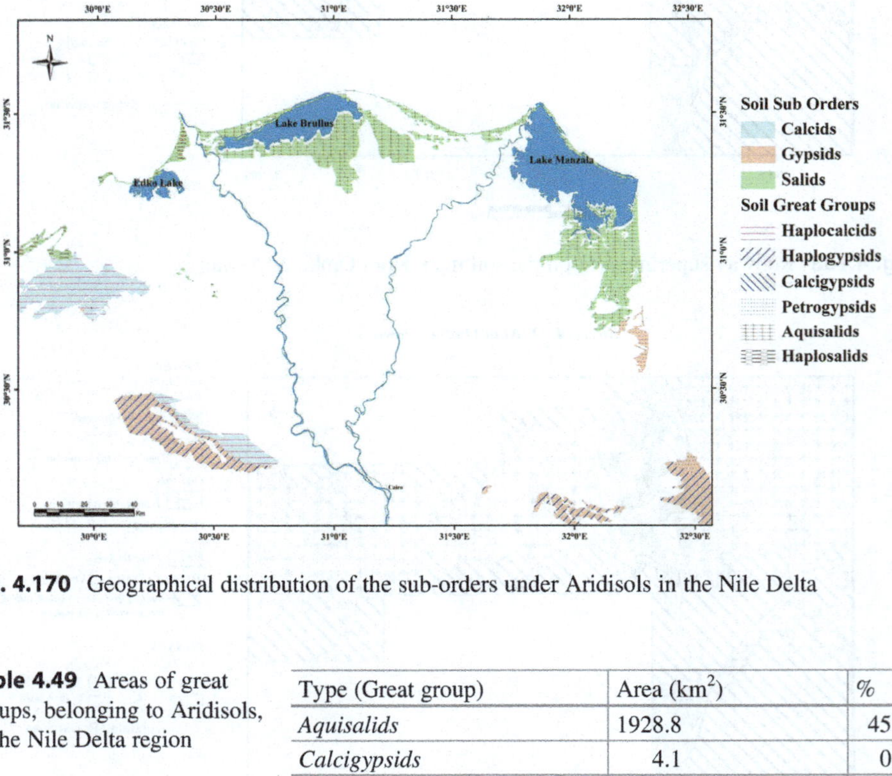

Fig. 4.170 Geographical distribution of the sub-orders under Aridisols in the Nile Delta

Table 4.49 Areas of great groups, belonging to Aridisols, in the Nile Delta region

Type (Great group)	Area (km^2)	%
Aquisalids	1928.8	45.2
Calcigypsids	4.1	0.1
Haplocalcids	1012.6	23.7
Haplogypsids	1007.5	23.6
Haplosalids	166.9	3.9
Petrogypsids	145.9	3.4
Total	4265.7	100.0

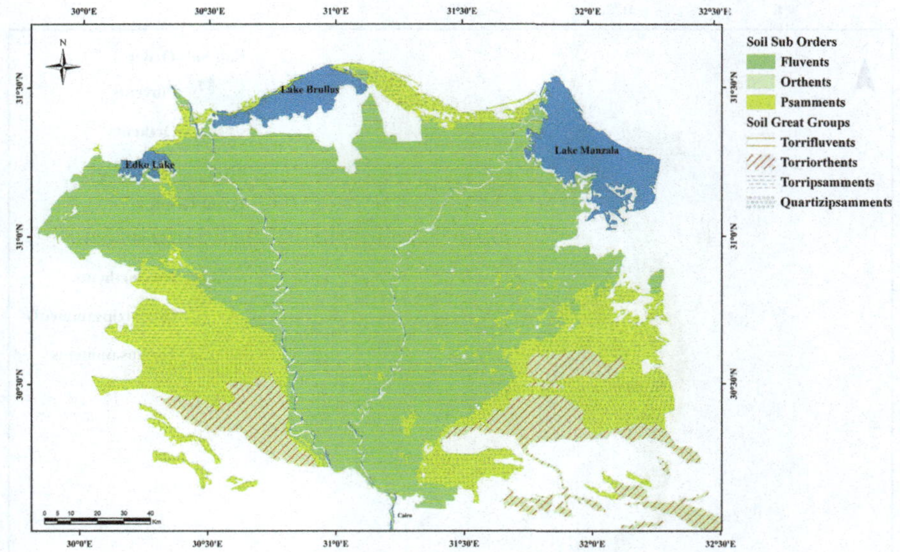

Fig. 4.171 Geographical distribution of the sub-orders under Entisols in the Nile Delta

Table 4.50 Areas of the great groups, belonging to Entisols, in the Nile Delta region

Type (Great group)	Area (km^2)	%
Torrifluvents	15893.3	66.9
Torriorthents	2390.9	10.1
Torripsamments	1385.0	5.8
Quartizpsamments	4082.0	17.2
Total	23751.3	100.0

Table 4.51 Areas of the great groups, belonging to Aridisols and Entisols, in the Nile Valley region

Soil order	Sub-order	Type (Great group)	Area (km^2)	%
Aridisols	*calcids*	*Haplocalcids*	17.4	1.0
	Gypsids	*Haplogypsids*	1.9	0.11
		Petrogypsids	24.3	1.4
	Salids	*Haplosalids*	3.9	0.2
Entisols	*Psamments*	*Quartizpsamments*	333.3	19.1
	Fluvents	*Torrifluvents*	910.2	52.1
	Orthents	*Torriorthents*	455.0	26.1
		Total area of valley soils	1746.0	100.0

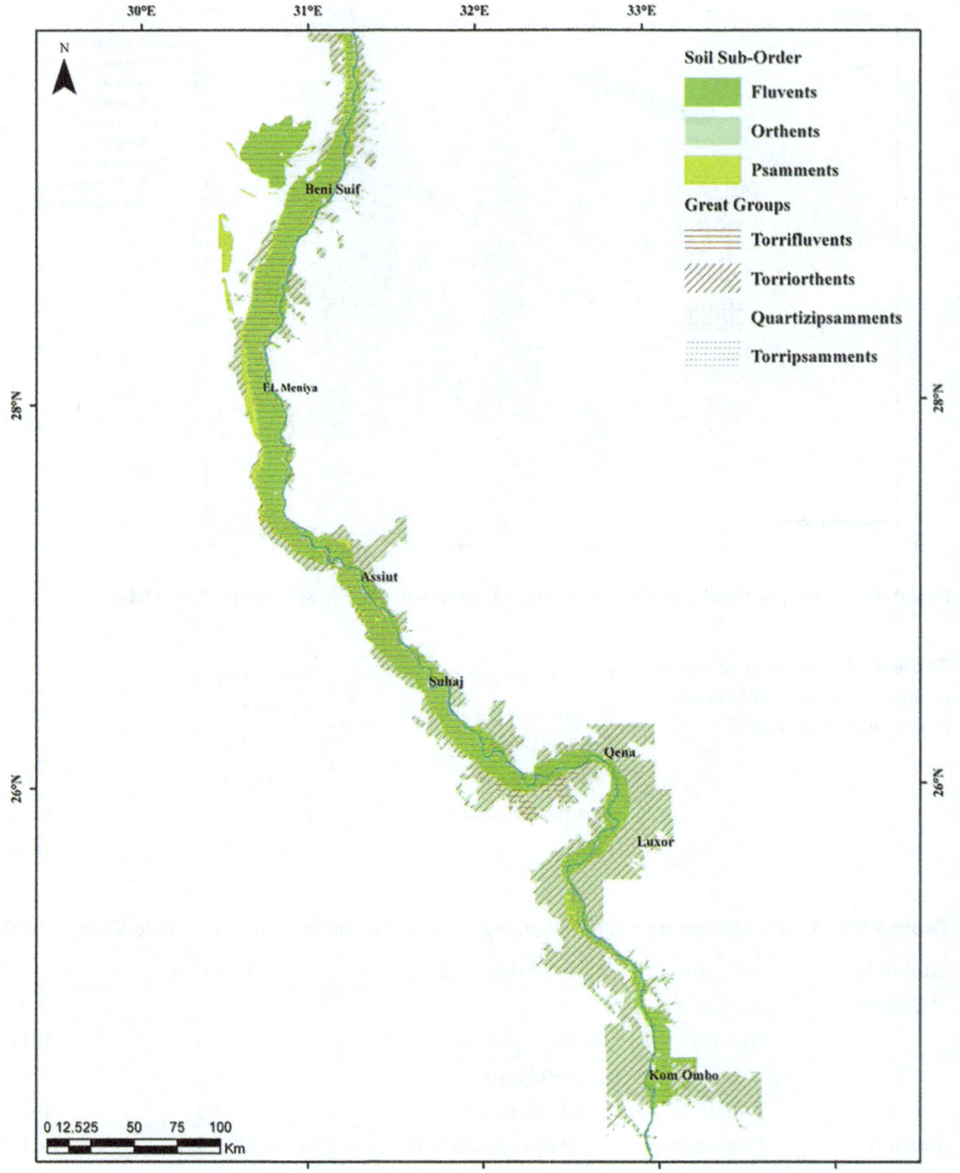

Fig. 4.172 Geographical distribution of Aridisols order in the Nile Valley Region

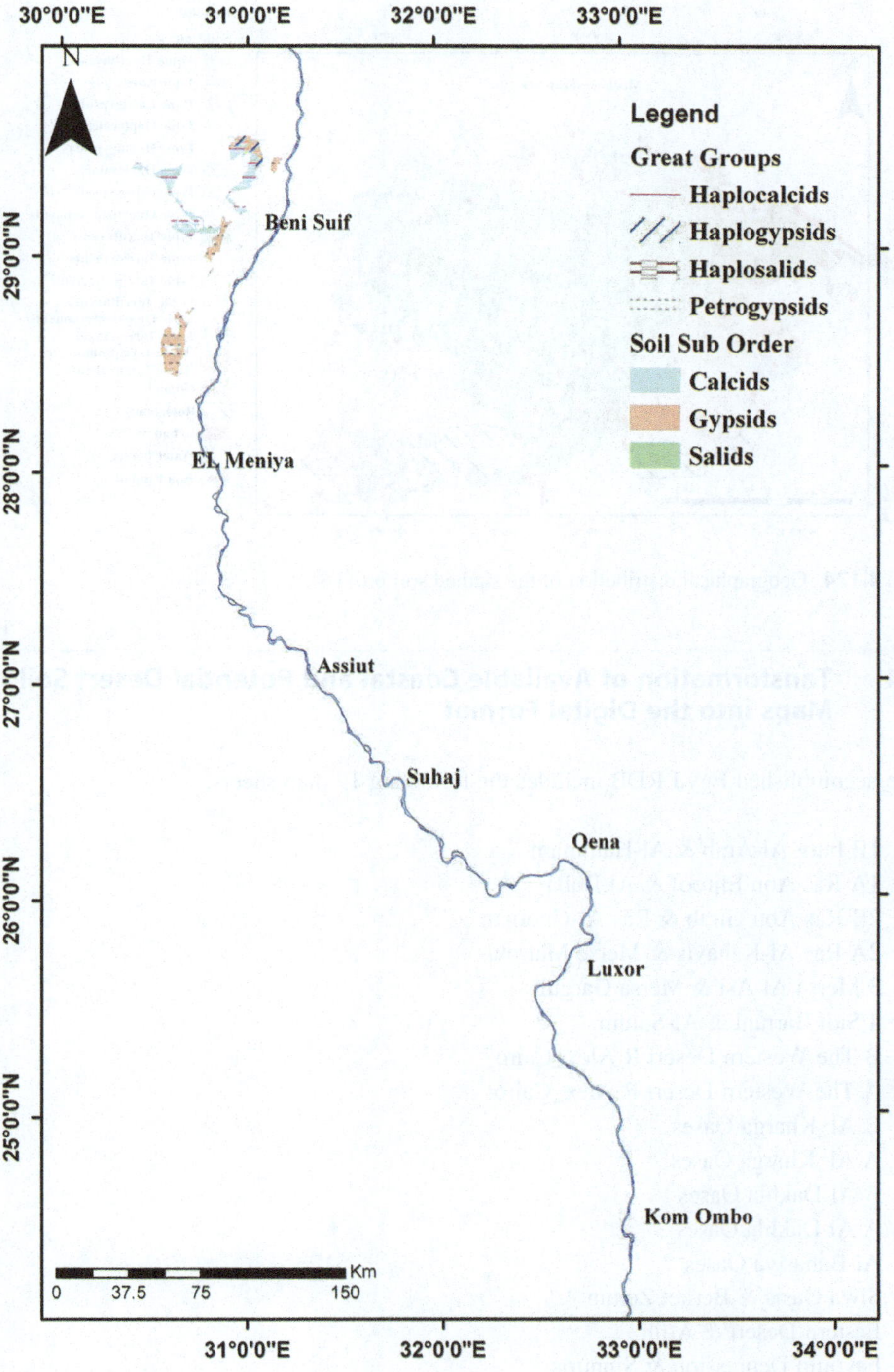

Fig. 4.173 Geographical distribution of Entisols order in the Nile Valley Regio

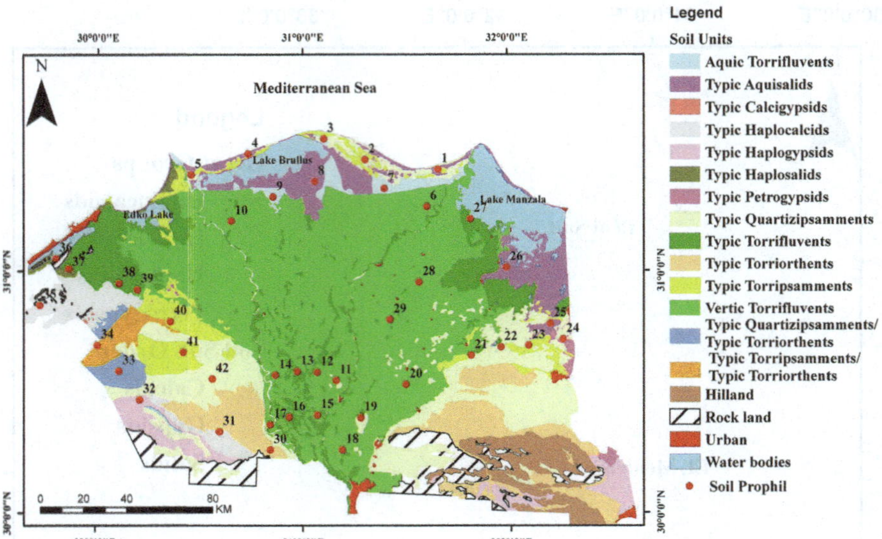

Fig. 4.174 Geographical distribution of the studied soil profiles

4.4 Tansformation of Available Coastal and Potential Desert Soil Maps into the Digital Format

The accomplished Egy.LRDB includes the following 19 map sheets;

1. 1B burg Al-Arab & Al-Hammam
2. 1A Ras Abu Eljroof & Al-Daba
3. 2B Ras Abu Girab & Ras Al-Ghergan
4. 2A Ras Al-Kanayis & Merse Matrouh
5. 3 Mersa Al-Asi & Mersa Gargub
6. 4 Sidi Barrani & Al-Salum
7. B The Western Desert R Alex_Cairo
8. A The Western Desert R Alex_Cairo
9. B Al_Kharga Oases
10. A Al_Kharga Oases
11. B Al-Dakhla Oases
12. A Al-Dakhla Oases
13. Al-Bahariya Oases
14. Siwa Oasis & Berket Zeitun
15. Eastern Desert & Atfih
16. Fayoum Depression & Sinnuris
17. Western Desert & Al-Nubaria

Table 4.52 Particle size distribution analysis of the representative Nile Delta soil profiles

Taxonomic unit	Representative profile no.	Depth in cm	Particle size %					Textural class
			Gravel	Coarse sand	Fine sand	Silt	Clay	
Aquic Torrifluvents	7	0–10	0.00	0.90	35.30	21.50	42.30	C
		10–25	0.00	1.10	34.90	19.80	44.20	C
		25–40	0.00	1.60	26.80	32.30	39.30	CL
		40–60	0.00	1.40	24.20	38.30	36.10	CL
		Mean	0.00	1.25	30.30	27.98	40.48	C
Typic Aquisalids	8	0–25	0.00	0.40	21.20	19.10	59.40	C
		25–50	0.00	0.20	16.70	17.50	65.60	C
		Mean	0.00	0.30	18.90	18.30	62.50	C
Typic Haplosalids	37	0–30	7.15	29.29	34.05	19.04	6.96	SL
		30–50	12.47	30.83	29.95	21.76	9.12	SL
		50–80	0.00	2.45	54.15	14.10	30.09	SCL
		80–110	0.00	5.62	42.15	20.50	36.26	CL
		Mean	4.90	17.05	40.05	18.80	20.80	SCL
Typic Torrifluvents	13	0–20	0.00	1.20	24.50	35.80	38.50	CL
		20–45	0.00	1.50	34.90	27.50	36.10	CL
		45–80	0.00	2.30	27.80	25.60	44.30	C
		80–110	0.00	1.20	24.10	24.90	49.80	C
		Mean	0.00	1.55	27.83	28.44	42.18	C
Vertic Torrifluvents	10	0–10	0.00	1.80	12.30	45.20	40.70	SiC
		10–35	0.00	1.00	11.80	40.50	46.70	SiC
		35–70	0.00	0.50	10.90	41.30	47.30	SiC
		70–120	0.00	0.70	20.40	35.60	43.30	C
		Mean	0.00	1.00	13.85	40.65	44.50	SiC

(continued)

Table 4.52 (continued)

Taxonomic unit	Representative profile no.	Depth in cm	Particle size %					Textural class
			Gravel	Coarse sand	Fine sand	Silt	Clay	
Typic Calcigypsids	35	0–15	0.00	3.46	15.07	31.68	49.79	C
		15–40	0.00	13.36	29.49	18.72	38.43	CL
		40–60	0.00	9.35	39.14	26.16	25.35	SCL
		60–90	0.00	23.65	59.91	5.78	10.66	LS
		Mean	0.00	12.46	35.90	20.58	31.06	SCL
Typic Haplocalcids	31	0–25	9.2	9.62	45.22	27.83	8.13	SL
		25–70	20.4	24.62	39.46	12.81	2.71	LS
		Mean	14.8	17.12	42.34	20.32	5.42	SL
Typic Haplogypsids	25	0–15	4.30	10.4	48.18	22.00	15.12	SL
		15–50	2.04	9.80	37.04	21.50	29.62	SCL
		50–85	9.50	40.50	12.88	9.02	29.10	SCL
		Mean	5.28	20.23	32.70	17.51	24.61	SCL
Typic Petrogypsids	24	0–20	9.10	0.79	33.10	31.00	26.10	CL
		20–75	8.33	0.74	13.54	25.07	52.32	CL
		Mean	8.66	0.77	23.32	28.04	39.21	CL
Typic Torriorthents	30	0–15	12.52	0.95	29.49	36.71	20.33	L
		15–35	8.74	0.60	31.52	37.13	22.01	L
		35–75	20.00	5.25	18.75	11.8	44.2	C
		75–130	11.96	5.75	26.27	14.2	41.82	C
		Mean	13.3	3.14	29.76	25.96	27.84	SCL
Typic Torripsamments	41	0–15	2.25	6.30	81.75	9.20	0.50	S
		15–40	0.70	5.33	82.03	11.72	0.22	S
		40–130	24.99	8.27	55.45	10.29	1.00	S
		Mean	9.31	6.63	73.08	10.40	0.57	S

Typic Quartizpsamments	42	0–25	3.00	18.69	67.19	3.36	7.76	S
		25–100	0.75	28.79	60.74	1.91	7.81	S
		Mean	1.28	24.50	69.96	2.25	2.01	S
Typic Quartizpsamments/Typic Torriorthents	33	0–15	9.10	22.3	51.50	10.7	6.4	LS
		15–50	7.90	26.1	54.40	7.2	4.4	S
		50–120	6.70	25.3	56.80	5.1	6.1	S
		Mean	7.90	24.57	54.23	5.67	7.63	LS
Typic Torripsamments/Typic Torriorthents	34	0–20	6.10	24	56.60	6.1	7.2	S
		20–110	6.50	26.6	55.30	5.1	6.5	S
		Mean	6.30	25.30	55.95	5.60	6.85	LS

Table 4.53 some chemicals analysis of the representative soil profiles

Taxonomic unit	Representative profile	Depth in cm	pH 1: 2.5	EC dS/m	$CaCO_3$%	OM %	ESP	CEC meq/ 100 g. soil	Soluble cations meq/L				Soluble anions meq/L			
									Na^+	K^+	Ca^{++}	Mg^{++}	CO_3^{--}	HCO_3^-	Cl^-	SO_4^{--}
Aquic Torrifluvents	7	0–10	7.73	3.68	2.51	1.39	12.5	46.5	19.70	0.64	3.50	2.28	0.00	0.26	17.73	8.13
		10–25	7.81	2.83	2.31	1.70	11.7	48.8	12.83	1.64	13.75	10.31	0.00	0.17	11.55	26.81
		25–40	7.88	2.56	1.62	0.31	14.3	40.3	25.43	0.95	3.50	2.45	0.00	0.37	24.67	7.29
		40–60	7.91	1.78	1.12	0.38	14.3	39.4	6.89	0.64	1.75	1.31	0.00	0.09	6.21	4.29
		Mean														
Typic Aquisalids	8	0–25	7.81	13.3	1.33	1.70	13.2	52.6	22.48	1.67	6.15	4.31	0.00	3.20	21.36	10.05
		25–50	7.88	11.04	2.05	0.3	12.6	43.2	23.48	0.85	7.05	4.23	0.00	3.35	22.31	9.95
		Mean														
Typic Haplosalids	37	0–30	7.95	11.70	5.55	0.39	16.2	14.5	16.96	2.26	4.25	2.97	0.00	0.25	16.45	9.74
		30–50	7.87	18.90	4.41	0.64	14.8	17.4	48.91	0.64	13.75	10.31	0.00	0.75	49.86	23.00
		50–80	8.05	30.60	7.22	0.41	15.6	32.4	63.04	3.33	15.00	11.25	0.00	0.94	62.37	29.31
		80–110	7.64	16.20	9.71	0.42	13.4	38.2	54.35	3.85	37.5	28.8	0.00	0.77	51.63	72.10
		Mean														
Typic Torrifluvents	13	0–20	8.15	3.20	3.15	2.86	16.5	40.6	11.21	0.26	4.65	3.56	0.00	3.03	10.09	6.56
		20–45	8.40	1.32	2.11	1.35	13.7	39.3	6.91	0.13	1.95	1.37	0.00	2.23	6.36	1.77
		45–80	8.01	1.80	1.60	1.91	14.5	44.3	2.39	0.13	1.25	0.87	0.00	0.43	2.15	2.06
		80–110	8.13	3.57	1.20	2.12	12.8	45.8	9.34	0.33	8.05	6.45	0.00	2.94	1.06	12.82
		Mean														
Vertic Torrifluvents	10	0–10	8.40	0.92	2.11	2.01	12.3	42.3	4.56	1.03	2.65	1.85	0.00	0.83	4.15	5.11
		10–35	8.50	0.79	1.03	0.70	14.5	47.6	3.96	0.15	1.65	1.15	0.00	0.75	3.56	2.60
		35–70	8.25	0.62	0.92	1.42	13.2	49.8	3.35	0.33	1.65	1.24	0.00	0.85	3.18	2.54
		70–120	8.34	0.80	0.88	1.77	15.8	45.1	4.61	0.18	1.30	0.98	0.00	1.25	4.15	1.67
		Mean														
Typic Calcigypsids	35	0–15	7.57	10.90	7.52	0.51	11.3	45.6	35.56	6.46	24.90	19.92	0.00	7.93	34.49	44.42
		15–40	7.74	6.10	8.61	0.57	14.7	37.4	26.61	5.15	16.90	13.52	0.00	5.81	25.28	31.09
		40–60	7.70	5.80	6.92	0.45	14.9	27.8	26.00	4.74	16.25	12.18	0.00	5.62	24.44	29.11
		60–90	7.86	2.52	9.52	0.23	15.7	14.8	10.22	2.51	7.00	5.21	0.00	2.11	9.19	13.64

		Mean														
Typic Haplocalcids	31	0–25	7.92	3.65	13.52	0.12	25.6	12.8	24.57	0.69	12.65	8.85	0.00	0.32	21.13	25.31
		25–70	8.10	4.22	14.23	0.15	26.4	8.6	23.91	5.51	10.40	7.81	0.00	20.00	21.52	6.11
		Mean														
Typic Haplogypsids	25	0–15	8.21	14.23	10.22	1.65	22.6	19.6	10.30	1.87	11.25	8.44	0.00	0.15	9.89	21.82
		15–50	8.41	11.65	8.65	1.11	24.2	30.5	10.22	1.18	9.75	7.31	0.00	3.32	14.46	10.86
		50–85	7.96	14.71	7.33	0.42	26.4	28.6	11.26	1.85	24.00	16.80	0.00	0.16	10.70	43.05
		Mean														
Typic Petrogypsids	24	0–20	8.21	7.51	9.12	0.51	16.84	28.3	27.17	0.64	13.75	8.25	0.00	9.03	25.81	14.97
		20–75	8.05	8.11	11.23	0.12	21.61	46.1	26.74	1.67	13.30	13.90	0.00	8.52	28.40	18.69
		Mean														
Typic Torriorthents	30	0–15	8.23	3.15	4.86	1.13	13.7	23.6	12.83	0.18	14.75	4.50	0.00	10.00	22.18	0.08
		15–35	8.41	4.33	5.19	1.52	11.9	25.4	23.48	0.67	8.85	4.88	0.00	7.81	22.31	7.76
		35–75	8.11	5.62	6.73	0.41	12.7	45.3	25.43	0.95	3.50	2.45	0.00	0.37	24.67	7.29
		75–130	8.05	7.11	7.12	0.10	18.6	39.5	23.48	1.72	4.55	3.64	0.00	3.78	16.44	13.17
		Mean														
Typic Torripsamments	41	0–15	7.60	7.62	1.51	0.41	14.8	8.9	53.69	0.54	28.61	12.20	0.00	0.60	86.01	8.43
		15–40	7.93	3.49	1.95	0.21	12.6	9.5	38.78	0.24	12.81	10.50	0.00	0.70	43.02	18.61
		40–130	8.00	3.51	2.03	0.02	13.5	9.3	32.61	0.19	5.65	16.13	0.00	0.60	8.01	45.97
		Mean	7.84	4.87	1.83	0.21	13.6	9.2	41.69	0.32	15.69	12.94	0.00	0.63	45.68	24.34
Typic Quartizpsamments	42	0–25	8.10	0.77	8.33	0.30	11.2	7.2	10.87	0.44	5.61	1.53	0.00	2.10	6.03	10.32
		25–100	8.57	0.20	10.23	0.32	10.3	8.3	3.04	0.23	2.55	2.25	0.00	1.60	1.50	4.97
		Mean	8.34	0.485	9.28	0.31	10.75	7.75	6.95	0.34	4.08	1.89	0.00	1.85	3.76	7.65
Typic Quartizpsamments/ Typic Torriorthents	33	0–15	7.72	5.03	7.74	0.36	12.5	12.3	32.81	0.75	25.65	16.03	0.00	1.70	45.51	28.06
		15–50	7.73	7.08	9.15	0.32	14.3	10.6	38.77	0.41	44.45	17.51	0.00	1.30	64.50	35.32
		50–120	7.61	5.92	3.65	0.05	9.8	11.4	29.82	0.29	39.35	16.05	0.00	0.90	54.51	29.66
		Mean	7.69	6.01	6.85	0.24	12.2	11.4	33.80	0.48	36.48	16.53	0.00	1.30	54.84	31.01
Typic Torripsamments/ Typic Torriorthents	34	0–20	8.32	1.55	1.44	0.30	11.9	9.6	27.39	0.30	9.45	8.51	0.00	0.80	16.52	28.32
		20–110	7.88	28.90	1.18	0.45	14.8	12.6	244.12	1.20	45.55	16.31	0.00	1.10	176.61	30.08
		Mean	8.10	15.23	1.31	0.38	13.3	11.1	135.76	0.75	27.50	12.41	0.00	0.95	96.57	29.20

Table 4.54 Land capability classes and limiting factors in the Nile Delta representative soil profiles

Factor		Representative soil profiles													
		7	8	37	13	10	35	31	25	24	30	41	42	33	34
Soil depth	D	3	3	1	1	1	2	2	2	2	1	1	2	1	1
Texture/Structure	T	1	1	1	1	1	2	1	1	1	1	3	3	2	2
% Coarse fragments	C	1	1	1	1	1	4	3	3	4	4	3	2	3	3
Drainage condition	L	1	3	1	1	1	1	1	2	2	1	1	1	1	1
% $CaCO_3$	K	1	1	1	1	1	2	3	3	1	1	2	1	1	1
CEC meq/100 g. soil	F	1	1	2	1	1	1	3	2	2	1	4	3	4	3
ECe $_{(1:1)}$ dS/m	E	2	4	4	2	1	3	2	4	3	3	2	3	3	4
ESP	S	2	2	1	2	1	2	3	3	2	2	1	1	1	1
Capability class		III	IV	IV	II	I	IV	III	IV	IV	IV	IV	III	IV	IV
Limiting factors	–	D	E, D, L	E	–	–	C, E	C, K, F, S	E, C, S	C, E	C, E	F, T, C	T, F, E	F, C, E	E, C, F

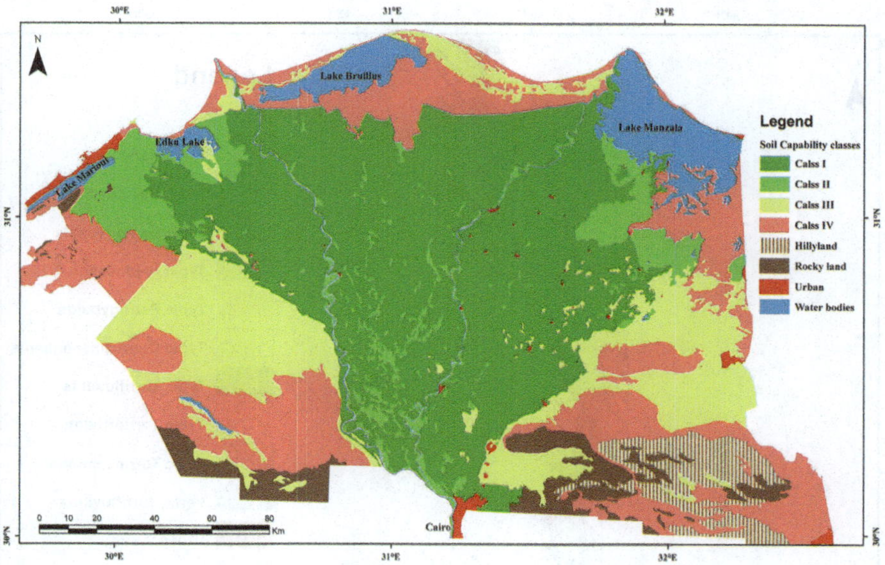

Fig. 4.175 Land capability classification map of the Nile Delta region

Table 4.55 Areas of the Nile Delta land capability classes

Capability class	Area km^2	Area %
I	12794.4	45.1
II	2510.8	8.9
III	4251.6	15.0
IV	5880.4	20.7
Other features		
R	573.8	2.0
U	208.1	0.7
W	2143.9	7.6
Total	28362.9	100.0

Table 4.56 Distribution of soil units in Burg Al-Arab–Al-Hammam

Soil Code	Area (km^2)	Area %
Aocat 3/3 a	110.9	3.7
Aogp 1/1b	39.5	1.3
Aosat 4/4a	99.9	3.3
Eott 1/1a	360.3	12.1
Eptl 2/1c	376.8	12.6
Eptt 2/1b	113.7	3.8
Eptt 2/1c	374.6	12.6
Plateau surface	2.4	0.1
Rock land	217.1	7.3
Water	1290.7	43.2
Total	2985.8	100.0

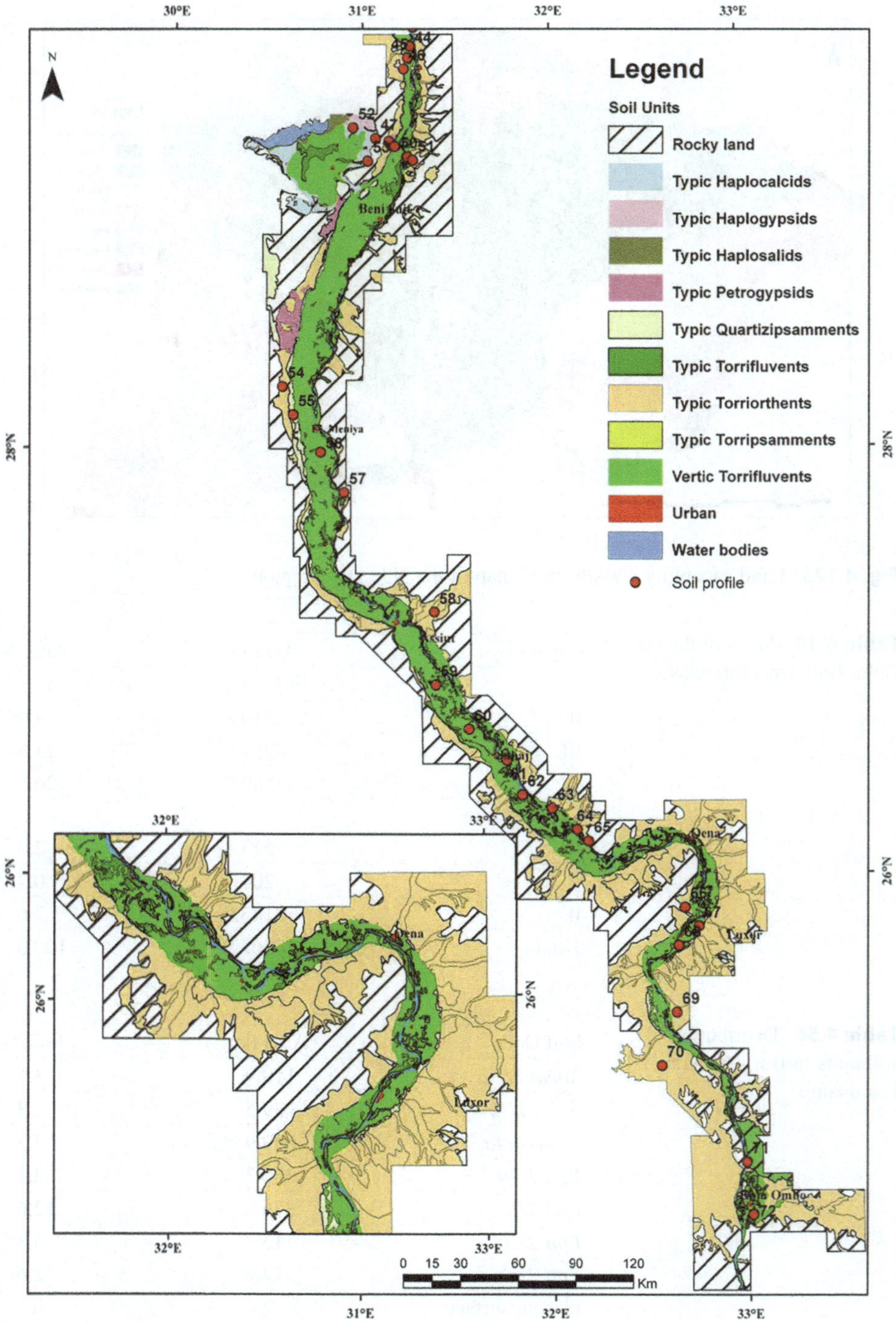

Fig. 4.176 Geographical distribution of the studied soil profiles of the Nile Valley

Table 4.57 Lengths of additional land resources, Burg Al-Arab–Sidi Abdel-Rahman

Layer	Length (km)	Area (km^2)
Urban		2.0
Roads	1309.9	
Railways	60.3	
Utilities	344.4	

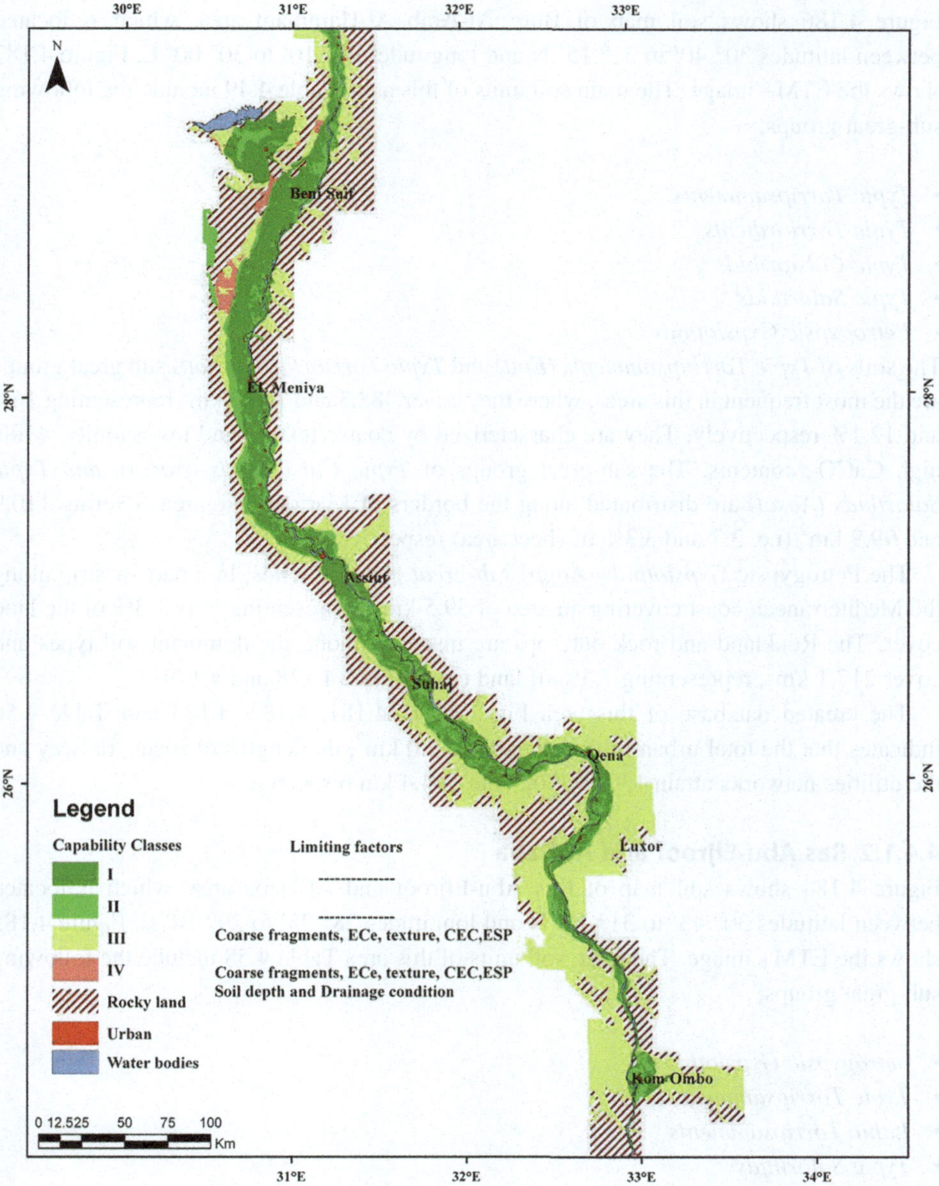

Fig. 4.177 Land capability classes and limiting factors of the Nile Valley region

18. B Al-Tumailat-Suez
19. A Al-Tumailat-Suez Al-Tell Al-Kabeir

4.4.1 Northwestern Coastal Zone Soil Maps

4.4.1.1 Burg Al-Arab–Al-Hammam

Figure 4.186 shows soil map of Burg Al-Arab–Al-Hammam area, which is located between latitudes 30° 40′ to 31° 15′ N and longitudes 29° 10′ to 30° 00′ E. Figure 4.187 shows the ETM+ image. The main soil units of this area, Table 4.49 include the following sub-great groups;

- *Typic Torripsamments*
- *Typic Torriorthents*
- *Typic Calsiorthids*
- *Typic Salorthids*
- *Petrogypsic Gypsiothids*

The soils of *Typic Torrbipsamments (Eptt)* and *Typic Torriorthents (Eott)* sub great groups are the most frequent in this areas, where they cover 488.3 and 360.3 km^2 representing 16.4 and 12.1% respectively. They are characterized by coarse texture and low salinity, while high $CaCO_3$ contents. The sub-great groups of *Typic Calsiorthids (Aocat), and Typic Salorthids (Aosat)* are distributed along the borders of Mariout lake area covering 110.9 and 99.9 km^2 (i.e. 3.7 and 3.3% of sheet area) respectively.

The Petrogypsic *Gypsiothids (Aogp) sub-great group* extends, in a narrow strip along the Mediterranean coast covering an area of 39.5 km^2, representing only 1.3% of the land cover. The Rockland and rock outcrops are inserted among the dominant soil types and cover 217.1 km^2, representing 7.3% of land cover (Figs. 4.178 and 4.179).

The created database of this area Figs. 4.180, 4.181, 4.182, 4.183 and Table 4.50 indicates that the total urban settlement area is 2.0 km^2, the lengths of roads, railway and the utilities networks attain 1309.9, 60.3 and 344.4 km respectively.

4.4.1.2 Ras Abu-Eljroof and Al-Daba

Figure 4.184 shows soil map of Ras Abu-Eljroof and Al-Daba area, which is located between latitudes 30° 43′ to 31° 15′ N and longitudes 28° 25′ to 29° 10′ E. Figure 4.185 shows the ETM+ image. The main soil units of this area Table 4.58 include the following sub-great groups;

- *petrogypsic Gypsiothids*
- *Typic Torripsamments*
- *Lithic Torripsamments*
- *Typic Salorthids*
- *Typic calciorthids*

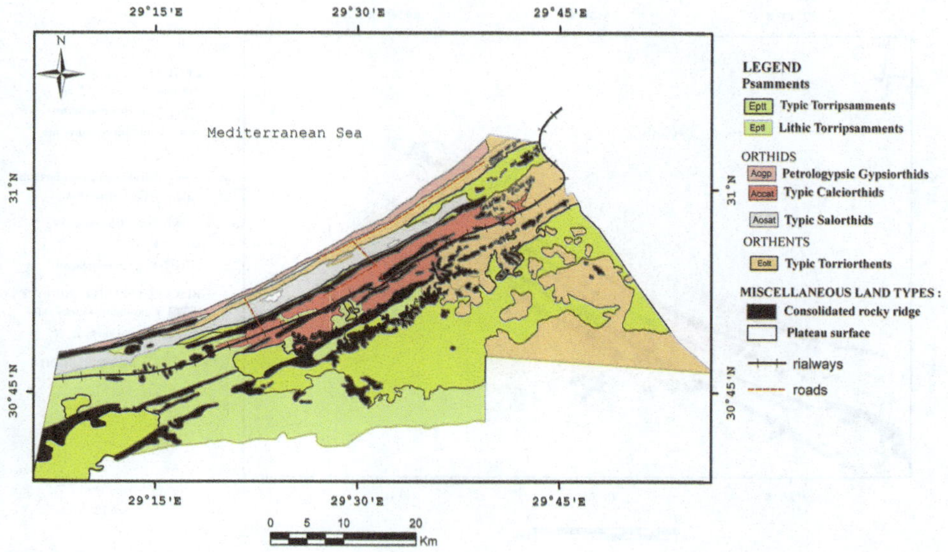

Fig. 4.178 Soil map of Burg Al-Arab–Al-Hammam

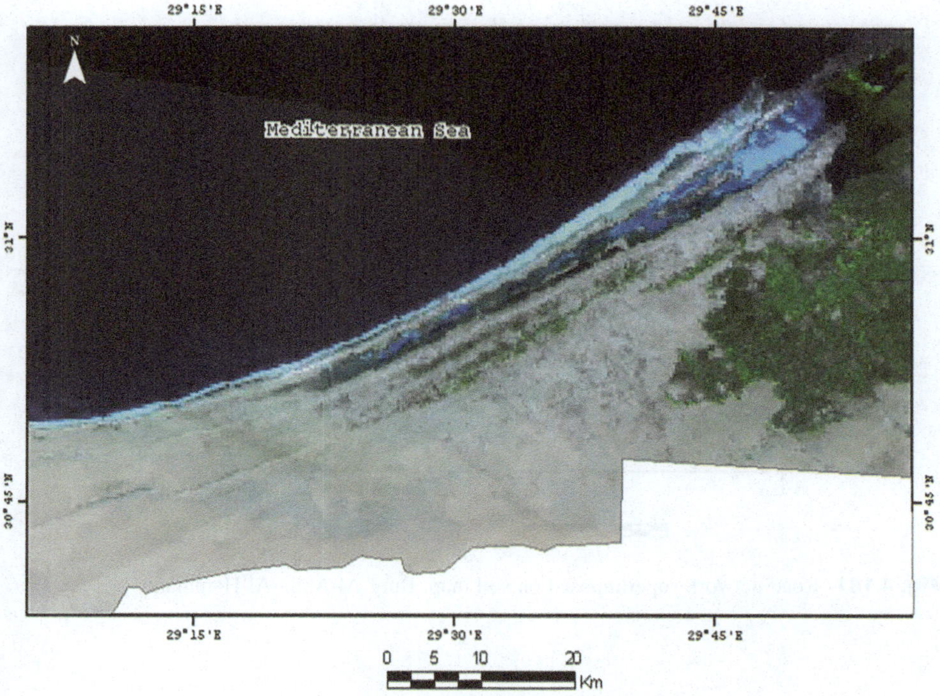

Fig. 4.179 ETM Landsat images of Burg Al-Arab–Al-Hammam

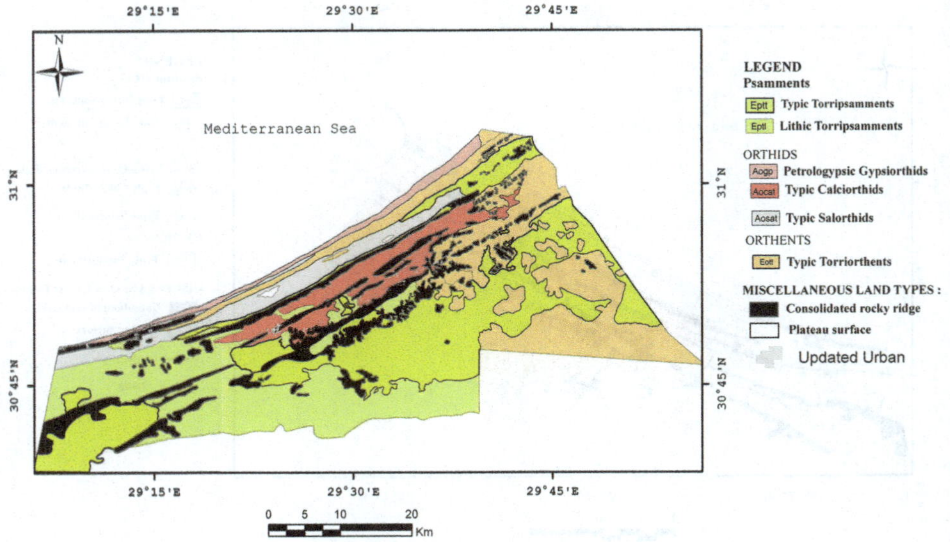

Fig. 4.180 Urban areas superimposed on the soil map, Burg Al-Arab–Al-Hammam

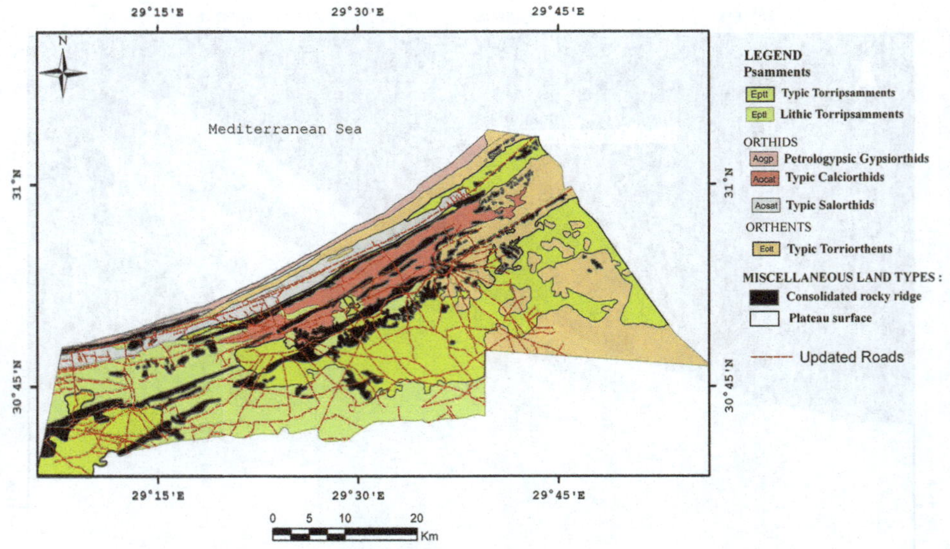

Fig. 4.181 Road network superimposed on soil map, Burg Al-Arab–Al-Hammam

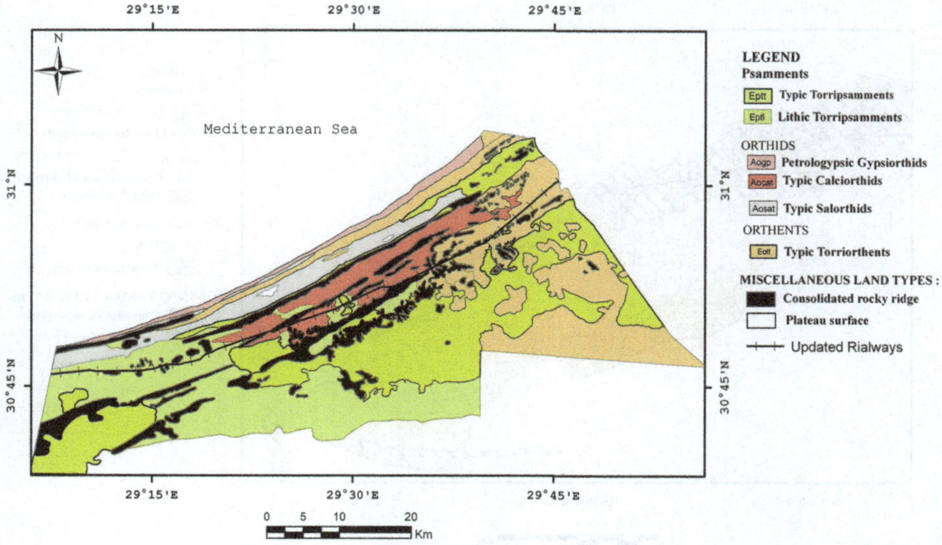

Fig. 4.182 Railway superimposed on the soil map, Burg Al-Arab–Al-Hammam

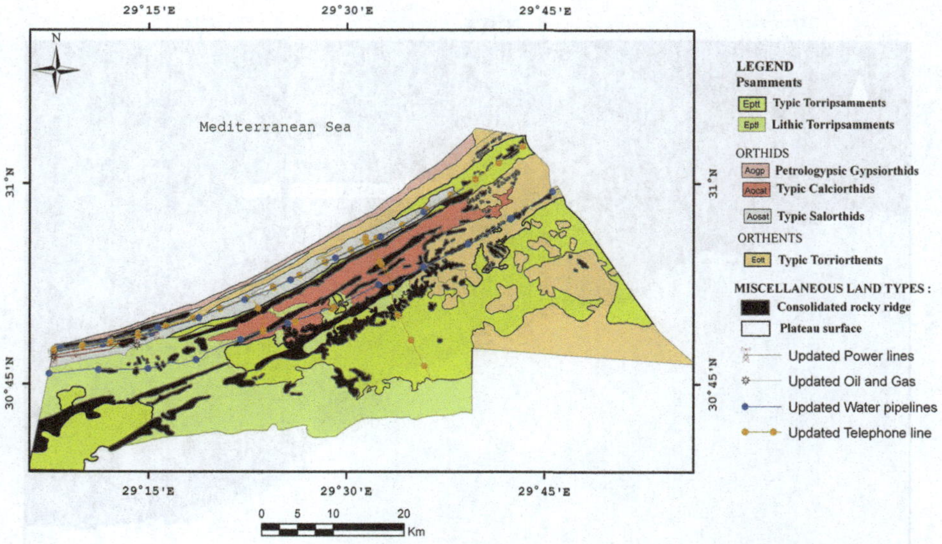

Fig. 4.183 Utilities, superimposed on the soil map, Burg Al-Arab–Al-Hammam Ras Abu-Eljroof and Al-Daba

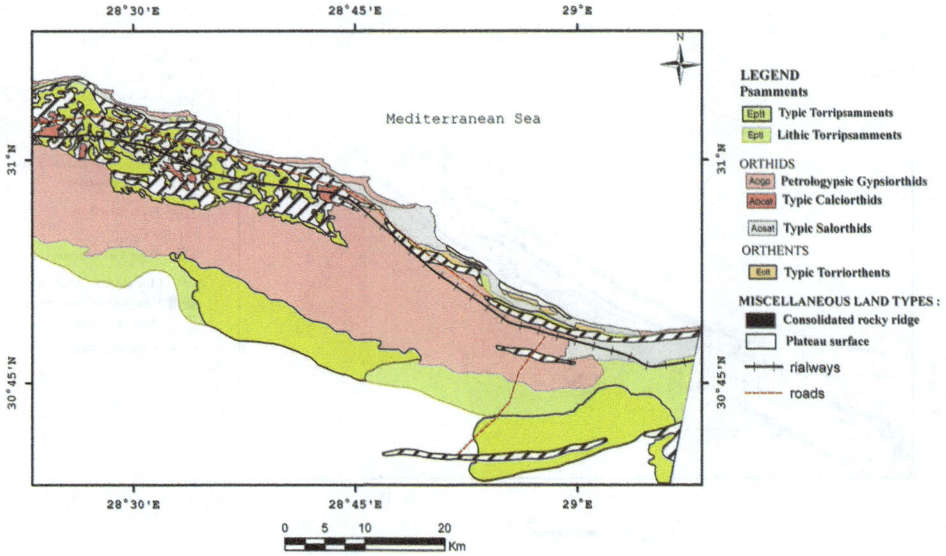

Fig. 4.184 Soil map of Ras Abu-Eljroof and Al-Daba

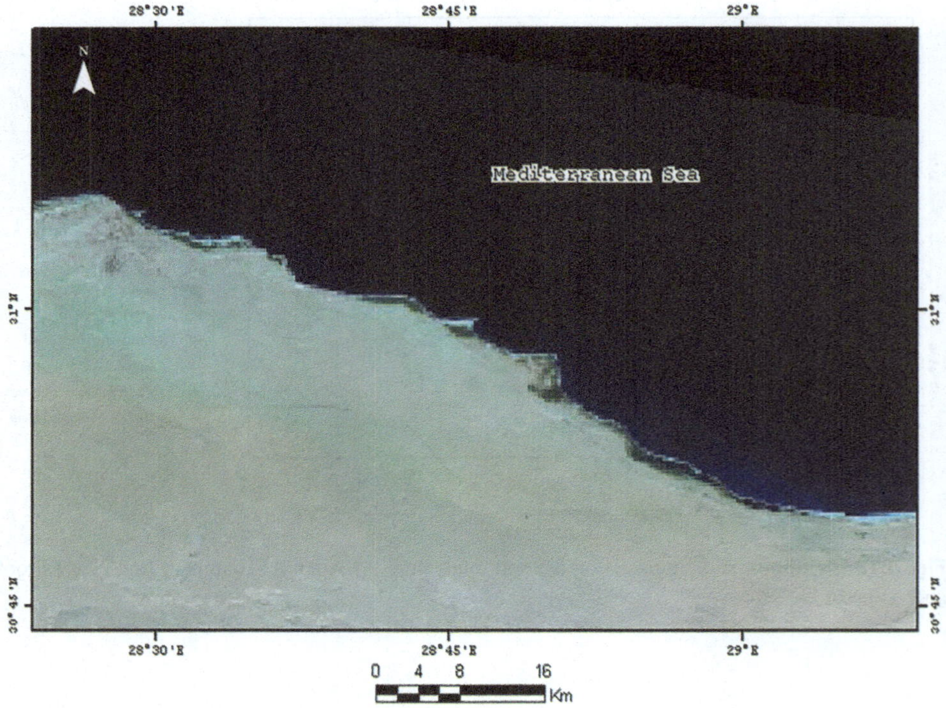

Fig. 4.185 ETM Landsat images of Ras Abu-Eljroof and Al-Daba

Table 4.58 Distribution of soil units in Ras Abu-Eljroof and Al-Daba

Soil Code	Area (km^2)	Area %
Aocat 3/3 a	12.2	0.3
Aogp 1/1b	682.8	15.1
Aosat 4/4a	87.3	1.9
Eott 1/1a	8.7	0.2
Eptl 2/1c	212.0	4.7
Eptt 2/1b	268.9	5.9
Eptt 2/1c	126.5	2.8
Plateau surface	6.9	0.2
Rock land	225.3	5.0
Water	2905.0	64.1
Total	4535.4	100.0

The soils of *Petrogypsic Gypsiothids (Aogp) and Typic Torripsamments (Eptt)* sub great groups are the most frequent in this areas, where they cover 682.8 and 395.4 km^2 representing 15.1 and 8.7% respectively. They are characterized by coarse texture and low salinity and high $CaCO_3$ contents. The sub-great groups of *Typic Calsiorthids (Aocat), and Typic Salorthids (Aosat)* are distributed along the sea shore, covering 12.2 and 87.3 km^2 (i.e. 0.3 and 1.9% of sheet area) respectively.

The *Typic Torriorthents (Eott)* sub-great group exist in patches of different areas distributed north and south of the area, covering 8.7 km^2 (i.e. 0.2%). The Rockland and rock outcrops are located parralel to the shoreline, covering an area of 225.3 km^2, representing 5.0% of land cover.

The created database of this area Figs. 4.186, 4.187, 4.188, 4.189 and Table 4.59 indicate that the total urban settlement area is 1.9 km^2, the lengths of roads, railway and the utilities networks attain 1740.3, 85.4 and 787.8 km respectively.

4.4.1.3 Ras Abu Girab & Ras Al-Ghergan

Figure 4.190 shows soil map of Ras Abu Girab & Ras Al-Ghergan area, which is located between latitudes 30° 15′ to 31° 15′ N and longitudes 27° 50′ to 28° 20′ E. Figure 4.191 shows the ETM+ image of the area. The main soil units of this area, Table 4.60 includes the following sub-great groups;

- *Typic Torriorthents*
- *Lithic Torripsamments*
- *Petrogypsic Gypsiothids*
- *Typic Calsiorthids*
- *Typic Torripsamments*

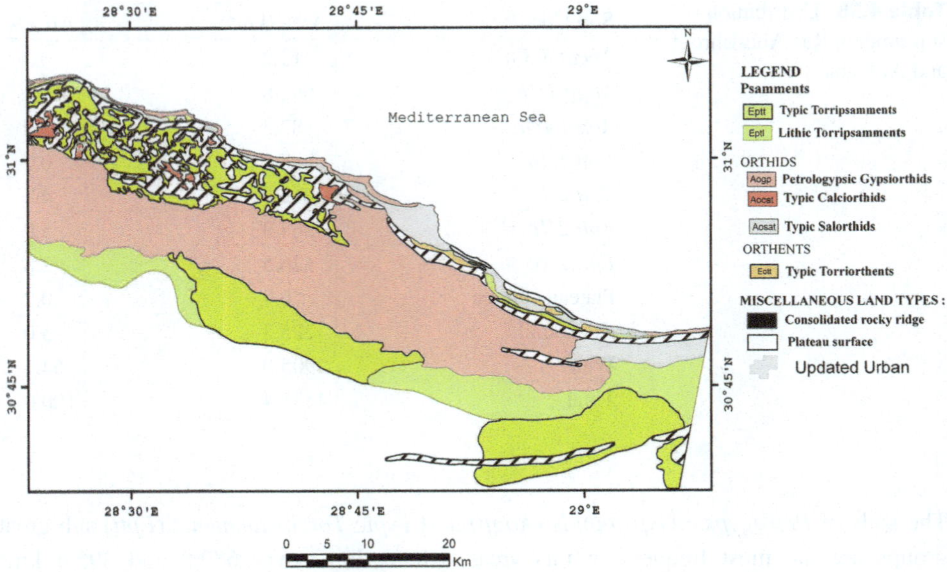

Fig. 4.186 Urban areas superimposed on the soil map, Ras Abu-Eljroof and Al-Daba

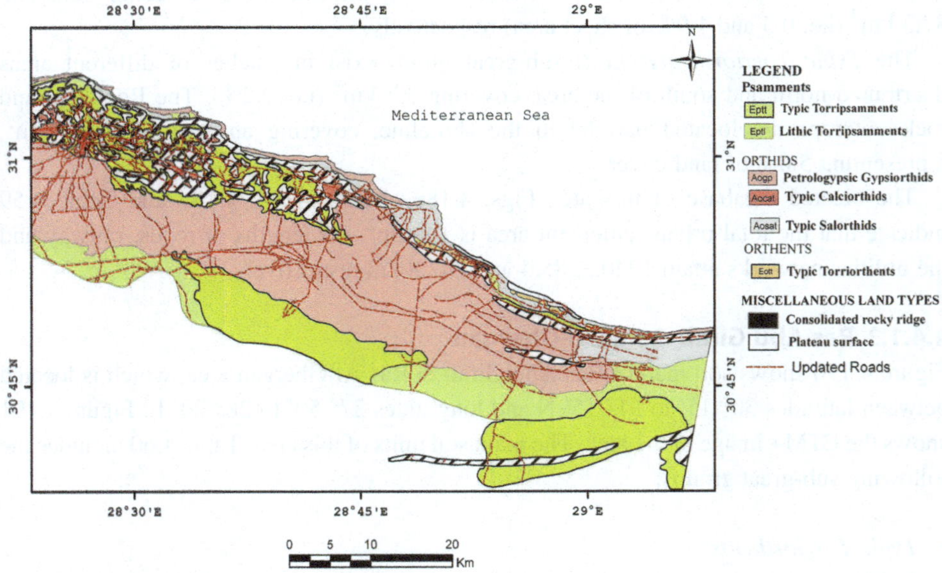

Fig. 4.187 Road network superimposed on soil map, Ras Abu-Eljroof and Al-Daba

The *Typic Torriorthents (Eott)* sub-great group is extensively distributed in patches along the wedth of the map sheet area. It covers 120.2 km^2, representing 6.9% of the area. The soils of *Lithic Torripsamments (Eptl)* sub-great group exists in few patches at the north and

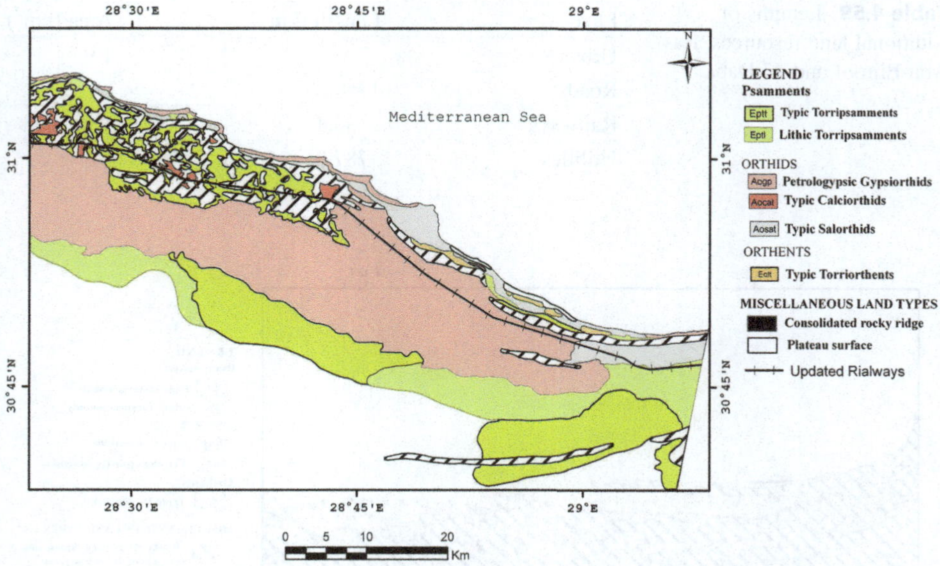

Fig. 4.188 Railway superimposed on the soil map, Ras Abu-Eljroof and Al-Daba

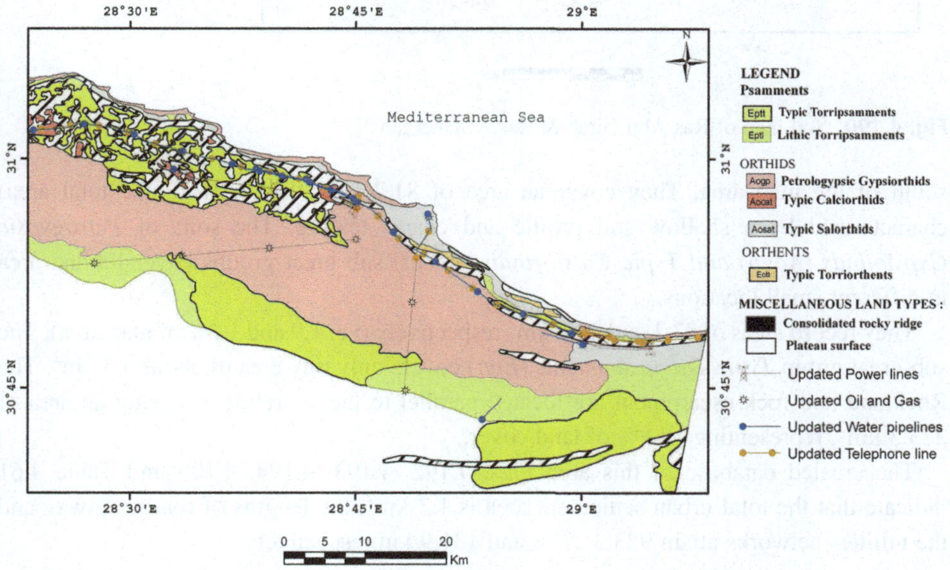

Fig. 4.189 Utilities, superimposed on the soil map, Ras Abu-Eljroof and Al-Daba

Table 4.59 Lengths of additional land resources, Ras Abu-Eljroof and Al-Daba

Layer	Length (km)	Area (km^2)
Urban		1.9
Roads	1740.3	
Railways	85.4	
Utilities	787.8	

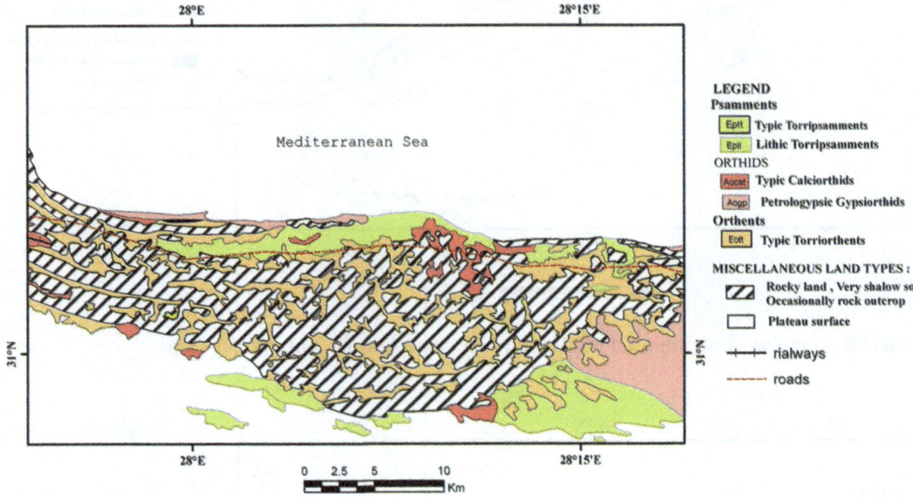

Fig. 4.190 Soil map of Ras Abu Girab & Ras Al-Ghergan

south of the map area. They cover an area of 81.2 km^2 (i.e. 4.7% of the total area) characterized by a shallow soil profile and coarse texture. The soils of *Petrogypsic Gypsiothids (Aogp) and Typic Calsiorthids (Aocat)* sub great groups cover limited area in different small locations.

They record areas of 33.1 and 22.0 km^2 respectively (i.e. 1.9 and 1.3% of map area). The sub-great group *Typic Torripsamments (Eptt)* covers only tiny area of about 0.7 km^2. The Rockland and rock escarpment are located parallel to the shoreline, covering an area of 273.3 km^2, representing 15.8% of land cover.

The created database of this area Figs. 4.192, 4.193, 4.194, 4.195 and Table 4.61 indicate that the total urban settlement area is 1.3 km^2, the lengths of roads, railway and the utilities networks attain 923.3, 52.8 and 131.9 km respectively.

4.4.1.4 Ras Al-kanaies–Matruh

Figure 4.196 shows soil map of Ras Al-kanaies—Matruh area, which is located between latitudes 30° 43′ to 31° 30′ N and longitudes 27° 00′ to 28° 00′ E. Figure 4.197 shows the

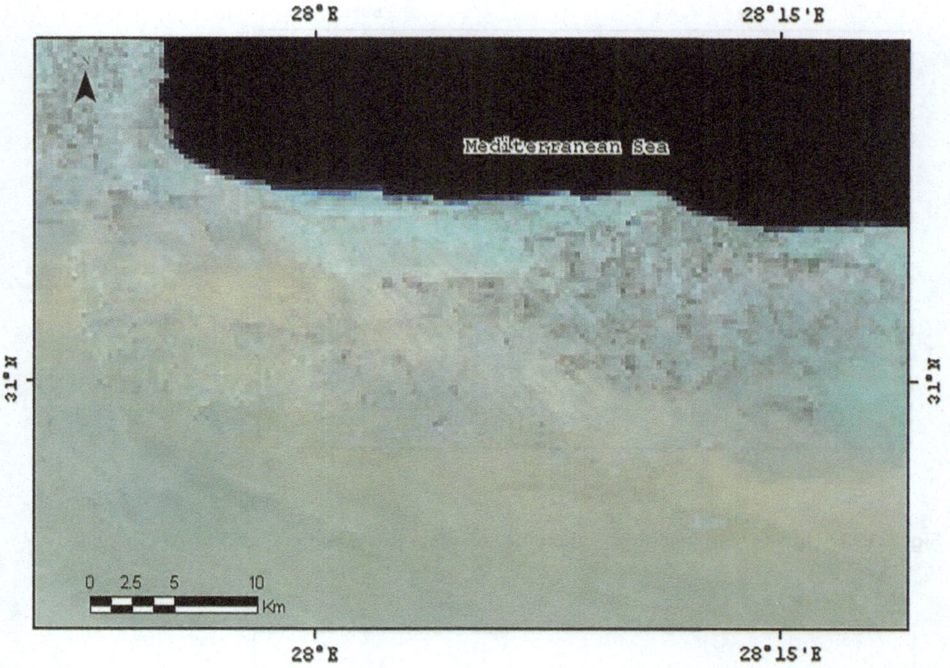

Fig. 4.191 ETM Landsat images of Ras Abu Girab & Ras Al-Ghergan

Table 4.60 Distribution of soil units in Ras Abu Girab & Ras Al-Ghergan area

Soil Code	Area (km^2)	Area %
Aocat 3/3 a	22.0	1.3
Aogp 1/1b	33.1	1.9
Eott 1/1a	120.2	6.9
Eptl 2/1c	81.2	4.7
Eptt 2/1c	0.7	0.0
Plateau surface	0.5	0.0
Rock escarpment	19.2	1.1
Rock land	254.1	14.7
Water	1200.4	69.3
Total	1731.5	100.0

ETM+ image. The main soil units of this area, Table 4.64 includes the following sub-great groups;

- *Lithic Torripsamments*
- *Typic Calsiorthids*

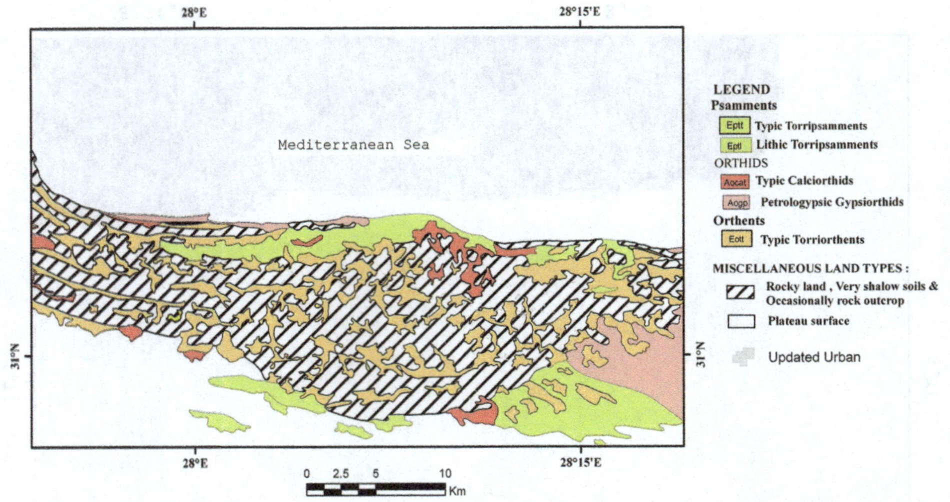

Fig. 4.192 Urban areas superimposed on the soil map, Ras Abu Girab & Ras Al-Ghergan

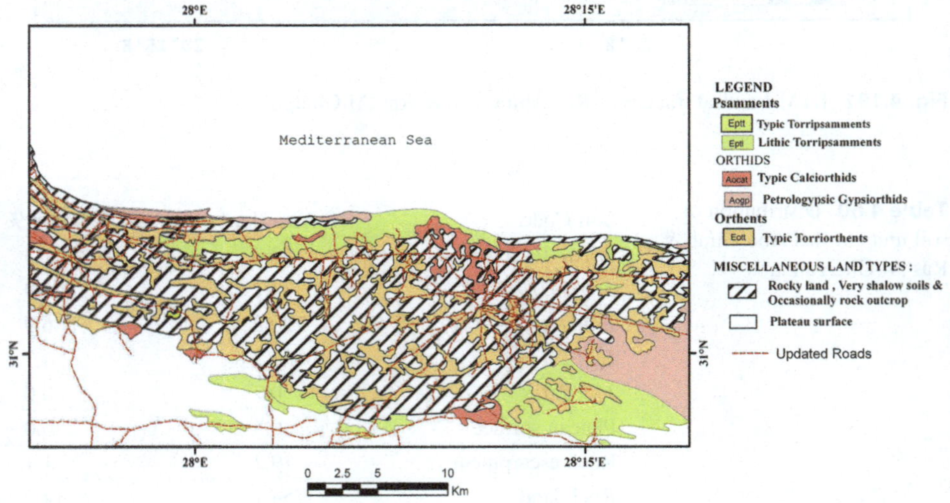

Fig. 4.193 Road network superimposed on soil map, Ras Abu Girab & Ras Al-Ghergan

- *Typic Torriorthents*
- *Petrogypsic Gypsiothids*

The soils of *Lithic Torripsamments (Eptl))* sub great groups are the most frequent in this areas, where they cover 228.0 km^2 representing 40.4% characterized by a shallow weakly developed soil profiles and coarse texture. The sub-great groups of Typic *Calsiorthids (Aocat), is* distributed, in patches, along the seashore, covering 83.0 km^2 (i.e. 14.7% of

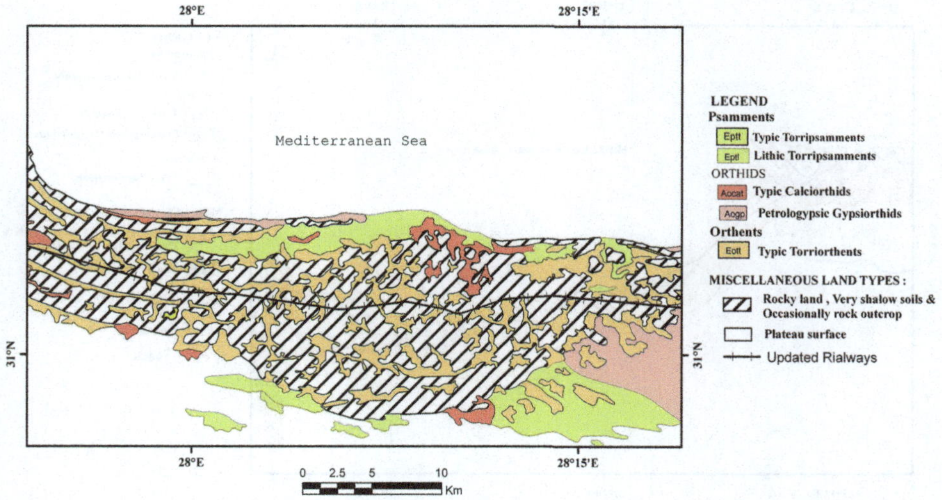

Fig. 4.194 Railway superimposed on the soil map, Ras Abu Girab & Ras Al-Ghergan

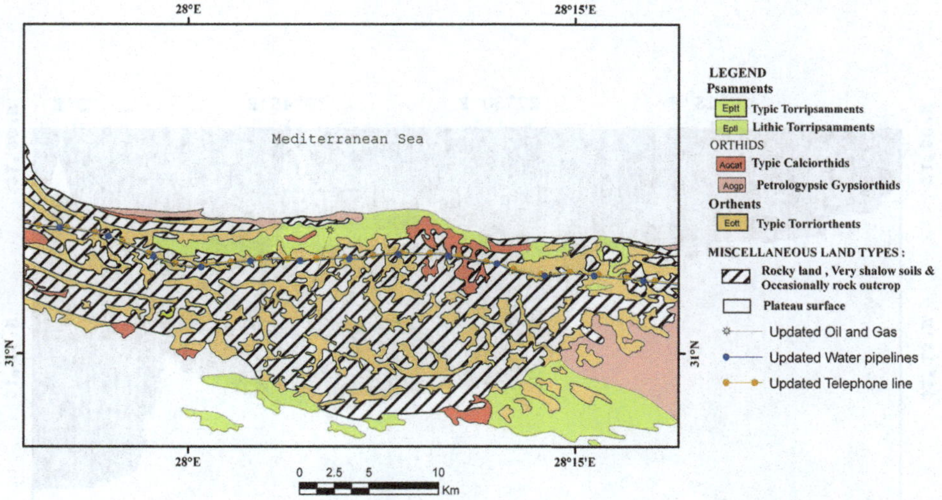

Fig. 4.195 Utilities, superimposed on the soil map, Ras Abu Girab & Ras Al-Ghergan

Table 4.61 Lengths of additional land resources, Ras Abu Girab & Ras Al-Ghergan

Layer	Length (km)	Area (km^2)
Urban		1.3
Roads	923.3	
Railways	52.8	
Utilities	131.9	

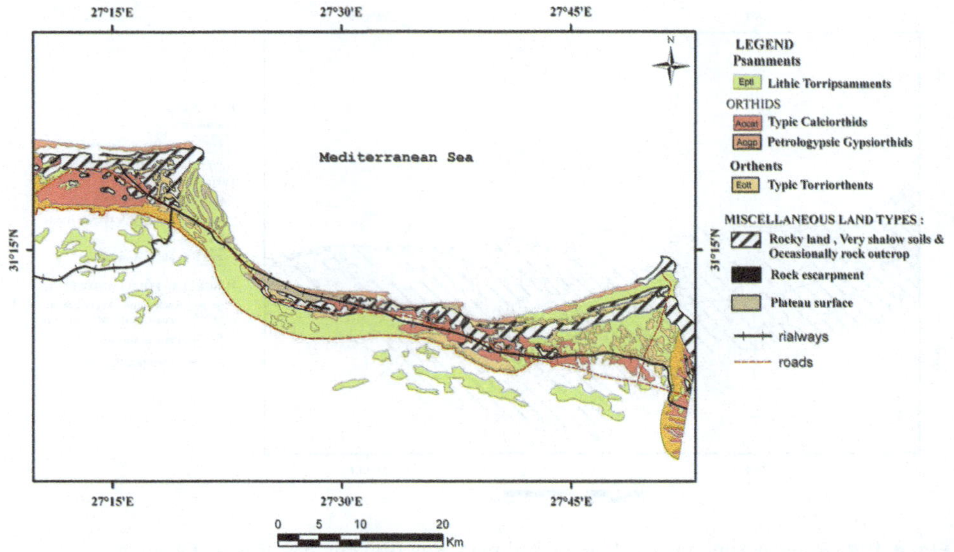

Fig. 4.196 Soil map of Ras Al-kanaies–Matruh

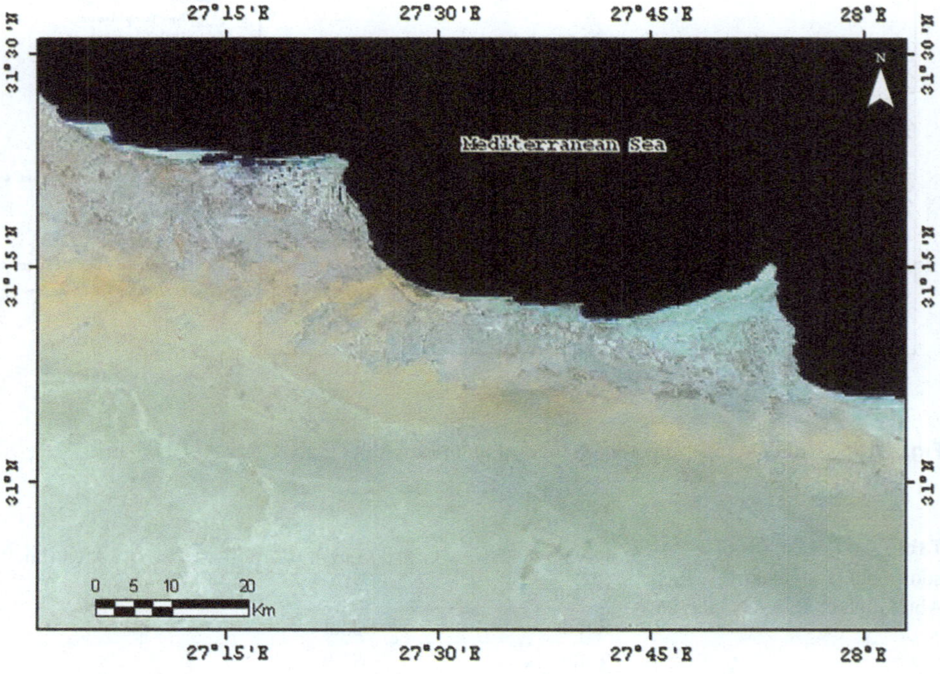

Fig. 4.197 ETM Landsat images of Ras Al-kanaies–Matruh

Table 4.62 Distribution of soil units in Ras Al-kanaies–Matruh

Soil Code	Area (km^2)	Area %
Aocat 3/3 a	83.0	14.7
Aogp 1/1b	22.0	3.9
Eott 1/1a	51.0	9.0
Eptl 2/1c	228.0	40.4
Plateau surface	19.0	3.4
Rock escarpment	32.0	5.7
Rock land	129.0	22.9
Total	564.0	100.0

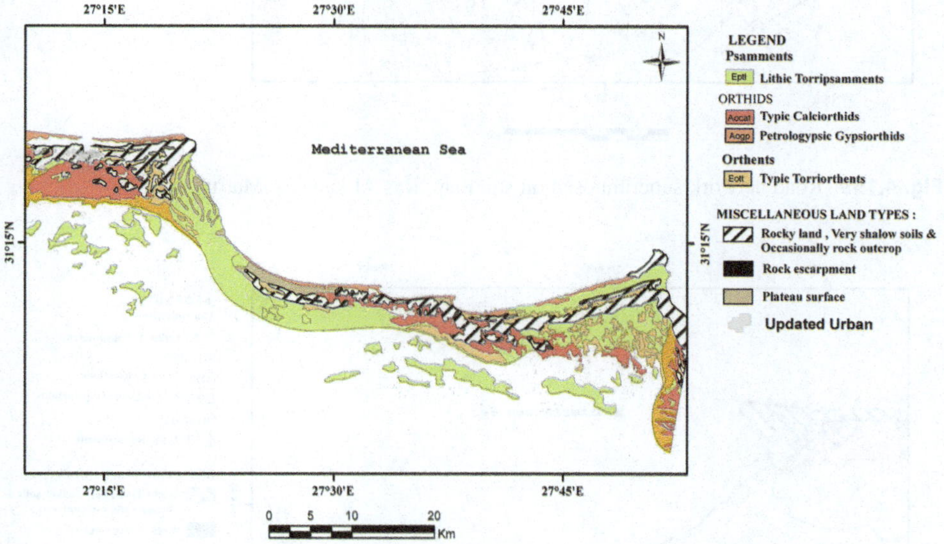

Fig. 4.198 Urban areas superimposed on the soil map, Ras Al-kanaies–Matruh

sheet area). The *Typic Torriorthents (Eott)* sub-great group exists mainly in two patches in the east and south west of the area, covering 51.0 km^2 (i.e. 9.0%). The *Petrogypsic Gypsiothids (Aogp) sub-great group* extends, in a narrow strip along the Mediterranean coast covering an area of 22.0 km^2, representing only 3.9% of land cover.

The Rockland and rock outcrops are located parallel to the shoreline, covering an area of 161.0 km^2, representing 28.6% of land cover (Table 4.62).

The created database of this area Figs. 4.198, 4.199, 4.200, 4.201 and Table 4.63 indicates that the total urban settlement area is 14.3 km^2, the lengths of roads, railway and the utilities networks attain 2847.0, 109.5 and 304.7 km respectively.

4.4.1.5 Mersa Al-Asi & Mersa Gargub

Figure 4.202 shows the soil map of Mersa Al-Asi & Mersa Gargub area, which is located between latitudes 31° 00′ to 32° 00′ N and longitudes 26° 10′ to 27° 10′ E. Figure 4.203

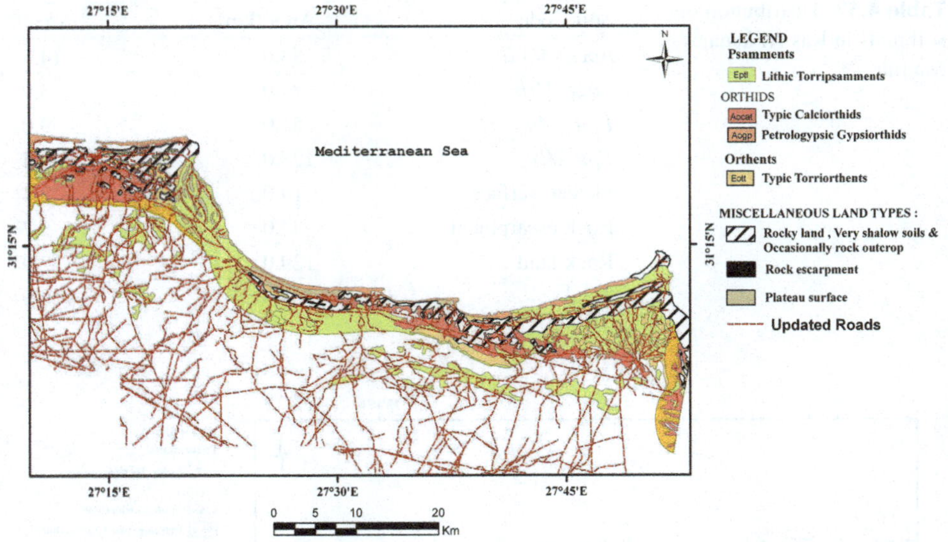

Fig. 4.199 Road network superimposed on soil map, Ras Al-kanaies–Matruh

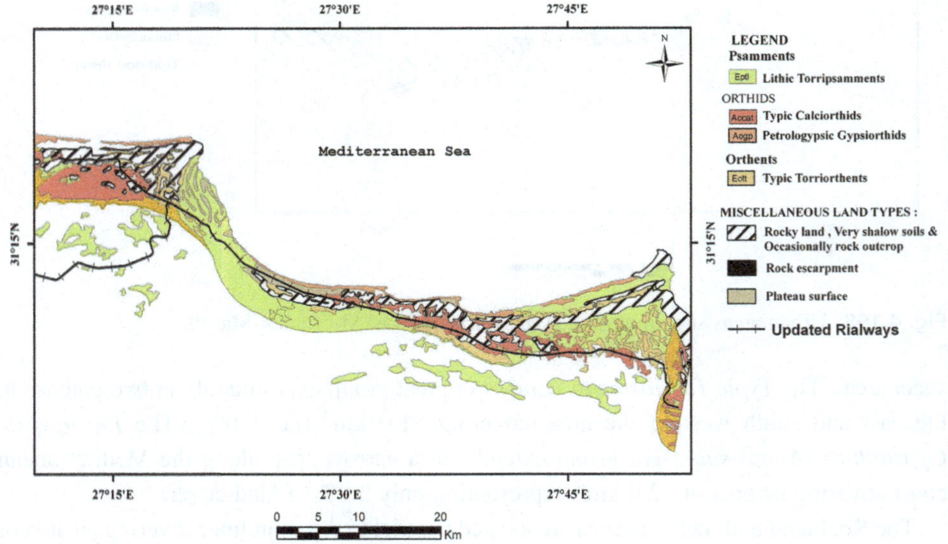

Fig. 4.200 Railway superimposed on the soil map, Ras Al-kanaies–Matruh

shows the ETM+ image of the area. The mature soils are rare in this area, as they don't represent more than 20% of sheet area. The main soil units of this area, Table 4.64 includes only three sub-great groups as the following;

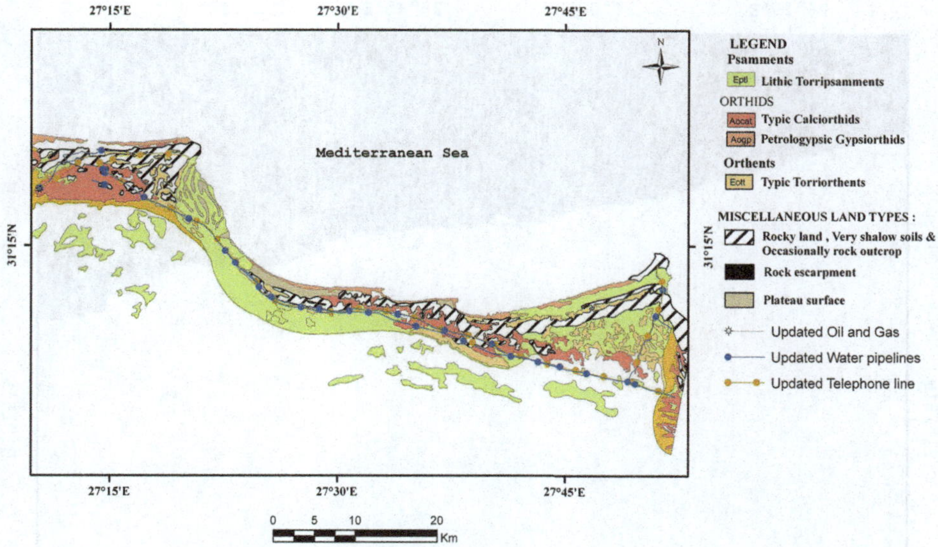

Fig. 4.201 Utilities, superimposed on the soil map, Ras Al-kanaies–Matruh

Table 4.63 Lengths of additional land resources, Ras Al-kanaies–Matruh

Layer	Length (km)	Area (km^2)
Urban		14.3
Roads	2847.0	
Railways	109.5	
Utilities	304.7	

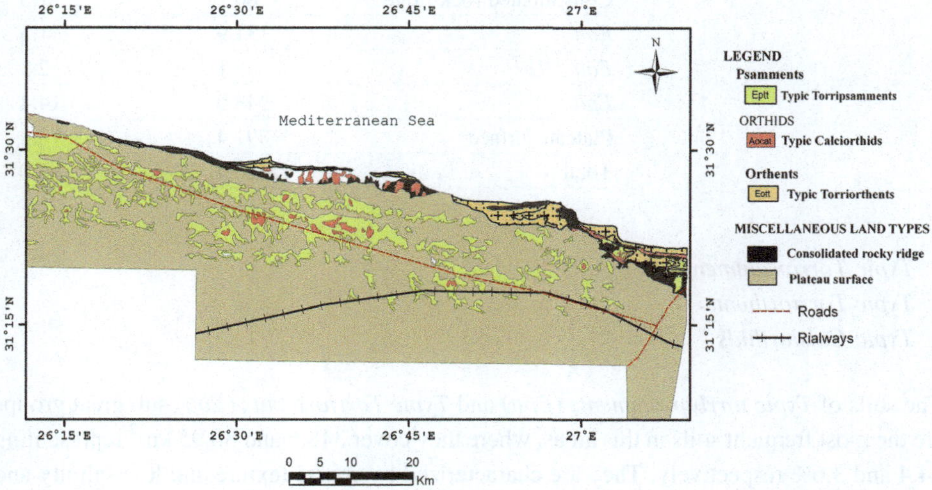

Fig. 4.202 Soil map of Mersa Al-Asi & Mersa Gargub

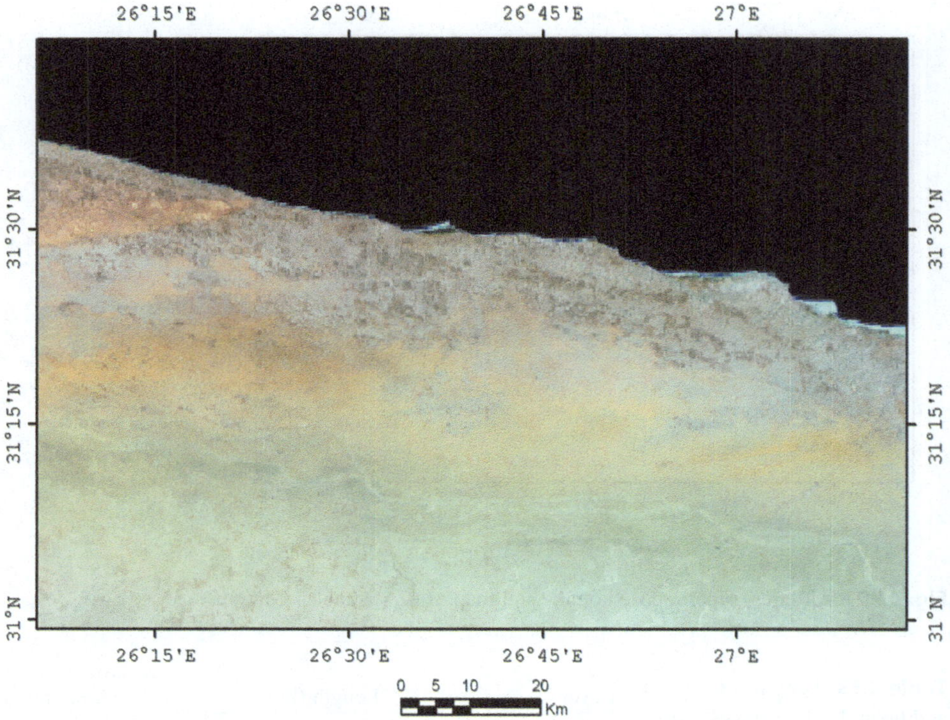

Fig. 4.203 ETM Landsat images of Mersa Al-Asi & Mersa Gargub

Table 4.64 Distribution of soil units in Mersa Al-Asi & Mersa Gargub

Soil Code	Area (km^2)	Area %
Aocat	33.4	1.4
Consolidated rock ridge	60.1	2.5
Eott	31.9	1.3
Eott 1/1b1	55.1	2.3
Eptt	348.6	14.4
Plateau surface	1891.4	78.1
Total	2420.4	100.0

- *Typic Torripsamments*
- *Typic Torriorthents*
- *Typic Calsiorthids*

The soils of *Typic torrbipsamments (Eptt)* and *Typic Torriorthents (Eott)* sub great groups are the most frequent soils in this areas, where they cover 348.6 and 86.95 km^2 representing 14.4 and 3.6% respectively. They are characterized by coarse texture and low salinity and

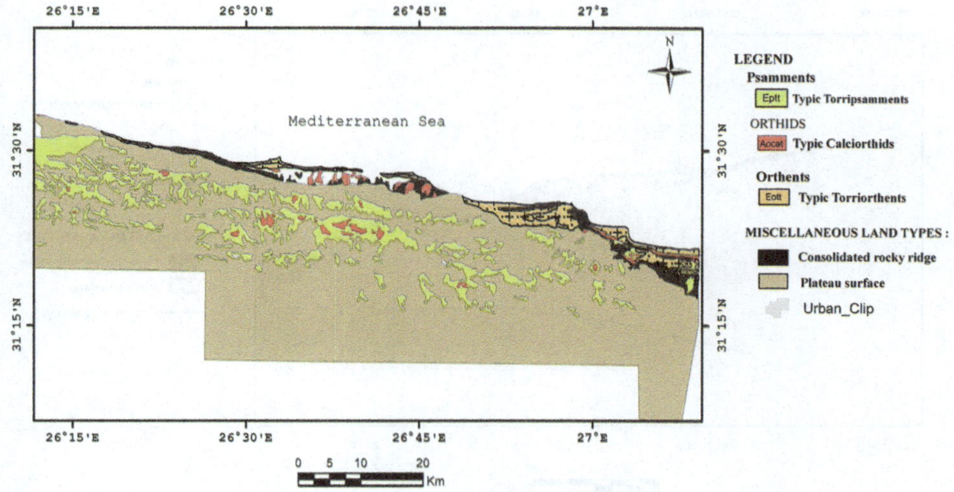

Fig. 4.204 Urban areas superimposed on the soil map, Mersa Al-Asi & Mersa Gargub

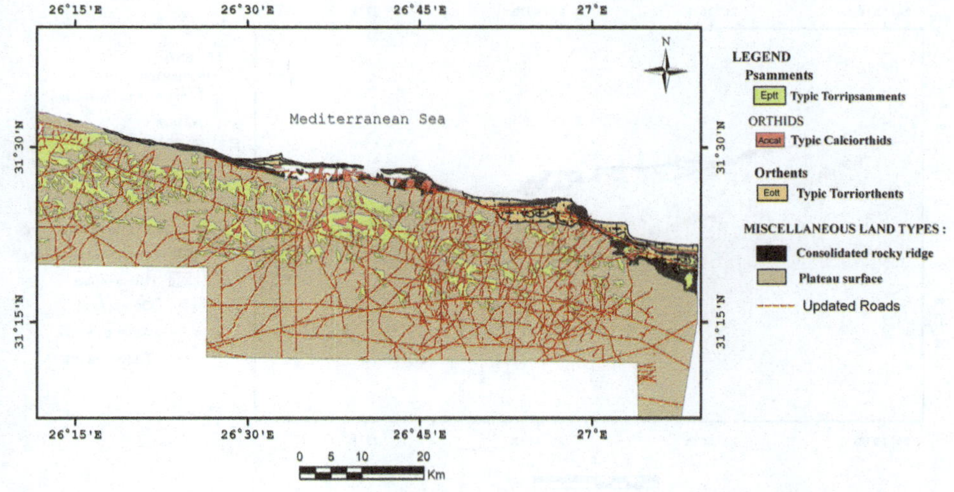

Fig. 4.205 Road network superimposed on soil map, Mersa Al-Asi & Mersa Gargub

high CaCO3 contents. The sub-great groups of Typic *Calsiorthids (Aocat), is* distributed, in patches, along the seashore, covering 17.6 km^2 (i.e. 0.7% of sheet area).

The rock lands area dominate in the middle and southern parts of the area, covering a total of an area of 1951.5 km^2, representing 80.6% of land cover.

The created database of this area Figs. 4.204, 4.205, 4.206, 4.207 and Table 4.65 indicate that the total urban settlement area is 2.6 km^2; the lengths of roads, railway and the utilities networks attain 2878.8, 97.4 and 292.7 km respectively.

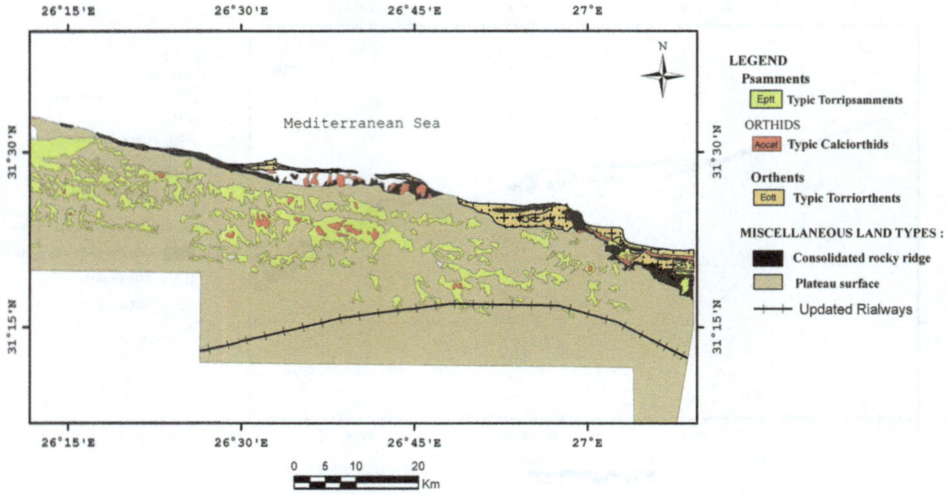

Fig. 4.206 Railway superimposed on the soil map, Mersa Al-Asi & Mersa Gargub

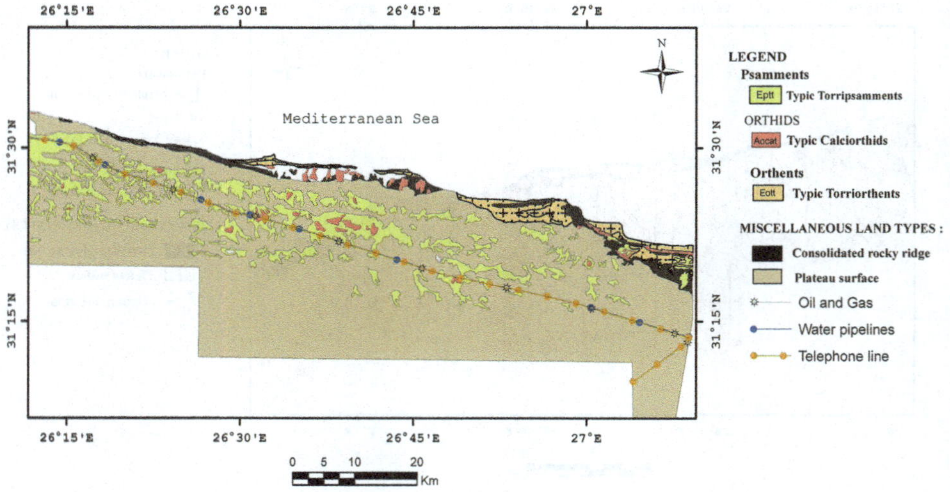

Fig. 4.207 Utilities, superimposed on the soil map, Mersa Al-Asi & Mersa Gargub

Table 4.65 Lengths of additional land resources, Mersa Al-Asi & Mersa Gargub

Layer	Length (km)	Area (km^2)
Urban		2.6
Roads	2878.8	
Railways	97.4	
Utilities	292.7	

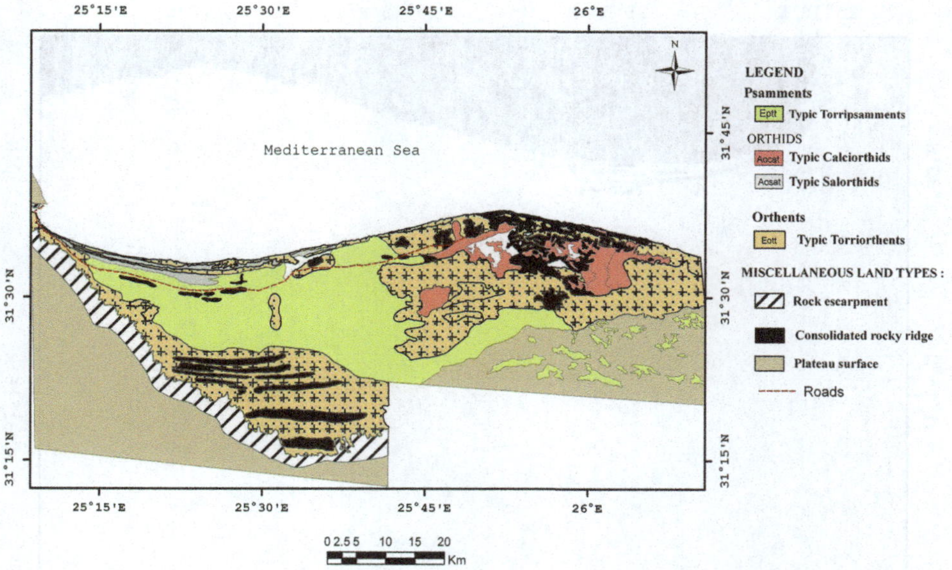

Fig. 4.208 Soil map of Sidi Barrani & Al-Salum

4.4.1.6 Sidi Barrani & Al-Salum

Figure 4.208 shows soil map of Sidi Barrani & Al-Salum area, which is located between latitudes 31° 00′ to 31° 40′ N and longitudes 25° 10′ to 26° 10′ E. Figure 4.209 shows the ETM+ image. The mature soils in this area represent only 32%, whereas the rock lands exhibit 68% of the land surface. The main soil units of this area, Table 4.59 include only three sub-great groups as follows:

- *Typic Torripsamments*
- *Typic Torriorthents*
- *Typic Calsiorthids*

The soils of *Typic Torrbipsamments (Eptt)* and *Typic Torriorthents (Eott)* sub great groups are relatively most frequent soils in this areas, where they cover 716.8 and 518.9 km^2 representing 17% and 12.3% respectively. They are characterized by coarse texture, low soil salinity and have a high $CaCO_3$ contents. The sub-great groups of *Typic Calsiorthids (Aocat)*, is distributed, in patches, along the seashore, covering an area of 35.7 km^2 (i.e. 0.8%). Nevertheless, the soils of subgreat group, Typic Salorthids cover an area of about 81.2 km^2 (i.e. 1.9%) and distributed close to the the sea as a small patches. The rock lands and escarpment cover a total area of 2875.6 km^2, representing 68% of land cover (Table 4.66).

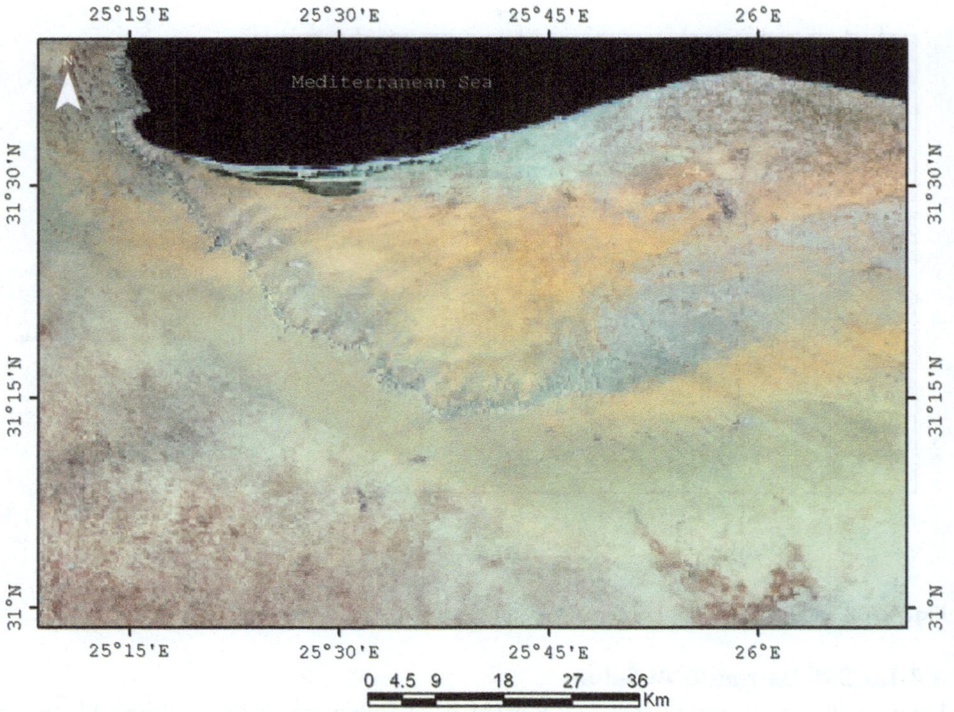

Fig. 4.209 ETM Landsat images of Sidi Barrani & Al-Salum

Table 4.66 Distribution of soil units in Sidi Barrani & Al-Salum

Soil Code	Area (km^2)	Area %
Aocat	35.7	0.8
Aosat	81.2	1.9
Eott	115.3	2.7
Eott 1/1a	403.6	9.6
Eptt	716.8	17.0
Plateau surface	2449.3	57.9
Rock escarpment	206.0	4.9
Consolidated rocky	220.3	5.2
Total	4228.2	100.0

The created database of this area Figs. 4.210, 4.211, 4.212, 4.213 and Table 4.67 indicate that the total urban settlement area is 1.4 km^2; the lengths of roads, railway and the utilities networks attain 3210.9, 84.0 and 292.7 km respectively.

4.4.1.7 Western Desert Road Alex Cairo (A)

Figure 4.214 shows soil map of Western Desert Road Alex Cairo (A) area, which is located between latitudes 29° 35′ to 30° 25′ N and longitudes 30° 05′ to 30° 50′ E. Figure 4.215

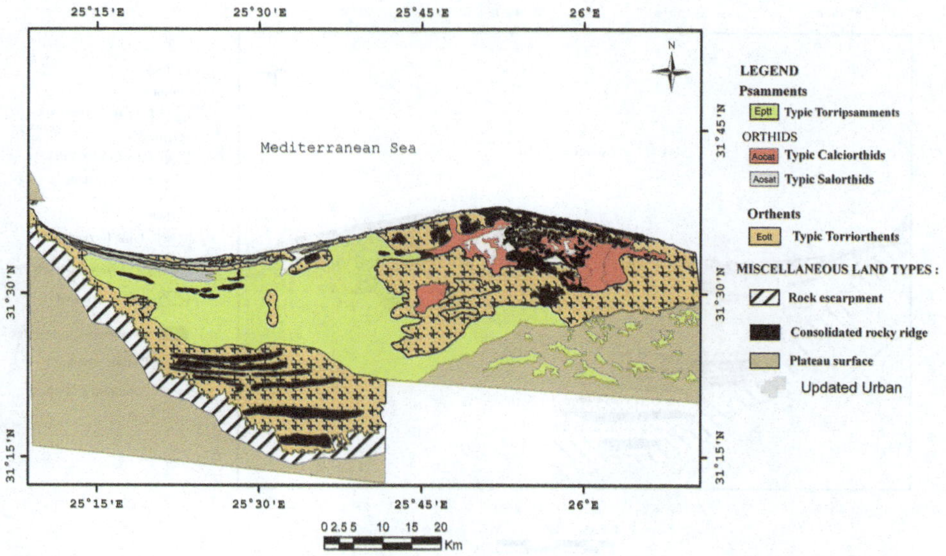

Fig. 4.210 Urban areas superimposed on the soil map, Sidi Barrani & Al-Salum

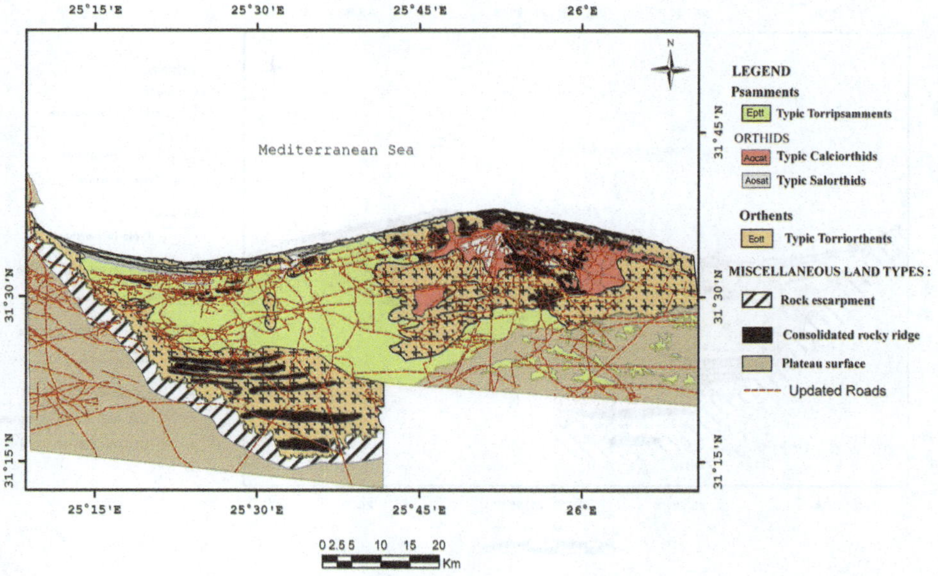

Fig. 4.211 Road network superimposed on soil map, Sidi Barrani & Al-Salum

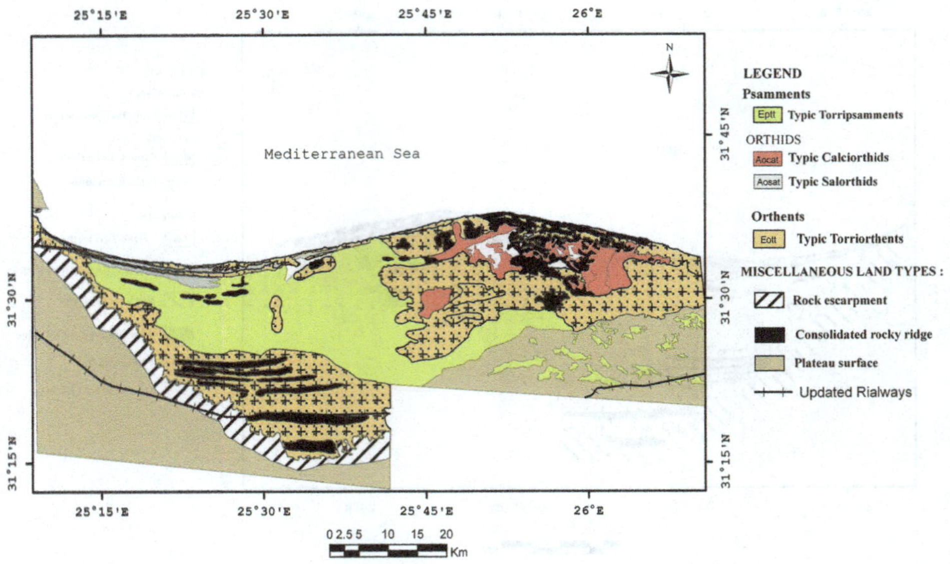

Fig. 4.212 Railway superimposed on the soil map, Sidi Barrani & Al-Salum

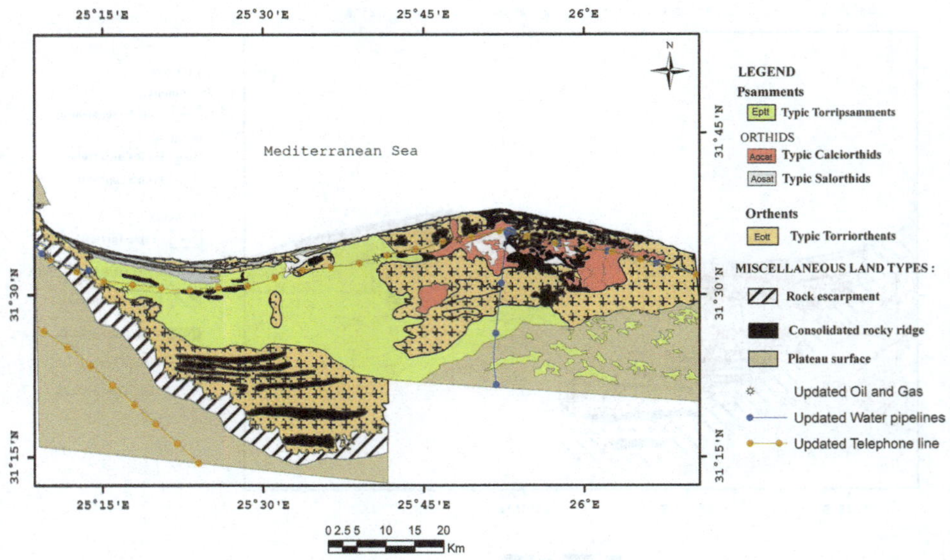

Fig. 4.213 Utilities, superimposed on the soil map, Sidi Barrani & Al-Salum

shows the ETM+ image. Two sub-great groups exhibit the soils in this map sheet area, as demonstrated in Table 4.68:

Table 4.67 Lengths of additional land resources, Mersa Al-Asi & Mersa Gargub

Layer	Length (km)	Area (km^2)
Urban		1.4
Roads	3210.9	
Railways	84.0	
Utilities	292.7	

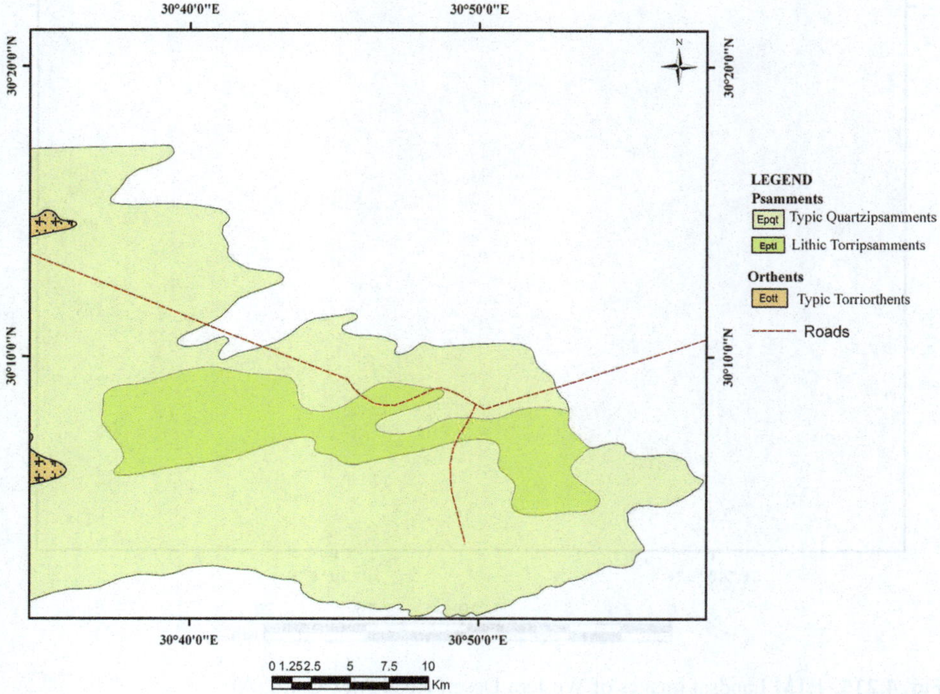

Fig. 4.214 Soil map of Western Desert Road Alex Cairo (A)

- *Typic Quartizpsamments*
- *Typic Torriorthents*

The soils of Typic *Quartizpsamments (Epqt)* dominate the soils in the map sheet area, where it coversn 3399.9 km^2 representing about 84% of total land cover. These soils are characterized by a coarse texture in both surface and sub-surface layers, while the salinity level exhibits a low class (i.e. 0–4 dS/m). Also, $CaCO_3$ content is mostly low (i.e. 4%). The sub-great group soils of *Typic Torriorthents (Eott)* exist in limited locations, covering an area of 606.3 km^2, representing less 15% of map sheet area. These soils are characterized by high content of $CaCO_3$ and a coarse texture class. The sand dunes zone is clear in both.

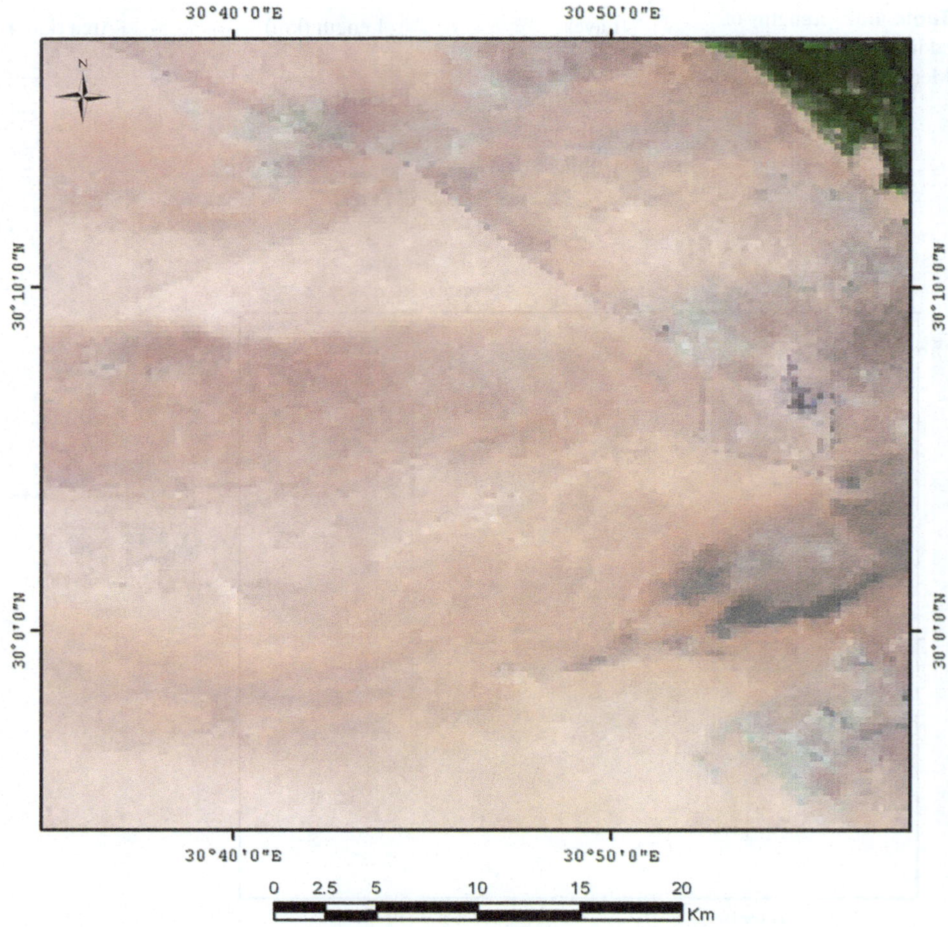

Fig. 4.215 ETM Landsat images of Western Desert Road Alex Cairo (A)

Table 4.68 Distribution of soil units in Western Desert Road Alex Cairo-A

Soil Code	Area (km^2)	Area %
Eott 1/1a	227.9	5.6
Eott 1/1b	378.4	9.4
Epqt 1/1 a	3399.9	84.0
Sand dune	39.4	0.9
Total	4045.6	100.0

4.4.1.8 Western Desert Road Alex Cairo (B)

Figure 4.216 shows soil map of Western Desert Road Alex Cairo (B) area, which is located between latitudes 29° 55′ to 30° 18′ N and longitudes 30° 35′ to 31° 00′ E. Figure 4.217 shows the ETM+ image. The main soil units of this area, Table 4.69 includes the following sub-great groups;

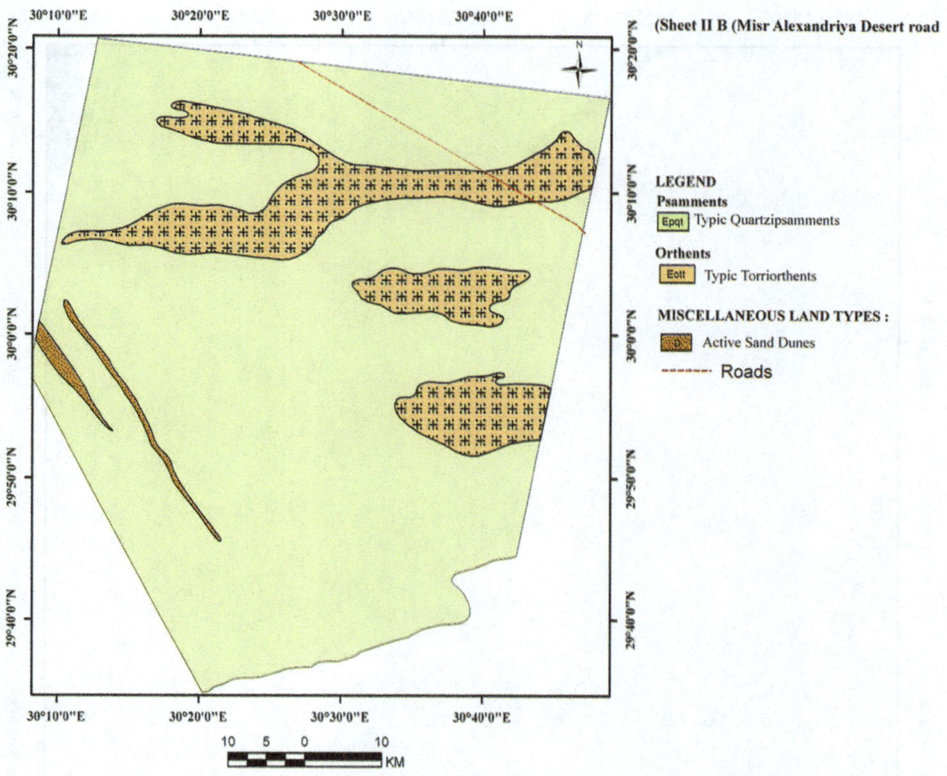

Fig. 4.216 Soil map of Western Desert Road Alex Cairo (B)

- *Typic Quartizpsamments*
- *Lithic Torripsamments*
- *Typic Torriorthents*

The soils of Typic *Quartizpsamments (Epqt)* dominates the soils in the map sheet area, where it covers 512.7 km^2, representing about 80.1% of total land cover. These soils are characterized by a medium to coarse texture in both surface and sub-surface layers, while the salinity level exhibits a low class (i.e. 0–4 dS/m). Also, $CaCO_3$ content is mostly low (i.e. 4%). The *Lithic Torripsamments (Eptl)* cover an area of 113.5 km^2, representing 17.7% of map sheet area. The sub-great group soils of *Typic Torriorthents (Eott)* exist in limited locations, covering an area of 5.7 km^2, representing less 1% of map sheet area.

4.4.1.9 Nubaryia

Figure 4.218 shows the soil map of Nubaryia area, which is located between latitudes 29° 30′ to 30° 20′ N and longitudes 30° 05′ to 30° 27′ E. Figure 4.219 shows the ETM+ image. The soils in the map sheet area include the following sub-great groups Table 4.70;

Fig. 4.217 ETM Landsat images of Western Desert Road Alex Cairo (B)

Table 4.69 Distribution of soil units in Western Desert R Alex Cairo-B

Soil Code	Area (km^2)	Area %
Eott 1/1a	5.7	0.9
Epqt 1/1b	512.7	80.1
Eptl 1/1b	113.5	17.7
River Nile	7.9	1.2
Total	639.9	100.0

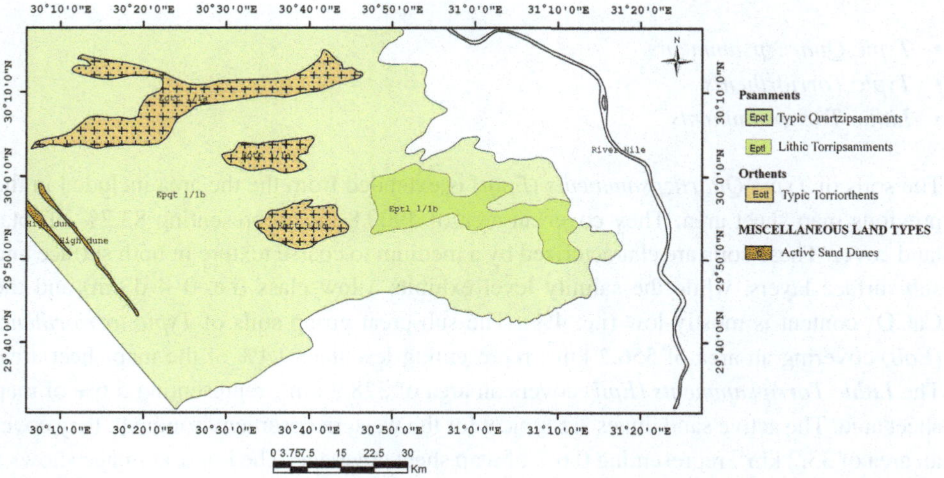

Fig. 4.218 Soil map of Nubaryia

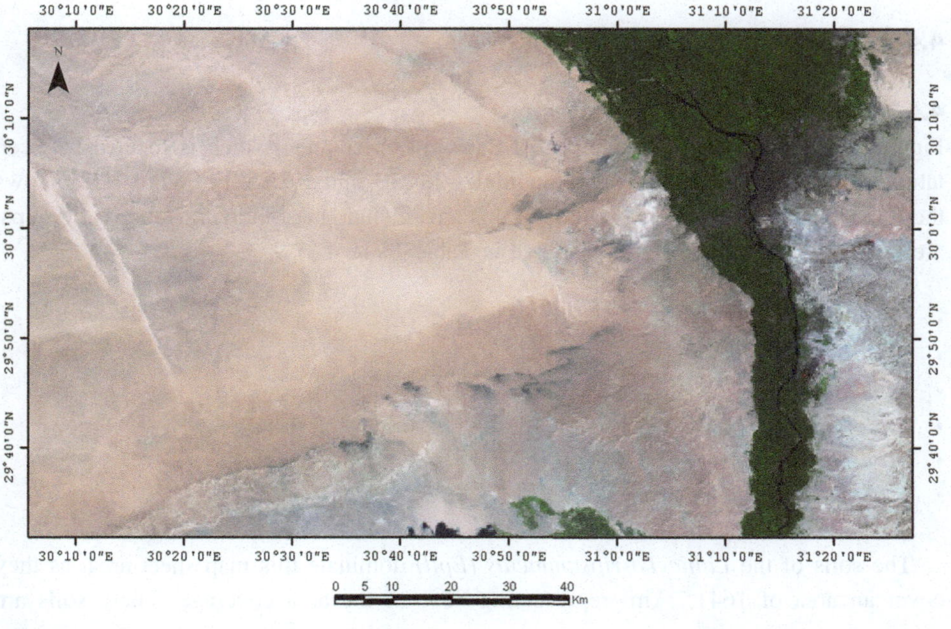

Fig. 4.219 ETM Landsat images of Nubaryia

Table 4.70 Distribution of soil units in Nubaryia area

Soil Code	Area (km^2)	Area %
Eott 1/1a	556.2	9.4
Epqt 1/1b	4900.8	83.2
Eptl 1/1b	328.9	5.6
High dune	33.2	0.6
River Nile	70.0	1.2
Total	5889.1	100.0

- *Typic Quartizpsamments*
- *Typic Torriorthents*
- *Lithic Torripsamments*

The soils of *Typic Quartizpsamments (Epqt)* is extended from the the area included in the previous map sheet area. They cover an area of 4900.8 km^2 representing 83.2% of total land cover. These soils are characterized by a medium to coarse texture in both surface and sub-surface layers, while the salinity level exhibits a low class (i.e. 0–4 dS/m) and the $CaCO_3$ content is mostly low (i.e. 4%). The sub-great group soils of *Typic torriorthents (Eott)* covering an area of 556.2 km^2, representing less than 9.4% of the map sheet area. The *Lithic Torripsamments (Eptl)* covers an area of 328.9 km^2, representing 5.6% of map sheet area. The active sand dunes are typical for the western desert environment, they cover an area of 33.2 km^2, representing 0.6% of map sheet coverage. The Landsat image shows a longitudinal pattern of moving, according to the main wind direction, from northwest to southeast direction.

4.4.2 Western Desert Potential Areas

4.4.2.1 Al-Baharyia Oases

Figure 4.220 shows the soil map of Al-Baharyia oases area, which is located between latitudes 28° 00′ and 28° 30′ N and longitudes 28° 30′ and 29° 15′ E. Figure 4.221 shows the ETM+ image. The main soil units of this area include the following sub-great groups, areas of different soil units are illustrated in Table 4.71:

- *Lithic Torripsamments*
- *Typic Torripsamments*
- *Typic Salorthids*
- *Typic Torriorthents*
- *Typic Torrifluvents*
- *Typic Quartizpsamments*

The soils of the *Lithic Torripsamments (Eptl)* dominate this map sheet area, as they cover an area of 1641.7 km^2 representing 85% of the area coverage. These soils are

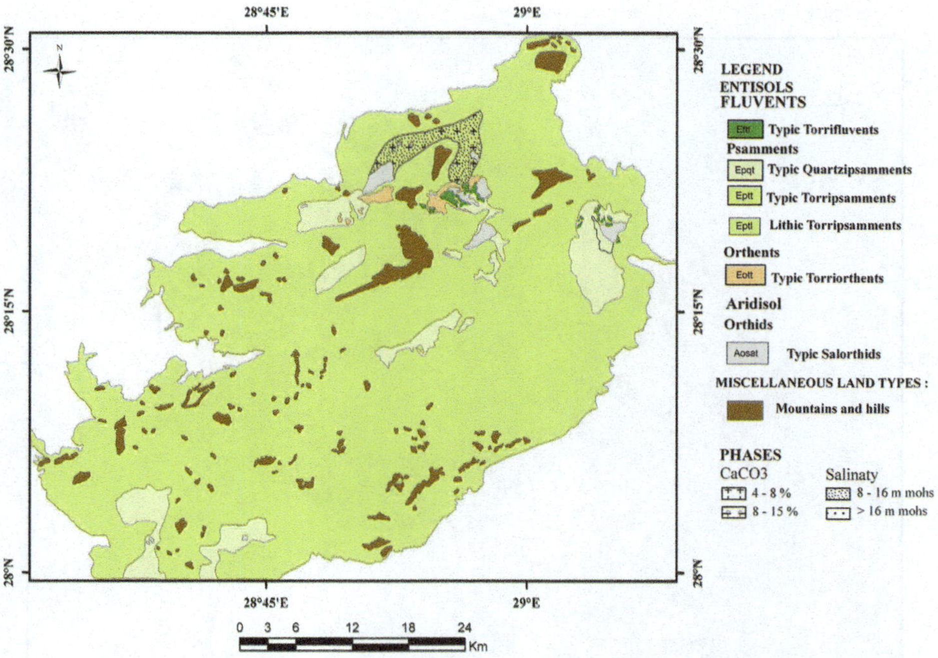

Fig. 4.220 Soil map of Al-Baharyia oases

characterized by a shallow soil profile and coarse texture. The soils of *Typic Torripsamments (Eptt)* occupy 161.5 km^2 representing 8.4% of land cover. The soils of sub-great group *Typic Salorthids (Aosat)* are restricted to the depressions area covering a total of 17.8, representing 0.9% of the area. The *Typic Torriorthents (Eott)* sub-great group covers 10 km^2 (0.5% of the total area).

The *Typic Torrifluvents* (*Eftt*) sub-great group covers 5.3 km^2 (0.3% of the total area) restricted in small patches concentrated at the northeastern corner. The soil of *Typic Quartizpsamments (Epqt)* sub-great group cover an area of 2.8 km^2 representing 0.1% of total sheet area. The rock land, referred as Mountains and hills, have an area of 91.3 km^2, representing 4.7%.

The created database of this area Figs. 4.222, 4.223, 4.224 and Table 4.72 indicates that the total urban settlement area is 2.3 km^2; the lengths of roads, railway and the utilities networks attain 1286.5, 114.6 and 260.7 km respectively.

4.4.2.2 Al-Dakhla Oases (A)

Figure 4.225 shows soil map of Al-Dakhla Oases (A) area, which is located between latitudes 25° 00′ and 25° 45′ N and longitudes 28° 30′ and 29° 30′ E. Figure 4.226 shows the ETM+ image. The main soil units of this area include the following sub-great groups Table 4.73:

Fig. 4.221 ETM Landsat images of Al-Baharyia oases

Table 4.71 Distribution of soil units in Al-Baharyia oases

Soil Code	Area (km^2)	Area %
Aosat	17.8	0.9
Eftt2/2a	5.3	0.3
Eott 1/1a	10.0	0.5
Epqt	2.8	0.1
Eptl 1/1a	1641.7	85.0
Eptt 1/1a	30.6	1.6
Eptt 1/1b	130.9	6.8
Mountains and hills	91.3	4.7
Total	1930.4	100.0

- *Palustollic Chromusterts*
- *Lithic Torripsamments*
- *Typic Torripsamments*
- *Palustollic Torrerts*

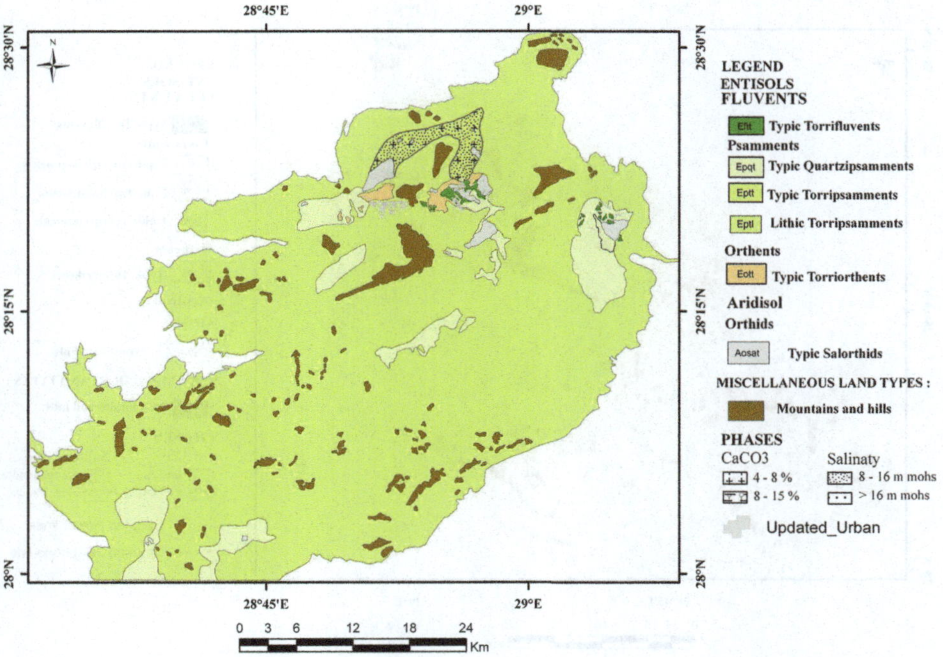

Fig. 4.222 Urban areas superimposed on the soil map Al-Baharyia oases

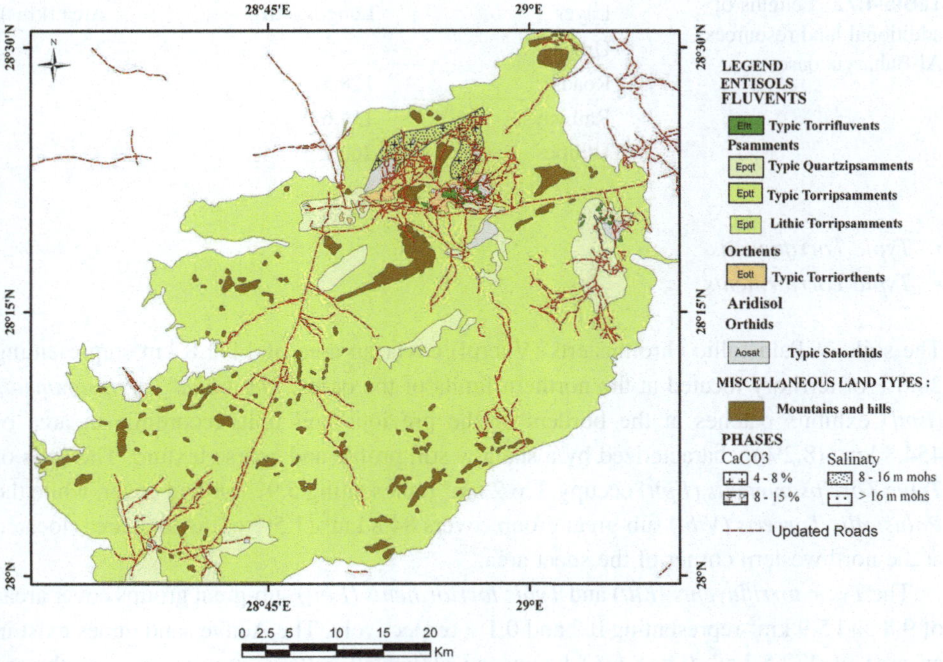

Fig. 4.223 Road network superimposed on soil map, Al-Baharyia oases

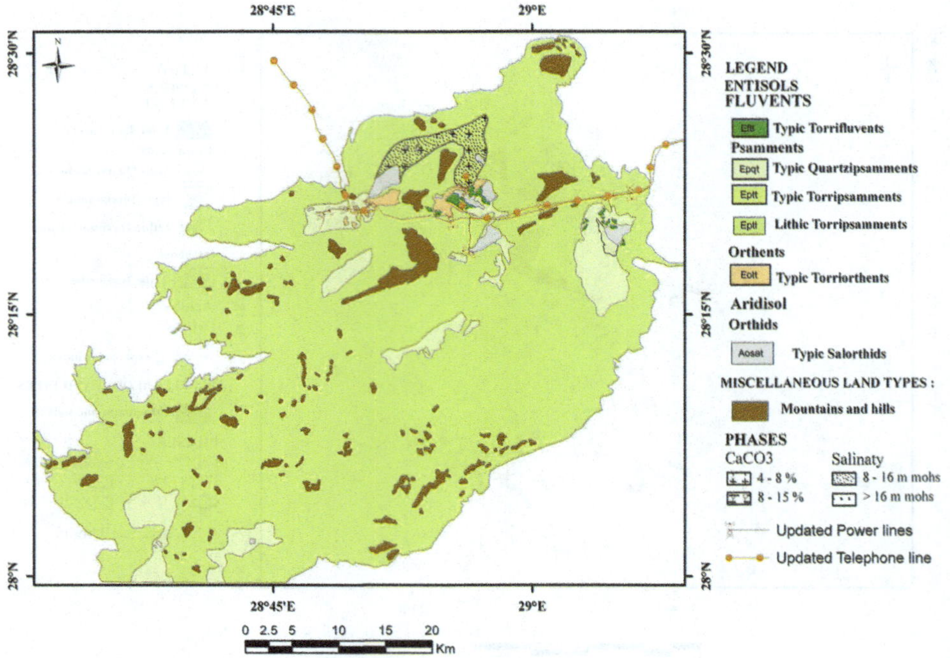

Fig. 4.224 Utilities, superimposed on the soil map, Al-Baharyia oases

Table 4.72 Lengths of additional land resources, Al-Baharyia oases

Layer	Length (km)	Area (km^2)
Urban		2.3
Roads	128.5	
Railways	114.6	
Utilities	260.7	

- *Typic Torrifluvents*
- *Typic Torriorthents*

The soils of Palustollic chromusterts (Vucrpl) cover an area of 1471.8 km^2 representing 26.6% of territory located at the northern limits of the oases. *The Lithic Torripsamments (Eptl)* exhibits patches at the borders of the previous soil unit, recording an area of 454.8 km^2 (8.2%), characterized by a shallow soil profile and coarse texture. The soils of *Typic Torripsamments (Eptt)* occupy 326.2 km^2 representing 5.9% of land cover, while the *Palustollic Torrerts (Vtbl)* sub-great group covers 84.8 km^2 (1.5% of the total area) located at the northwestern corner of the sheet area.

The *Typic torrifluvents* (*Eftt*) and *Typic torriorthents (Eott)* sub-great groups cover areas of 9.8 and 5.9 km^2 representing 0.2 and 0.1% respectively. The Active sand dunes exist in an area of 472.5 km^2 (i.e. 8.6%) having an elongated pattern, moving in a southward

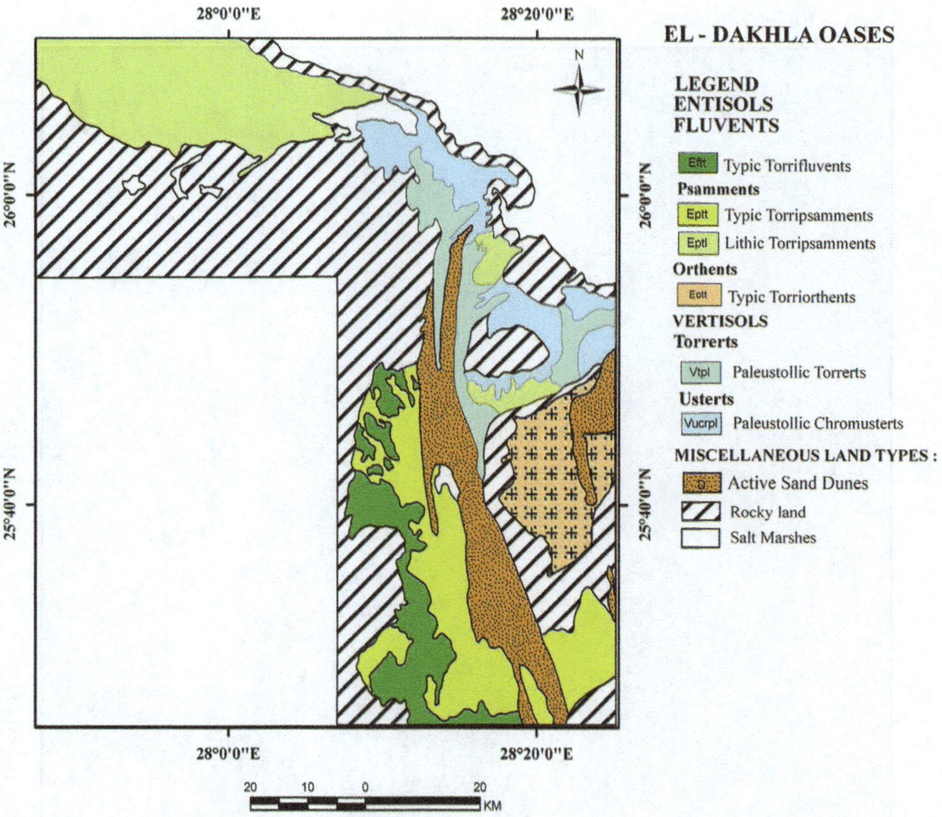

Fig. 4.225 Soil map of Al-Dakhla Oases (A)

direction. The rock land exhibits an extensive coverage of 2612.6 km ^{2} area (i.e. 47.2% of the land cover). The occurrence of salt marshes, characterizes such depression area covering a patch of 94.5 km^2 representing 1.7% of the land cover.

4.4.2.3 Dakhla Oases3

Figure 4.227 shows soil map of Al-Dakhla Oase B area, which is located between latitudes 25° 00′ and 25° 55′ N and longitudes 28° 00′ and 28° 50′ E. Figure 4.228 shows the ETM+ image. The main soil units of this area, Table 4.74 include the following sub-great groups;

- *Lithic Torripsamments*
- *Typic Torripsamments*
- *Typic Torrifluvents*
- *Palustollic Chromusterts*
- *Typic Torriorthents*
- *Palustollic Torrerts*

Fig. 4.226 ETM Landsat images of Al-Dakhla Oases (A)

The soils of *Lithic Torripsamments (Eptl)* sub-great group are found in the northwestern parts of Al-Dakhla oases, covering an area of 745.1 km^2 (i.e. 13.8%). They are characterized by a shallow soil profile and coarse texture. The soils of *Typic Torripsamments (Eptt)* exhibit the southern parts of the oases covering an area of 573.7 km^2 as it representing 10.7% of the total area. The *Typic Tifluvents* (*Eftt*) sub-great group is found in the southestern part as small patch, covering an area of 312.9 km^2 (i.e. 5.8% of the total area). The *Typic Torriorthents (Eott)* soils covers an area of 281.8 km^2 (i.e. 5.2% of the total area), distributed maily in the western part of the map. The soils of *Palustollic Torrerts (Vtbl)* and *Palustollic Chromusterts (Vucrpl)* sub-great

Table 4.73 Distribution of soil units in Al-Dakhla (A) Oases

Soil Code	Area (km^2)	Area %
Eftt 1/1b	2.3	0.1
Eftt 3/4a	7.5	0.1
Eott 1/1a	5.9	0.1
Eptl 1/1b	138.1	2.5
Eptl 1/1c	316.7	5.7
Eptt 1/1b	13.1	0.2
Eptt 1/1c	313.1	5.7
Vtpl 4/4a	84.8	1.5
Vu cr pl 4/4a	1471.8	26.6
Active sand dune	472.5	8.6
Rock land	2612.6	47.2
Salt marshes	94.5	1.7
Total	5532.9	8.5

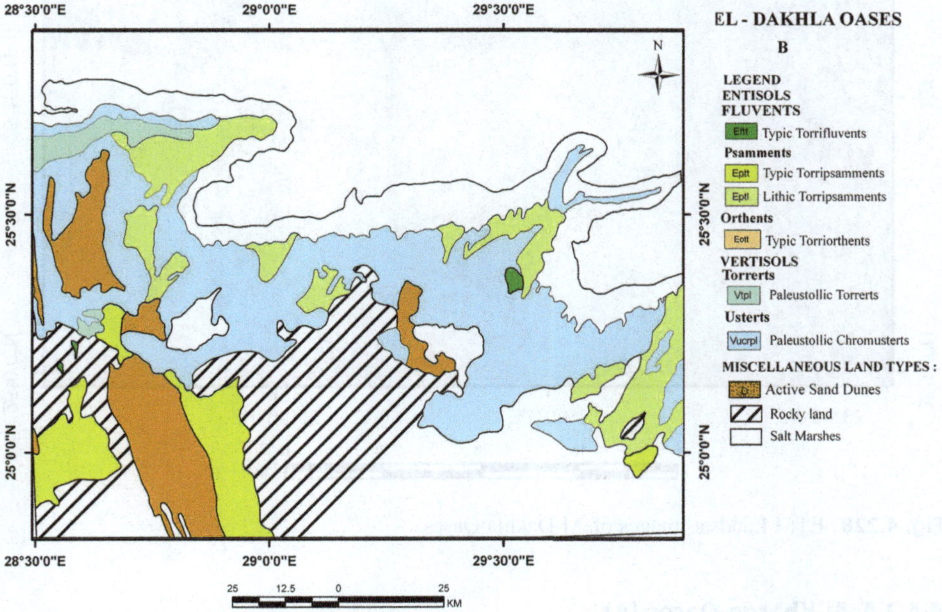

Fig. 4.227 Soil map of Al-Dakhla Oases

groups are found in the northeast parts of the map area covering an area of 246.2 and 310.3 km^2 respectively. Distributed as a small parts at the north of the oases and are dominated by salt marches (75.1 km^2). The active sand dunes (515.2 km^2) cross the map area towards the south direction. The rock land covers large area around the oases i.e. 2321.9 km^2 representing 43.1% of the total area.

Fig. 4.228 ETM Landsat images of Al-Dakhla Oases

4.4.2.4 Al-Kharga Oases (A)

Figure 4.229 shows the soil map of Al-Kharga Oases (A) area, which is located between latitudes 25° 15′ and 26° 05′ N and longitudes 30° 05′ and 30° 45′ E. Figure 4.230 shows the ETM+ image. The main soil units of this area, Table 4.75 include the following sub-great groups:

- *Lithic Torripsamments*
- *Palustollic Chromusterts*
- *Typic Torripsamments*

Table 4.74 Distribution of soil units in Al-Dakhla Oases

Soil Code	Area (km^2)	Area %
Eftt 1/1b	58.9	1.1
Eftt 1/1c	254.0	4.7
Eott 1/1c	281.8	5.2
Eptl 1/1c	745.1	13.8
Eptt 1/1c	573.7	10.7
Vtpl 4/4a	246.2	4.6
Vu cr pl 4/4a	310.3	5.8
Active sand dune	515.2	9.6
Rock land	2321.9	43.1
Salt marshes	75.1	1.4
Total	5382.0	100.0

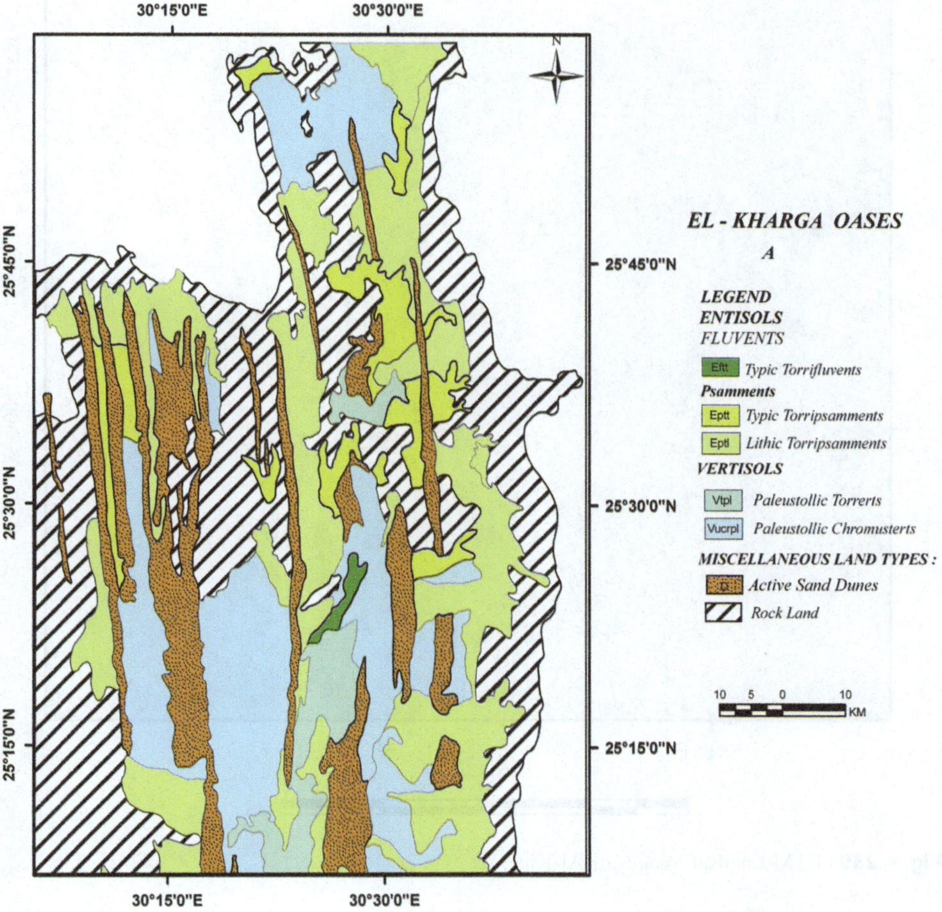

Fig. 4.229 Soil map of Al-Kharga Oases (A)

Fig. 4.230 ETM Landsat images of Al-Kha

Table 4.75 Distribution of soil units in Al-Kharga Oas

Soil Code	Area (km^2)	Area %
Active sand dune	665.0	16.0
Eftt 3/2a	15.0	0.4
Eptl 1/1c	11.0	0.3
Eptl 1/1d	108.0	2.6
Eptl 2/1b	14.0	0.3
Eptl 2/1c	775.0	18.6
Eptl 2/1d	43.0	1.0
Eptt 1/1c	42.0	1.0
Eptt 2/1b	24.0	0.6
Eptt 2/1c	157.0	3.8
Rock land	1400.0	33.6
Vtpl 4/4a	145.0	3.5
Vu cr pl 4/4a	768.0	18.4
Total	4166.0	100.0

Eott 2/1c	140.4	2.4
Eptl 2/1a	52.1	0.9
Eptl 2/1c	2294.4	38.6
Eptt 2/1a	41.4	0.7
Vtpl 4/4a	473.7	8.0
Vu cr pl 4/4a	1047.6	17.6
Active sand dune	1244.0	20.9
Rock land	647.0	11.0
Total	5940.6	100.0

- *Palustollic Torrerts*
- *Typic Torrifluvents*

The soils of the *Lithic Torripsamments (Eptl)* dominate this map sheet area, as they cover an area of 951.0 km^2 (i.e. 22.8% of the total area). They are characterized by a shallow soil profile and coarse texture. The *Palustollic Chromusterts (Vucrpl)* sub-great group exhibits the middle parts of the oases covering an area of 768.0 km^2, representing 18.4% of the land cover. The *Palustollic Torrerts (Vtbl)* sub-great group is restricted in 145 km^2 on the extension of the Palustollic chromusterts, representing 3.5%. The soils of *Typic Torripsamments (Eptt)* cover an area of 223.0 km^2 (5.3%), while the *Typic Torrifluvents (Eftt)* sub-great group covers 15.0 km^2 (0.4% of the total area). The Active sand dunes cover an area of 665.0 km^2 (i.e. 16% of the map sheet area), moving in a southward direction. The rock land exhibits different location in this area, covering a total of 1400.0 km^2 area (i.e. 33.6% of the land cover).

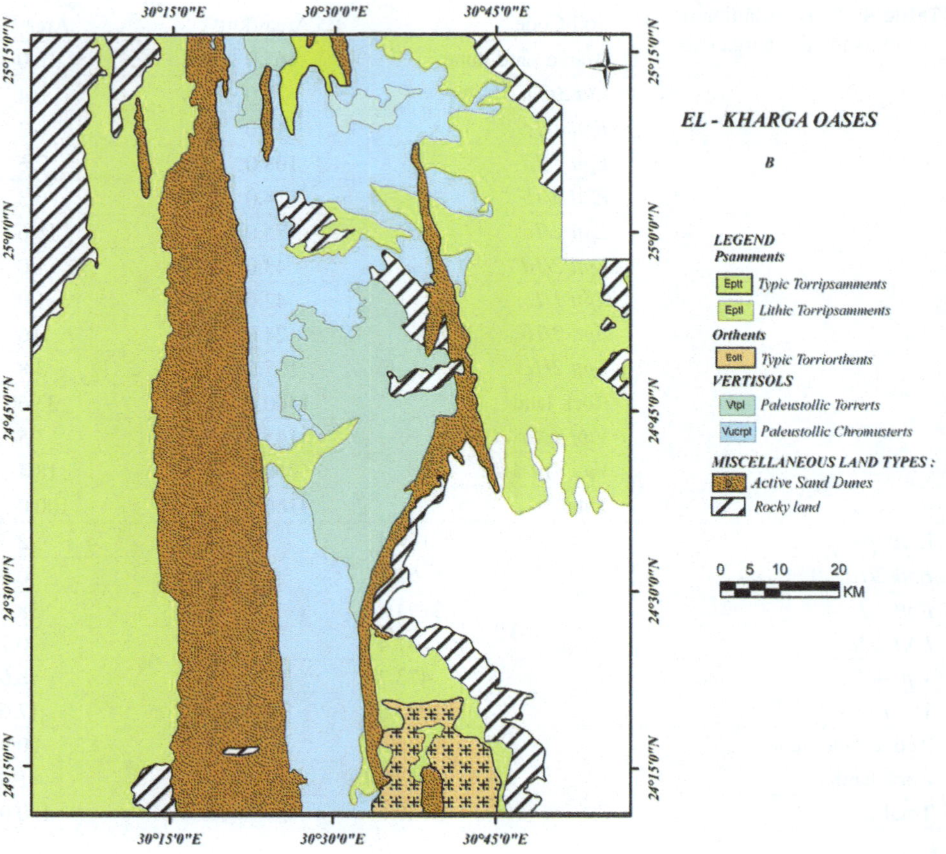

Fig. 4.231 Soil map of Al-Kharga Oases

4.4.2.5 Al-Kharga Oases

Figure 4.231 shows the soil map of Al-Kharga Oases area, which is located between latitudes 24° 20′ and 25° 20′ N and longitudes 30° 15′ and 30° 50′ E. Figure 4.232 shows the ETM+ image. The main soil units of this area Table 4.69 include the following sub-great groups:

- *Lithic Torripsamments*
- *Palustollic Chromusterts*
- *Palustollic Torrerts*
- *Typic Torriorthents*
- *Typic Torripsamments*

The soils of *Lithic Torripsamments (Eptl)* sub-great group dominate this map sheet area, as they cover an area of 2346.5 km^2 (i.e. 39.5% of the total area). These soils are found in both eastern and western sides of the oases, characterized by a shallow soil profile and coarse

Fig. 4.232 ETM Landsat images of Al-Kharga Oases

texture. The *Palustollic Chromusterts (Vucrpl)* and *Palustollic Torrerts (Vtbl)* exhibit the middle parts of the oases covering an area of 1047.6 and 473.7 km^2 respectively. Small parts (41.4 km^2) in the north of the oases are exhibited by the *Typic Torripsamments (Eptt)* sub-great group. The soils of *Typic Torriorthents (Eott)* cover an area of 140.0 km^2 representing 2.4% of the total area. Active sand dunes cover an area of 1244.0 km^2 (i.e. 20.9% of the map sheet area), moving in a southward direction. The rock land exhibits different location in this area, covering a total of 647.0 km^2 area (i.e. 10.9% of the land cover).

4.4.2.6 Siwa Oasis & Berket Zeitun

Figure 4.233 shows soil map of Siwa Oasis & Berket Zeitun area, which is located between latitudes 29° 05′ and 29° 30′ N and longitudes 25° 15′ and 25° 46′ E. Figure 4.234 shows the ETM+ image. The main soil units of this area include the following sub-great groups (Table 4.76):

- *Typic Salorthids*
- *Typic Torripsamments*
- *Calcic Salorthids*
- *Typic Calciorthids*
- *Typic Torriorthents*
- *Typic Haplaquents*
- *Calcic Gypsiorthids*
- *Hemists Salorthids*

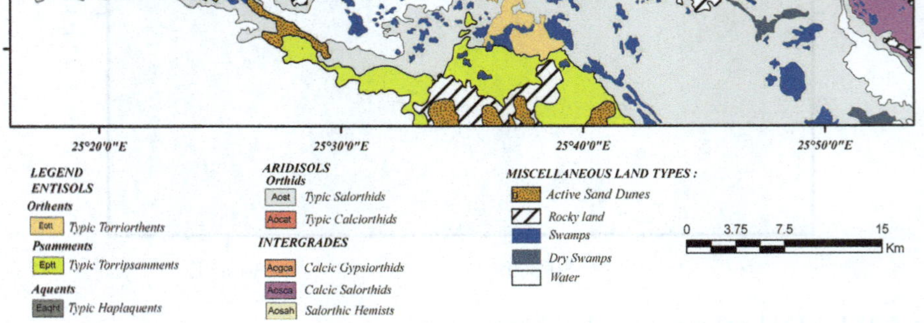

Fig. 4.233 Soil map of Siwa Oasis & Berket Zeitun\

Fig. 4.234 ETM Landsat images of Siwa Oasis & Berket Zeitun

Table 4.76 Distribution of soil units in Siwa Oasis & Berket Zeitun

Soil Code	Area (km^2)	Area %
Aocat	9.3	2.5
Aogca	0.8	0.2
Aosah	0.5	0.1
Aosca	21.2	5.7
Aost	196.6	52.9
Dry swamps	6.2	1.7
Eaght	4.8	1.3
Eott	7.5	2.0
Eptt	23.7	6.4
Rock land	19.6	5.3
Sand dune	12.4	3.3
Swamps	26.5	7.1
Water	42.9	11.6
Total	371.8	100.0

The soils of *Typic Salorthids (Aost)* sub-great group dominates the existing soil units as it covers an area of 196.6 km^2, representing 52.9% of the map sheet area. Other soil sub-great groups exist in this area with low occurrences, however representing the particular environment of the Siwa oases, related to the occurrence of salinity and Gypsum. Particular features include the swamps, sand dunes and rock land areas which cover 26.5, 12.4 and 19.6 km^2 representing 7.1, 3.3 and 5.3% of map sheet area. The *Typic Torripsamments (Eptt)* sub-great group covers an area of 23.7 km^2, representing 6.4% of map sheet area. The sub-great group of *Calcic Salorthids (Aosca)* cover an area of 21.2, representing 5.7%, while the *Typic Calciorthids (Aocat)* exhibits 9.3 (i.e. 2.5%) of map sheet area. The soils sub-great group *Typic Torriorthents (Eott)* cover an area of 7.5 km^2, representing 2%. The *Typic Haplaquents (Eaght)* covers an area of 4.8 km^2, representing 1.3% of the map sheet

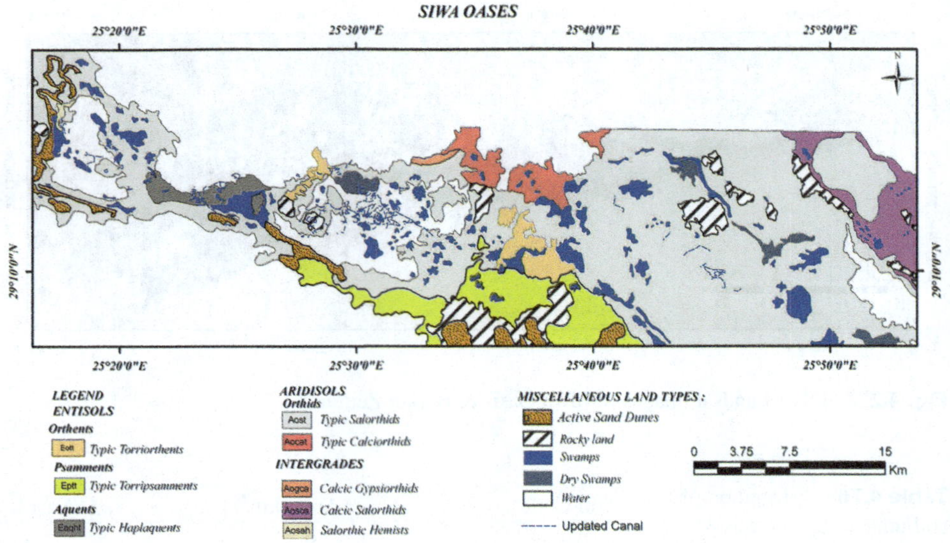

Fig. 4.235 Urban areas superimposed on the soil map, Siwa Oasis & Berket Zeitun

area. The sub-great groups *Calcic Gypsiorthids (Aogca)* and *Hemists Salorthids (Aosah)* cover areas of 0.8 and 0.5 km^2, representing 0.2 and 0.1% respectively.

The created database of this area Figs. 4.235, 4.236, 4.237, 4.238 and Table 4.77 indicates that the total urban settlement area is 2.3 km^2; the lengths of roads and the utilities networks attain 840.3 and 47.7 km respectively.

4.4.3 Eastern Desert Frngs Soils

4.4.3.1 AL-Tumailat (A)-Suez Al-Tell Al-Kabeir

Figure 4.239 shows soil map of AL-Tumailat (A)-Suez Al-Tell Al-Kabeir area, which is located between latitudes 29° 50′ to 30° 35′ N and longitudes 32° 00′ to 32° 40′ E. Figure 4.240 shows the ETM+ image. The main taxonomic units of the soils in this area are the following sub-great groups Table 4.78:

- *Typic Quartzipsamments*
- *Typic Torriorthents*
- *Typic Gypsiorthids*
- *Calcic Gypsiorthids*
- *Petrogypsic Gypsiorthids*

The sub-great group *Typic Quartzipsamments (Epqt)* exhibits the northern part of map sheet area, covering 790.7 km^2 representing 26.6% of the area.

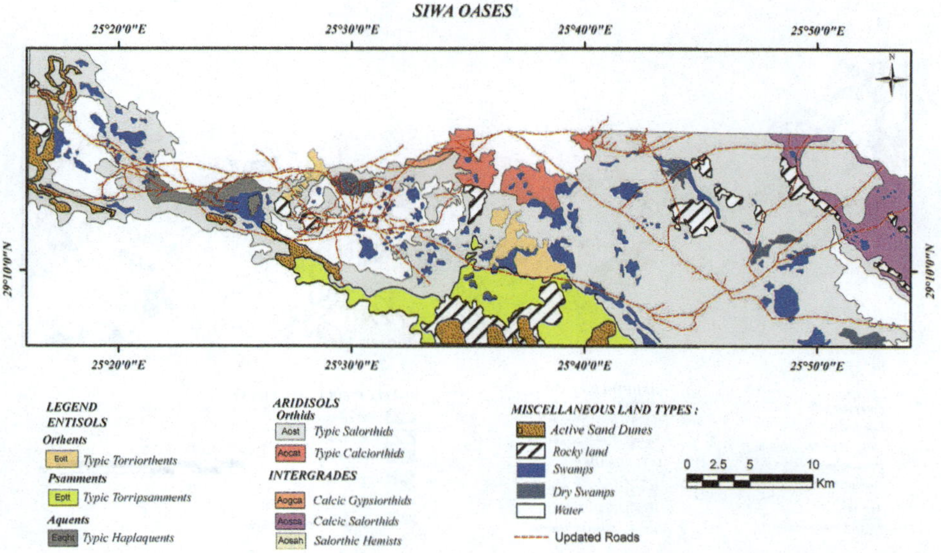

Fig. 4.236 Irrigation and drainage canals superimposed on soil map, Siwa Oasis & Berket Zeitun

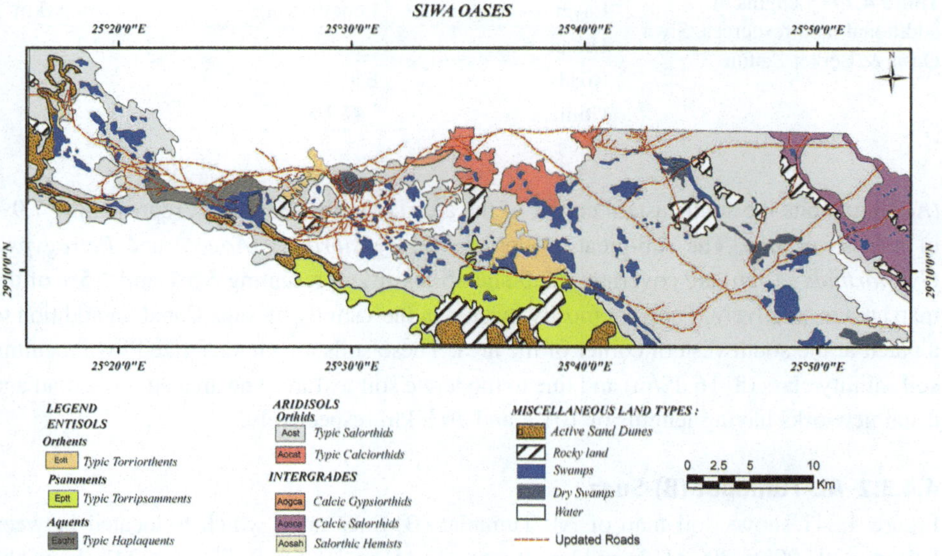

Fig. 4.237 Road network superimposed on soil map, Siwa Oasis & Berket Zeitun

These soils are characterized by coarse soil texture and medium salinity (8–16 dS/m) is expressed in most of its coverage. The *Typic Torriorthents (Eott)* covers an area of 548.8 $km^{2,}$ representing 18.5% of the area. The soils in this area attain 4–8% of $CaCO_3$ and salinity range between 8 and 16 dS/m. The sub-great group of *Typic Gypsiorthids*

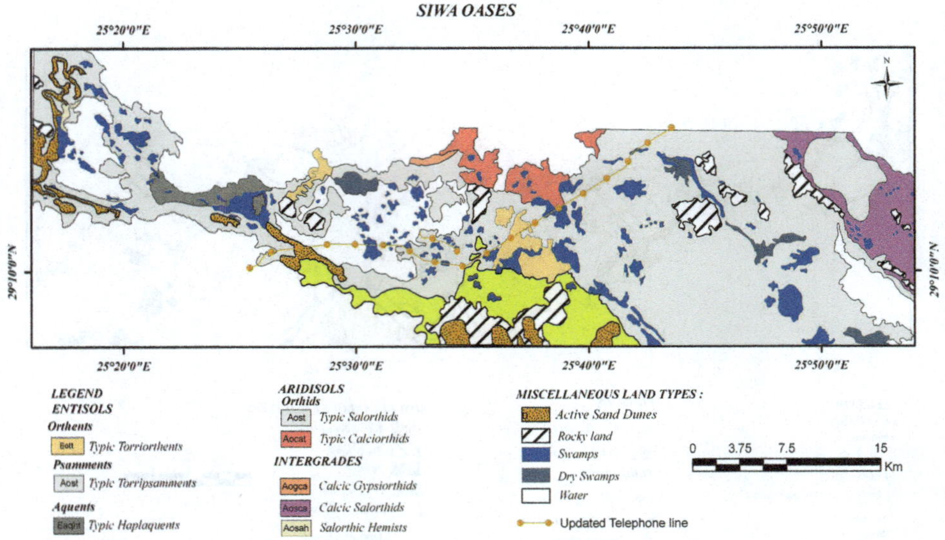

Fig. 4.238 Utilities, superimposed on the soil map, Siwa Oasis & Berket Zeitun

Table 4.77 Lengths of additional land resources, Siwa Oasis & Berket Zeitun

Layer	Length (km)	Area (km^2)
Urban		2.3
Roads	840.3	
Utilities	47.7	

(Aogt) exhibits the southeastern corner of the area covering 235.5 km^2, representing 7.9% of map sheet area. The sub-great groups *Calcic Gypsiorthids (Aogca)* and *Petrogypsic Gypsiorthids (Aogp)* are covering 162.5 and 45.5 km^2, representing 5.5% and 1.5% of the map area respectively. They are mostly located in the vicinity of Suez Canal, in addition to a patch at the southwestern corner of the area. These soils are characterized by a medium soil salinity class (8–16 dS/m) and fine to moderate soil texture. The area attains a road and canal networks having lengths of 101.7 and 70.8 km respectively.

4.4.3.2 AL-Tumailat (B)-Suez

Figure 4.241 shows soil map of AL-Tumailat (B)-Suez area, which is located between latitudes 30° 00′ to 30° 31′ N and longitudes 31° 15′ to 31° 55′ E. Figure 4.242 shows the ETM+ image. The main taxonomic units of the soils in this area are the following sub-great groups, Table 4.79:

- *Typic Torriorthents*
- *Typic Quartizpsamments*
- *Calcic Gypsiorthids*

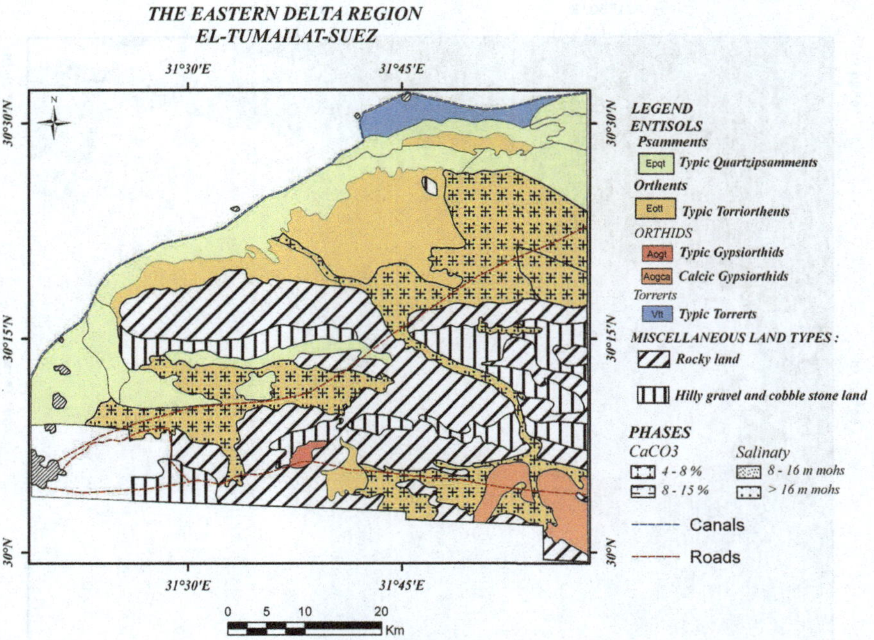

Fig. 4.239 Soil map of AL-Tumailat (A)-Suez Al-Tell Al-Kabeir

- *Typic Torrerts*
- *Typic Gypsiorthids*

The sub-great group *Typic Torriorthents (Eott)* dominates the soil units in the current map sheet, as it covers an area of 923.0 km^2 representing 36.5% of the area. In this area, the soils of *Typic Torriorthents* are characterized by low $CaCO_3$ contents and moderate salinity. The *Typic Quartizpsamments (Epqt)* sub-great group is mostly exhibiting the northern borders of the area, covering 477.6 km^2 (i.e. 18.8% of map sheet area).

The sub-great groups *Calcic Gypsiorthids (Aogca)* cover an area of 69.3 km^2 (2.7% of mapped area) at the north characterized by coarse soil texture and non-salinity conditions. The soils of *Typic Torrerts (Vtt)* cover almost similar area as the previous one (55.0 km^2) but located at the northeastern corner on the fringes of eastern Delta region characterized. It is characterized by fine soil texture, non-salinity conditions and non-to slight $CaCO_3$ contents. The rock land dominates the area, as it covers 982.8 km^2 representing 38.8% of the mapped area.

4.4.3.3 Eastern Desert-Atfih

Figure 4.243 shows soil map of Eastern Desert-Atfih area, which is located between latitudes 29° 05′ to 29° 30′ N and longitudes 31° 10′ to 31° 30′ E. Figure 4.244 shows

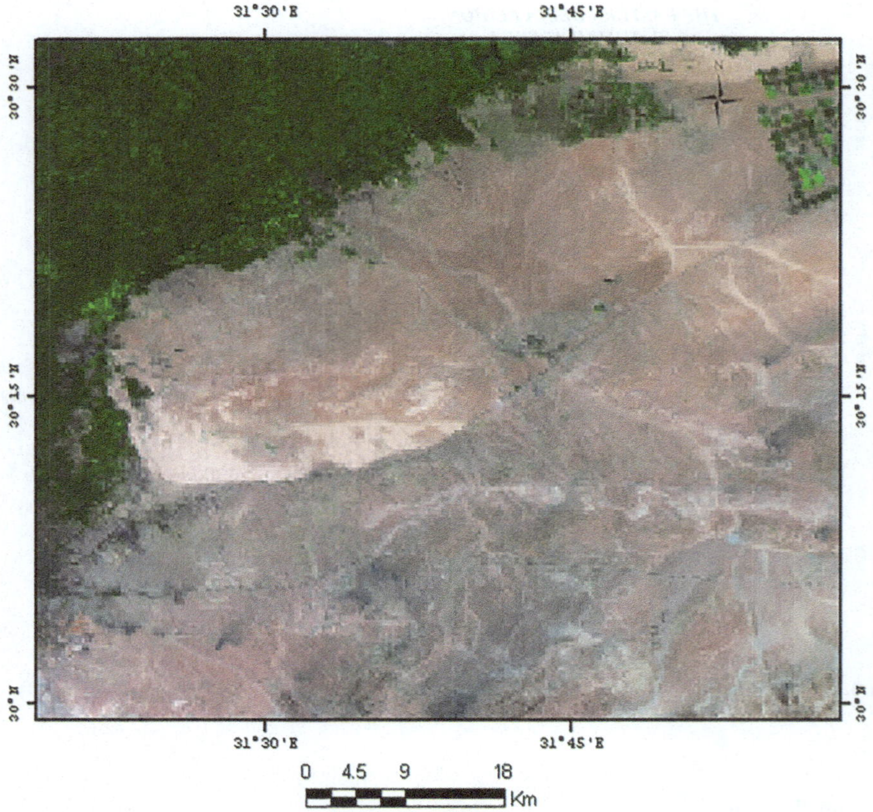

Fig. 4.240 ETM Landsat images of AL-Tumailat (A)-Suez Al-Tell Al-Kabeir

the ETM+ image. Only three sub-great soil groups are found in this area, Table 4.80 including the following;

- *Typic Quartizpsamments*
- *Typic Torriorthents*
- *Typic Torripsamments*
- *Lithic Torripsamments*

The sub-great group *Typic Quartizpsamments (Epqt)* dominates the map sheet area as it represents 52.8% of the total area (i.e. 422.0 km^2), these soils are found in the middle and north parts of the map sheet. The soils of *Typic Torriorthents (Eott)* represents 21.5% of the total area (i.e. 171.5 km^2), Thesse soils in this area are characterized by a high salinity (EC more than 16 dS\m) and moderate $CaCO_3$ content (8–15). The *Typic Torripsamments*

Table 4.78 Distribution of soil units in AL-Tumailat (A)-Suez Al-Tell Al-Kabeir

Soil Code	Area (km^2)	Area %
Aogca 1/1a1	162.5	5.5
Aogp 1/3a1	45.5	1.5
Aogt 1/1a1	29.1	1.0
Aogt 1/1b1	51.8	1.7
Aogt 1/3a3	154.6	5.2
Eott 1/1 a4	188.4	6.3
Eott 1/1a	125.2	4.2
Eott 1/1a3	79.5	2.7
Eott 1/1b	69.8	2.3
Eott 1/2a1	85.9	2.9
Epqt 1/1a	64.9	2.2
Epqt 1/1a1	176.1	5.9
Epqt 1/1b	111.1	3.7
Epqt 1/1b-c	101.9	3.4
Epqt 1/1b1	274.2	9.2
Epqt 1/1c	62.5	2.1
Rock escarpment	312.7	10.5
Rock land	830.0	27.9
Urban	44.9	1.5
Total	2970.5	100.0

(Eptt) sub great group are found in the eastern south of the area covering an area of 107.0 km^2 (i.e. 13.4% of the mapped area), this sub great group are characterized by a high salinity in this area as the EC ranges between 8 and 16 dS\m. The soils of a shallow soil profile *Lithic Torripsamments (Eptl)* are found in the old river deposits at the eastern desert fringes, it covers an area of 48.0 km^2 (i.e. 6.0% of the mapped area).

4.5 Soil Taxonomy of Northwestern Costal Region and Desert Areas

4.5.1 Northwestern Costal Region

Table 4.81 and Fig. 4.245 show the geogeaphical distribution of soil units in the Northwestern coastal zone. Both *Arididsols* and Entisols soil orders are found covering 42.1 and 57.9% of the mapped soils respectively.

The Calcids sub-order is mostly clustered in areas of Burg Al-Arab, Marsa Matrouh and Sidi Barani exihibiting 14.5% of the area. The Salids sub-order exists around both Matiout and Salum lagoons representing 6.1% of soils. The Gypsids sub-order soil is restricted in the area between Al-Hammam and Sedi Heneish covering 21.5% of the maped soils.

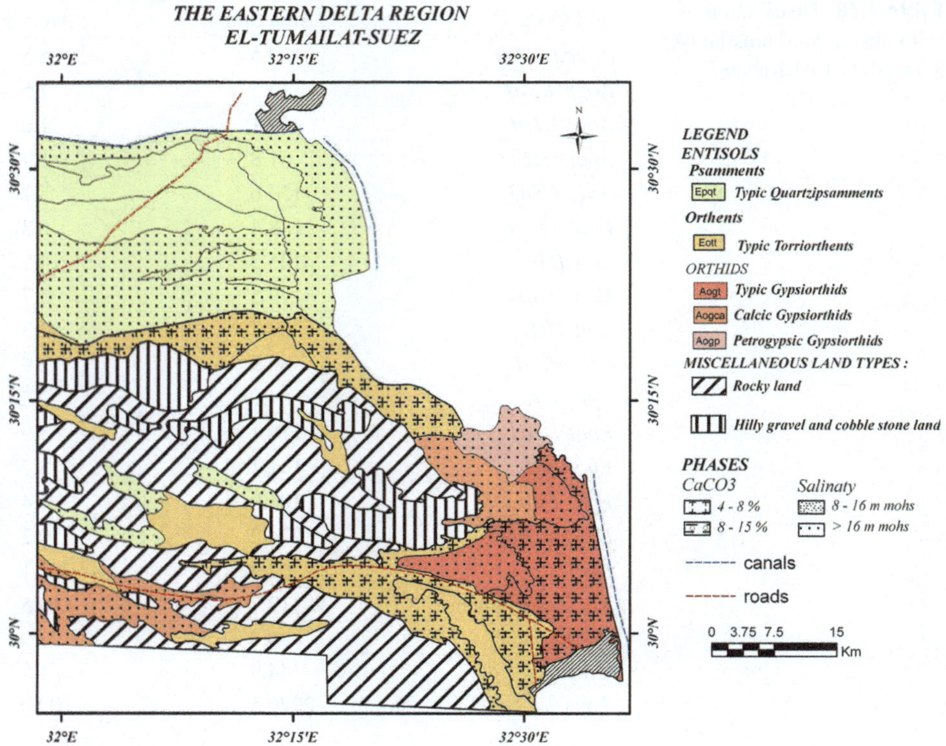

Fig. 4.241 Soil map of AL-Tumailat (B)-Suez

The *Entisols* soil order includes the sub-orders *Orthents* and *Psamments*, Fig. 4.246 representing 34.0 and 23.9% of the mapped soils respectively. It can be noticed that the area from Burg Al-Arab to Matrouh is characterized by variability of soil units (e.g. *Torripsamments, Torriorthents*), while the area from Matrouh to Al-Saloum is occupied by the *Torriorthents* great group soils.

4.5.2 Western Desert Oases

Table 4.82 and Figs. 4.247, 4.248, 4.249 and 4.250 show the geographical distribution of the soil units in the western desert oases. Both *Aridisols* and *Entisols* soil orders exist in the four studied oases, however the *Vertisols* soil order only exist in the Kharga oases. In the Baharya oases, the most common soil unit is the *Torripsamments* great group covering an area of 1803.2 km^2 representing 98.1%. *Aridisols*, including only one sub-order (i.e. *Salids*) cover an area of 17.8 km^2 representing 0.9%. The Siwa oasis is considered as a closed basin, thus significantly affected by salt accumulation. The *Haplosalids* great group exists in the oases covering 64.9% of its soils spread all over the oases coverage.

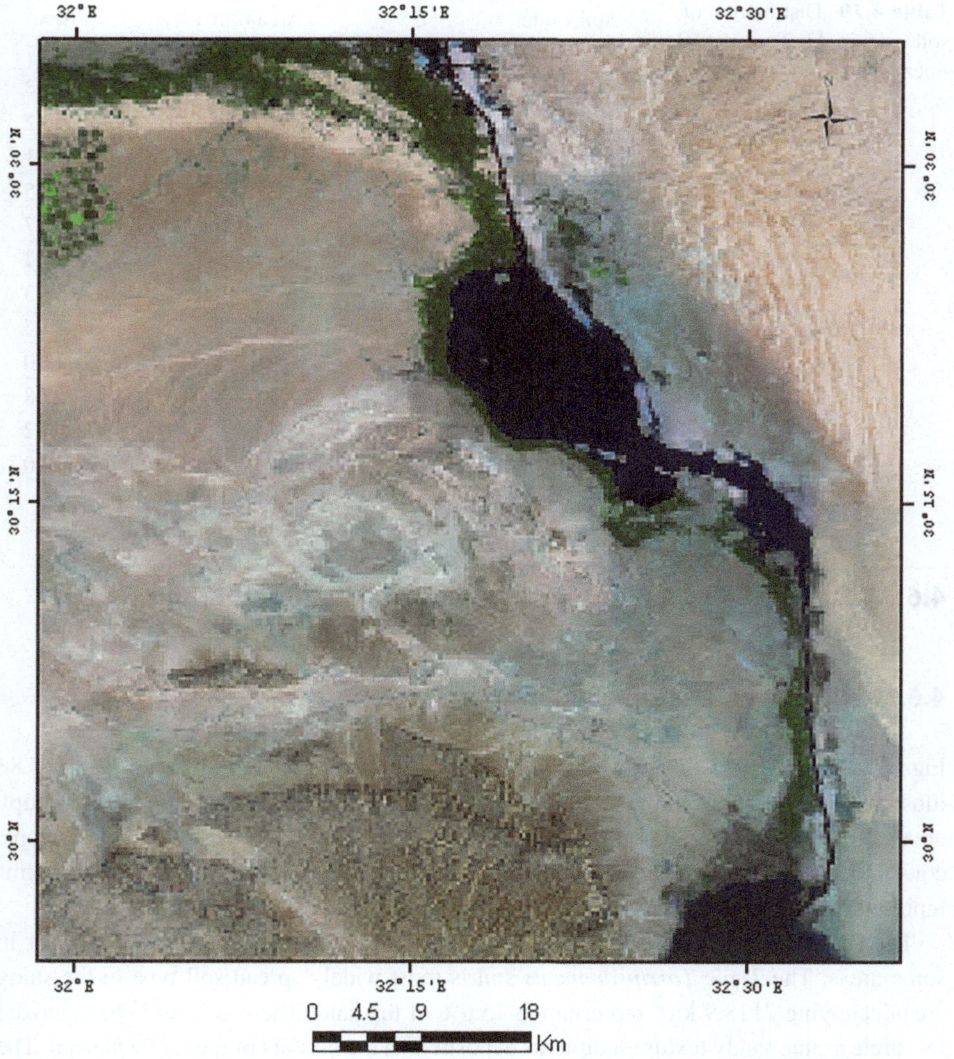

Fig. 4.242 ETM Landsat images of AL-Tumailat (B)-Suez

The suborder "*Psamments*" exists in the south of the oases covering an area of 57.3 km^2 representing 22.7% of the mapped soils. The only oases having the soil order *Vertisols* is Kharga it is covered by 40.2% of those soils. The *Psamments* sub-order covers an area of 4313.2 km^2 representing 59.6% of the Karga oases. A percentage of 57.9% of Dakhla oases is exhibited by *Entisols* soil order, where the *Torripsamments* great group represents 44.6% of the mapped soils. The *Aridisols* soil order, represented by the *Petrogypsids* great group covers an area of 1816.1 km^2 representing 42.1% of the soils.

Table 4.79 Distribution of soil units in AL-Tumailat (B)–Suez

Soil Code	Area (km^2)	Area %
Aogca 1/1a	69.3	2.7
Aogt2/1a	7.3	0.3
Eott 1/1a	510.4	20.2
Eott 1/1b	305.7	12.1
Eott 1/2a1	38.9	1.5
Eott 2/1a	68.0	2.7
Epqt 1/1a	334.7	13.2
Epqt 1/1b-c	34.0	1.3
Epqt 1/1c	108.9	4.3
Hilly gravel and cob	278.2	11.0
Rock land	704.6	27.8
Vtt 4/4 a	55.0	2.2
Total	2531.3	100.0

4.6 Digital Soil Map of Northwestern Costal Region and Desert Areas

4.6.1 Northwestern Coastal Zone

Figure 4.251 shows soil map mosaic of the Northwestern coastal zone, Table 4.83 illustrates that the Rockland exhibit most of the study region, as the Libyan plateau occupy an area of 6878.8 km^2, representing 48.7% of the area. Moreover, the Rockland outcrops, consolidated rocky ridges and rock escarpments cover areas of 824.8, 299.4 and 243.0 km^2 representing 5.8, 2.1 and 1.7% respectively.

The consolidated rocky ridges are situated parallel to the shoreline, and disappear in some areas. The *Typic Torripsamments* soil is most widely spread soil type in the study area occupying 2118.9 km^2 representing 15.0% of the land. These soils are characterized by single graine sandy texture sediments deposited on the borders of the rocky plateau. The *Typic Torriorthents* sub-great group covers an area of 1275.6 km^2, representing 9.0% of the mapped area. The sub-great group *Lithic Torripsamments* exhibit 897.1 km^2 representing 6.4% of land. The morphological description and laboratory analyses of the soil profiles representing this unite indicate that the soil depth is shallow (< 60 cm) and almost rest on a rock layer. The *Petrogypsic gypsiorthids* cover an area of 808.4 km^2 representing 5.7% of the (Figs. 4.252, 4.253, 4.254, 4.255, 4.256, 4.257).

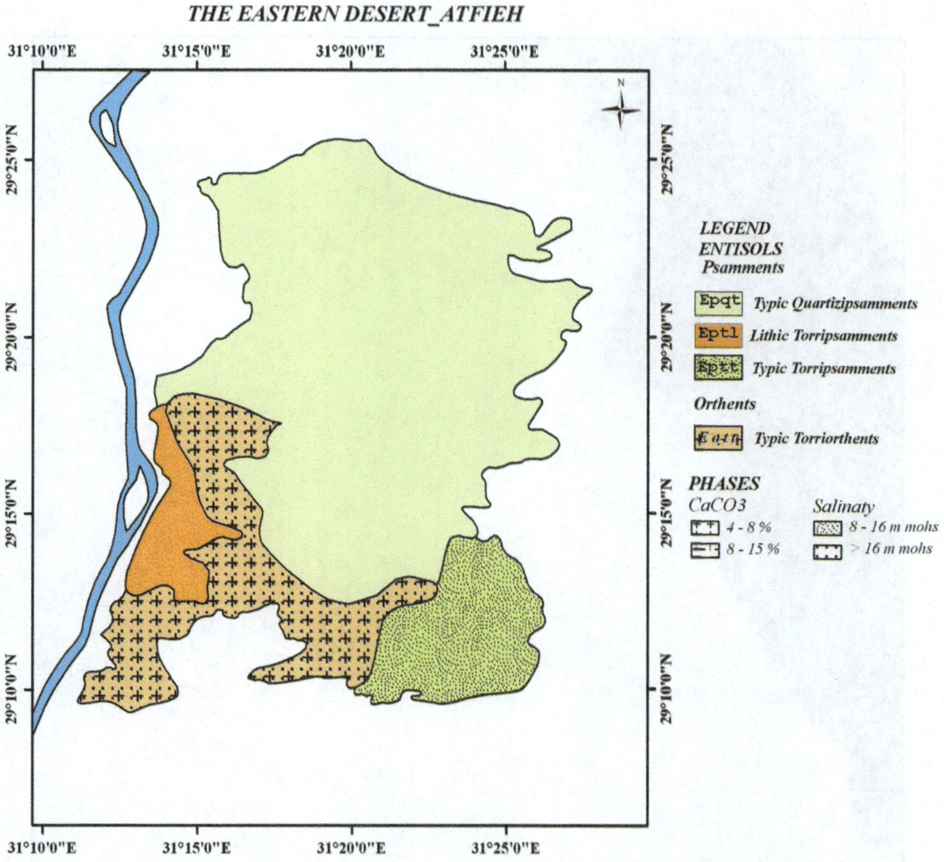

Fig. 4.243 Soil map of Eastern Desert-Atfih

4.6.2 Western Desert Oases

A number of four oases, located in the western desert of Egypt, are included in this study (i.e. Siwa, Baharyia, Dakhla and Kharga). Each of the oases represents a particular community characterized by its culture, land use pattern and environment. The distance from Dakhla oases, south, to Farafra, north, is about 230.0 km while the distance to Kharga, East, is about 150.0 km. The following paragraphs include the analytical studies of each oasis. Concerning the morphological description of the representative soil profile and laboratory analysis, Annexes II and III are referred.

4.6.2.1 Siwa Oases

Siwa Oases is located in the western desert of Egypt with 15,000 inhabitants, near the border with Libya, at a depression of 18 meters below sea level. The Oases is 82 km long and has a width varying between 2 and 20 km. It sites in ancient times of the famous oracle

Fig. 4.244 ETM Landsat images of Eastern Desert-Atfih

Table 4.80 Distribution of soil units in Eastern Desert-Atfih

Soil Code	Area (km^2)	Area %
Eott 1/1a	171.5	21.5
Epqt 1/1a	422.0	52.8
Eptl 1/1a	48.0	6.0
Eptt 1/1a	107.0	13.4
River Nile	50.1	6.3
Total	798.6	100.0

Table 4.81 Areas of the great groups, belonging to Aridisols and Entisols, in the Northwestern coast of Egypt

Order	Sub-Order	Type (Great group)	Area (km^2)	%
Aridisols	*Calcids*	*Haplocalcids*	542.2	14.5
	Salids	*Haplosalids*	229.4	6.1
	Gypsids	*Petrogypsids*	808.5	21.5
Entisols	*Orthents*	*Torriorthents*	1275.6	34.0
	Psamments	*Torripsamments*	897.1	23.9
Total area of the coast			3752.7	100.0

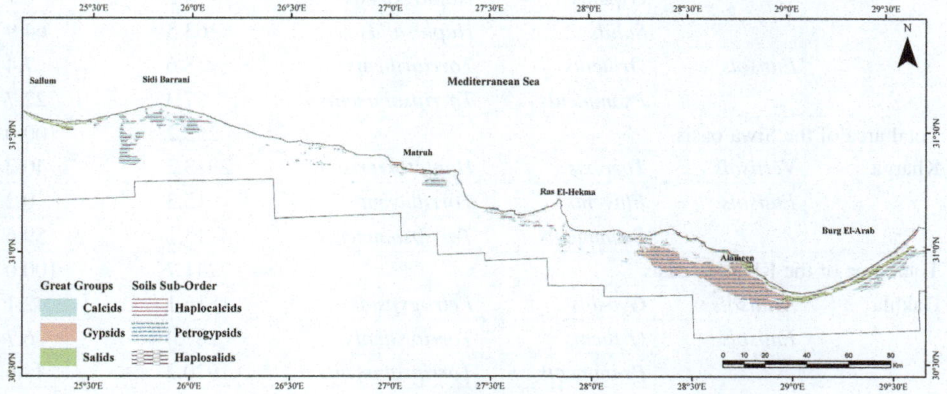

Fig. 4.245 Geographical distribution of Aridisols order in North Western Coast region

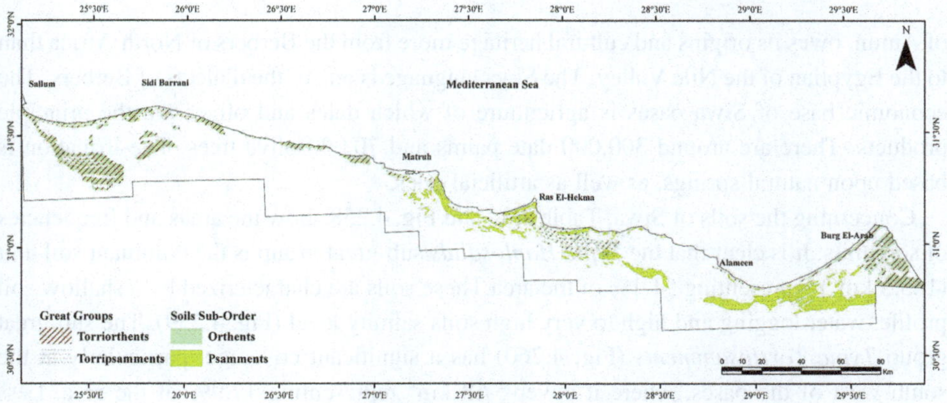

Fig. 4.246 Geographical distribution of Entisols order in North Western Coast region

Table 4.82 Areas of sub-great groups, belonging to Aridisols, Entisols, and Vertisols in the Western Desert Oases

Oases	Order	Sub-Order	Great group	Area (km^2)	%
Baharya	*Aridisols*	*Salids*	*Haplosalids*	17.8	0.9
	Entisols	*Psamments*	*Quartizpsamments*	2.8	0.2
			Torripsamments	1803.2	98.1
		Fluvents	*Torrifluvents*	5.3	0.3
		Orthents	*Torriorthents*	9.9	0.5
Total area of the Baharya oasis				1839.0	100.0
Siwa	*Aridisols*	*Calcids*	*Haplocalcids*	8.6	3.4
		Gypsids	*Haplogypsids*	4.2	1.7
		Salids	*Haplosalids*	163.5	64.9
	Entisols	*Orthents*	*Torriorthents*	18.6	7.4
		Psamments	*Torripsamments*	57.3	22.7
Total area of the Siwa oasis				252.2	100.0
Kharga	*Vertisols*	*Torrerts*	*Haplotorrerts*	2913.2	40.2
	Entisols	*Fluvents*	*Torrifluvents*	15.3	0.2
		Psamments	*Torripsamments*	4313.2	59.6
Total area of the Kharaga oasis				7241.7	100.0
Dakhla	*Aridisols*	*Gypsids*	*Petrogypsids*	1816.1	42.1
	Entisols	*Orthents*	*Torriorthents*	287.4	6.7
		Psamments	*Torripsamments*	1920.2	44.6
		Fluvents	*Torrifluvents*	285.7	6.6
Total area of the Dakhla oasis				4309.5	100.0

of Amun, owes its origins and cultural heritage more from the Berbers of North Africa than to the Egyptian of the Nile Valley. The Siwa language is one of the dialects of Berbers. The economic base of Siwa oases is agriculture of which dates and olives are the principle products. There are around 300,000 date palms and 70,000 olive trees. The irrigation is based upon natural springs, as well as artificial ones.

Concerning the soils of Siwa, Table 4.84 and Fig. 4.258 show the areas and frequencies of soil units. It is clear that the *Typic Haplosalids* sub-great group is the dominant soil unit (163.5 km^2) representing 51.1% of the area.These soils are characterized by a shallow soil profile, water logging and high to very high soils salinity level (Fig. 4.259). The sub-great group *Typic Torripsamments* (Fig. 4.260) has a significant coverage, particularly at the south west of the oases, where it cover 57.3 km^2 representing 17.9% of the area. Less coverage of sub-great groups *Typic Torriorthents* (Fig. 4.261), *Typic Calciorthids* and *Typic Haplogypsids* (Fig. 4.262) exist in patches clustered in the north and middle of the oases. They cover areas of 18.6, 8.6 and 4.2 km^2, representing 5.8, 2.7 and 1.3% of the area

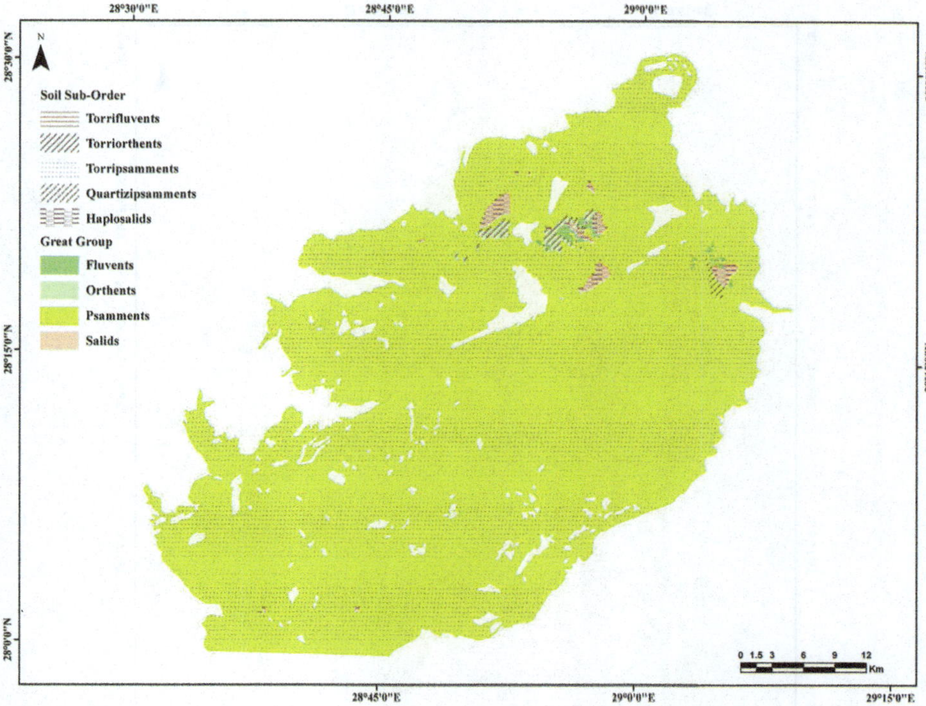

Fig. 4.247 Geographical distribution of soil orders in Bahariya Oasis

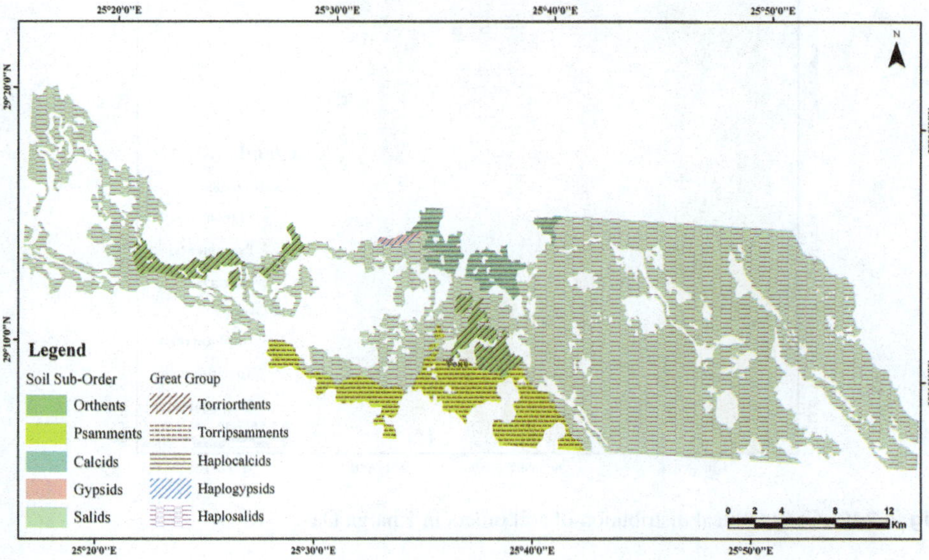

Fig. 4.248 Geographical distribution of soil orders in Siwa Oasis

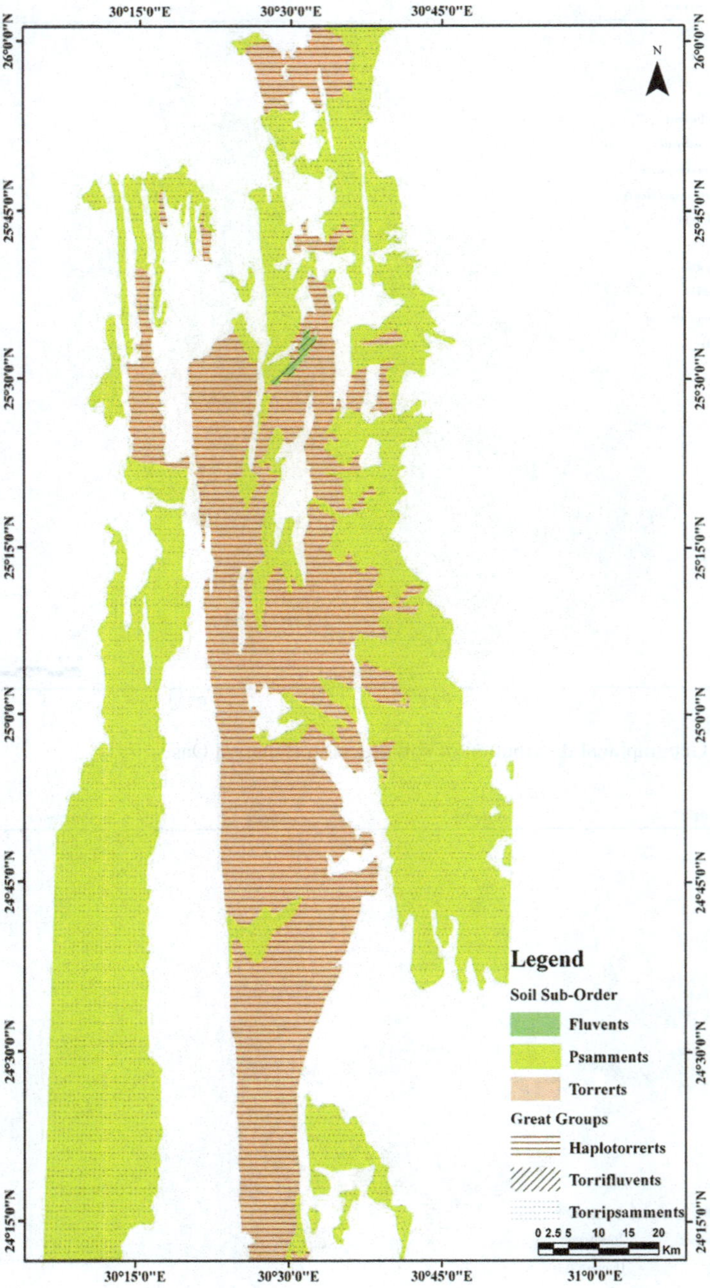

Fig. 4.249 Geographical distribution of soil orders in Kharga Oases

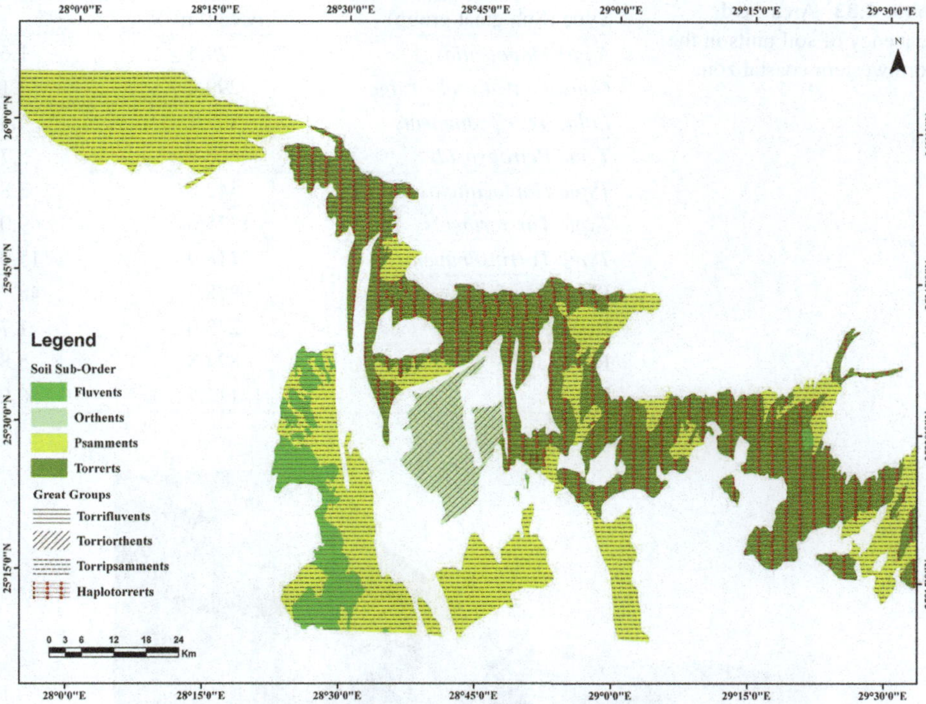

Fig. 4.250 Geographical distribution of soil orders in Dakhla Oasis

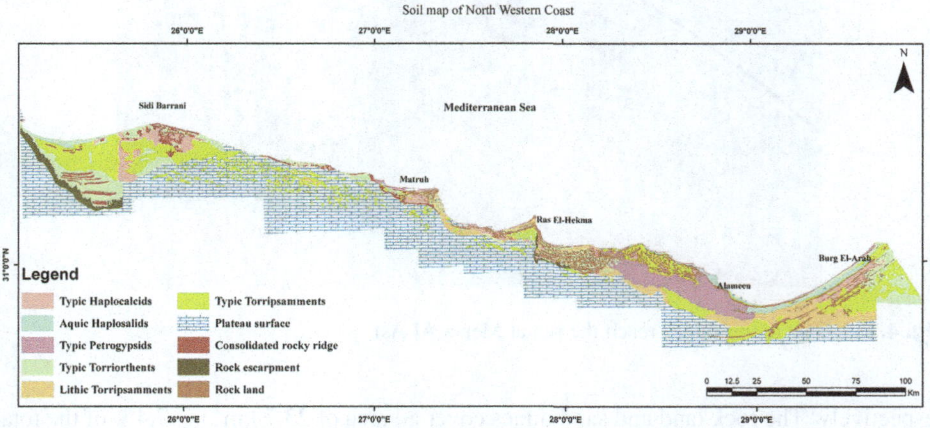

Fig. 4.251 Soil map of the North Western Coast region

Table 4.83 Areas and frequency of soil units in the Northwestern coastal zone

Type (Sub great group)	Area (km^2)	%
Aquic Hapolsalids	229.4	1.6
Consolidated rocky ridge	299.4	2.1
Lithic Torripsamments	897.1	6.4
Typic Petrogypsids	808.4	5.7
Typic Haplocalicids	542.2	3.8
Typic Torriorthents	1275.6	9.0
Typic Torripsamments	2118.9	15.0
Plateau surface	6878.8	48.7
Rock escarpment	243.0	1.7
Rock land	824.8	5.8
Total	14117.7	100.0

Fig. 4.252 The rocky ridges reach the sea at Mersa Al Asi

respectively. The rock land and sand dunes cover an area of 23.7 km^2 i.e. 7.4% of the total area.

4.6.2.2 Baharyia Oasis

Baharyia oasis is situated in a depression, 325.0 km southwest of Cairo. The depression is characterized by about 100 km length and an average of 40 km width. It is completely surrounded by high escarpment. Al-Bawiti is the modern administrative town, where many

Fig. 4.253 The rock outcrops on the plateau surface

Fig. 4.254 *Typic Torripsamments* landscape

ancient monuments, springs and gardens are found. The oases possess several natural springs; some of them have slight iron contents while others have sulphurous. Hot springs exist in the district Ain Al-Bishmu, on the western edge of the oases, where water temperature reaches 40 °C. The fortunes of the oases have changed through the history, in times of decline and poverty many of the population migrated to more wealthy regions, especially Cairo, in search of work. In 1958, the population of the oases was only around 6000 inhabitants. The development of the desert began with the Government plans of the New Valley, where many immigrants returned to the oases and population raised now to as 30,000 inhabitants. The oases conditions have flourished since the road from Cairo was

Fig. 4.255 *Typic Torriorthents* soils

Fig. 4.256 *Lithic Torripsamments* on the plateau surface with rock out crop

Fig. 4.257 Petrogypsic Gypsiorthids soils

Table 4.84 Areas and frequency of soil units in Siwa Oasis

Soil class	Area (km^2)	%
Typic Haplosalids	163.5	51.1
Typic Torripsamments	57.3	17.9
Typic Torriorthents	18.6	5.8
Typic Calciorthids	8.6	2.7
Typic Haplogypsids	4.2	1.3
Rock land	10.9	3.4
Sand dunes	12.8	4.0
Dry swamps	24.0	7.5
Swamps	20.2	6.3
Total	320.0	100.0

paved in 1967 and mineral mines at El-Managim were developed, together with the modern paved road connecting the oases with Farafra. Concerning the soils of Baharia oases, Fig. 4.263 and Table 4.85 show the distribution of the sub-great groups. The data indicate that the *Typic Torripsamments* is the dominant soil sub great group, as it covers an area of 1803.2 km^2, representing 93.4% of the Oases soils. The *Typic Haplosalids* Fig. 4.264 sub great group exhibits an area of 17.8 km^2, at the north of the oases representing 0.9% of soils. Salic characteristic horizon distinct this soil unit, where EC is >30 dS/m and the depth of the horizon is generally >15 cm.

It can be noticed that the north of the oases is characterized by soil type variability, where sub-great groups of *Typic Torriorthents* Fig. 4.265, *Typic Torrifluvents* and *Typic*

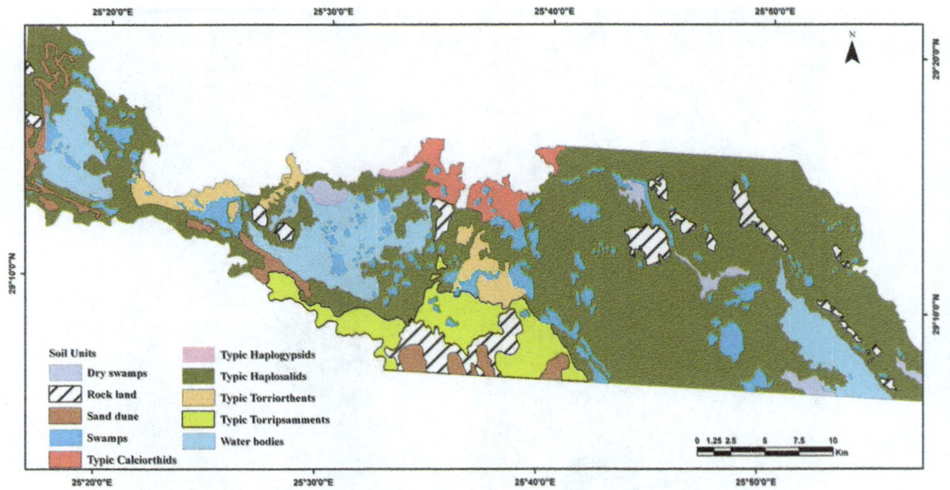

Fig. 4.258 Soil map of Siwa Oasis

Fig. 4.259 Salt marches and salt crust at Siwa Oasis

Quartizpsamments soils exist in small patches. This distribution is like the old cultivation in the oases. The Rock land spreads allover the oases covering an area of 91.3 km^2 representing 4.7% of land.

4.6.2.3 Dakhla Oases

Dakhla oasis was known in ancient times as "Zeszes", the place of two swords because it is divided into two distinct areas. It is an area of around 2000 km^2 bounded on the west by the

Fig. 4.260 *Typic Torripsamments* landscape at Siwa Oasis

Fig. 4.261 *Typic Torriorthents* at Siwa Oasis

Great Sand Sea, on the north by a high limestone escarpment and on the east by Abu-Tartur Plateau. Dakhla is the second most highly populated region in the New Valley region. There is evidence that Dakhla, like other desert regions, has been inhabited since Prehistoric times. When the region gradually became more arid, the people began to move closer towards the sources of water.

Fig. 4.262 *Typic Haplocalcids* Landscape at Siwa Oasis

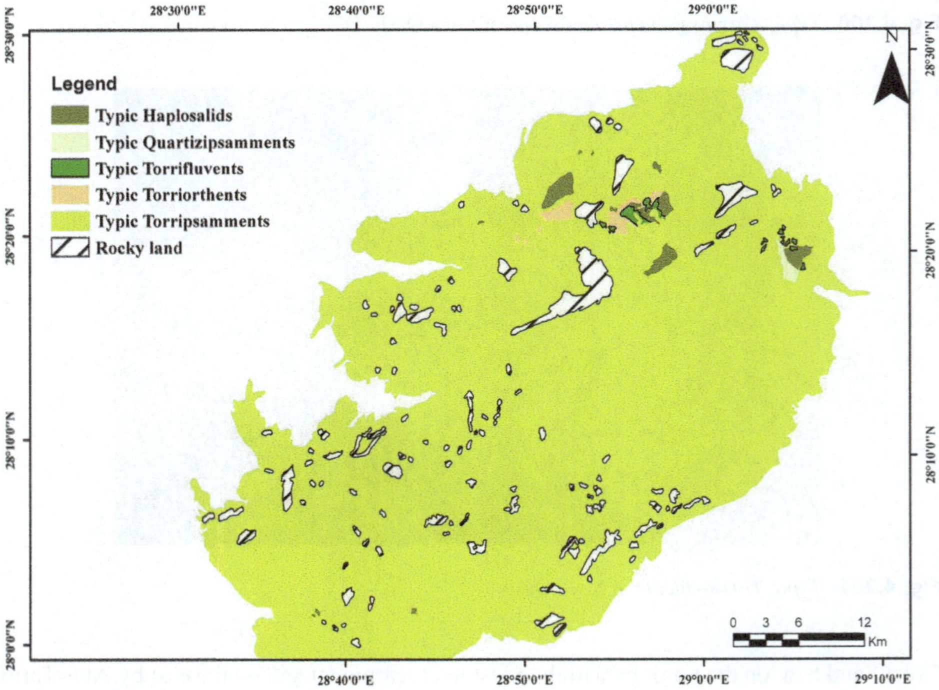

Fig. 4.263 Soil map of Baharyia oasis

Concerning the land resources database of Dakhla oases, Fig. 4.266, and Table 4.86 demonstrate the distribution of different soil units within the oasis.

Table 4.85 Areas and frequency of soil units in the Baharyia Oasis

Soil class	Area (km^2)	%
Typic Torripsamments	1803.2	93.4
Typic Haplosalids	17.8	0.9
Typic Torriorthents	9.9	0.5
Typic Torrifluvents	5.3	0.3
Typic Quartizpsamments	2.8	0.1
Rock land	91.3	4.7
Total	1930.4	100.0

Fig. 4.264 *Typic Haplosalids* and water logged areas at Bahariya Oasis

Fig. 4.265 *Typic Torriorthents* soils in Bayariya Oasis

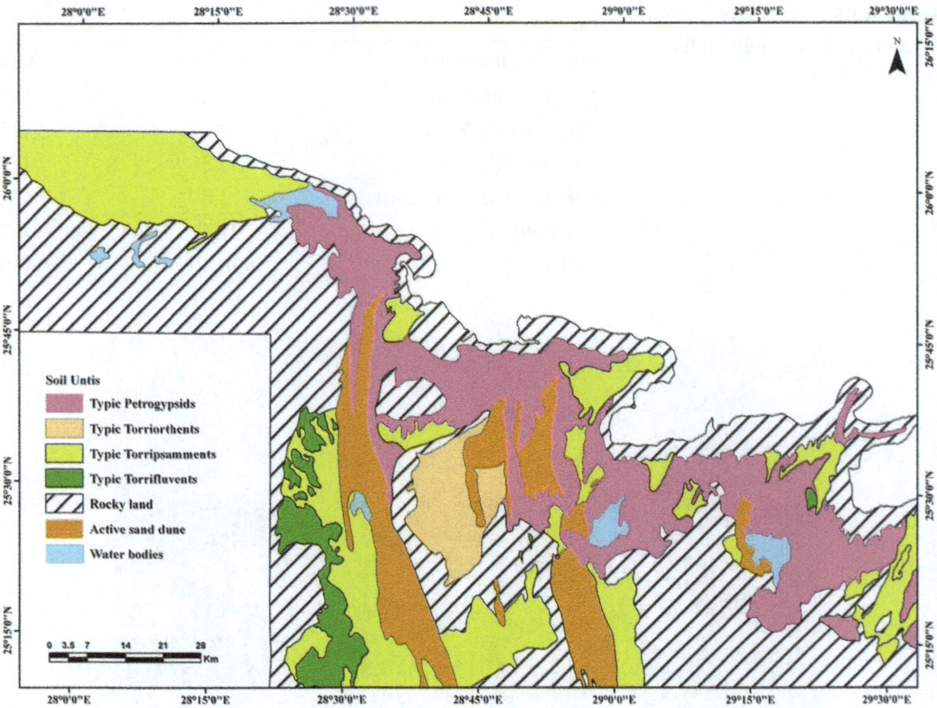

Fig. 4.266 Soil map of Dakhla Oasis

Table 4.86 Areas and frequency of soil units in the Dakhla Oasis

Soil class	Area (km^2)	%
Typic Torripsamments	1920.2	20.3
Typic Petrogypsids	1816.1	19.2
Typic Torriorthents	287.4	3.0
Vertic Torrifluvents	285.7	3.0
Rock land	4271.7	45.1
Active sand dunes	881.4	9.3
Total	9462.6	100.0

It is clear that the sub-great group *Typic Torripsamments* Fig. 4.267 dominates the oases, where it covers an area of 1920.2 km^2 representing 20.3% of its soils. The spatial distribution of this soil unit is mostly linked with the distribution of active sand dune zones, which represent 9.3% of land surface. This might explain the aeolian origin of the soils. The *Typic Petrogypsids* Fig. 4.268 sub-great group, covering an area of 1816.1 km^2 (19.2% of land surface) is restricted in the northern part of the oasis. The existence of such soil refers to a fact that most of these areas were submerged by water for long periods in the past time. Both sub-great groups *Typic Torriorthents* Fig. 4.269 and *Vertic Torrifluvents*

Fig. 4.267 *Typic Torripsamments* landscape at Dakhla Oasis

Fig. 4.268 *Typic Petrogypsids* landscape at Dakhla Oasis

Fig. 4.269 *Typic Torriorthents* soils at Dakhla Oasis

Fig. 4.270 *Vertic Torrifluvents* landscape at Dakhla Oasis

Fig. 4.270 corresponds with cultivated lands in the oases. They cover areas of 287.4 and 285.7 km^2 representing 3.0 and 3.0% respectively. These soils indicate the fluvial origins of

Table 4.87 Areas and frequency of soil units in Kharga Oasis

Soil class	Area (km^2)	%
Typic Torripsamments	4313.2	36.6
Chromic Haplotorrerts	2144.3	18.2
Typic Haplotorrerts	768.9	6.5
Typic Torrifluvents	15.3	0.1
Active sand dunes	2292.9	19.4
Rock land	2262.7	19.2
Total	11797.4	100.0

their sediments. The Rock land surround the oases, forming its skeletal covering an area of 4271.7 km^2 representing 9.3% of land surface.

4.6.2.4 Kharga Oases

Kharga oases, known to the ancient Egyptians as the 'Southern Oasis' is the largest oases of the wetern Desert and consist of a depression about 160 km long and from 20 km to 80 km wide. Today it is often referred to as the 'Great Oasis' and considered as the capital of the New Valley Governorate. The oasis is inhibited by sixty thousand inhabitants, including one thousand Nubians who were settled there after the building of the High Dam in Asswan. In ancient times a lake occupied a large part of the depression and the thick deposits of sandy clay laied down and formed the cultivated land. Although the oasis gets little rain, water is obtained from wells dug in the porous sandstone which underlies much of the desert area and extensive land reclamation work has been undertaken since the beginning of the twentieth century. Throughout its history Kharga seems to have been the place to where undesirable inhabitants of the Nile Valley were punished—the fierce summer heat, devastating winds and remote location makes it an ideal place of exile.

Concerning the established digital land resources data base, Table 4.87 and Fig. 4.271 show the distribution of the soil sub-great groups. *Typic Torripsamments* sub great group dominates the soils of the oasis Fig. 4.272, showing a spatial relation with sand dune distribution, covers an area of 4313.2 km^2 representing 36.6% of land. The *Chromic Haplotorrerts* Fig. 4.273 soils are distinctive for the Kharga oasis, covering an area of 2144.3 km^2, representing 18.2% of its soils. These soils are characterized by a high amount of iron oxides, Reddish Brown to dark reddish brown colors, clay to silty clay texture and wide (5–12 cm) & deep (30–60 cm) surface cracks. The sub-great group *Typic Haplotorrerts* Fig. 4.274 is found associated with the *Chromic Haplotorrerts* having the same characteristics except for the high amount of iron oxides; it is dominated by a brown to dark. These soils cover an area of 768.9 km^2 representing 6.5% of the soils. A small patch of the sub-great group *Typic Torrifluvents* Fig. 4.275 is found in the north of the oasis depression covering an area of 15.3 km^2 representing 0.1% of land surface. The evidences of wide and deep surface cracks in these soils are absent due to its inclusion of relatively high percentage of sand fraction. One of the distinct features of the Kharga Oases is the dense existence of active sand dunes, which exhibit an area of 2292.9 km^2 representing

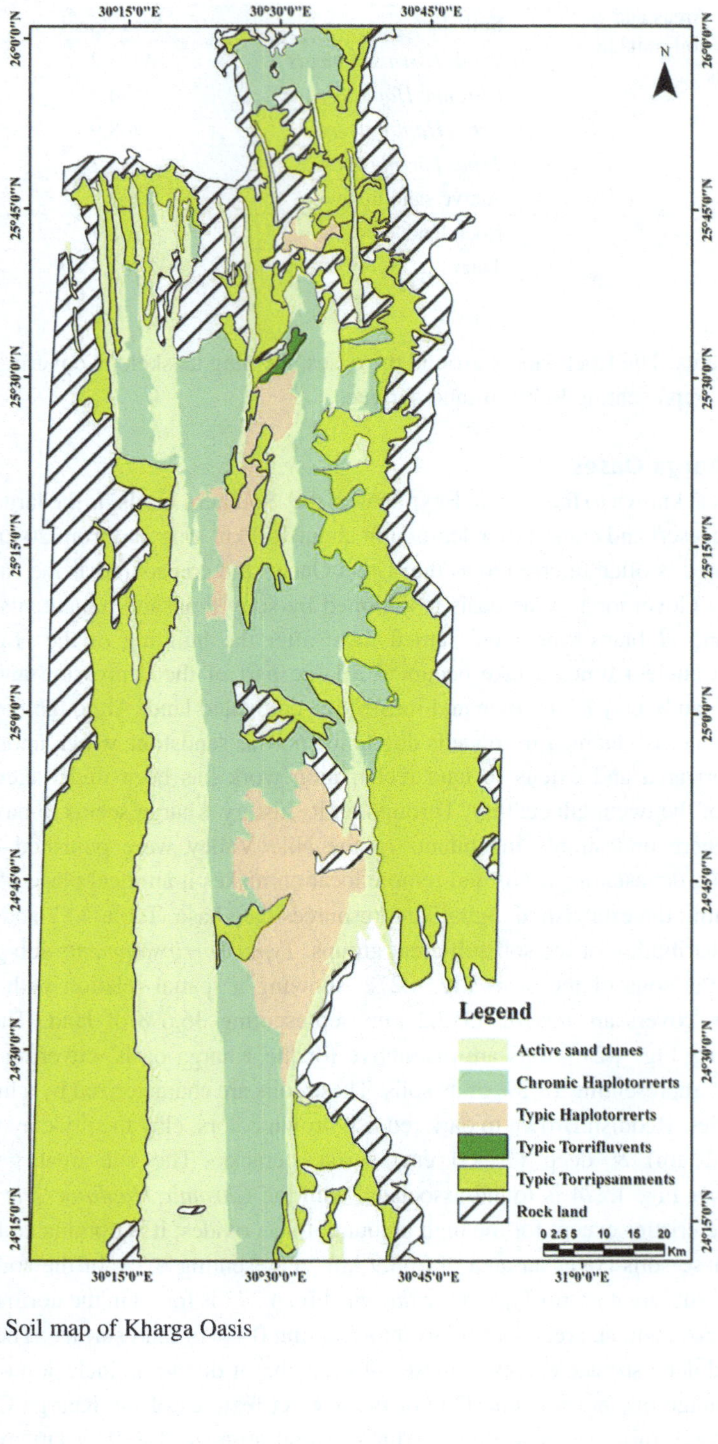

Fig. 4.271 Soil map of Kharga Oasis

Fig. 4.272 *Typic Torripsamments* landscape at Kharga Oasis

Fig. 4.273 *Chromic Haplotorrerts* soils at Kharga Oasis

Fig. 4.274 *Vertic Torripfluvents* soils at Kharga Oasis

Fig. 4.275 *Typic Torripfluvents* soils at Kharga Oasis

19.4% of land. The dune pattern is elongated strips invading the oases from the north. The Rock Land surround the oasis from the north, west and east covering an area of 2262.7 km^2 representing 19.2% of land surface.

4.7 Land Capability Classification of Northwestern Costal Region and Desert Areas

4.7.1 Northwestern Coastal Zone

Twelve soil profiles, representing the different soil units of the Northwestern Coastal zone, (profiles no. 73 to 84), Fig. 4.276 were selected to evaluate the land capability of the region. The morphological description and the laboratory analyses representing these profiles are included in Annex II, and III. These data were used to assess the land capability for each sub-great group. The assessment of land quality was based on several factors (i.e. drainage condition, texture/structure, % coarse fragments, soil depth, % $CaCO_3$, CEC, soil salinity (EC) and exchangeable sodium percentage (ESP) as shown in Table 4.88.

The results indicate that the soil salinity, soil texture, soil depth, $CaCO_3$ content, CEC and coarse fragments are the main limiting factors in the soil of the Northwestern region. Figure 4.277 shows the land capability classification map and the main land qualities of the studied area, the different classes and their areas are shown in Table 4.88. It is clear that the high capable soil (class II) represents about 12.9% of the total area, it is associated with the *Typic Torriorthents* sub-great group. It dominates the eastern and western borders of the region in addition to a narrow strip along the coast. The moderately capable soils (class III) represent 27.1% of the total area; associated with the soils of *Typic Torripsamments*. It is found in different places between the coastal plain and the plateau. Low capability classes represent 1.6% geographically distributed in patches along the coast and mostly associated

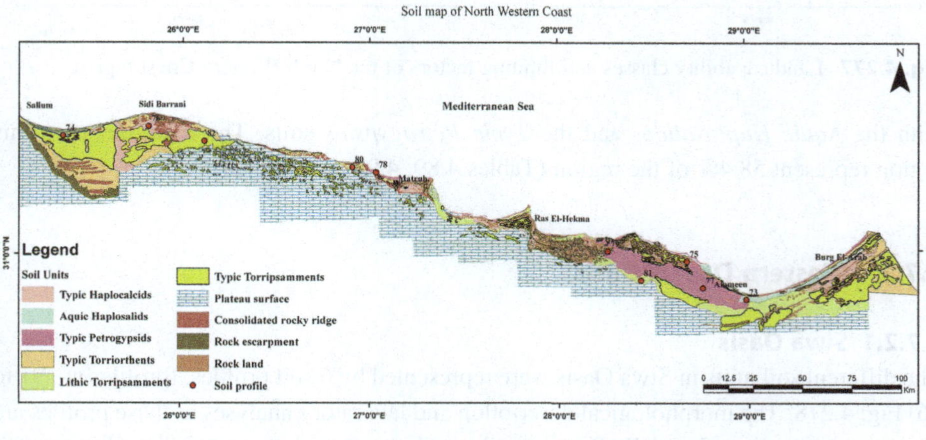

Fig. 4.276 Geographical distribution of the studied soil profiles of North Western Coast

Table 4.88 Land capability classes and limiting factors of the Nile-Valley representative soil profile

	Representative soil profiles										
Factor	46	48	52	65	71	47	72	62	73	53	
Soil depth	D	2	3	2	1	1	1	2	2	3	2
Texture/Structure	T	3	1	2	2	1	1	1	1	1	3
% Coarse fragments	C	1	1	2	2	1	2	1	1	1	1
Drainage condition	L	1	3	2	1	1	1	2	2	3	1
% $CaCO_3$	K	1	2	3	3	1	2	1	1	1	2
CEC (meq/100 g. soil)	F	3	4	3	1	1	3	1	1	1	4
ECe (1:1) (dS/m)	E	1	2	3	3	1	2	4	1	4	1
ESP	S	2	2	1	2	1	2	3	2	2	2
Capability class		III	IV	III	III	I	III	IV	II	IV	IV
Limiting factors		T, F	D, L, F	K,F, E	K, E	–	F, E	E, S	–	C, E	T, F

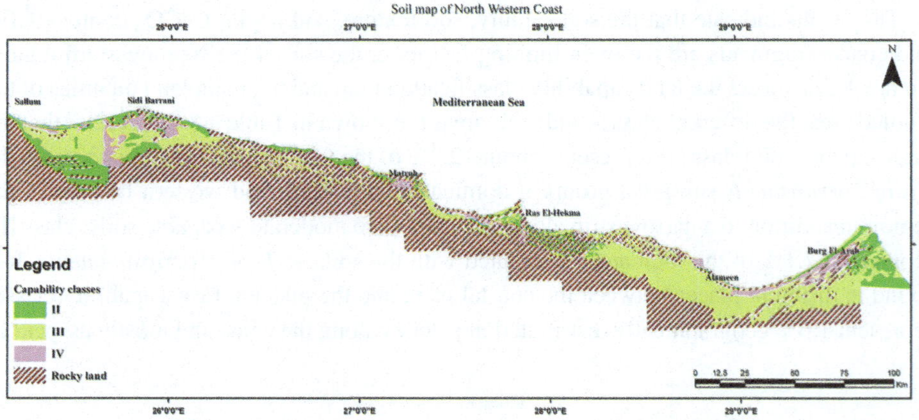

Fig. 4.277 Land capability classes and limiting factors of the North Western Coast region

with the *Aquic Haplosalides* and the *Typic Petrogypsids* soils. The rocky areas in this region represent 58.4% of the region (Tables 4.89, 4.90, 4.91, 4.92).

4.7.2 Western Desert Oasis

4.7.2.1 Siwa Oasis

The different soil units in Siwa Oasis were represented by 6 soil profiles, (profiles no. 91 to 96) Fig. 4.278. The morphological description and laboratory analyses of these profiles are given in Appendixes I and II. Based on the soil parameters the capability classes were

Table 4.89 Areas of Nile Valley land capability classes

Capability class	Area km^2	Area %
I	3058.3	10.4
II	1510.9	5.1
III	16432.6	55.8
IV	3058.3	1.7
Other features		
R	7506.3	25.5
U	190.3	0.7
W	239.3	0.8
Total	29451.7	100.0

worked out, the results indicate that soil salinity, soil depth, texture, CEC and drainage condition are the main limiting factors in Siwa Oasis. Figure 4.279 shows the land capability classification and the main land qualities of the Oasis, different classes and their areas are shown in Table 4.93. The data reveals that there are no high capable soils (class II) in the Oases due to the high severity of limiting factors specially salinity and drainage condition. The moderately capable soils (class III) represent 11.9% of the total area; it is associated with the soils of *Typic Torripsamments*. The low capable soils dominate the Oasis as they represent 75.5% of the total area. The low capable soils are associated with the soils located under the order *Aridisols*. The non capable soils represent 12.6% of the Oases. These soils are mainly affected by the sand dunes or the intrusion of saline water of Siwa lakes.

4.7.2.2 Bahariya Oasis

The different soil units in Bahariya Oasis were represented by 6 soil profiles, (profiles no. 85 to 90), Fig. 4.280 the morphological description and the laboratory analyses of these profiles are given in Annexes II and III.

The results indicate that soil salinity, soil depth, texture, CEC and coarse fragments are the main limiting factors in the Oasis. Figure 4.281 shows the land capability classification of Bahariya Oasis, Table 4.94. The high capable soil (class II) represents small areas (i.e. 0.8%) of the Oasis. It is associated with the *Typic Torrifluvents* sub-great group, and found mainly in the northern parts of the Oasis. The moderately capable soils (class III) dominate the Oasis as it represents 93.6% of the total area, associated with the soils of *Typic Torripsamments*. The rest of the Oases is considered as non capable soils or rock land, these units represent 5.7% of the total area.

4.7.2.3 Kharga Oasis

The different soil units in Kharga Oasis were represented by 5 soil profiles, (profiles no. 97 to 101) Fig. 4.282. The morphological description and laboratory analyses of these profiles are given in Appendixes I and II. Figure 4.283 shows the land capability classification and the main land qualities of the Kharga Oasis, the different classes and their

Table 4.90 Land capability classes and limiting factors of the Northwestern Coast soil profiles

Factor		Representative soil profiles											
		73	74	75	76	77	78	79	80	81	82	83	84
Soil depth	D	3	3	1	1	3	2	1	1	1	1	4	4
Texture/Structure	T	1	1	1	1	1	3	3	3	3	3	3	3
% Coarse fragments	C	1	1	1	1	1	3	3	3	2	2	2	2
Drainage condition	L	2	2	1	1	2	1	1	2	2	1	3	1
% $CaCO_3$	K	2	2	2	2	2	3	2	3	1	1	2	1
CEC meq/100 g. soil	F	1	1	1	1	1	1	3	3	3	3	3	3
ECe $_{(1:1)}$ dS/m	E	4	4	2	2	3	1	1	3	2	2	2	1
ESP	S	3	3	1	3	1	1	1	1	1	1	1	1
Capability class		IV	IV	II	II	II	III	III	III	III	III	IV	IV
Limiting factors	–	D	E, D, L	E	–	–	C, E	C, K, F, S	E, C, S	C, E	C, E	F, T, C	T, F, E

Table 4.91 Areas of Northwestern Oasis land capability classes

Capability	Area (km^2)	Area %
II	1817.7	12.9
III	3824.6	27.1
IV	229.4	1.6
Rock	8246.0	58.4
Total	14117.7	100.0

Table 4.92 Areas of western desert Oasis land capability classes

oases	Capability classes	Area km^2	Area %
Bahariya Oasis	II	15.2300	0.79
	III	1805.9900	93.55
	N	17.8100	0.92
	Rock	91.4300	4.74
	Total	1930.4600	100.00
Siwa Oasis	III	35.9700	11.89
	IV	228.3500	75.49
	N	38.1600	12.62
	Total	302.4800	100.00
Kharga Oasis	II	2928.5000	24.45
	III	182.2800	1.52
	IV	4313.2200	36.00
	N	4555.6900	38.03
	Total	11979.6900	100.00
Dakhla Oasis	II	1816.1200	19.19
	III	573.0900	6.06
	IV	1920.2400	20.29
	N	5153.1200	54.46
	Total	9462.5700	100.00

areas are shown in Table 4.95. The obtained data indicate that the high capable soils (class II) represent 24.5% of the Oasis; it is associated with the *Typic Haplotorrerts* and *Typic Torrifluvents* sub-great groups, and found mainly in the middle parts of the Oasis. The moderately capable soils (class III) represent 1.5% of the total area; it is associated with the soils of *Typic Torriorthents*, and exhibit the southwestern corner of the depression. The low capable soils represent 36.0% of the total area; this class is associated with the soils of *Torripsamments* great group and found mainly closed to the depression margins. The rest of the Oasis is considered as non capable soils or rock land, representing 38.0% of the total area.

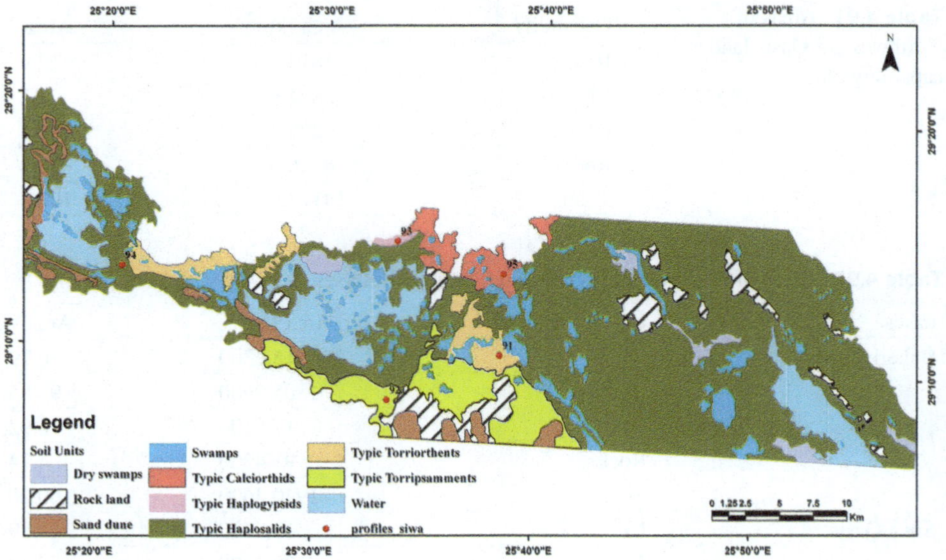

Fig. 4.278 Geographical distribution of the studied soil profiles of Siwa oasis

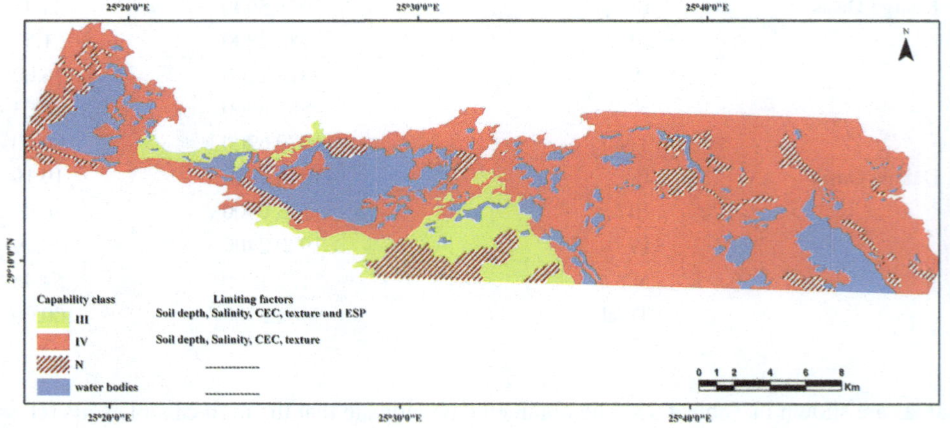

Fig. 4.279 Land capability classes and limiting factors of Siwa Oasis

Table 4.93 Areas of Siwa Oasis land capability classes

Capability class	Area (km^2)	Area %
III	35.9	11.9
IV	228.4	75.5
N	38.1	12.6
Total	302.5	100.0

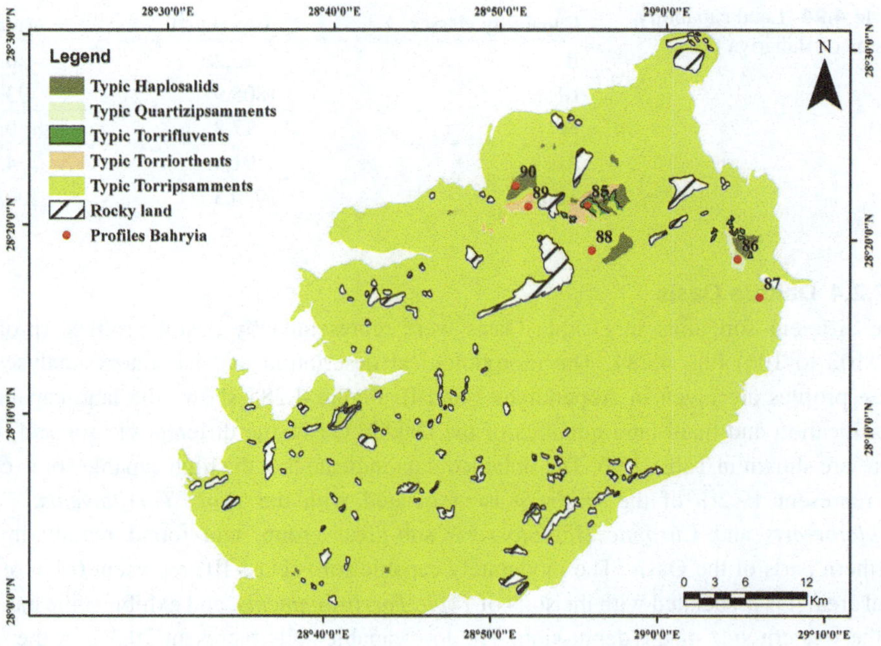

Fig. 4.280 Geographical distribution of the studied soil profiles of Baharyia Oasis

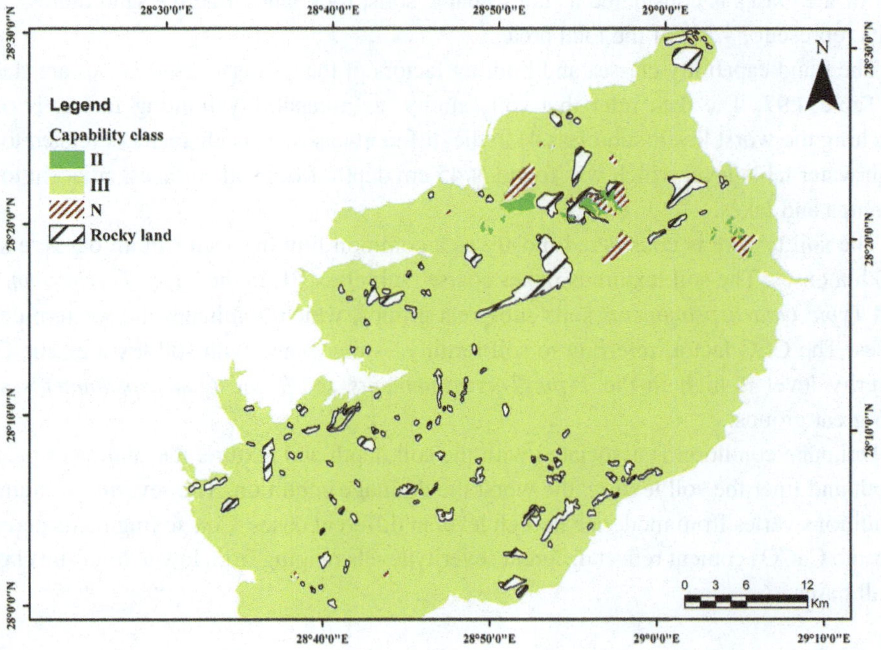

Fig. 4.281 Land capability classes and limiting factors of Bahariya Oasis

Table 4.94 Land capability classes of Bahariya Oasis

Capability classes	Area (km^2)	Area %
II	15.2	0.8
III	1805.9	93.6
N	17.8	0.9
Rock	91.4	4.7
Total	1930.5	100.0

4.7.2.4 Dakhla Oasis

The different soil units in Dakhla Oasis were represented by 5 soil profiles, (profiles no. 102 to 106) Fig. 4.284. The morphological description and laboratory analyses of these profiles are given in Appendixes I and II. Figure 4.285 shows the land capability classification and main land qualities of the Dakhla Oasis, the different classes and their areas are shown in Table 4.96. The obtained data indicate that the high capable soils (class II) represent 19.2% of the Oasis; it is associated with the *Typic Torrifluvents*, *Typic Haplotorrerts* and *Chromic Haplotorrerts* sub-great group, and found mainly in the northern parts of the Oasis. The moderately capable soils (class III) represent 6.1% of the total area; it is associated with the soils of *Typic Torripsamments*, and exhibit some patches in the western side of the depression. The low capable soils represent 20.3% of the total area; this class is associated with the soils of *Lithic Torripsamments* sub-great group and found mainly closed to the depression margins or adjacent to the active sand dunes. The rest of the Oases is considered as non capable soils, rock land or active sand dunes, these units represent 54.5% of the total area.

The Land capability classes and limiting factors of the western desert Oases are shown in Table 4.97. The data refer that soil salinity, as a capability limiting factor, is often reaching the worst level (sub-class 4) in the different oases. This character is related to the high water table level which was found at 45 cm depth, from soil surfaces, adjacent to the swamps and lakes.

The soil texture is considered mostly as a common limiting factor in all oases, except Dakhla oases. The soil texture is rather coarse (sub-class III) in the *Typic Torripsamments* and *Typic Quartizpsamments* soils sub-great groups, which dominates the western desert oases. The CEC factor, referring to soil fertility, is associated with soil texture. The CEC severity level is high in the *Typic Torripsamments* and *Typic Quartizpsamments* soils sub-great groups.

Drainage condition is associated with the soil depth and texture, the shallower the soil depth and finer the soil texture, the worst the drainage condition. The severity of drainage conditions varies from moderate to high level in different oases. Coarse fragments percentage and $CaCO_3$ content reflect different severity levels ranging from low to high sub classes in all oases.

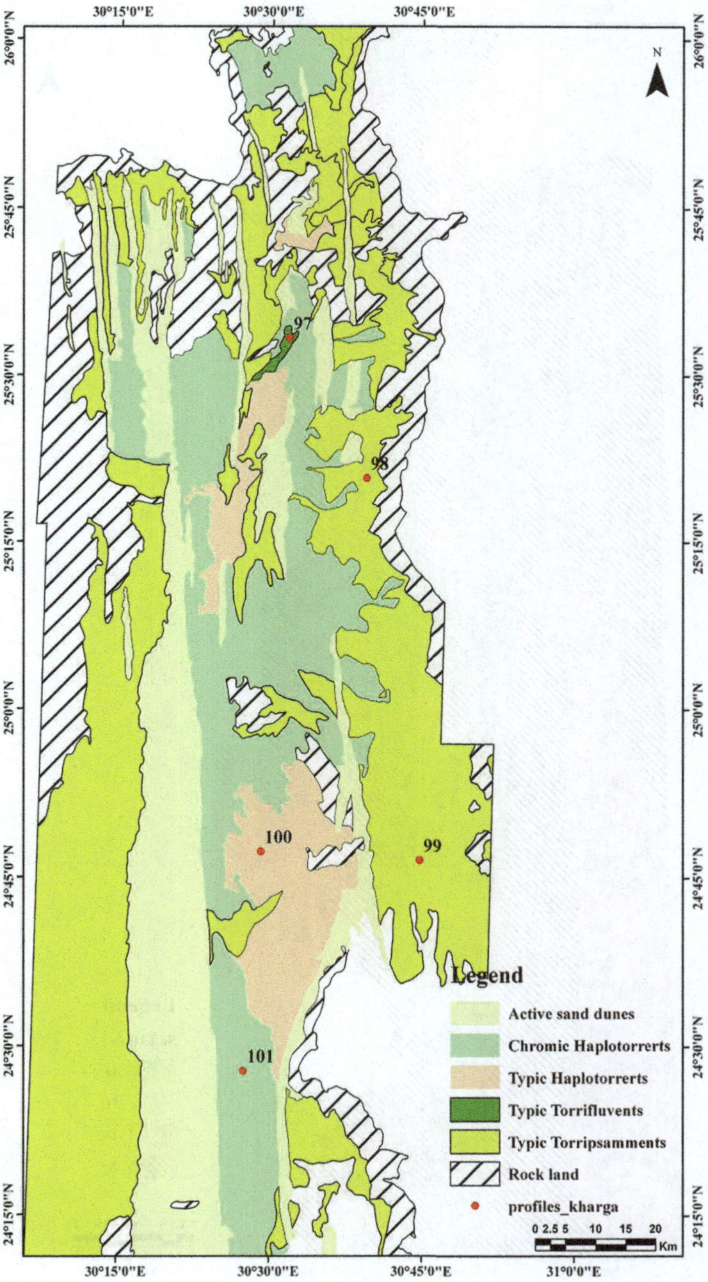

Fig. 4.282 Geographical distribution of the studied soil profiles of Kharga Oasis

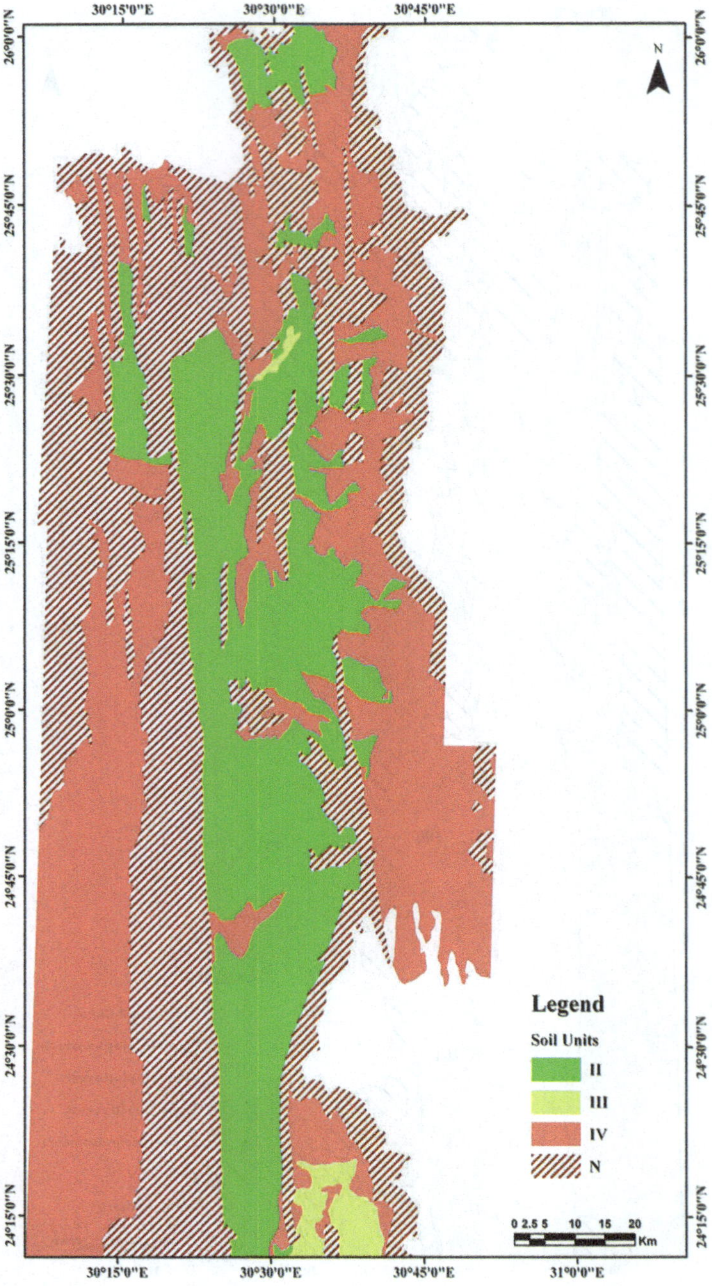

Fig. 4.283 Land capability classes and limiting factors of Kharga Oasis

Table 4.95 Areas of Kharga Oasis land capability classes

Capability class	Area (km^2)	Area %
II	2928.5	24.5
III	182.3	1.5
IV	4313.2	36.0
N	4555.7	38.0
Total	11979.7	100.0

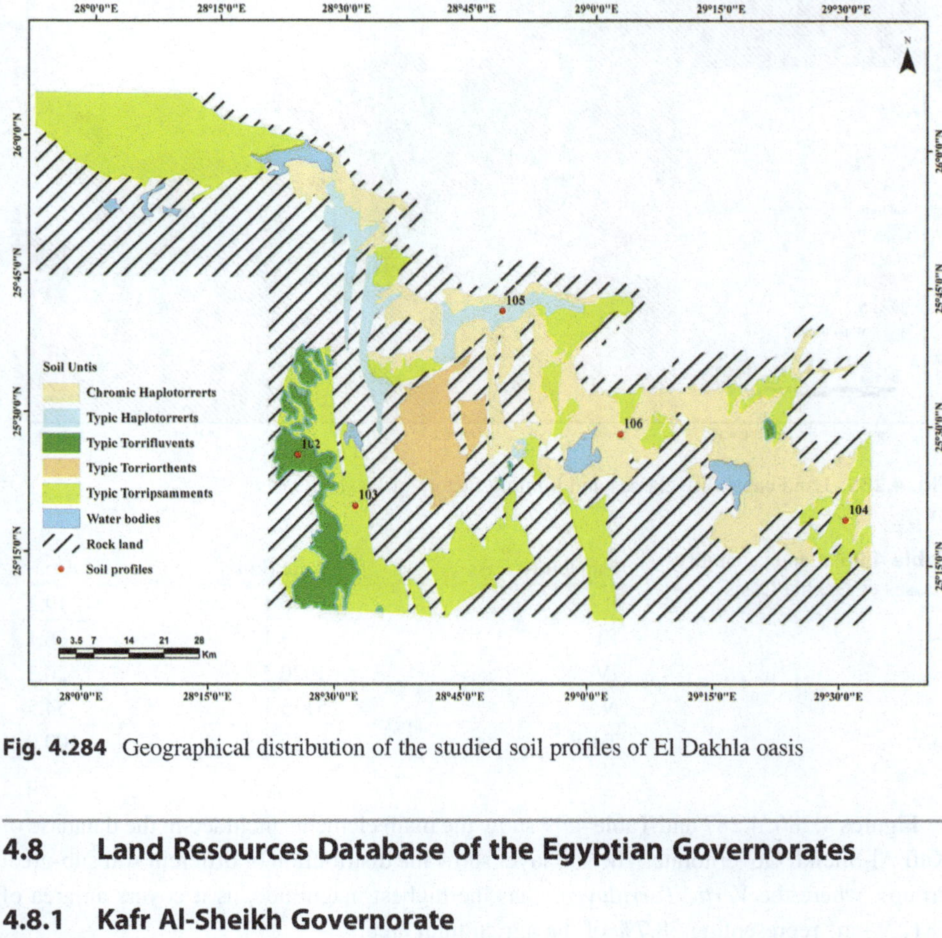

Fig. 4.284 Geographical distribution of the studied soil profiles of El Dakhla oasis

4.8 Land Resources Database of the Egyptian Governorates

4.8.1 Kafr Al-Sheikh Governorate

The Kafr Al-Sheikh Governorate is administratively sub-divided into 10 administrative districts, locally called Markaz (Table 4.98). Sidi Salim district is the largest area as it covers 525.1 km^2, representing 15.9% of the total area. Al-Burllus and Qillin districts are covering the smallest districts, as they occupy areas of 156.4 and 161.5 km^2 representing 4.7 and 4.9% respectively.

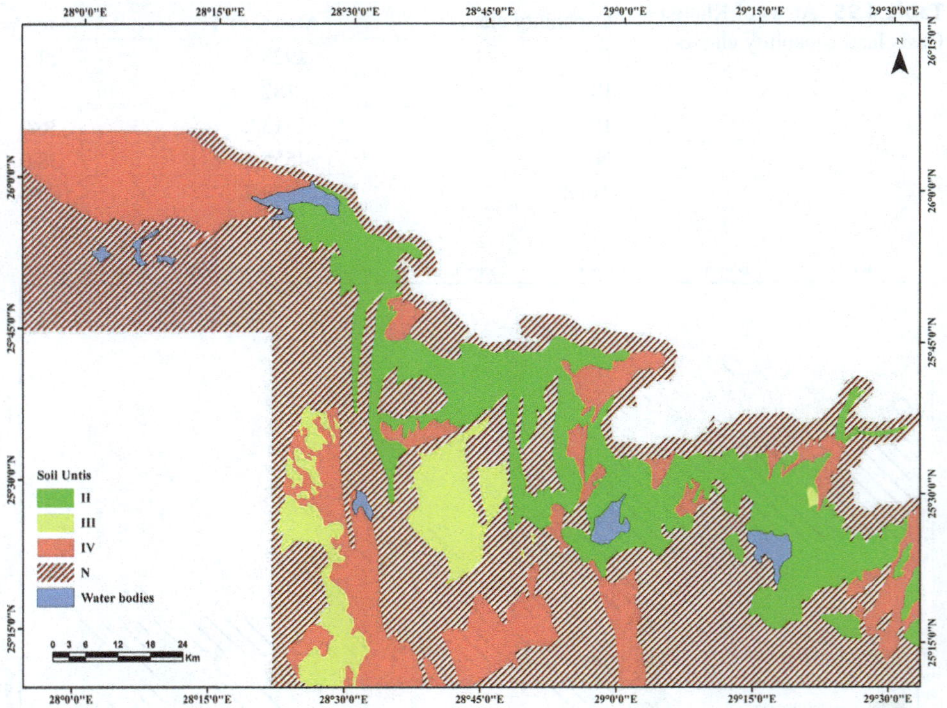

Fig. 4.285 Land capability classes and limiting factors of Dakhla Oasis

Table 4.96 Land capability classes of Dakhla Oasis

Capability class	Area (km^2)	Area %
II	1816.1	19.2
III	573.1	6.1
IV	1920.2	20.3
N	5153.1	54.5
Total	9462.6	100.0

Figures 4.286, 4.287 and Table 4.99 show the main elements included in the database of Kafr Al-Sheikh Governorate. The soil layer show the distribution of different soil sub-great groups, where the *Vertic Torrifluvents* has the highest magnitude, as it covers an area of 1812.9 km^2 representing 58.7% of the agricultural area.

These soils are located to the south of the *Aquollic Salorthids*, which surrounds the Lake Brullus area. The later sub great group covers an area of 810.2 km^2, representing 26.2% of cultivated land of the governorate.

Other sub-great groups have lower frequencies compared with the previous two, as the *Typic Haplosalids* exist in patches around the lake covering an area of 244.4 km^2, representing 7.9% of agricultural land. Also, the *Typic Quartizpsamments* exist in the

Table 4.97 Land capability classes and limiting factors of the Western Desert Oases representative profiles

		Bahariya Oasis						Siwa Oasis						Kharga Oasis					Dakhla Oasis				
Factor		85	86	87	88	89	90	91	92	93	94	95	96	97	98	99	100	101	102	103	104	105	106
Soil depth	D	1	1	1	3	1	1	1	1	3	2	1	2	1	3	1	1	1	1	3	3	1	1
Texture/ Structure	T	3	1	3	3	2	3	3	3	1	2	3	1	1	3	3	1	1	1	1	1	1	1
% Coarse fragments	C	1	1	1	1	1	3	3	3	1	2	2	2	2	2	2	2	1	1	1	1	1	1
Drainage condition	L	1	1	1	3	1	1	1	1	2	2	1	2	1	3	1	1	1	1	3	3	1	1
% $CaCO_3$	K	1	1	1	1	1	2	3	1	1	1	3	2	1	1	1	1	1	1	1	1	2	1
CEC meq/100 g. soil	F	1	3	2	2	2	2	2	3	1	2	3	2	1	2	3	1	1	1	1	1	1	2
ECe $_{(1:1)}$ dS/m	E	2	2	3	2	2	4	2	1	2	4	2	4	3	1	2	2	3	2	2	2	4	1
ESP	S	2	2	1	2	1	2	1	2	2	1	2	1	2	2	3	1	1	1	3	2	3	2
Capability class		III	III	III	III	II	IV	III	III	III	IV	III	IV	III	III	III	II	III	II	III	III	IV	II
Limiting factors	–	T	F	T, E	D, T, L	--	T, C, E	T, C, K	T, C, F	D	E	T, K, F	E	E	D, T, L	T, F	--	E	--	D, L,S	D, L	E,S	--

Table 4.98 Areas and frequencies of different administrative districts within Kafr Al-Sheikh Governorate

Name	Area (km^2)	%
Al-Burllus	156.4	4.7
Al-Hamul	657.6	19.9
Ar-Riyad	298.6	9.1
Biyala	324.9	9.8
Disuq	319.5	9.7
Fowa	127.7	3.9
Kafr el Sheikh	430.7	13.1
Metobas	297.0	9.0
Qillin	161.5	4.9
Sidi Salim	525.1	15.9
Total	3298.5	100.0

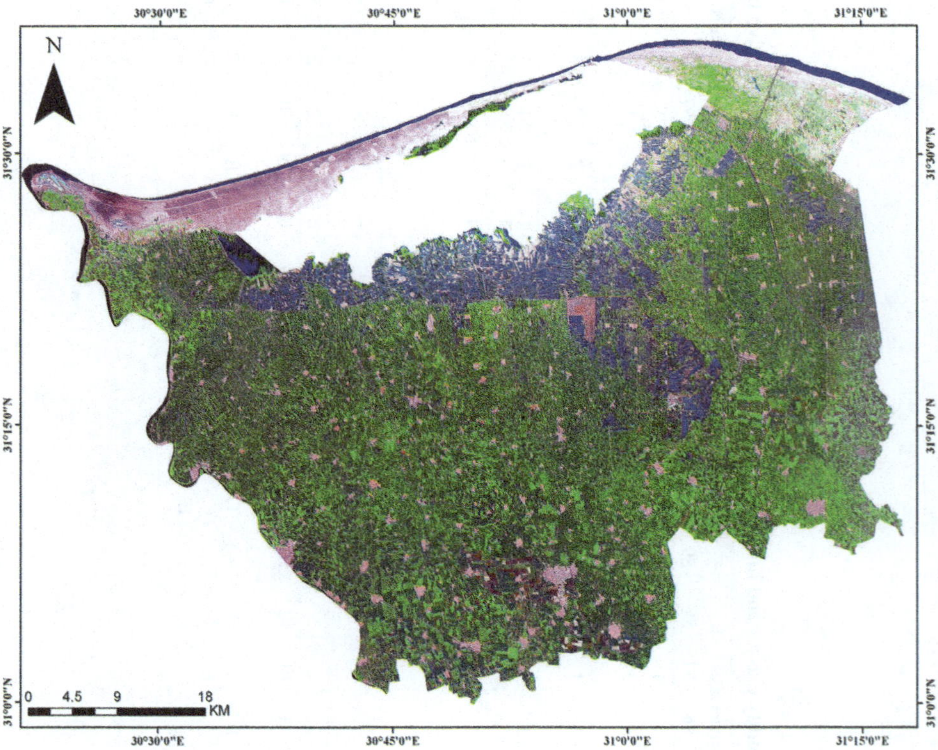

Fig. 4.286 Landsat ETM+ 2002 of Kafr Al-Sheikh Governorate

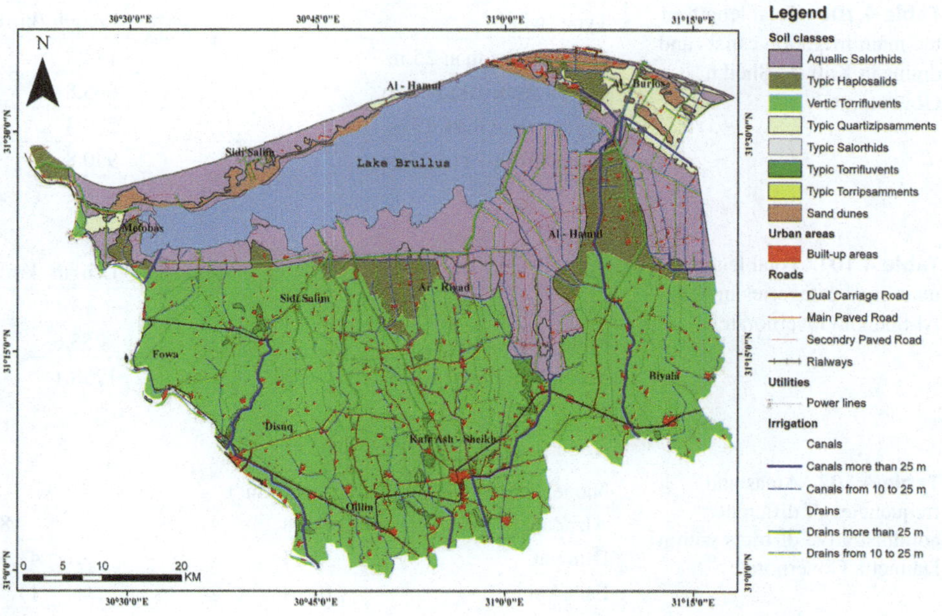

Fig. 4.287 Main elements included in the database of Kafr Al-Sheikh Governorate

Table 4.99 Areas and frequencies of soil sub-great groups in Kafr Al-Sheikh Governorate

Type (Sub great group)	Area (km^2)	%
Aquallic Salorthids	810.2	26.2
Typic Haplosalids	244.4	7.9
Typic Quartzipsamments	94.1	3.0
Typic Torrifluvents	40.4	1.3
Typic Torripsamments	0.1	0.0
Vertic Torrifluvents	1812.9	58.0
Sand dunes	85.7	2.8
Total	3087.9	100.0

northeastern corner of the Governorate and seem mostly to be originated from the sand dunes running parallel to the coastal line. Very minor frequencies of the *Typic Quartizpsamments, Typic Salorthids, Typic Torrifluvents* and *Typic Torripsamments* area are found in the Governorate. Sand dunes are frequent in narrow strips along the northern borders of the Governorate covering an area of 85.7 km^2, representing 2.8% of the agricultural land.

Two main irrigation canals and drain categories are shown in the database layout of the Governorate, lengths of each category are presented in Table 4.100. The irrigation canals of more than 25 m width have a total length of 174.8 km, while those of 10–25 m width attain a total length of 665.8 km. The drainage canals of more than 25 m width have a total length of 269.3 km, while those having 10 to 25 m width attains 940.0 km length.

Table 4.100 Total length of the main irrigation canals and drains in Kafr Al-Sheikh Governorate

Type (width)	Length (km)
Canals more than 25 m	174.8
Canals from 10 to 25 m	665.8
Drains more than 25 m	269.3
Drains from 10 to 25 m	940.8

Table 4.101 Total lengths of main road categories in Kafr Al-Sheikh Governorate

Type	Length (km)
Dual Carriage Road	13.1
Main Paved Road	253.6
Secondary Paved Road	1258.1

Table 4.102 Areas and frequencies of different administrative districts within Damietta Governorate

Name	Area (km^2)	%
Alz-Zarqa	70.8	7.8
Dumyat	376.4	41.3
Faraskour	106.9	11.7
Kafr Saad	357.6	39.2
Total	911.8	100.0

Table 4.101 reveals the length of different road categories within the Governorate. Three main roads are illustrated in the land resources map of the govenorate identified as dual carriage, main paved and secondary paved attaining lengths of 13.1, 253.6 and 1258.1 km respectively.

The land resources database moreover includes the layers of urban areas and utilities. The urban area reaches 89.8 km^2 and the total length of utilities is 765 km.

4.8.2 Damietta Governorate

The Damietta Governorate is one of the smallest Governorates in the Nile Delta. It is administratively sub-divided into 4 administrative districts, locally called Markaz Table 4.102. Damietta district is the largest area as it covers 70.8 km^2, representing 41.3% of the total area. Kafr Saad district covers an area of 357.6 km^2, representing 39.2% of Governorate area, whereas Faraskour covers an area of 106.9 km^2. El-Zarqa is the smallest district which covers only 70.8 km^2, representing 7.8% of the Governorate.

Figures 4.288, 4.289 and Table 4.103 show the main elements included in the database of Damietta Governorate. The soil map shows the distribution of the different soil sub-great groups, where the *Typic Haplocalsids* has the largest area, as it covers an area of 283.2 km^2 representing 32.4% of the agricultural area. These soils are located at the northeastern corner of the Governorate, to the south and west of the *Typic Aquisalids* sub-great group

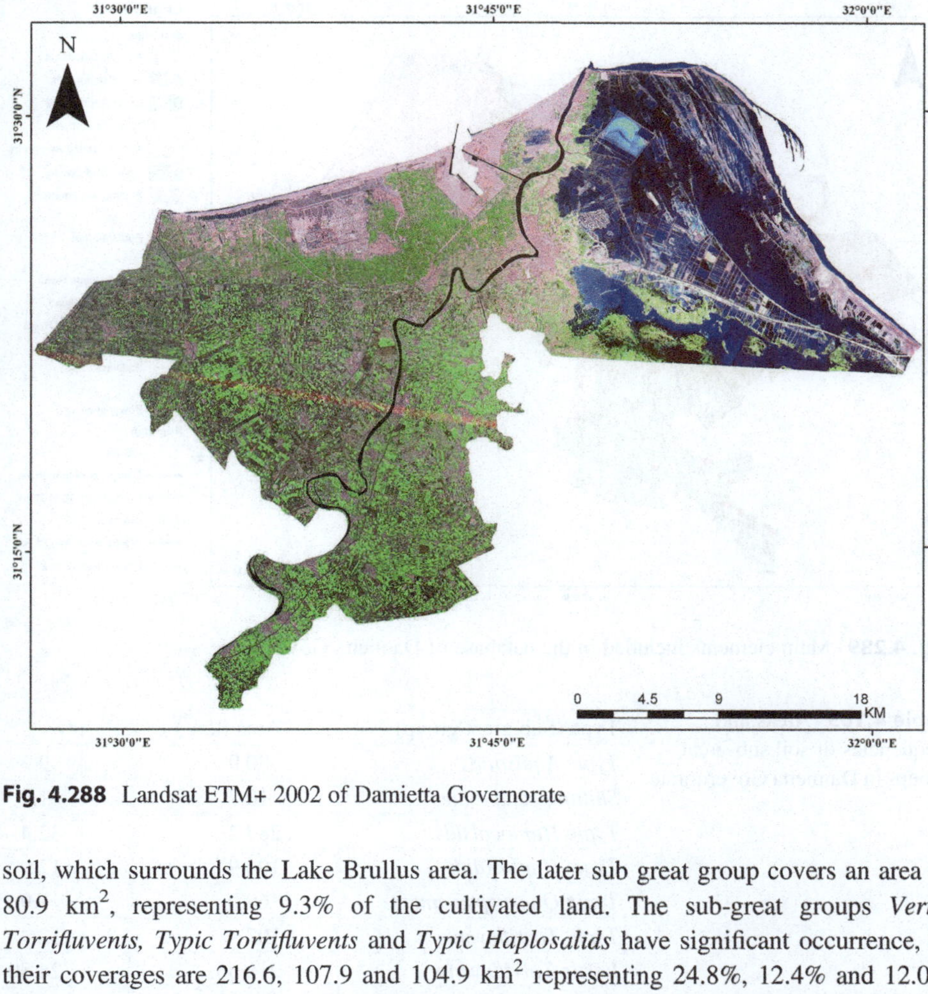

Fig. 4.288 Landsat ETM+ 2002 of Damietta Governorate

soil, which surrounds the Lake Brullus area. The later sub great group covers an area of 80.9 km^2, representing 9.3% of the cultivated land. The sub-great groups *Vertic Torrifluvents, Typic Torrifluvents* and *Typic Haplosalids* have significant occurrence, as their coverages are 216.6, 107.9 and 104.9 km^2 representing 24.8%, 12.4% and 12.0% respectively. Also, the *Typic Quartizpsamments* exist in the northwestern corner of the Governorate covering an area of 69.0 km^2, representing 7.9% of agricultural land. Patches of shifting sand dunes are immersed within the sandy soils, covering an area of 10.7 km^2.

Two main irrigation and drain canals are shown in the database layout of the Governorate, lengths of each element are presented in Table 4.104. The irrigation canals of more than 25 m width have a total length of 15.5 km, while those of 10–25 m width attain a total length of 207.9 km. The drainage canals of more than 25 m width have a total length of 32.8 km, while those having 10 to 25 m width attains 131.2 km length.

The road network is included in the land resources database Table 4.105, where the main roads are illustrated as dual carriage, main paved and secondary paved. They attain lengths of 93.8, 56.2 and 730.2 km respectively.

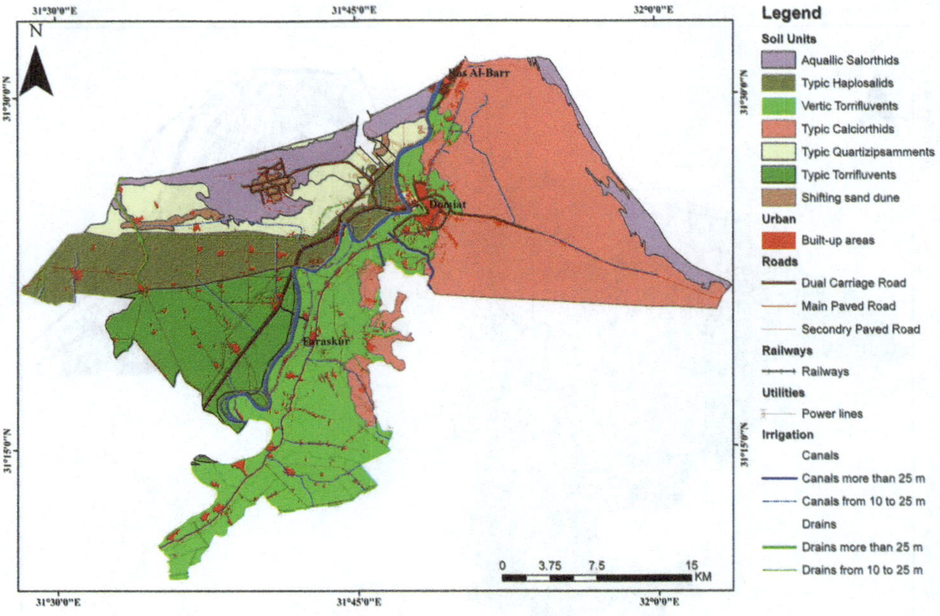

Fig. 4.289 Main elements included in the database of Damietta Governorate

Table 4.103 Areas and frequencies of soil sub-great groups in Damietta Governorate

Type (Sub great group)	Area (km^2)	%
Typic Aquisalids	80.9	9.3
Shifting sand dune	10.7	1.2
Typic Haplocalsids	283.2	32.4
Typic Haplosalids	104.9	12.0
Typic Quartizpsamments	69.0	7.9
Typic Torrifluvents	107.9	12.4
Vertic Torrifluvents	216.56	24.80
Total	873.3	100.0

Table 4.104 Total length of the main irrigation canals and drains in Damietta Governorate

Type (width)	Length (km)
Canals more than 25 m	15.5
Canals from 10 to 25 m	207.9
Drains more than 25 m	32.8
Drains from 10 to 25 m	131.2

Table 4.105 Total lengths of main roads in Damietta Governorate

Type	Length (km)
Dual Carriage Road	93.8
Main Paved Road	56.2
Secondary Paved Road	730.2

Table 4.106 Areas of different administrative districts within Al-Dakahlia Governorate

Name	Area (km^2)	%
Belqas	761.7	19.6
Al-Manzalah	763.1	19.6
Shirbin	252.4	6.5
Menyet Al-Nasr	186.1	4.8
Talkha	307.4	7.
Al-Mansourah	350.3	8.9
Dekernes	360.7	9.3
Aga	237.9	6.1
Al-Sinbelaween	426.9	10.9
Mit Ghamr	250.3	6.4
Total	3896.7	100.0

The database includes also the layers of urban areas, railways and utilities, where the total urban area reach 34.2 km^2, while the lengths of railways and utilities are 45.5 km and 181.3 km respectively.

4.8.3 Al-Dakahlia Governorate

The Al-Dakahlia Governorate is one of the North Delta Governorates. It is administratively sub-divided into 10 administrative districts or Markaz Table 4.106.

Al-Manzalah and Belqas districts are the largest ones as they cover 763.1 km^2 and 761.7 km^2, representing 19.6 and 19.6% of the total Governorate area. Al-Sinbelaween district covers an area of 426.9 km^2, representing 10.9% of the Governorate area, whereas Dekernes covers an area of 360.7 km^2 representing 9.3%. The districts of Al-Mansourah, Talkha, Aga, Mit Ghamr, Shirbin and Menyet AL-Nasr cover 350.3, 307.4, 237.9, 250.3, 252.4, 186.01 km^2, representing 8.9%, 7.9%, 6.1%, 6.4%, 6.5% and 4.8% of the Governorate respectively.

Figures 4.290, 4.291 and Table 4.107 show the main elements included in the database of Al-Dakahlia Governorate. The soil layer shows the distribution of different soil sub-great groups, where the *Vertic Torrifluvents* has the highest occurrence, as it covers an area of 2070.6 km^2 representing 53.8% of the agricultural area. The *Typic Torrifluvents* sub-great group also exhibits an area of 849.2, representing 22.1% of cultivated land. The *Typic Haplocalsids* and *Typic Aquisalids* sub-great groups exist on both sides of Manzala Lake area; covering 461.2 km^2 and 202.2 km^2 representing 12.0% and 5.3% of the agricultural land respectively. The *Typic Quartizpsamments* and *Typic Haplosalids* exist in the northwestern corner of the Governorate covering an areas 132.6 and 86.3 km^2, representing 3.5% and 2.2% of agricultural land respectively. Patches of shifting sand dunes exist along the costal area exhibiting 44.6 km^2.

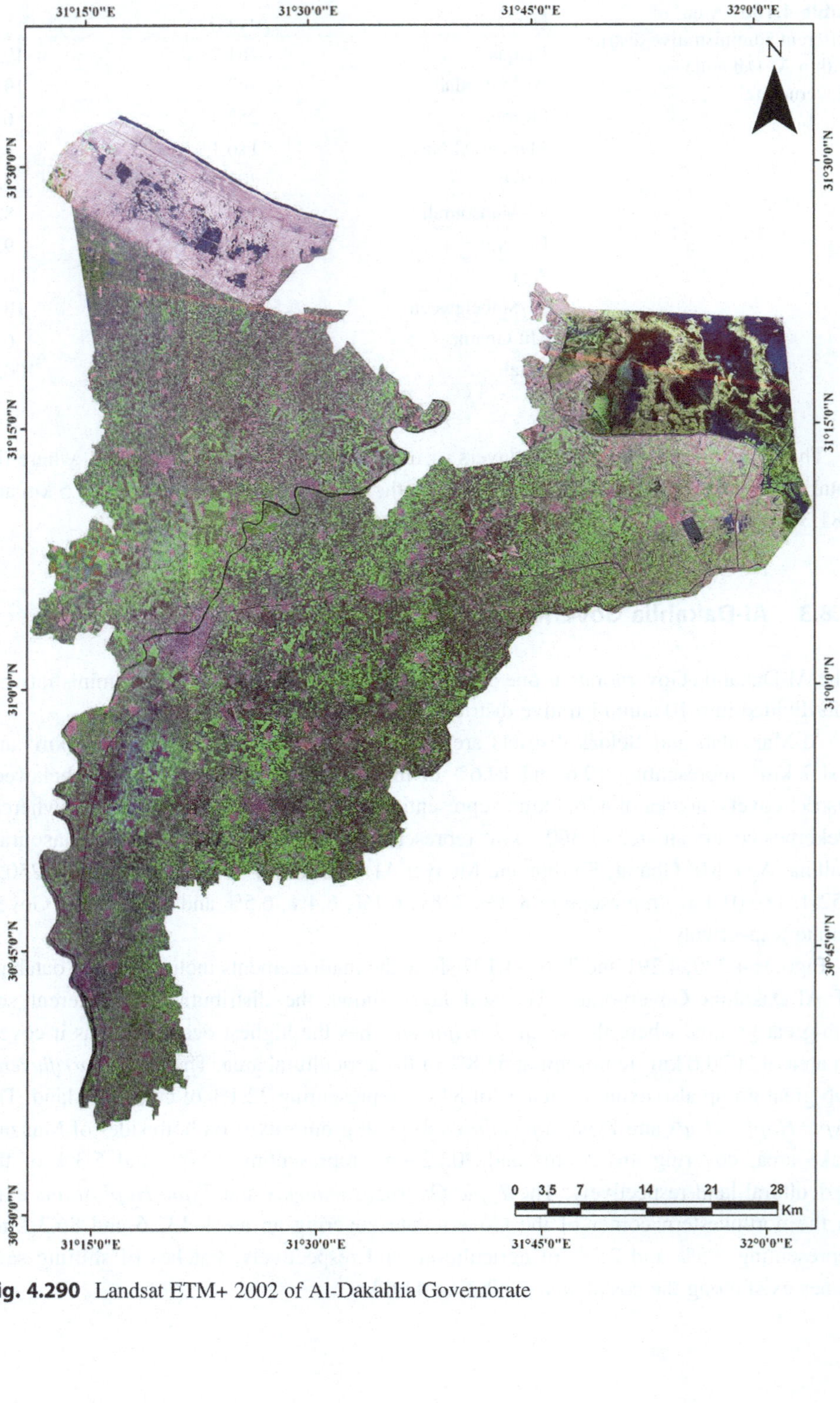

Fig. 4.290 Landsat ETM+ 2002 of Al-Dakahlia Governorate

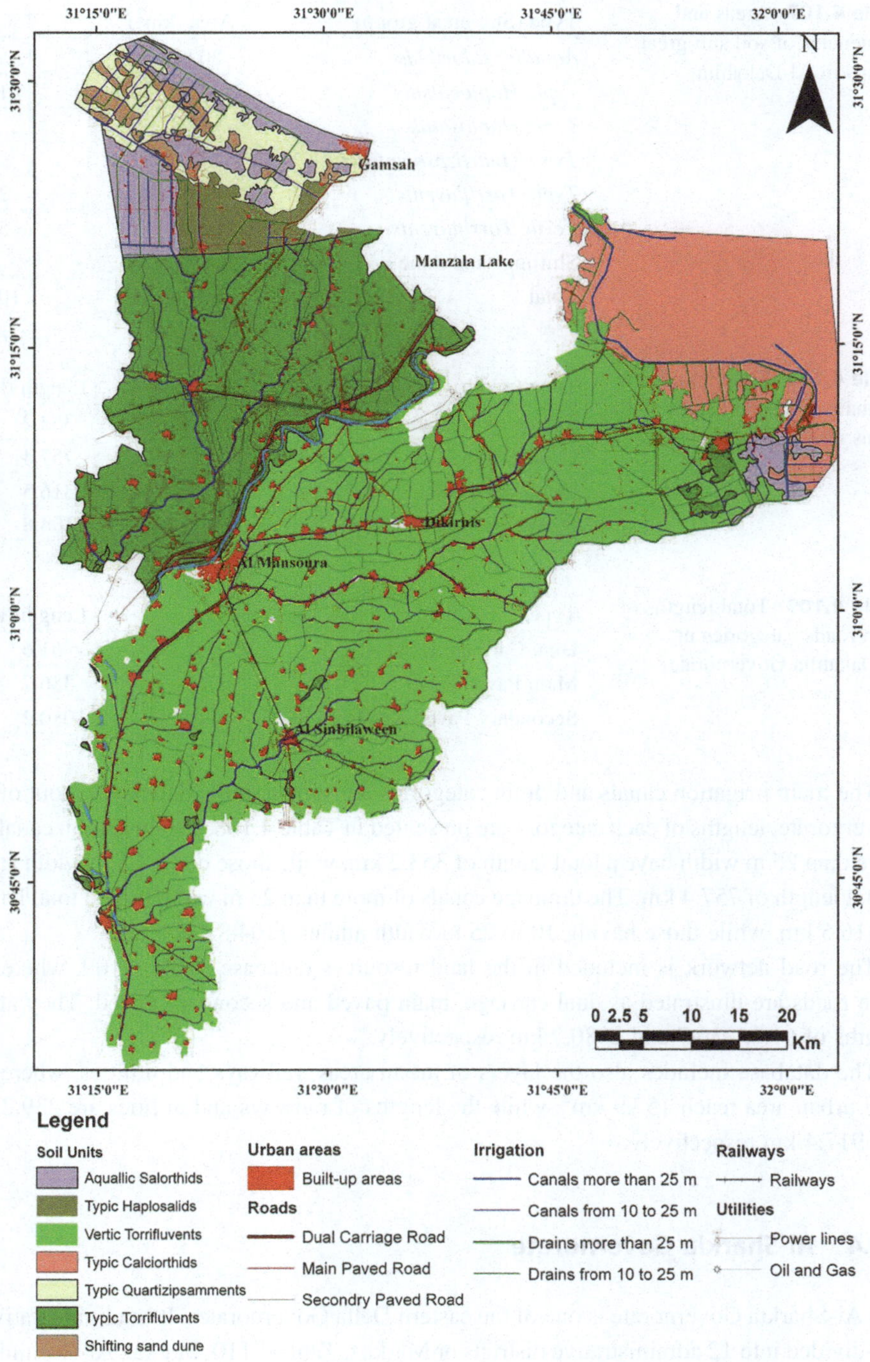

Fig. 4.291 Main elements included in the database of Al-Dakahlia Governorate

Table 4.107 Areas and frequencies of soil sub-great groups in Al-Dakahlia

Type (Sub great group)	Area (km^2)	%
Aquallic salorthids	202.2	5.3
Typic Haplocalsids	461.2	12.0
Typic Haplosalids	86.3	2.2
Typic Quartizpsamments	132.6	3.5
Typic Torrifluvents	849.2	22.1
Vertic Torrifluvents	2070.6	53.8
Shifting sand dune	44.6	1.2
Total	3846.6	100.0

Table 4.108 Total lengths of the main irrigation canals and drains in Al-Dakahlia

Type (width)	Length (km)
Canals more than 25 m	353.2
Canals from 10 to 25 m	757.4
Drains more than 25 m	316.5
Drains from 10 to 25 m	1104.8

Table 4.109 Total lengths of main roads categories in Al-Dakahlia Governorate

Type	Length (km)
Dual Carriage Road	61.6
Main Paved Road	336.7
Secondary Paved Road	1080.2

The main irrigation canals and drain categories are shown in the database layout of the Governorate, lengths of each category are presented in Table 4.108. The irrigation canals of more than 25 m width have a total length of 353.2 km, while those of 10–25 m width attain a total length of 757.4 km. The drainage canals of more than 25 m width have a total length of 316.5 km, while those having 10 to 25 m width attains 1104.8 km length.

The road network is included in the land resources database Table 4.109, where the main roads are illustrated as dual carriage, main paved and secondary paved. They attain lengths of 61.6, 336.7 and 1080.2 km respectively.

The database includes also the layers of urban areas, railways and utilities, where the total urban area reach 157.9 km^2, while the lengths of railways and utilities are 239.2 km and 917.4 km respectively.

4.8.4 Al-Sharkia Governorate

The Al-Sharkia Governorate is one of the eastern Delta Governorates. It is administratively sub-divided into 12 administrative districts or Markaz, Table 4.110. Markaz Al-Husainiyah district is the largest one as it covers 1523.0 km^2, representing 34.9% of the total Governorate area. Markaz Belbes covers an area of 454.6 km^2, representing 10.4% of

Table 4.110 Areas and frequencies of different administrative districts within Al-Sharkia Governorate

Name	Area (km^2)	%
Markaz Al-Husainiyyah	1523.0	34.9
Markaz Awlad Saqr	257.6	5.9
Markaz Kafr Saqr	176.0	4.0
Markaz Faqus	430.5	9.9
Markaz Abu Kabir	189.9	4.4
Markaz Derab Nijm	211.7	4.9
Markaz Al-Ibrhimiyyah	72.0	1.7
Markaz Hehya	129.3	2.9
Markaz Al-Zakazik	332.6	7.6
Markaz Abu Hamad	289.2	6.6
Markaz Minia Al-Kamh	295.4	6.8
Markaz Belbes	454.6	10.4
Total	4361.9	100.0

Governorate area, whereas the other districts exhibit small homogenous areas comparatively.

Figures 4.292, 4.293 and Table 4.111 show the main elements included in the database of Al-Sharkia Governorate. The soil map shows that the *Vertic Torrifluvents* has the highest occurrence, as it covers an area of 2720.2 km^2 representing 62.6% of the alluvial soils. The *Typic Quartizpsamments* sub-great group is found at the eastern part of the Governorate with an area of 778.3 km^2, representing 17.9% of the Governorate. This sub-great group represents the recently reclaimed soils, which are irrigated by Pivot and Sprinkler irrigation systems. The *Aquollic Salorthids* and *Typic Torriorthents* sub-great groups exhibit areas at the northeast of the Governorate covering 328.6 km^2 and 241.6 km^2 representing 7.6% and 5.6% of the total area.

Patches of *Typic Calciorthids, Typic Torrifluvents and Typic Torripsamments* are scattered in the Governorate covering areas of 84.94, 49.73 and 106.72 km^2, representing 1.9%, 1.1% and 2.5% respectively. Also, limited coverage of rockland and rock escarpments exist in the southeastern corner exhibiting 34.7 km^2 and 3.0 km^2 representing 0.8% and 0.1% respectively.

Two main irrigation and drain canals categories are shown in the database layout of the Governorate, lengths of each category are presented in Table 4.112. The irrigation canals of more than 25 m width have a total length of 240.6 km, while those of 10–25 m width attain a total length of 375.5 km. The drainage canals of more than 25 m width have a total length of 285.9 km, while those having 10 to 25 m width attains 874.1 km length.

The road network is included in the land resources database Table 4.113, where the main roads are illustrated as dual carriage, main paved and secondary paved. They attain lengths of 28.7, 237.5 and 1533.2 km respectively.

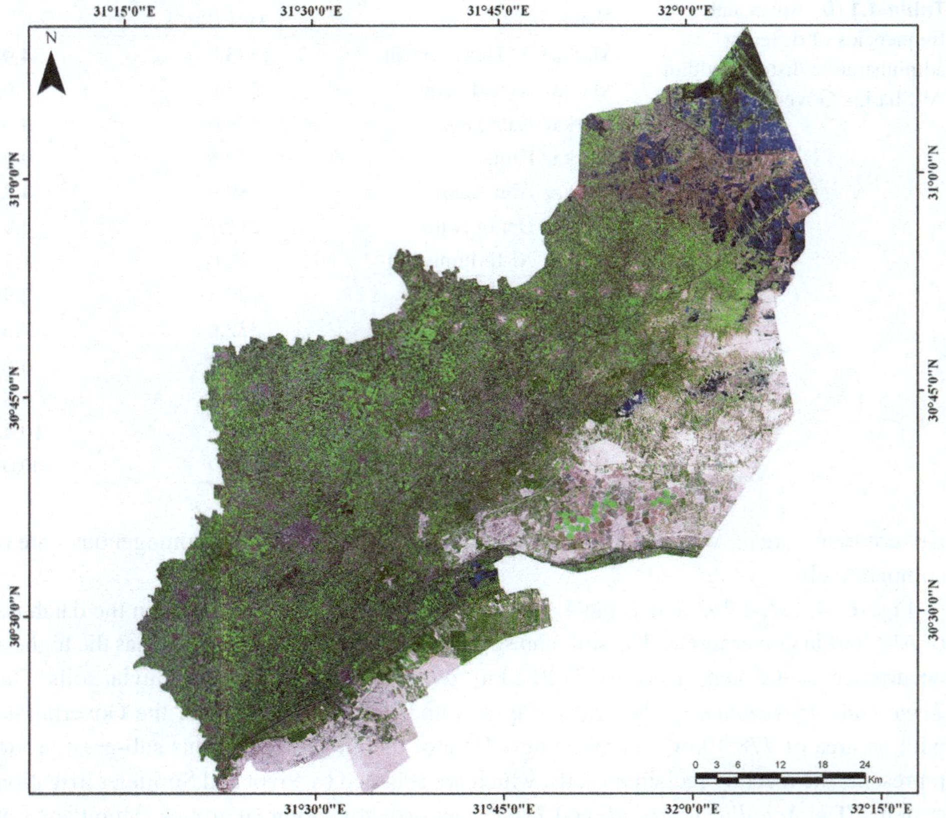

Fig. 4.292 Landsat ETM+ 2002 of Al-Sharkia Governorate

The database includes also the layers of urban areas, railways and utilities, where the total urban area reach 379.9 km^2, while the lengths of railways and utilities are 215.7 km and 489.7 km respectively.

4.8.5 Al-Gharbia Governorate

The Al-Gharbia Governorate is located in the Middle of the Nile Delta Governorates. It is administratively sub-divided into 8 administrative districts or Markaz, Table 4.114. Markaz Al-Mahalah Al-Kobra and Kafr Al–Zayat districts, known by their famous textile industry, cover areas of 441,9km^2 and 204.9 km^2 representing 22.7% and 10.5% respectively. Tanta district, including the government capital, covers 331.2 km^2, representing 16.98%.

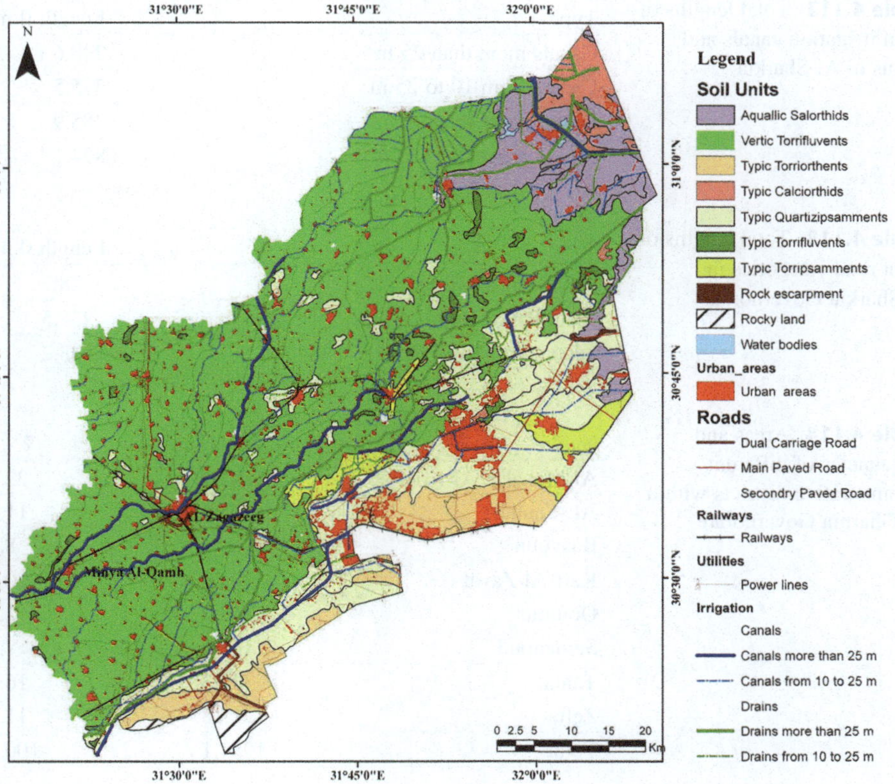

Fig. 4.293 Main elements included in the database of Al-Sharkia Governorate

Table 4.111 Areas and frequencies of soil sub-great groups in Al-Sharkia

Type (Sub great group)	Area (km^2)	%
Aquollic Salorthids	328.6	7.6
Typic Calciorthids	84.9	1.9
Typic Quartizpsamments	778.3	17.9
Typic Torrifluvents	49.7	1.1
Typic Torriorthents	241.6	5.6
Typic Torripsamments	106.7	2.5
Vertic Torrifluvents	2720.2	62.6
Rock land	34.7	0.8
Rock escarpment	3.0	0.1
Total	4347.9	100.0

The districts of AL-Santah, Qutour and Zefta have amost simlar areas ranging from 229.0 km^2 to 215.8 km^2. Basyoun and Samanoud districts represent the smallest ones covering areas of 161.1 km^2 and 148.6 km^2, representing 8.3% and 7.6% respectively.

Table 4.112 Total lengths of main irrigation canals and drains in Al-Sharkia

Type	Length (km)
Canals more than 25 m	240.6
Canals from 10 to 25 m	375.5
Drains more than 25 m	285.9
Drains from 10 to 25 m	874.1

Table 4.113 Total lengths of main road categories in El-Sharkia Governorate

Type	Length (km)
Dual Carriage Road	28.7
Main Paved Road	237.5
Secondary Paved Road	1533.2

Table 4.114 Areas and frequencies of different administrative districts within Al-Gharbia Governorate

Name	Area (km^2)	%
Al-Mahalah Al-Kobra	441.9	22.7
AL-Santah	218.6	11.2
Basyoun	161.1	8.3
Kafr Al-Zayat	204.9	10.5
Qoutour	229.0	11.7
Samanoud	148.6	7.6
Tanta	331.2	16.9
Zefta	215.8	11.1
Total	1951.1	100.0

Figures 4.294, 4.295 and Table 4.115 show the main elements included in the database of Al-Gharbia Governorate. The soil layer shows that only two sub-great soil groups exist in the Governorate (i.e. *Vertic Torrifluvents and Typic Torrifluvents*). Most of the Governorate soils are exhibited by the *Vertic Torrifluvents sub-great group,* as they cover an area of 1889.1 km^2 representing 96.9% of arable land. The *Typic Torrifluvents* sub-great group covers 60.6 km^2, representing 3.1% of the soils aligned along the eastern and western borders of the Governorate.

The main irrigation canals and drain network are shown in the database layout of the Governorate; total lengths of each category are presented in Table 4.116. The irrigation canals of more than 25 m width exhibit a total length of 212.8 km, while those of 10–25 m width attain a total length of 350.7 km. The drainage canals of more than 25 m width have a total length of 18.9 km, while those having 10 to 25 m width attains 662.3 km length.

The road network is included in the land resources database Table 4.117, where the main roads are illustrated as dual carriage, main paved and secondary paved roads. They attain lengths of 95.9, 164.6 and 572.2 km respectively.

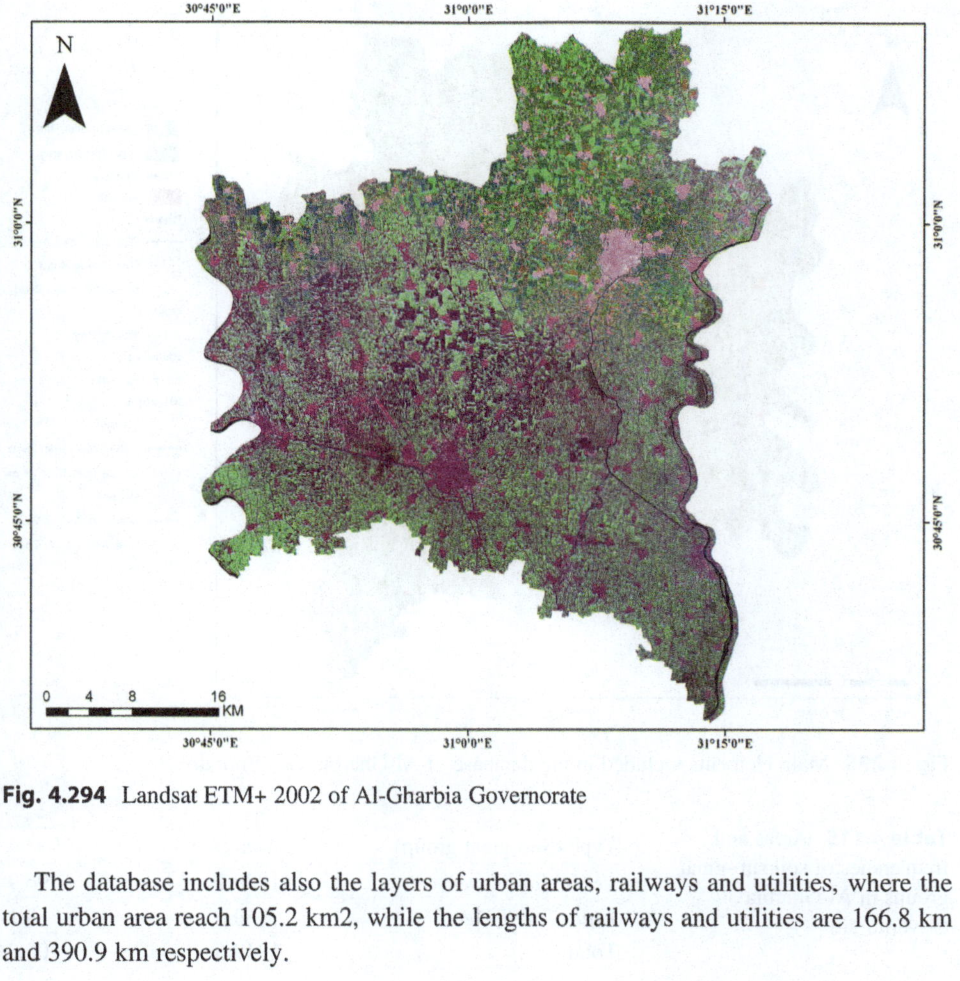

Fig. 4.294 Landsat ETM+ 2002 of Al-Gharbia Governorate

The database includes also the layers of urban areas, railways and utilities, where the total urban area reach 105.2 km2, while the lengths of railways and utilities are 166.8 km and 390.9 km respectively.

4.8.6 Al-Menoufia Governorate

Al-Menoufia Governorate is one of the Middle Delta Governorates. It is administratively sub-divided into 8 administrative districts or Markaz Table 4.118. Markaz Al-Menoufia is the largest district, as it covers an area of 1091.1 km^2 and representing 44.9% of the Governorate total area. Ashmoun district is the second in its coverage, covering an area of 316.6 km^2 and representing 13.1%. The Qewasna district covers 203.9 km^2, representing 7.7% of the soils. The districts of Shibin Al-Kawm, Tala, Al-Bagour and Ash-Shohada have almost similar areas ranging from 185.9 km^2 to 155.4 km^2, representing 7.7% to 6.4% respectively. The smallest markz is attained by the Birkat As-Sabea district which covers an area of 116.7 km^2, representing 4.8% of total Governorate area.

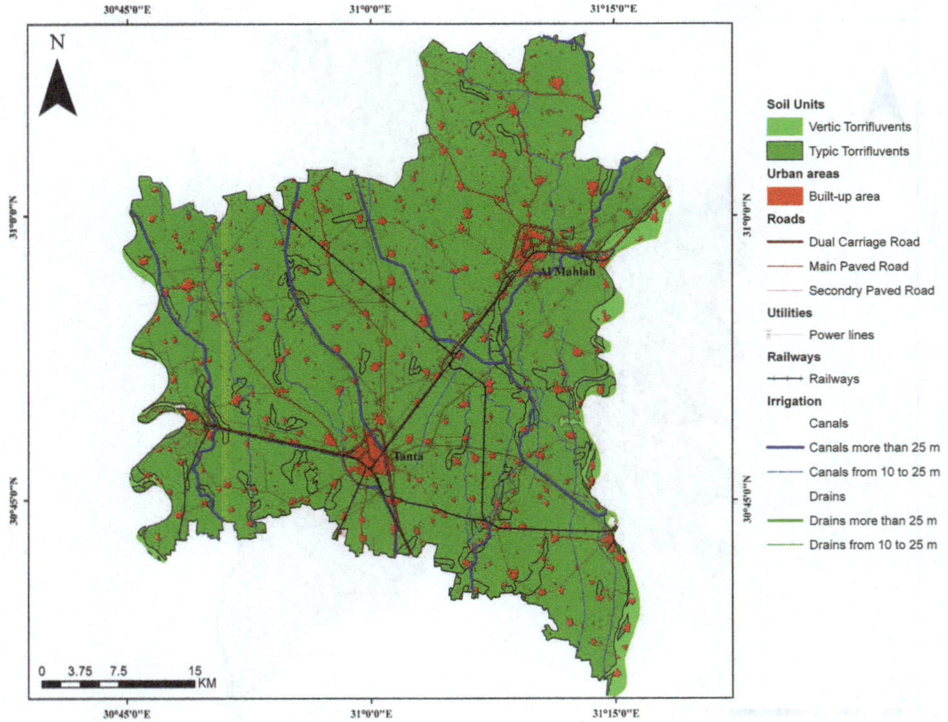

Fig. 4.295 Main elements included in the database of Al-Gharbia Governorate

Table 4.115 Areas and frequencies of soil sub-great groups in Al-Gharbia Governorate

Type (Sub great group)	Area (km^2)	%
Vertic Torrifluvents	1889.1	96.9
Typic Torrifluvents	60.6	3.1
Total	1949.6	100.0

Table 4.116 Total length of the main irrigation canals and drains in Al-Gharbia Governorate

Type	Length (km)
Canals more than 25 m	212.8
Canals from 10 to 25 m	350.7
Drains more than 25 m	18.9
Drains from 10 to 25 m	662.3

Table 4.117 Total lengths of main road categories in Al-Gharbia Governorate

Type	Length (km)
Dual Carriage Road	95.9
Main Paved Road	164.6
Secondary Paved Road	572.2

Table 4.118 Areas and frequencies of different administrative districts within Al-Menoufia Governorate

Name	Area (km^2)	%
Al-Bagour	169.0	6.9
Ash-Shohada	155.4	6.4
Ashmoun	316.6	13.1
Birkat As-Sabea	116.7	4.8
Menouf	1091.1	44.9
Qewasna	203.9	8.4
Shibin Al-Kawm	185.9	7.7
Tala	187.2	7.7
Total	2425.8	100.0

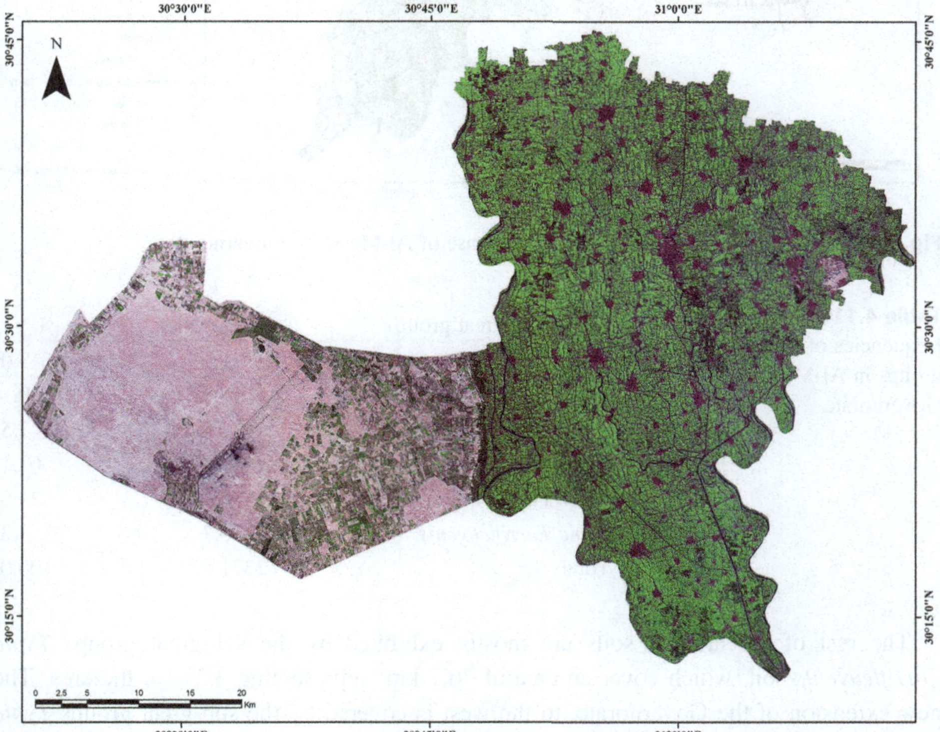

Fig. 4.296 Landsat ETM+ 2002 of Al-Menoufia Governorate

Figures 4.296, 4.297 and Table 4.119 show the main elements included in the database of Al-Menoufia Governorate. The soil layer shows the distribution of different soil sub-great groups, where the *Vertic Torrifluvents* has the highest occurrence covering most of the old alluvial soils. It covers an area of 1473.0 km^2 representing 62.1% of the soils.

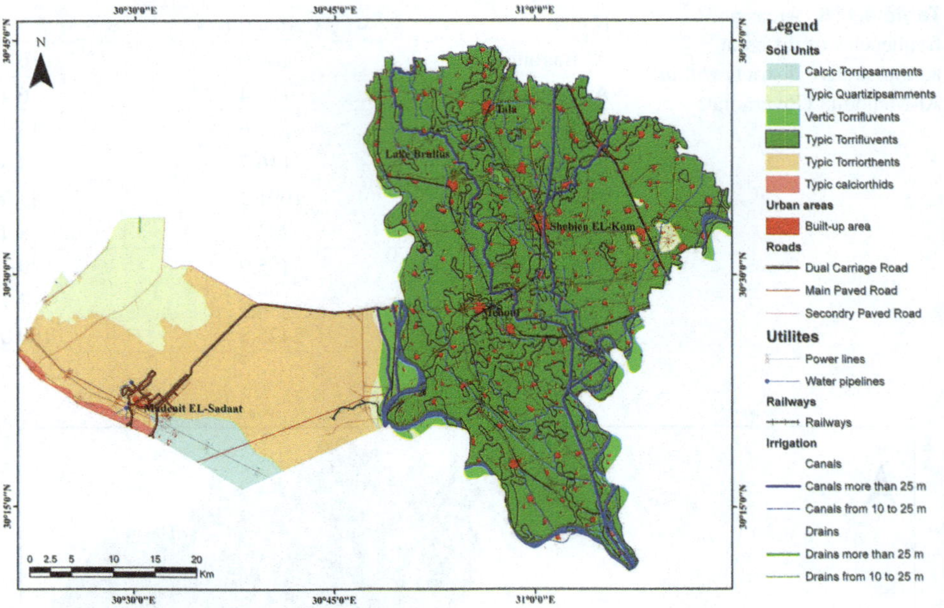

Fig. 4.297 Main elements included in the database of Al-Menoufia Governorate

Table 4.119 Areas and frequencies of soil sub-great groups in Al-Menoufia Governorate

Type (Sub great group)	Area (km²)	%
Typic Torripsamments	71.9	3.0
Typic Calciorthids	15.0	0.
Typic Quartizpsamments	178.6	7.5
Vertic Torrifluvents	1473.0	62.1
Typic Torriorthents	556.3	23.5
Typic Torrifleuvents	76.7	3.2
Total	2371.57	100.0

The rest of the alluvial soils are mostly exhibited by the sub-great groups *Typic Torrifleuvents* soil, which cover an area of 76.7 km^2 representing 3.2% of the area. The new extension of the Governorate, to the west is covered by the sub-great groups *Typic Torriorthents, Typic Quartizpsamments and Typic Torripsamments*. These sub-great groups cover areas of 556.3 km^2, 178.6 km^2 and 71.9km^2 representing 23.5%, 7.5% and 3.0% of total area respectively. The two sub-great groups *Calcic Torripsamments* and *Typic Calciorthids* soils exist in two distinctive spots at the west of the Governorate, covering areas of 71.9 km^2 and 15.0 km^2 representing and 3.0% and 0.6% respectively.

The main irrigation canal and drain are shown in the database layout of the Governorate, total lengths of each category are presented in Table 4.120. The irrigation canals of more than 25 m width exhibit a total length of 275.8 km, while those of 10–25 m width attain a

Table 4.120 Total length of the main irrigation canals and drains in Al-Menoufia Governorate

Type	Length (km)
Canals more than 25 m	275.8
Canals from 10 to 25 m	406.2
Drains more than 25 m	42.3
Drains from 10 to 25 m	486.5

Table 4.121 Total lengths of main road categories in Al-Menoufia Governorate

Type	Length (km)
Dual Carriage Road	104.9
Main Paved Road	76.4
Secondary Paved Road	999.1

total length of 406.2 km. The drainage canals of more than 25 me width have a total length of 42.3 km, while those having 10 to 25 m width attains 486.5 km length.

The road network is included in the land resources database Table 4.121, where the main roads are illustrated as dual carriage, main paved and secondary paved. They attain lengths of 104.9, 76.4 and 999.1 km respectively.

The database includes also the layers of urban areas, railways and utilities, where the total urban area reach 81.1 km2, while the lengths of railways and utilities are 287.1 km and 426.7 km respectively.

4.8.7 Al-Beheira Governorate

Al-Beheira Governorate is one of the largest governorate in Egypt, located in the north Delta region south of Alexandria. It is administratively sub-divided into 13 administrative districts or Markaz Table 4.122. Kowm Hamadah and Abu Al-Matamir are the largest two districts in the Governorate, covering 1683.5 and 1278.6 km^2, representing 26.4% and 20.1% of the total area respectively. The medium size districts are located in the alluvial landscape area (i.e. Kafr Al-Dawar, Abu Homos and Damanhour) ranging between 582.4 km^2 and 389.9 km^2, representing 9.1% to 6.1% respectively. The other districts (i.e. Rashid, Idko, Al-Mahmoideyah, Al-Rahmaneyah, Shubrakhit, Hosh Eisa, Itay Al-Baroud and Ad-Dilinjat) cover relatively small areas. Their areas range between 379.9 km^2 in Al-Dilinjat and 197.7 km^2 in Rashid district representing 5.9% and 3.1% respectively.

Figures 4.298, 4.299 and Table 4.123 show the main elements included in the database of Al-Beheira Governorate. The soil map shows the distribution of different soil sub-great groups, where the *Vertic Torrifluvents* is dominating the soil types and concentrated in the alluvial region. It covers an area of 3026.2 km^2, representing 47.8% of the soils. The *Typic Torripsamments* and *Typic Torrifluvents* sub-great groups interfere with the previous one

Table 4.122 Areas and frequencies of different administrative districts within Al-Beheira Governorate

Name	Area (km^2)	%
Rashid	197.7	3.1
Idko	239.9	3.8
Al-Mahmoudeyah	174.7	2.7
Kafr Ad-Dawar	582.4	9.1
Abu Homos	529.1	8.3
Ar-Rahmaneyah	92.9	1.5
Damanhour	389.9	6.1
Shubrakhit	233.2	3.7
Abu Al-Matamir	1278.6	20.1
Hosh Eisa	278.6	4.4
Itay Al-Baroud	308.7	4.9
Ad-Dilinjat	379.9	5.9
Kawm Hamadah	1683.5	26.4
Total	6368.9	100.0

within the alluvial landscape covering area of 550.0 km^2 and 428.8 km^2 representing 8.7% and 6.8% respectively. The sub-great group *Typic Haplocalsids,* covers significant areas around Idko and Mariout lakes, with a total of 536.7 km^2 representing 8.5% of total Governorate.

Limited patches of sub-great groups *Typic Aquisalids, Typic Haplosalids and Typic Calcigypsids* soil interfere with *Typic Haplocalsids* soils covering an areas of 28.0, 72.3 and 1.8 km^2 representing 0.4%, 1.1% and 0.03% respectively.

The *Typic Quartizpsamments* sub-great group, covers an area of 994.4 km^2 (15.7% of the soils) found at the southwestern part of the Governorate, adjacent to the areas of sand dunes. It is also interfered by the sub-great group *Typic Torriorthents,* covers limited patches having an area of 107.3 km^2 representing 1.7% of total soil area.

The main irrigation and drain canal categories are shown in the database layout of the Governorate, total lengths of each kind are presented in Table 4.124. The irrigation canals of more than 25 m width exhibit a total length of 439.3 km, while those of 10–25 m width attain a total length of 1077.9 km. The drainage canals of more than 25 m width have a total length of 246.8 km, while those having 10 to 25 m width attains 1634.7 km length.

The road network is included in the land resources database Table 4.125, where the main roads are illustrated as dual carriage, main paved and secondary paved. They attain lengths of 112.9, 169.5 and 2481.9 km respectively.

The database includes also the layers of urban areas, railways and utilities, where the total urban area reach 128.8 km^2, while the lengths of railways and utilities are 337.4 km and 1200.9 km respectively.

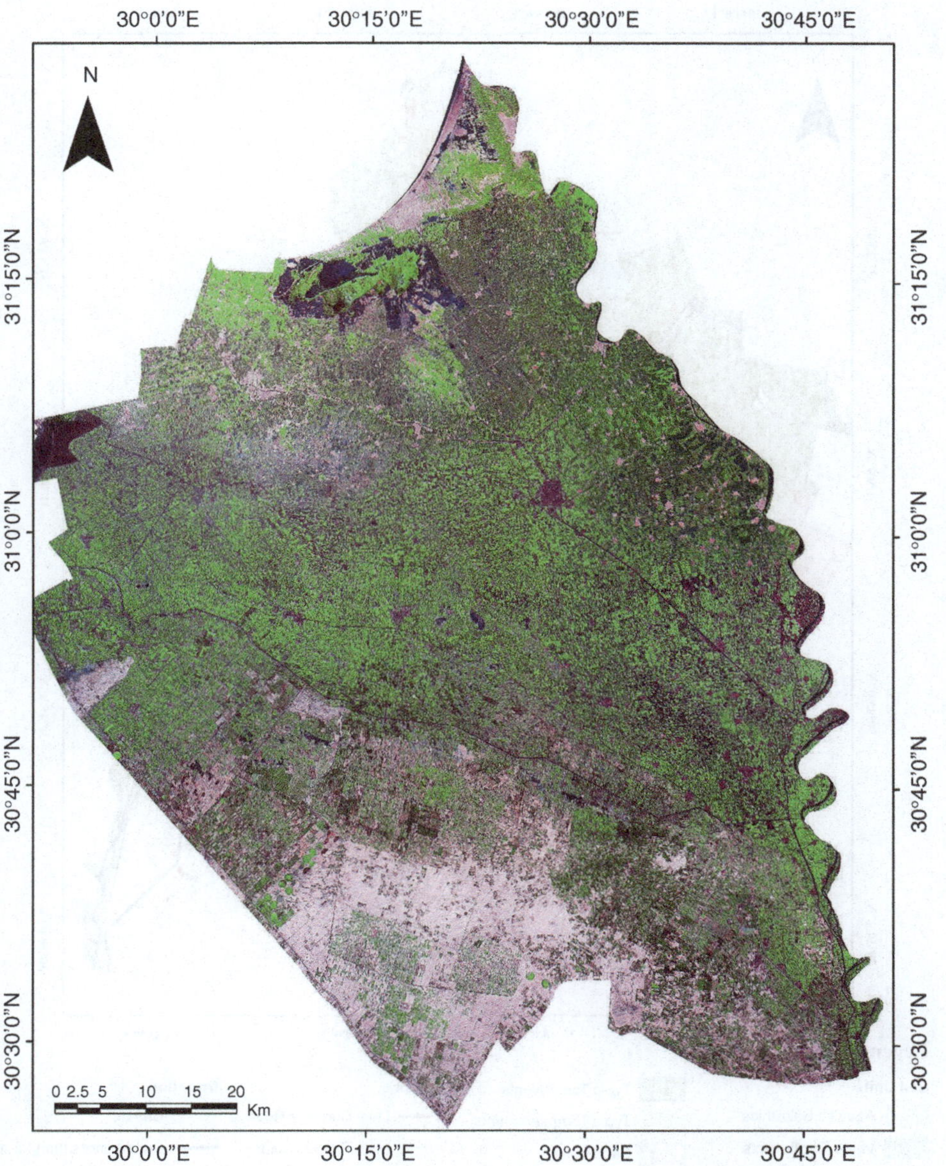

Fig. 4.298 Landsat ETM+ 2002 of Al-Beheira Governorate

4.8.8 Alexandria Governorate

Alexandria Governorate is known by its distinctive history and is considered as the main Egyptian harbour on the Mediterranean Sea. It is administratively sub-divided into 15 administrative districts or Markaz Table 4.126. Some administrative districts combine

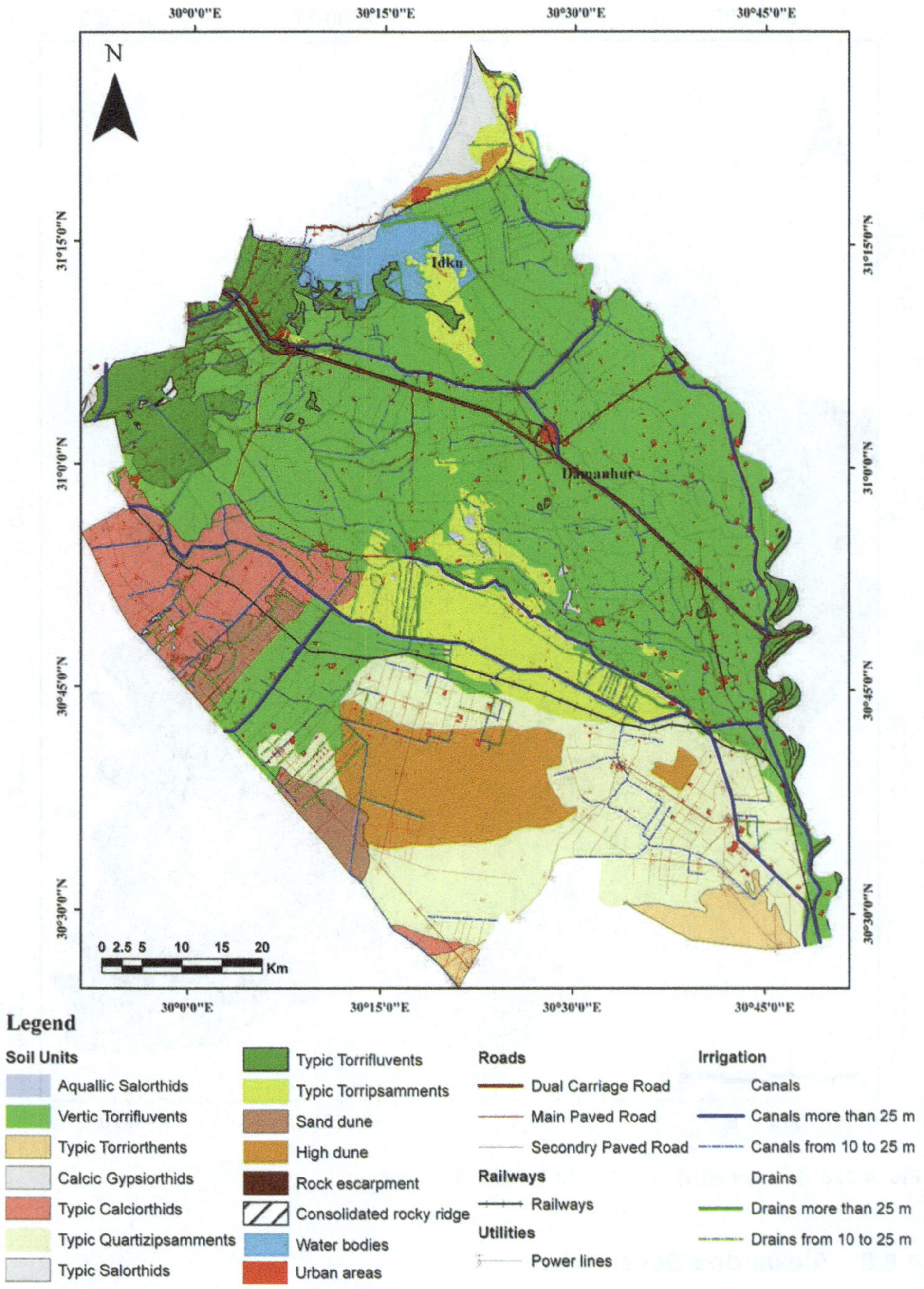

Fig. 4.299 Main elements included in the database of Al-Beheira Governorate

Table 4.123 Areas and frequencies of soil sub-great groups in Al-Beheira Governorate

Type (Sub great group)	Area (km^2)	%
Typic Aquisalids	28.0	0.4
Typic Calcigypsids	1.8	0.03
Typic Haplocalsids	536.7	8.5
Typic Quartizpsamments	994.4	15.7
Typic Haplosalids	72.3	1.1
Typic Torrifluvents	428.8	6.8
Typic Torriorthents	107.3	1.7
Typic Torripsamments	550.0	8.7
Vertic Torrifluvents	3026.2	47.8
Consolidated rocky ridge	7.8	0.1
High dune	430.9	6.8
Sand dune	149.7	2.4
Total	6333.9	100.0

Table 4.124 Total length of the main irrigation and drain canals in Al-Beheira Governorate

Type (width)	Length (km)
Canals more than 25 m	439.3
Canals from 10 to 25 m	1077.9
Drains more than 25 m	246.8
Drains from 10 to 25 m	1634.7

Table 4.125 Total lengths of main road categories in Al-Beheira Governorate

Type	Length (km)
Dual Carriage Road	112.9
Main Paved Road	169.5
Secondary Paved Road	2481.9

both urban and rural pattern, while others are almost urban. New agriculture sustainable projects are found is particular districts as Borg Al-Arab and Mina Al-Basel. Al-Montazah and Al-Amereya districts include limited areas of successful orchards. The Al-Dekheila (29.1 km^2) is known as one of the industrial areas as Iron-steel and cement factories exist. Regarding the geographical extension, Borg Al-Arab and Al-Amereya are the largest districts covering areas of 951.5 km^2 and 421.1 km^2 representing 60.8% and 26.9% of the total Govenorates area respectively. The rest of the administrative districts are almost urban covering areas ranging from 1.1 km^2, for Al-Laban districts, to 87.4 km^2, for Al-Montazah district.

Figures 4.300, 4.301 and Table 4.127 show the main elements included in the database of Alexandria Governorate. The rock land type exhibits limited patches, with a total area of 49.1 km^2, representing 7.8% of land at the middle of the Governorate boundaries. The soil layer shows the distribution of different soil sub-great groups, where the *Typic*

Table 4.126 Areas and frequencies of different administrative districts within Alexandria Governorate

Name	Area (km^2)	%
Al-Dekheila	29.1	1.9
Al-Amereya	421.1	26.9
Al-Attarin	1.9	0.1
Al-Gomrok	3.5	0.2
Al-Laban	1.1	0.1
Al-Mansheya	0.6	0.0
Al-Montazah	87.4	5.6
Al-Raml	31.5	2.0
Bab Sharqi	5.7	0.4
Borg Al-Arab	951.5	60.8
Karmouz	3.6	0.2
Mina Al-Basel	10.3	0.7
Moharam Bek	5.5	0.4
Shortet Mina Al-Iskandariyyah	0.5	0.0
Sidi Gaber	11.7	0.8
Total	1564.9	100.0

Haplocalsids is dominating the soil types and concentrated in the southwestern region around Mariout Lake. These soils exhibits an area of 330.2 km^2 representing 52.3% of the Governorate soils. The *Typic Torrifluvents* and *Typic Haplosalids* sub-great groups have a significant occurrence in the northeast and middle of the Governorate, as they cover areas of 147.2 and 97.5 km^2 representing 23.3% and 15.4% respectively. The sub-great groups *Typic Aquisalids* and *Typic Torripsamments* cover very small areas (4.7 km^2 and 3.2 km^2 respectively) on the coast, respectively 0.8% and 0.5% resepectively.

4.8.9 Al-Qaliyubia Governorate

Al-Qaliyubia Governorate is one of the largest ones in Egypt, located in the southern Delta region north of Cairo. It is administratively sub-divided into 8 administrative districts or Markaz Table 4.128. Tokh is the largest districts in the Governorate, covering 234.8 km^2, representing 23.9% of the whole Governorate. Banha district is the second in its area covering 169.8 km^2, representing 17.3% of total coverage and including Banha city government capital. Shibin Al-Qanatir and Al-Khankah are important districts located at the middle of the Governorate, where transportation is available to most of the other towns. They cover areas of 141.5 and 121.1 km^2 representing 14.4% and 12.3% of the governorte. Also, having the famous barrages and a vegetated resort place for the residence of Cairo Al-Qanatir Al-Khayreyah district is well known area which covers an area of 100.7 km^2 representing 10.2% of the total area of the Governorate. Kafr Shokr and Awal Shobra

Fig. 4.300 Landsat ETM+ 2002 of Alexandria Governorate

Al-Khaymah are small districts at the borders of adjacent southern Governorates, having areas of 65.5 km^2 and 28.3 km^2 representing 6.7% and 2.9% respectively.

Figures 4.302, 4.303 and Table 4.129 show the main elements included in the database of Al-Qaliyubia Governorate. The soil layer demonstrates that only three sub-great groups exhibit the whole Governorate, where *Vertic Torrifluvents* is dominating the soil types covering an area of 696.1 km^2, representing 84.1% of the region.

The Typic Torrifluvents sub-great group cover limited patches within the alluvial area, exhibiting 87.9 km^2 representing 10.6% of the soils. The *Typic quartzipsamments* and *Vertic Torrifluvents* sub-great groups interfere with the previous one in the alluvial area covering area of 43.3 km^2 and 696.1 km^2 representing 5.2% and 84.1% respectively.

The main irrigation and drain canal are shown in the database layout of the Governorate, total lengths of each category are presented in Table 4.130. The irrigation canals of more than 25 m width exhibit a total length of 128.1 km, while those of 10–25 m width attain a total length of 135.9 km. The drainage canals of more than 25 m widths have a total length of 19.7 km, while those having 10 to 25 m width attains 146.5 km length.

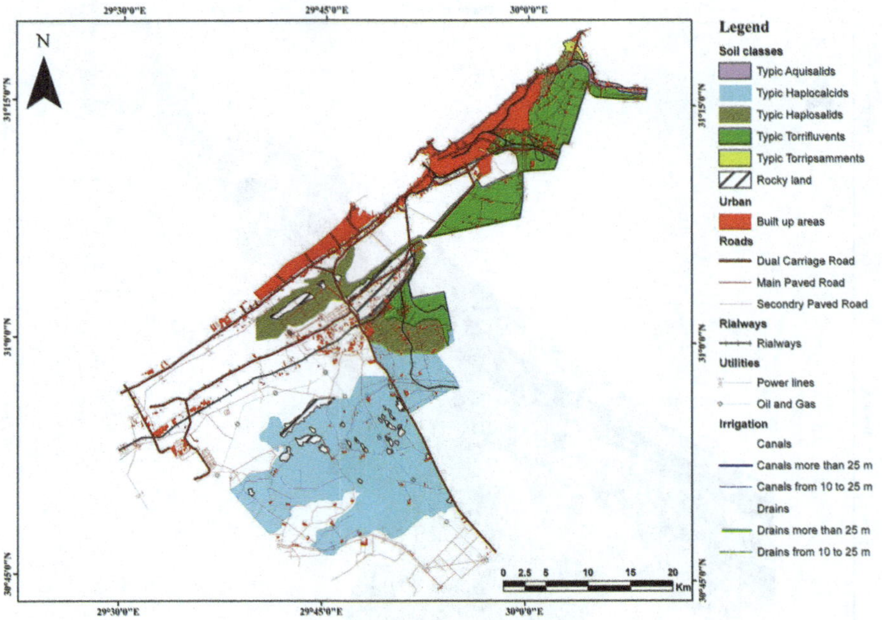

Fig. 4.301 Main elements included in the database of Alexandria Governorate

Table 4.127 Areas and frequencies of soil sub-great groups in Alexandria Governorate

Type (Sub great group)	Area (km^2)	%
Typic Aquisalids	4.7	0.8
Typic Haplocalcids	330.2	52.3
Typic Haplosalids	97.5	15.4
Typic Torrifluvents	147.2	23.3
Typic Torripsamments	3.2	0.5
Rock land	49.1	7.8
Total	631.9	100.0

Table 4.128 Areas and frequencies of different administrative districts within Al-Qaliyubia Governorate

Name	Area (km^2)	%
Al-Khankah	121.1	12.3
Al-Qanatir Al-Khayreyah	100.7	10.2
Awal Shobra Al-Khaymah	28.3	2.9
Banha	169.8	17.3
Kafr Shokr	65.5	6.7
Qelyoub	121.9	12.4
Shibin Al-Qanatir	141.5	14.4
Tokh	234.8	23.9
Total	983.6	100.0

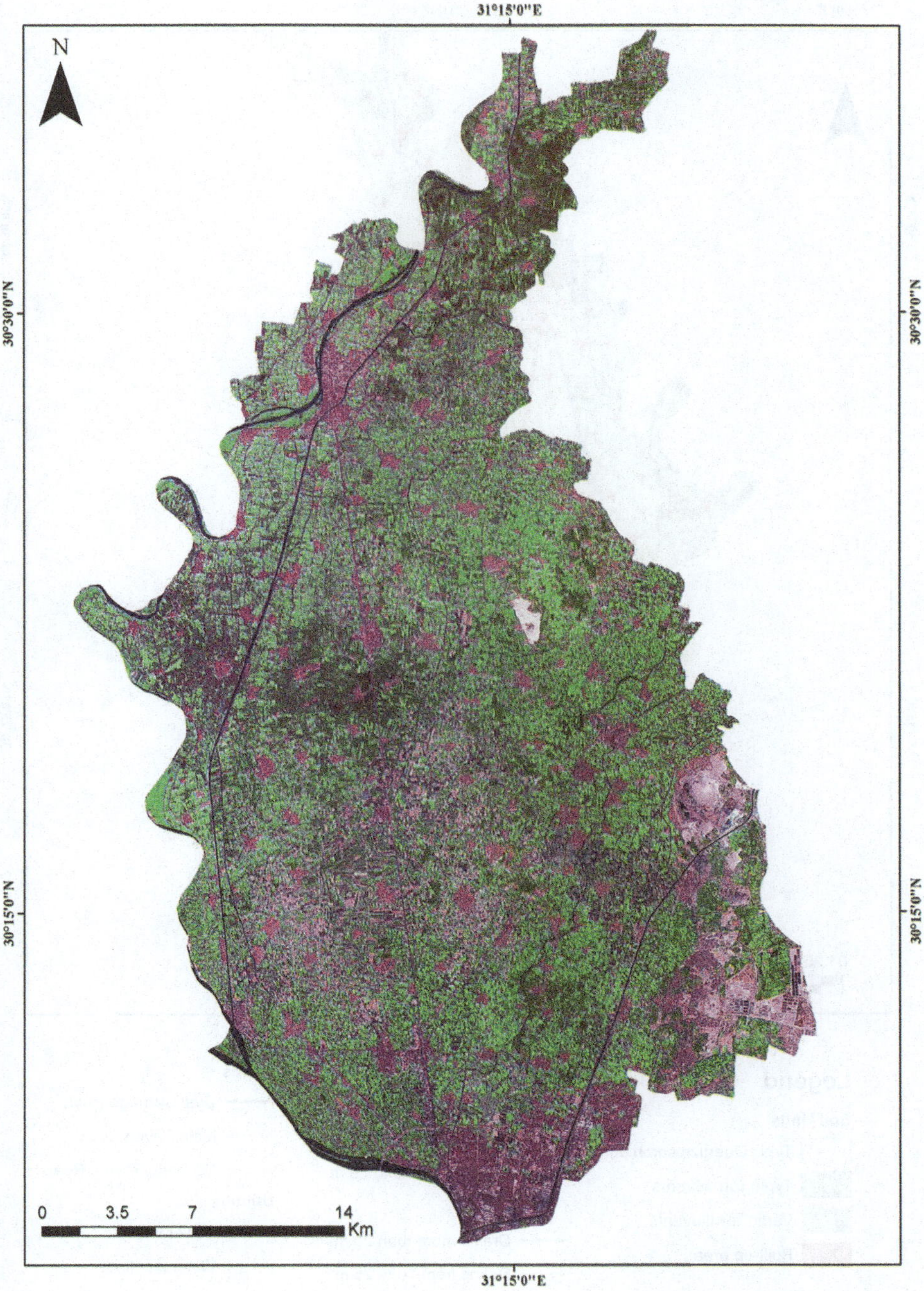

Fig. 4.302 Landsat ETM+ 2002 of Al-Qaliyubia Governorate

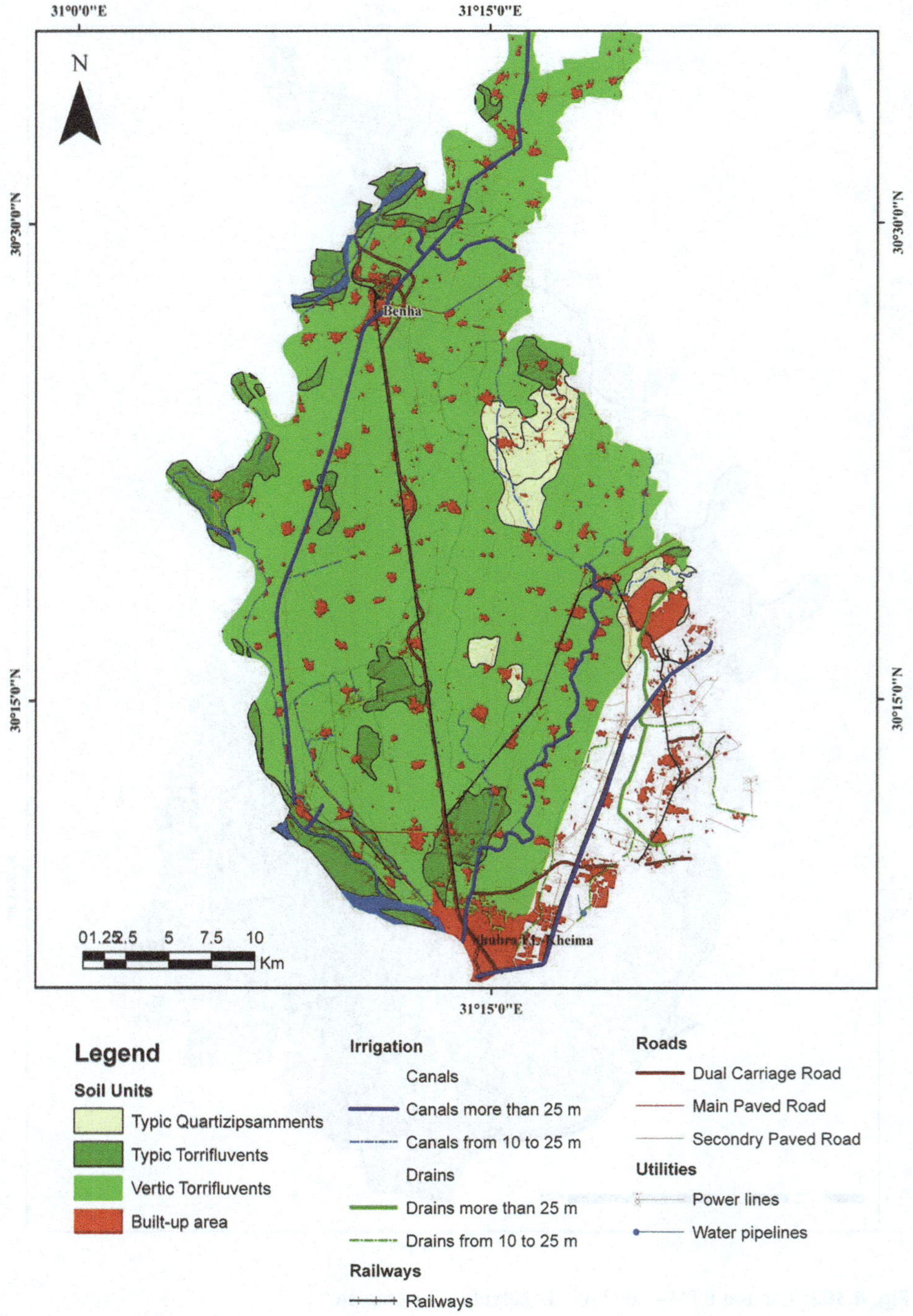

Fig. 4.303 Main elements included in the database of Al-Qaliyubia Governorate

Table 4.129 Areas and frequencies of soil sub-great groups in Al-Qaliyubia Governorate

Type (Sub great group)	Area (km^2)	%
Typic Quartizpsamments	43.3	5.2
Typic Torrifluvents	87.9	10.6
Vertic Torrifleuvents	696.1	84.1
Total	827.3	100.0

Table 4.130 Total length of the main irrigation and drain canals in Al-Qaliyubia Governorate

Type	Length (km)
Canals more than 25 m	128.1
Canals from 10 to 25 m	135.9
Drains more than 25 m	19.7
Drains from 10 to 25 m	146.5

Table 4.131 Total lengths of the main roads categories in Al-Qaliyubia Governorate

Type	Length (km)
Dual Carriage Road	93.2
Main Paved Road	81.4
Secondary Paved Road	783.9

The road network is included in the land resources database Table 4.131, where the main roads are illustrated as dual carriage, main paved and secondary paved road. They attain lengths of 93.2, 81.4 and 783.9 km respectively.

The database includes also the layers of urban areas; railways and utilities, where the total urban area reaches 75.2 km^2, while the lengths of railways and utilities are 92.4 km and 260.3 km respectively.

4.8.10 Cairo Governorate

Al–Qahirah is the local name of Cairo, which is the capital of Egypt. Cairo Governorate includes 35 administrative units Table 4.132. All the administrative regions are characterised by its urban nature, with a total area is 282.5 km^2.

Figures 4.304, 4.305 and Table 4.133 show the main elements included in the database of Cairo Governorate. The soil map demonstrates that the rock land has an extensive occurrence within the administrative boundaries of the Governorate, extending 1172.5 km^2 in the eastern desert fringes. Cultivable soils are rare except in areas at the extreme east and south of the Governorate. The sub-great groups *Typic Torriorthents and Vertic Torrifluvents* are the dominant soils, where they cover 334.6 km^2 and 106.5 km^2 representing 20.6% and 6.5% of the cultivated area respectively. Small patches of the sub-great groups *Typic Haplogypsids, Typic Quartizpsamments and Typic Torrifluvents*

Table 4.132 Areas and frequencies of different administrative districts within Cairo Governorate

Name	Area (km^2)	%
15 Mayou	19.9	7.0
Abdin	1.7	0.6
Al-Darb Al-Ahmar	1.4	0.5
Ain Shams	8.9	3.2
Al-Azbakeia	1.3	0.5
Al-Basatin	14.1	5.0
Al-Gameleya	1.9	0.7
Al-Khelifa	11.6	4.1
Al-Maadi	6.5	2.3
Al-Marg	14.4	5.1
Al-Matareya	6.4	2.3
Al-Mouski	0.8	0.3
Al-Wayli	10.3	3.6
Al-Nozha	21.0	7.4
Al-Sahel	5.9	2.1
Al-Sayeda Zeinab	3.8	1.3
Al-Selam	20.4	7.2
Al-Sharabeya	3.3	1.2
Al-Tebin	12.5	4.4
Al-Zaher	1.9	0.7
Al-Zamelek	3.5	1.3
Al-Zawya Al-Hamra	4.6	1.6
Al-Zaytoun	7.6	2.7
Bab Al-Shaiereya	1.1	0.4
Boulaq	3.1	1.1
Hadaeq El Qoba	5.2	1.8
Helwan	33.2	11.8
Misr Al-Gadida	30.6	10.8
Misr Al-Qadima	12.3	4.3
Monshat Naser	3.3	1.2
Qasr Al-Nil	1.5	0.5
Rod Al-Farag	2.7	0.9
Shobra	1.5	0.5
Madinat Nasr	0.6	0.2
Tora	3.8	1.3
Total	282.5	100.0

exist in the extension of Cairo city to the northeast and southwest direction. They cover areas of 6.1, 5.9 and 3.0 km^2 representing 0.4%, 0.4% and 0.2% respectively.

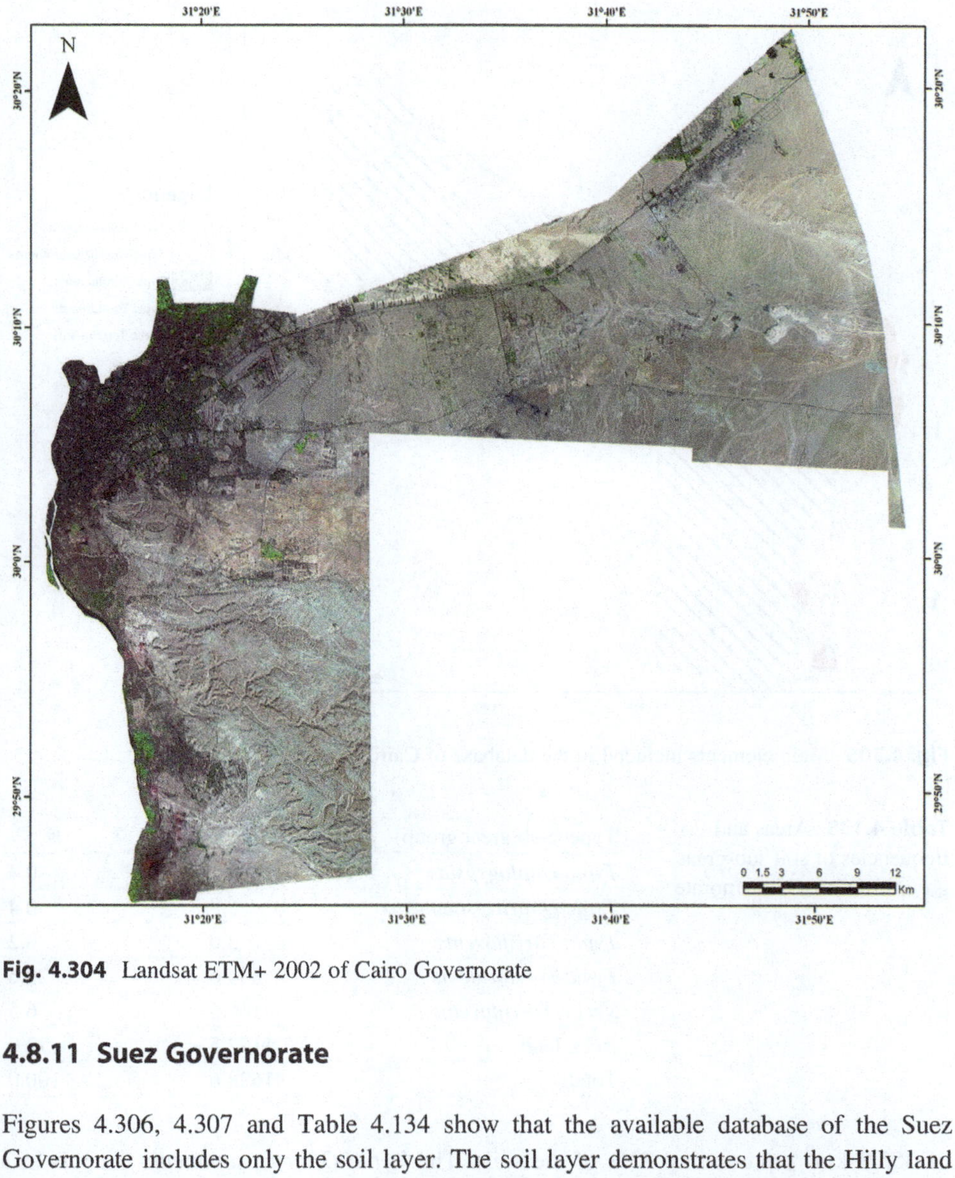

Fig. 4.304 Landsat ETM+ 2002 of Cairo Governorate

4.8.11 Suez Governorate

Figures 4.306, 4.307 and Table 4.134 show that the available database of the Suez Governorate includes only the soil layer. The soil layer demonstrates that the Hilly land has an extensive occurrence within the administrative boundaries of the Governorate, covering an area of 564.5 km^2 representing 36.5% of the soils. The rock land also covers an area of 138.8 km^2, representing 8.9% within the hilly land cover.

The *Typic Haplogypsids* and *Typic Petrogypsids* sub-great groups cover significant areas specially along the Suez canal. The total area of the *Typic Haplogypsids* soils amount 431.5 km^2, representing 27.9% while the *Typic Petrogypsids* exhibit 11.4 km^2 representing only 0.7%. The *Typic Torriorthents* sub-great group exist along the wadis covering an area

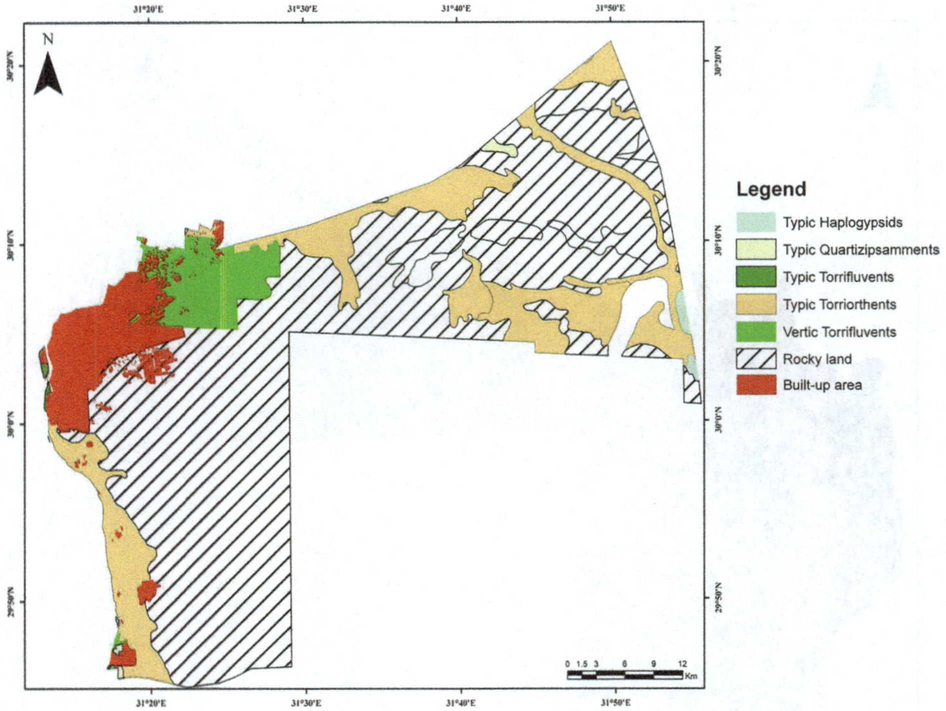

Fig. 4.305 Main elements included in the database of Cairo Governorate

Table 4.133 Areas and frequencies of soil sub-great groups of Cairo Governorate

Type (Sub great group)	Area (km^2)	%
Typic Haplogypsids	6.1	0.4
Typic Quartizpsamments	5.9	0.4
Typic Torrifluvents	3.0	0.2
Typic Torriorthents	334.6	20.6
Vertic Torrifluvents	106.5	6.5
Rock land	1172.5	72.0
Total	1628.6	100.0

of 355.3 km^2 representing 23% of the soils. The *Typic Quartizpsamments* extends in a limited area (43.3 km^2) along desert plains, representing 2.8% of the existent soils.

4.8.12 Ismailia Governorate

Figures 4.308, 4.309 and Table 4.135 show that the available database of the Ismailia Governorate includes only the soil layer. The soil layer demonstrates that the Hilly land has an extensive occurrence within the administrative boundaries of the Governorate,

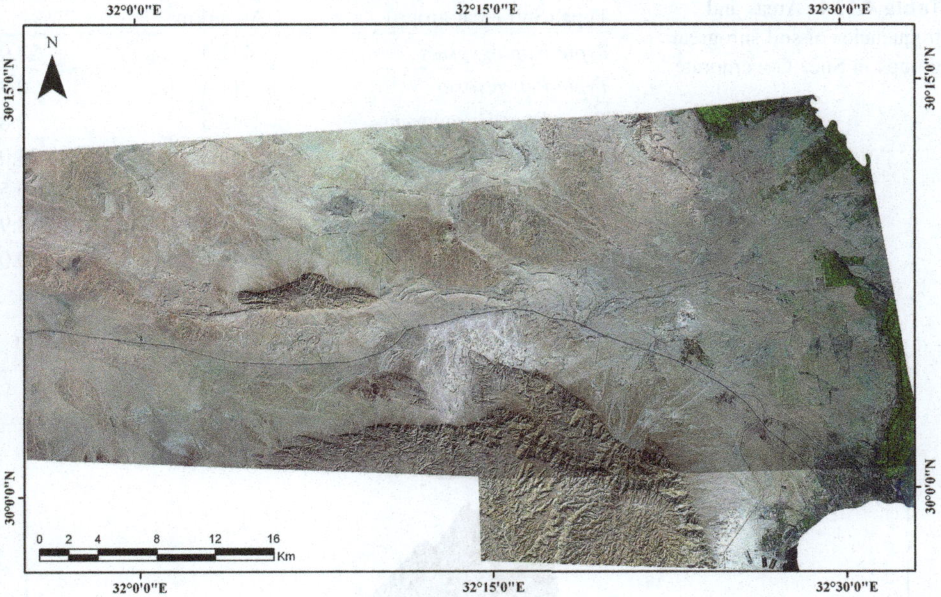

Fig. 4.306 Landsat ETM+ 2002 of Suez Governorate

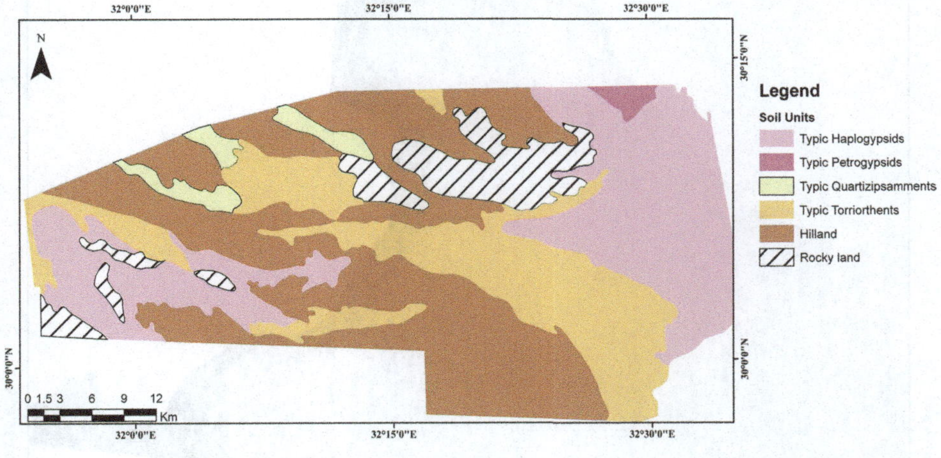

Fig. 4.307 Main elements included in the database of Suez Governorate

extending in the eastern part 531.4 km^2 representing 25.0% of the soils. The rock land also covers an area of 88.8 km^2, representing 4.2% within the hilly land coverage. The *Typic Quartizpsamments* have the widest soil coverage (788.4 km^2) at the north of the Governorate, representing 37.1% of the soils. The Typic Torriorthents sub-great group is the main soil type having alluvial nature, covering 385.5 km^2, and representing 18.2% of the

Table 4.134 Areas and frequencies of soil sub-great groups in Suez Governorate

Type (Sub great group)	Area (km^2)	%
Typic Haplogypsids	431.5	27.9
Typic Petrogypsids	11.4	0.7
Typic Quartizpsamments	43.3	2.8
Typic Torriorthents	355.3	23.0
Hilland	564.5	36.5
Rock land	138.8	8.9
Total	1544.8	100.0

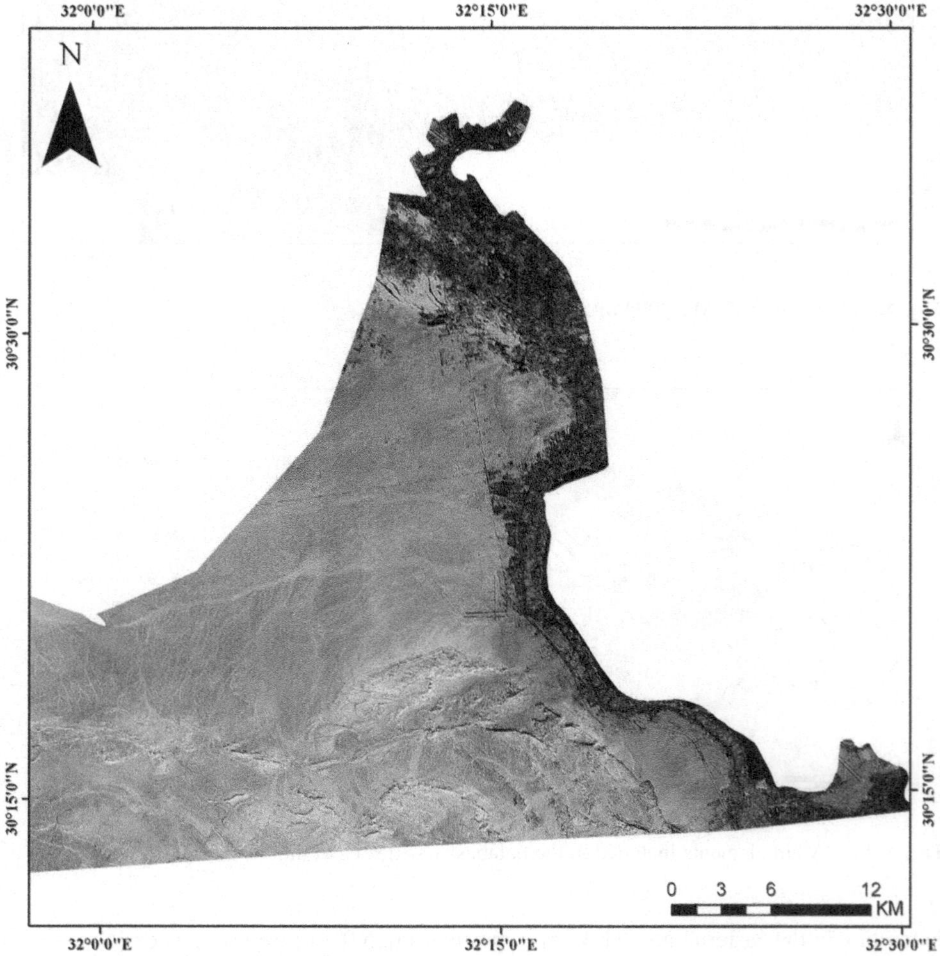

Fig. 4.308 Landsat ETM+ 2002 Ismailia Governorate

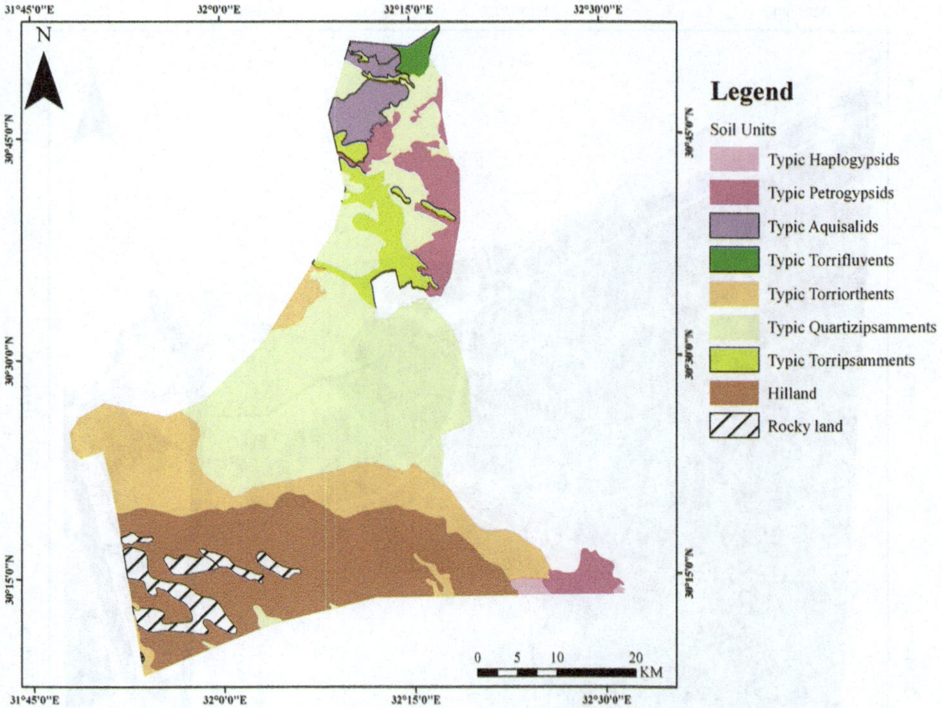

Fig. 4.309 Main elements included in the database of Ismailia Governorate

Table 4.135 Areas and frequencies of soil sub-great groups in Ismailia Governorate

Type (Sub great group)	Area (km^2)	%
Typic Aquisalids	68.9	3.3
Typic Haplogypsids	9.5	0.5
Typic Petrogypsids	134.1	6.3
Typic Quartizpsamments	788.4	37.1
Typic Torrifluvents	17.2	0.8
Typic Torriorthents	385.5	18.2
Typic Torripsamments	99.7	4.7
Hilly land	531.4	25.0
Rock land	88.8	4.2
Total	2123.7	100.0

Governorate soils. These soils are bordering the hilly land at the north, and might have been formed through the fluvial conditions due to torrential rain on the plateau. The sub-great groups *Typic Aquisalids, Typic Haplogypsids, Typic Petrogypsids, Typic Torripsamments and Typic Torrifluvents* cover small patches along the Suzie canal area at the north and south of the Governorate. They cover areas of 68.9, 9.5, 134.1, 99.7 and 17.2 km^2 representing 3.3%, 0.5%, 6.3%, 4.7% and 0.8% of the area respectively.

Fig. 4.310 Landsat ETM+ 2002 of Port Said Governorate

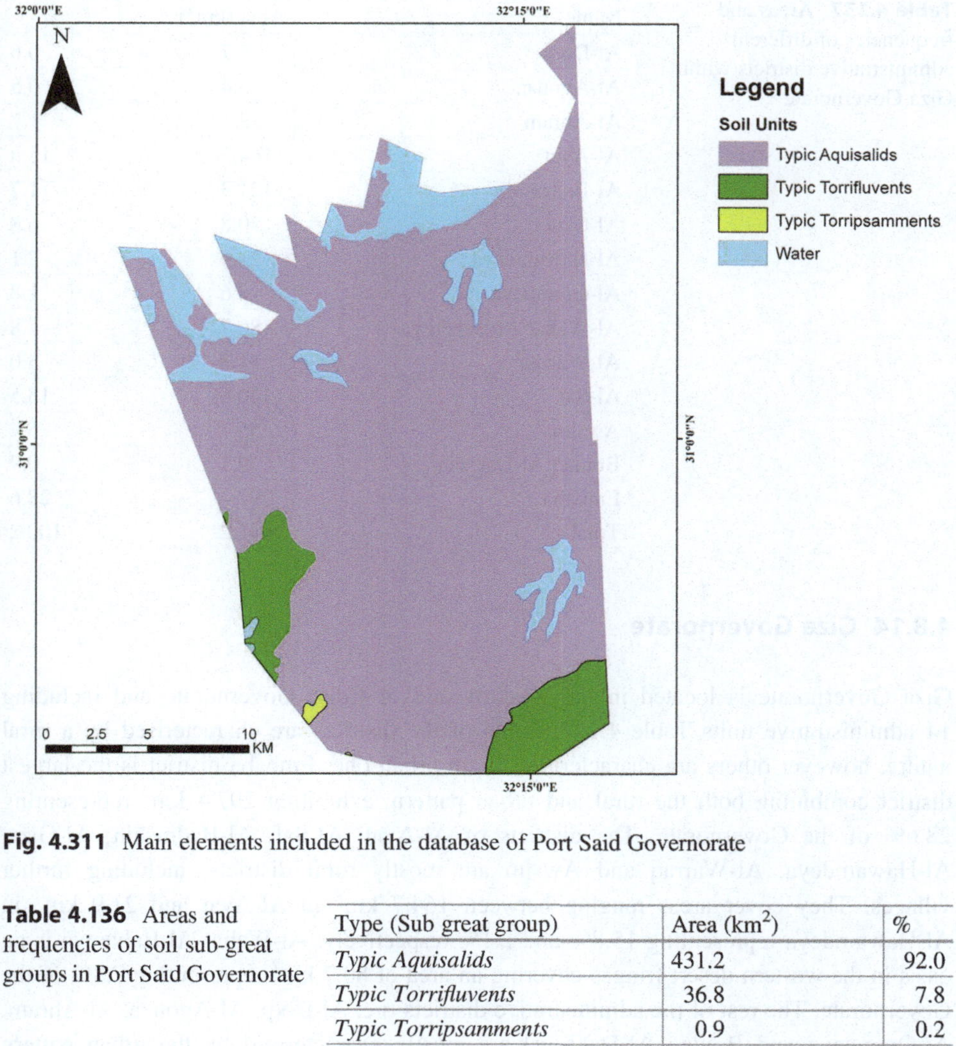

Fig. 4.311 Main elements included in the database of Port Said Governorate

Table 4.136 Areas and frequencies of soil sub-great groups in Port Said Governorate

Type (Sub great group)	Area (km^2)	%
Typic Aquisalids	431.2	92.0
Typic Torrifluvents	36.8	7.8
Typic Torripsamments	0.9	0.2
Total	468.9	100.0

4.8.13 Port Said Governorate

Figures 4.310, 4.311 and Table 4.136 show that the available database of Port Said Governorate includes only the soil layer. The soil layer shows that only three soil sub-great groups exist within the Governorate's boundaries. The sub-great group *Typic Aquisalids* is most extensive soils, covering an area of 431.2 km^2 and representing 92.0% of the soils area. The *Typic Torrifluvents* and *Typic Torripsamments* exhibit small patches of 36.8 and 0.9 km^2 representing 7.8% and 0.2% of the area respectively.

Table 4.137 Areas and frequencies of different administrative districts within Giza Governorate

Name	Area (km^2)	%
Al-Doqi	5.7	0.6
Al-Agouza	5.4	0.5
Al-Ahram	18.1	1.7
Al-Ayat	164.7	15.8
Al-Badrashin	121.3	11.7
Al-Giza	70.8	6.8
Al-Hawamdeya	21.9	2.1
Al-Omraneya	18.6	1.8
Al-Wahat Al-Bahareya	80.7	7.8
Al-Warraq	31.3	3.0
Al-Saf	140.6	13.5
Awsim	55.7	5.4
Boulaq Al-Dakrour	9.1	0.9
Embaba	297.4	28.6
Total	1041.2	100.2

4.8.14 Giza Governorate

Giza Governorate is located in the western side of Cairo Governorate and including 14 administrative units Table 4.137. Some of the districts are characterised by a rural nature, however others are characterised by an urban one. Embaba district is the largest district combining both the rural and urban pattern, exhibiting 297.4 km^2 representing 28.6% of the Governorate. The districts of Al-Ayat, Al-Saf, Al-Badrashin, Al-Giza, Al-Hawamdeya, Al-Warraq and Awsim are mostly rural districts, including further villages. They cover areas ranging between 164.7 km^2 in Al-Ayat and 21.9 km^2 in Al-Hawamdeya representing 15.8% and 2.1% respectively. Al-Wahat Al-Bahareya is an oasis in the western desert fringes covering an area of 80.7 km^2 representing 7.8% of the Governorate. The rest of the administrative districts (i.e. Al-Doqi, Al-Agouza, Al-Ahram, Al-Omraneya and Boulaq Al-Dakrour) are mostly characterised by the urban pattern having relatively smaller areas compared with the previous districts.

Figures 4.312, 4.313 and Table 4.138 show the main elements included in the database of Giza Governorate. The soil layer demonstrates that the rock land has an extensive occurrence within the administrative boundaries of the Governorate, extending 1674.0 km^2 in the western desert fringes. The sub-great groups *Typic Torriorthents* and *Vertic Torrifluvents* are the dominant soils, where they cover 867.3 km^2 and 551.6 km^2 representing 24.5% and 15.6% respectively. The *Typic Torriorthents* is associated with old deltaic deposits in the western desert fringes, while the *Vertic Torrifluvents* is associated with the recent terraces of the flood plain. The sub-great group *Typic Torrifluvents* associates the previous ones in the Nile alluvial plain, covering relatively a smaller area (114.2 km^2) representing 3.2% of the total. Small patches of Typic *Calciorthids, Typic*

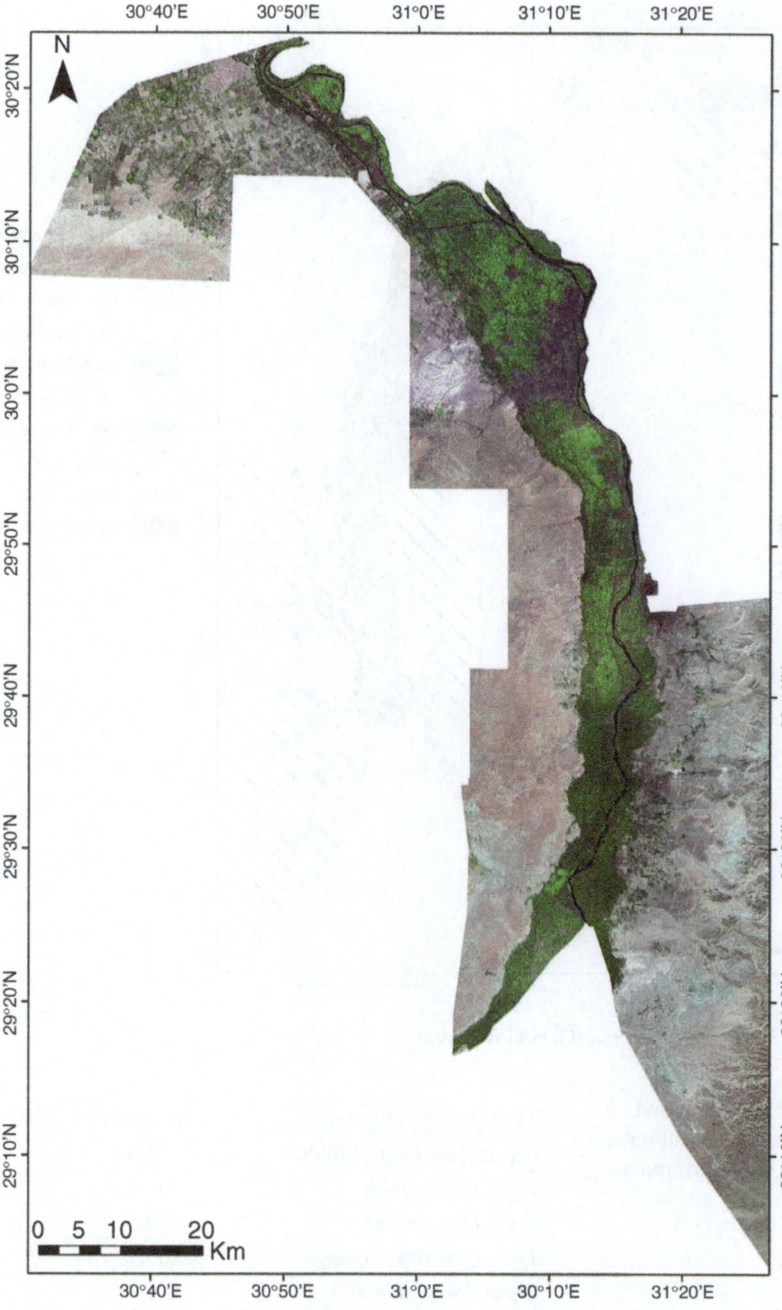

Fig. 4.312 Landsat ETM+ 2002 of Giza Governorate

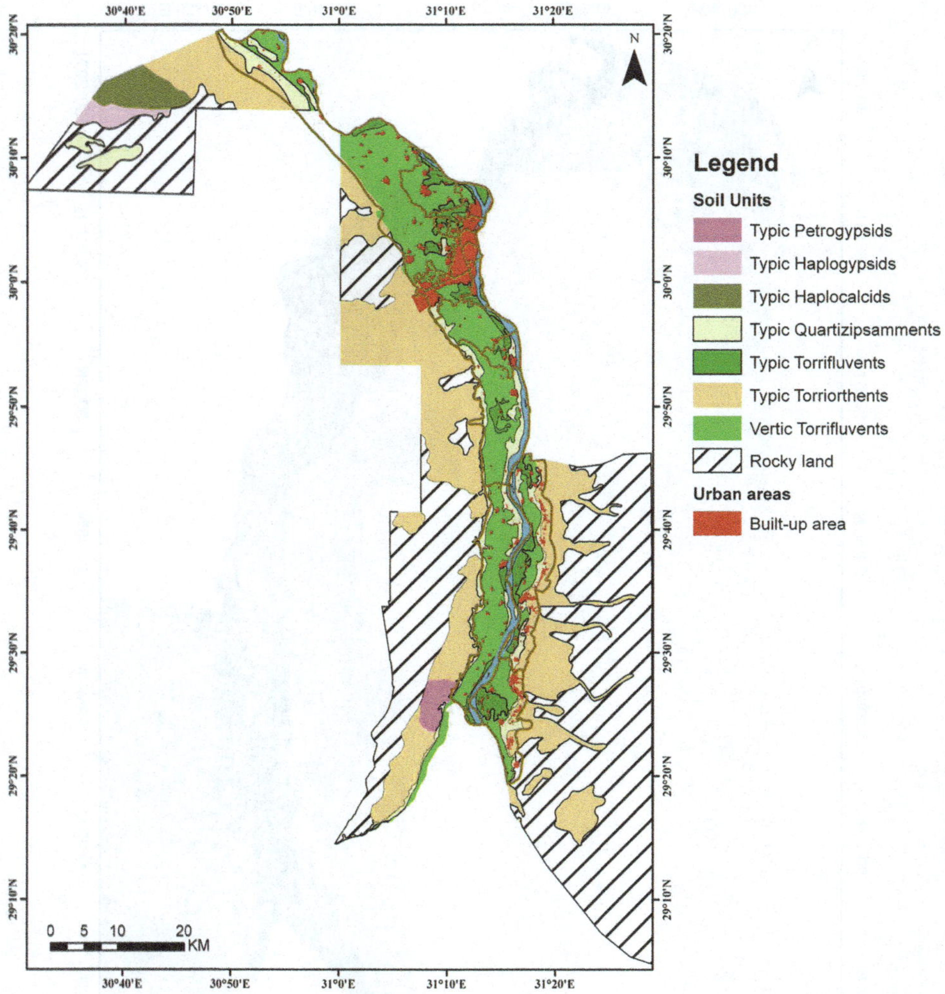

Fig. 4.313 Database layers ofGiza Governorate

Table 4.138 Areas and frequencies of soil sub-great groups in Giza Governorate

Type (Sub great group)	Area (km^2)	%
Petrogypsic Gypsiorthids	28.0	0.8
Typic Calciorthids	61.7	1.7
Typic Gypsiorthids	30.3	0.9
Typic Quartizpsamments	209.3	5.9
Typic Torrifluvents	114.2	3.2
Typic Torriorthents	867.3	24.5
Vertic Torrifluvents	551.6	15.6
Rock land	1674.4	47.3
Total	3536.8	100.0

Table 4.139 Areas and frequencies of different administrative districts within Bani Suef Governorate

Name	Area (km^2)	%
Ahnasya	205.5	14.7
Al-Fashn	249.7	17.8
Al-Wasta	205.1	14.7
Bani Swaif	204.7	14.6
Beba	192.9	13.8
Naser	212.9	15.2
Samasta	128.9	9.2
Total	1399.8	100.0

Gypsiorthids and *Petrogypsic Gypsiorthids* spread on the west of the alluvial plain covering areas of 61.7, 30.3 and 28.0 km^2 representing 1.7%, 0.9% and 0.8% respectively.

4.8.15 Bani Suef Governorate

Bani Suef Governorate is located to the south of Giza Governorate, including 7 administrative units Table 4.139. Al-Fashn district is the largest one, covering an area of 249.7 km^2, representing 17.8% of the total area. Ahnasya, Al-Wasta, Beni Swaif and Beba have almost similar coverage ranging between 13.8and 14.7% of the Governorate area. Samasta district occupy the smallest area, as it covers 128.9 km^2 representing 9.2% of the whole area.

Figures 4.314, 4.315 and Table 4.140 show the main elements included in the database of Bani Suef Governorate.

The soil layer demonstrates that the rock land has an extensive occurrence within the administrative boundaries of the Governorate, extending 1730.3 km^2 in both western and eastern desert fringes of the Nile valley. The sub-great group *Vertic Torrifluvents* is the dominant soil sub-great group, it covers an area of 1065.4 km^2 representing 30.0% of the mapped soils. Inclusion of *Typic Torrifluvents* within *Vertic Torrifluvents* sub-great groups occurs within the Nile flood plain covering an area of 167.4 km^2 representing 4.7% of soil units area. These soils are associated with the recent terraces of the flood plain. The *Typic Torriorthents* associated with the *Typic Quartizpsamments* exhibit the wadies of the eastern desert and the Nile Valley interference zone with the western desert. They cover areas of 286.0 and 239.1 km^2 representing 8.1 and 6.7% respectively. Small occurrence of the *Typic Pertrogypsids* and *Typic Haplocalcids* exist in the western borders of the flood plain and the eastern borders of Al-Fayoum depression. They cover areas of 57.2 and 2.2 km^2 representing 1.6 and 0.1% respectively.

Two main irrigation canals and drain categories are shown in the database layout of the Governorate, lengths of each category are presented in Table 4.141. The irrigation canals of more than 25 m width have a total length of 177.5 km, while those of 10–25 m width attain

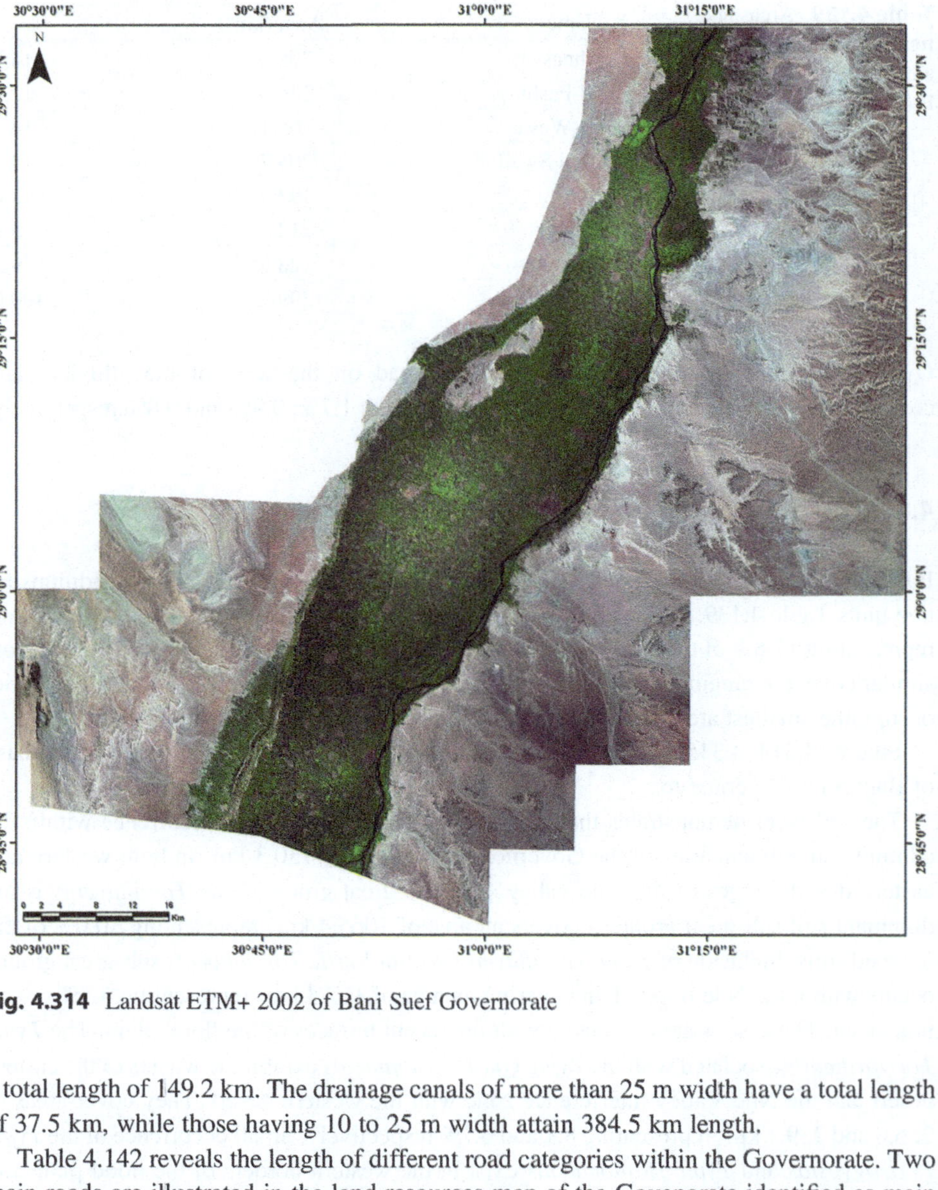

Fig. 4.314 Landsat ETM+ 2002 of Bani Suef Governorate

a total length of 149.2 km. The drainage canals of more than 25 m width have a total length of 37.5 km, while those having 10 to 25 m width attain 384.5 km length.

Table 4.142 reveals the length of different road categories within the Governorate. Two main roads are illustrated in the land resources map of the Govenorate identified as main paved and secondary paved attaining lengths of 100.7 and 900.8 km respectively.

The land resources database, moreover, includes the layers of urban areas and utilities. The urban area amount 53.5 km^2 and the total length of utilities is 324.9 km.

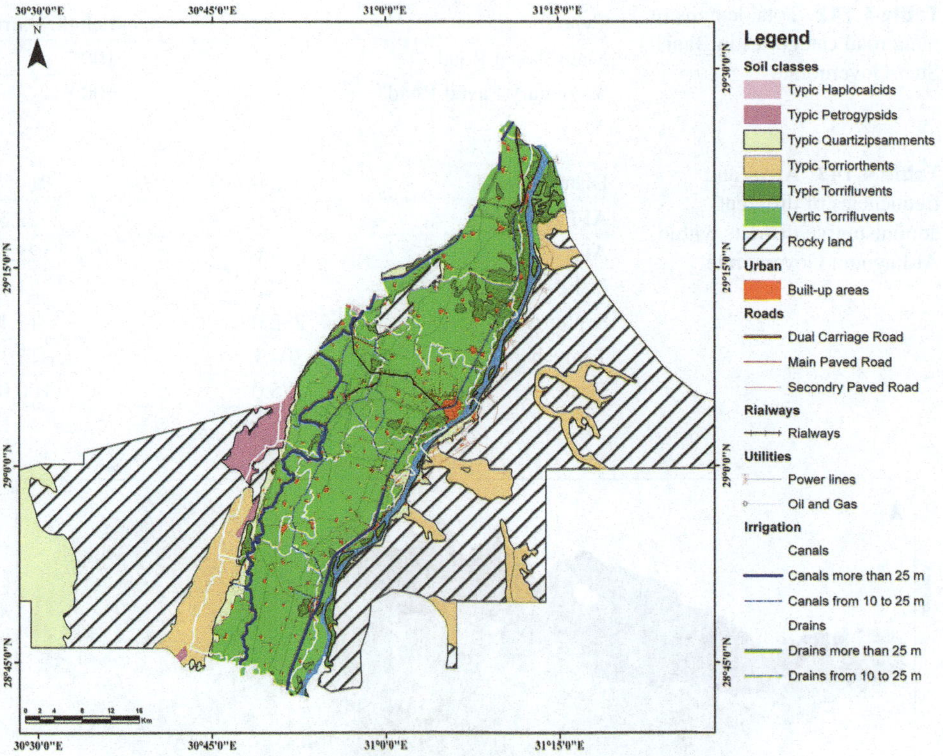

Fig. 4.315 Database layers of Bani Suef Governorate

Table 4.140 Areas and frequencies of soil sub-great groups in Bani Suef Governorate

Sub-Great Group	Area (km^2)	%
Vertic Torrifluvents	1065.4	30.0
Typic Torriorthents	286.0	8.1
Typic Quartizpsamments	239.1	6.7
Typic Torrifluvents	167.4	4.7
Typic Petrogypsids	57.2	1.6
Typic Haplocalcids	2.2	0.1
Rock land	1730.3	48.8
Total	7463.6	100.0

Table 4.141 Total length of the main irrigation and drain canals in Bani Suef Governorate

Type	Length (km)
Canals more than 25 m	177.5
Canals from 10 to 25 m	149.2
Drains more than 25 m	37.5
Drains from 10 to 25 m	384.5

Table 4.142 Total lengths of main road categories in Bani Suef Governorate

Type	Length (km)
Main Paved Road	100.7
Secondary Paved Road	900.8

Table 4.143 Areas and frequencies of different administrative districts within Al-Fayoum Governorate

District Name	Area (km^2)	%
Al-Fayoum	395.9	22.3
Atsa	454.6	25.6
Ibsheway	388.3	21.9
Senouris	232.0	13.1
Tamyah	304.1	17.1
Total	1775.0	100.0

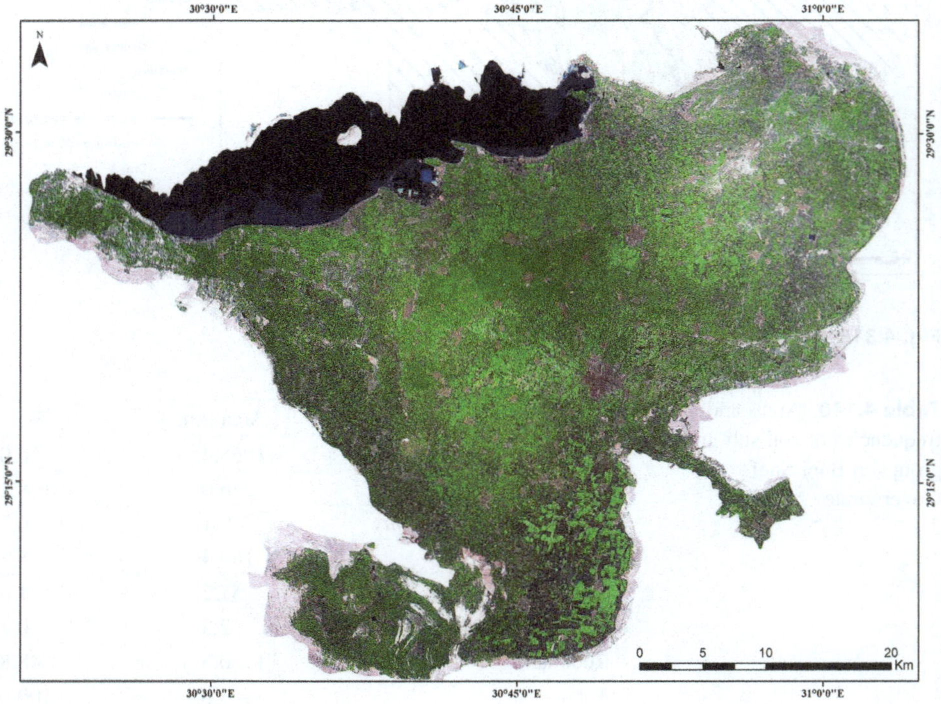

Fig. 4.316 Landsat ETM+ 2002 of Al-Fayoum Governorate

4.8.16 Al-Fayoum Governorate

Al-Fayoum Governorate is administratively sub-divided into 5 districts, Table 4.143. The areas of the districts can be arranged as follows Atsa > Al-Fayoum > Ibsheway > Tamyah > Senouris districts having areas of 395.9, 454.6, 388.3, 304.1 and 232.0 km^2 respectively.

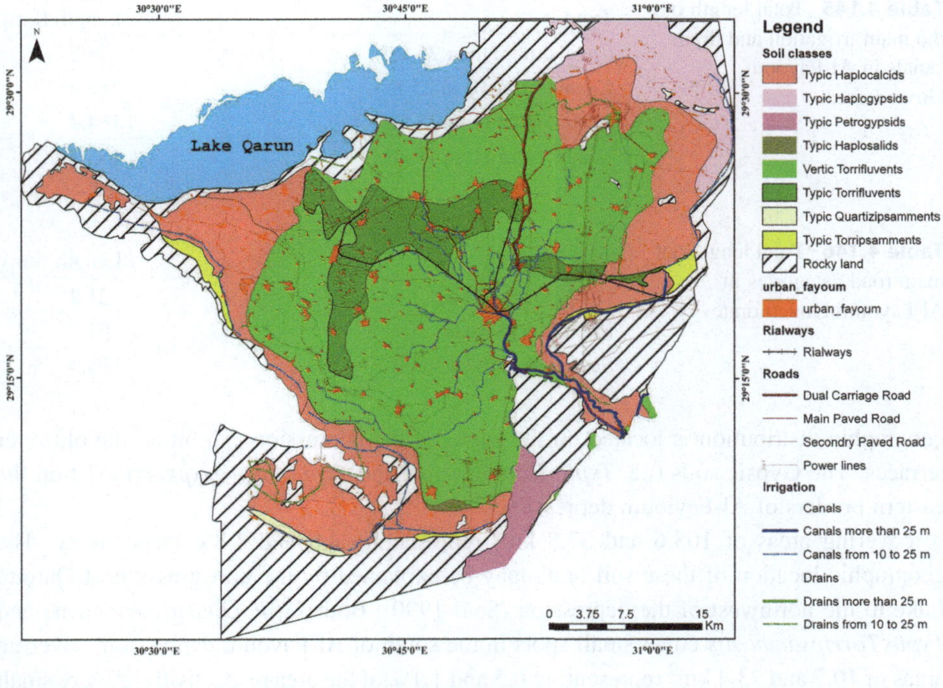

Fig. 4.317 Database layers of Al-Fayoum Governorate

Table 4.144 Areas and frequencies of soil sub-great groups in Al-Fayoum Governorate

Soil class	Area (km^2)	%
Vertic Torrifluvents	792.6	37.7
Typic Haplocalcids	444.1	21.1
Typic Torrifluvents	148.0	7.0
Typic Haplogypsids	103.5	4.9
Typic Petrogypsids	37.8	1.8
Typic Torripsamments	23.4	1.1
Typic Quartizpsamments	10.7	0.5
Typic Haplosalids	3.6	0.2
Rock land	539.8	25.7
Total	2103.4	100.0

Figures 4.316, 4.317 and Table 4.144 show the main elements included in the database of Al-Fayoum Governorate. The soil layer demonstrates that the sub-great group *Vertic Torrifluvents* is the dominant soil sub-great group, it covers an area of 792.6 km^2 representing 37.7% of the mapped soils. Inclusion of *Typic Torrifluvents* occurs within the depression, covering an area of 148.0 km^2 representing 7.0% of the mapped soils. These soils are associated with the recent terraces of the flood plain. The sub-great group of *Typic Haplocalcids* covers an area of 444.1 km^2 representing 21.1% of soil units area. Its

Table 4.145 Total length of the main irrigation and drain canals in Al-Fayoum Governorate

Type	Length (km)
Canals more than 25 m	75.2
Canals from 10 to 25 m	282.4
Drains more than 25 m	184.4
Drains from 10 to 25 m	135.7

Table 4.146 Total lengths of main road categories in Al-Fayoum Governorate

Type	Length (km)
Dual carriage road	21.2
Main Paved Road	97.7
Secondary Paved Road	598.4

geographic distribution is located on the edges of the depression exhibiting the old river terraces. The Gypsic soils (i.e. *Typic Haplogypsids* and *Typic Pertrogypsids*) exist on the eastern borders of Al-Fayioum depression.

covering areas of 103.6 and 37.8 km^2 representing 4.9 and 1.8% respectively. The geographic location of these soil units may be explained by the transgression of Qaroun Lake to the northwest of the depression (Said 1990). Both *Typic Quartizpsamments* and *Typic Torripsamments* cover small spots in the south of Al-Fayoum depression, covering areas of 10.7 and 23.4 km^2 representing 0.5 and 1.1% of the area respectively. A very small patch of the sub-great group *Typic Haplosalids* exist in the south east of the depression exhibiting an area of 3.6 km^2 representing 0.2% of the mapped soils. It should be noticed that such variability of sub-great groups is unique for Al-Fayoum Governorate due to its location, altitude, and formation processes pattern and agricultural practices.

Two main irrigation and drain canals are shown in the database layout of Al-Fayoum Governorate, lengths of each canal are presented in Table 4.145. The irrigation canals of more than 25 m width have a total length of 75.2 km, while those of 10–25 m width attain a total length of 282.4 km. The drainage canals of more than 25 m width have a total length of 184.4 km, while those having 10 to 25 m width attain 135.7 km length.

Table 4.146 reveals that three road categories are spread within the Governorate. Total lengths of Dual carriage, Main paved and secondary paved roads are 21.2, 97.7 and 598.4 km respectively.

Urban areas, utilities and railways are included in the land resources database of Al-Fayoum Governorate. The urban area covers 46.2 km^2, while the length of utilities and railways are 532.8 and 77.2 km respectively.

4.8.17 Al-Minya Governorate

Al-Minya Governorate is located to the south of Bani Suef Governorate including 9 administrative units Table 4.147. Samalout district is the largest one, covering an area of

Table 4.147 Areas and frequencies of different administrative districts within Al-Minya Governorate

Name	Area (km^2)	%
Abu Qurqas	288.8	12.4
Al-Adwah	143.8	6.2
Al-Minya	351.3	15.0
Bani Mazar	282.5	12.1
Dayr Mowas	179.6	7.7
Maghagha	230.2	9.8
Matay	171.8	7.4
Melawi	327.6	14.0
Samalout	362.5	15.5
Total	2338.2	100.0

362.5 km^2, representing 15.5% of the total area. Al-Minya district extends in an area of 351.3 km^2 representing 15.0%. Al-Minya district is the Governorate capital have Al-Minya University and South Valley radio and TV broadcast. Melawi district exhibits a coverage of 327.6 km^2 representing 14.0% of the whole area. Abu Qurqas and Bani Mazar districts have mostly similar area coverage (i.e 288.8 and 282.5 km^2 respectively). The districts of Maghagha, Dayr Mowas, Matay and Al-Adwah extend for relatively small areas figuring 230.2, 179.6, 171.8 and 143.8 km^2 representing 9.9, 7.7, 7.4 and 6.2% of the Governorate area respectively.

Figures 4.318, 4.319 and Table 4.148 show the main elements included in the database of the Al-Minya Governorate. The soil layer demonstrates that the rock land has an extensive occurrence within the administrative boundaries of the Governorate, extending 1852.1 km^2 in the both western and eastern desert fringes. The sub-great group *Vertic Torrifluvents*, with inclusion of *Typic Torrifluvents,* is the main soil unites in the Nile flood plain in the governorate. These two soil units cover areas of 1632.7 and 270.7 km^2 representing 31.9% and 5.3% of the mapped soils respectively.

The *Typic Torriorthents* associated with the *Typic Quartizpsamments* sub-great groups exhibit the interference zones between the Nile Valley and the desert fringes. They cover areas of 689.1 and 434.1 km^2 representing 13.5 and 8.5% of the area respectively. The *Typic Pertrogypsids* occupy an area of 241.0 km^2 representing 4.7% of the mapped soils at the Northwestern part of the Governorate.

Two main irrigation and drain canals are included in the database layout of Al-Minya Governorate, lengths of each type are presented in Table 4.149. The irrigation canals of more than 25 m width have a total length of 342.5 km, while those of 10–25 m width attain a total length of 218.2 km. The drainage canals of more than 25 m width have a total length of 115.5 km, while those having 10 to 25 m width attains 318.2 km length.

Table 4.150 reveals that two road categories are spread within the Governorate. Total lengths of main paved and secondary paved roads are 142.8 and 821.7 km respectively.

Fig. 4.318 Landsat ETM+ 2002 of Al-Minya Governorate

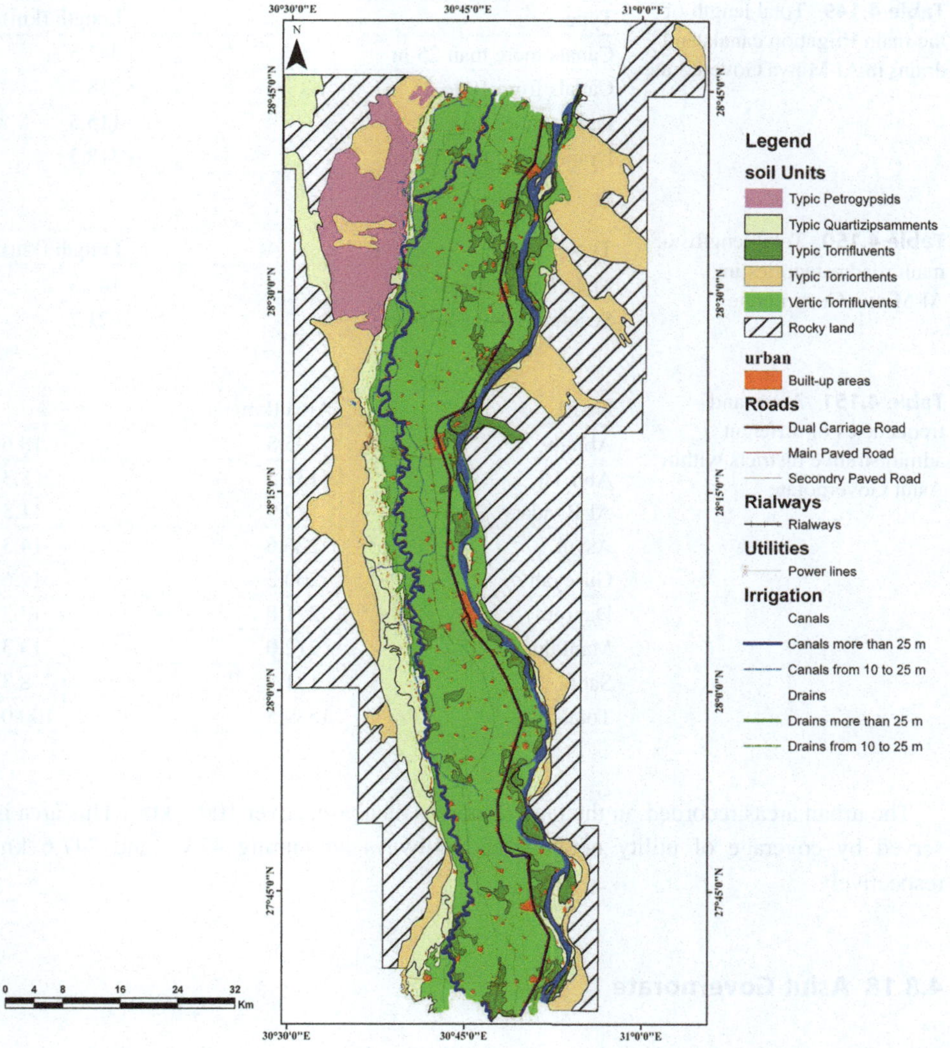

Fig. 4.319 Database layers of Al-Minya Governorate

Table 4.148 Areas and frequencies of soil sub-great groups in Al-Minya Governorate

Soil class	Area (km^2)	%
Vertic Torrifluvents	1632.7	31.9
Typic Torriorthents	689.1	13.5
Typic Quartizpsamments	434.1	8.5
Typic Torrifluvents	270.7	5.3
Typic Petrogypsids	241.0	4.7
Rock land	1852.1	36.2
Total	5119.6	100.0

Table 4.149 Total length of the main irrigation canals and drains in Al-Minya Governorate

Type	Length (km)
Canals more than 25 m	342.5
Canals from 10 to 25 m	218.2
Drains more than 25 m	115.5
Drains from 10 to 25 m	318.2

Table 4.150 Total lengths of main road categories in Al-Minya Governorate

Type	Length (km)
Main Paved Road	142.8
Secondary Paved Road	821.7

Table 4.151 Areas and frequencies of different administrative districts within Asiut Governorate

Name	Area (km^2)	%
Abnub	295.8	18.6
Abu Tij	131.5	8.3
Al-Bbadary	179.9	11.3
Asyut	226.6	14.3
Guseyah	200.2	12.6
Dayrout	211.8	13.3
Manfalut	212.0	13.3
Sadfa	131.7	8.3
Total	1589.5	100.0

The urban areas recorded, in the land resources database, cover 100.1 km^2. This area is served by coverage of utility network and railways amounting 423.3 and 247.6 km respectively.

4.8.18 Asiut Governorate

Asiut Governorate is located to the south of Al-Minya Governorate including 8 administrative units Table 4.151. Abnub district is the largest one, covering an area of 295.8 km^2, representing 18.6% of the total area. Asiut district extends in an area of 226.6 km^2 representing 14.3%. The importance of Asiut district is related to its inclusion of Asiut capital and Asiut University. Melawi district exhibits a coverage of 327.6 km^2 representing 14.0% of the whole area. Manfalut and Dayrout districts have mostly similar area coverage (i.e 212.0 and 211.8 km^2 respectively). The districts of Guseyah, Al-Badarry, Sadfa and Abu Tij extend for relatively smaller areas figuring 200.2, 179.9, 131.7 and 131.5 km^2 representing 12.6, 11.3, 8.3 and 8.3% of the Governorate area respectively.

Figures 4.320, 4.321 and Table 4.152 show the main elements included in the database of Asiut Governorate. The soil layer demonstrates that the rock land has an extensive

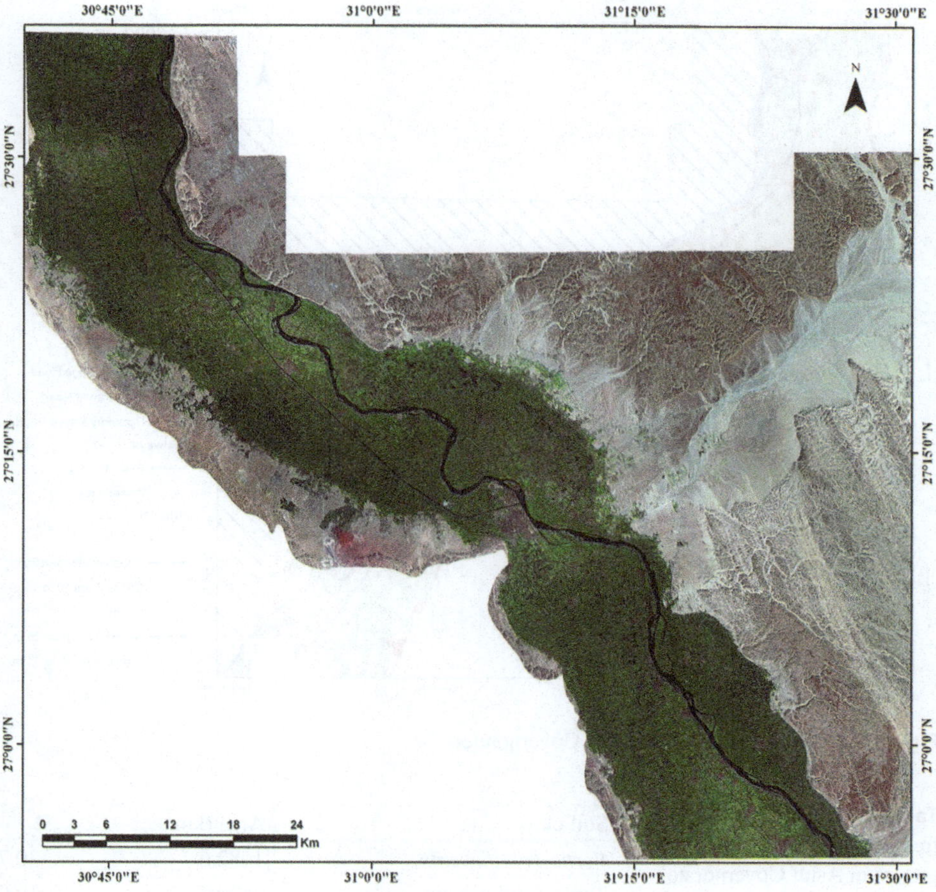

Fig. 4.320 Landsat ETM+ 2002 of Asiut Governorate

occurrence within the administrative boundaries of the Governorate, reaching 1359.5 km^2 representing 38.6% of the mapped area. The sub-great groups Vertic Torrifluvents, with inclusion of Typic Torrifluvents, are the main soil unites in the Nile alluvial plain.

They cover areas of 1082.6 and 231.6 km^2 representing 30.7 and 6.6% of the mapped soils respectively. The *Typic Torriorthents* sub-great group exhibits the wadis in the eastern desert, covering an area of 670.2 km^2. The *Typic Quartizpsamments* sub-great group exhibits the eastern desert wadi mouths, in addition to the interference zone between the Valley and western desert. It covers an area of 179.4 km^2 representing 5.1% of the total soils area.

The irrigation and drain canals are included in the database layout of Asiut Governorate; lengths of each category are presented in Table 4.153. The irrigation canals of more than 25 m width have a total length of 161.6 km, while those of 10–25 m width attain a total length of 206.4 km. The drainage canals of more than 25 m width have a total length of 136.5 km, while those having 10 to 25 m width attains 275.2 km length.

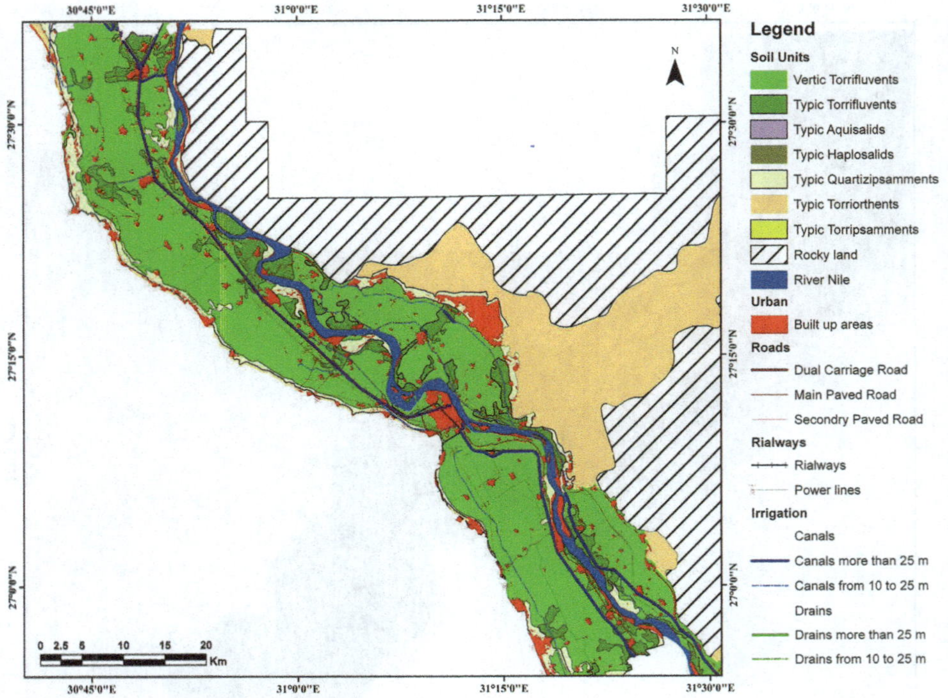

Fig. 4.321 Database layers of Asiut Governorate

Table 4.152 Areas and frequencies of soil sub-great groups in Asiut Governorate

Soil class	Area (km^2)	%
Vertic Torrifluvents	1082.6	30.7
Typic Torriorthents	670.3	19.0
Typic Torrifluvents	231.6	6.6
Typic Quartizpsamments	179.4	5.1
Rock land	1359.5	38.6
Total	3523.4	100.0

Table 4.153 Total length of the main irrigation and drain canals in Asiut Governorate

Type	Length (km)
Canals more than 25 m	161.6
Canals from 10 to 25 m	206.4
Drains more than 25 m	136.5
Drains from 10 to 25 m	275.2

Table 4.154 reveals that two road categories are spread within the Governorate. Total lengths of the main paved and the secondary paved roads are 134.3 and 855.5 km respectively.

Table 4.154 Total lengths of main road categories in Asiut Governorate

Type	Length (km)
Main Paved Road	134.3
Secondary Paved Road	855.5

Table 4.155 Areas and frequencies of different administrative districts within Suhag Governorate

Name	Area (km^2)	%
Akhmim	97.3	6.1
Ghohayna Al-gharbeiyyah	81.2	5.1
Jirja	149.7	9.4
Saqaltah	82.2	5.2
Suhaj	187.2	11.8
Tahta	171.6	10.8
Tama	148.7	9.4
Al-Belina	163.6	10.3
Al-Maraghah	130.7	8.2
Al-Monshaah	181.4	11.4
Dar Al-Salam	194.4	12.2
Total	1588.1	100.0

The urban areas of Asiut Governorate exhibit 215 km^2, served by 855.5 km utility network. The railways network records 120.9 km in the land resources database of the Governorate.

4.8.19 Suhag Governorate

Suhag Governorate is located to the south of Asiut Governorate and including 11 administrative units Table 4.155. The five districts (Dar Al-Salam, Suhaj, Al-Monshaah, Tahta and Al-Belina) are relatively larger districts than the others, covering areas ranging between 194.4 and 171.6 km^2 representing 12.2 to 10.3% of the total Governorate area respectively. The districts Jirja, Tama, and Al-Maraghah are considered as medium area coverage, as they extend 149.7, 148.7 and 130.7 km^2 representing 9.4, 9.4 and 8.2% respectively. The three relatively smaller districts area are Akhmim, Saqaltah and Ghohayna Al-Gharbiyyah cover areas of 97.3, 82.2 and 81.2 km^2 representing 6.1, 5.1 and 5.1% of total urban area respectively.

Figures 4.322, 4.323 and Table 4.156 show the main elements included in the database of Suhag Governorate. The soil map demonstrates that the rock land has an occurrence of 804.0 km^2, representing 28.7% of the mapped soils. Four soil sub-great groups exist in the Governorate. The soils of *Vertic Torrifluvents* with *Typic Torrifluvents* exhibit the Nile alluvial plain part in the Governorate and covering areas of 926.2 and 507.3 km^2 representing 33.0 and 18.1% respectively. The sub-great group *Typic Torriorthents*, with

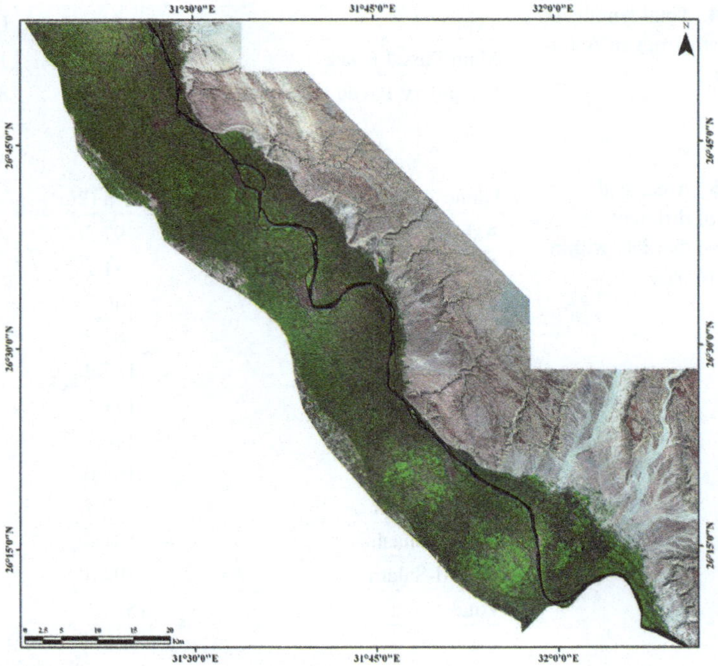

Fig. 4.322 Landsat ETM+ 2002 of Suhag Governorate

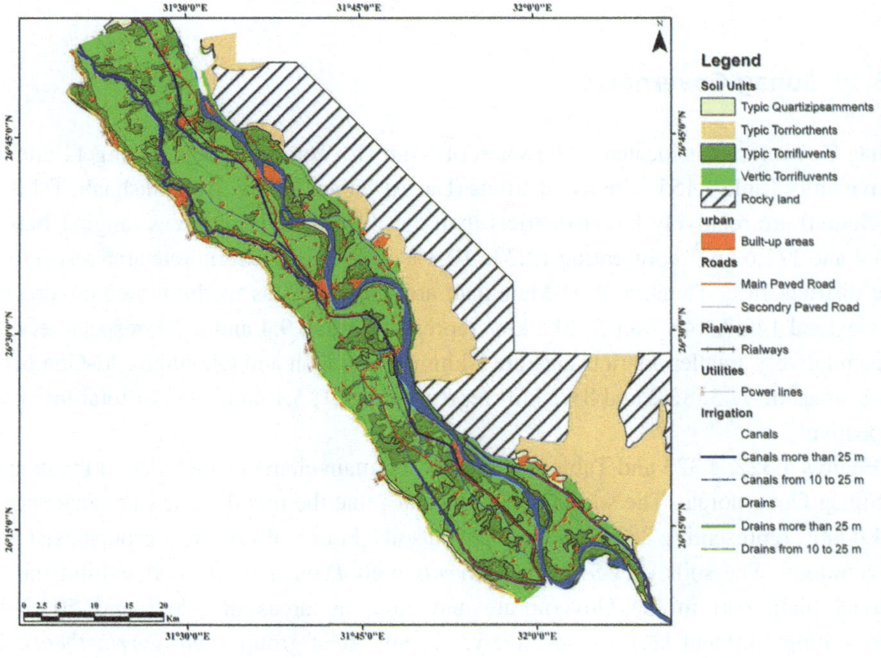

Fig. 4.323 Database layers of Suhag Governorate

Table 4.156 Areas and frequencies of soil sub-great groups in Sohag Governorate

Soil class	Area (km^2)	%
Vertic Torrifluvents	926.2	33.0
Typic Torriorthents	513.0	18.3
Typic Torrifluvents	507.3	18.1
Typic Quartizpsamments	53.1	1.9
Rock land	804.0	28.7
Total	2803.5	100.0

Table 4.157 Total length of the main irrigation and drain canals in Sohag Governorate

Type	Length (km)
Canals more than 25 m	258.8
Canals from 10 to 25 m	163.8
Drains more than 25 m	209.5
Drains from 10 to 25 m	284.6

Table 4.158 Total lengths of main road categories in Sohag Governorate

Type	Length (km)
Main Paved Road	180.5
Secondary Paved Road	839.2

inclusion of patches exhibited the *Typic Quartizpsamments* sub-great groups, exist in the desert plains and interference zone between the alluvial soils and desert soils. These soils cover areas of 513.0 and 53.1 km^2 representing 18.3 and 1.90% of the mapped soils respectively.

The irrigation and drain canals are shown in the database layout of Sohag Governorate, lengths of each type are presented in Table 4.157. The irrigation canals of more than 25 m width have a total length of 258.8 km, while those of 10–25 m width attain a total length of 163.8 km. The drainage canals of more than 25 m width have a total length of 209.5 km, while those having 10 to 25 m width attains 248.6 km length.

Table 4.158 reveals that two road categories are spread within the Governorate. Total lengths of Main paved and secondary paved roads are 180.5 and 839.2 km respectively.

The urban areas of Sohag Governorate exhibit 221.3 km^2, served by 191.4 km utility network. The railways network records 109.2 km in the land resources database of the Governorate.

4.8.20 Qena Governorate

The Qena Governorate is located to the south of Souhag Governorate including 12 administrative units Table 4.159. Isna district is the largest one, covering an area of 292.9 km^2, representing 15.2% of the total area. Luxor extends in an area of 241.0 km^2 representing

Table 4.159 Areas and frequencies of different administrative districts within Qena Governorate

Name	Area (km^2)	%
Abu Tesht	217.9	11.3
Al-Uqsur	241.0	12.5
Al-Waqf	45.3	2.4
Arment	112.9	5.9
Dishna	194.8	10.1
Farshout	50.2	2.6
Isna	292.9	15.2
Naj Hammadi	210.4	10.9
Naqada	91.1	4.7
Qaft	60.4	3.1
Qina	226.6	11.7
Qus	186.0	9.6
Total	1929.6	100.0

12.5%. It is considered as one of the greatest capitals of the ancient world. Charming and evocative, with the Nile banks lined with modern hotels and the feluccas sailing along the quiet waters of the river. The tourists are fascinated by such great, ancient city of Thebes, capital of the Egyptian empire for almost one thousand years. The districs of Qena, Abu Tesht, Naj Hammadi, Dishna and Qus have almost similar coverages ranging between 11.7 and 9.6 km^2. Naj Hammadi district has a special economic importance, as it includes both Aluminium and sugar industries, in addition to the existence of Naj Hammadi Barrages that regulate the Nile stream water. The districts Arment, Naqada, Qaft, Farshout and Al-Waqf extend for small area coverages ranging from 45.3 to 112.9 km^2 representing 2.4 to 5.9% of the Governorate.

Regarding the soil layer in the land resources database, Figs. 4.324 and 4.325 and Table 4.160 show that the *Typic Torriorthents* soil sub-great group has the most extensive coverage extending for an area of 4152.0 km^2 representing 58.5% of the soil. These soils exhibit the wadi bottoms and plains as the Wadi Qena on the eastern side of the Nile Valley. The Valley is dominated by the soil sub-great group Vertic Torrifluvents, associated with Typic Torrivlufents which extend for areas of 1043.3 km^2 and 359.1 km^2 representing 14.7% and 5.1% of the Governorate soils respectively. The sub-great group *Typic Quartizpsamments* spreads on desert fringes adjacent to the Nile plain, covering an area of 115.4 km^2 representing 1.6% of total soil coverage. The rock land on is very near from the western edges of the Nile Valley. In the mapped area, the rock land covers 1422.4 km^2 representing 20.1% of the mapped coverage.

Two main irrigation canals and drains categories are shown in the database layout of the Governorate, lengths of each category are presented in Table 4.161. The irrigation canals of more than 25 m width have a total length of 180.2 km, while those of 10–25 m width attain a total length of 341.6 km. The drainage canals of more than 25 m width have a total length of 195.1 km, while those having 10 to 25 m width attains 470.5 km length.

Fig. 4.324 Landsat ETM+ 2002 of Qena Governorate

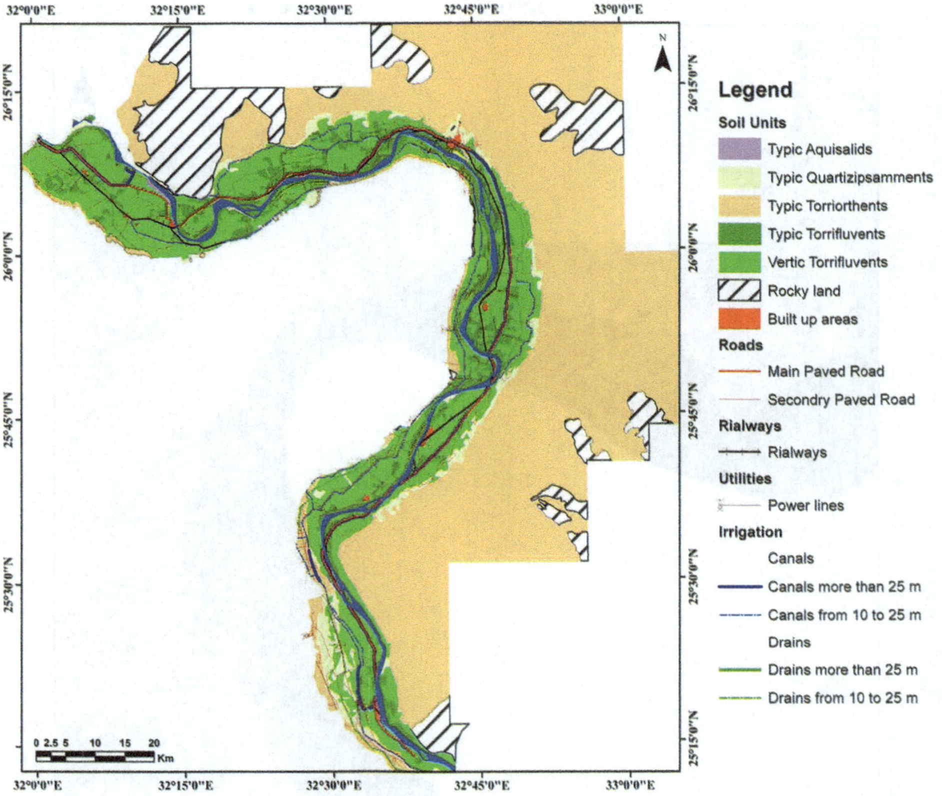

Fig. 4.325 Database layers of Qena Governorate

Table 4.160 Areas and frequencies of soil sub-great groups in Qena Governorate

Soil class	Area (km^2)	%
Typic Torriorthents	4152.0	58.5
Vertic Torrifluvents	1043.3	14.7
Typic Torrifluvents	359.1	5.1
Typic Quartizpsamments	115.4	1.6
Rock land	1422.4	20.1
Total	7092.1	100.0

Table 4.161 Total lengths of main road categories in Qena Governorate

Type	Length (km)
Canals more than 25 m	180.2
Canals from 10 to 25 m	341.6
Drains more than 25 m	195.1
Drains from 10 to 25 m	470.5

Table 4.162 Total length of the main irrigation canals and drains in Qena Governorate

Type	Length (km)
Main Paved Road	197.4
Secondary Paved Road	1307.2

Table 4.163 Areas and frequencies of different administrative districts within Aswan Governorate

Name	Area (km^2)	%
Kawm Umbu	484.4	47.4
Aswan	48.4	4.7
Idfu	303.3	29.7
Nasr	186.3	18.2
Total	1022.4	100.0

Table 4.162 reveals the length of different road categories within the Governorate. Two main roads are illustrated in the land resources database map of Qena Govenorate identified as main paved and secondary paved, attaining lengths of 197.4 and 1307.2 km respectively.

The urban areas recorded, in the land resources database, cover 238.4 km^2. This area is served by coverage of utility network and railways amounting 694.8 and 365.7 km respectively.

4.8.21 Aswan Governorate

Aswan Governorate is located in extreme south of the Nile Valley. In addition to its historic background, it is linked with the existence of both Aswan reservoir and the High Dam. The Governorate includes 4 administrative units Table 4.165. Kawm Umbu district, north of the Governorate exhibits almost half of the total area. Also, Idfu district covers an area of 303.3 km^2, representing 29.7% of the Governorate. Nasr district occupies an area of 186.0 km^2, representing 18.2%. Aswan is the smallest district in the Governorate extending for only 48.4 km^2, representing 4.7% of total area (Table 4.163).

Figures 4.326, 4.327 and Table 4.164 show the main elements included in the database of the Aswan Governorate. The soil layer demonstrates that the rock land extends for an extensive coverage within the Governorate boundaries.

It exhibits an area of 2401.3 km^2 representing 39.5% of total mapped area, distributed on both the western and eastern sides of the valley. Four soil sub-great groups are found in the Governorate, the most extensive one is *Typic Torriorthents* covering an area of 2796.3 km^2 representing 46.0% of the mapped area. These soils spread in the wadi plains and wadi bottoms existing in the western and eastern desert fringes. The sub-great groups Vertic Torrifluvents, with inclusion of Typic Torrifluvents, are the main soil unites in the Nile alluvial plain. They cover areas of 626.9 and 218.1 km^2 representing 10.3 and 3.6% of the mapped soils respectively. The *Typic Quartizpsamments* sub-great group exhibits

Fig. 4.326 Landsat ETM+ 2002 Aswan Governorate

patches in the interference zones between the Nile valley and desert areas. These soils cover an area of 35.6 km^2 representing 0.6% of the total soils area.

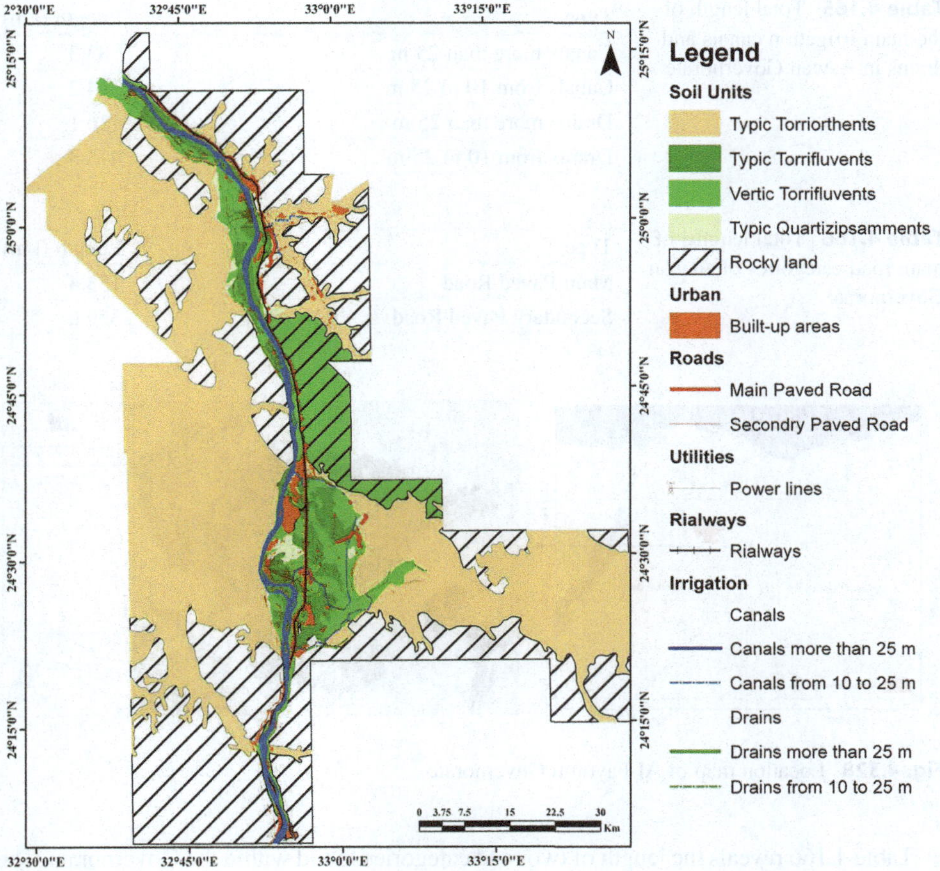

Fig. 4.327 Database layers of Aswan Governorate

Table 4.164 total lengthof the main irrigation and drainage canals in Aswan Governorate

Soil class	Area (km^2)	%
Typic Torriorthents	2796.3	46.0
Vertic Torrifluvents	626.9	10.3
Typic Torrifluvents	218.1	3.6
Typic Quartizpsamments	35.6	0.6
Rock land	2401.3	39.5
Total	6078.2	100.0

Two main irrigation and drain canals are shown in the database layout of the Governorate, lengths of each type are presented in Table 4.165. The irrigation canals of more than 25 m width have a total length of 83.3 km, while those of 10–25 m width attain a total length of 344.2 km. The drainage canals of more than 25 m width have a total length of 116.1 km, while those having 10 to 25 m width attains 478.8 km in length.

Table 4.165 Total length of the main irrigation canals and drains in Aswan Governorate

Type	Length (km)
Canals more than 25 m	83.3
Canals from 10 to 25 m	344.2
Drains more than 25 m	116.1
Drains from 10 to 25 m	478.8

Table 4.166 Total lengths of main road categories of Aswan Governorate

Type	Length (km)
Main Paved Road	153.4
Secondary Paved Road	359.6

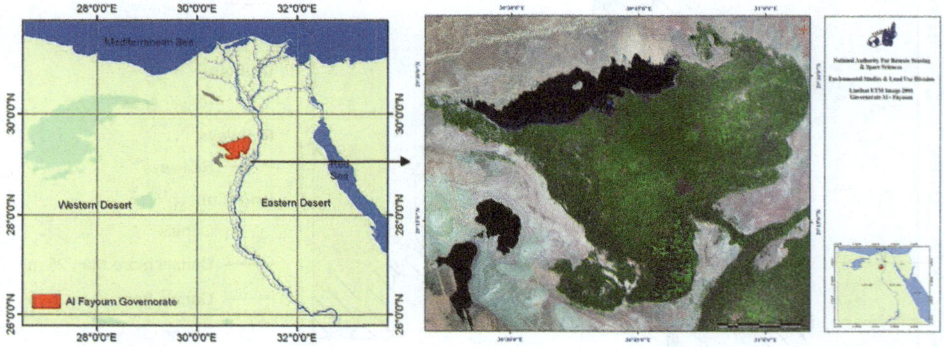

Fig. 4.328 Location map of Al-Fayoum Governorate

Table 4.166 reveals the length of two road categories found within the Governorate. The main paved road network extends for 153.4 km, while the secondary paved one exhibits a total length of 359.6 km.

The urban areas of Aswan Governorate exhibit 277.3 km^2, served by 174.8 km utility network. The railways network records 112.9 km in the land resources database of the Governorate.

4.9 Detailed Land Resources Database of Al-Fayoum Governorates

As mentioned earlier in the project proposal, the final phase is dedicated for developing the detailed land resources database of the Al Fayoum and Al Qalubia Governorates. The two Governorates were undertaken as examples of the Nile Valley and Delta provinces, aiming to proceed later with the other Governorates of whole Egypt.

Al-Fayoum Governorate is located, in a depression, at the southwest of Cairo, where the distance to Al-Fayoum town is 90 km (Fig. 4.328). It is connected to the Nile Valley by the Hawara canal, through which Bahr Yousef is transporting the Nile water. Al-Fayoum

Governorate was chosen for this study, as an example representing the alluvial soils of the Nile Valley, and will be regarded as an example to be followed in other Middle and Upper Egypt Governorates. Al Fayoum Governorate is distinguished by its temperate climate throughout the year, in addition to its long history extending back millions of years. The civilisation started at Al-Fayoum since the Stone Age, where it had a prestigious position in the era of the central state and family second session. Its importance emerged during the Greek, Roman, Coptic and Islamic eras. Al-Fayoum Governorate is the only one, where lakes, vegetation and desert, in their diverse features and unique combination, exist. It is the only Governorate where a salt lake (i.e. Qaroun) and a fresh one (Al-Rayan) are found.

The comprehensive database of Al-Fayoum Governorate was developed, including a number of 13 thematic layers. Three satellite images (i.e. TM of 1992, ETM+ of 2002 and SPOT of 2006) were used to extract the situation of urban areas and urban encroachment. Distribution of land resources in each of the administrative units (Markaz) is issued and discussed. Some applications were implemented, using the created database (i.e. land capability classification, urban encroachment).

4.9.1 Environmental Conditions of Al-Fayoum Depression

Al Fayoum Governorate is occupying a depression west of the Nile between latitudes 29° 02′ and 29° 35′ N and longitudes 30° 23′ and 31° 05′ E. The climatic data of Al-Fayoum district indicate that the total rainfalls does not exceed 7.2 mm/year and the mean minimum and maximum annual temperatures are 14.5 and 31.0 °C respectively. The evaporation rates are coinciding with temperatures, where the lowest evaporation rate (1.9 mm/day) was recorded in January while the highest value (7.3 mm/day) was recorded in June (CLAC 2004). According to the aridity index classes (Hulme and Marche 1990) the depression is located under arid climatic condition. The depression has a particular nature, differing from the Delta and Upper Egypt and from the Oases as well. The differences are not limited to agriculture, they extend to geographical and topographical features, as the environment vary between agricultural, desert and coastal areas. It occupies a portion of the western desert Eocene limestone plateau, while the subsurface lithology consists of marine sedimentary strata, which has undergone alternating periods of erosion and deposition.

The present depression has been formed when the basin was subsided relative to the Nile River, allowing it to break through and to flood the area. This led to the formation of a thick fertile alluvium (Euroconsult 1992). The main identified landforms in Al-Fayoum depression are fans, recent and old lake terraces, depression, plain, and basins (Abo El-Einein 1999). With the present regime of irrigation provided to Al-Fayoum province, rising of water level in Qaroun Lake would endanger surrounding lands. The total area of Al-Fayoum depression is 6068.7 km^2; the land use in this area includes only 1849.6 km^2 (i.e. 30.5% of the total area). The agricultural land in the depression amount to 1609.3 km^2 classified into five grades according to their production capacity. The first and second

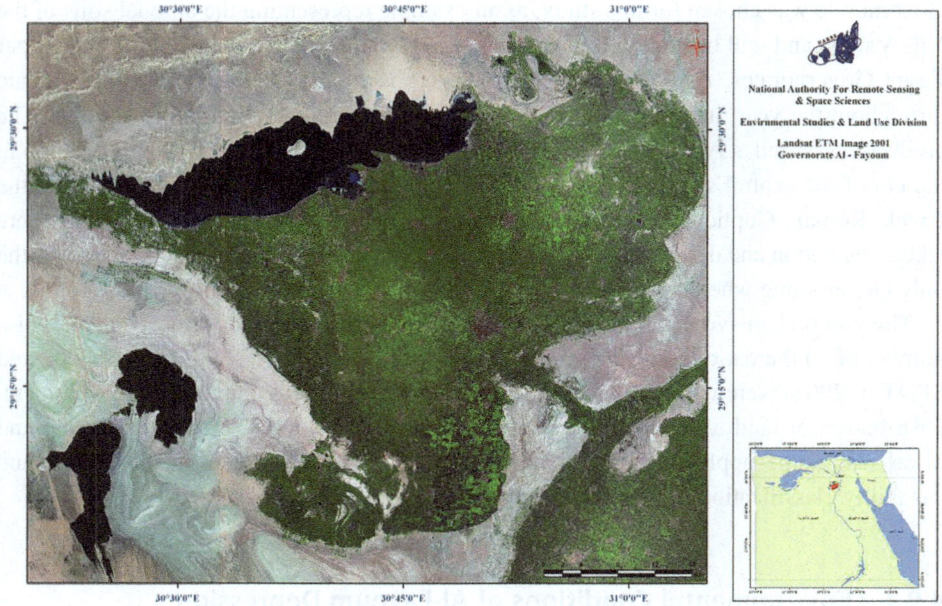

Fig. 4.329 Landsat ETM+ image of Al Fayoum Governorate (2001)

grades do not exceed 20.2% of the total cultivated area, whereas about 70.2% accommodate the third and fourth classes (ASRT 1996).

4.9.2 Physiography of Al-Fayoum Governorate

Field survey data, Landsat ETM+ (Fig. 4.329) and SPOT images (Fig. 4.330) and digital elevation model (DEM–Fig. 4.331) were used to define the physiographic units in Al-Fayoum Governorate (Al-Fayoum Depression—Fig. 4.332, Wadi Al-Rayian Depressions—Fig. 4.333, and north Qaroun lake area–Fig. 4.334). The obtained data refer that the studied areas include the following physigraphic units:

4.9.3 Detailed Geographic Database of Al-Fayoum Governorate

4.9.3.1 Al-Fayoum depression

(a) Lacustrine plain:

This landscape occupies an area of 194.2 km^2 and is subdivided into high, moderately high and low terraces (Fig. 4.335). The digital elevation model of the depression shows that the

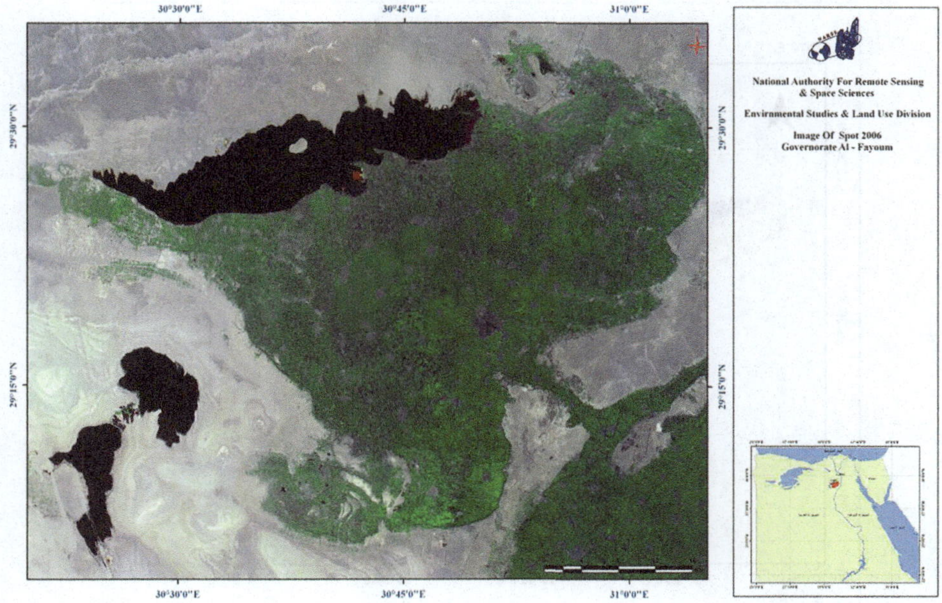

Fig. 4.330 SPOT image of Al Fayoum Governorate (2006)

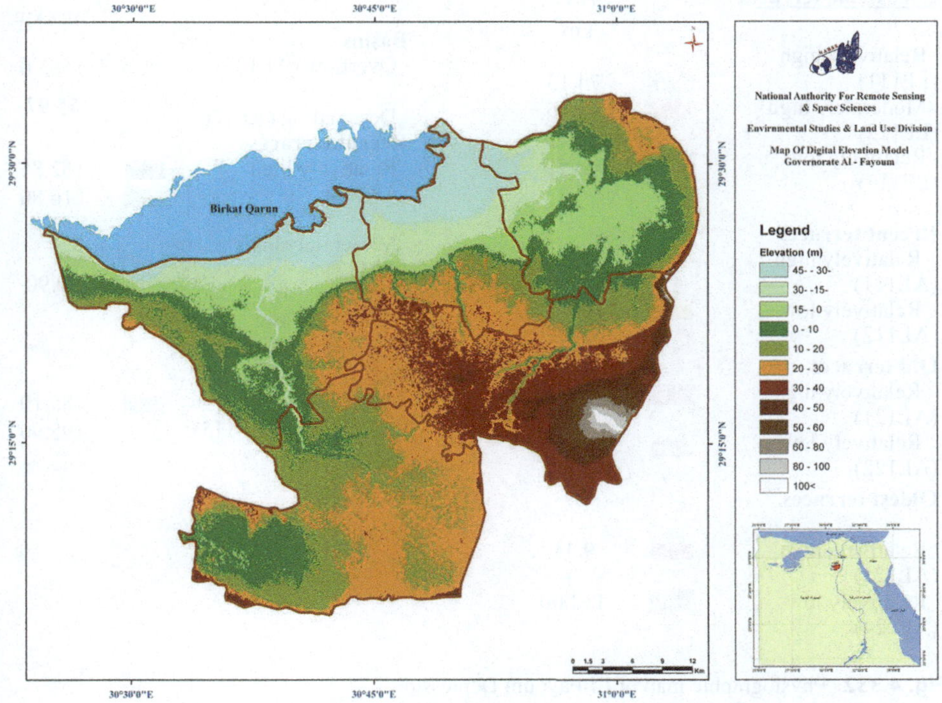

Fig. 4.331 Digital Elevation Model (DEM) of Al Fayoum Governorate

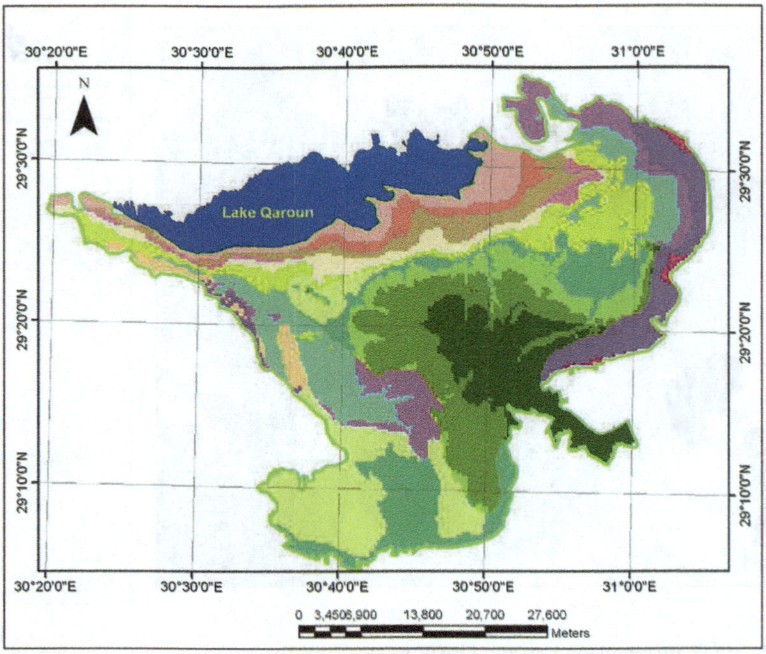

Lacustrine terraces	Area km^2
Relatively high (LP111)	74.13
Moderately high (LP112)	44.36
Relatively low (LP113)	75.71
Recent terraces	
Relatively high (AL111)	38.16
Relatively low (AL112)	147.70
Old terraces	
Relatively high (AL121)	0.17
Relatively low (AL122)	20.17
Oldest terraces	
Relatively high (AL123)	9.43
Relatively low (AL124)	113.00

	Area km^2
Basins	
Overflow (AL125)	158.70
Decantation (AL126)	55.97
Recent terraces	
Relatively high (AP111)	162.85
	216.80
Moderately high (AP112)	
	90.90
Relatively low (AP113)	
Basins	
Overflow (AP114)	185.10
Decantation (AP115)	195.90

Fig. 4.332 Physiographic map of El-Fayoum Depressio

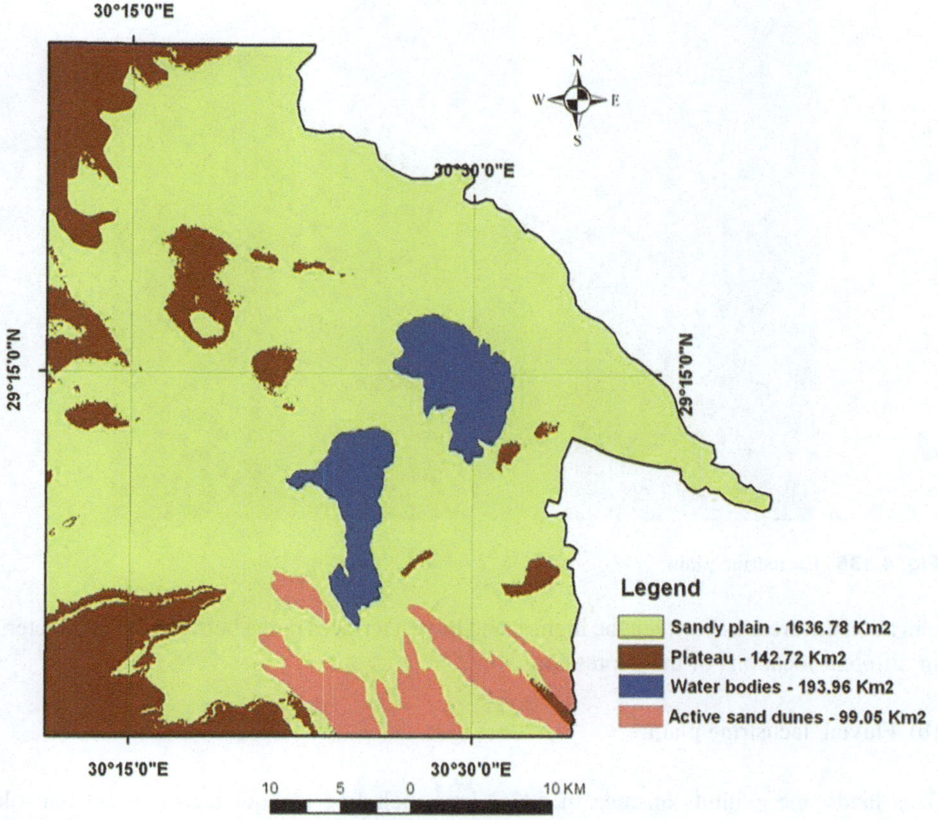

Fig. 4.333 Physiographic map of Wadi Al-Rayian Depression

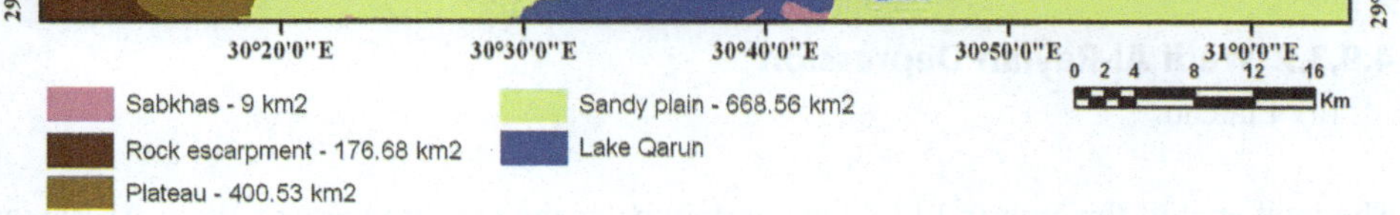

Fig. 4.334 Physiographic map of north Qaroun lake area

Fig. 4.335 Lacustrine plain

elevation differences between the highest and lowest terraces range between 3 and 5 meters, in different locations of the depression.

(b) Fluvial–lacustrine plain:

This landscape exhibits an area of 543.3 km^2 including a sequence of recent and old terraces which occupy the border of the depression (Fig. 4.336). The terraces vary widely in elevation, as the difference between the highest and lowest terraces ranges between 5 and 25 meters. Also the landforms of overflow and decantation basins exist, dominating the middle parts of the depression, covering an area of 214.7 km^2.

(c) Alluvial plain:

This landscape covers an area of 851.6 km^2 including a sequence of recent terraces and basins dominating the middle and southern parts of the depression (Fig. 4.337).

4.9.3.2 Wadi Al-Rayian Depression

(a) Plateau:

The total area of this unit is 142.7 km^2; it dominates the western parts of Wadi Al-Rayian depression. It is characterized by a surface elevation ranging from 62 to 94 meters above sea level.

Fig. 4.336 Fluvial–lacustrine plain

Fig. 4.337 Alluvial plain

Fig. 4.338 Active sand dunes

(b) Active sand dunes:

The active sand dunes dominate the southern parts of Wadi Al-Rayian depression, covering an area of about 99.1 km^2. The general movement direction of the dunes is to the south east, as they extend to Al-Minyea governorate at Samalout district (Fig. 4.338).

(c) Lakes:

Wadi Al-Rayian lakes exhibit the middle parts of the depression, covering an area of 194.0 km^2.

(d) Sandy plain:

The total area of this unit is about 1636.8 km^2, combining sand sheets of various elevations ranging between −21 to +33 meters (Fig. 4.339).

4.9.3.3 North Qarun Lake area

Plateau

This unit dominates the north western part of the area, covering about 400.53 km^2. The surface elevation of the plateau ranges between 205 to 297 meters above sea level.

Fig. 4.339 Sabkhas

(a) Rock escarpment:

This unit dominates an area of about 176.68 km^2, the surface elevation of this unit is sharply decreased from 94 m to 297 m above sea level, whereas its width does not exceed more than 1.5 km.

(b) Sabkhas:

The sabkhas land form is found close to the Shore of Qarun Lake, covering an area of about 9 km^2.

(c) Sandy plain:

The sandy plain dominates the area extending from the lake to the plateau escarpment, including sand sheets of various elevations. The total area of these units is 668.56 km^2 and the surface elevation differs widely from −20 to +62 meter above sea level (Fig. 4.340).

4.9.4 Detailed Database of Al-Fayoum Governorate

4.9.4.1 Al-Fayoum Depression Area

Al-Fayoum Governorate is administratively sub-divided into 5 districts, (Table 4.167). The areas of the districts can be arranged as follows Atsa > Al Fayoum > Ibsheway > Senouris

Fig. 4.340 Sandy Plain

Table 4.167 Areas and frequencies of different administrative districts, Al-Fayoum Governorate

District Name	Area (km^2)	%
Atsa	454.6	25.6
Al-Fayoum	395.9	22.3
Ibsheway	388.3	21.9
Senouris	232.0	13.1
Tamyah	304.1	17.1
Total	1775.0	100.0

> Tamyah districts having areas of 454.6, 395.9, 388.3, 232.0 and 304.1 km^2 respectively (Fig. 4.341).

An updated digital soil map of Al-Fayoum Governorate was elaborated using the SPOT image of 2006 and GIS techniques, as well as the land surveying where 30 soil profiles were observed (Fig. 4.342). The soils were classified according to the USDA soil Taxonomy (2006). The updated map includes also the areas of North Lake Qarun Lake and Wadi Al-Rayian areas which were not included in the soil map of Egypt, 1982.

Figure 4.343 and Table 4.168 indicate that the sub-great group *Vertic Torrifluvents* dominates the soils, covering an area of 701.0 km^2 representing 39.5% of the mapped soils. The *Typic Torrifluvents* occurs within the depression, covering an area of 133.0 km^2, representing 7.5% of the mapped soils. These soils are associated with the recent terraces of the flood plain. The sub-great group of *Typic Calciorthids* soil covers an area of 390.0 km^2 representing 22.0% of soil units area. Its geographic distribution is located on the edges of the depression, exhibiting the old river terraces. The Gypsic soils (i.e. *Typic Haplogypsids*)

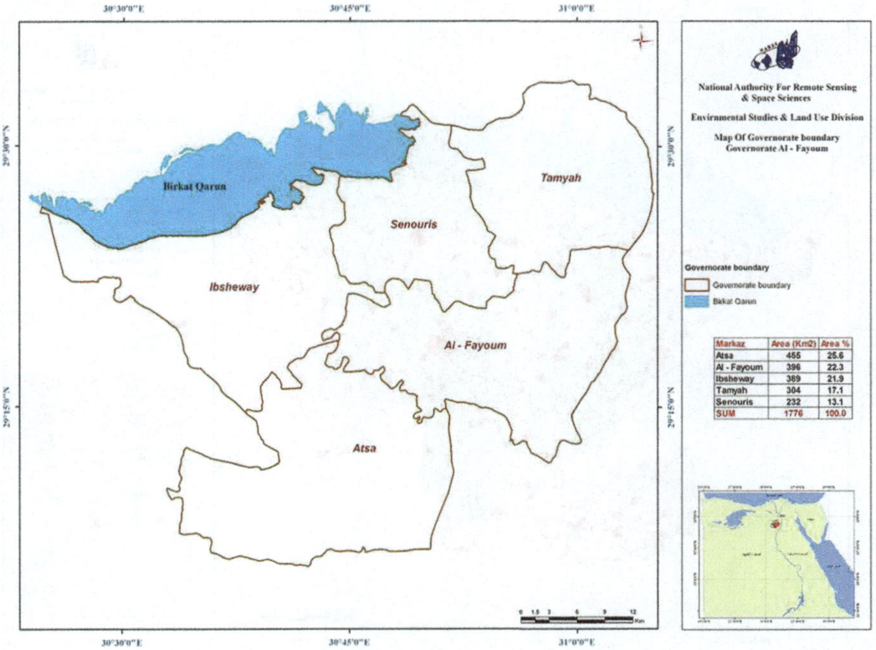

Fig. 4.341 Administrative boundaries of Al Fayoum Governorate districts

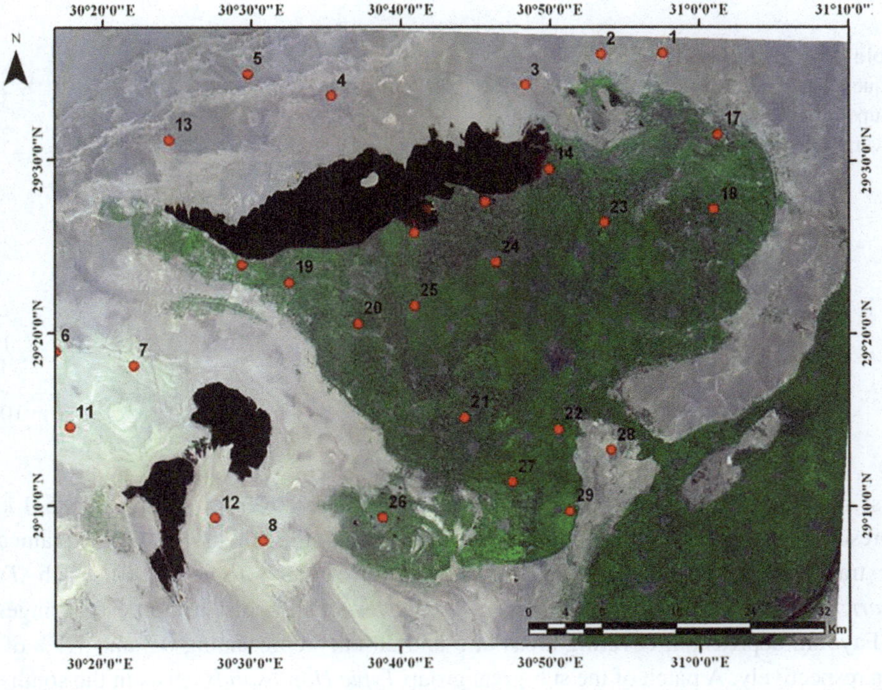

Fig. 4.342 Distribution of the studied soil profiles in Al-Fayoum Governorate

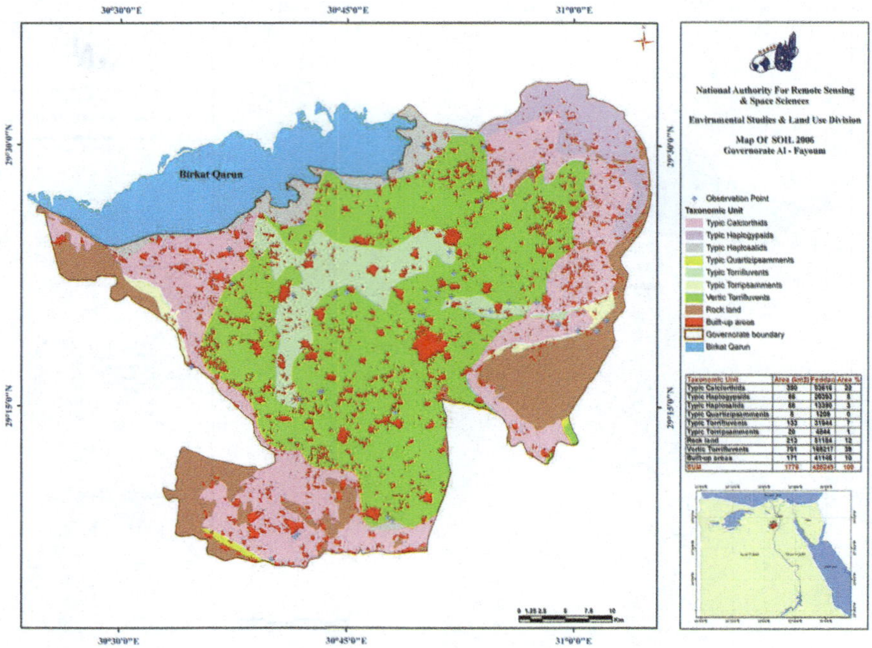

Fig. 4.343 Soil map of Al-Fayoum Governorate

Table 4.168 Areas and frequencies of soil sub-great groups in Al-Fayoum Governorate

Soil class	Area (km^2)	%
Vertic Torrifluvents	701.0	39.5
Typic calciorthids	390.0	22.0
Typic Torrifluvents	133.0	7.5
Typic Haplogypsids	85.0	4.8
Typic Torripsamments	20.0	1.1
Typic Quartizipsamments	5.0	0.3
Typic Haplosalids	56.0	3.2
Rock land	213.0	12.0
Built up area	171.0	9.6
Total	1774.0	100.0

exist on the eastern borders of Al-Fayioum depression, covering an area of 85.0 km^2, representing 4.8% of the area. The geographic location of this soil unit may be explained by the transgression of Qaroun Lake to the northwest area (Said 1990). Both *Typic Quartzipssaments* and *Typic Torripsamments* cover small spots in the desert fringes of Al-Fayoum depression, covering areas of 5 and 20 km^2 representing 0.3 and 1.1% of the area respectively. A patch of the sub-great group *Typic Haplosalids* exists in the south east

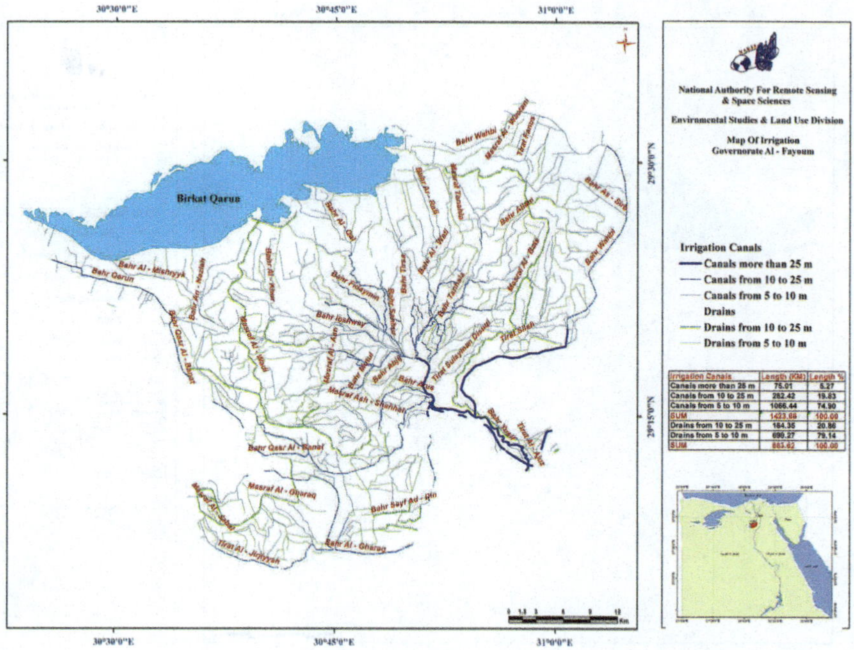

Fig. 4.344 Canals and drainage network layer of Al Fayoum Governorate

Table 4.169 Total length of the main irrigation canals and drains in Al-Fayoum Governorate

Type	Length (km)
Canals more than 25 m	75.2
Canals from 10 to 25 m	282.4
Drains more than 25 m	184.4
Drains from 10 to 25 m	135.7

of the depression, exhibiting an area of 56.0 km^2 representing 3.2% of the mapped soils. It should be noticed that such variability of sub-great groups is unique for Al-Fayoum Governorate due to its location, altitude, and formation processes.

Two main irrigation canals and drains categories are shown in the database layout of Al-Fayoum Governorate (Fig. 4.344); lengths of each category are presented in Table 4.169. The irrigation canals of more than 25 m width have a total length of 75.2 km, while those of 10–25 m width attain a total length of 282.4 km. The drainage canals of more than 25 m width have a total length of 184.4 km, while those having 10 to 25 m width attains 135.7 km length. This represent a fairly good balance between irrigation and drainage network.

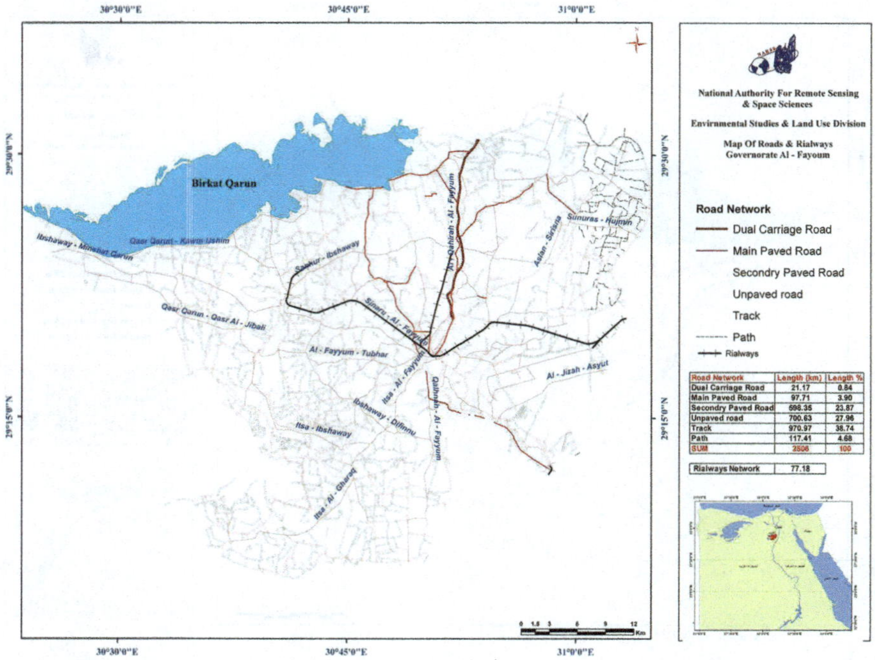

Fig. 4.345 Roads network layer of Al Fayoum Governorate

Table 4.170 Total lengths of main road categories in Al-Fayoum Governorate

Type	Length (km)
Dual carriage road	21.2
Main Paved Road	97.7
Secondary Paved Road	598.4

Figure 4.345 and Table 4.170 reveal three roads categories spread within Al-Fayoum Governorate. Total lengths of Dual carriage, Main paved and secondary paved roads are 21.2, 97.7 and 598.4 km respectively.

Built areas, utilities and railways (Fig. 4.346) are included in the land resources database of Al-Fayoum Governorate. The urban area cover 64.2 km^2, while the length of utilities and railways are 532.8 and 77.2 km respectively.

Change detection technique resulted in following the evolution of urban areas and its implication with soil and land capability units. The built-up areas in the Governorate (Figs. 4.347, 4.348, 4.349, 4.350) increased from 64.0 km^2 (i.e. 4%) in the year 1992 to 107.0 km^2 (i.e. 6%) in 2001 and to 171.0 km^2 (i.e. 10%) in the year 2006.

Studying the urban expansion from 1992 to 2006 (Figs. 4.351, 4.352, 4.353) revealed that urbanization mainly occurs on the account of the most fertile soils of Al-Fayoum

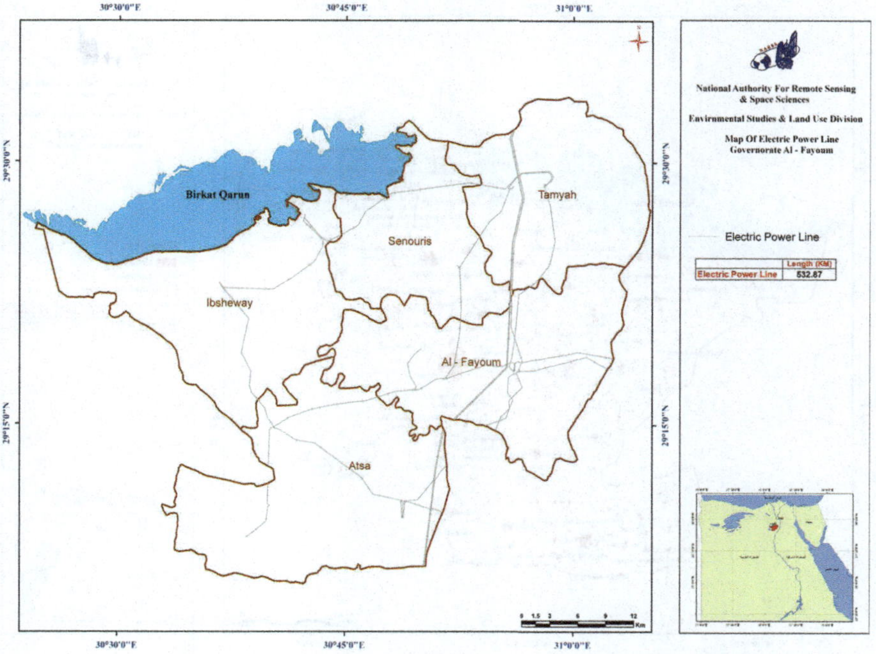

Fig. 4.346 Utility network layer of Al Fayoum Governorate

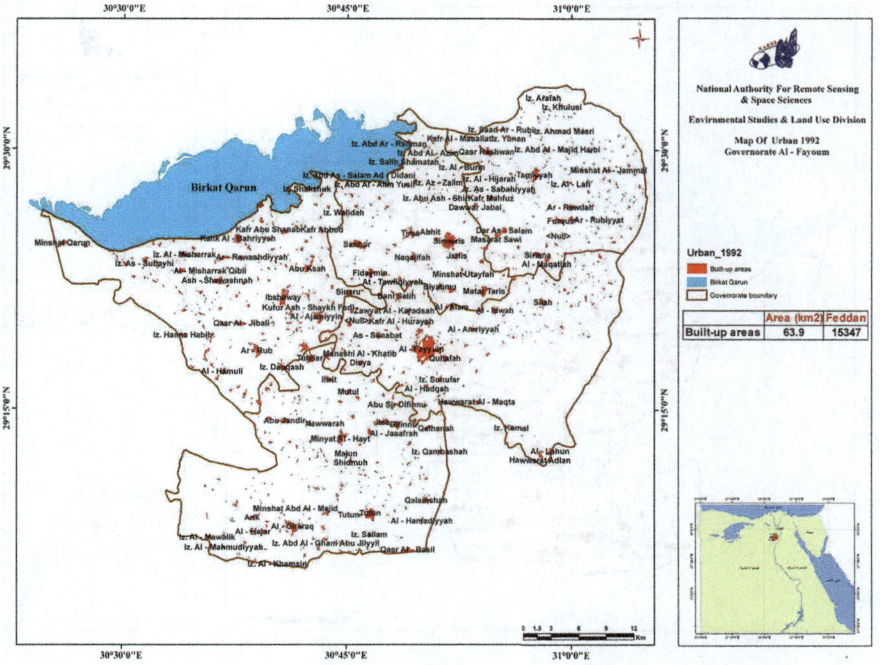

Fig. 4.347 Built-up areas in Al Fayoum Governorate in 1992

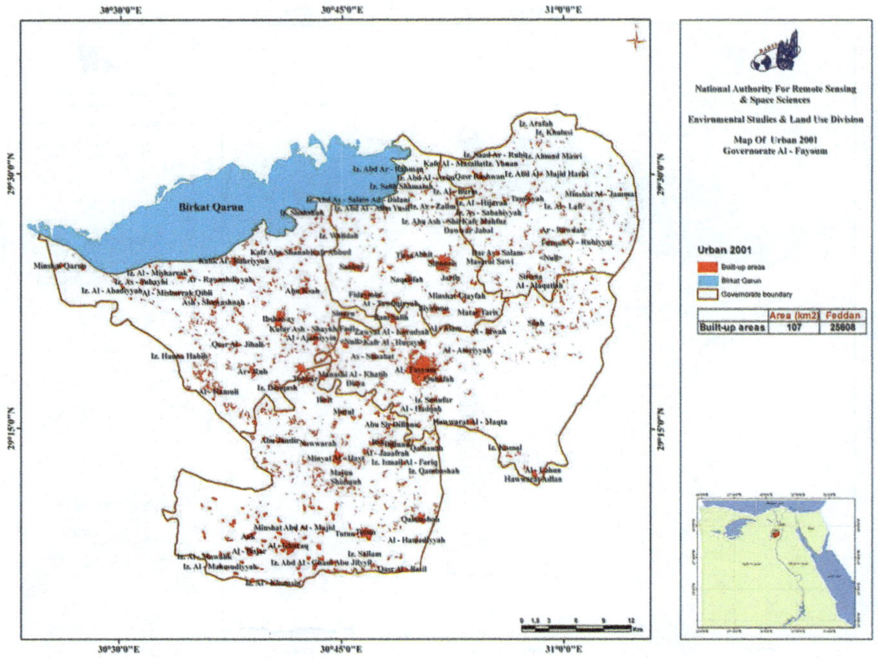

Fig. 4.348 Built-up areas in Al Fayoum Governorate in 2001

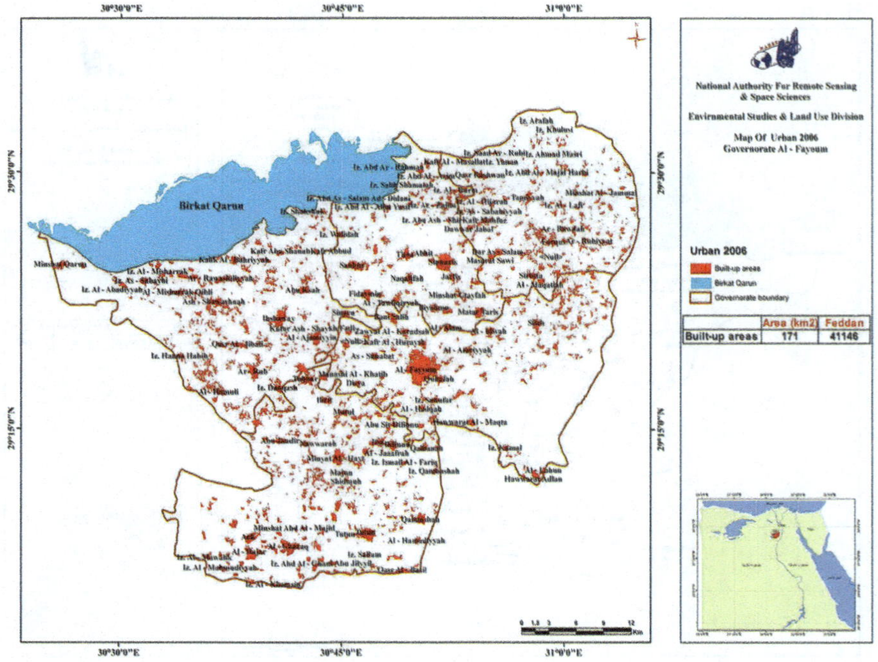

Fig. 4.349 Urban areas layer of Al Fayoum Governorate (2006)

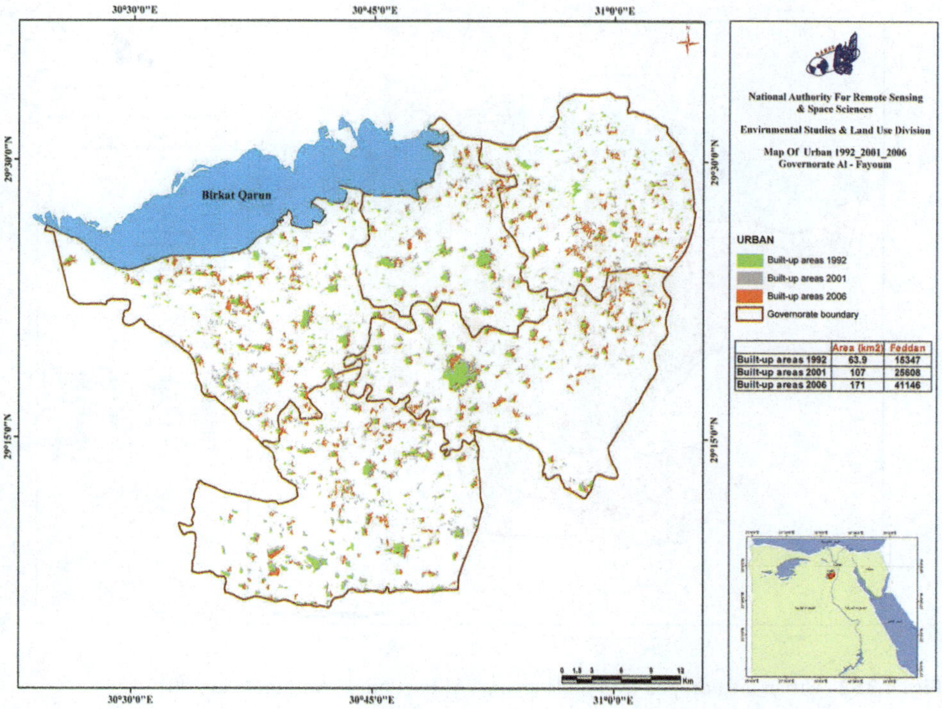

Fig. 4.350 Changes in the Built-up areas, Al Fayoum Governorate, 1992, 2001 and 2006

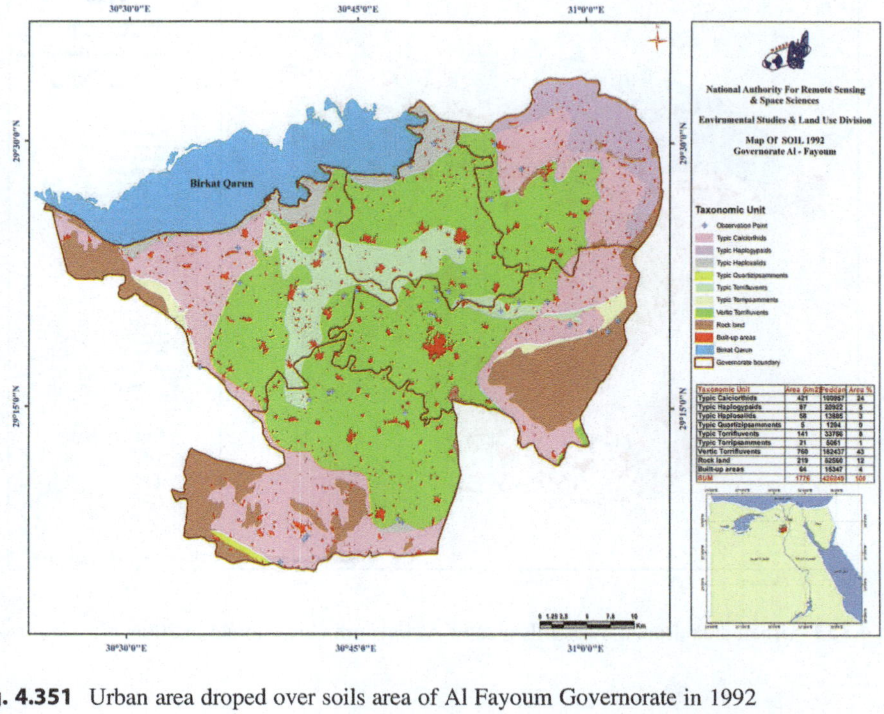

Fig. 4.351 Urban area droped over soils area of Al Fayoum Governorate in 1992

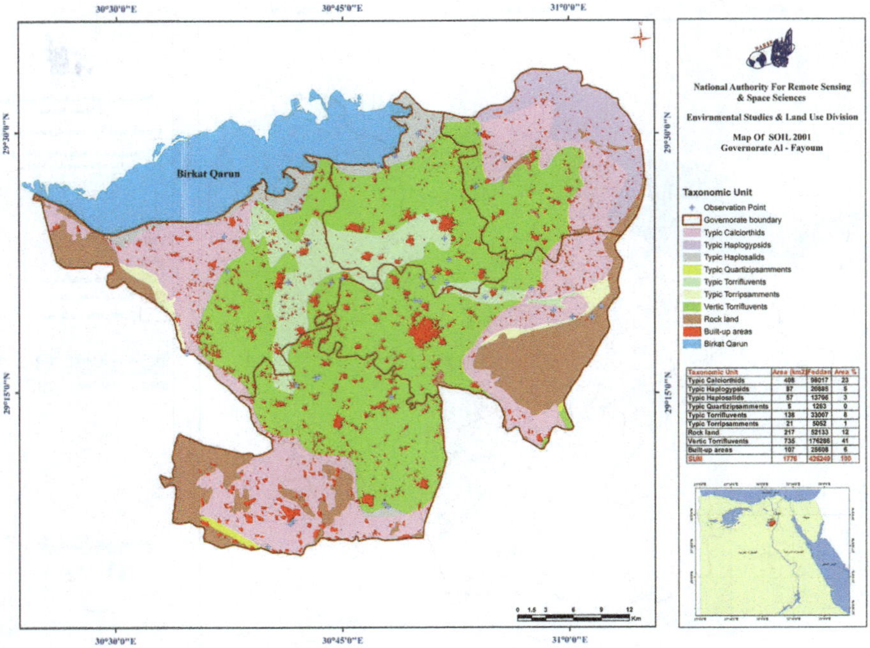

Fig. 4.352 Urban area droped over soils area of Al Fayoum Governorate in 2001

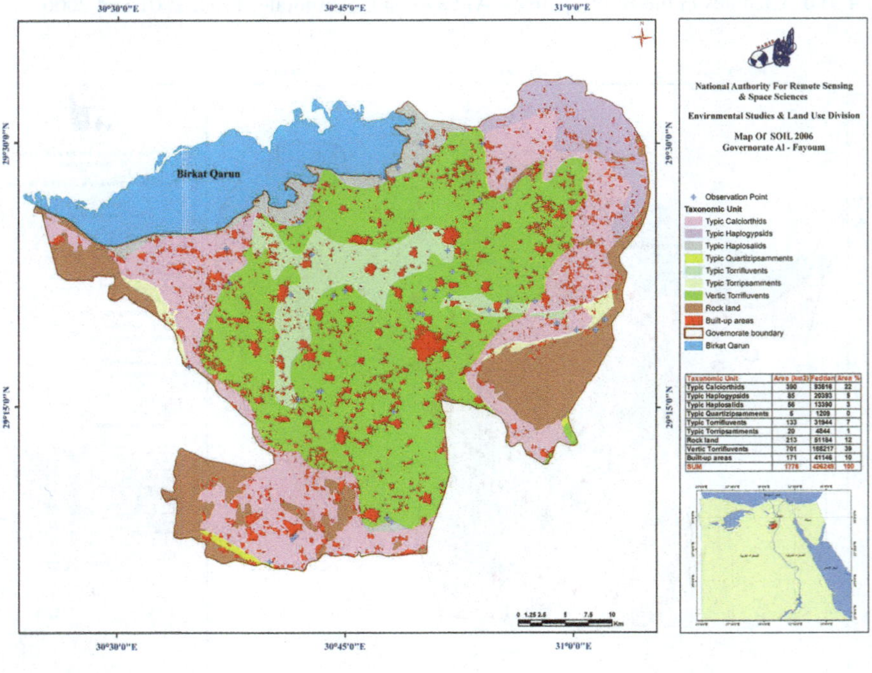

Fig. 4.353 Urban area droped over soils area of Al Fayoum Governorate in 2006

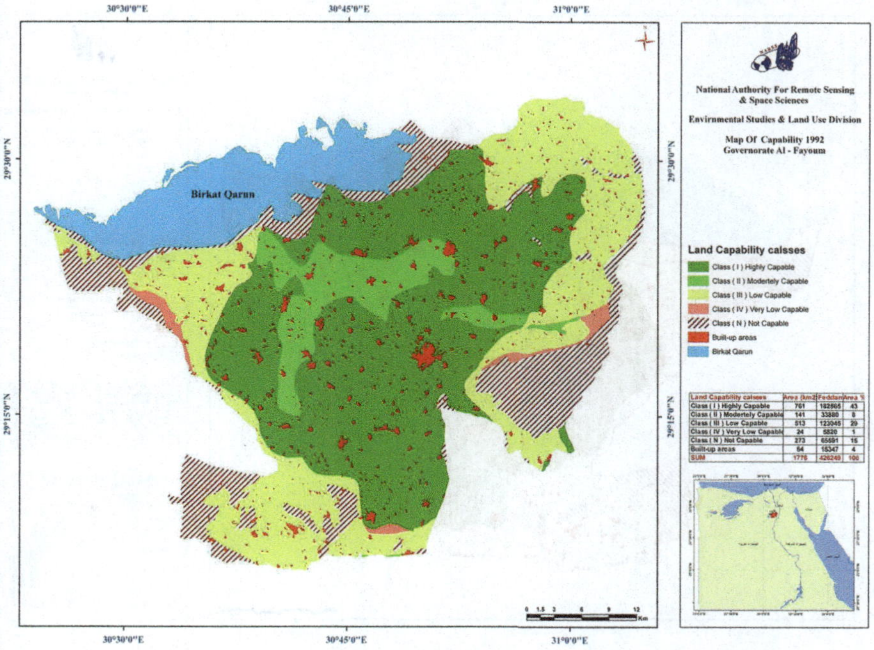

Fig. 4.354 Urban area droped over areas of soil capability classes of Al Fayoum Governorate in 1992

depression. The remained cultivable land exhibited by the soil sub-great group "*Vertic Torrifluvents*", decreased from 760.0 km^2 to 701.0 km^2, where the area of Typic *Torrifluvents* soils decreased from 141.0 km^2 to 133.o km^2 and *Haplocalcids* area decreased from 421.0 km^2 to 390.0 km^2 during the period 1992–2006.

In terms of land capability classification (Figs. 4.354, 4.355, 4.356), the high capable soils (Class I) decreased from 761.0 km^2 in 1992 to 701.0 km^2 in 2006. The moderate capable soils decreased from 141.0 to 133.0 km^2 between 1992 and 2006, while the marginally capable soils decreased from 513.0 km^2 to 480.0 km^2 during the same period.

4.9.4.2 Soil Mapping of North Qarun Lake and Wadi Al-Rayian Areas

The soil maps of North of Qarun lake and Wadi Al-Rayian areas were not included in the outcomes of the project "Soil Map of Egypt, 1982". Thus, it was thought to accomplish these maps to integrate the land resources database of the whole Al-Fayoum Governorate. It should be referred that the areas of wadi-Al-Rayian and north Qaroun Lake have a good potentiality in the sustainable development programs of the Governorate.

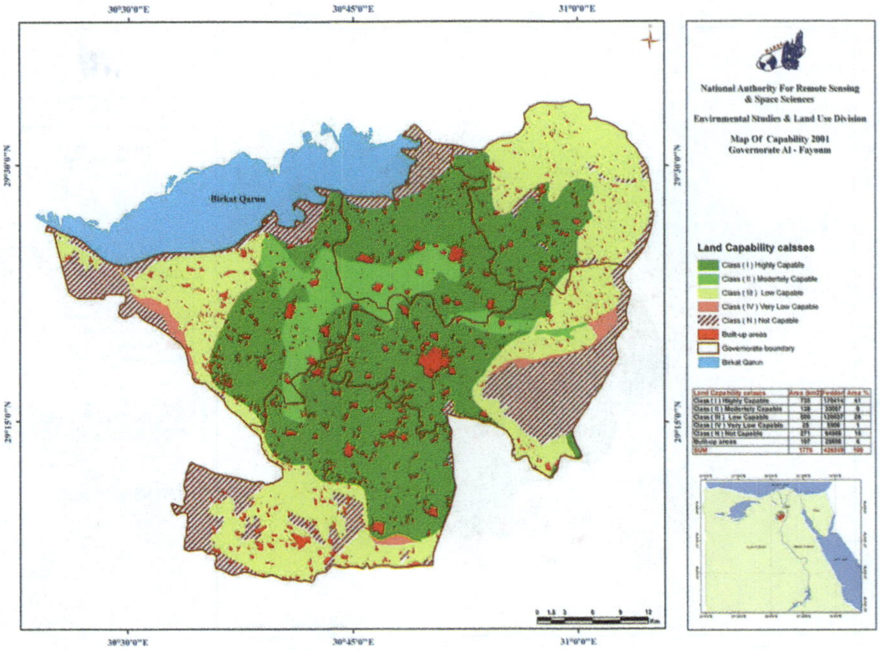

Fig. 4.355 Urban area droped over areas of soil capability classes of Al Fayoum Governorate in 2001

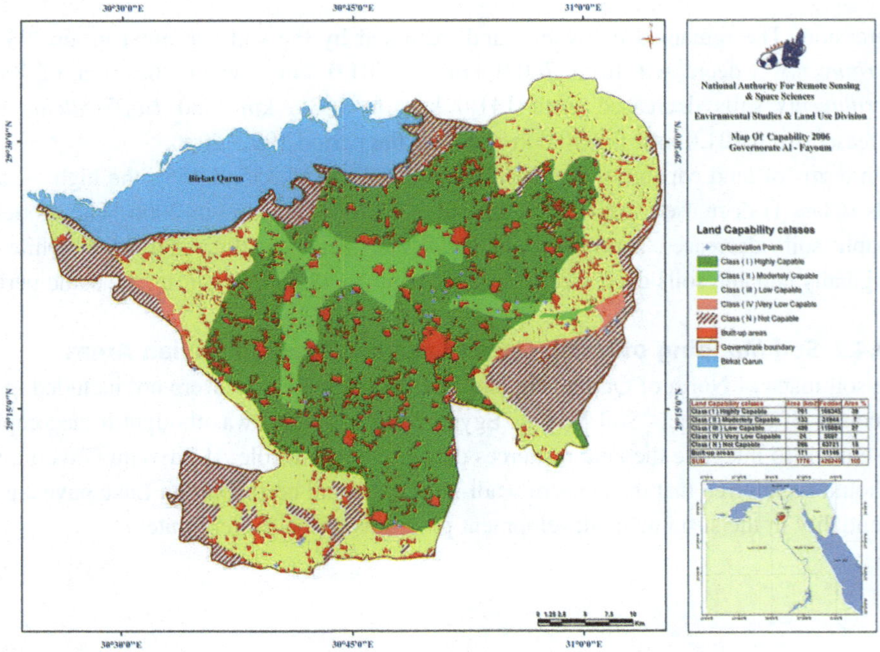

Fig. 4.356 Urban area droped over areas of soil capability classes of Al Fayoum Governorate in 2006

Table 4.171 Soil units of North Lake Qarun area

Soil unit	Area (km^2)	Area (fed)	Area (%)
Lithic Calciorthids	156.3	3750	10.1
Typic Calciorthids	35.9	8628	2.3
Typic Haplosalids	91.4	21,924	5.9
Typic Quartizipsamments	161.6	38781.6	10.4
Typic Torriorthents	306.5	73,548	19.8
Typic Torripsamments	220.1	52833.6	14.2
Rock escarpment	176.7	42403.2	11.4
Rock land	400.5	96127.2	25.9
Total	1548.9	371745.6	100.0

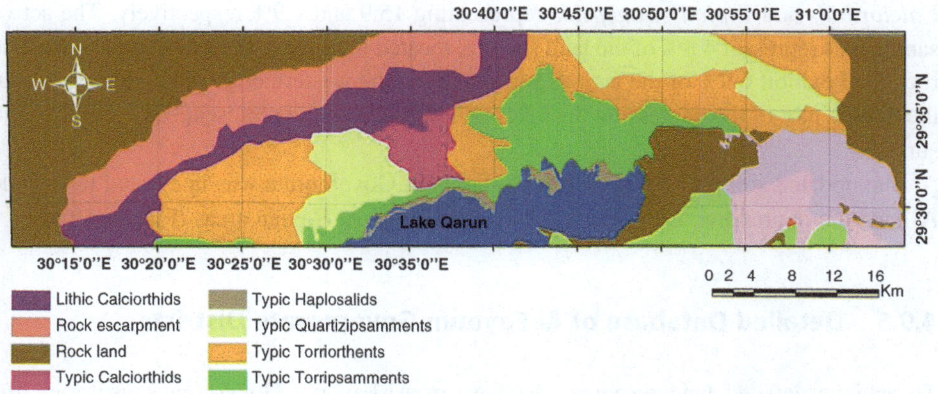

Fig. 4.357 Soil layer of the North Lake Qarun area in 2006

North Lake Area

The total mapped area of Northern Qarun lake is 1548.9 km^2, Table 4.171 and Fig. 4.357. The soil units include the sub great groups of *Lithic Calciorthids, Typic Calciorthids, Typic Haplosalids, Typic Quartzipssamentss, Typic Torriorthents, Typic Torripsamments* which represent 156.3, 35.9, 91.4, 161.6, 306.5, and.

220.1 km^2 respectively. The rest of the area is exhibited by rock escarpment and rock land covering areas of 176.7 and 400.5 km^2 respectively.

4.9.4.3 Wadi Al-Rayian Area

The total mapped area of Wadi Al-Rayian is 2072.5 km^2, Table 4.172 and Fig. 4.358. The soils of this area belong mainly to the soil Order *Entisols* and *Aridisols*. The *Entisols* soil order includes the sub-great groups *Typic Quartzipssamentss*, *Typic Torriorthents* and *Typic Torripsamments*, representing 15.8, 16.4 and 22.7% of tthe area respectively. These soils classified under the soil order "*Aridisols*" include the sub-great groups *Typic*

Table 4.172 soil units of Wadi Al-Rayian area

Soil unit	Area (km^2)	Area (fed)	Area (%)
Active sand dunes	99.1	23770.9	4.8
Lake	193.9	46550.1	9.4
Rock land	142.7	34251.8	6.8
Typic Calciorthids	330.1	79209.4	15.9
Typic Petrogypsids	164.9	39569.0	7.9
Typic Quartizipsamments	327.1	78508.4	15.8
Typic Torriorthents	339.4	81464.1	16.4
Typic Torripsamments	475.3	114077.3	22.9
Total	2072.5	497400.9	100.00

Calciorthids and *Typic Petrogypsids* representing 15.9 and 7.9% respectively. The active sand dunes represent 4.8% of the land surface, located in the southern parts of the area. The rock land exhibit 6.9% of the area and is located in the western edge of the depression. In the middle part of the depression the lakes of Wadi Al-Rayian represent 9.4% of total land surface.

The updated soil map of the whole Al-Fayoum Governorate was integrated to include Al-Fayoum depression, north Qaroun Lake and Wadi Al-Rayian areas (Fig. 4.359).

4.9.5 Detailed Database of Al Fayoum Governorate Districts

To achieve detailed land resources database of Al-Fayoum Governorate territories, the districts were considered as a geographical baseline. The layer of administrative boundary of Al-Fayoum Governorate was used for this objective. Moreover, the areas of Northern Qarun Lake and Wadi Al-Rayian were based on their physiographic features. The established database for each district include the same pervious thematic layers of the whole Governorate, the data included in the database reveal the following:

4.9.5.1 Al-Fayoum District

Al-Fayoum district includes the Al-Fayoum town which is the capital of Al-Fayoum Governorate, it incorporate an area of 396.0 km^2 extending between longitude 30° 45′ to 31° 05′ and latitudes 29° 10′ to 29° 25′. The database of the district includes the Landsat ETM+ and SPOT images taken during the years 2001 and 2006 respectively (Figs. 4.360 and 4.361). The satellite data were used to update the thematic layers. The digital elevation model of the district indicates that the surface elevation varies widely from −45 meter below the sea level to 100 meter above sea level, where the lowest areas are dominating the northern parts of the depression (Fig. 4.362). Regarding the soil layer of the district (Fig. 4.363), the soils of *Vertic Torrifluvents* sub great group are common in the study area, covering an area of 152.0 km^2 (i.e. 38.4% of the district area). The middle of the

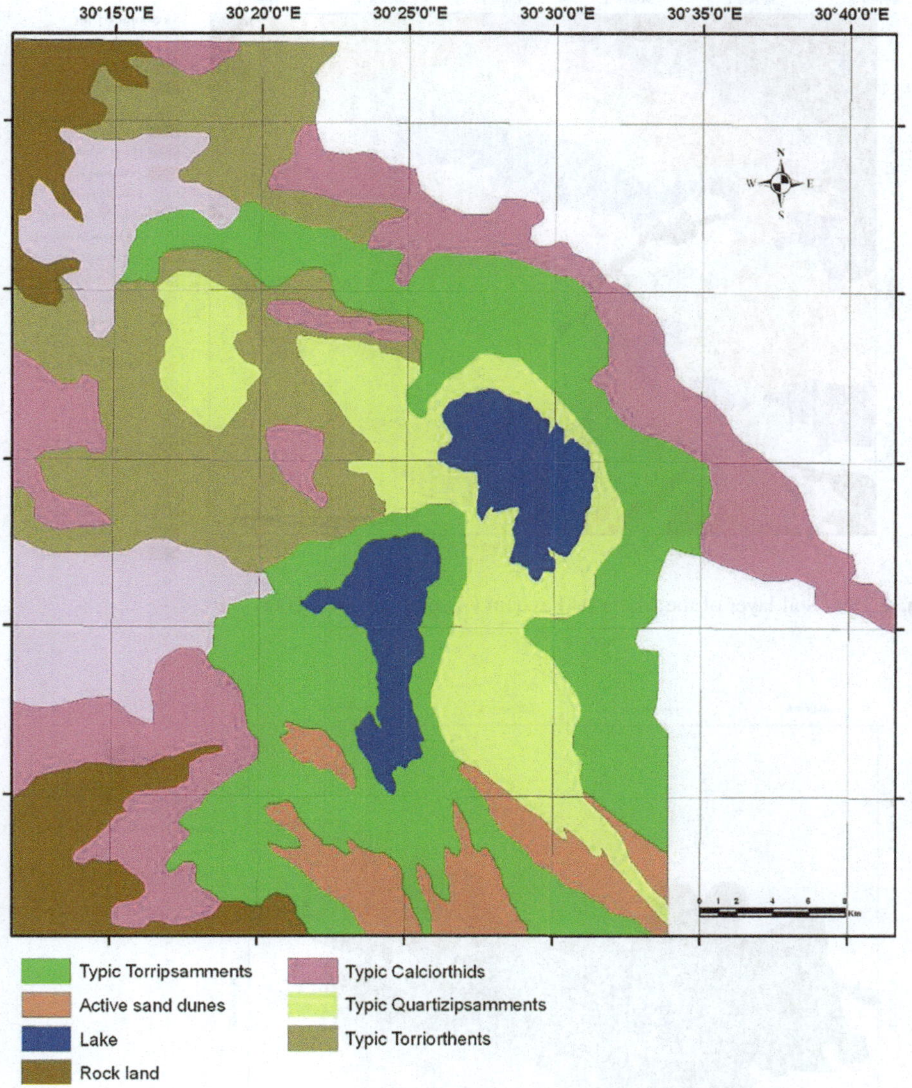

Fig. 4.358 Soil layer of Wadi Al-Rayian area in 2006

district is exhibited by the soils of *Typic calciorthids* (84.0 km^2) which represent 21.2% of the whole district area. The southern parts of the district are dominated by rock land, covering an area of 98.0 km^{2,} representing 24.9% of the total district area. The rest of the district is covered by the soils of *Typic Torripsamments* (2.6%), *Typic Quartizipsamments* (0.1%) and *Typic Torrifluvents* (3.3%).

The irrigation and drainage canals thematic layer of the district (Fig. 4.364) revels that the total length of irrigation canals is 524.8 km, while the total length of drainage network

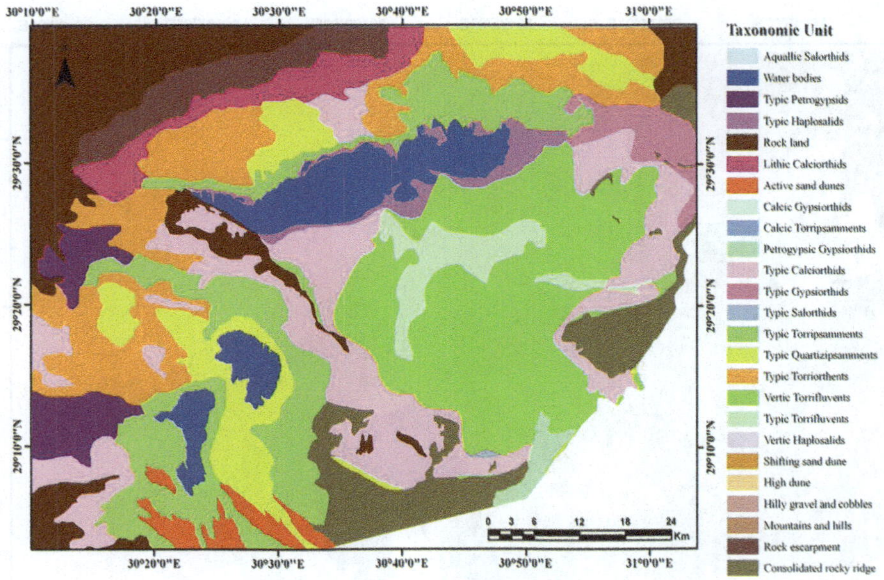

Fig. 4.359 Soil layer of the whole Al Fayoum Governorate area, 2006

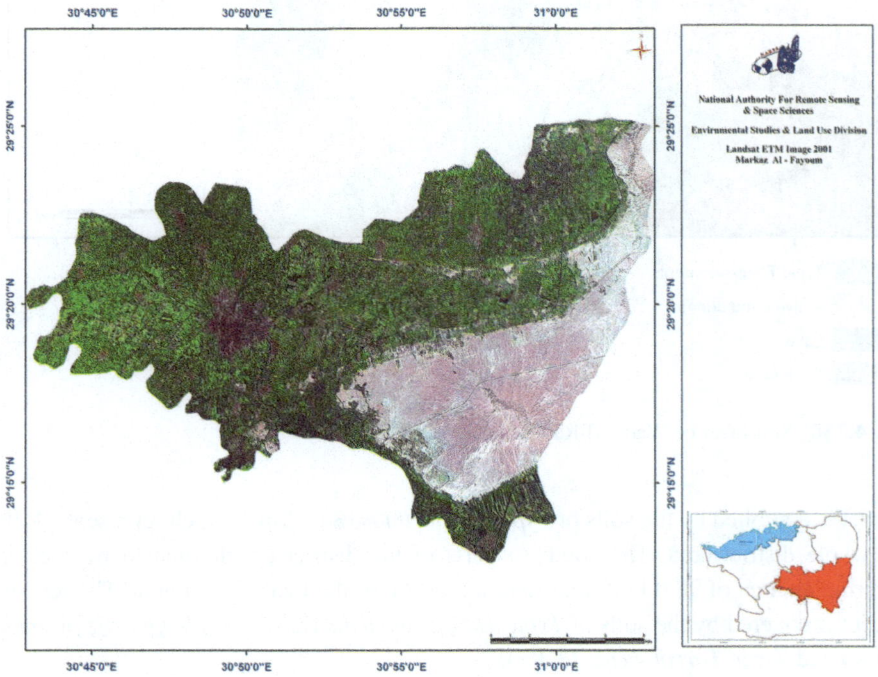

Fig. 4.360 Landsat ETM+ image of Al Fayoum district (2001)

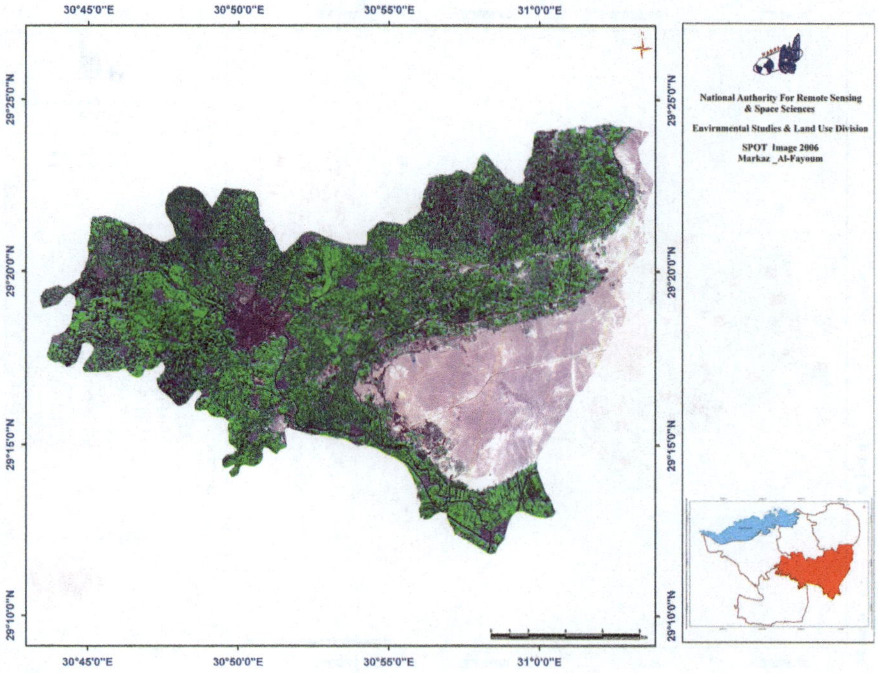

Fig. 4.361 Spot image of Al Fayoum district (2006)

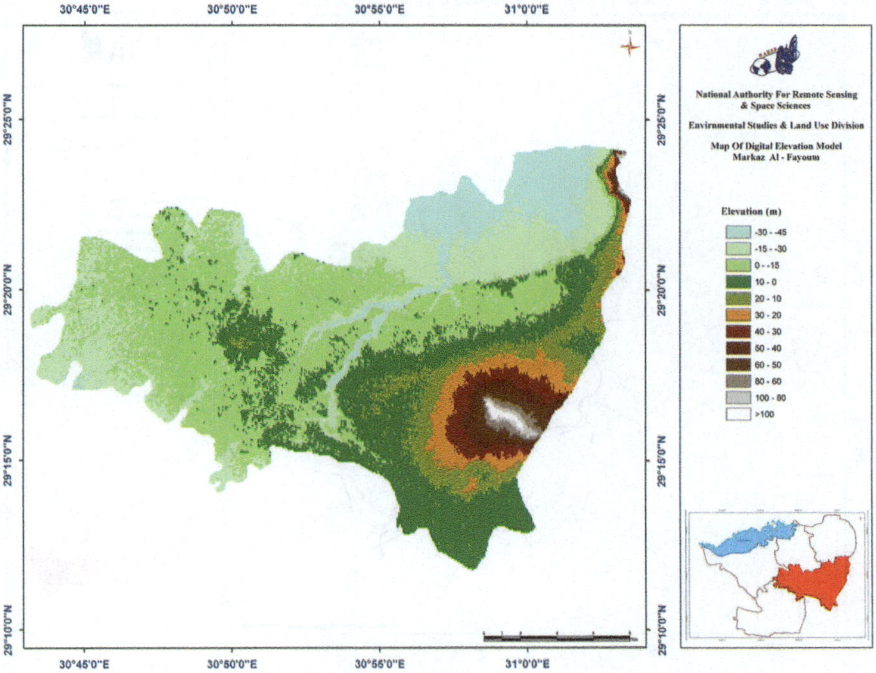

Fig. 4.362 Digital Elevation Model (DEM) of Al Fayoum district

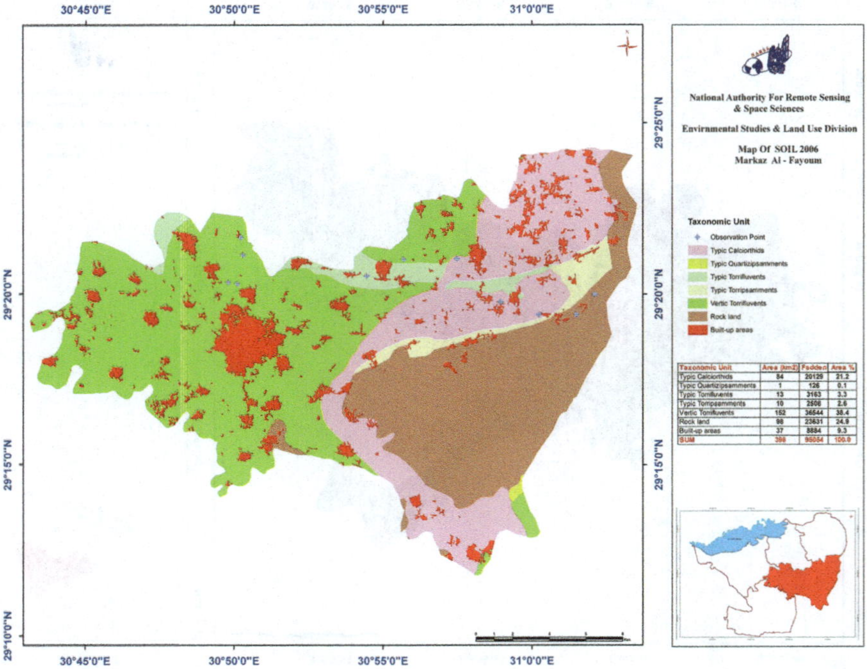

Fig. 4.363 Soil map layer of Al Fayoum district

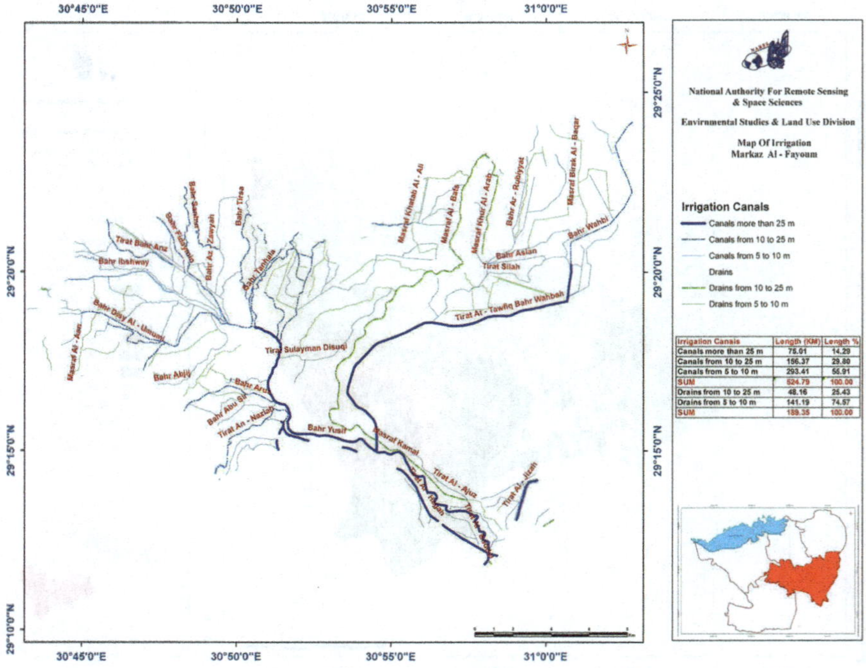

Fig. 4.364 Irrigation and drainage canals network of Al Fayoum district

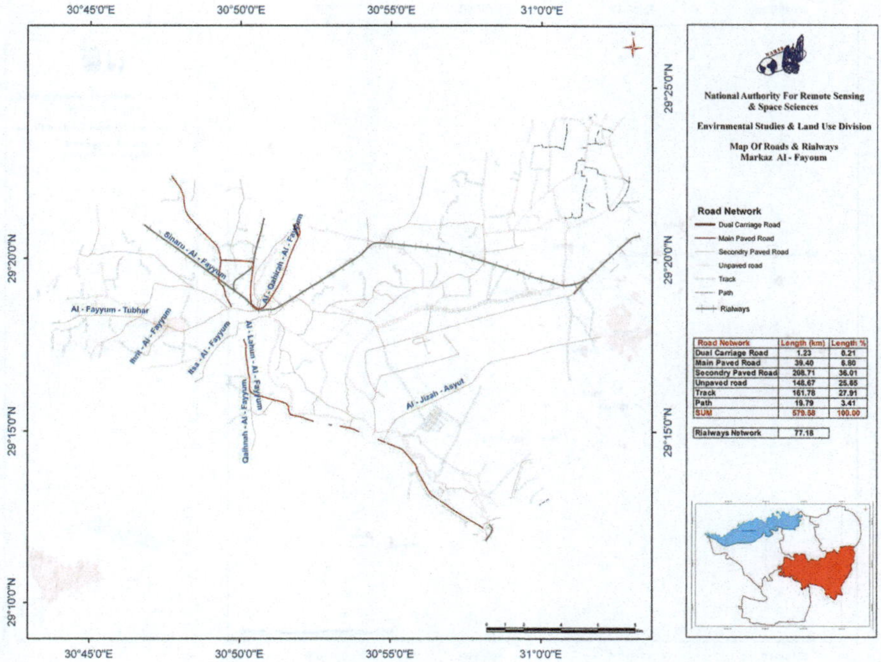

Fig. 4.365 Roads network of Al Fayoum district

is 189.4 km. Here, one may conclude that Al-Fayoum district is characterized by a wide gap between the open drainage and irrigation network. However, tile drainage system may be compensating such gap.

The roads network layer (Fig. 4.365) indicates that the district in a good and satisfactory state. The total length of the roads network is 579.6 km, indicating that each 1 km^2 of the district attains 1.5 km road network length.

Urban areas were monitored during the 1992–2006 period and the thematic layers were attached to the database (Figs. 4.366, 4.367 and 4.368). The obtained data reveal that the urban areas increased from 14.0 km^2 in the year 1992 to 23.0 km^2 in the year 2001 and to 37.0 km^2 in 2006 with annual rate of 1.6 km^2 per year. Results of change detection technique (Fig. 4.369) reveal that urban encroachment spreads in the whole district, however mostly dominates the northern parts where most fertile soils are found.

Interrelationship between soil sub-great groups and urban encroachment was analyzed to find out which type of soils is most endangered by urbanization (Figs. 4.370, 4.371, 4.372). The area of the fertile soils *Vertic Torrifluvents* decreased from 165.0 to 158.0 to 152.0 km^2 in the years 1992, 2001 and 2006 respectively. The soils of *Typic calciorthids* also decreased from 91.0 to 89.0 to 84.0 km^2 in the years 1992, 2001 and 2006 respectively.

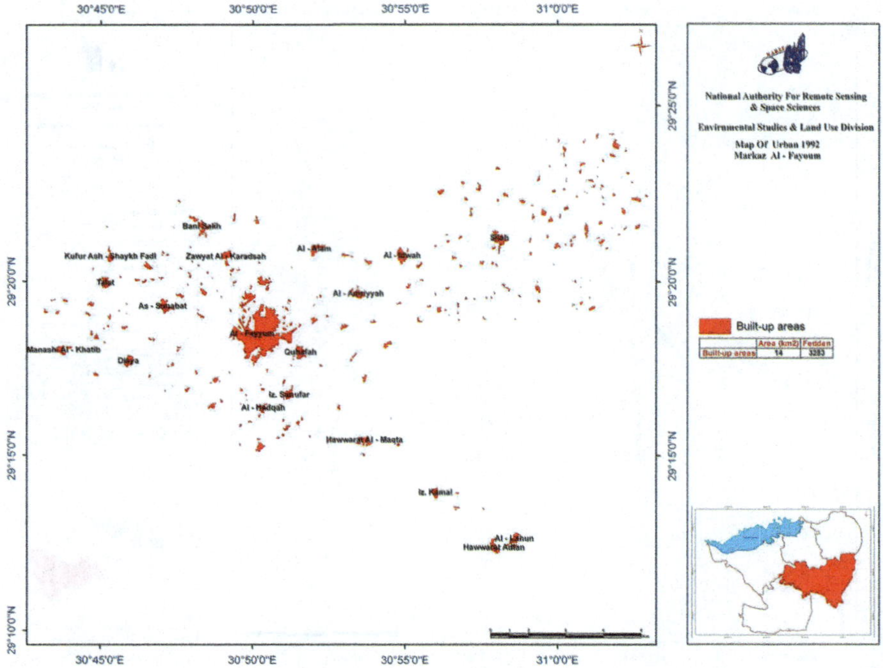

Fig. 4.366 Urban areas of Al Fayoum district (1992)

The soil capability was determined and matched with the urban layer of Al-Fayoum district (Figs. 4.373, 4.374, 4.375). The output layers indicate that the highly capable soils lost an area of 13.0 km^2 during the 1992 to 2006 period (i.e. 0.9 km^2/year). The moderately capable soils lost only an area of 1.0 km^2 with an annual rate of 0.1 km^2, while the soils of low capability lost 7.0 km^2 with an annual rate of 0.5 km^2 during the 1992–2006 periods,. The marginal and non capable soils decreased slightly with 2.1% and annual rate of 0.8% during the same period.

4.9.5.2 Atsa District

Atsa is the largest district in Al-Fayoum Governorate, it incorporates an area of 455.0 km^2, extending between longitudes 30° 30′ and 31° 55′ and latitudes 29° 05′ and 29° 20′. The database of the district includes the Landsat ETM+ and SPOT images taken during the years 2001 and 2006 respectively (Figs. 4.376 and 4.377). The satellite data were used to update the land resources thematic layers. The digital elevation model of the district (Fig. 4.378) indicates that the surface elevation values differ widely from −30 meter below the sea level, in small spots at the northern parts of the district, to 100 meter above sea level. The common surface elevation, within the district, ranges from 10–60 meter above sea level.

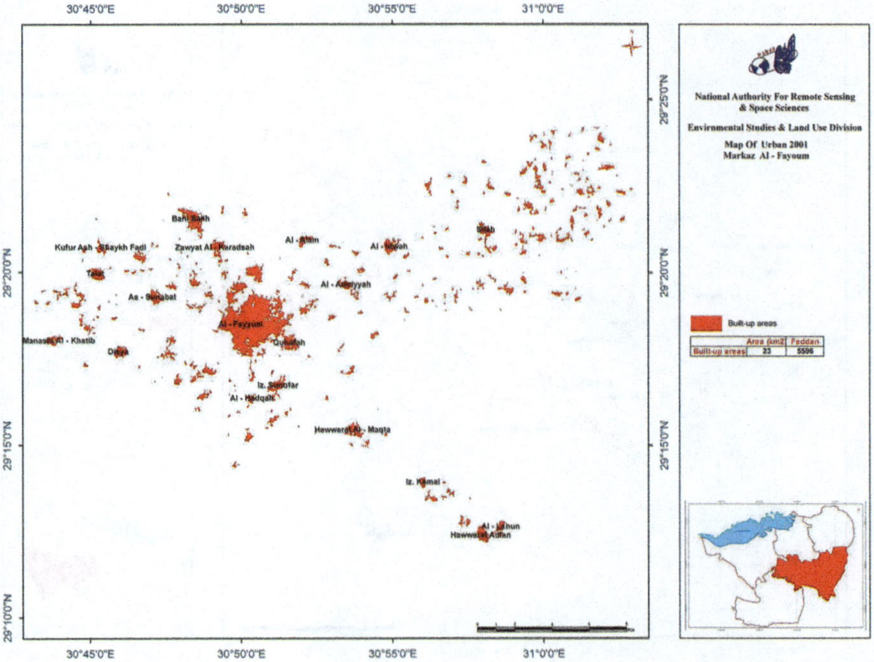

Fig. 4.367 Urban areas of Al Fayoum district (2001)

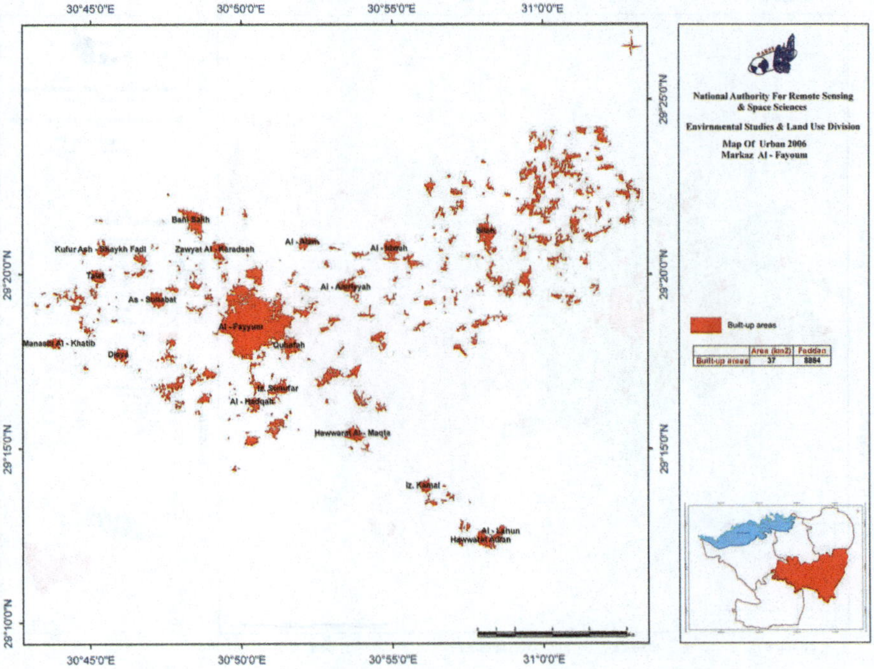

Fig. 4.368 Urban areas of Al Fayoum district (2006)

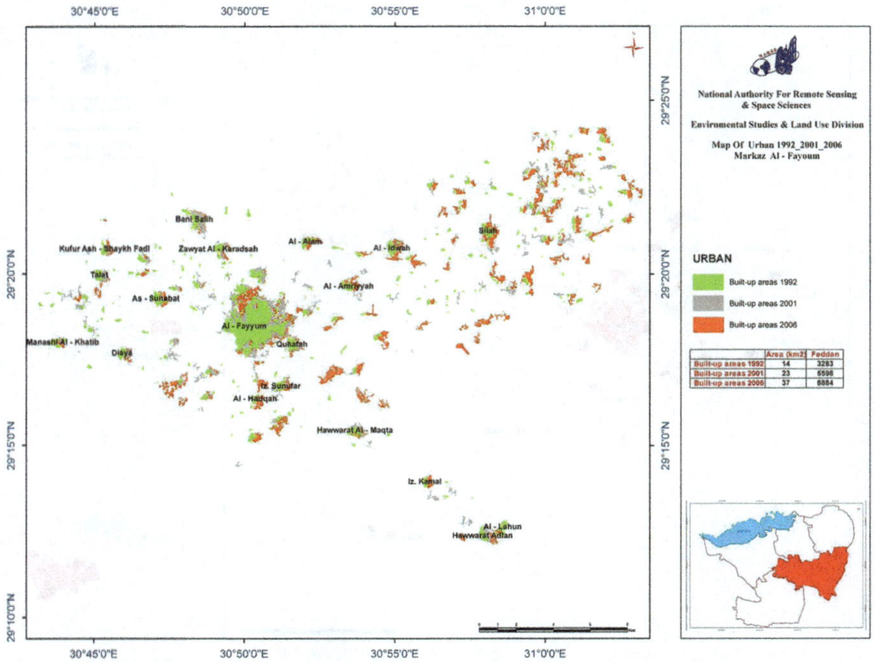

	Area (km2)	Feddan
Built-up areas 1992	14	3283
Built-up areas 2001	23	5598
Built-up areas 2006	37	8884

Fig. 4.369 Urban areas of Al Fayoum district (1992, 2001, 2006)

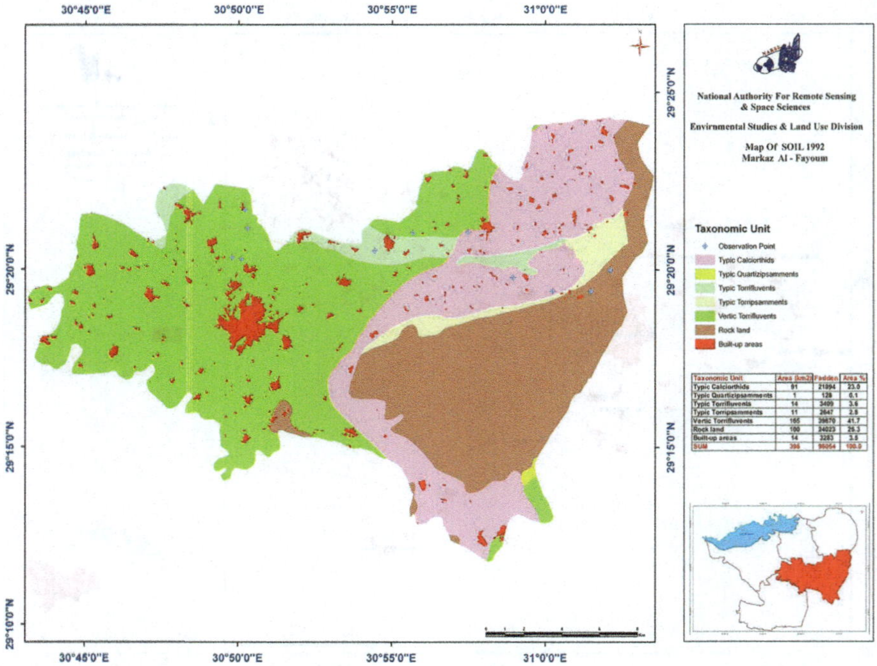

Taxonomic Unit	Area (km2)	Feddan	Area %
Typic Calciorthids	91	21894	23.0
Typic Quartizipsamments	1	128	0.1
Typic Torrifluvents	14	3409	3.6
Typic Torripsamments	11	2647	2.8
Vertic Torrifluvents	165	39670	41.7
Rock land	100	24023	25.3
Built-up areas	14	3283	3.5
SUM	396	95054	100.0

Fig. 4.370 Urban areas draped over the soil layer of Al Fayoum district (1992)

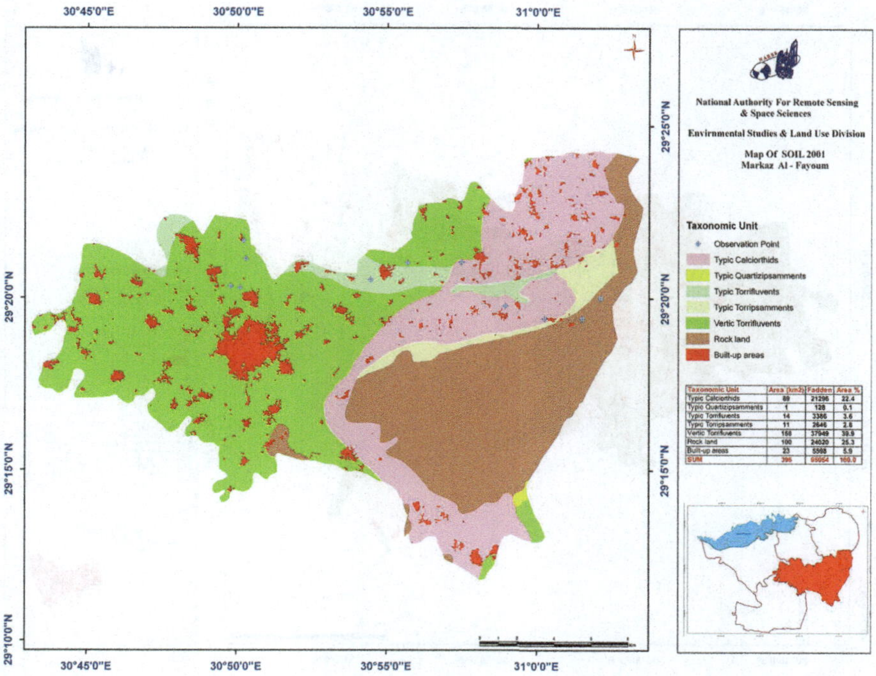

Fig. 4.371 Urban areas draped over the soil layer of Al Fayoum district (2001)

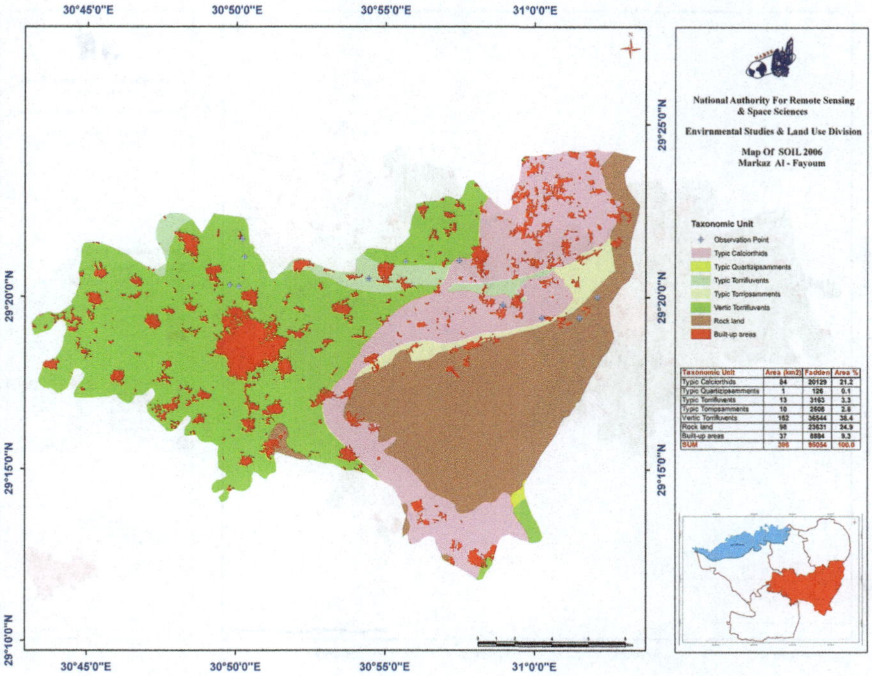

Fig. 4.372 Urban areas draped over the soil layer of Al Fayoum district (2006)

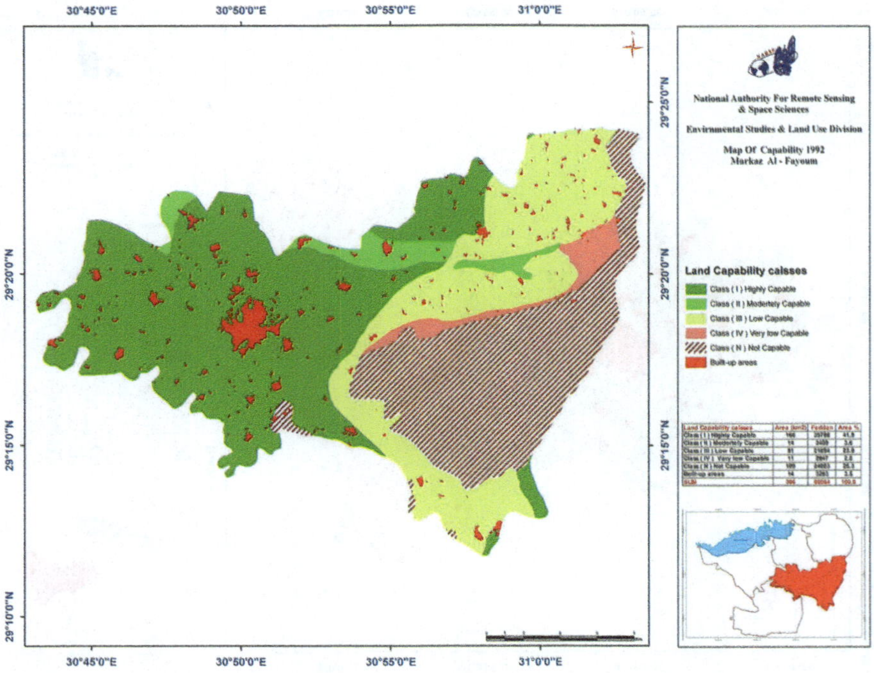

Fig. 4.373 Urban areas draped over the soil capability of Al Fayoum district (1992)

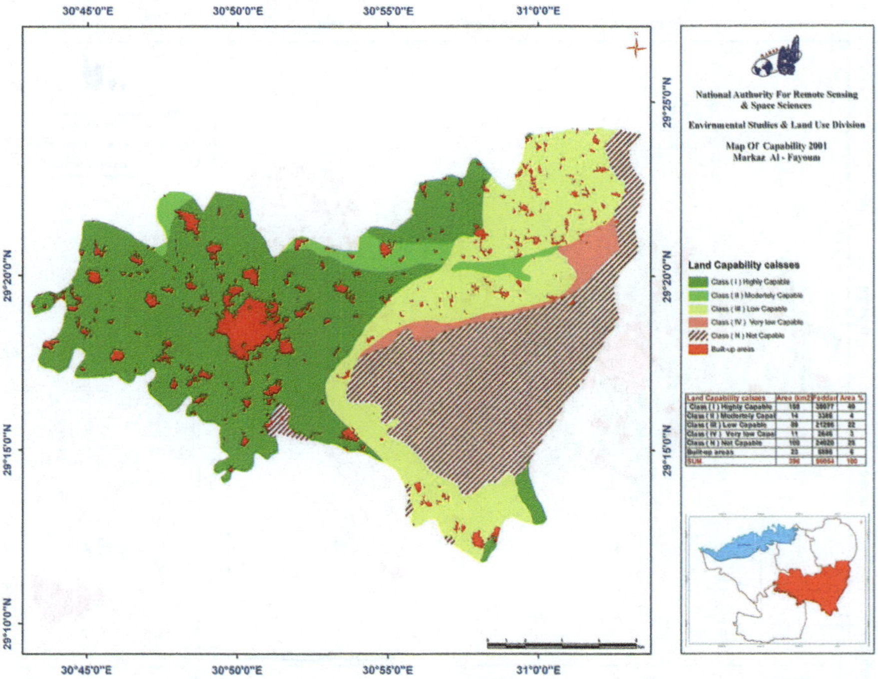

Fig. 4.374 Urban areas draped over the soil capability of Al Fayoum district (2001)

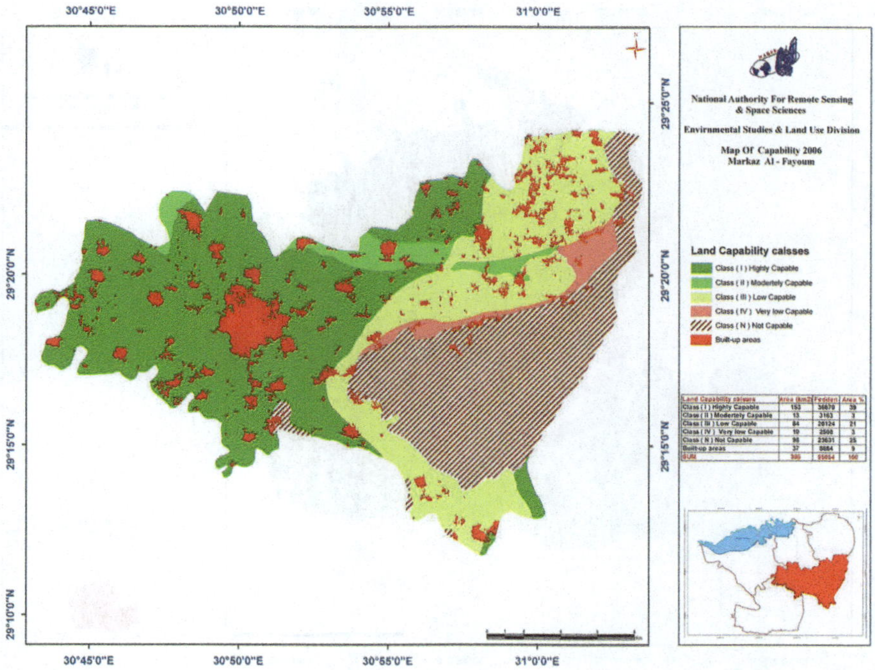

Fig. 4.375 Urban areas draped over the soil capability of Al Fayoum district (2006)

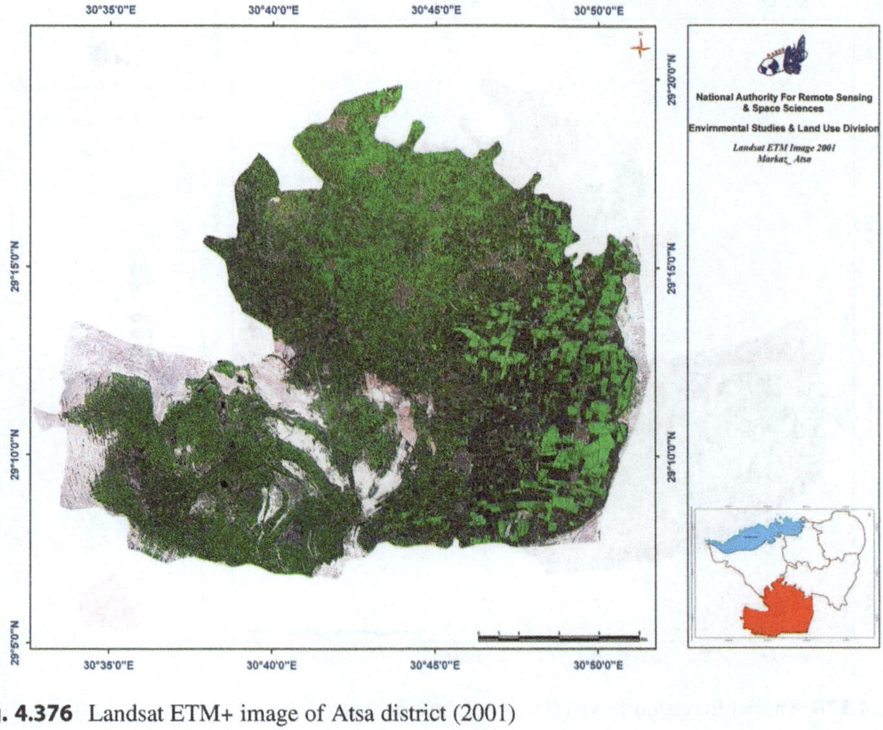

Fig. 4.376 Landsat ETM+ image of Atsa district (2001)

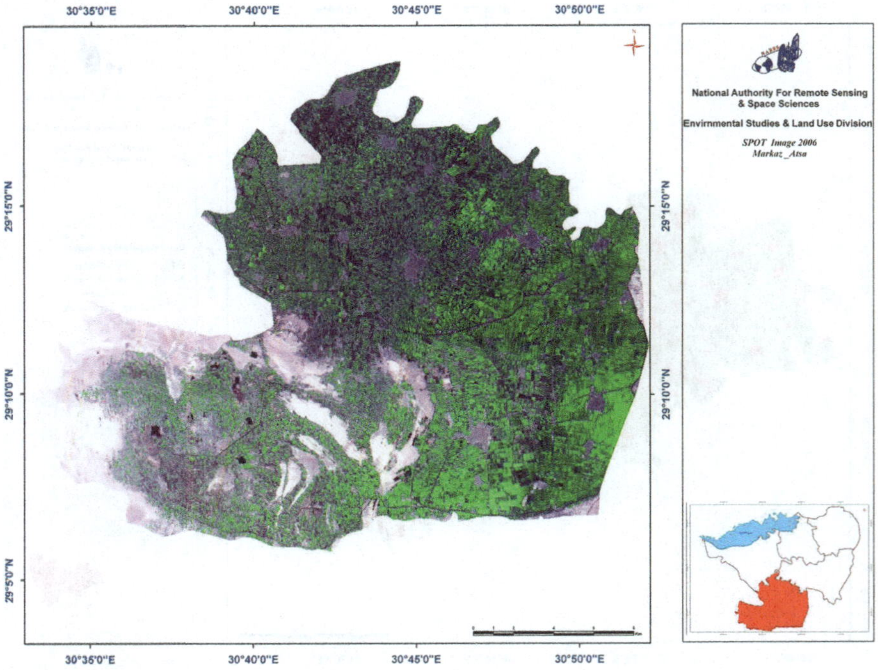

Fig. 4.377 SPOT image of Atsa district (2006)

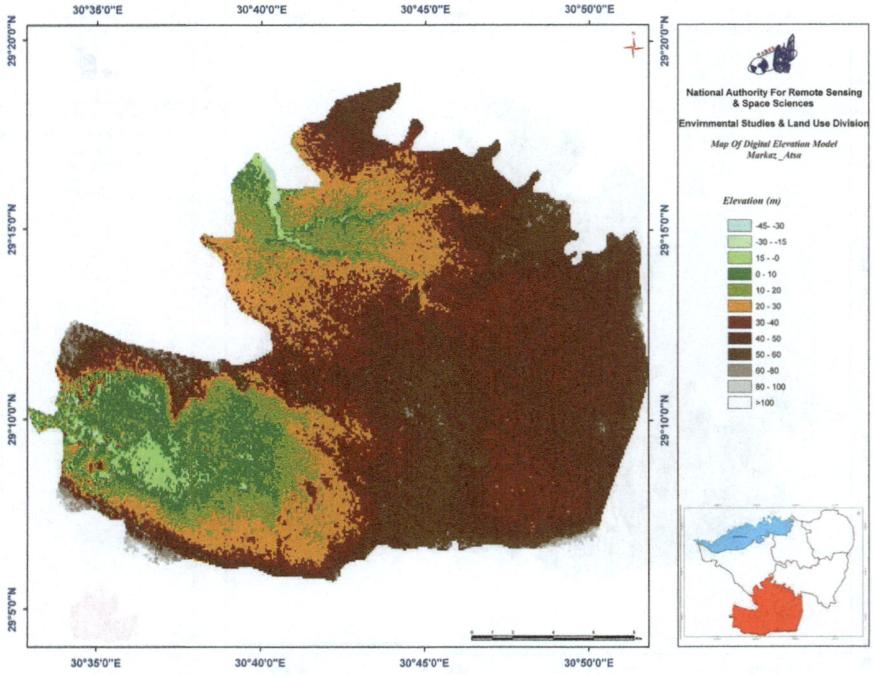

Fig. 4.378 Digital Elevation Model (DEM) of Atsa district

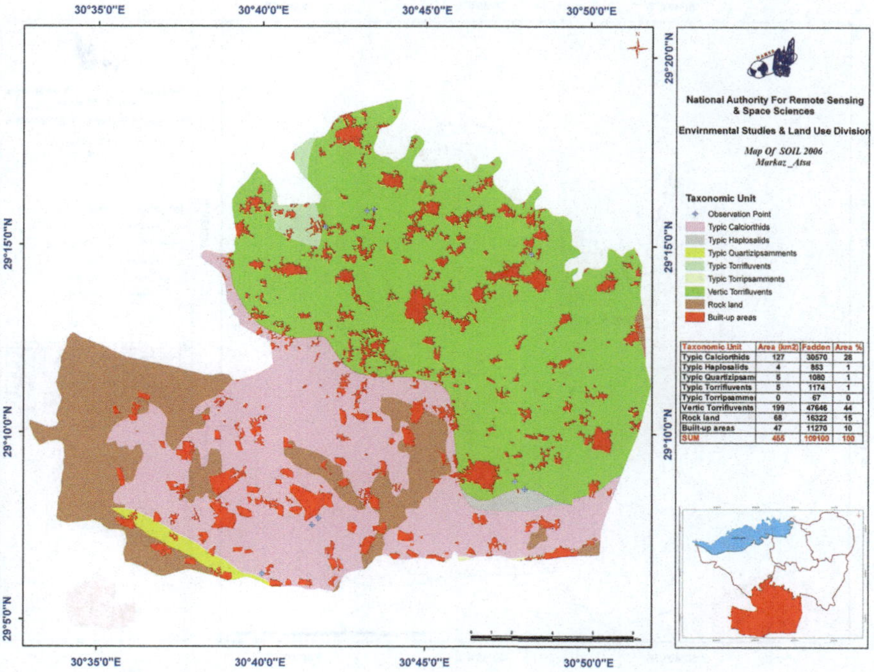

Fig. 4.379 Soil map layer of Atsa district

Regarding the soil layer (Fig. 4.379) of the district, the North East parts are dominated by the *Vertic Torrifluvents* sub great group, having an area of 199.0 km^2 (i.e. 44.0% of the district area). The southwestern parts are exhibited by the sub-great group soil of *Typic calciorthids* covering an area 127.0 km^{2,} representing 28.0% of the total district area. The rock land represents 68.0 km^2, representing 15.0% of the total area, dominating the southern part of the district. The rest of the district is dominated by the soils of *Typic Torripsamments* (0.04%), *Typic Quartzipssamentss* (1.0%), *Typic Torrifluvents* (1.0%) and *Typic Haplosalids* (0.9%).

The irrigation and drainage canals thematic layer of the Atsa district (Fig. 4.380) reveals that the total length of irrigation canals is 363.7 km, while the total length of drainage network is 320.0 km.

Figure 4.381 shows the different types of roads network and its total length. The data indicates that the district have a well distributed network. The total length of the network is 750.8 km, this indicate that each 1 km^2 of the district have a length of 1.7 km of roads.

Progress of urban areas was monitored during the 1992–2006 period (Figs. 4.382, 4.383, 4.384). Change detection of settlement areas (Fig. 4.385) resulted in pointing out the location and magnitudes of urban encroachment in Atsa district. The obtained data reveal that the urban area increased from 15.0 km^2 in the year 1992 to 32.0 km^2 in the year 2001 and to 47.0 km^2 in 2006 with an annual rate of 1.6 km^2 per year.

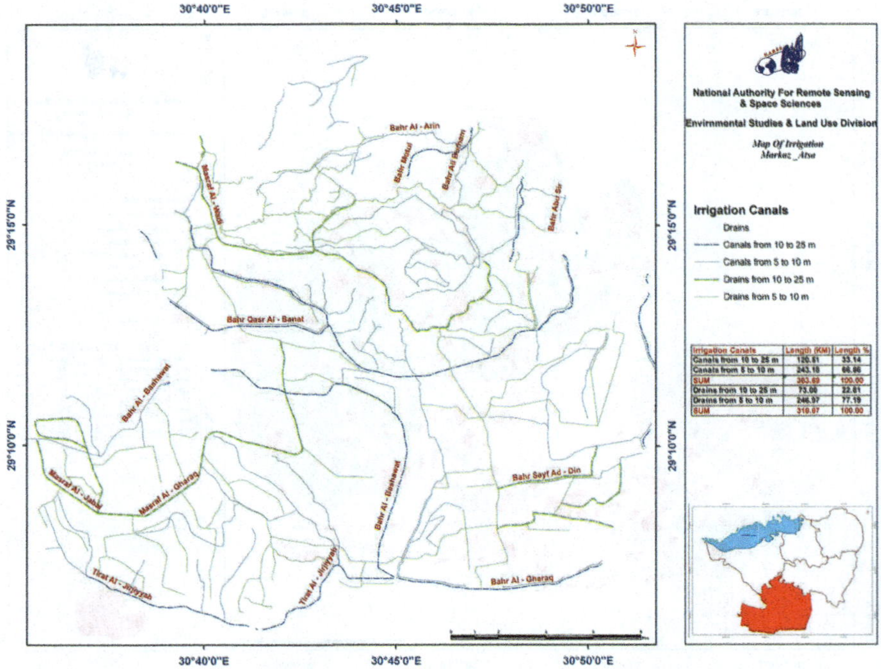

Fig. 4.380 Irrigation and drainage canals network of Atsa district

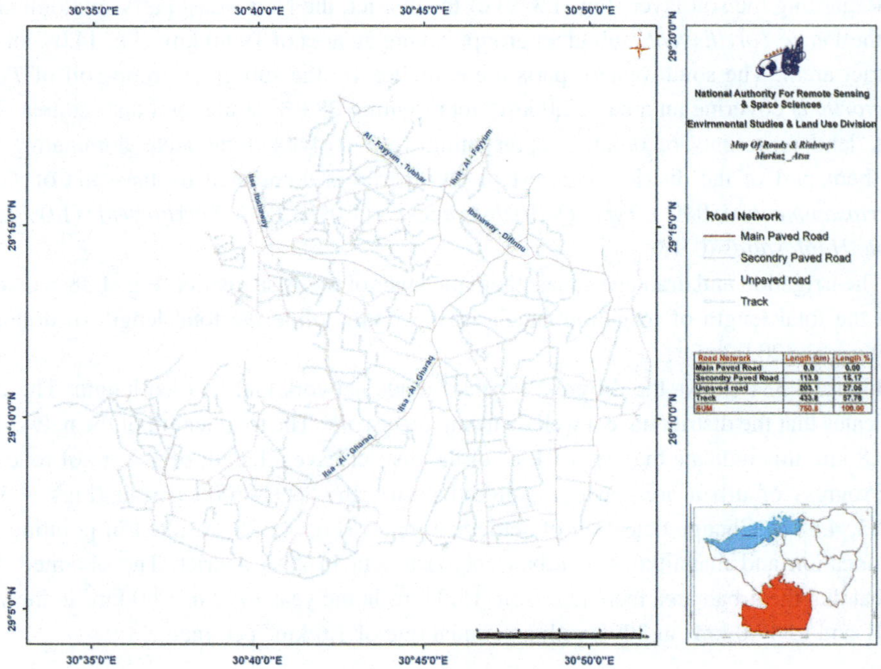

Fig. 4.381 Roads network of Atsa district

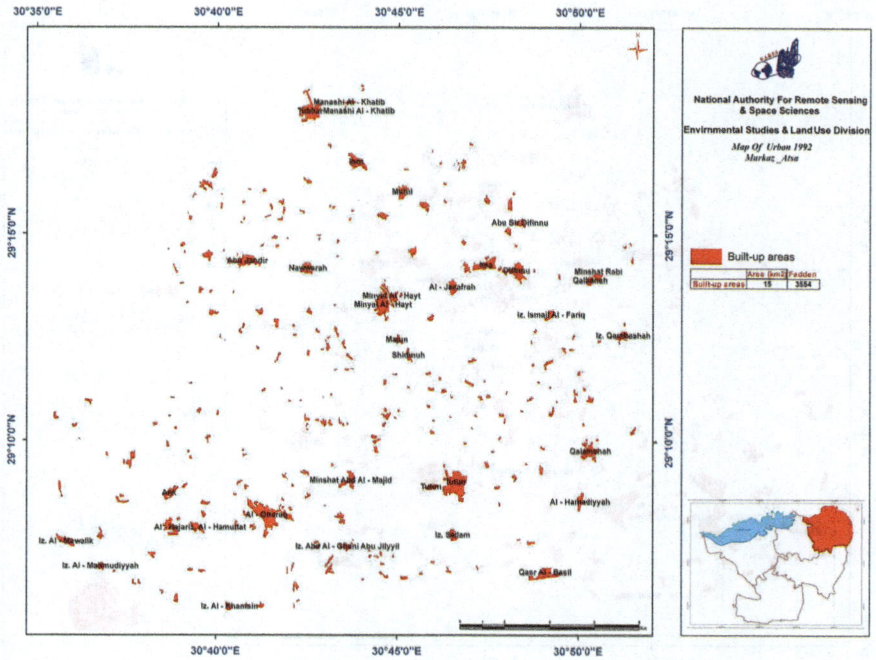

Fig. 4.382 Urban areas of Atsa district (1992)

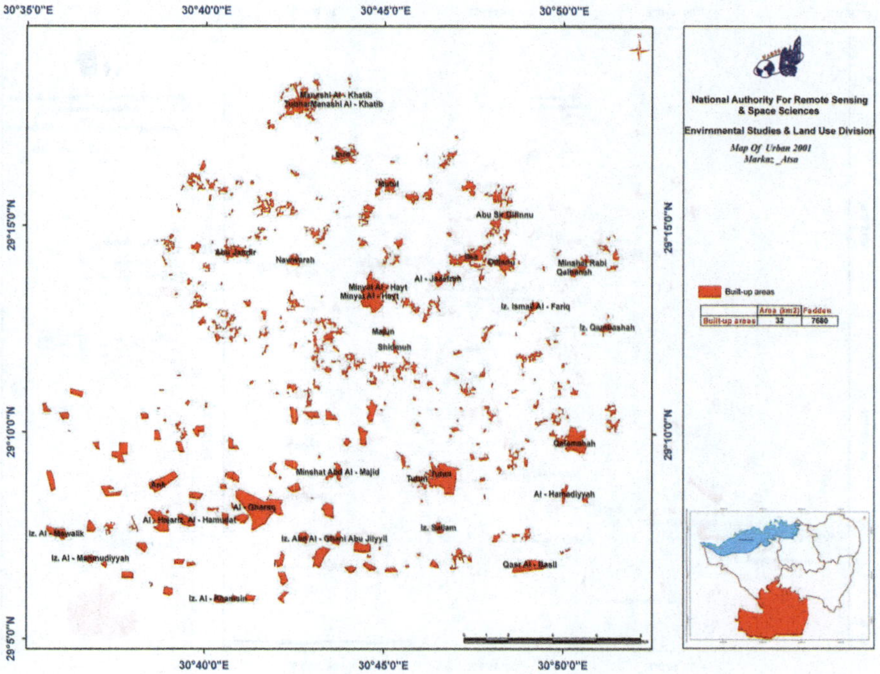

Fig. 4.383 Urban areas of Atsa district (2001)

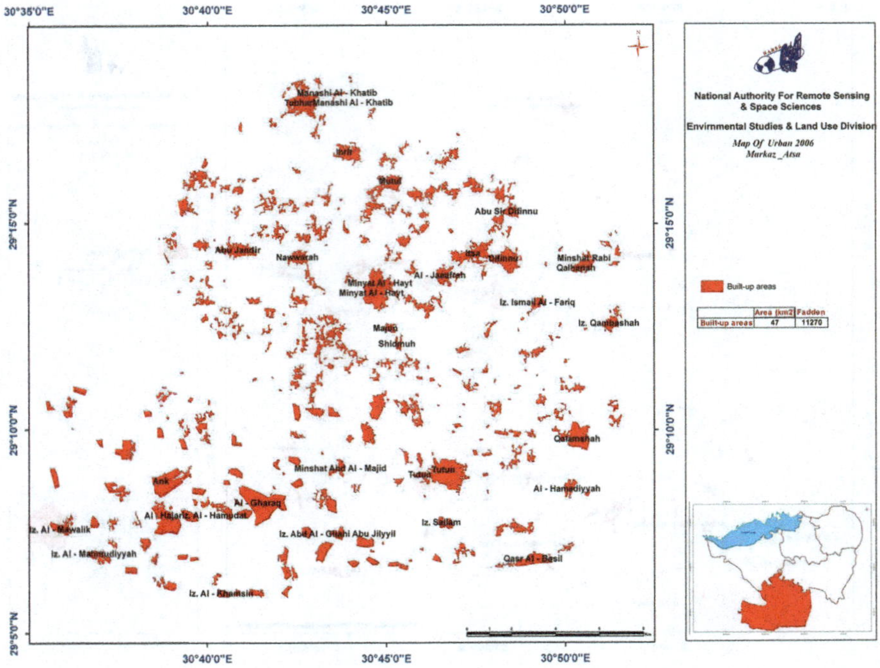

Fig. 4.384 Urban areas of Atsa district (2006)

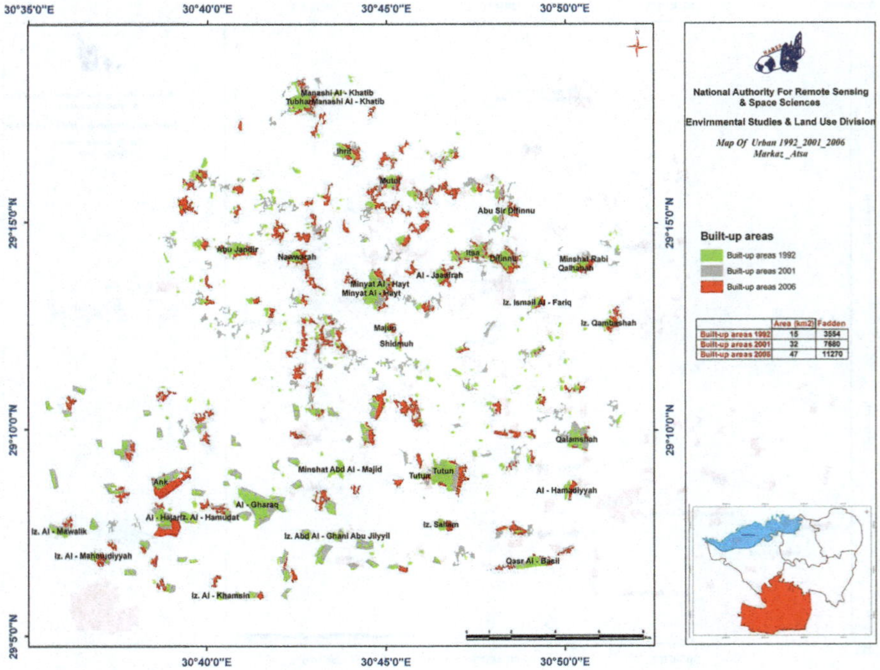

Fig. 4.385 Urban areas of Atsa district (1992, 2001, 2006)

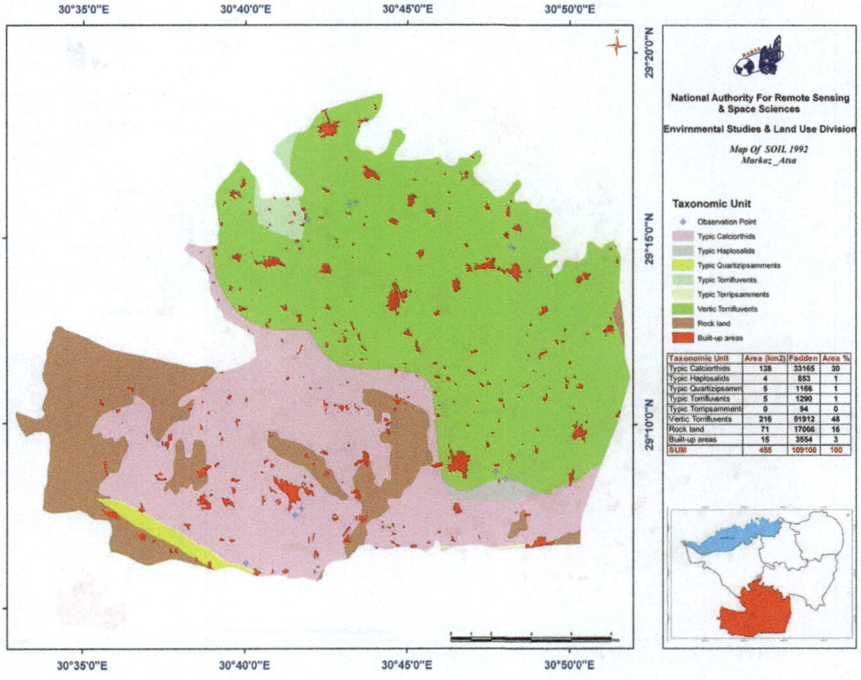

Fig. 4.386 Urban areas draped over the soil layer of Atsa district (1992)

Attaching the soil and land capability layers, in one hand, and multi-temporal urban areas, in another hand, was performed. Figures 4.386, 4.387 and 4.388 show that the urban encroachment was almost on the account of most fertile soils. The area of the cultivable soils of *Vertic Torrifluvents* sub-great group decreased from 216.0 to 208.0 to 199.0 km^2 in the years 1992, 2001 and 2006 respectively. The soils of *Typic calciorthids* also decreased from to 131.0 km^2 in 2001 and to 127.0 km^2 in the year 2006.

The soil capability was established and matched with the urban layer of the district (Figs. 4.389, 4.390 and 4.391), the output layers indicate that the highly capable soils lost 17.0 km^2 during the 1992–2006 period (i.e. 1.2 km^2/year), the moderately capable soils lost only an area of 1.0 km^2 with annual rate of 0.1 km^2. The soils of low capability lost 11.0 km^2 during the 1992–2006 period, with an annual rate of 0.8 km^2, while the marginal and non capable soils lost a very small area during the same period with an annual rate of 0.2 km^2.

4.9.5.3 Ibsheway District

Ibsheway district incorporate an area of 389.0 km^2 and extending between longitudes 30° 25′ and 30° 50′ and latitudes 29° 15′ and 29° 30′. The land resources database of the district include the Landsat ETM+ and SPOT images taken during the years 2001 and 2006 respectively (Figs. 4.392 and 4.393), used in updating the thematic layers. The digital

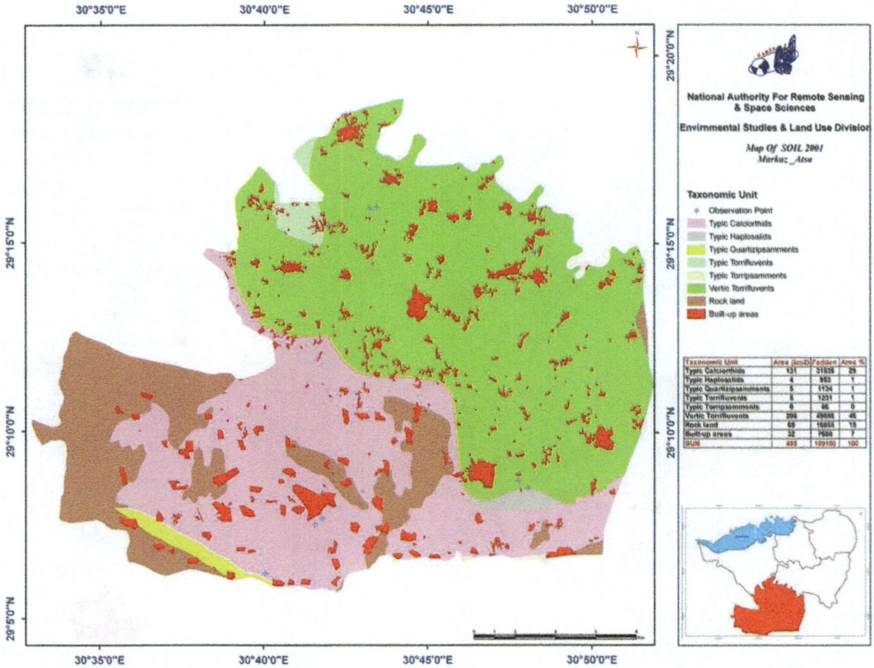

Fig. 4.387 Urban areas draped over the soil layer of Atsa district (2001)

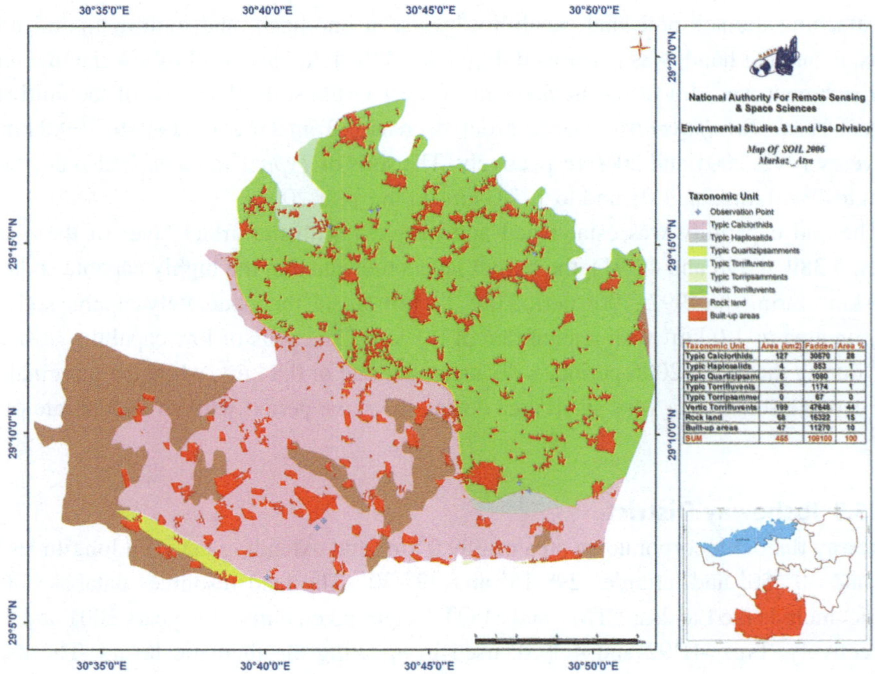

Fig. 4.388 Urban areas draped over the soil layer of Atsa district (2006)

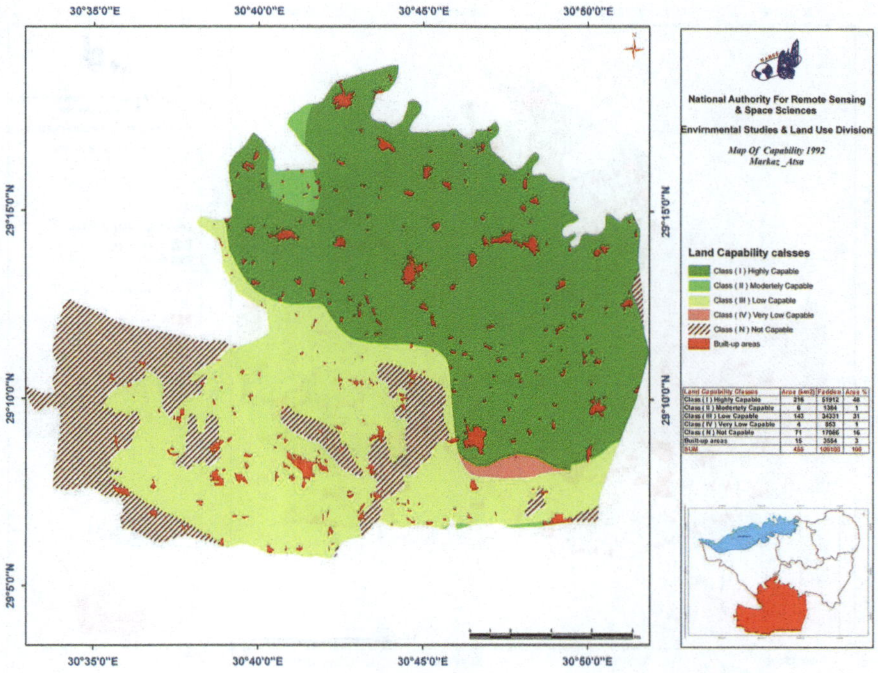

Fig. 4.389 Urban areas draped over the soil capability of Atsa district (1992)

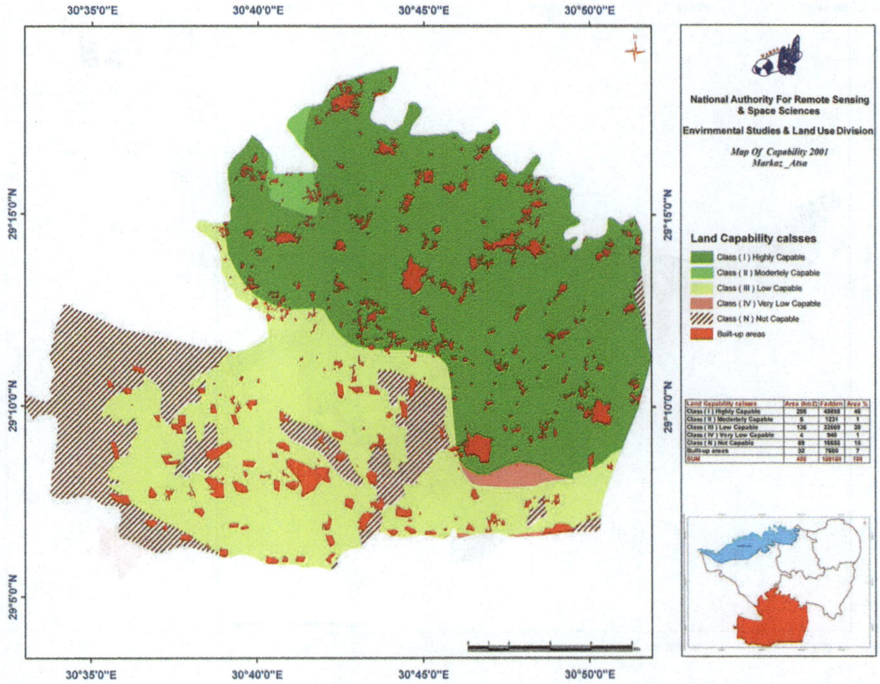

Fig. 4.390 Urban areas draped over the soil capability of Atsa district (2001)

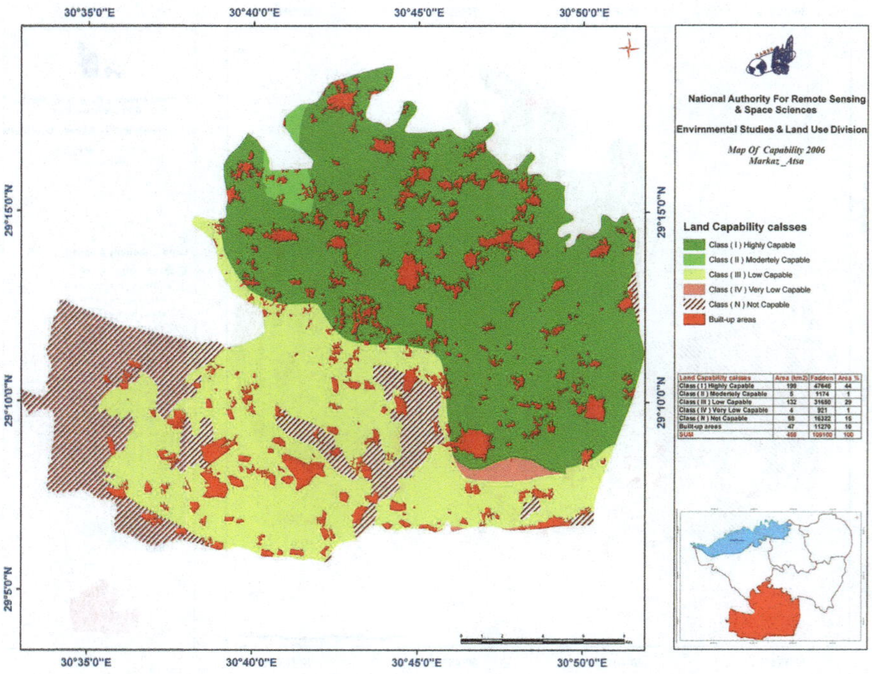

Fig. 4.391 Urban areas draped over the soil capability of Atsa district (2006)

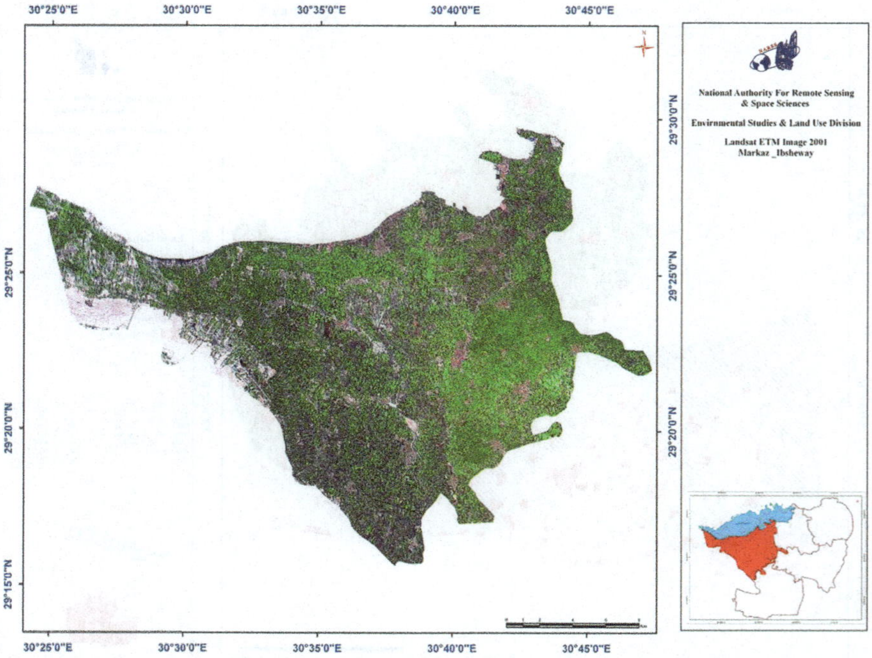

Fig. 4.392 Landsat ETM+ image of Ibsheway district (2001)

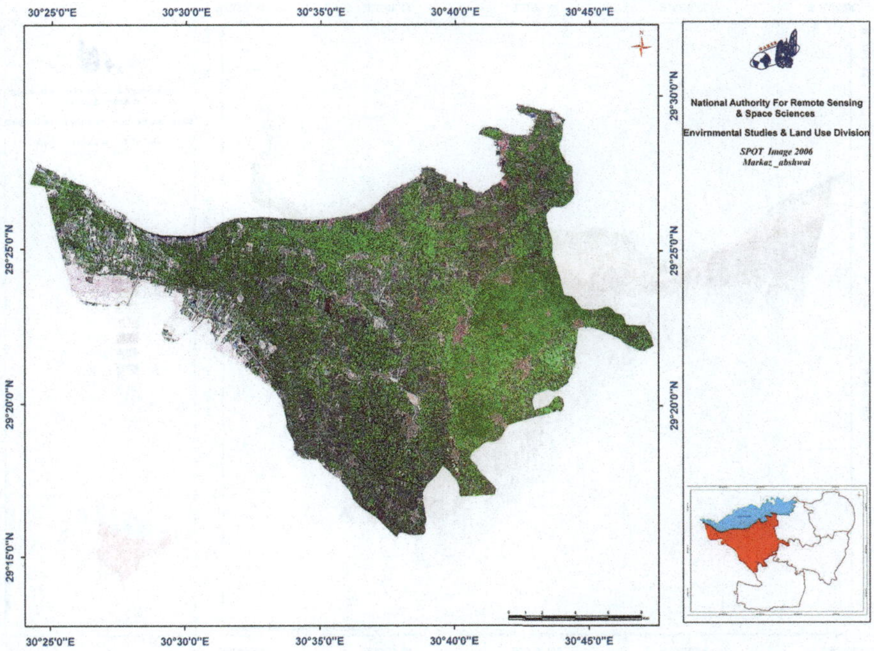

Fig. 4.393 SPOT image of Ibsheway district (2006)

elevation model of the district (Fig. 4.394) indicates that the surface elevation differs widely from −45 meter below sea level, in small spots at the northern parts, to 80 meter above sea level. However, the common surface elevation, within the district, ranges from 0 to 50 meter above sea level.

Regarding the soil layer (Fig. 4.395), the middle and northeastern parts of the district are dominated by the *Vertic Torrifluvents* sub great group with an area of 119.0 km^2 (i.e. 31.0% of the district area). The northwestern parts of the district are exhibited by the *Typic calciorthids* soil sub great group, extending for 97.0 km^2, representing 25.0% of the total district area. The soils of *Typic Torrifluvents* sub great group dominate the eastern parts of the district, having an area of 65.0 km^2 (i.e. 17% of total soils). The rock land covers an area of 34.0 km^2, representing 9.0% of the whole territory. The rest of the district is dominated by small spots characterized by the soils of Typic *Torripsamments* (2.0%), and *Typic Haplosalids* (6.0%).

The irrigation and drainage canals thematic layer of the district (Fig. 4.396) reveal that the total length of irrigation canals is 348.5 km while the total length of drainage network is 199.2 km. The situation indicates a large gap between the drainage and irrigation network in the district.

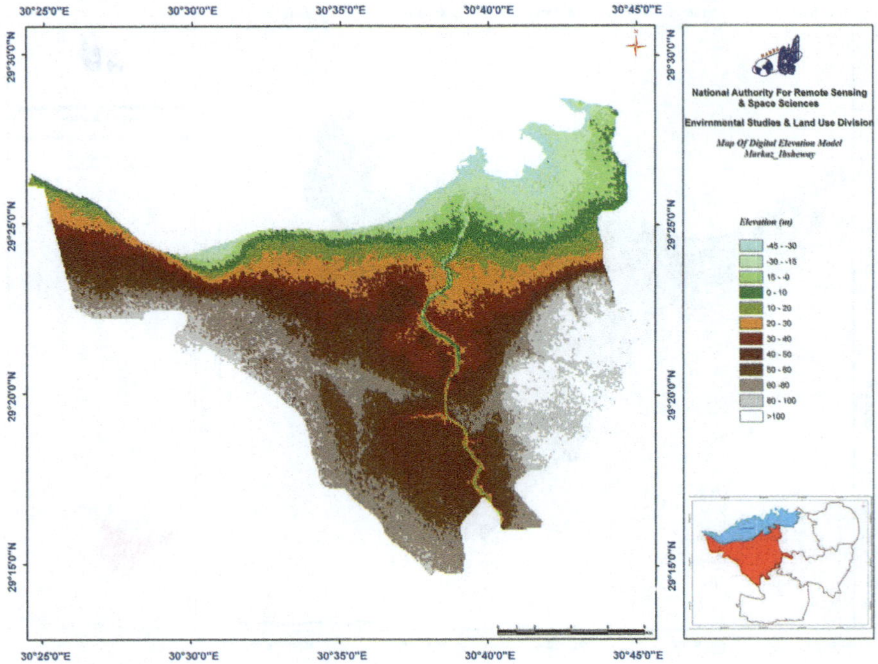

Fig. 4.394 Digital Elevation Model (DEM) of Ibsheway district

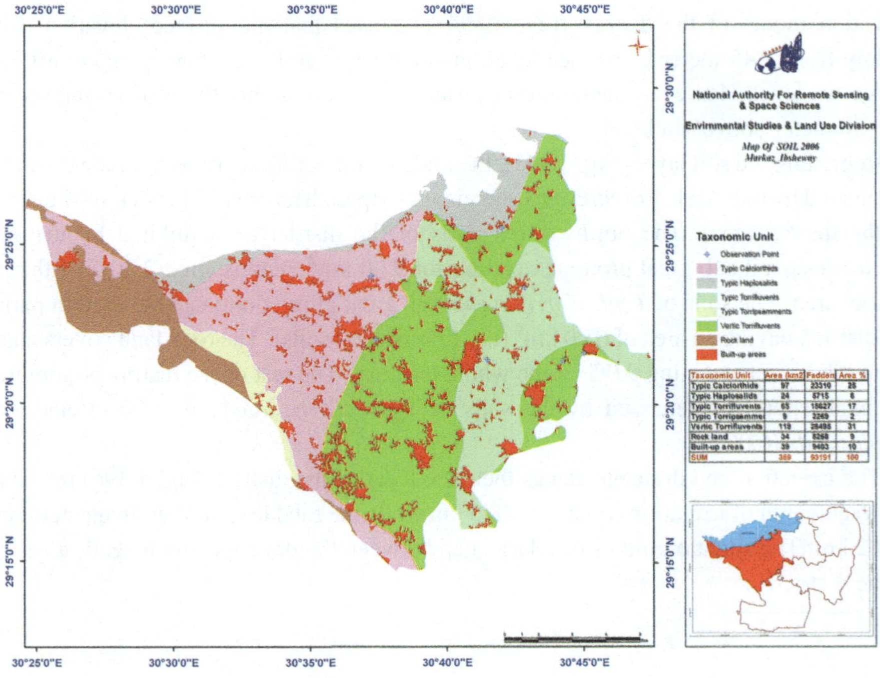

Fig. 4.395 Urban areas draped over the soil layer of Ibsheway district (2006)

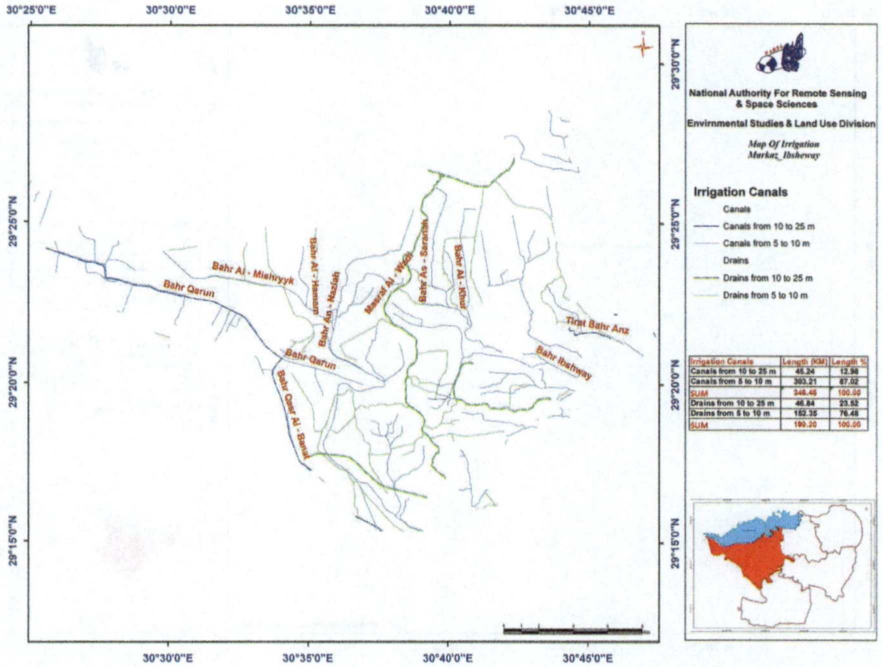

Fig. 4.396 Irrigation and drainage canals network of Ibsheway district

The roads network layer (Fig. 4.397) indicates that the district has a good and satisfactory network, characterized by a total roads length of 617.0 km. That demonstrates that each 1 km^2 of the district is served by 1.6 km of roads.

Monitoring the urban areas during the years 1992–2006 period resulted in obtaining the settlements thematic layers, which were attached to the created land resources database (Figs. 4.398, 4.399 and 4.400). Change detection of urban growth demonstrates (Fig. 4.401) reveal that the urban areas have increased from 16.0 km^2 in the year 1992 to 24.0 km^2 in the year 2001 and to 39.0 km^2 in 2006 with an annual rate of 1.6 km^2 per year (the same rate of Atsa district). The urban encroachment dominates the whole parts of the district where most fertile soils are found.

Combining the soil and urban thematic layers shows that the cultivable soils of *Vertic Torrifluvents* decreased from 129.0 km^2 (year 1992) to 125.0 km^2 (year 2001) and to 119.0 km^2 (year 2006). The soils of *Typic calciorthids* also decreased from 105.0 km^2 (year 1992) to 103.0 km^2 (year 2001) and to 97.0 km^2 (year 2006). The cultivable *Typic Torrifluvents* soil sub great group decreased from 69.0 km^2 (year 1992) to 67.0 km^2 (year 2001) and to 65.0 km^2 (year 2006) with an annual rate of 0.3 km^2 per year (Figs. 4.402, 4.403 and 4.404).

Matching the soil capability classification with the multi-temporal urban layers of the district (Figs. 4.405, 4.406 and 4.407), show that the highly capable soils lost 10.0 km^2

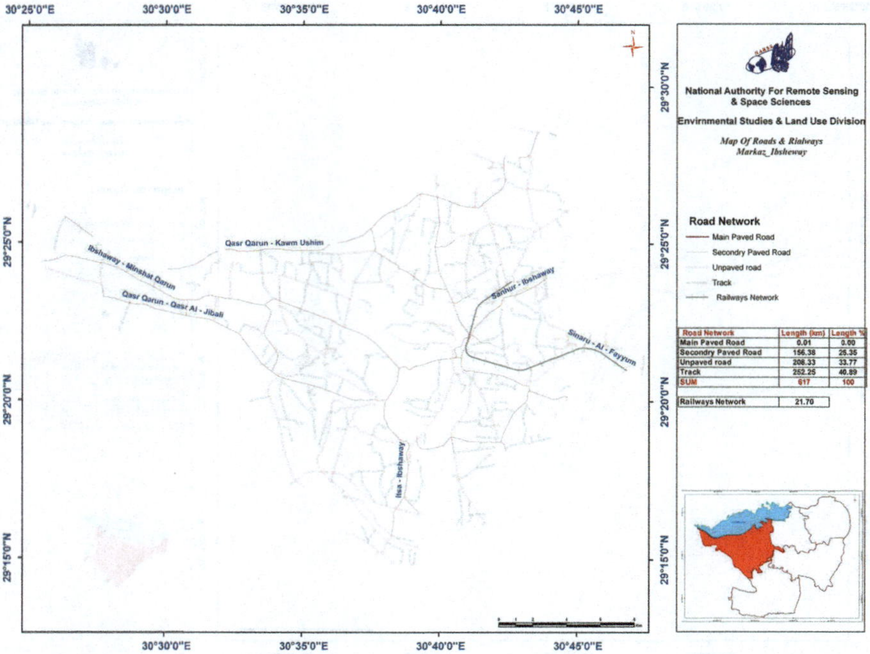

Fig. 4.397 Roads network of Ibsheway district

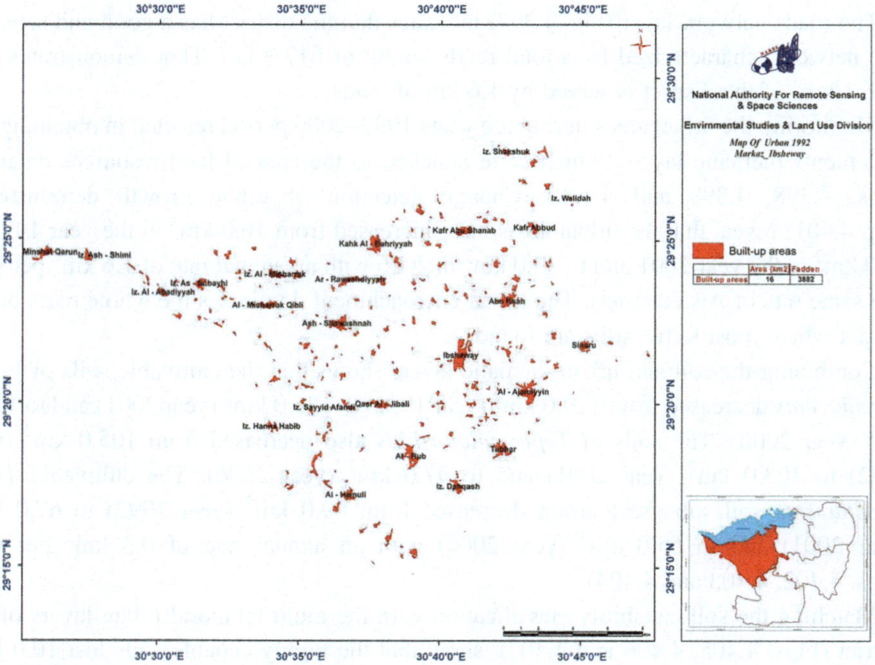

Fig. 4.398 Urban areas of Ibsheway district (1992)

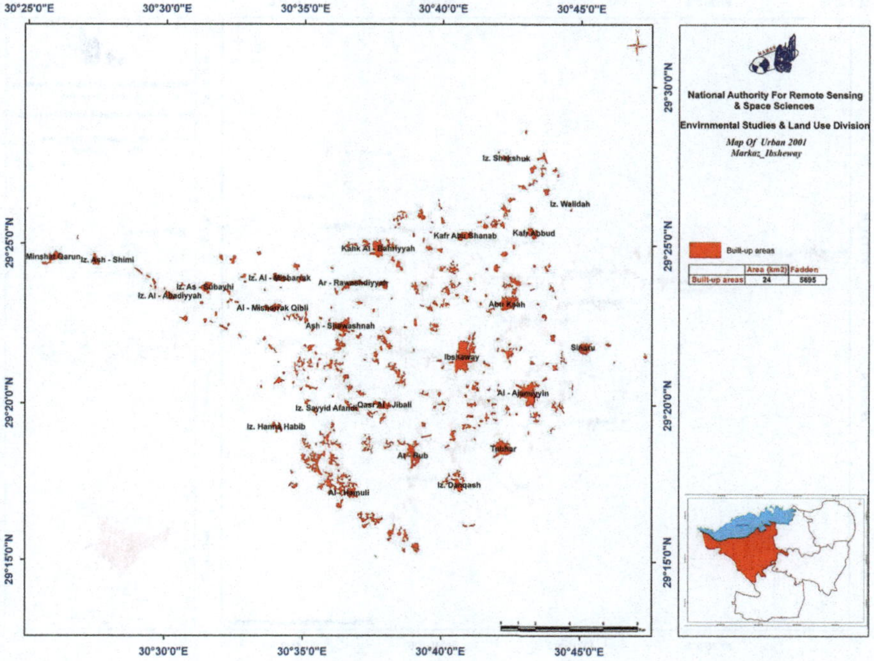

Fig. 4.399 Urban areas of Ibsheway district (2001)

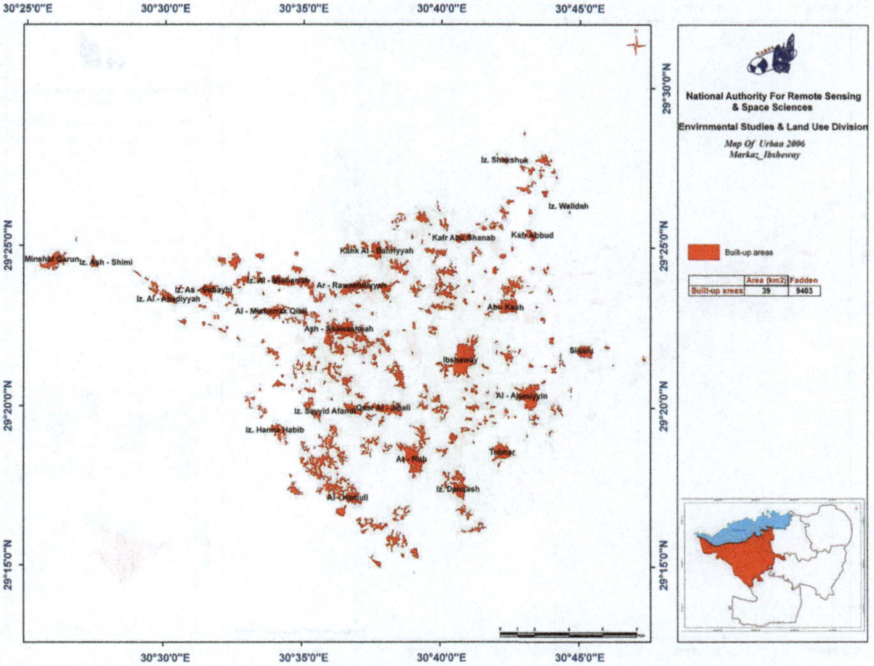

Fig. 4.400 Urban areas of Ibsheway district (2006)

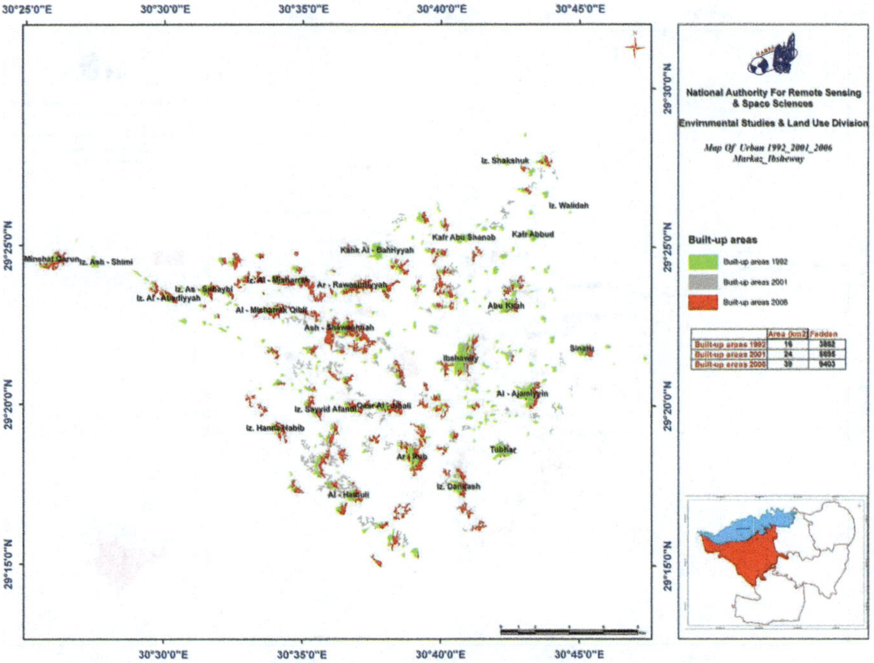

Fig. 4.401 Urban areas of Ibsheway district (1992, 2001, 2006)

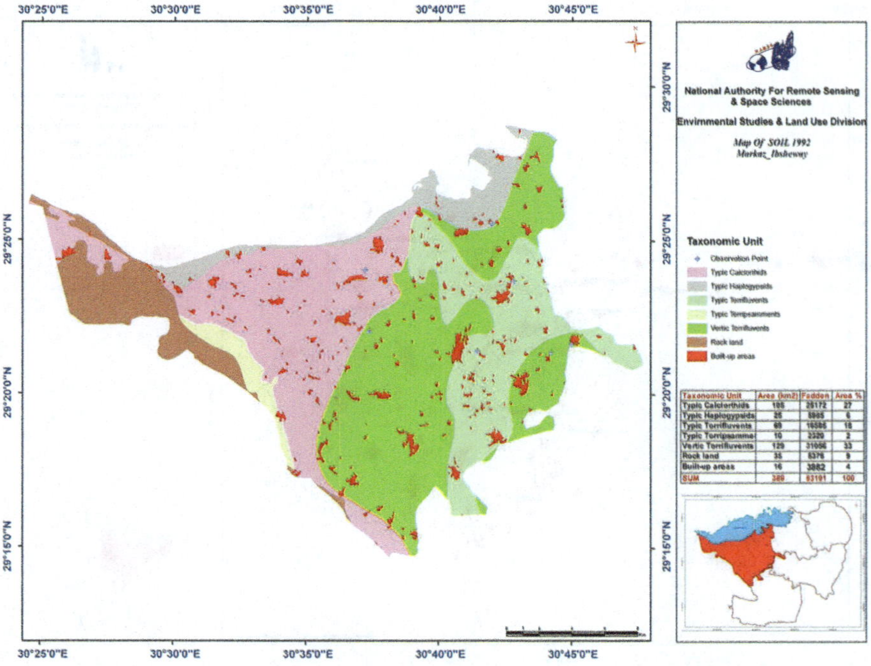

Fig. 4.402 Urban areas draped over the soil layer of Ibsheway district (1992)

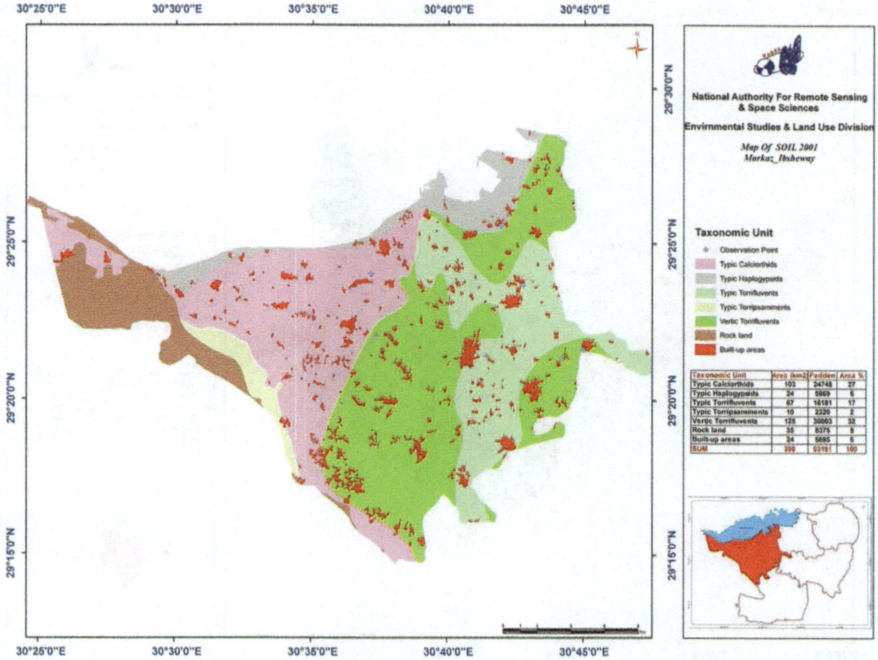

Fig. 4.403 Urban areas draped over the soil layer of Ibsheway district (2001)

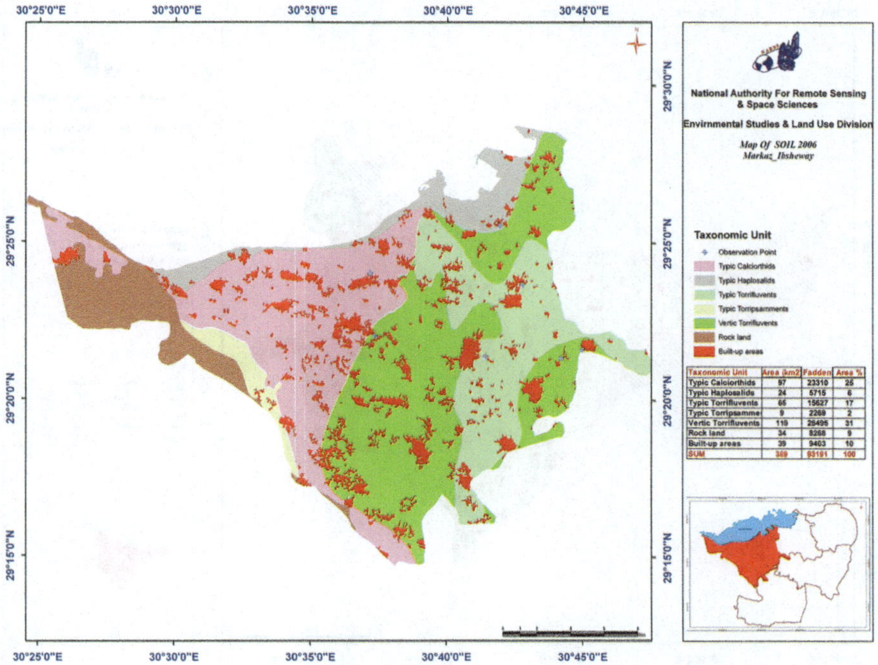

Fig. 4.404 Urban areas draped over the soil layer of Ibsheway district (2006)

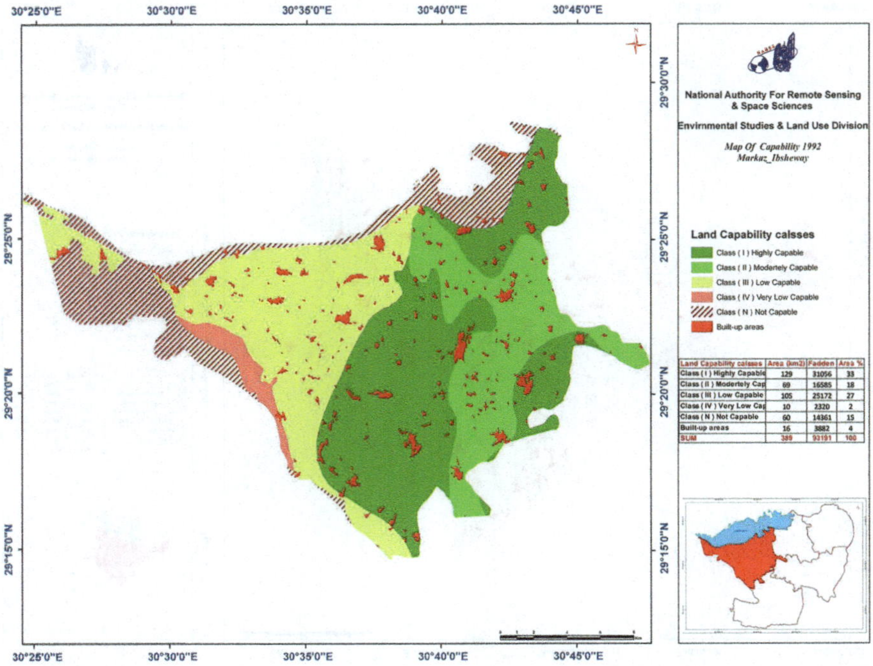

Fig. 4.405 Urban areas draped over the soil capability of Ibsheway district (1992)

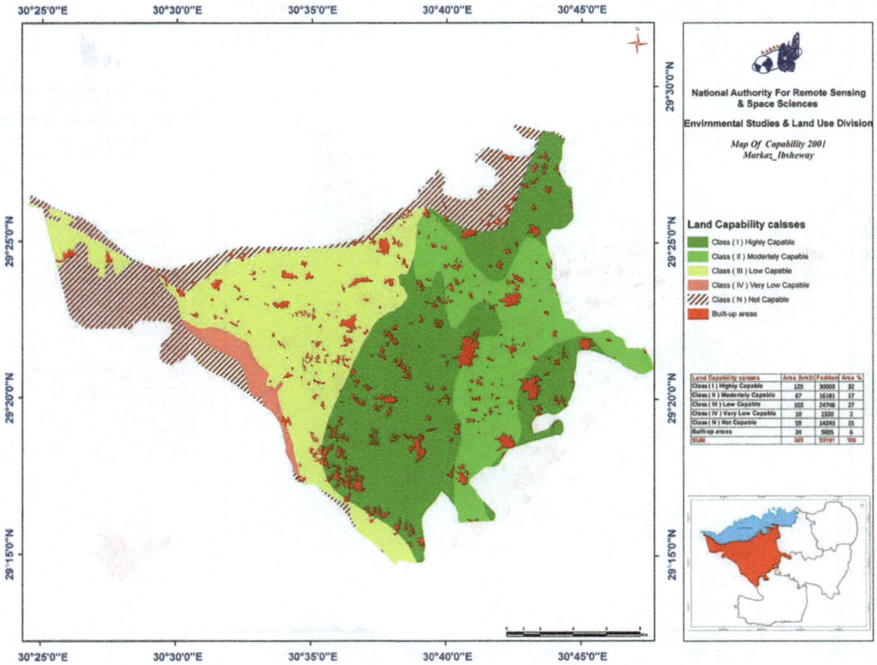

Fig. 4.406 Urban areas draped over the soil capability of Ibsheway district (2001)

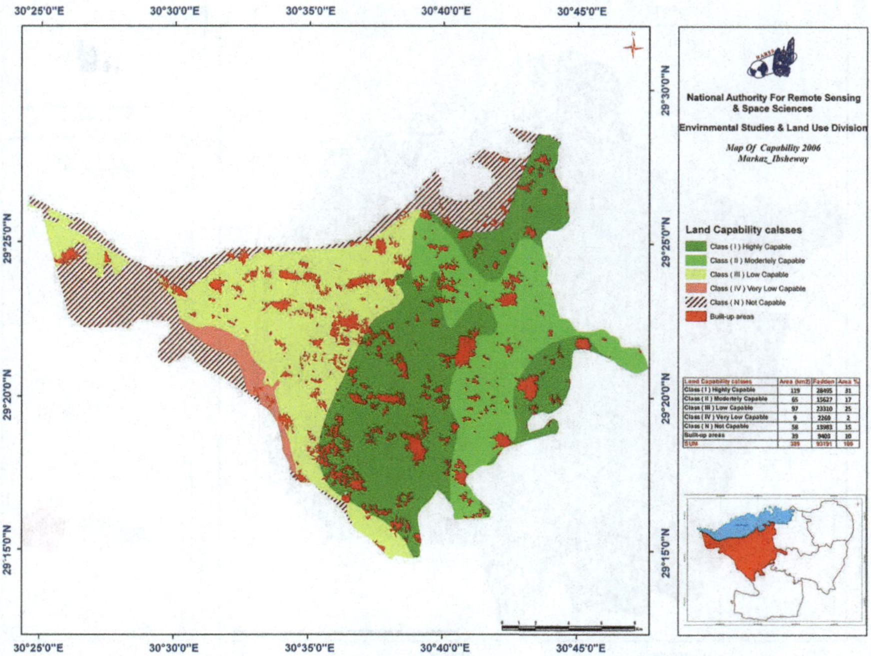

Fig. 4.407 Urban areas draped over the soil capability of Ibsheway district (2006)

during the 1992–2006 period (i.e. 0.7 km^2/year). The moderately capable soils lost only an area of 4.0 km^2, with an annual rate of 0.3 km^2. The soils of low capability lost 8.0 km^2 during the same period, with an annual rate of 0.6 km^2, while the marginal and non capable soils lost a very small area during the same period with an annual rate of 0.2 km^2.

4.9.5.4 Tamyah District

Tamyah district incorporates an area of 304.0 km^2, extending from longitudes 30° 50′ to 31° 50′ N and latitudes from 29° 20′ to 29° 40′ E. The land resources database of the district includes the Landsat ETM+ and SPOT images taken during the years 2001 and 2006 respectively (Figs. 4.408 and 4.409), which were used for updating the thematic layers. The digital elevation model of the district (Fig. 4.410) indicates that the surface elevation differs widely from −45 meter below the sea level, characterizing limited spots in the northern parts, to 60 meter above sea level. The common surface elevation, characterizing the district, ranges from 0 to 20 meters above sea level.

Regarding the soil layer (Fig. 4.411) of the district, the soils of *Vertic Torrifluvents* sub great group cover an area of 98.0 km^2 (i.e. 32.0% of the district area) mainly distributed in the south western part of the study area. The northern parts of the district are exhibited by the soils of *Typic Haplogypsids* sub-great group covers an area of 85.0 km^2 (i.e. 28.0% of the total district area). While, the soils of *Typic calciorthids,* covering an area of 82.0 km^2 representing 27.0% of the total area. The *Typic haplosalids* sub-great group is found in a

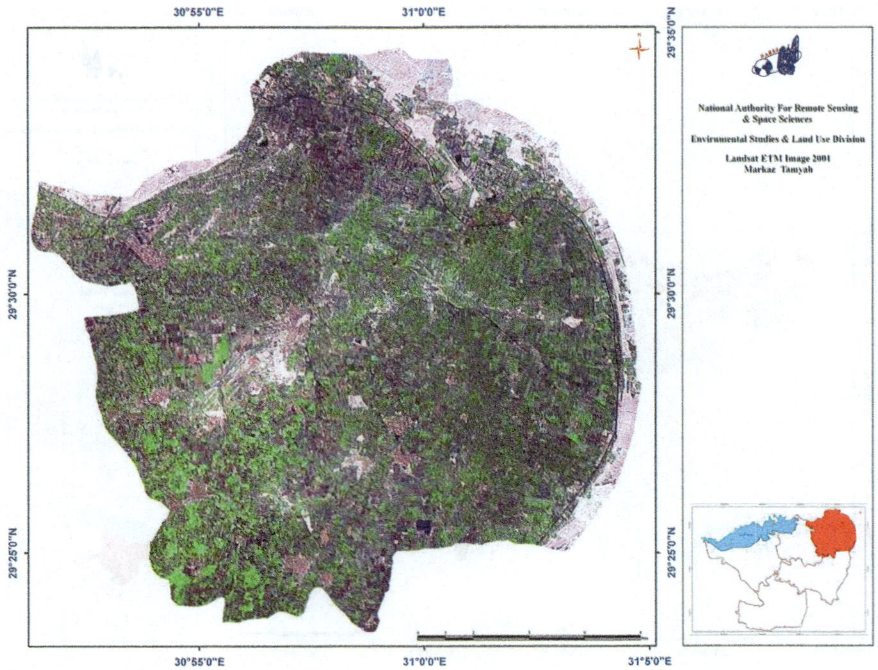

Fig. 4.408 Landsat ETM+ image of Tamyah district (2001)

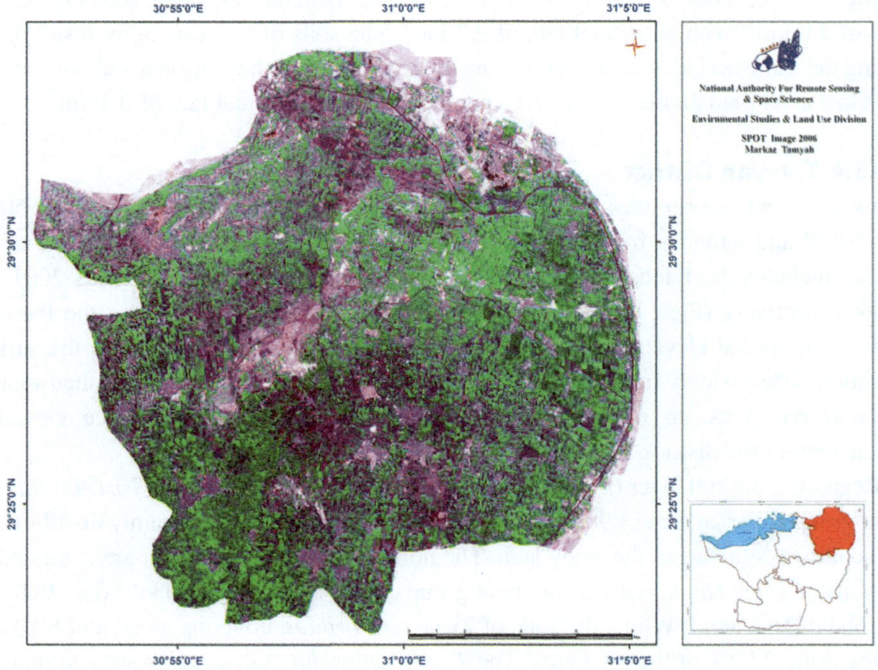

Fig. 4.409 SPOT image of Tamyah district (2006)

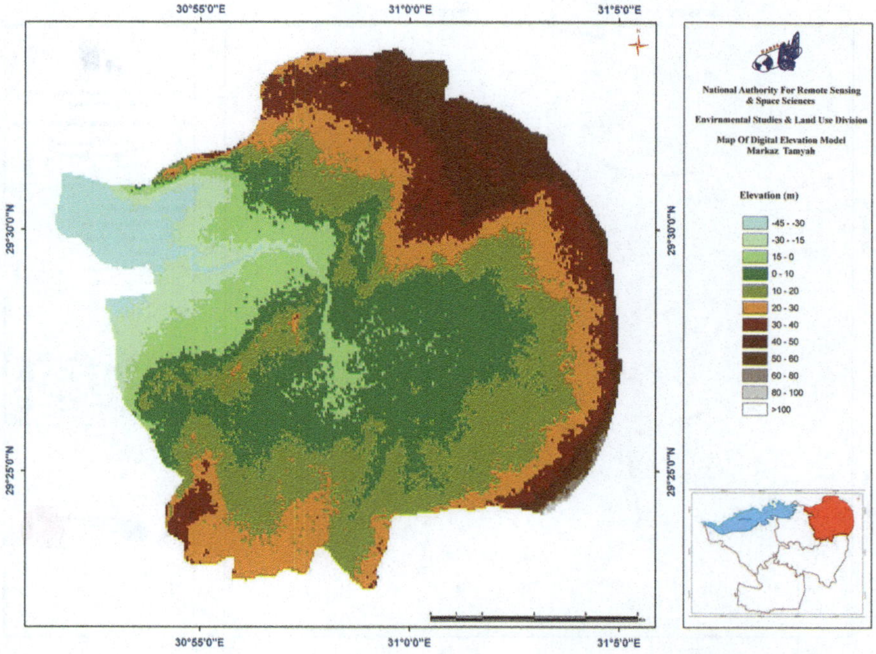

Fig. 4.410 Digital Elevation Model (DEM) of Tamyah district

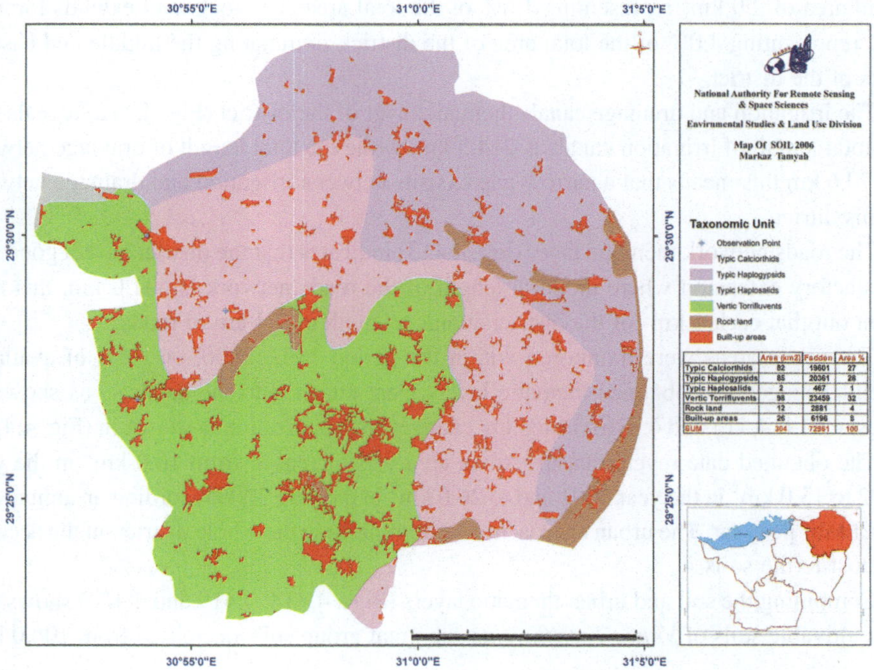

Fig. 4.411 Urban areas draped over the soil layer of Tamyah district (2006)

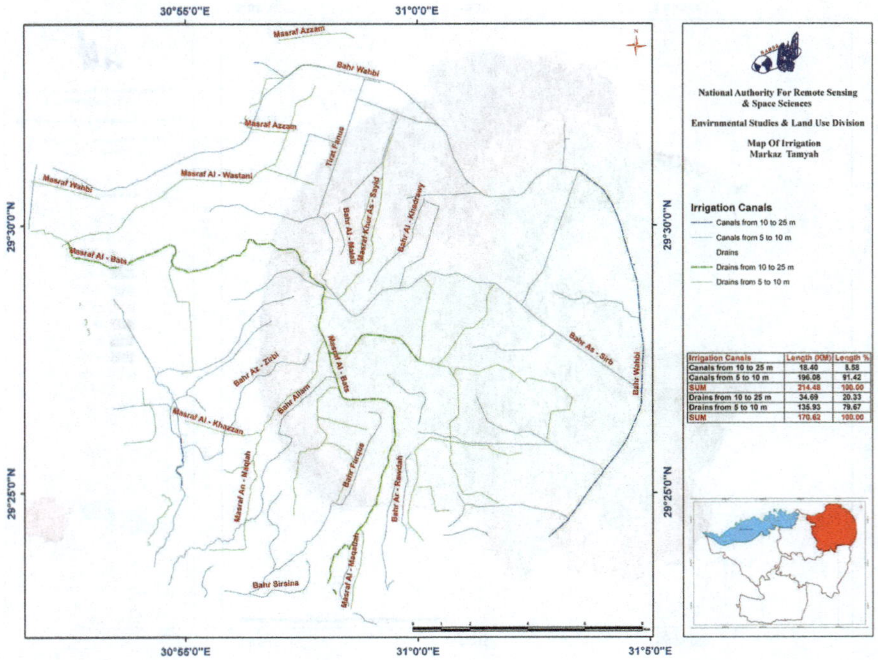

Fig. 4.412 Irrigation and drainage canals network of Tamyah district

small area of 2.0 km^2 representing 1.0% of the total area. The rock land extends for 12.0 km^2 representing 4.0% of the total area of the district, dominating the middle and eastern parts of the district.

The irrigation and drainage canals thematic layer of the district (Fig. 4.412) reveals that the total length of irrigation canals is 214.5 km, while the total length of drainage network is 170.6 km this means that a narrow gap exsisits between irrigation and drainage network in this ditrict.

The roads network thematic layer (Fig. 4.413) indicates that the district have a good and satisfactory extension where the total length of the roads network is 431.0 km. this may point out that each 1 km^2 of the district attains a length of 1.4 km of roads.

The urban areas were monitored within the period 1992–2006, on basis of available satellite images, the obtained thematic layers were attached to the database as shown in (Figs. 4.414, 4.415 and 4.416). Detecting changes of urbanization is shown in (Fig. 4.417).

The obtained data reveal that the urban area was increased from 10.0 km^2 in the year 1992 to 13.0 km^2 in the year 2001 and to 26.0 km^2 in the year 2006 recording an annual rate of 1.1 km^2 per year. The urban encroachment dominates in the whole district on the account of most fertile soils.

Combining the soil and urban thematic layers (Figs. 4.418, 4.419 and 4.420) shows that the cultivable soils of *Vertic Torrifluvents* sub great group soils decreased from 106.0 km^2

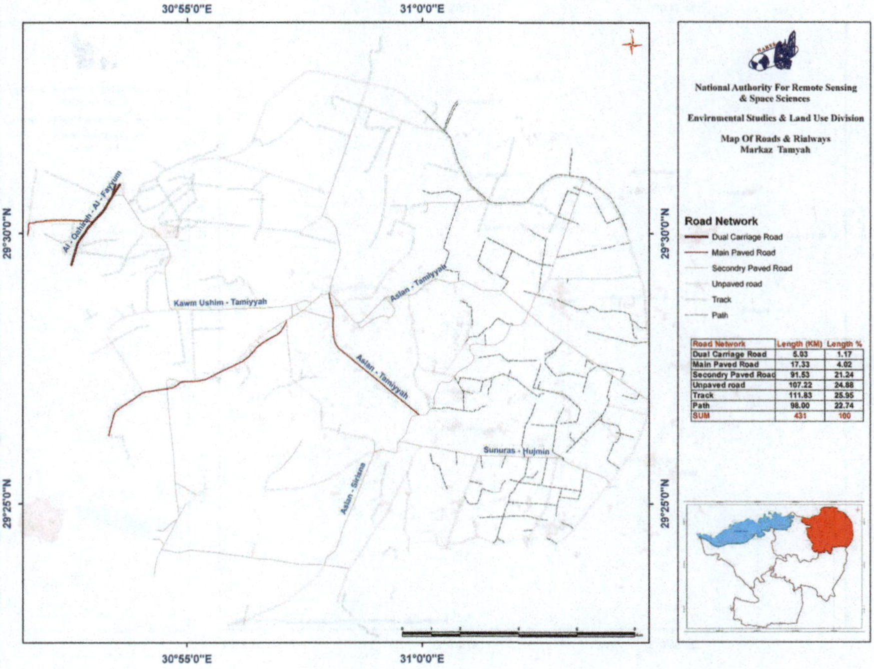

Fig. 4.413 Roads network of Tamyah district

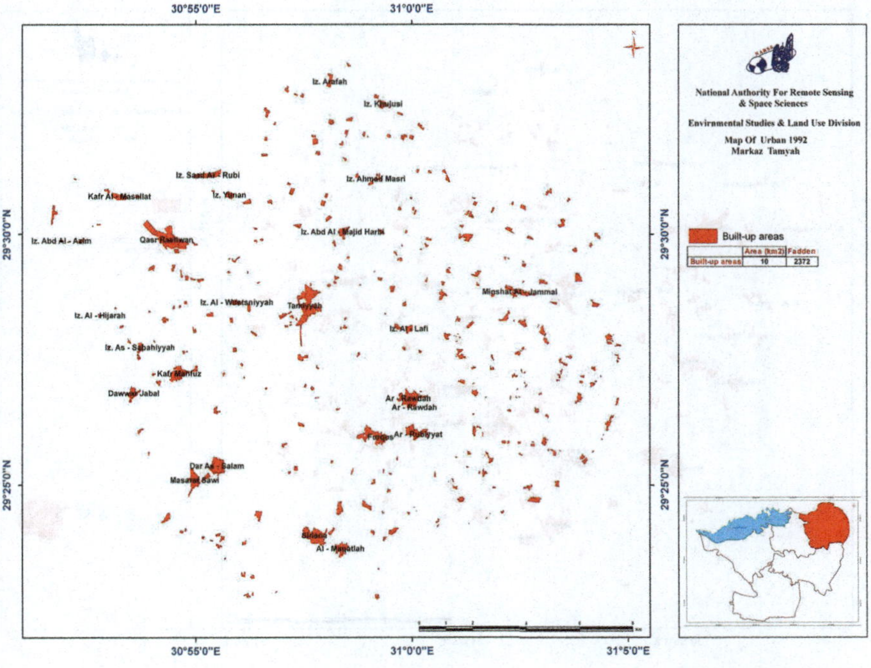

Fig. 4.414 Urban areas of Tamyah district (1992)

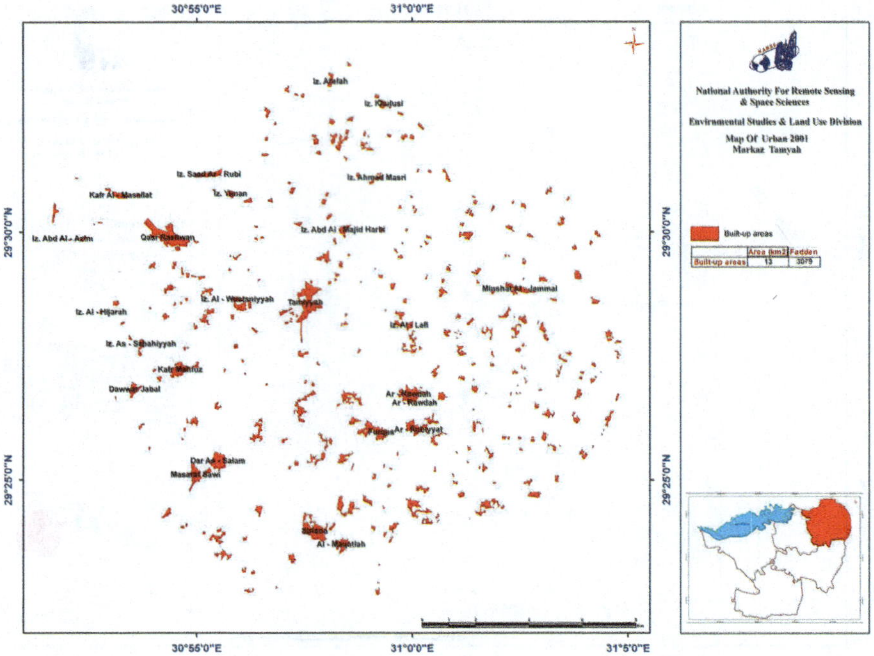

Fig. 4.415 Urban areas of Tamyah district (2001)

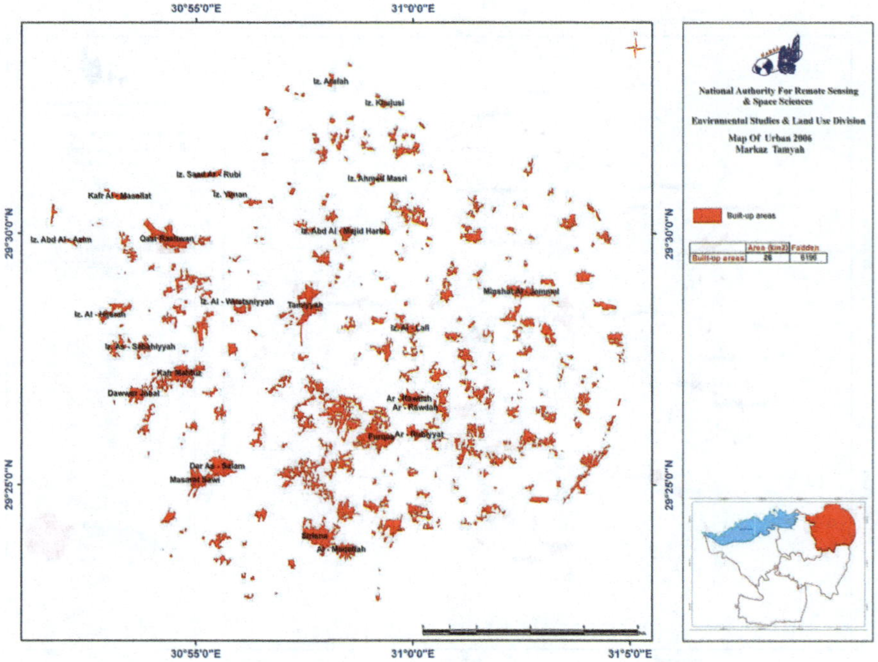

Fig. 4.416 Urban areas of Tamyah district (2006)

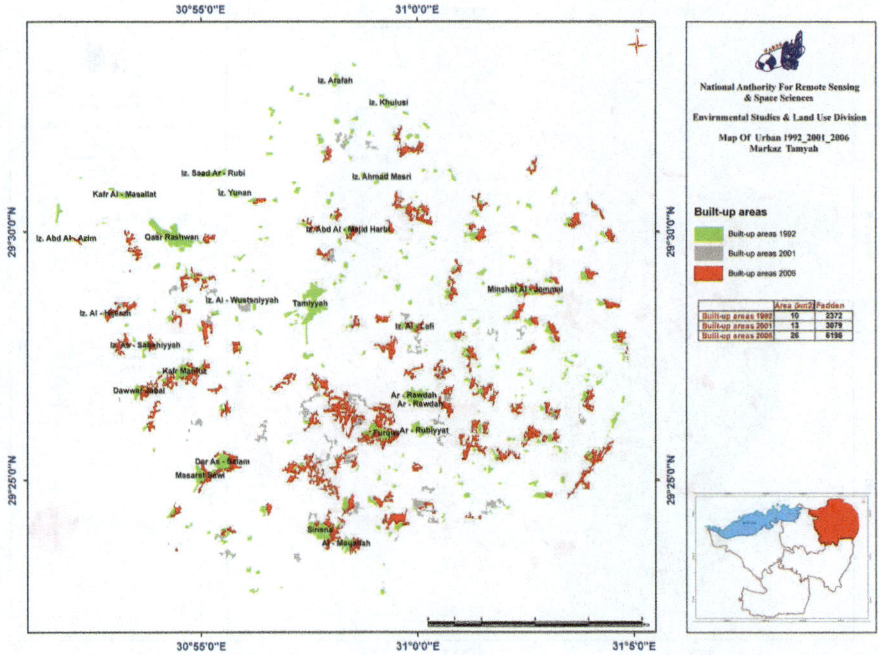

	Area (km2)	Fadden
Built-up areas 1992	10	2372
Built-up areas 2001	13	3079
Built-up areas 2006	26	6196

Fig. 4.417 Urban areas of Tamyah district (1992, 2001, 2006)

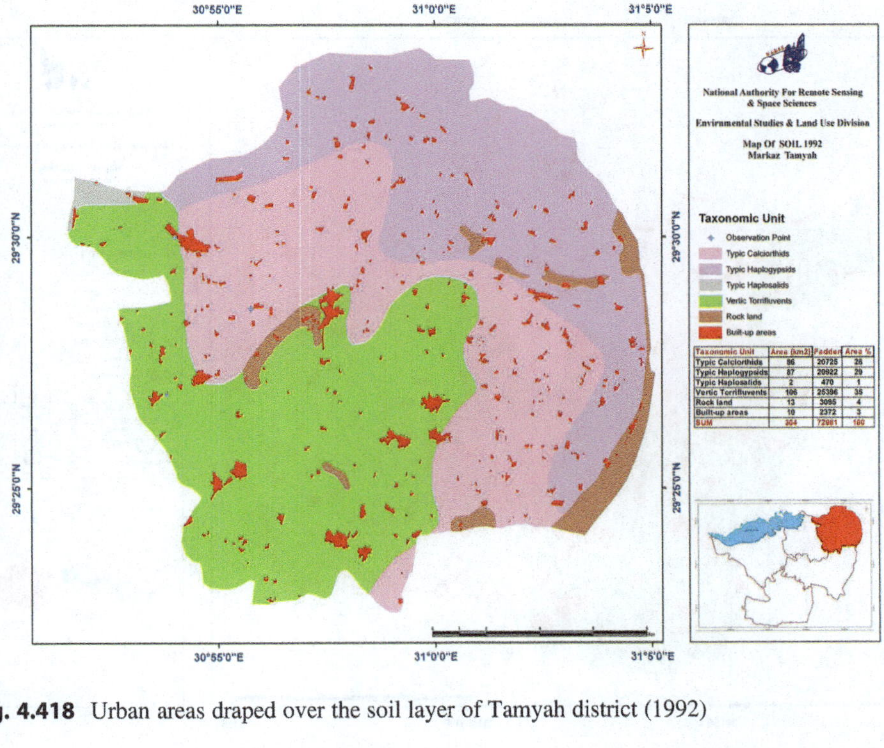

Taxonomic Unit	Area (km2)	Fadden	Area %
Typic Calciorthids	86	20725	28
Typic Haplogypsids	87	20922	29
Typic Haplosalids	2	470	1
Vertic Torrifluvents	106	25396	35
Rock land	13	3095	4
Built-up areas	10	2372	3
SUM	304	72981	100

Fig. 4.418 Urban areas draped over the soil layer of Tamyah district (1992)

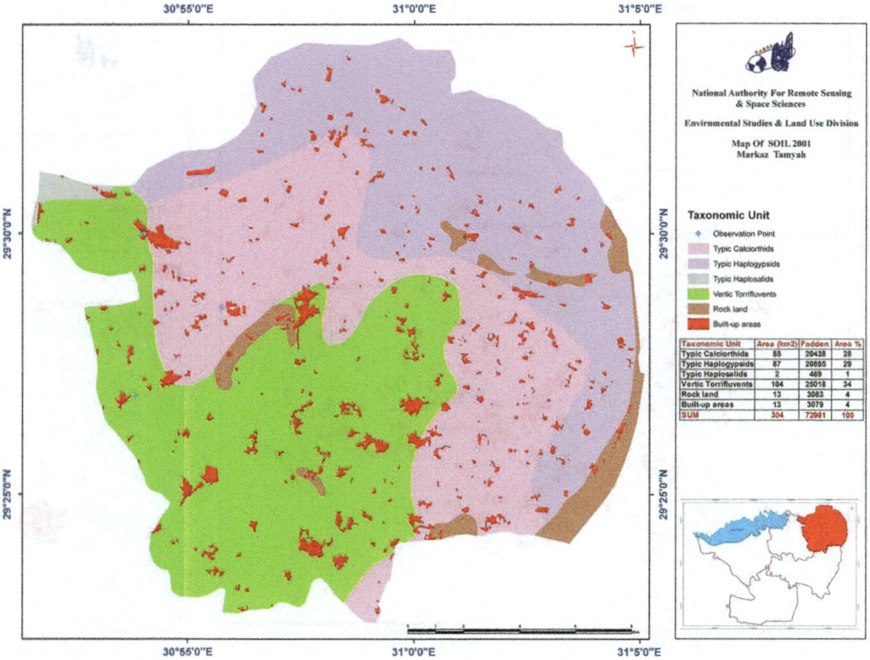

Fig. 4.419 Urban areas draped over the soil layer of Tamyah district (2001)

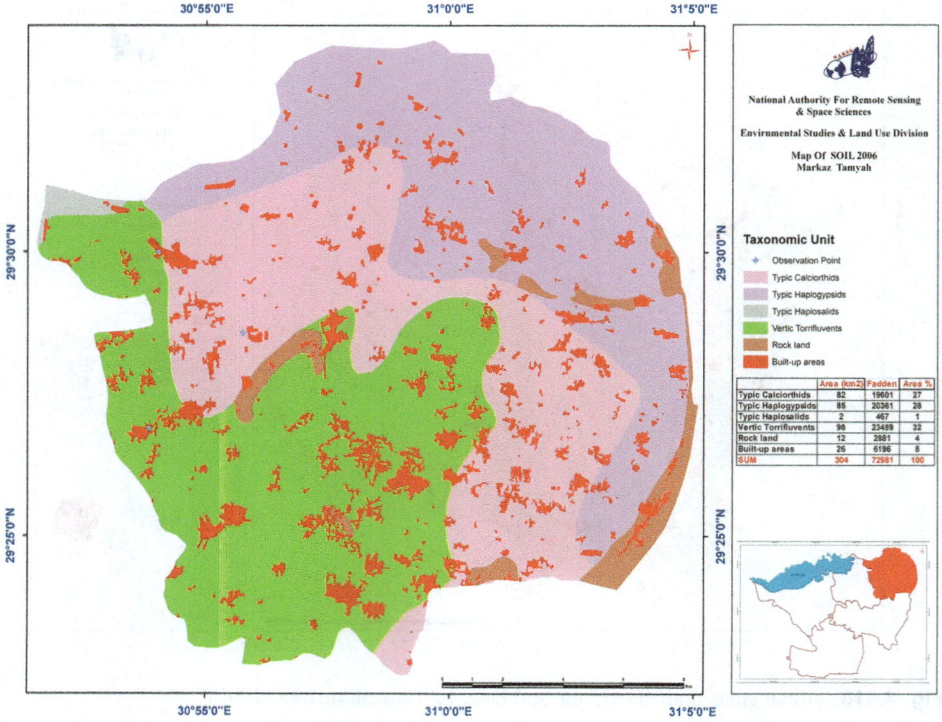

Fig. 4.420 Urban areas draped over the soil layer of Tamyah district (2006)

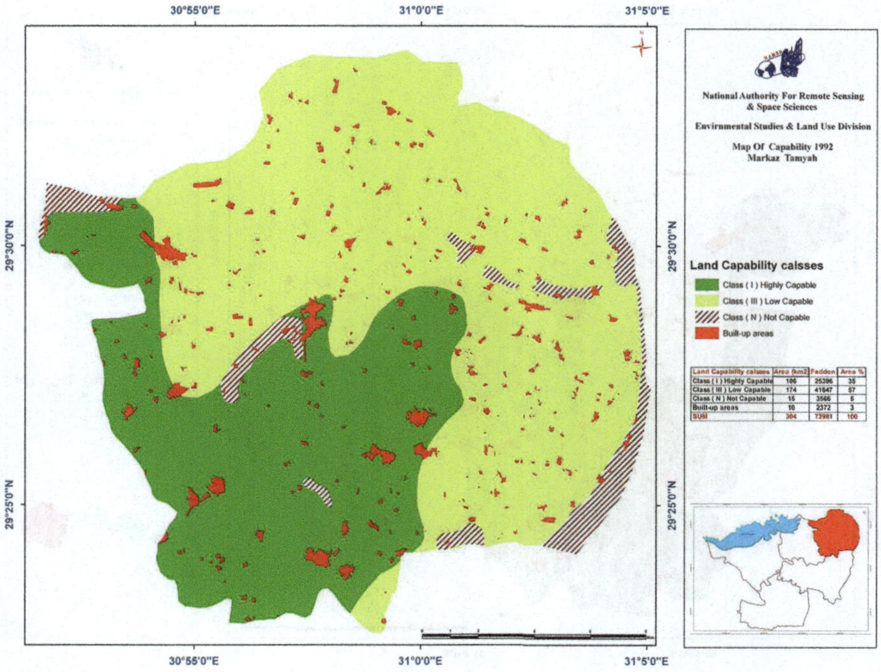

Fig. 4.421 Urban areas draped over the soil capability of Tamyah district (1992)

in year 1992 to 104.0 km^2 in year 2001 and to 98.0 km^2 in the year 2006, while the soils of sub great group *Typic calciorthids* also decreased from 86.0 km^2 in 1992 to 85 km^2 in 2001 and to 82.0 km^2 in the year 2006. The sub great group *Typic Haplogypsids* soils decreased from 87.0 km^2 to 85.0 km^2 during the period 2001–2006, recording an annual rate of 0.1 km^2 per year.

The land capability was established and matched with the multi-temporal urban layers of the district (Figs. 4.421, 4.422 and 4.423), the output layers indicate that the high capable soils lost 8.0 km^2 during the years 1992 to 2006 (i.e. 0.6 km^2/year). However, the low capable soils lost an area of 7.0 km^2 with annual rate of 0.5 km^2 per year, the areas of non capable soils are stable during this period.

4.9.5.5 Senouris District

Senouris is the smallest district in Al-Fayoum Governorate, exhibiting an area of 232.0 km^2 and extending from longitudes 30° 43′ and 30° 58′ N and latitudes 29° 20′ and 29° 35′E. The land resources database of the district includes the Landsat ETM+ and SPOT images taken during the years 2001 and 2006 respectively (Figs. 4.424 and 4.425), which were used for updating the thematic layers. The digital elevation model of the district (Fig. 4.426) indicates that the surface elevation differs widely from −45 meter below sea level, recorded in limited small patches at the northern parts, to 100 meters above sea level.

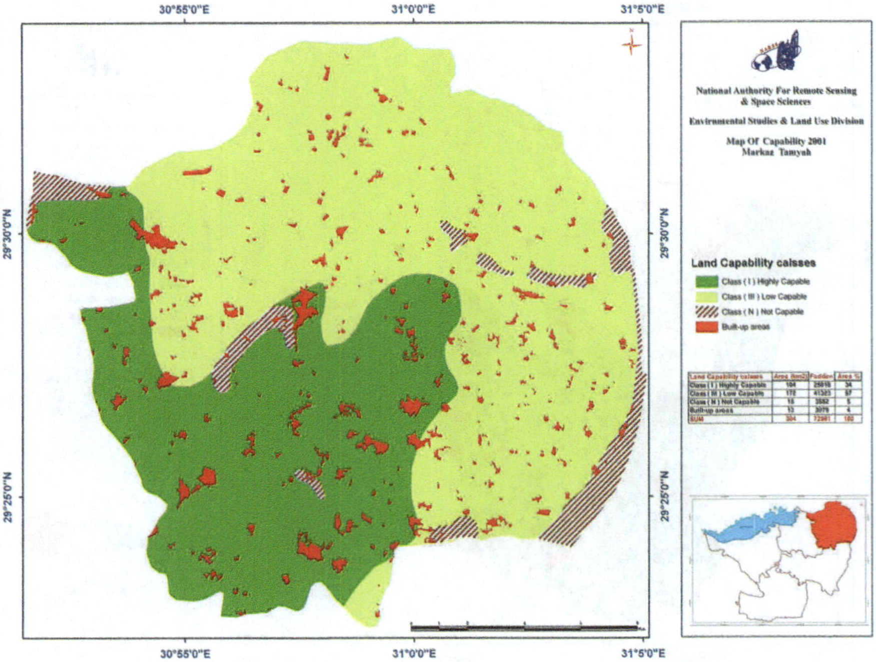

Fig. 4.422 Urban areas draped over the soil capability of Tamyah district (2001)

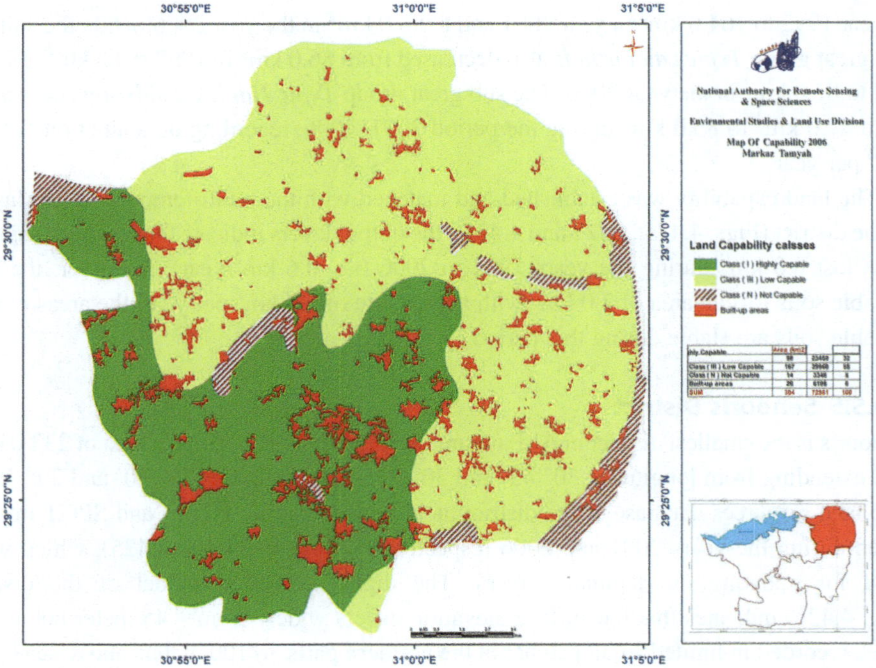

Fig. 4.423 Urban areas draped over the soil capability of Tamyah district (2006)

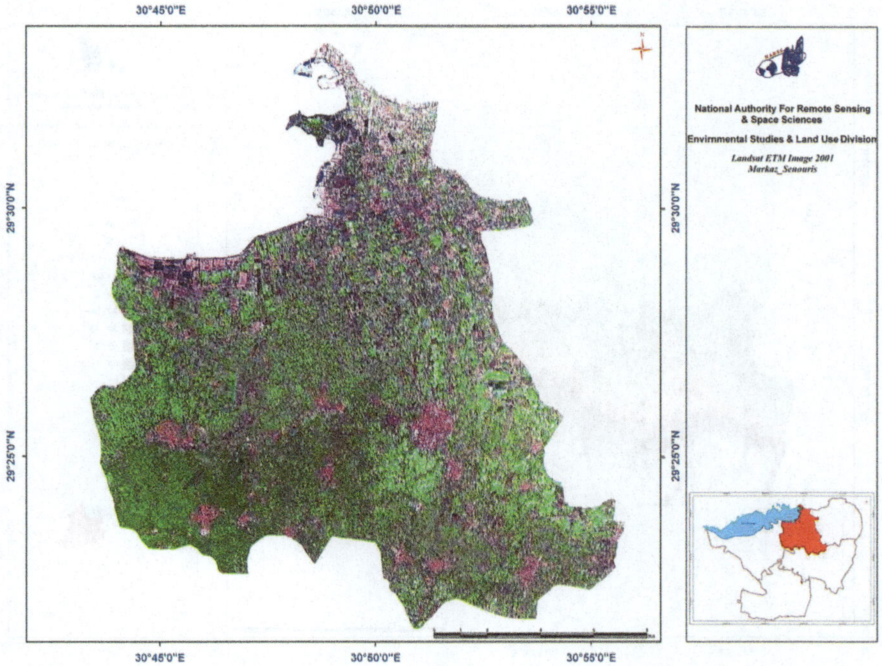

Fig. 4.424 Landsat ETM+ image of Senouris district (2001)

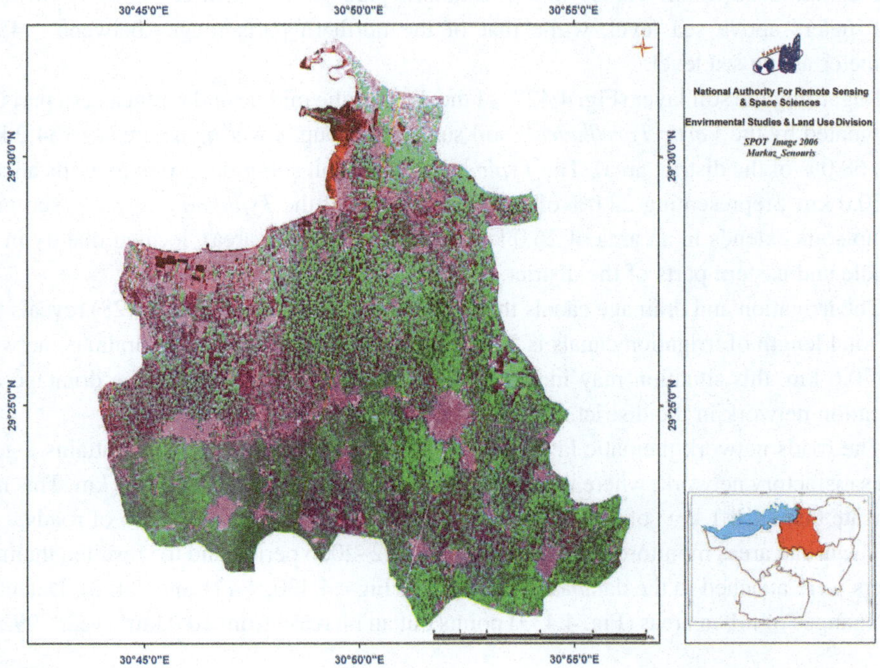

Fig. 4.425 Spot image of Senouris district (2006)

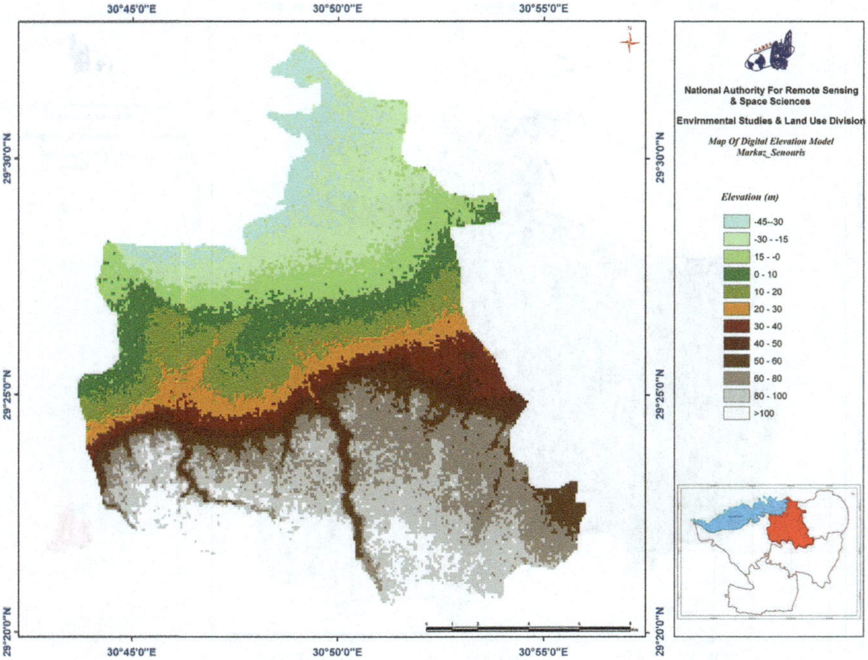

Fig. 4.426 Digital Elevation Model (DEM) of Senouris district

The common elevation values in the southern parts of the district varies from 50 to 100 meters above sea level, while that of the northern parts ranges between −45 to 50 meter above sea level.

Regarding the soil layer (Fig. 4.427) of the district, the middle and northeastern parts are dominated by the *Vertic Torrifluvents* soil sub great group, covering an area of 134.0 km^2 (i.e. 68.0% of the district area). The *Typic torrifluvent* soil sub-great group exhibits an area of 50.0 km^2, representing 22.0% of soil area. Moreover the *Typic haplogypsids* sub great group soils extends in an area of 26.0 km^2 (i.e. 10.0% of soil area), located mainly in the middle and eastern parts of the district.

The irrigation and drainage canals thematic layer of the district (Fig. 4.428) reveals that the total length of irrigation canals is 214.5 km, while the total length of drainage network is 170.6 km. this situation may indicate that there a large gap between the drainage and irrigation network in the district.

The roads network thematic layer (Fig. 4.429) indicates that the district attains a good and satisfactory network, where the total length of the roads network is 431.0 km. This may indicate that each 1 km^2 of the district area is served by a length of 1.4 km of roads.

The urban areas monitored during the years 1992–2006 period and the resulted thematic layers were attached to the database as shown in (Figs. 4.430, 4.431 and 4.432). Detecting the changes of urban areas (Fig. 4.433) points out an increase from 10.0 km^2 (year 1992) to

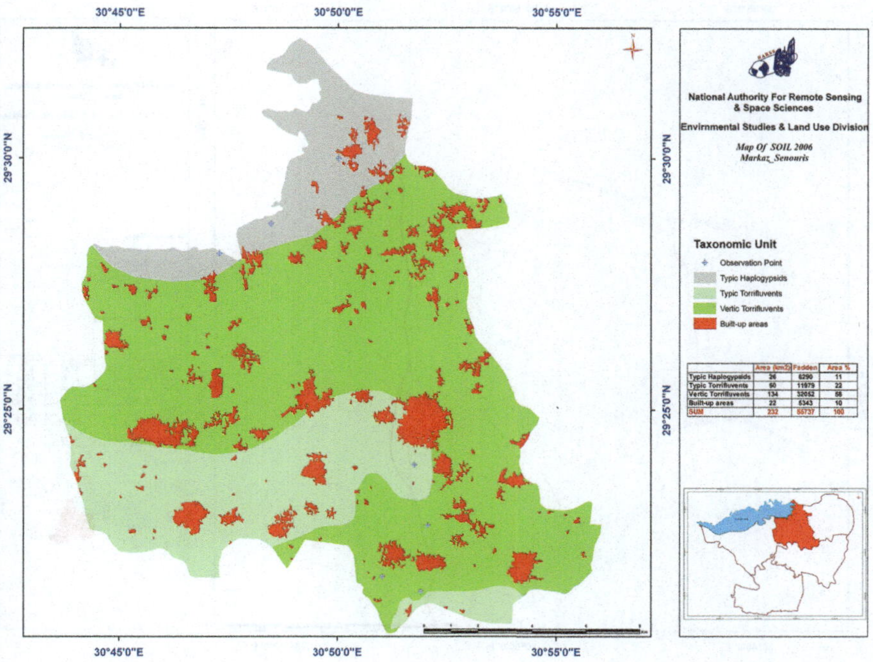

Fig. 4.427 Urban areas draped over the soils of Senouris district (2006)

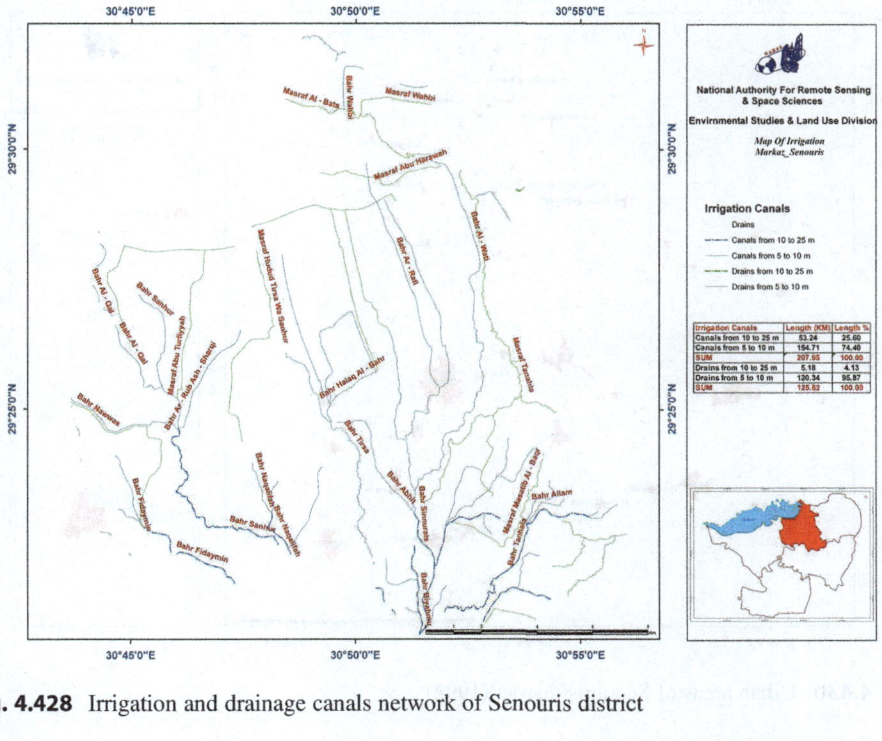

Fig. 4.428 Irrigation and drainage canals network of Senouris district

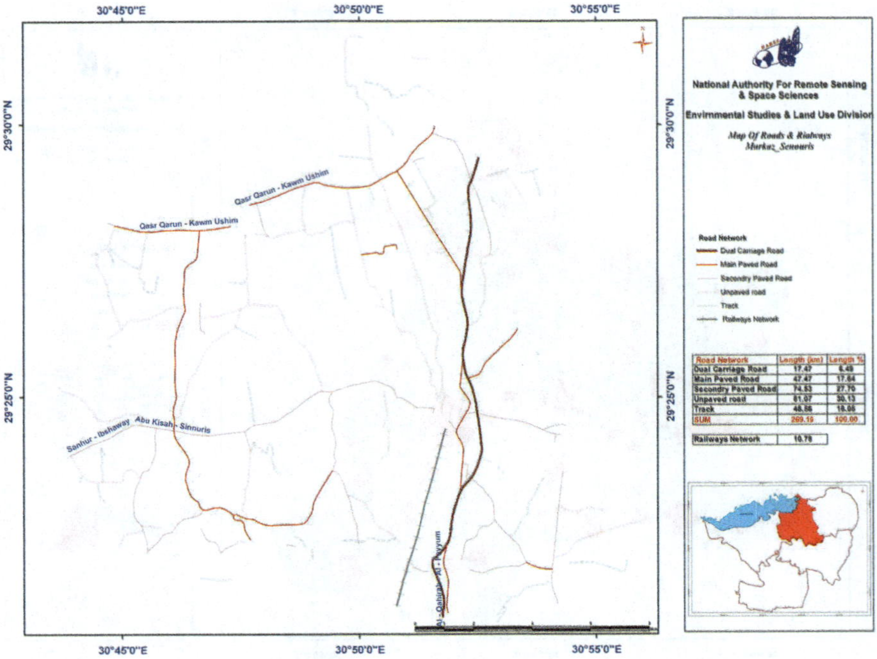

Fig. 4.429 Roads network of Senouris district

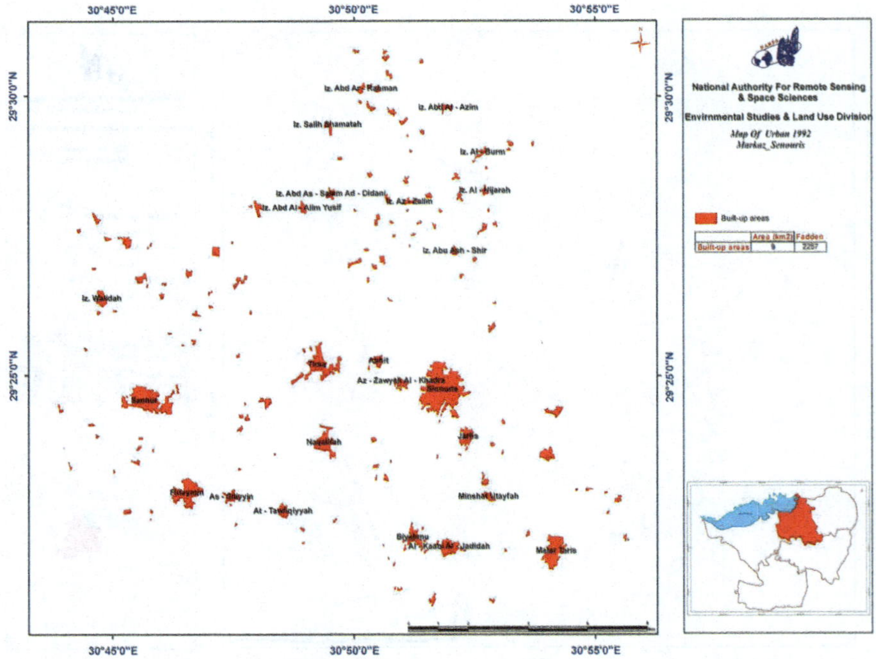

Fig. 4.430 Urban areas of Senouris district (1992)

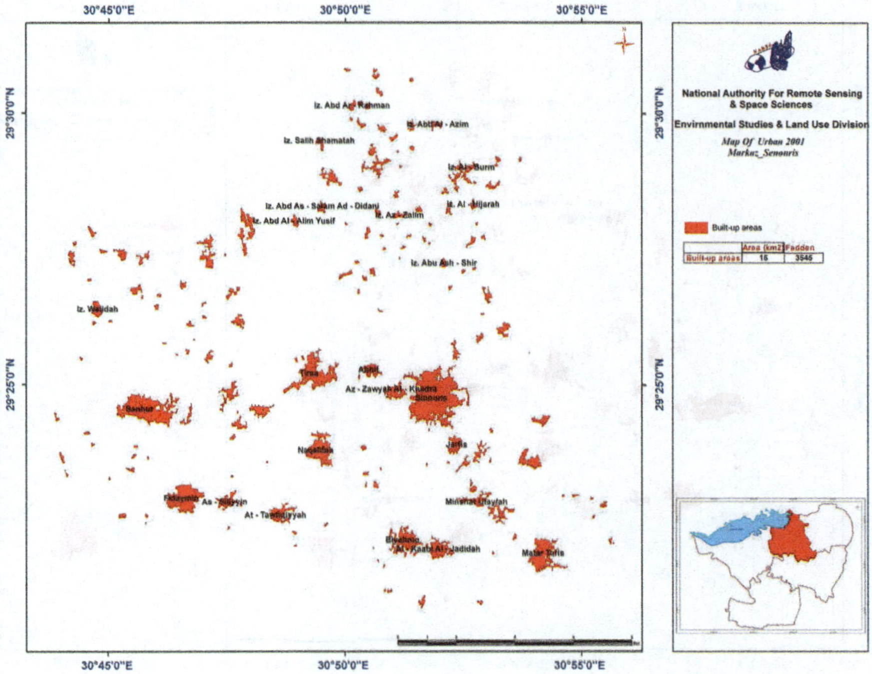

Fig. 4.431 Urban areas of Senouris district (2001)

13.0 km^2 in (year 2001) and to 26.0 km^2 (year 2006). The data refers to an annual urban expansion rate of 1.1 km^2 per year, dominating the whole district.

Change detection technique made it possible to demonstrate the interrelationship between the soil units and urban expansion. Figs. 4.434, 4.435 and 4.436 show that the cultivable *Vertic Torrifluvents* sub great group soils decreased from 143.0 km^2 (year 1992) to 139 km^2 (year 2001) and to 134.0 km^2 (year 2006).

The soil capability was established and matched with the urban layer of the district (Figs. 4.437, 4.438 and 4.439). The output layer indicates that the high capable soils lost 8.0 km^2 during the years 1992–2006 period (i.e. 0.6 km^2/year), while the low capable soils lost an area of 7.0 km^2 with annual rate of 0.5 km^2 per year. The areas of non capable soils show a stability during the same period.

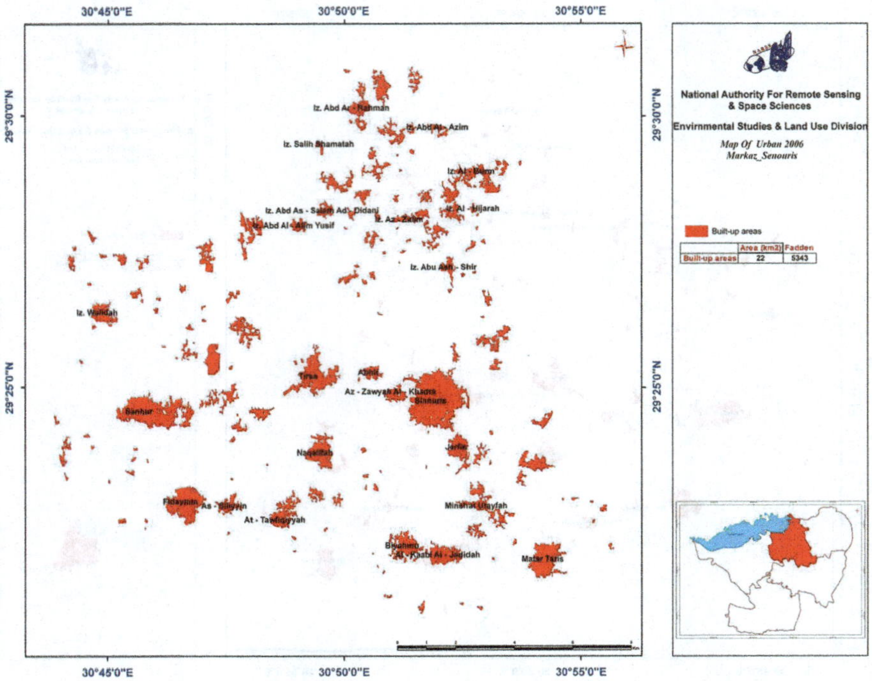

Fig. 4.432 Urban areas of Senouris district (2006)

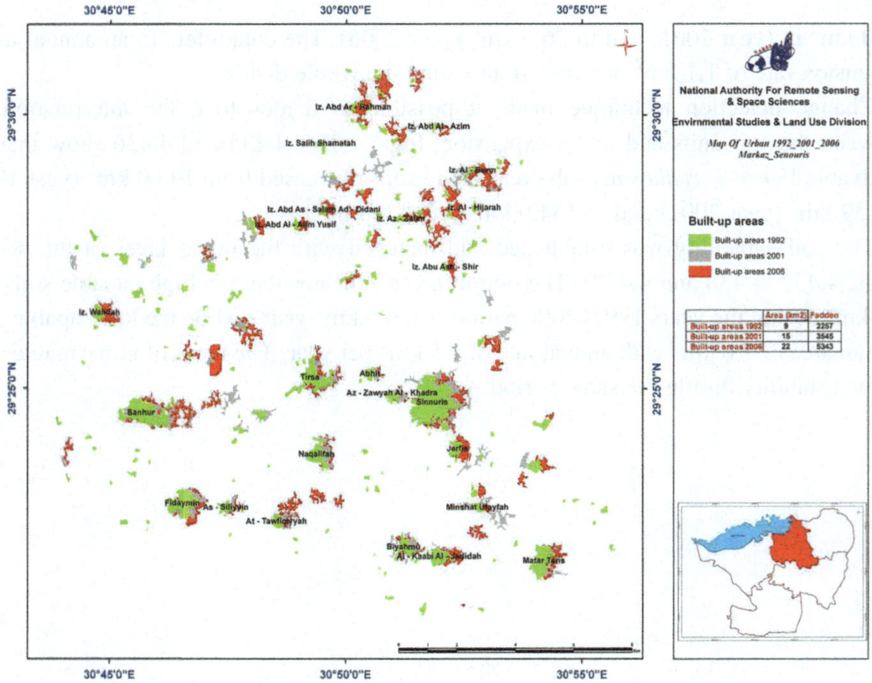

Fig. 4.433 Urban areas of Senouris district (1992, 2001, 2006)

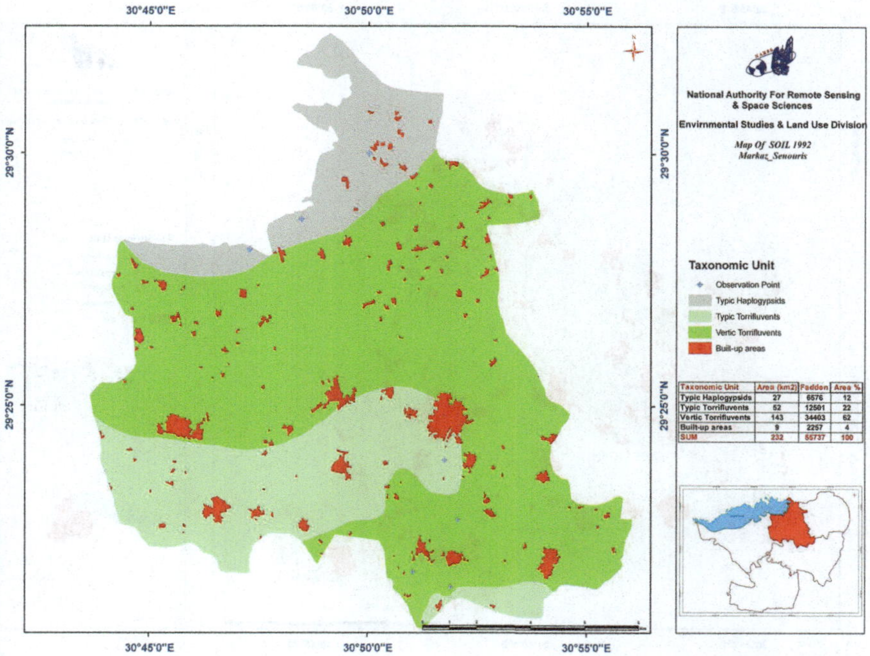

Fig. 4.434 Urban areas draped over the soils of Senouris district (1992)

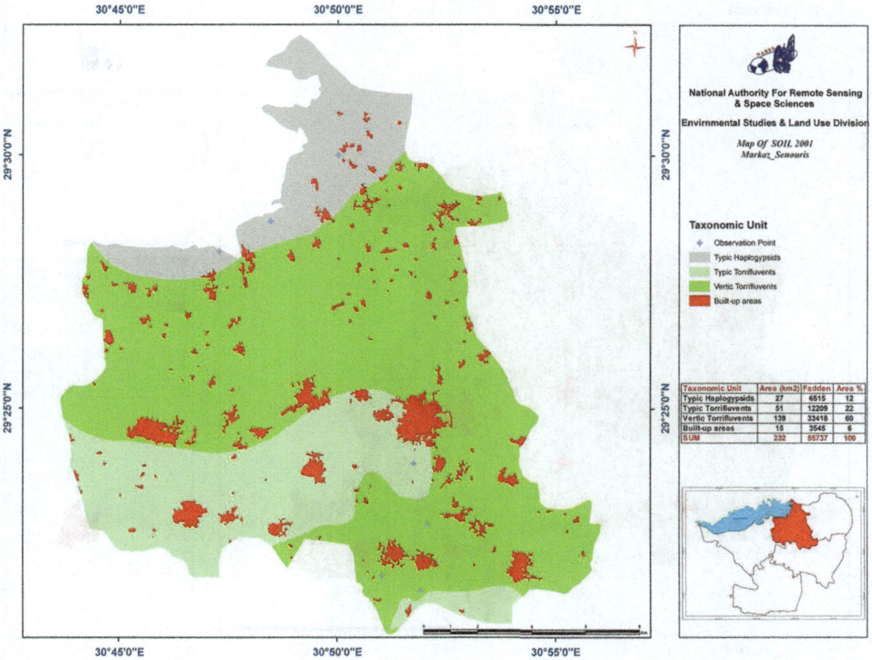

Fig. 4.435 Urban areas draped over the soils of Senouris district (2001)

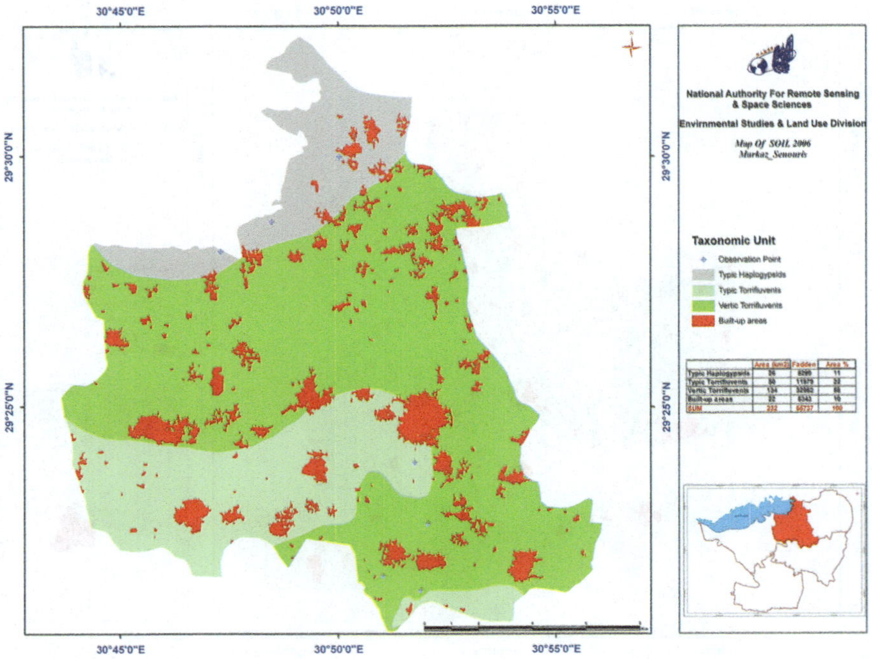

Fig. 4.436 Urban areas draped over the soils of Senouris district (2006)

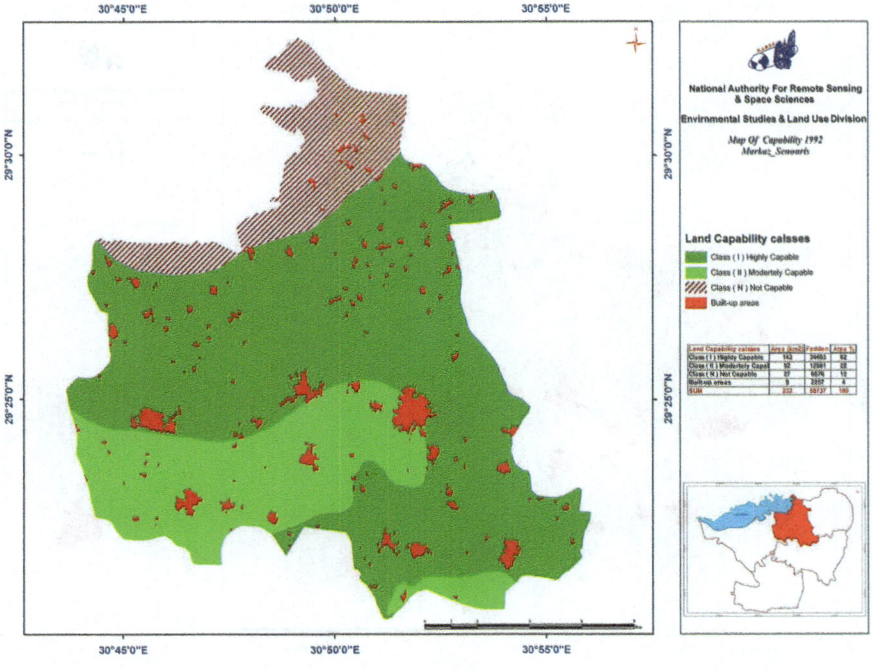

Fig. 4.437 Urban areas draped over the soil capability of Senouris district (1992)

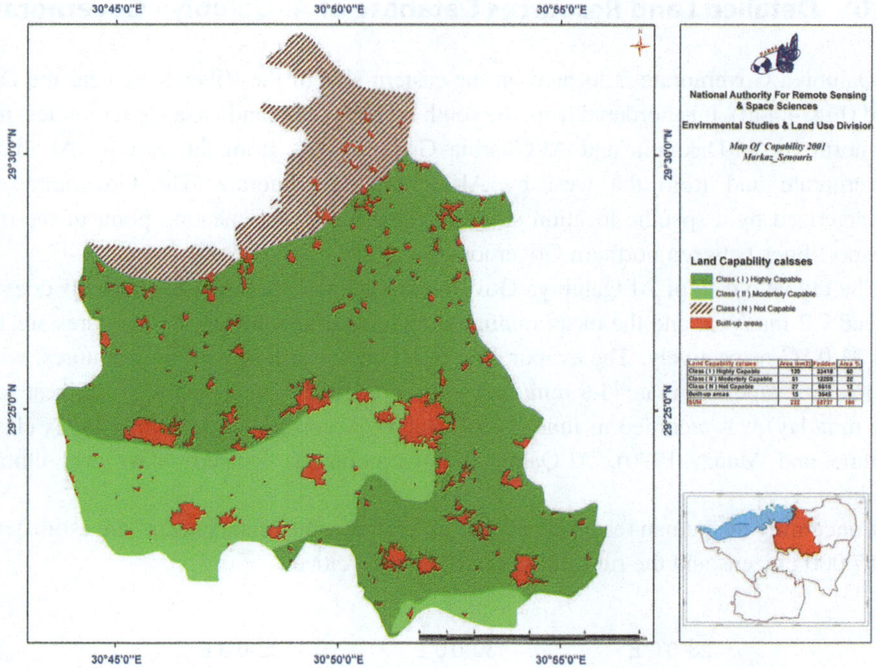

Fig. 4.438 Urban areas draped over the soil capability of Senouris district (2001)

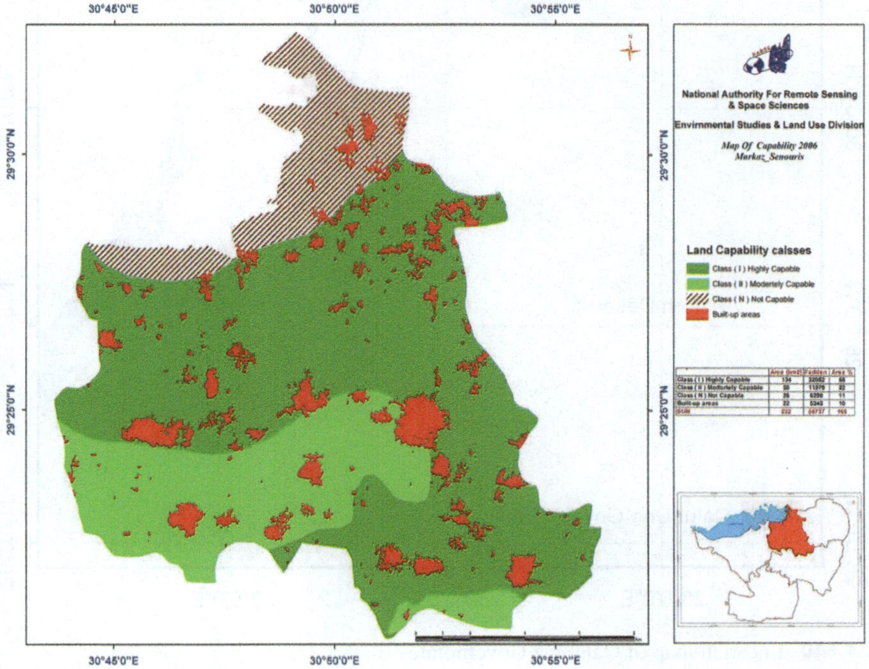

Fig. 4.439 Urban areas draped over the soil capability of Senouris district (2006)

4.10 Detailed Land Resources Database of Al-Qalubiya Governorates

Al-Qalubiya Governorate is located on the eastern side of the River Nile, near the Delta head (Fig. 4.440). It is bordered from the south by both Cairo and Giza Governorates, from the north by Al-Dakahlia and Al-Gharbia Governorates, from the east by Al-Sharkia Governorate and from the west by Al-Menofia Governorate. The Governoratse is characterized by a specific location since it is considered the meeting point of the main transport lines between northern Governorates.

The climatic data of Al-Qalubiya Governorate indicate that the total rainfalls does not exceed 7.2 mm/year and the mean minimum and maximum annual temperatures are 16.5 and 31.0 °C respectively. The evaporation rates are coinciding with temperatures, where the lowest evaporation rate (1.9 mm/day) was recorded in January while the highest value (7.6 mm/day) was recorded in June (CLAC 2004). According to the aridity index classes (Hulme and March 1990), Al-Qalubiya Governorate is located under arid climatic conditions.

Concerning the human resources of the Governorate, the urban citizens are estimated as 1,457,000 citizens and the rural citizen as 2,130,000 citizens.

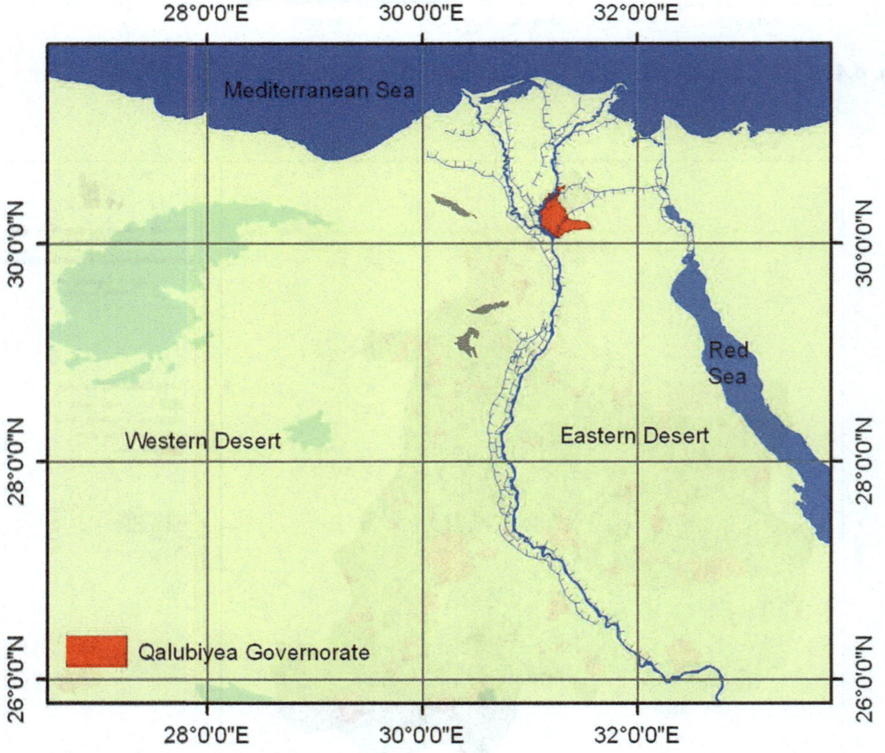

Fig. 4.440 Location map of Qalubiya Governorate

Al-Qalubiya Governorate consists of 8 districts, locally named Markaz (i.e. Al-Khankah, Al-Qanatir Al-Khayreyah, Awal Shobra Al-Khaymah, Benha, Kafr Shokr, Qelyoub, Shibin Al-Qanatir and Tokh. The total area of the Governorate is estimated to be 1140.7 km^2. The cultivated area is 810.0 km^2 (71% of total area).

The Governorate celebrates its national day on the thirtieth of August which was the same date of opening its famous barrages named "Al-Kanater Al-Khairia" in 1868. Al-Kanater Al-Khairia is characterized by its beautiful natural scenery, parks and orchards, extending mostly for 500 feddans, playing an important role for the governmental income.

The Governorate is famous for its agriculture production of crops, fruits and vegetables such as maize, cotton, wheat, citrus, banana, apricot etc. Moreover, the Governorate attains some major industrial regions such as Shoubra AL-Kheima, where activities of spinning, Electrical cables, plastic and oil refinery factories, in addition to food industries, packing, metallurgical industries exist. Abo Zaabel area is famous by fertilizers and chemical industries. Recently, the Governorate constructed an integral industrial region in Al-Khanka city.

4.10.1 Physiography of Al-Qalubiya Governorate

Field survey data, Landsat ETM+ (Fig. 4.441) and SPOT image (Fig. 4.442) and digital elevation model (DEM—Fig. 4.443) were used to define the physiographic units of Al-Qalubiya Governorate. The obtained data (Fig. 4.444) presents the physiographic features of the Governorate; the studied area includes the following physiographic units:

(a) River terraces:

This landscape dominates the middle parts as well as the borders of the Governorate, it occupies an area of 169.6 km^2 (i.e. 14.9% of whole territory), consisted of high, moderately high and low terraces. The digital elevation model shows that the elevation difference between the highest and lowest terraces ranges between 1 and 3 meter.

(b) Overflow basins:

This landscape dominates the middle parts of the Governorate; it represents 29.5% of its total territory (i.e.336.9 $km^{2)}$. It varies in elevation as the difference between the highest and lowest parts ranges between 14 and 20 meter above sea level.

(c) Decantation basins:

This landscape cover an area of 364.0 km^2 (i.e. 31.9% of the total area) exhibiting the lowest parts of the studied areas. The common elevation of these physiographic units is less than 12 metre above sea level.

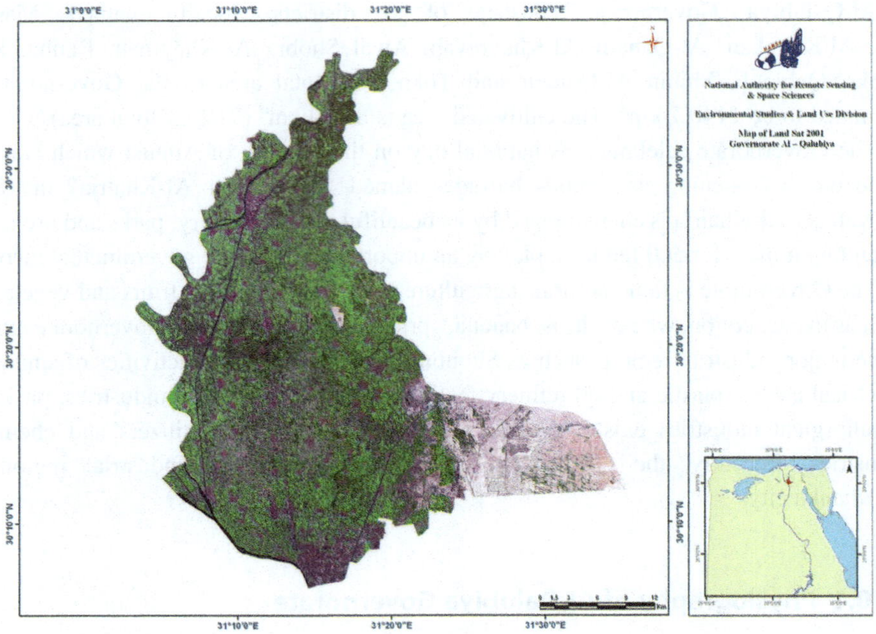

Fig. 4.441 Landsat ETM+ image of Qalyobiya Governorate (2001)

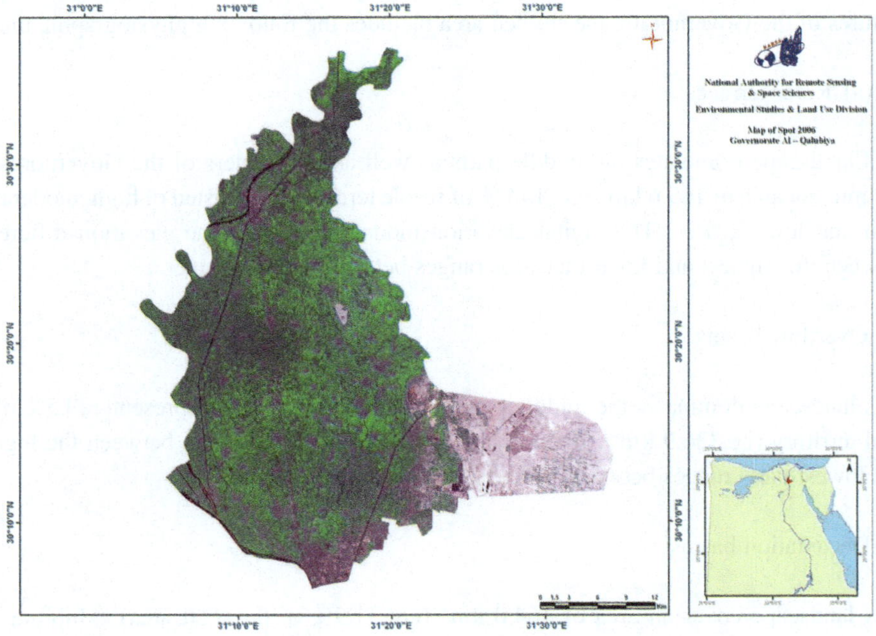

Fig. 4.442 SPOT image of Qalubiya Governorate (2006)

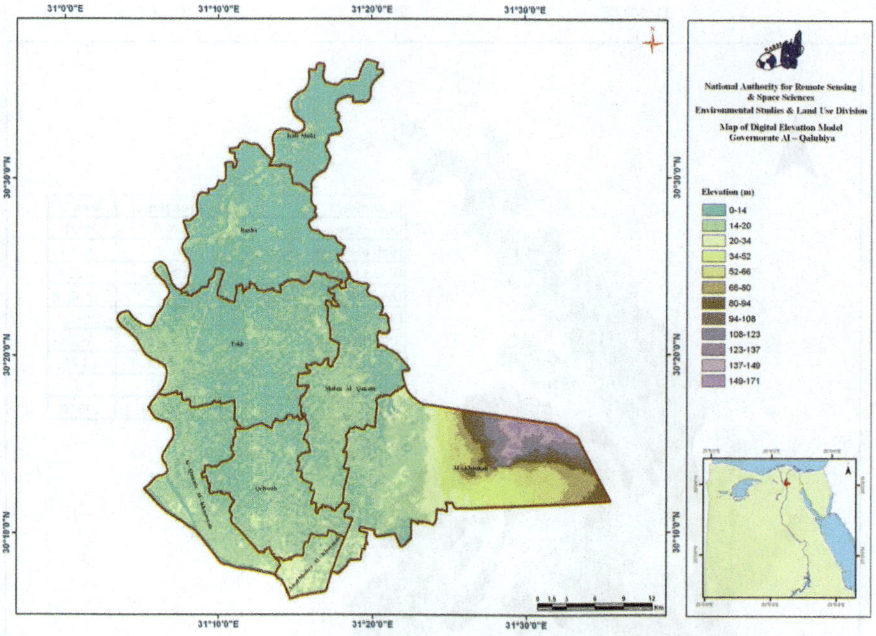

Fig. 4.443 Digital Elevation Model (DEM) of Qalubiya Governorate

(d) River levees:

This landscape is found mainly close to the river course, having a gentle slope and moderately higher elevation than the adjacent features, it ranges between 14 to 22 meters above sea level. The river levees of Al-Qalubiya governorate represent 5.7% of the total area (ie.64.6 km^2).

(e) High and low elevated sand sheets:

These units dominate great parts of Al-Khanka district; representing 13.6% of the total area. The surface elevation differs widely from 34 meters in the low level areas to 123 meters above sea level in the highly elevated areas.

(f) Sand ripples:

This landscape represents 3.6% of the total area, existing mainly in the middle parts of Al-Khanka district. The surface elevation of these units varies widely between 54 to 150 meters above sea level.

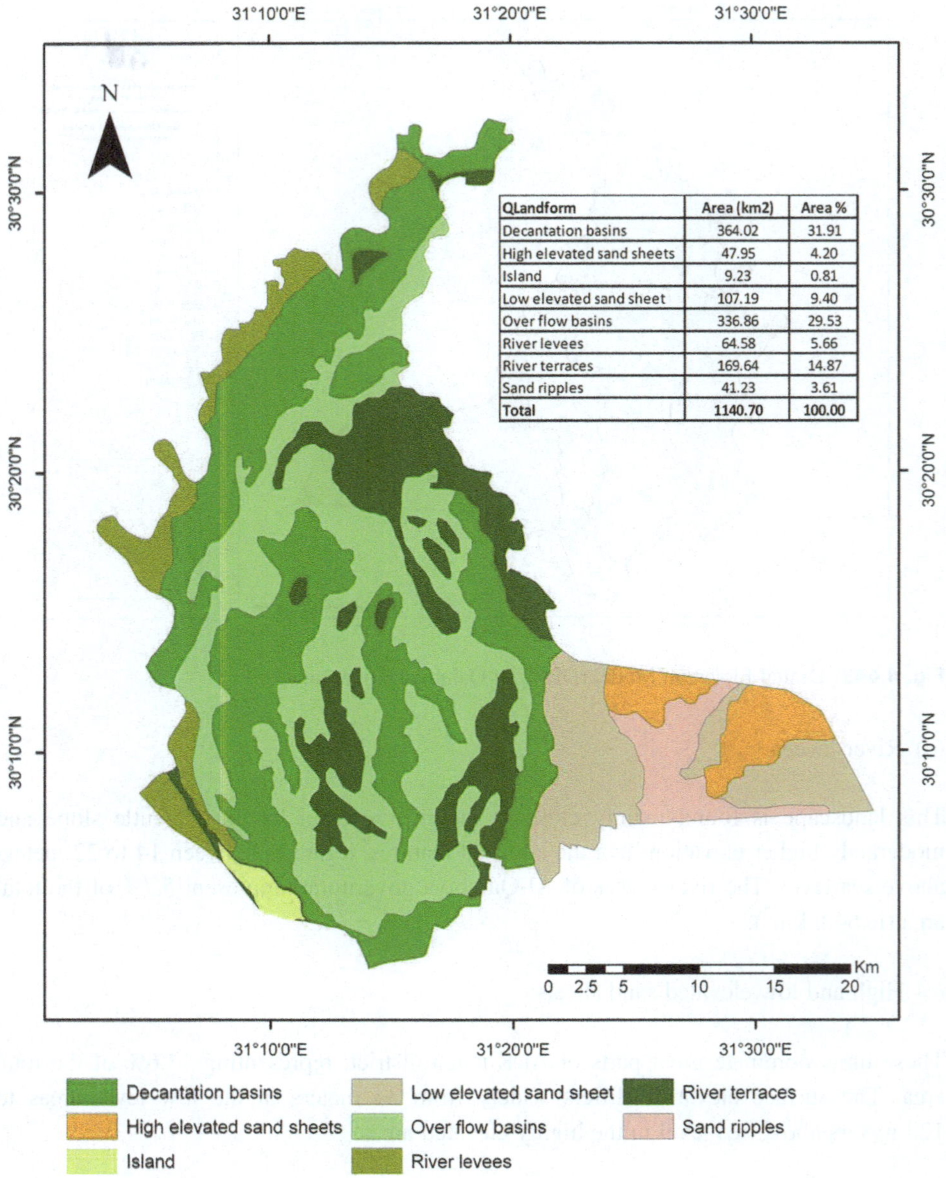

QLandform	Area (km2)	Area %
Decantation basins	364.02	31.91
High elevated sand sheets	47.95	4.20
Island	9.23	0.81
Low elevated sand sheet	107.19	9.40
Over flow basins	336.86	29.53
River levees	64.58	5.66
River terraces	169.64	14.87
Sand ripples	41.23	3.61
Total	**1140.70**	**100.00**

Fig. 4.444 Physiographic units of Qalubiya Governorate

4.10.2 Detailed Database of Al-Qalubiya Governorate

Al-Qalubiya Governorate is one of the largest governorate in Egypt, located in the southern Nile Delta region, north of Cairo. It is administratively sub-divided into 8 administrative districts or Markaz (Table 4.173 and Fig. 4.445).

Table 4.173 Areas and frequencies of different administrative districts within Al-Qalubiya Governorate

Name	Area (km^2)	%
Khankah	278.2	24.0
Al-Qanatir Al-Khayreyah	112.8	10.0
Awal Shobra Al-Khaymah	28.8	3.0
Benha	170.4	15.0
Kafr Shokr	60.6	5.0
Qelyoub	122.0	11.0
Shibin Al-Qanatir	138.4	12.0
Tokh	229.7	20.0
Total	1140.7	100.0

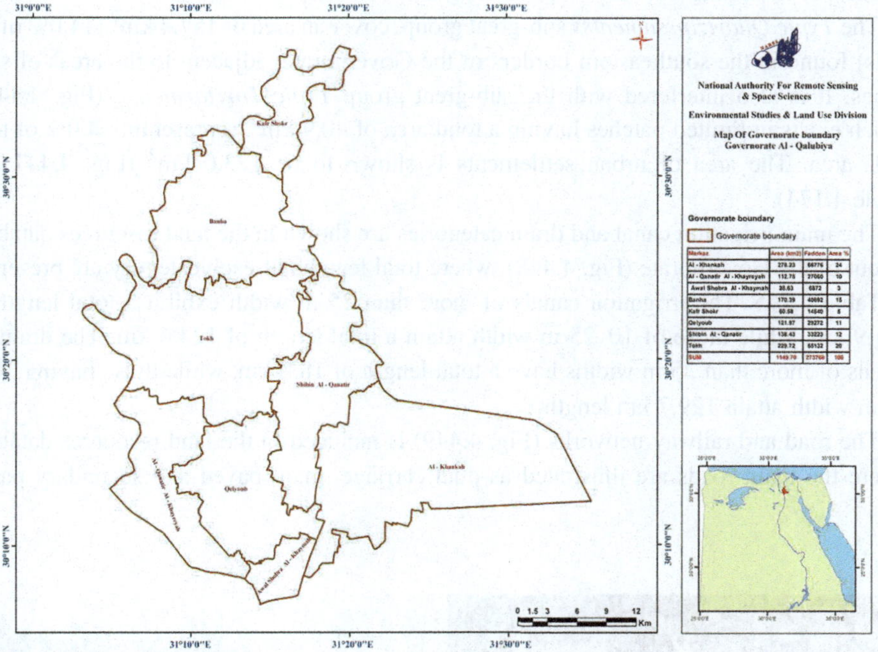

Fig. 4.445 The administrative boundary of Qalubiya Governorate

Tokh is the largest district in the governorate, covering an area of 229.7 km^2, representing 20.0% of the whole Governorate. Benha district is the second in its area coverage (170.4 km^2), representing 15.0% of total Governorate area. It includes Benha city, the Governorate's capital town where most of administrations are located. Shibin Al-Qanatir and Al-Al-Khankah are important districts located in the middle part of the Governorate where transportation is available to most of other towns. They cover areas of 138.4 and 278.2 km^2, representing 12.0% and 24.0% of the Governorate. Also, having the famous barrages and being a vegetated resort place for the residence of Cairo, Al-Qanatir

Al-Khayreyah district is Well known. It covers an area of 112.8 km^2 representing 10.0% of total area. Kafr Shokr and Awal Shobra Al-Khaymah are small districts at the borders of adjacent other Governorates, covering areas of 60.6 and 28.8 km^2 representing 5.0% and 3.0% respectively.

An updated digital soil map of Al-Qalubiya Governorate was elaborated using the SPOT image of 2006 and GIS techniques, as well as ground survey. A number of 30 soil profiles were selected covering the governorate. The American Soil Taxonomy (USDA 2006) were followed for classifying the soil units. The soil layer demonstrate that only four sub-great groups exhibit the whole governorate, where *Vertic Torrifluvents* (Fig. 4.446a) dominates the soil types, covering an area of 618.5 km^2, representing 54.0% of the soils. The *Typic Torrifluvents* sub-great group (Fig. 4.446b) covers limited patches within the alluvial landscape, exhibiting 93.0 km^2 representing 8.0% of the soils.

The *Typic Quartzipssamentss* sub-great group, cover an area of 157.4 km^2 (14.0% of the soils) found in the southeastern borders of the Governorate, adjacent to the areas of sand dunes. It is also interfered with the sub-great group *Typic Torriorthents,* (Fig. 4.446c) which exists in limited patches having a total area of 40.9 km^2, representing 4.0% of total soils area. The area of urban settlements is shown to be 173.0 km^2 (Fig. 4.447 and Table 4.174).

The main irrigation canal and drain categories are shown in the land resources database layout of the Governorate (Fig. 4.448), where total lengths of each category are presented in Table 4.175. The irrigation canals of more than 25 m width exhibit a total length of 111.9 km, while those of 10–25 m width attain a total length of 124.6 km. The drainage canals of more than 25 m widths have a total length of 18.0 km, while those having 10 to 25 m width attain 129.7 km length.

The road and railway networks (Fig. 4.449) is included in the land resources database where the main roads are illustrated as dual carriage, main paved and secondary paved

Fig. 4.446 (**a**) Vertic Torrifluvents soil sub-great group. (**b**) Typic Torriorthents soil sub-great group. (**c**) Typic Torriorthents soil sub-great group

Fig. 4.446 (continued)

Fig. 4.446 (continued)

(Table 4.176). They attain lengths of 82.9, 76.9 and 745.9 km respectively, the railway network has a length of 204.0 km.

The land resources database includes also the layer of utilities (Fig. 4.450) where the total length of the utilities is 260.3 km.

Change detection technique resulted in following the evolution of urban areas and its implication with soil and land capability units. The built-up areas in the Governorate (Figs. 4.451, 4.452, 4.453) increased from 90.0 km^2 (i.e. 8.0%) in 1992 to 133.5 km^2 (i.e. 12.0%) in 2001 and to 173.0 km^2 (i.e. 15.0%) in the year 2006. Figure 4.454 shows the outcome of performing the change detection technique; it shows that urbanization spread in the whole Governorate, and especially practiced near Cairo (i.e. Shoubra Al-Khema and Al-Khanka areas).

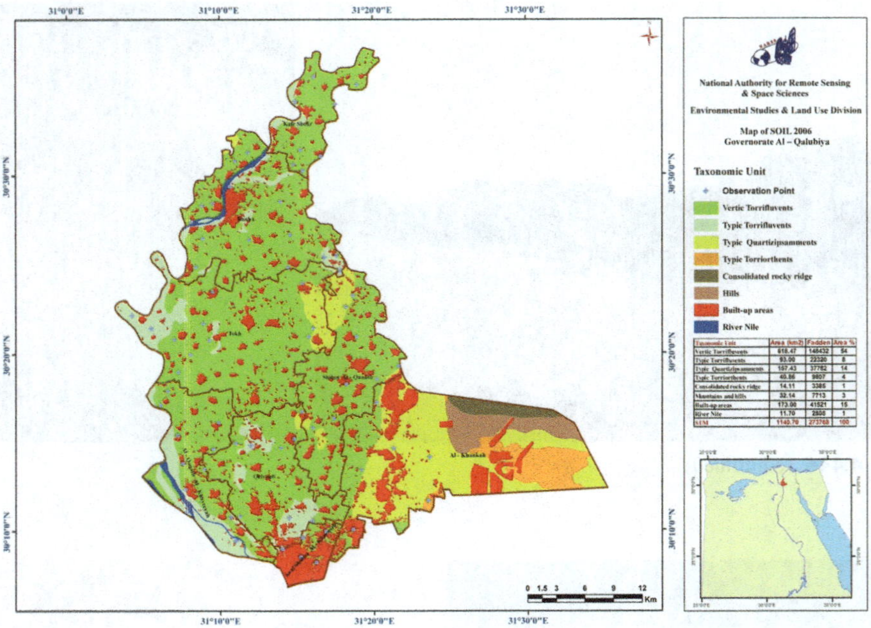

Fig. 4.447 Soil layer of Qalyopiya Governorate (2006)

Table 4.174 Areas and frequencies of soil sub-great groups in Al-Qaliyubia Governorate

Type (Sub great group)	Area (km^2)	%
Vertic Torrifleuvents	618.47	54
Typic Quartizipsamments	157.43	14
Typic Torrifluvents	93.00	8
Typic Torriorthents	40.86	4
Others	230.95	20
Total	1140.7	100

Studying the urban expansion from 1992 to 2006 (Figs. 4.455, 4.456, 4.457) revealed that urbanization mainly occurs on the account of most fertile soils of Al-Qalubiya Governorate. The remained cultivable land of the soil sub-great group "*Vertic Torrifluvents*", decreased from 683.2 km^2 in 1992 to 647.0 km^2 in 2001 and to 618.7 km^2 in 2006. The sub-great group *Typic Torrifluvents* areas decreased from 100.5 to 96.5 and to 93.0 km^2 in the years 1992, 2001 and 2006 respectively. During the same period, the areas of cultivable *Typic Quartzipssamentss* decreaesed from 168.0 to 163.8 and to 153.4 km^2. Less changes, due to urbanization was noteced for the soil sub-great group *Typic Torriorthents*.

In terms of land capability classification (Figs. 4.458, 4.459, 4.460), the high capable soils (Class I) decreased from 683.2 km^2 in 1992 to 647.6 km^2 in 2001 and to 618.5 in

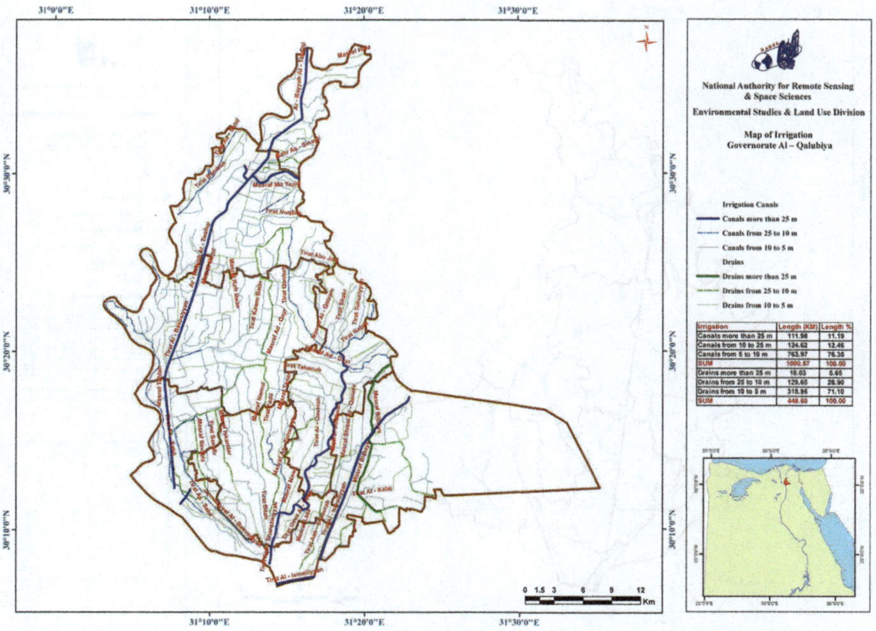

Fig. 4.448 Irrigation and drainage canals network of Qalubiya Governorate

Table 4.175 Total length of the main irrigation canals and drains in El-Qaliyubia Governorate

Type	Length (km)
Canals more than 25 m	111.9
Canals from 10 to 25 m	124.6
Drains more than 25 m	18.0
Drains from 10 to 25 m	129.7

2006. The moderately capable soils decreased from 100.5 to 93.8 km^2 from 1992 to 2006, while the marginally capable soils decreased from 209.1 km^2 to 198.3 km^2 during the same period. It is noticed that urban encroachment over the non capable soils are very limited, as their coverage was found stable during the period 1992–2006.

4.10.2.1 Benha District

Benha district includes Benha city, the capital of Al-Qalubiya Governorate. It incorporates an area of 170.4 km^2 extending from longitudes 31° 18′ and 31° 07′ E and latitudes 30° 32′ and 30° 23′ N. The database of the district includes the Landsat ETM+ and SPOT images taken during the years 2001 and 2006 respectively (Figs. 4.461 and 4.462), these images were used to update the thematic layers. The digital elevation model of the district indicates that the surface elevation ranges from 0 sea level to 34 meters above sea level.

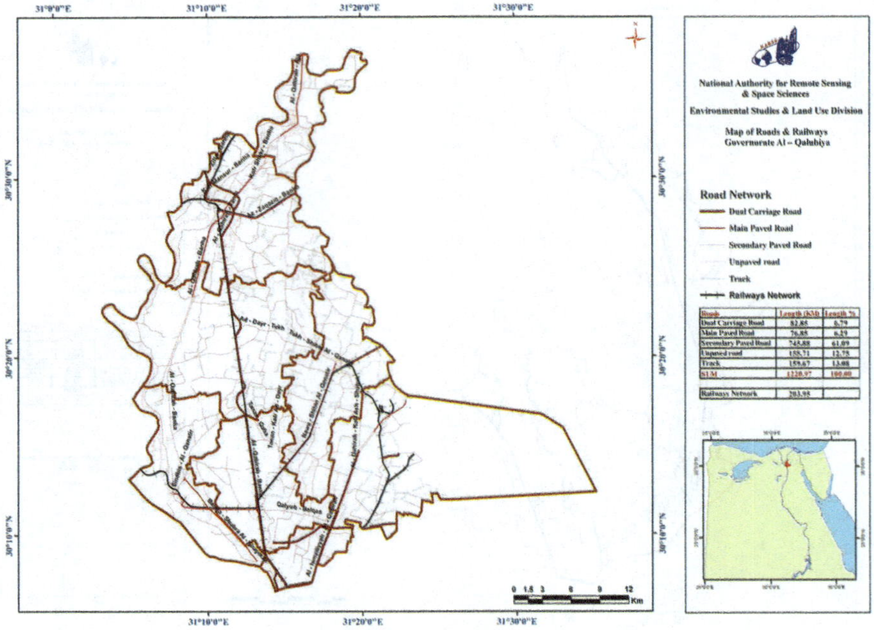

Fig. 4.449 Roads network of Qalubiya Governorate

Table 4.176 Total lengths of main road railway networks of Qaliyubia Governorate

Type	Length (km)
Dual Carriage Road	82.9
Main Paved Road	76.9
Secondary Paved Road	745.9
Railway	204.0

Regarding the soil layer (Fig. 4.463) of the district, it is found the *Vertic Torrifluvents* soil sub great group dominates the soil units, covering an area of 122.4 km^2, representing 72.0% of the district soil area. The *Typic Torrifluvents* covers 13.4 km^2, representing 8% of the district total area. The soil sub-great group Typic *Quartzipssamentss* exhibits an area of 1.5 km^2, representing only 1% of the total area.

The irrigation and drainage canals thematic layer of the district (Fig. 4.464) reveals that the total length of irrigation canals is 190.3 km, while that one for drainage network is 67.1 km. This fact may indicate the unbalance between the lengths of drainage and irrigation networks in the Benha district.

The roads network thematic layer (Fig. 4.465) indicates that the district attains a good and satisfactory network, where the total length of roads is 265.8 km. The roads network is characterized by a road density of 1.6 km roads for each 1 km^2 area.

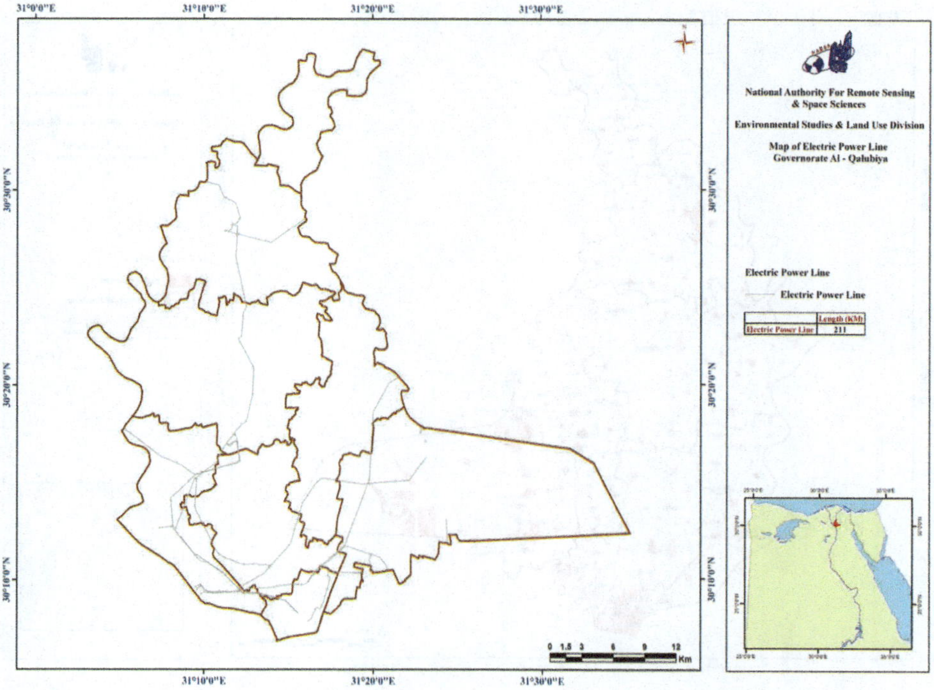

Fig. 4.450 Utility network layer of Qalubiya Governorate

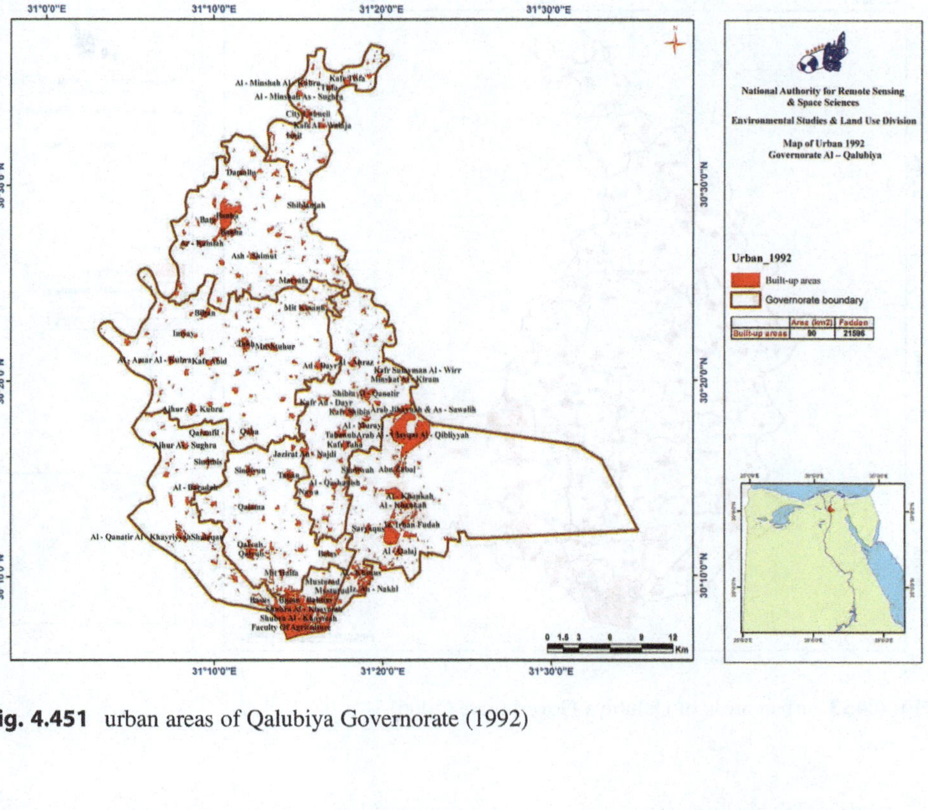

Fig. 4.451 urban areas of Qalubiya Governorate (1992)

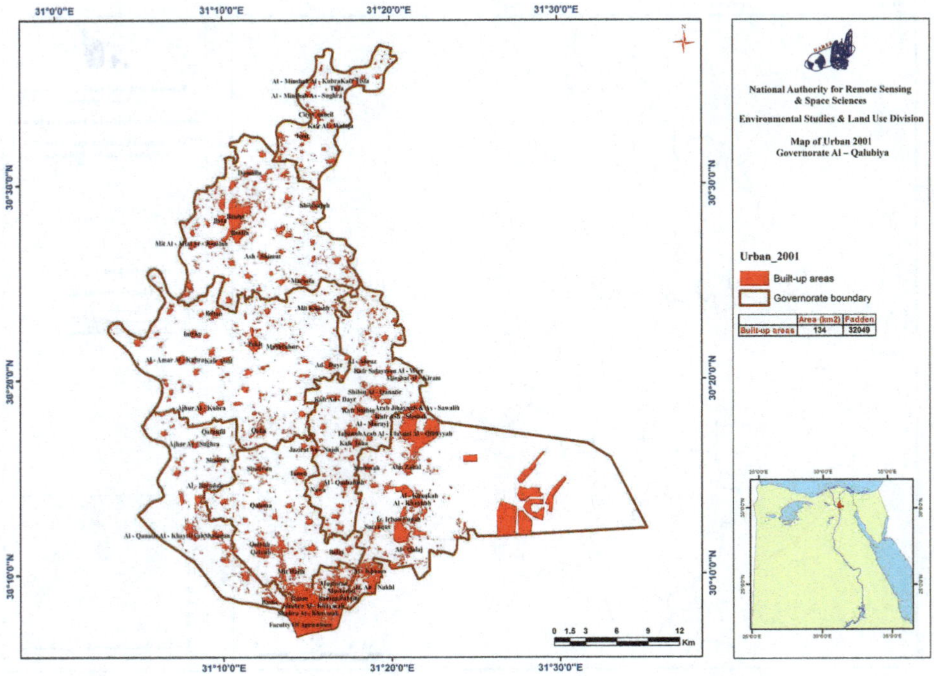

Fig. 4.452 urban areas of Qalubiya Governorate (2001)

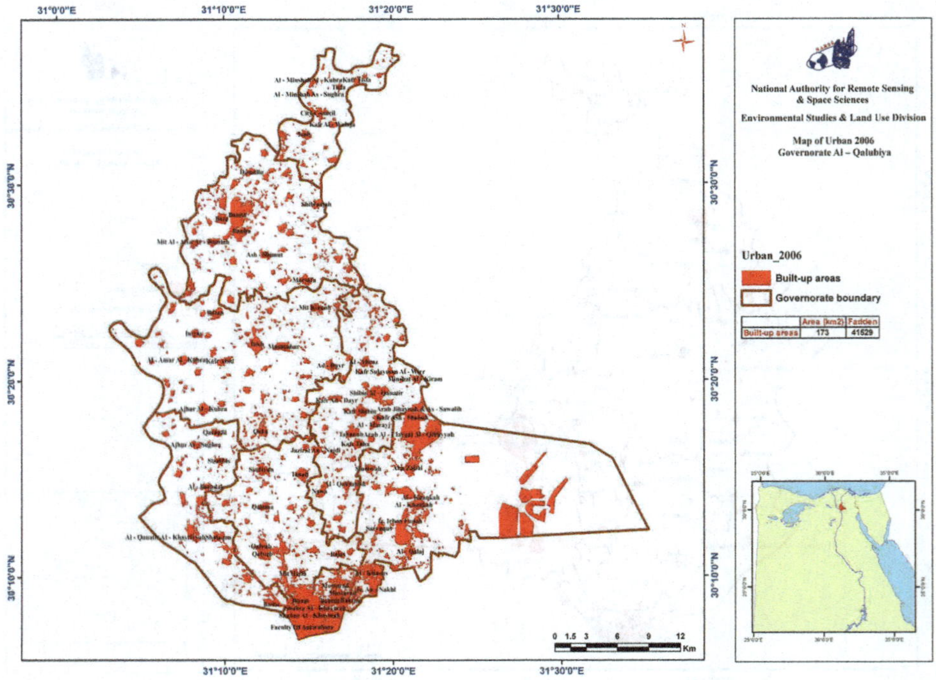

Fig. 4.453 urban areas of Qalubiya Governorate (2006)

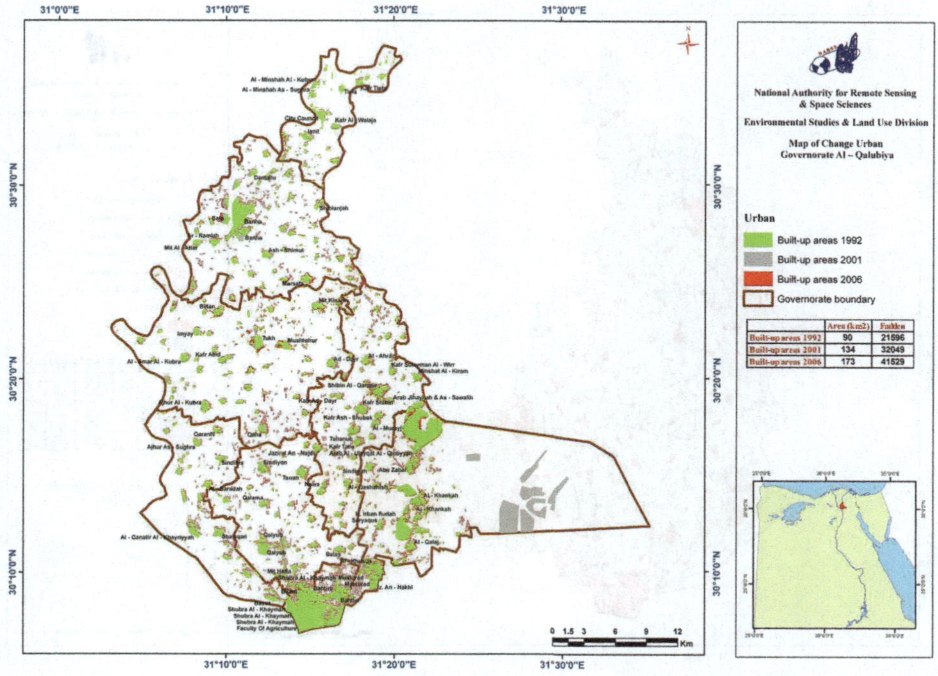

	Area (km2)	Fadden
Built-up areas 1992	90	21596
Built-up areas 2001	134	32049
Built-up areas 2006	173	41529

Fig. 4.454 Urban areas of Qalubiya Governorate (1992, 2001, 2006)

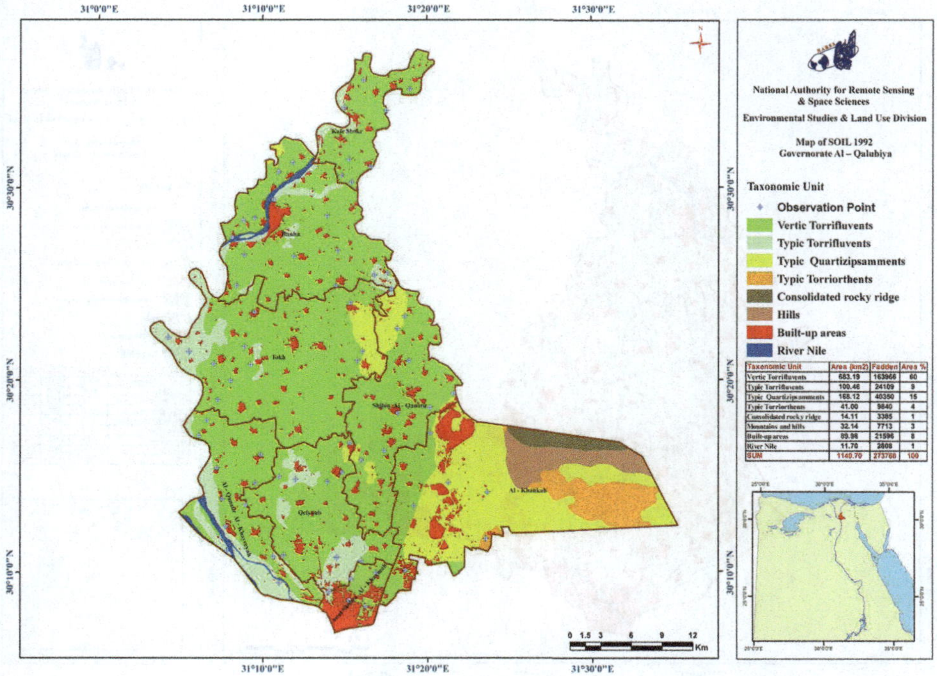

Taxonomic Unit	Area (km2)	Fadden	Area %
Vertic Torrifluvents	683.19	163966	60
Typic Torrifluvents	100.46	24109	9
Typic Quartizipsamments	168.12	40350	15
Typic Torriorthents	41.00	9840	4
Consolidated rocky ridge	14.11	3385	1
Mountains and hills	32.14	7713	3
Built-up areas	89.98	21596	8
River Nile	11.70	2808	1
SUM	1140.70	273768	100

Fig. 4.455 Urban areas draped over the soil layer of Qalubiya Governorate (1992)

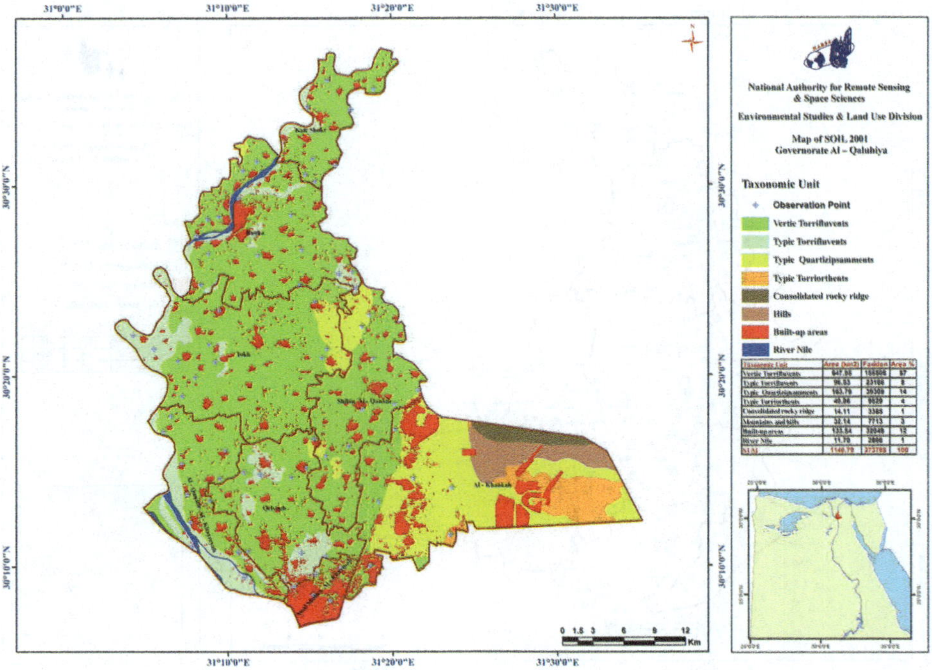

Fig. 4.456 Urban areas draped over the soil layer of Qalubiya Governorate (2001)

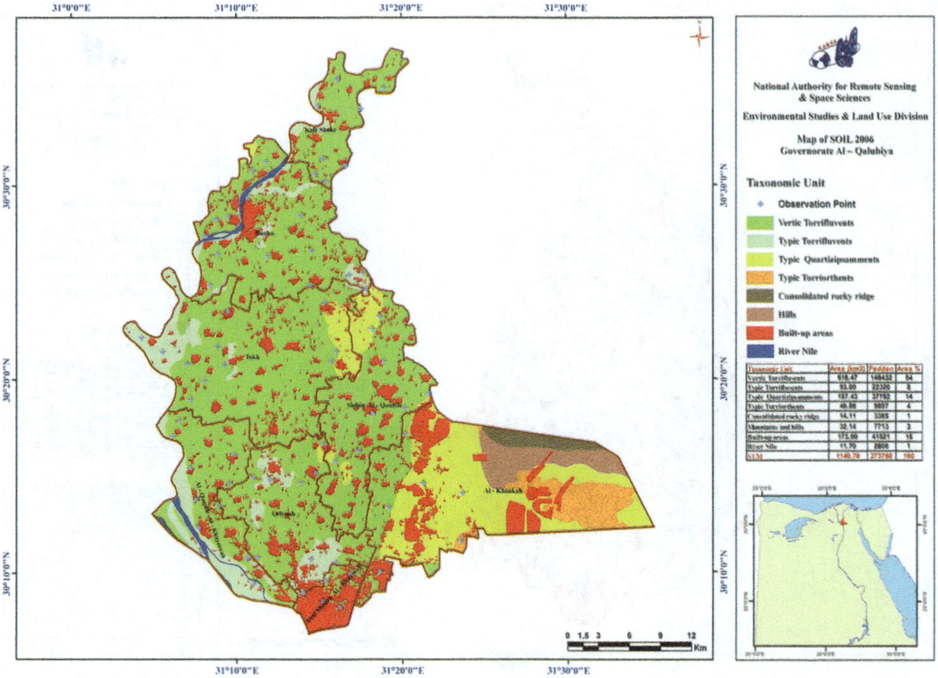

Fig. 4.457 Urban areas draped over the soil layer of Qalubiya Governorate (2006)

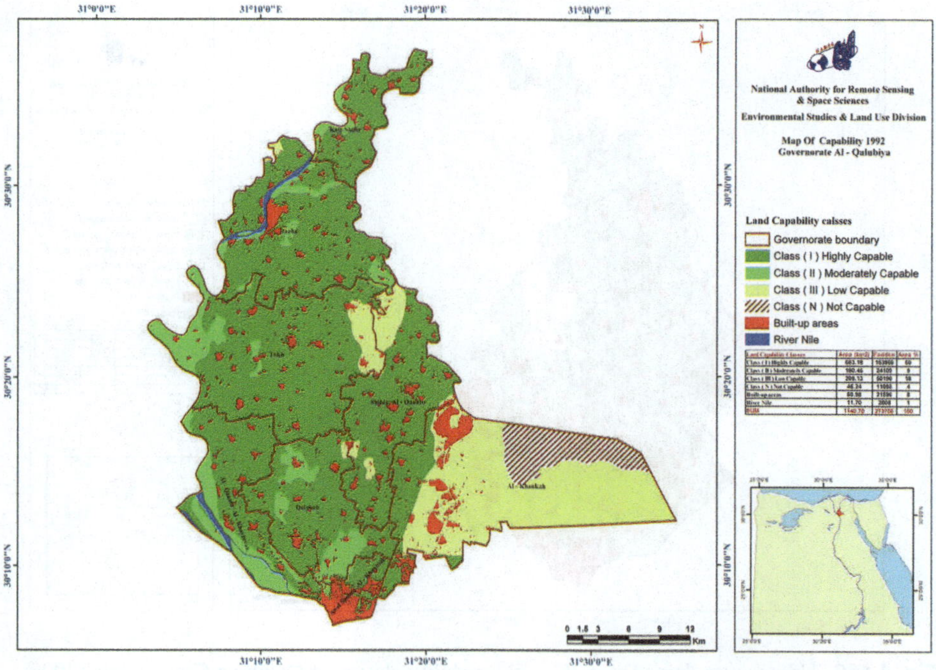

Fig. 4.458 Urban areas draped over the soil capability classes of Qalubiya Governorate (1992)

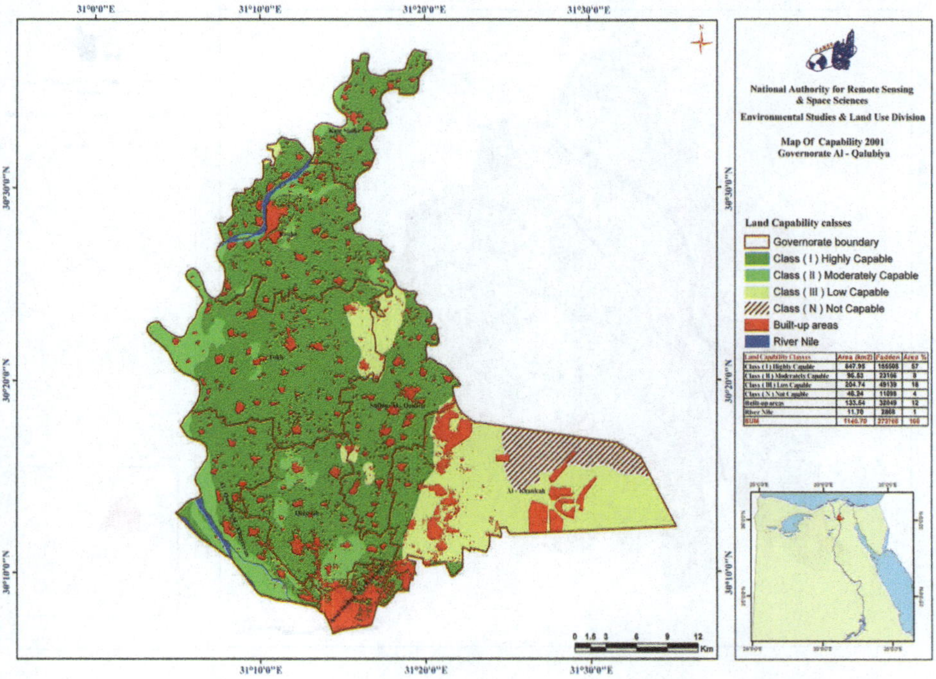

Fig. 4.459 Urban areas draped over the soil capability classes of Qalubiya Governorate (2001)

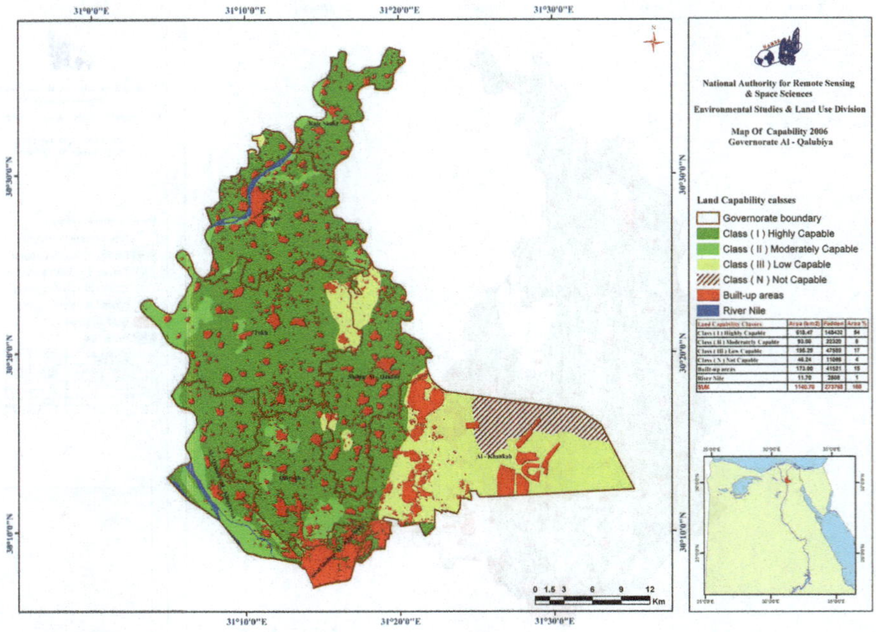

Fig. 4.460 Urban areas draped over the soil capability classes of Qalubiya Governorate (2006)

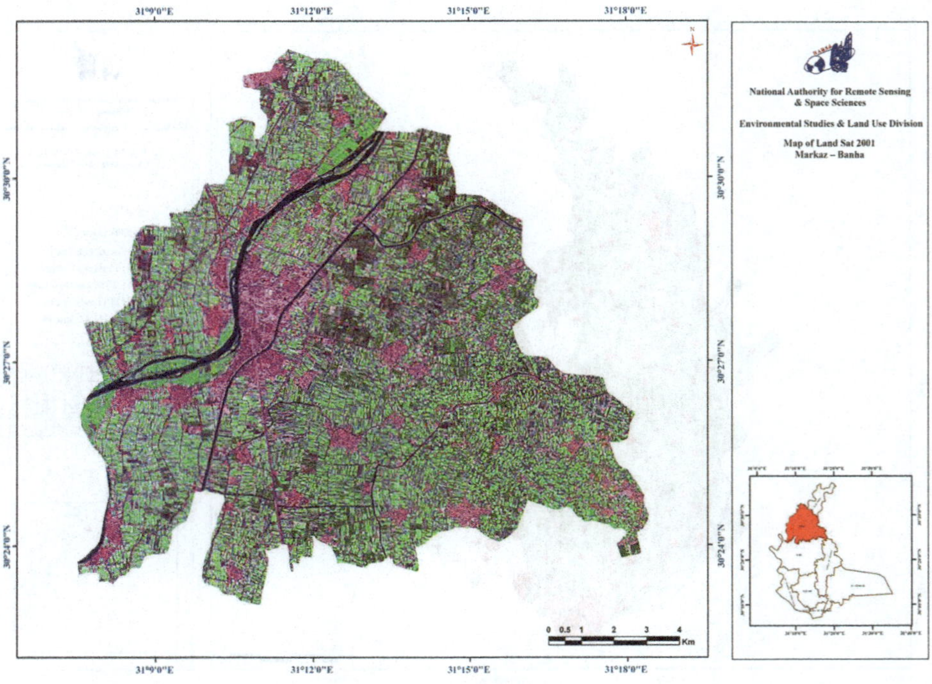

Fig. 4.461 Landsat ETM+ image of Benha district (2001)

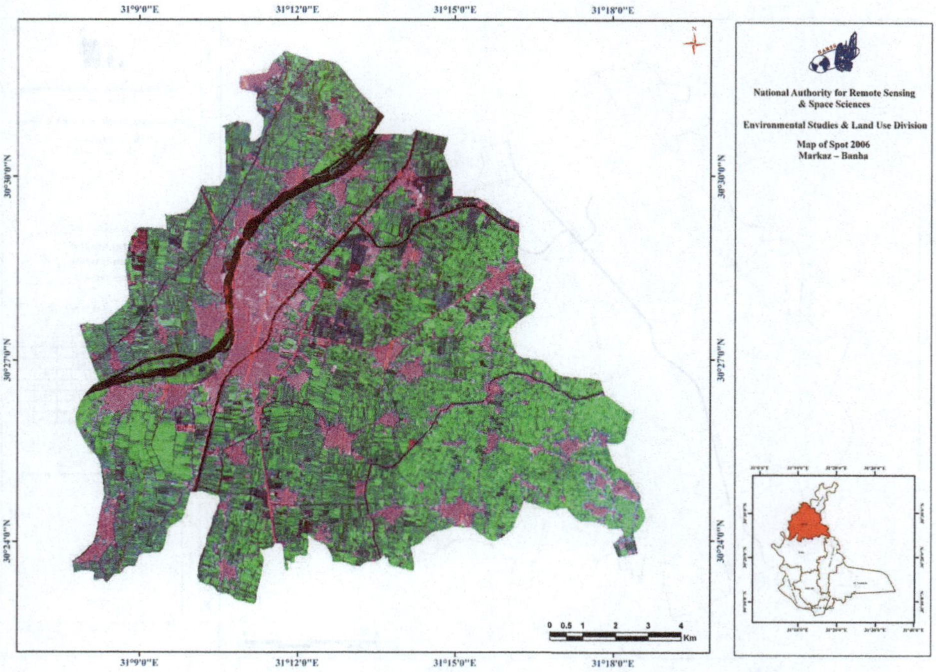

Fig. 4.462 SPOT image of Benha district (2006)

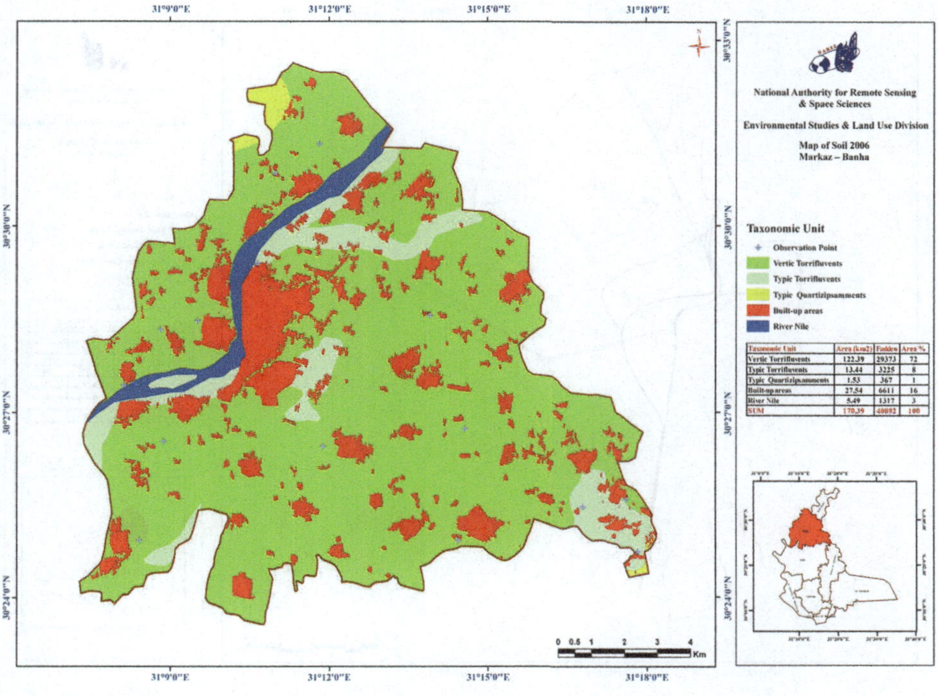

Fig. 4.463 Soil layer of Benha district

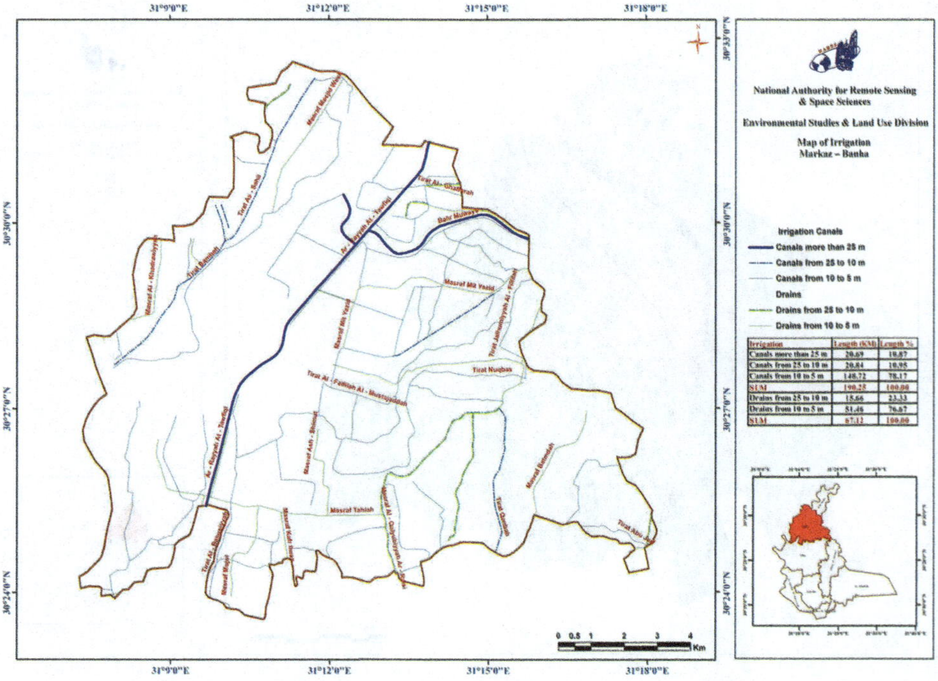

Fig. 4.464 Irrigation and drainage canals network of Benha district

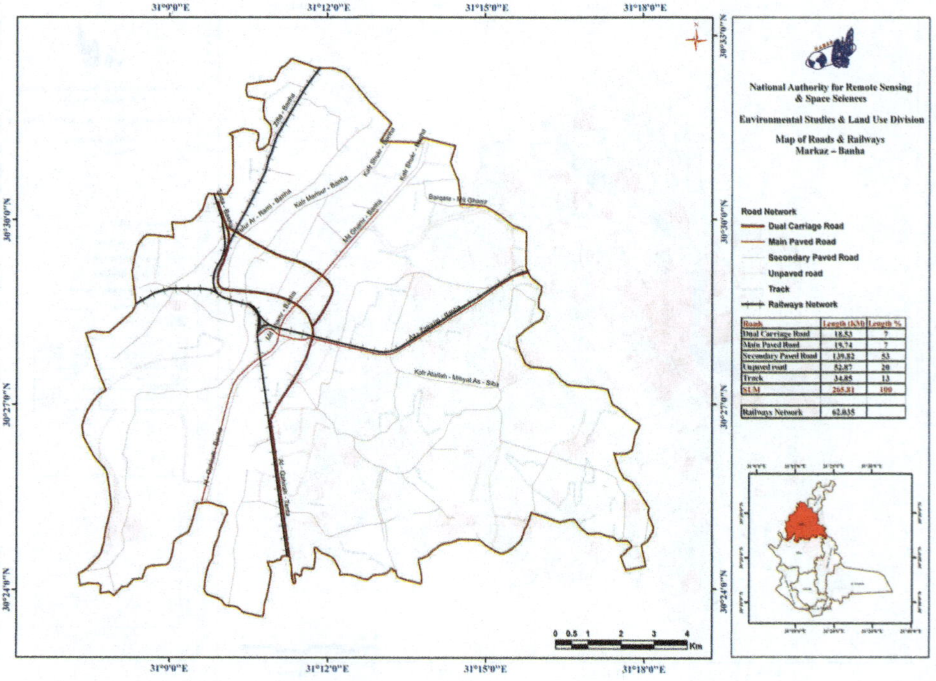

Fig. 4.465 Roads network of Benha district

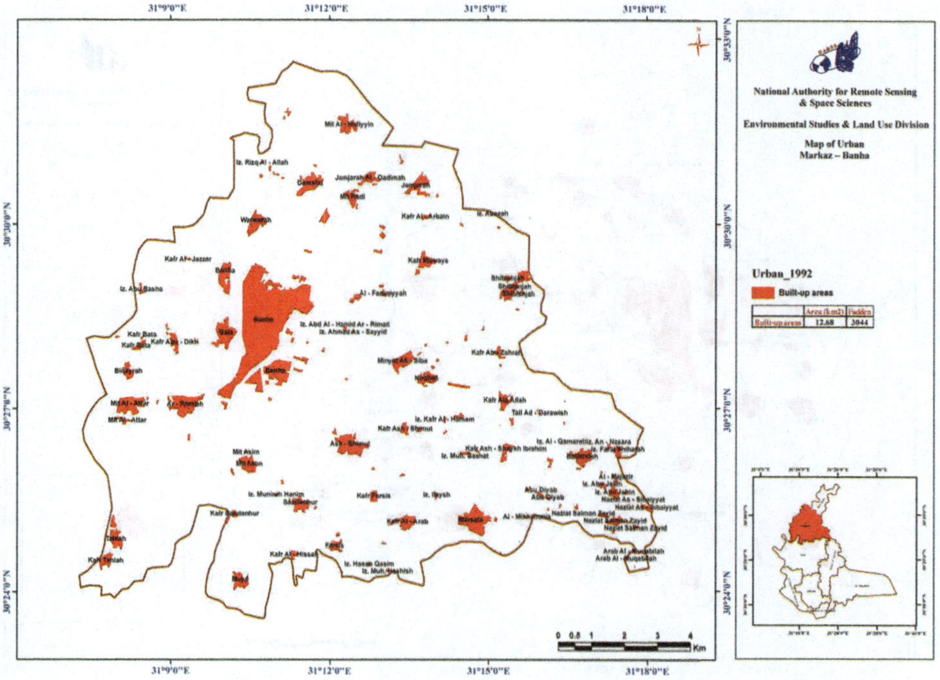

Fig. 4.466 Urban areas of Benha district (1992)

The multi-temporal monitoring of urban areas during the 1992–2006 period resulted in obtaining the thematic layers shown in (Figs. 4.466, 4.467 and 4.468). The results show an increase in the district settlement area from 12.9 km^2 in the year 1992 to 21.1 km^2 in the year 2001 and to 27.5 km^2 in 2006. The urban expansion (Fig. 4.469) records an average annual increasing rate of 1.1 km^2 per year, meaning that 249.7 feddans are excluded annually from the agricultural practices.

Combining the soil and land capability maps, in one hand, and the urban thematic layers, in another hand, made it possible to point out the risk of urban expansion, Figs. 4.470, 4.471, 4.472, 4.473, 4.474, 4.475 show that the cultivable soils of *Vertic Torrifluvents* sub great group, which coincides with the highly capability class, was the most negatively affected by urban expansion. It decreased from 135.7 km^2 in year 1992 to 128.0 km^2 in year 2001 and to 122.4 km^2 in year 2006. The soils of sub great group *Typic Torrifluvents,* coinciding with moderately capable soil class, decreased from 14.9 km^2 in 1992 to 14.2 km^2 in 2001 and to 13.4 km^2 in the year 2006. The sub great group *Typic Quartzipssamentss,* characterized as low capable soils, decreased from 1.6 km^2 to 1.5 km^2 during the period 1992–2006.

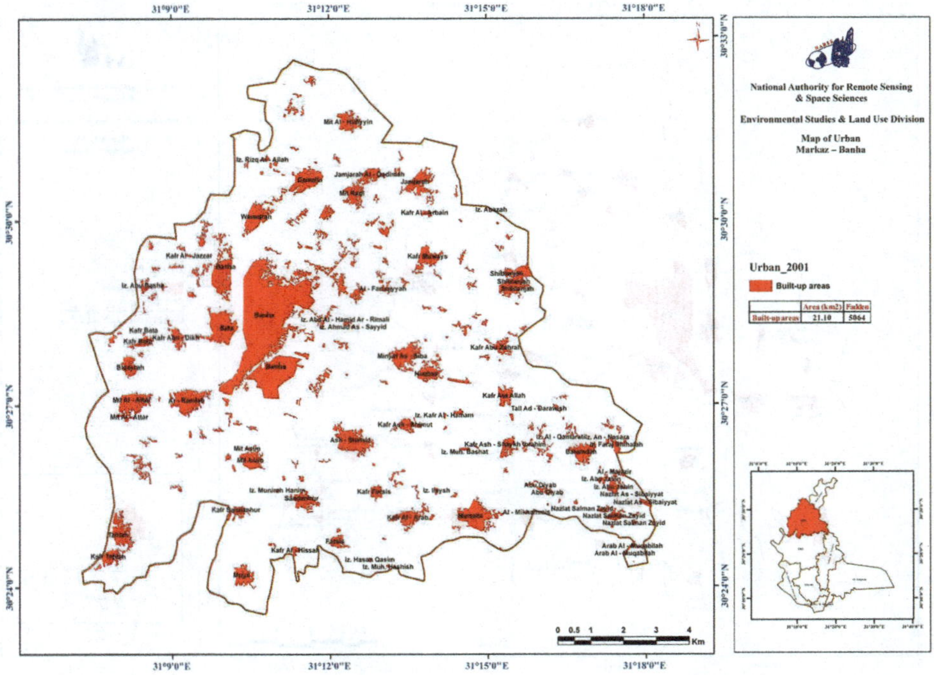

Fig. 4.467 Urban areas of Benha district (2001)

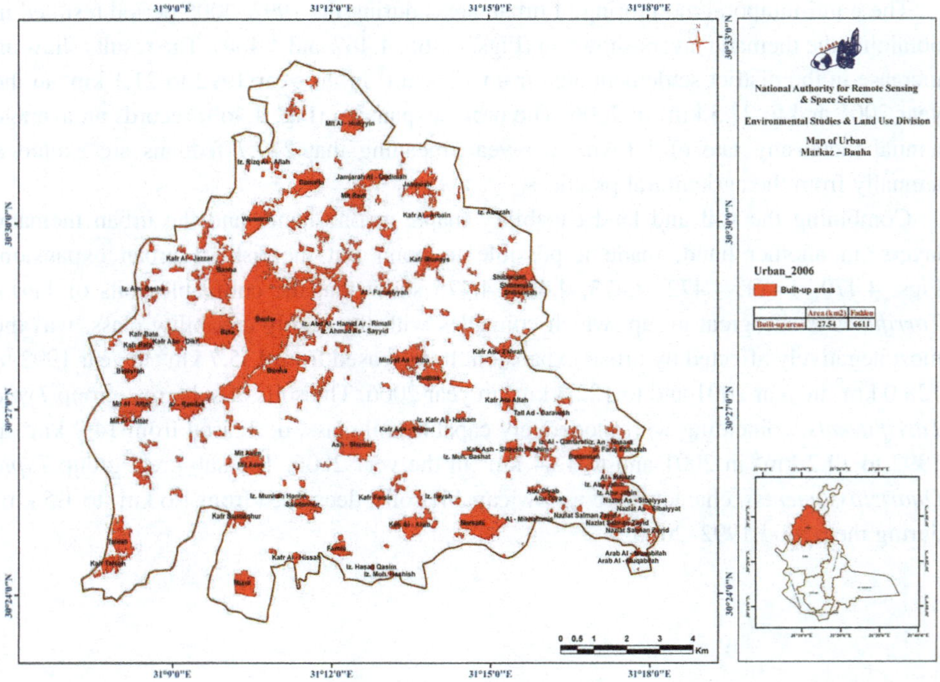

Fig. 4.468 Urban areas of Benha district (2006)

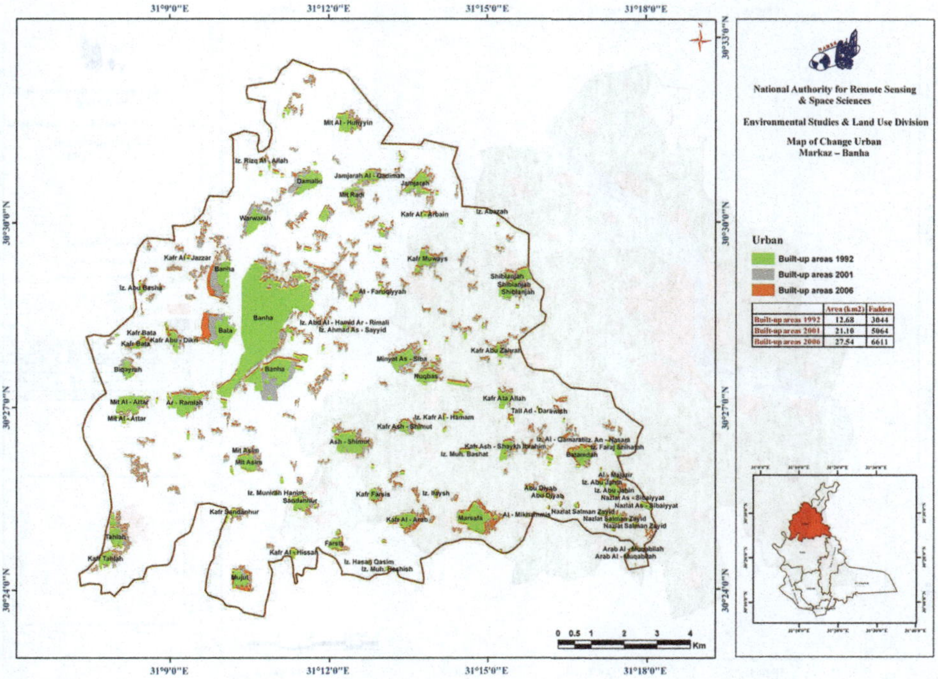

Fig. 4.469 Urban areas of Benha district (1992, 2001, 2006)

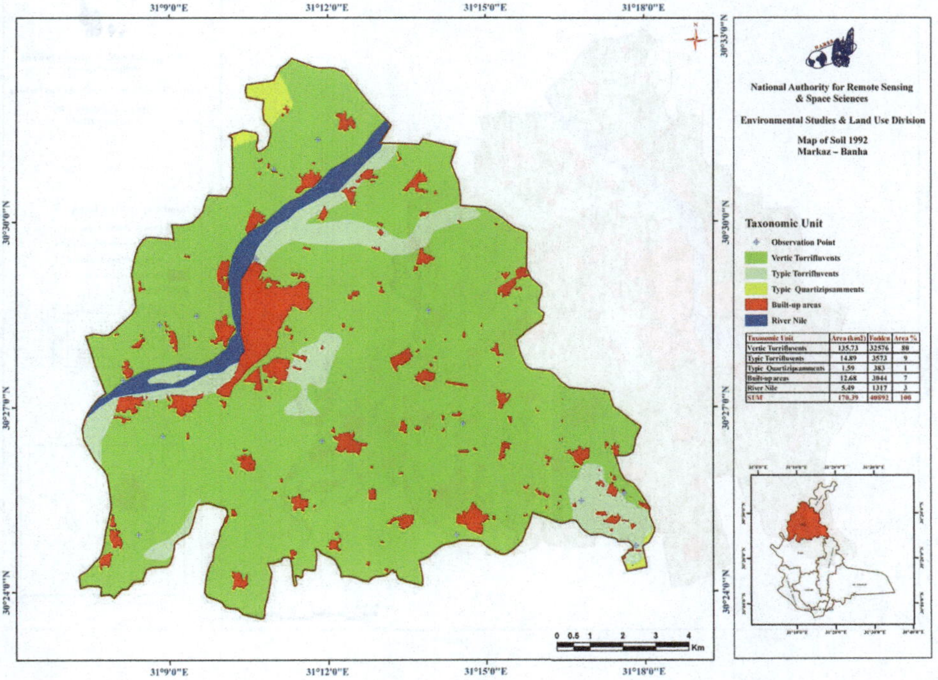

Fig. 4.470 Urban areas draped over the soil layer of Benha district (1992)

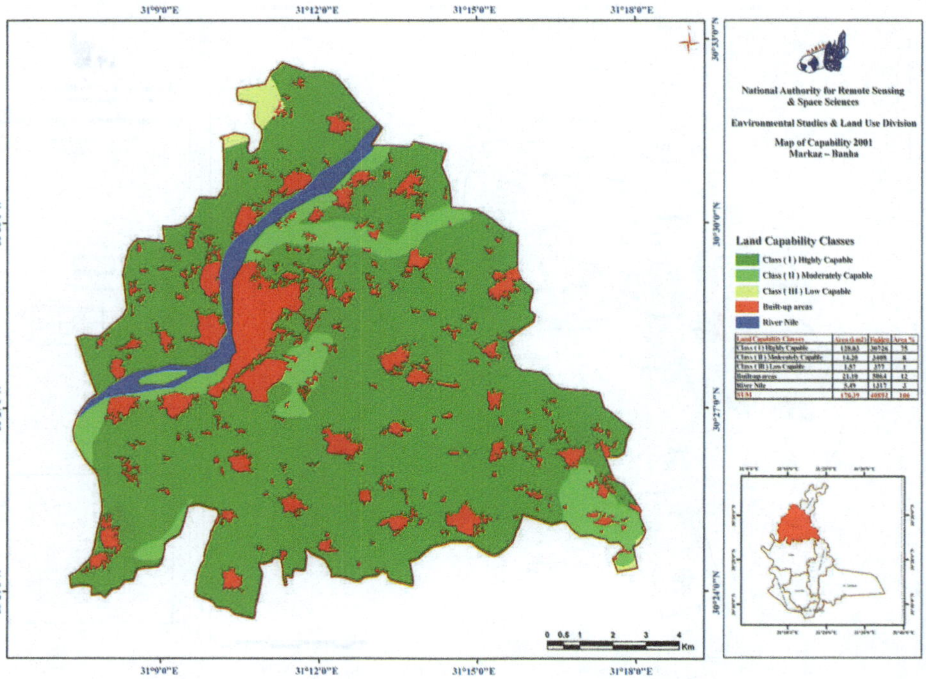

Fig. 4.471 Urban areas draped over the soil layer of Benha district (2001)

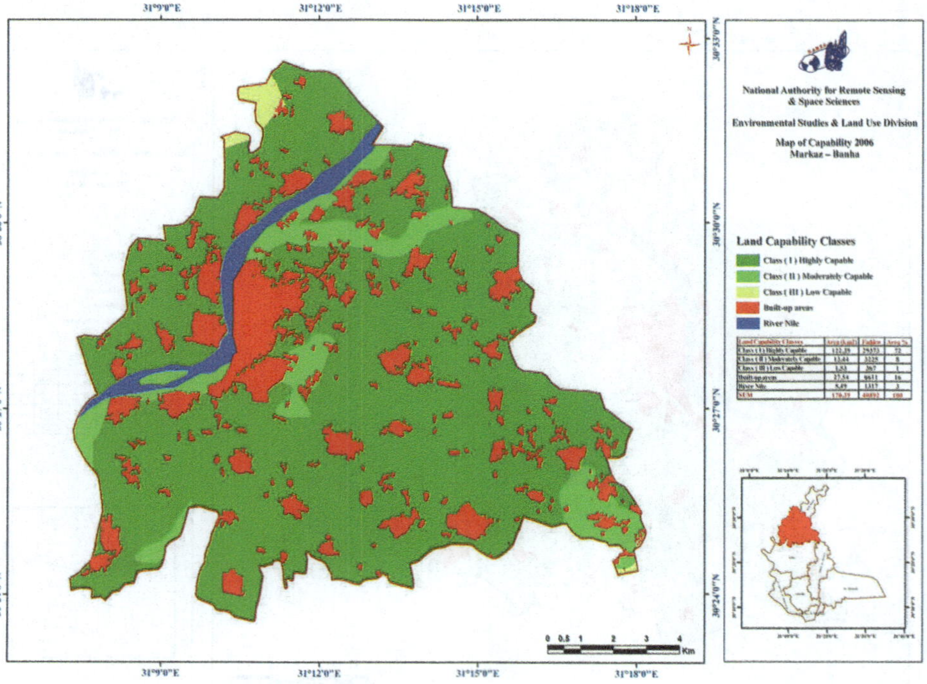

Fig. 4.472 Urban areas draped over the soil layer of Benha district (2006)

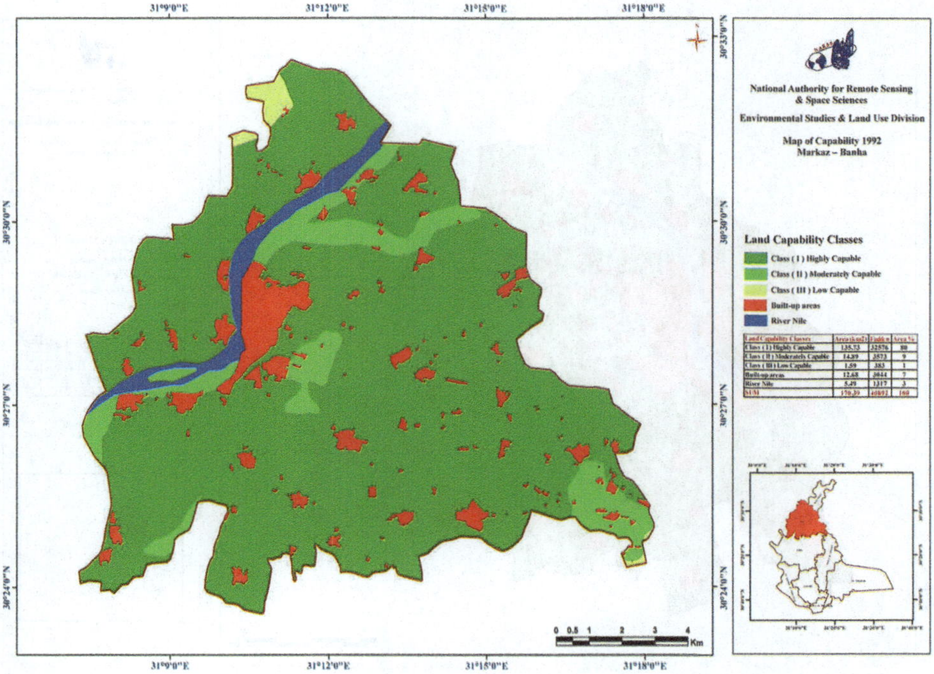

Fig. 4.473 Urban areas draped over the soil capability of Benha district (1992)

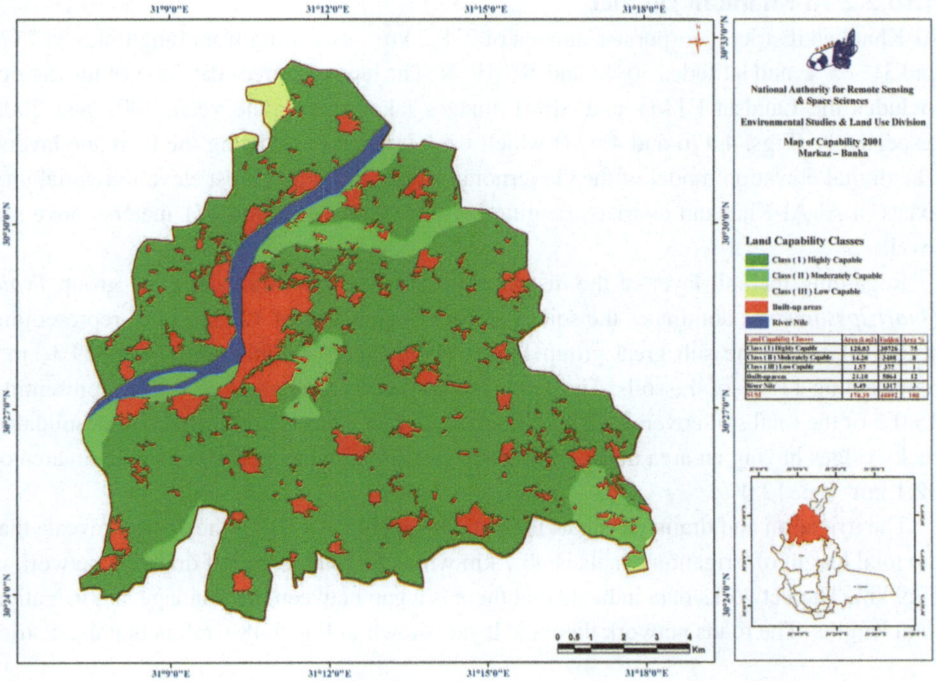

Fig. 4.474 Urban areas draped over the soil capability of Benha district (2001)

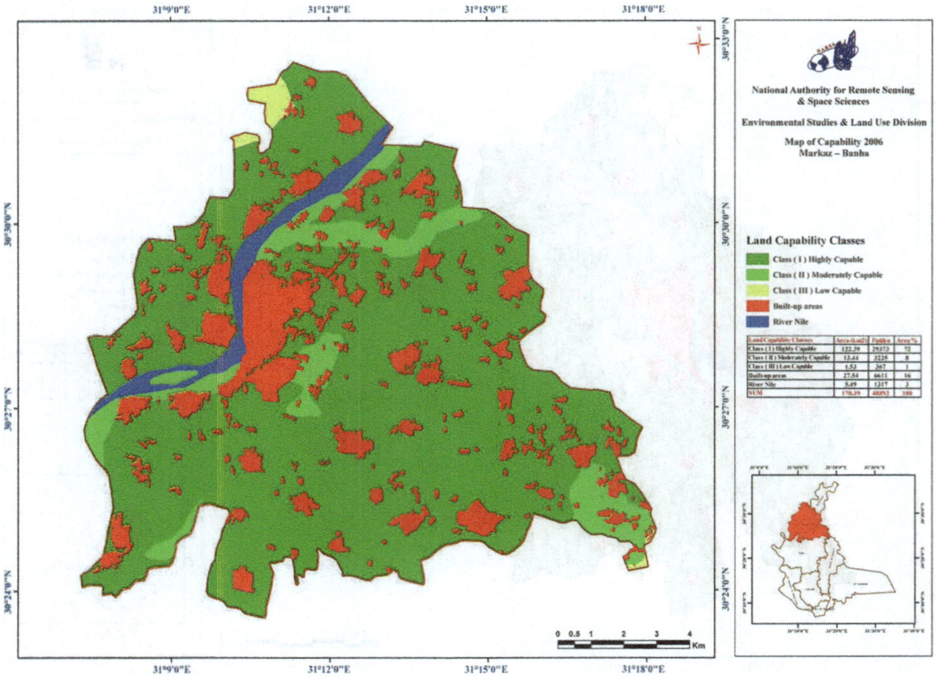

Fig. 4.475 Urban areas draped over the soil capability of Benha district (2006)

4.10.2.2 Al-Khankah District

Al-Khankah district incorporates an area of 278.2 km^2, extending from longitudes 31° 17′ and 31° 35′ E and latitudes 30° 8′ and 30° 19′ N. The land resources database of the district includes the Landsat ETM+ and SPOT images taken during the years 2001 and 2006 respectively (Figs. 4.476 and 4.477) which were utilized in updating the thematic layers. The digital elevation model of the Governorate area, shows that most elevation variability exists in Al-Al-Khankah districts, ranging between 0 sea level and 171 meters above sea level.

Regarding the soil layer of the district (Fig. 4.478) shows that sub-great group *Typic Quartzipssamentss* dominates the soil units covering an area of 125.4 0 km^2, representing 45.0% of soils. The sub great group *Vertic Torrifluvents* exhibits an area of 29.3 km^2, representing 11.0% of the soils. The *Typic Torriorthents* cover an area of 40.9 representing 15.0% of the total soil coverage. The rest of the district areas is dominated by consolidated rocky ridges having an area of 14.1 km^2 (i.e. 5%) and mountains and hills with an area of 32.1 km^2 (i.e. 12.0%).

The irrigation and drainage canals thematic layer (Fig. 4.479) of the district reveals that the total length of irrigation canals is 90.7 km while the total length of drainage network is 57.9 km. The networks data indicate that there is a gap between the drainage and irrigation total lengths. The roads network thematic layer, shown in Fig. 4.480, refers that the district

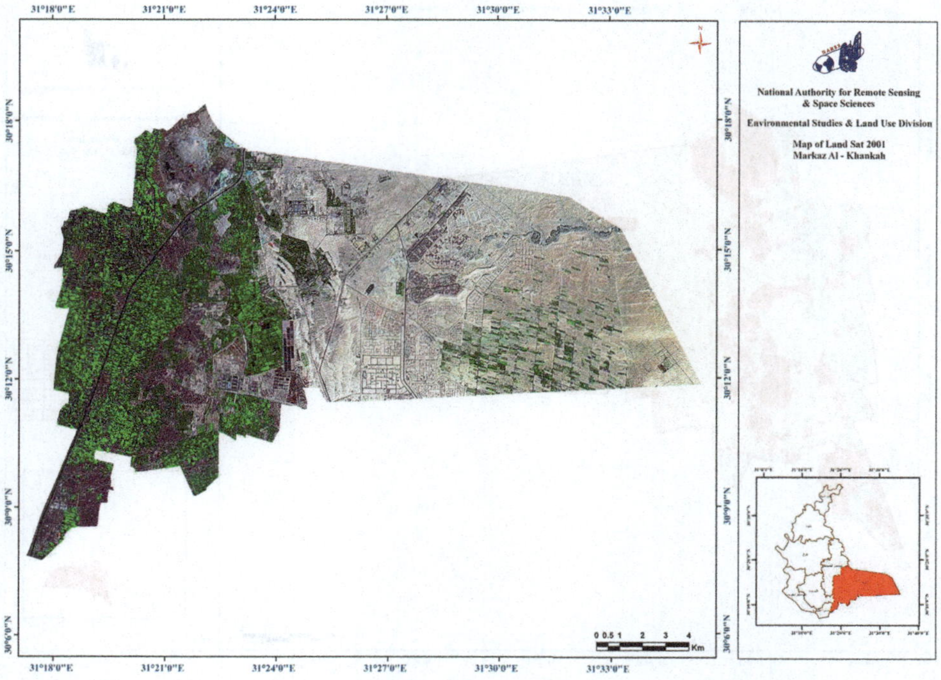

Fig. 4.476 Landsat ETM+ image of Al-Khankah district (2001)

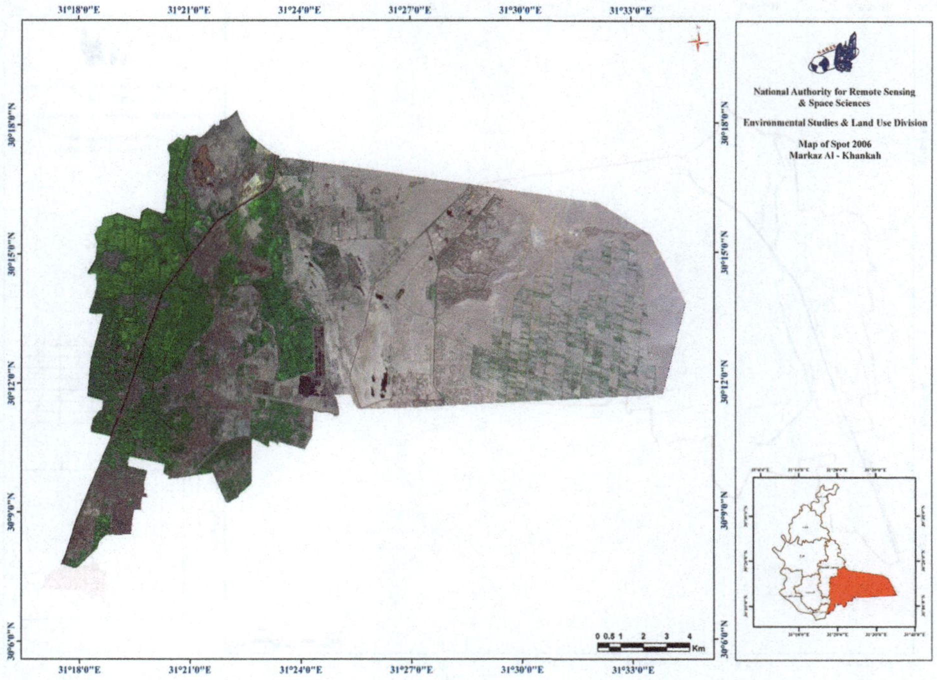

Fig. 4.477 Spot image of Al-Khankah district (2006)

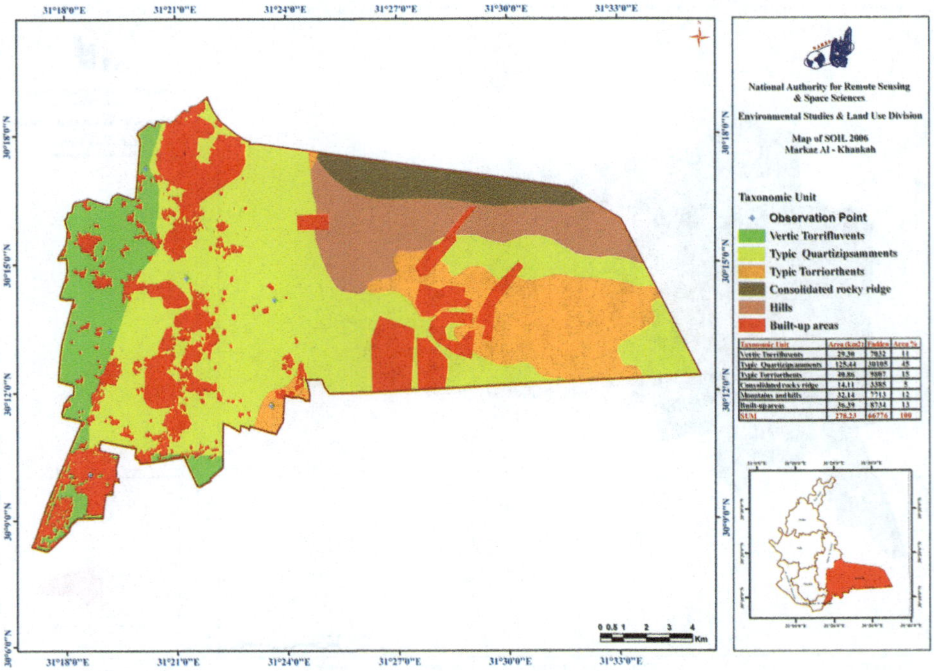

Fig. 4.478 Soil thematic layer of Al-Khankah district (2006)

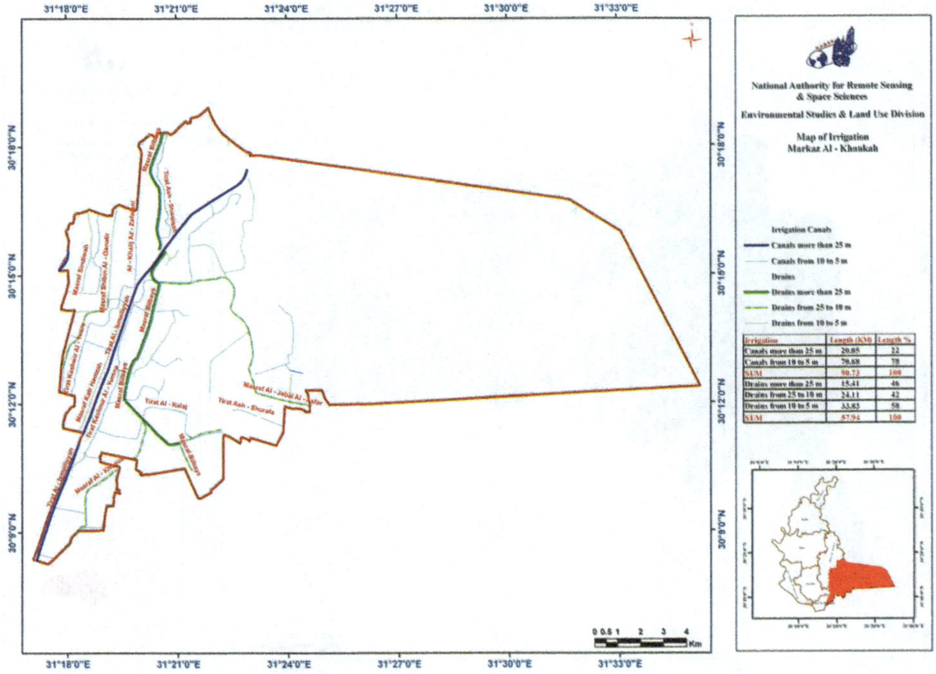

Fig. 4.479 Irrigation and drainage canals network of Al-Khankah district

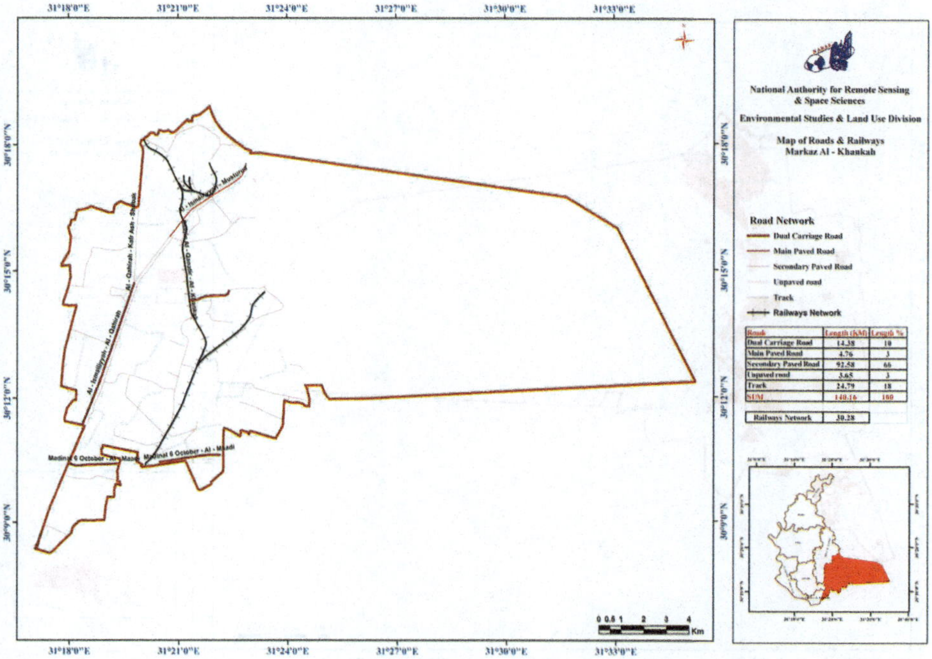

Fig. 4.480 Roads network of Al-Khankah district

attains unsatisfactory network where the total length of the roads is 140.6 km, this indicate that each 1 km^2 of the district is served only by 0.5 km of roads.

The progress in urban areas was monitored during the period 1992–2006 and the thematic layers were attached to the database, Figs. 4.481, 4.482 and 4.483. Detecting the changes in urban areas (Fig. 4.484) reveals that the urban area has increased from 22.4 km^2 in the year 1992 to 29.2 km^2 in the year 2001 and to 36.4 km^2 in 2006 with an almost annual rate of 1 km^2 per year.

Attaching the soil and land capability layers, in one hand, and multi-temporal urban areas, in another hand, was performed. Figures 4.485, 4.486 and 4.487 show that the urban encroachment was almost on the account of most fertile soils. The area of the cultivable soils of *Vertic Torrifluvents* sub-great group is the most influenced by the urban expansion, it decreased from 34.7 to 31.5 to 29.3 km^2 in the years 1992, 2001 and 2006 respectively. The soils of *Typic Torrionthents* and *Typic Quartzipssamentss* were slightly influenced in comparison with their actual areas. The interrelationship between urban growth and land capability classification is shown in Figs. 4.488, 4.489 and 4.490. The output layers indicate that the loos high capable soils was 5.4 km^2 (1288 feddans) during the period 1992–2006 (i.e. 92 feddan/year). The slightly low capable soils lost an area of 8.6 km^2 (20,357 feddans) with an annual rate of 1454 feddans. The non capable soils show some recent changes, caused by urban expansion, however not proportional with its large area.

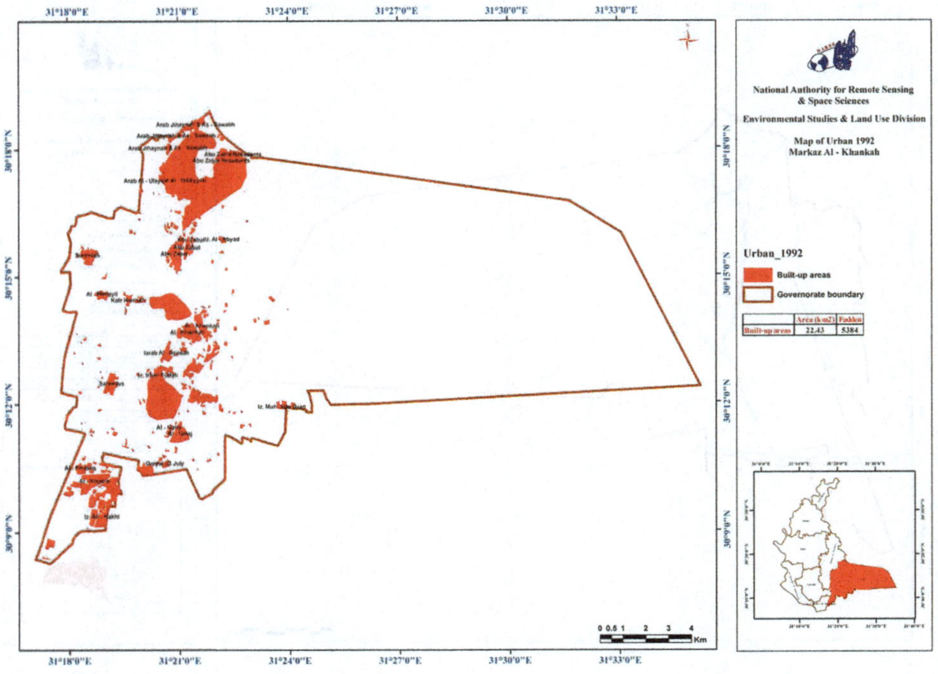

Fig. 4.481 Urban areas of Al-Khankah district (1992)

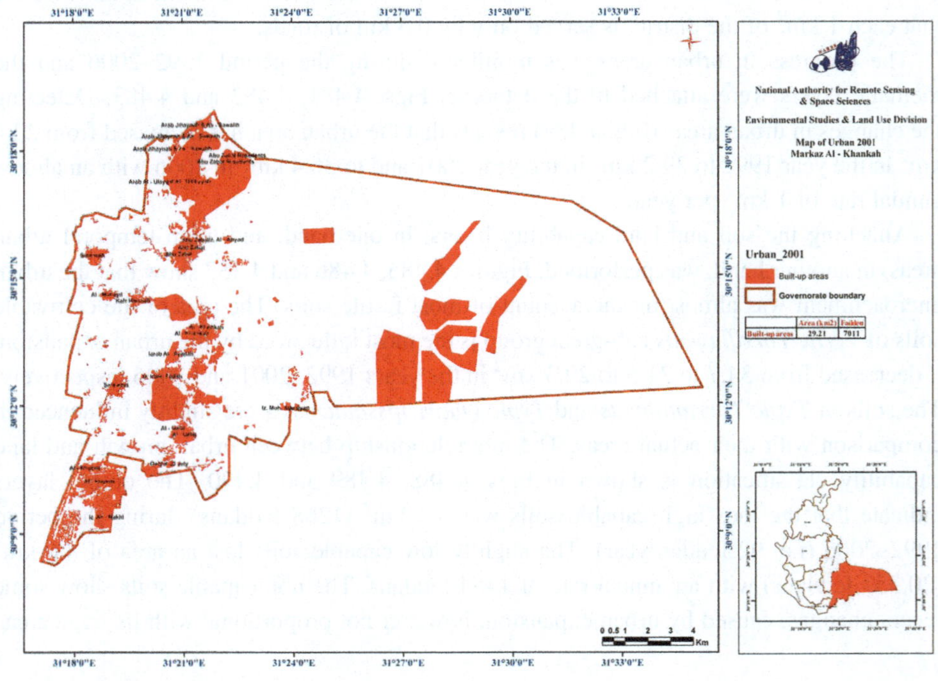

Fig. 4.482 Urban areas of Al-Khankah district (2001)

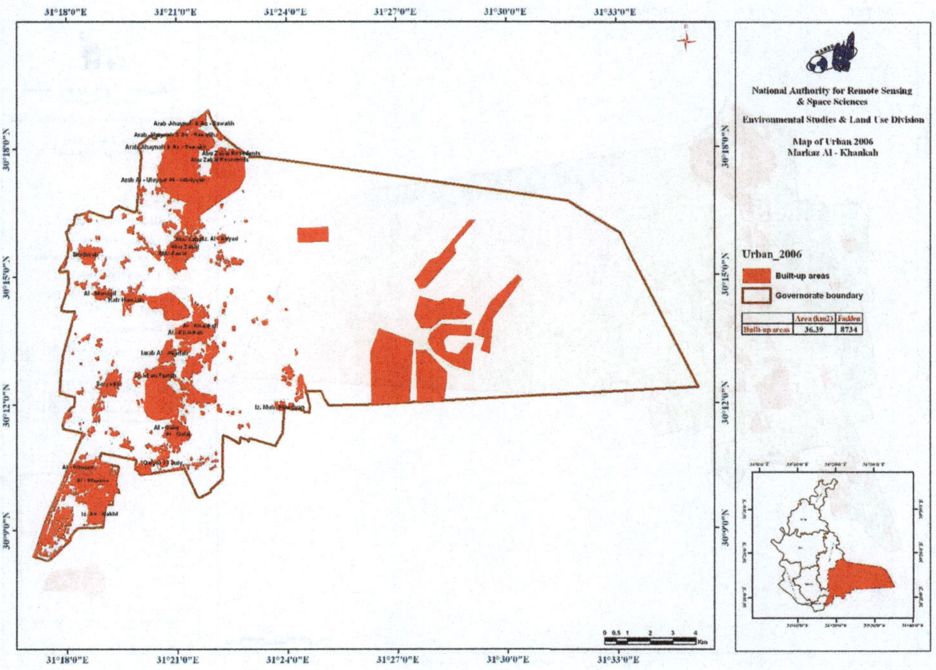

Fig. 4.483 Urban areas of Al-Khankah district (2006)

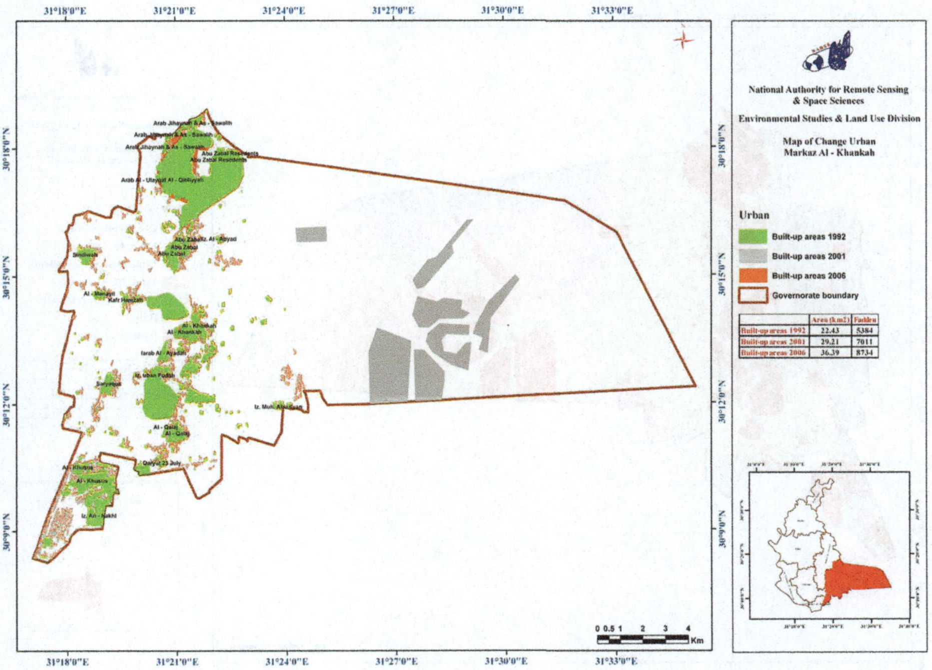

Fig. 4.484 Urban areas of Al-Khankah district (1992, 2001, 2006)

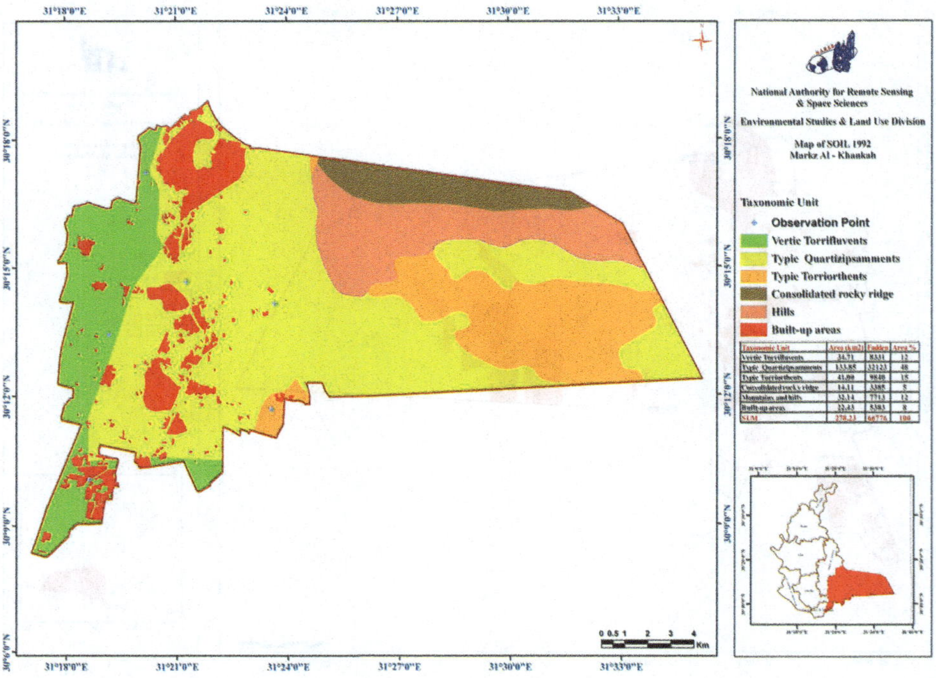

Fig. 4.485 Urban areas draped over the soil layer of Al-Khankah district (1992)

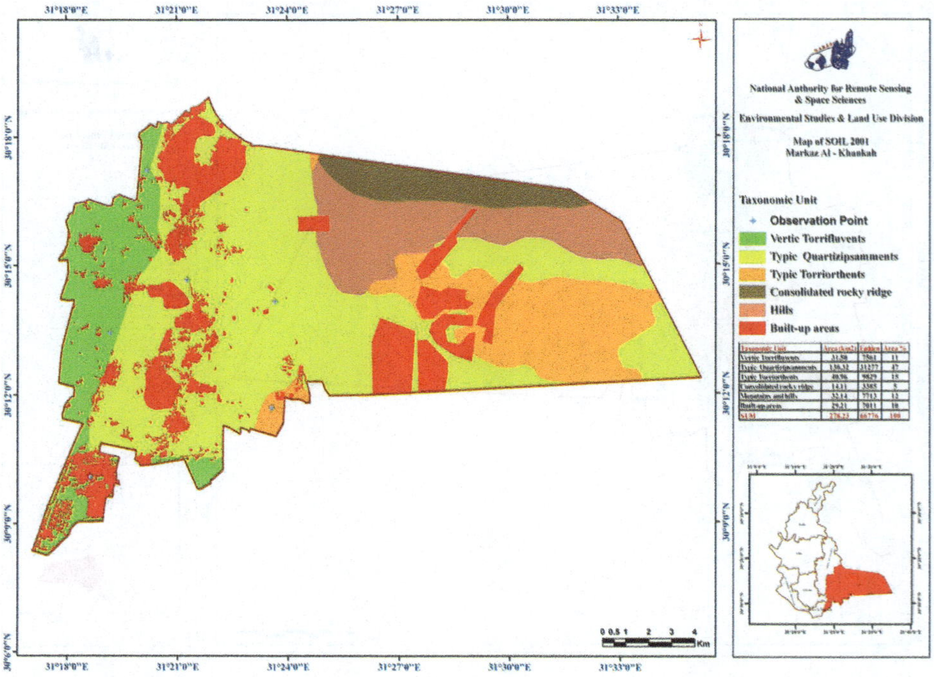

Fig. 4.486 Urban areas draped over the soil layer of Al-Khankah district (2001)

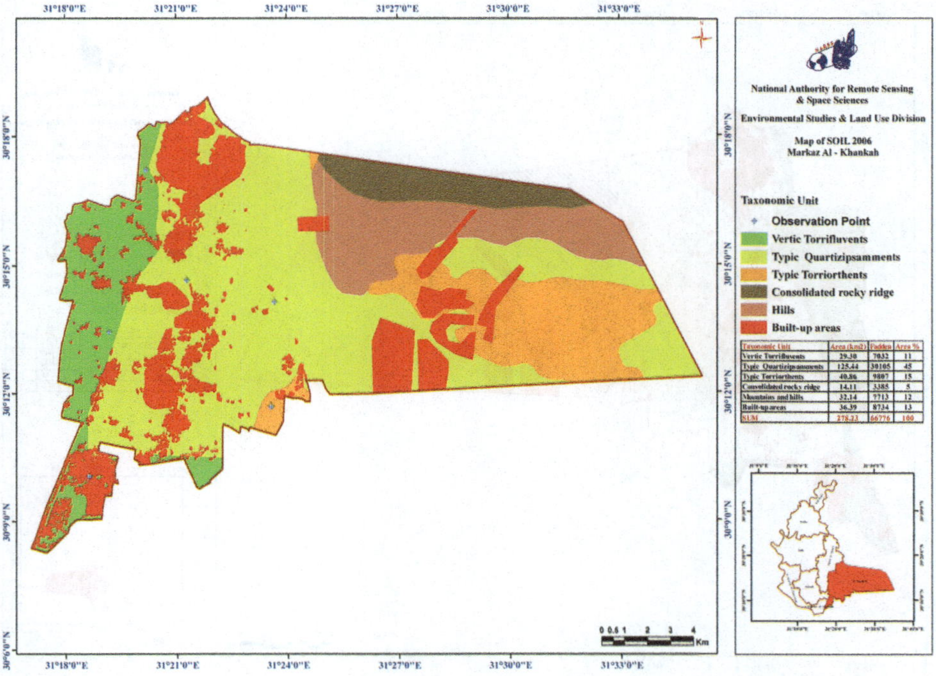

Fig. 4.487 Urban areas draped over the soil layer of Al-Khankah district (2006)

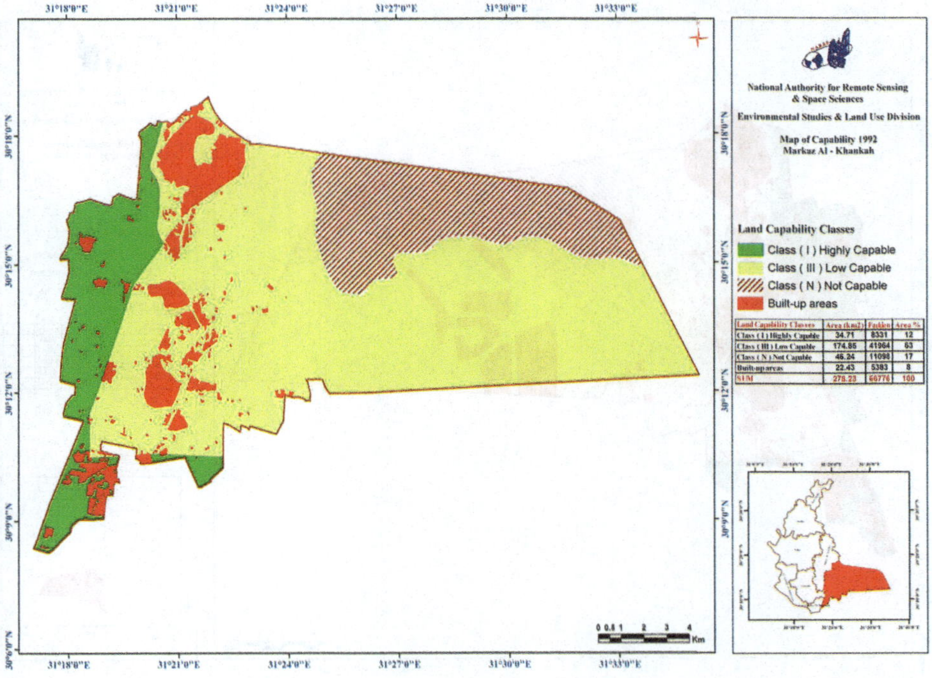

Fig. 4.488 Urban areas draped over the soil capability of Al-Khankah district (1992)

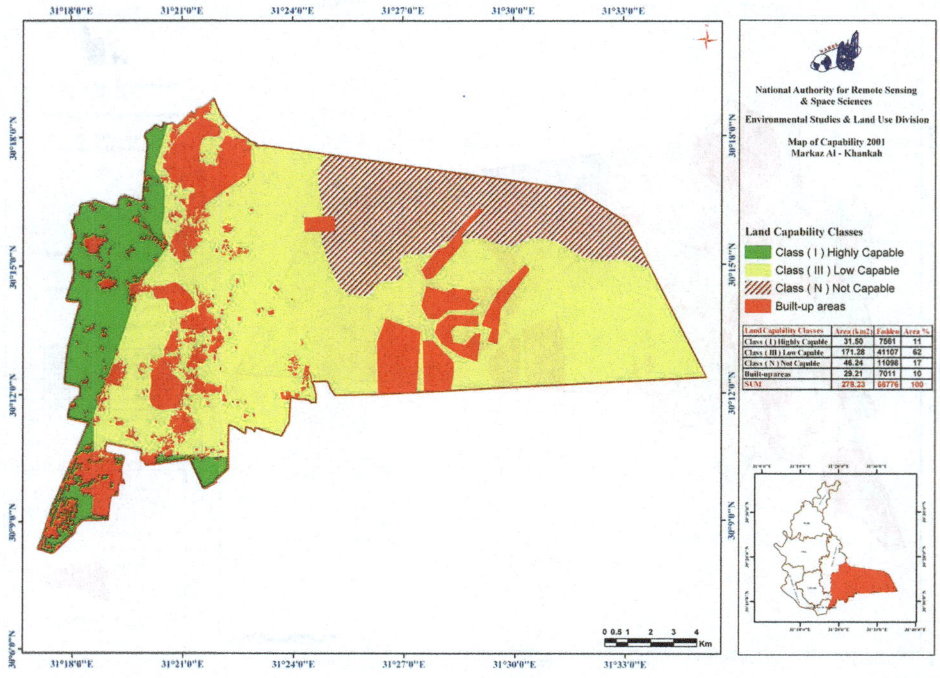

Fig. 4.489 Urban areas draped over the soil capability of Al-Khankah district (2001)

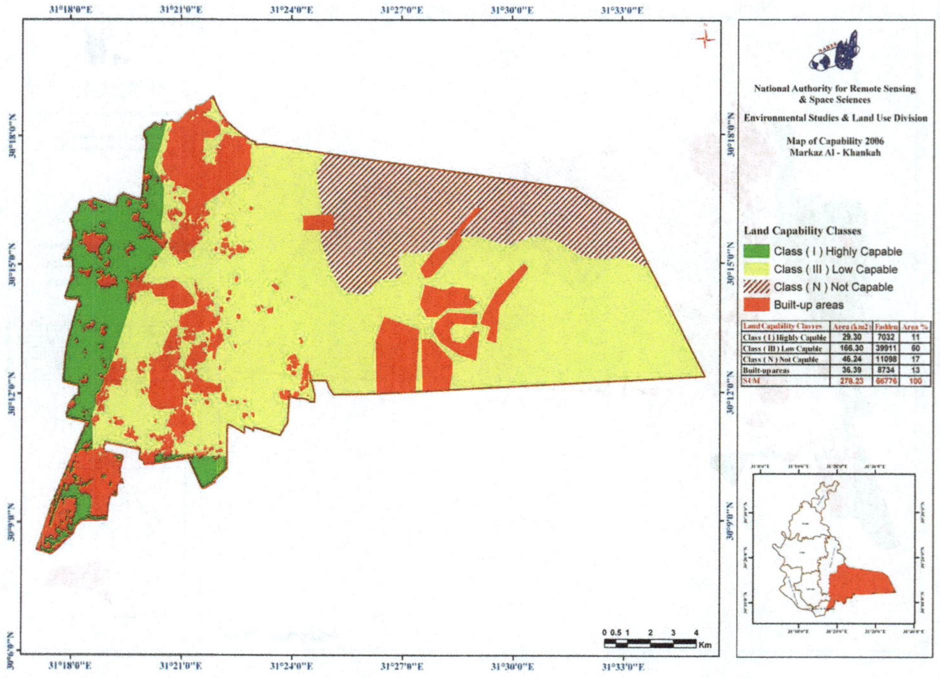

Fig. 4.490 Urban areas draped over the soil capability of Al-Khankah district (2006)

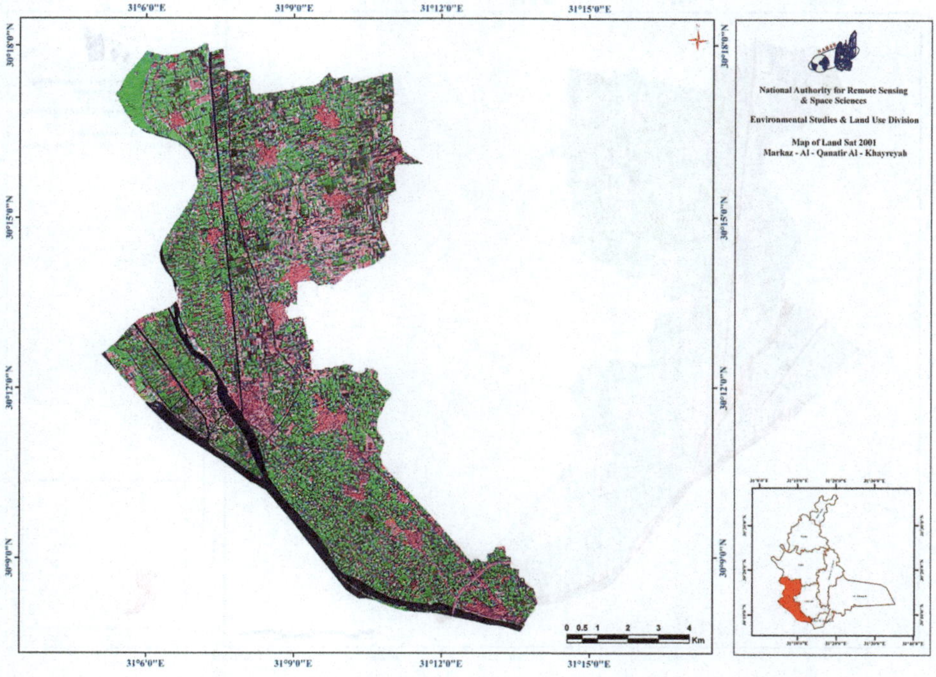

Fig. 4.491 Landsat ETM+ image of Al-Qanatir Al-Khayreyah district (2001)

4.10.2.3 Al-Qanatir Al-Khayreyah District

Al-Qanatir Al-Khayreyah district incorporates an area of 112.8 km^2, extending from longitudes 31° 04′ and 31° 14′ E and latitudes 30° 08′ and 30° 19′ N. The land resources database of the district includes the Landsat ETM+ and SPOT images taken during the years 2001 and 2006 respectively (Figs. 4.491 and 4.492) which were utilized in updating the thematic layers. The digital elevation model of the district indicates that the surface elevation ranges from 0 to 34 meters above sea level.

Regarding the soil layer (Fig. 4.493) of the district, it is found the *Vertic Torrifluvents* soil sub great group dominates the soil units, covering an area of 65.9 km^2, representing 58.4% of the district soil area. The *Typic Torrifluvents* covers 27.7 km^2, representing 24.6% of the district total area.

The irrigation and drainage canals thematic layer of the district (Fig. 4.494) reveals that the total length of irrigation canals is 122.0 km while the total length of drainage network is 49.1 km. This fact may indicate that there is a wide gap between the drainage and irrigation network in the district.

The roads network layer indicates that the district have a satisfactory network (Fig. 4.495) where the total length of the roads network is 111.6 km. the roads network is characterized by a road density of 1 km road length for each 1 km^2 area.

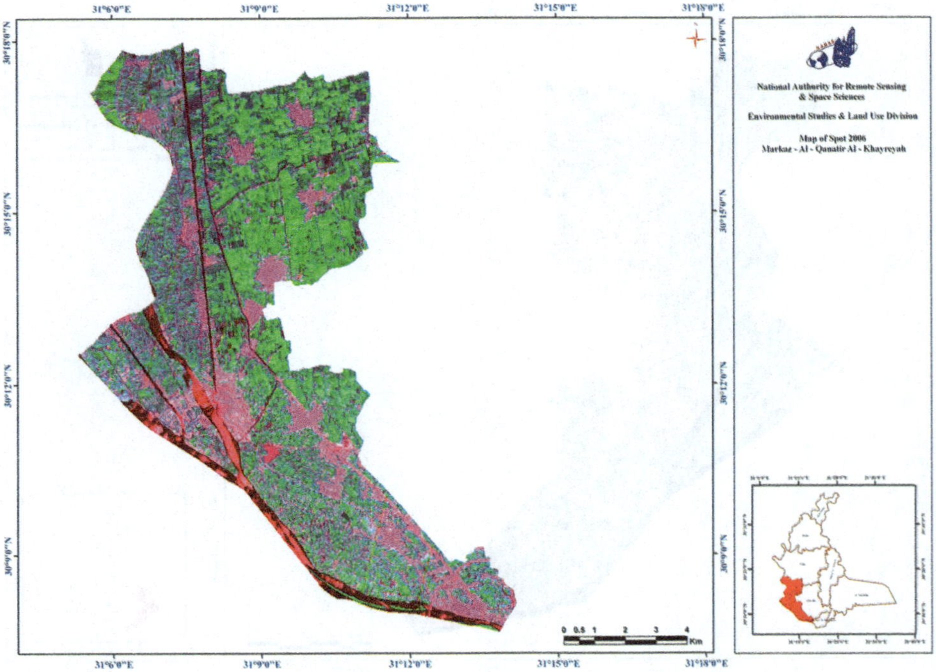

Fig. 4.492 SPOT image of Al-Qanatir Al-Khayreyah district (2006)

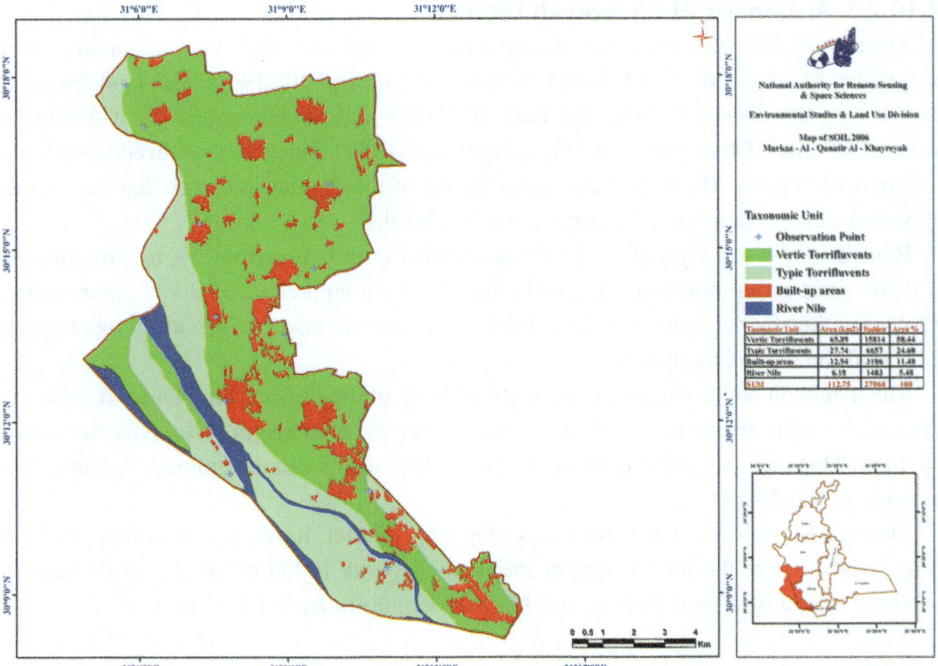

Fig. 4.493 Soil layer of Al-Qanatir Al-Khayreyah district

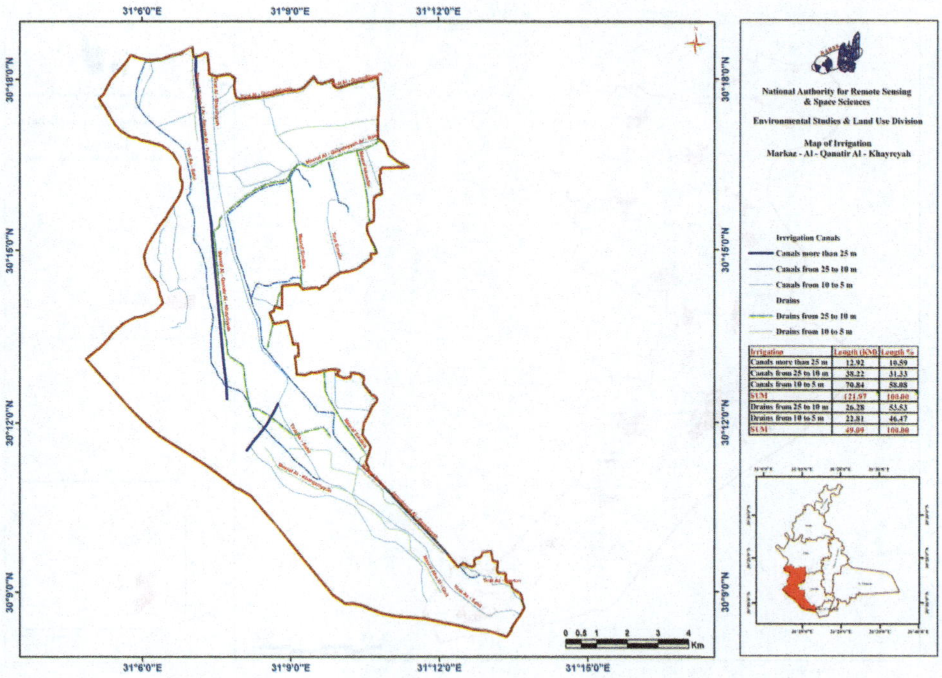

Fig. 4.494 Irrigation and drainage canals network of Al-Qanatir Al-Khayreyahdistrict

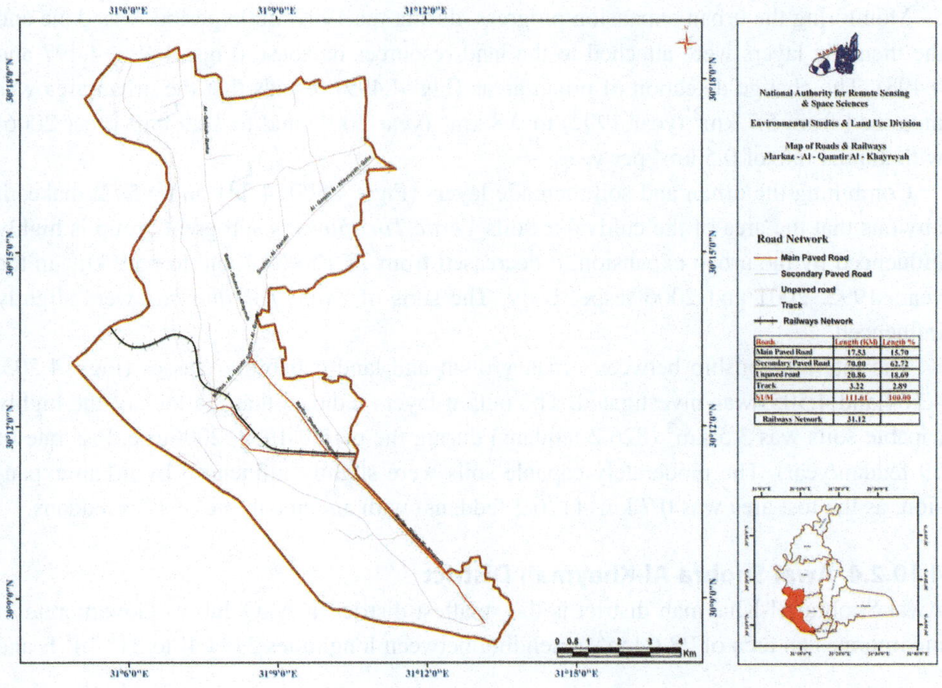

Fig. 4.495 Roads network of Al-Qanatir Al-Khayreyah district

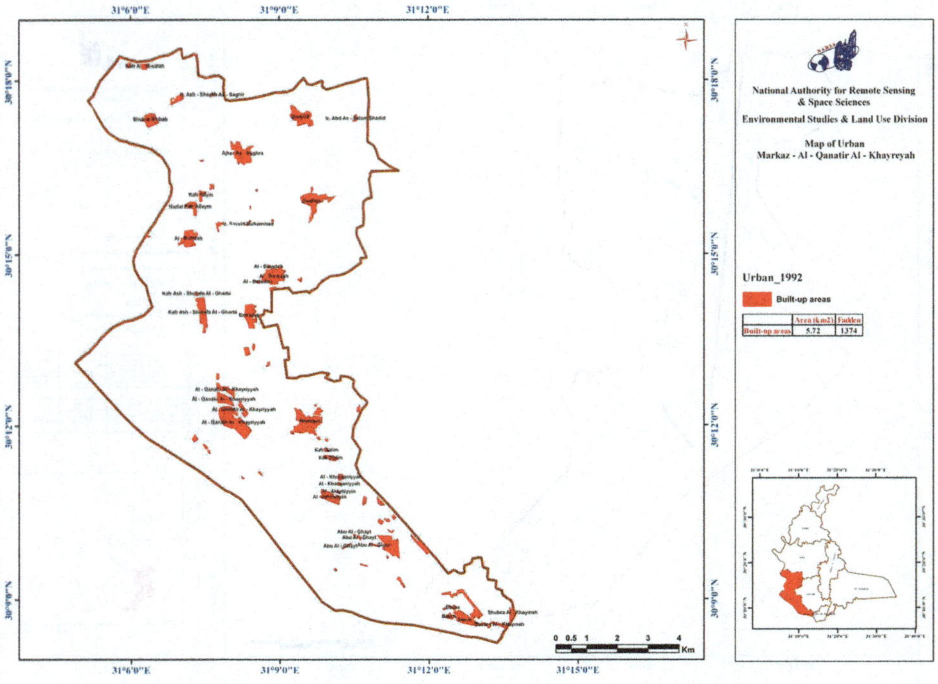

Fig. 4.496 Urban areas of Al-Qanatir Al-Khayreyah district (1992)

Monitoring the urban expansion progress, during the 1992–2006 period, was done and the thematic layers were attached to the land resources database (Figs. 4.496, 4.497 and 4.498). The change detection of urban areas (Fig. 4.499) reveals that the urban area was increased from 5.7 km^2 (year 1992) to 9.8 km^2 (year 2001) and to 12.9 km^2 (year 2006) with annual rate of 0.5 km^2 per year.

Combining the urban and soil thematic layers (Figs. 4.500, 4.501 and 4.502) makes it obvious that the area of the cultivable soils *Vertic Torrifluvents* sub-great group is highly influenced by the urban expansion, it decreased from 72.4 to 68.7 and to 65.9 km^2 in the years 1992, 2001 and 2006 respectively. The soils of *Typic Torrifluvents* were slightly influenced.

The interrelationship between urban growth and land capability classes (Figs. 4.503, 4.504 and 4.505) was investigated. The output layers indicate that the loos of the highly capable soils was 3.5 km^2 (826.2 feddans) during the period 1992–2006 (i.e. loss rate of 59 feddan/year). The moderately capable soils were slightly influenced by urban expansion, as the lost area was 0.7 km^2 (176.2 feddans) with an annual rate of 12.6 feddans.

4.10.2.4 Awal Shobra Al-Khaymah District

Awal Shobra Al-Khaymah district is the smallest district in Al-Qalubiya Governorate, it incorporates an area of 28.6 km^2, extending between longitudes 31° 04′ to 31° 14′ E and

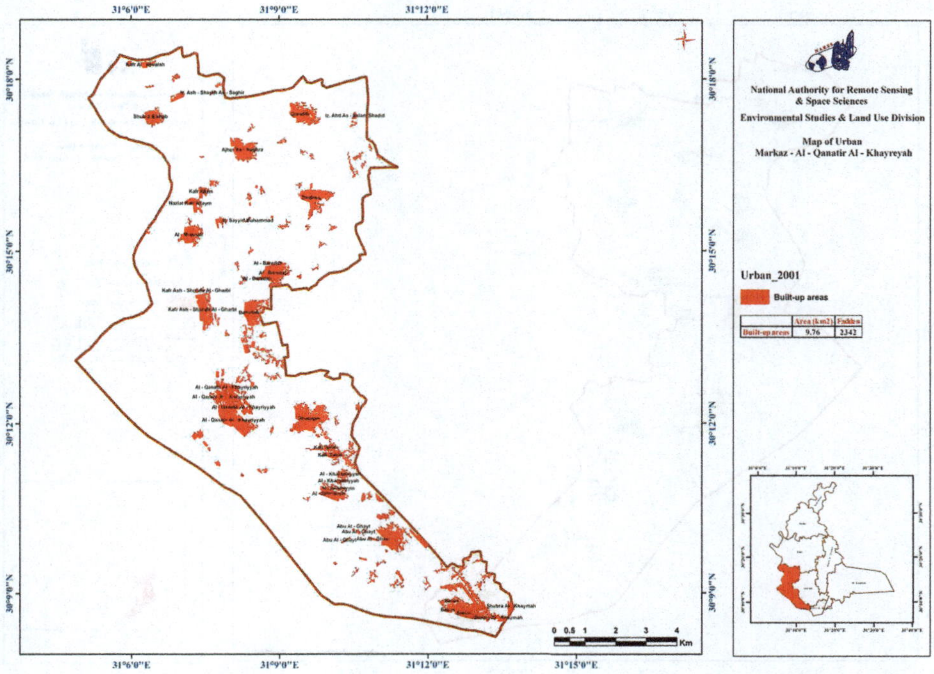

Fig. 4.497 Urban areas of Al-Qanatir Al-Khayreyah district (2001)

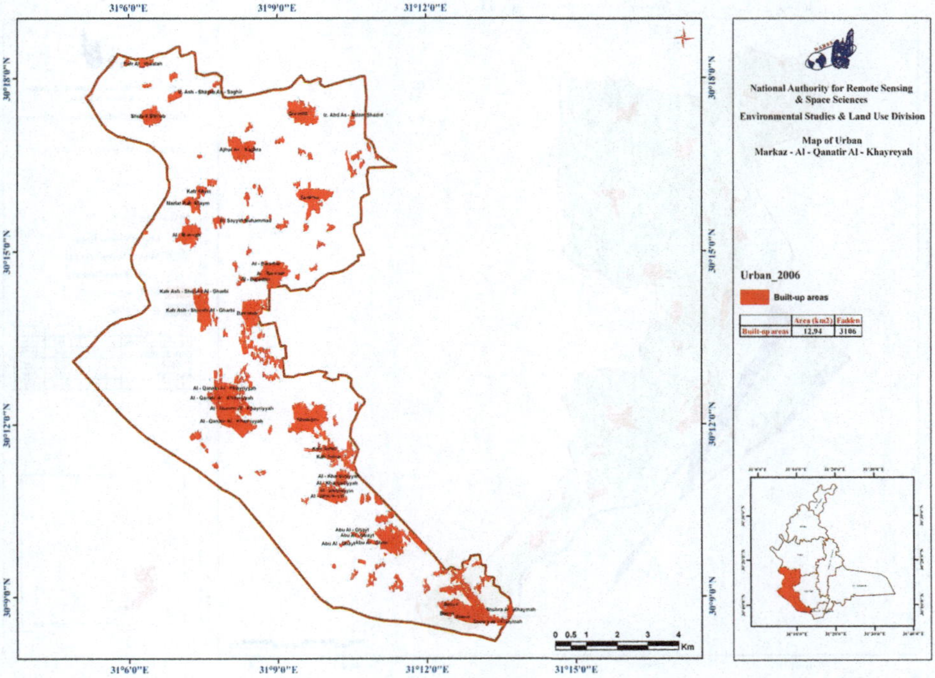

Fig. 4.498 Urban areas of Al-Qanatir Al-Khayreyah district (2006)

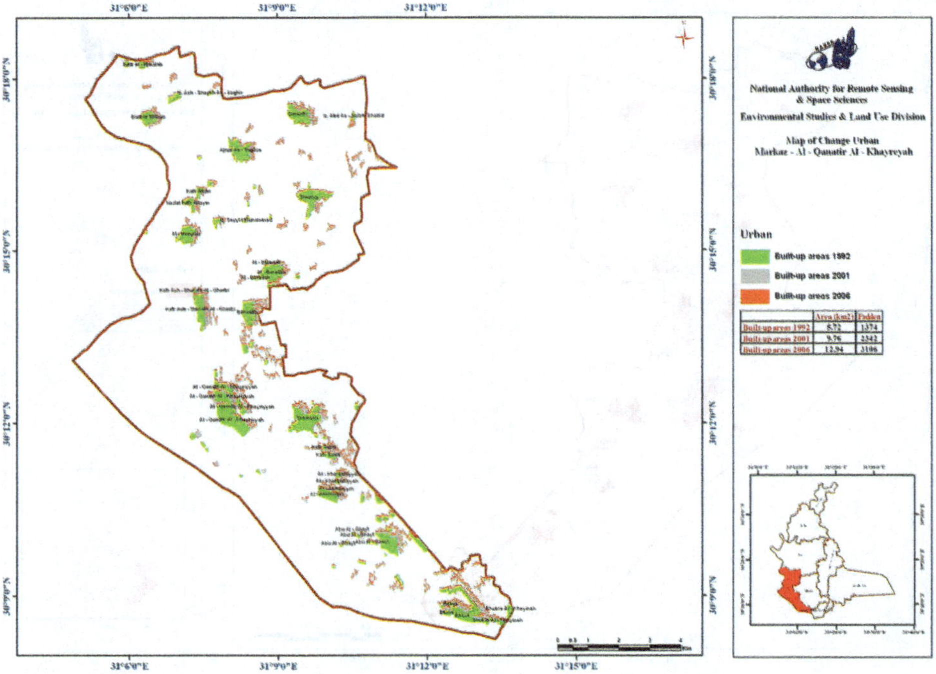

Fig. 4.499 Urban areas of Al-Qanatir Al-Khayreyahdistrict (1992, 2001, 2006)

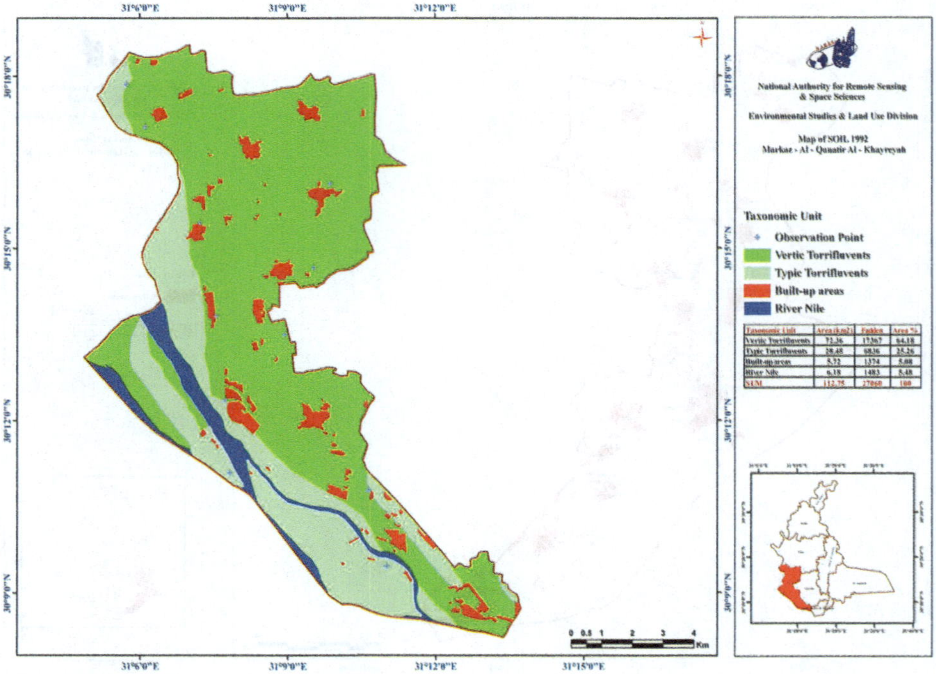

Fig. 4.500 Urban areas draped over the soil layer of Al-Qanatir Al-Khayreyah district (1992)

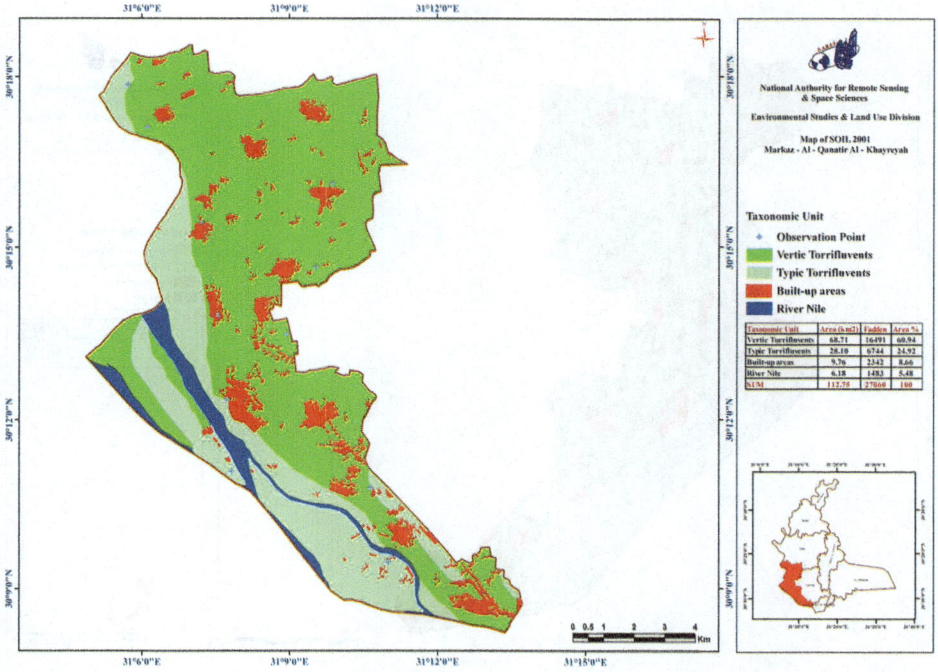

Fig. 4.501 Urban areas draped over the soil layer of Al-Qanatir Al-Khayreyah district (2001)

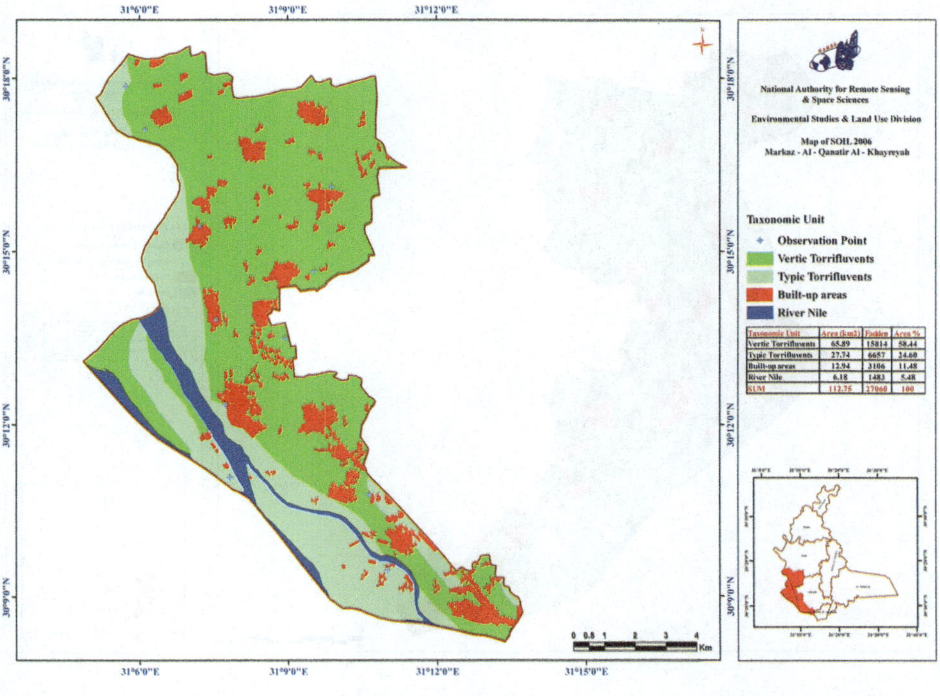

Fig. 4.502 Urban areas draped over the soil layer of Al-Qanatir Al-Khayreyah district (2006)

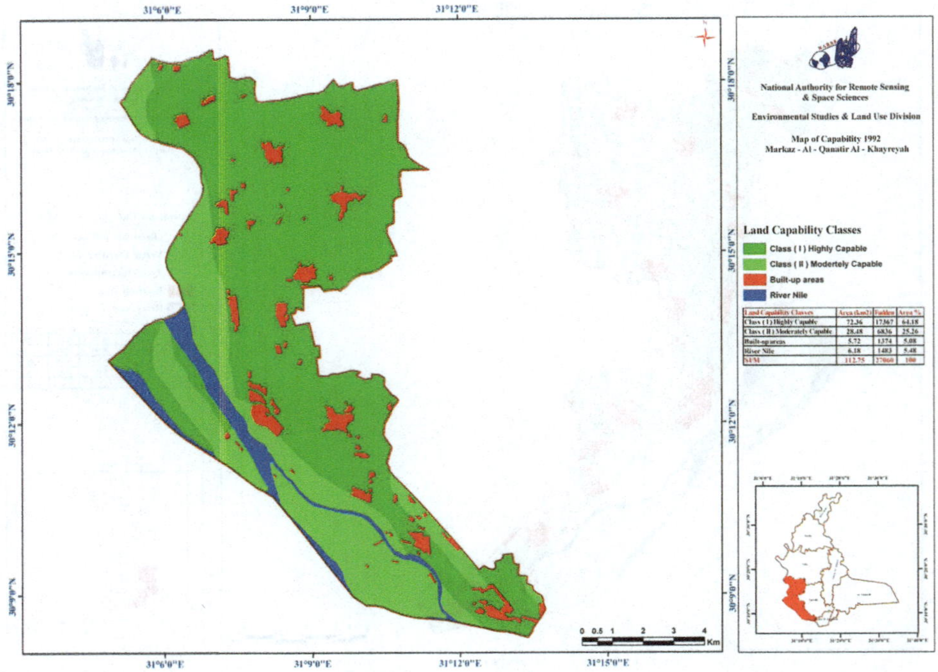

Fig. 4.503 Urban areas draped over the soil capability of Al-Qanatir Al-Khayreyah district (1992)

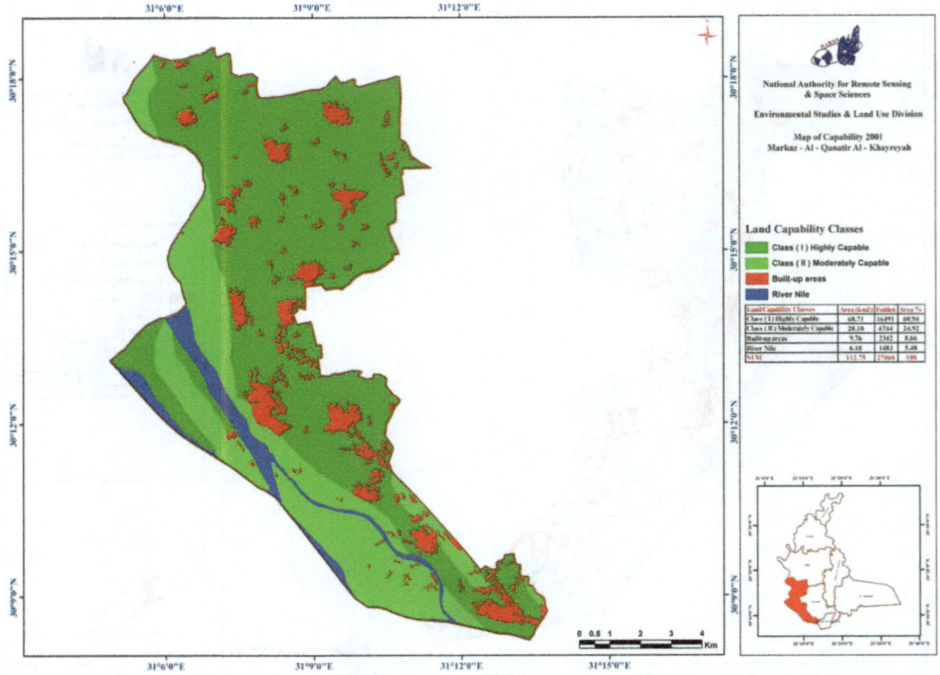

Fig. 4.504 Urban areas draped over the soil capability of Al-Qanatir Al-Khayreyah district (2001)

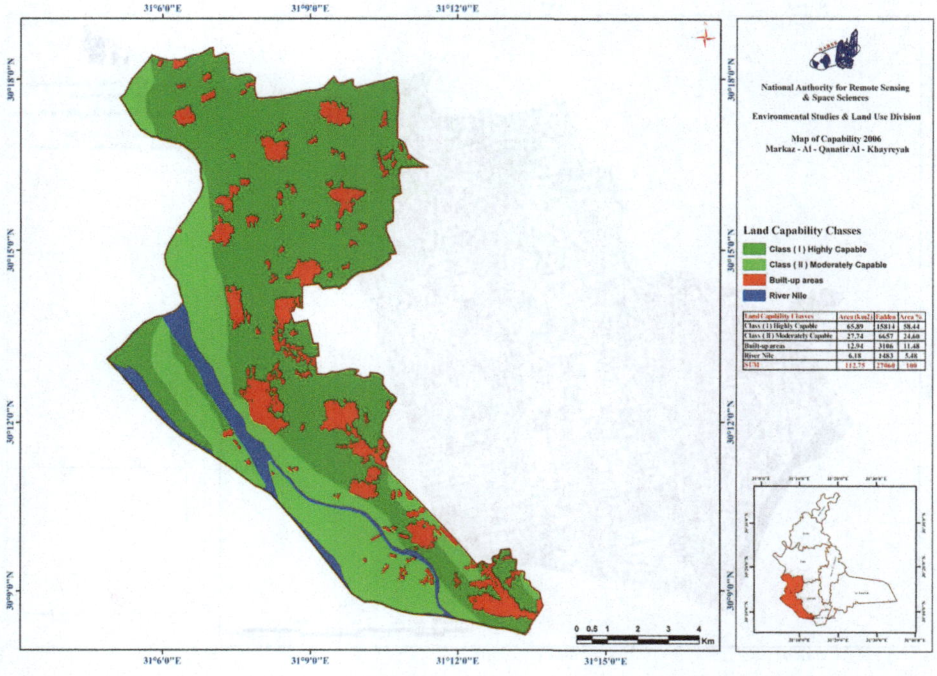

Fig. 4.505 Urban areas draped over the soil capability of Al-Qanatir Al-Khayreyah district (2006)

latitudes 30° 08′ to 30° 19′ N. The land resources database of the district includes the Landsat ETM+ and SPOT images taken during the years 2001 and 2006 respectively (Figs. 4.506 and 4.507), which were used to update the thematic layers. The digital elevation model of the district indicates that the surface elevation ranges from 0 to 34 meters above sea level.

Regarding the soil layer (Fig. 4.508) of the district, it is found that the *Vertic Torrifluvents* soil sub great group dominates the soil units, covering an area of 5.9 km^2, representing 21.0% of the district soil area. The *Typic Torrifluvents* covers 0.7 km^2, representing 2.0% of the district total area.

The irrigation and drainage canals thematic layer of the district (Fig. 4.509) reveals that the total length of irrigation and drainage networks are 16.7 km and 14.6 km respectively. This mostly indicates a balance between the drainage and irrigation networks in the district.

The roads network thematic layer (Fig. 4.510) indicates that the district have a satisfactory network where the total length of the roads is 35.0 km. The roads density of the district is 1 km road length for each 1 km^2 area.

The urban areas expansion was monitored, during the 1992–2006 period, and the thematic layers were attached to the database Figs. 4.511, 4.512 and 4.513. The change detection of urban areas (Fig. 4.514) reveals that an increase from 14.9 km^2 (year 1992) to 19.85 km^2 (year 2001) and to 22.0 km^2 (year 2006) with an average annual rate of 0.5 km^2.

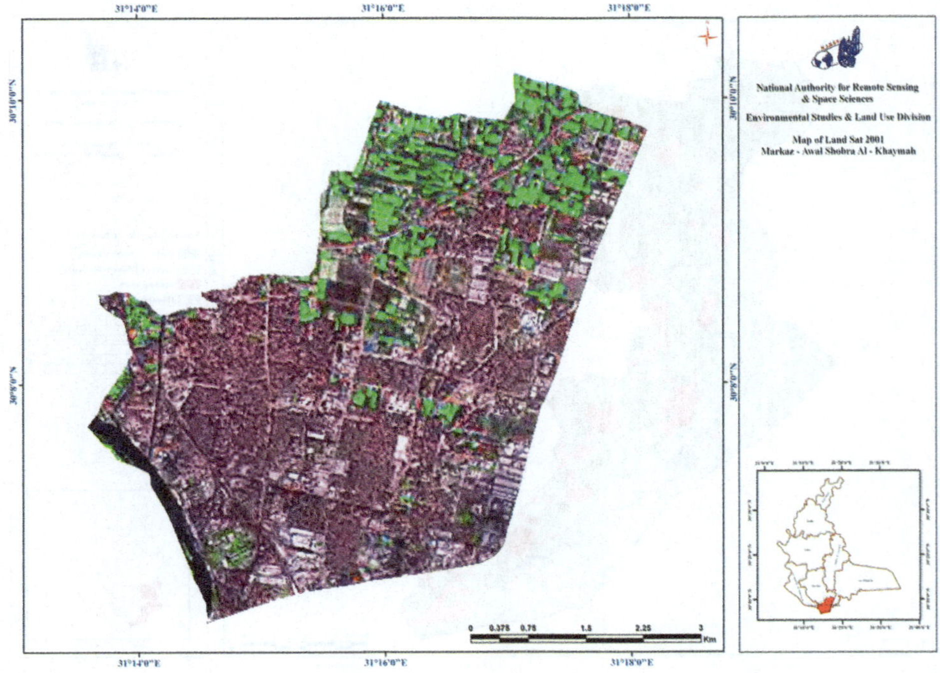

Fig. 4.506 Landsat ETM+ image of Awal Shobra Al-Khaymah district (2001)

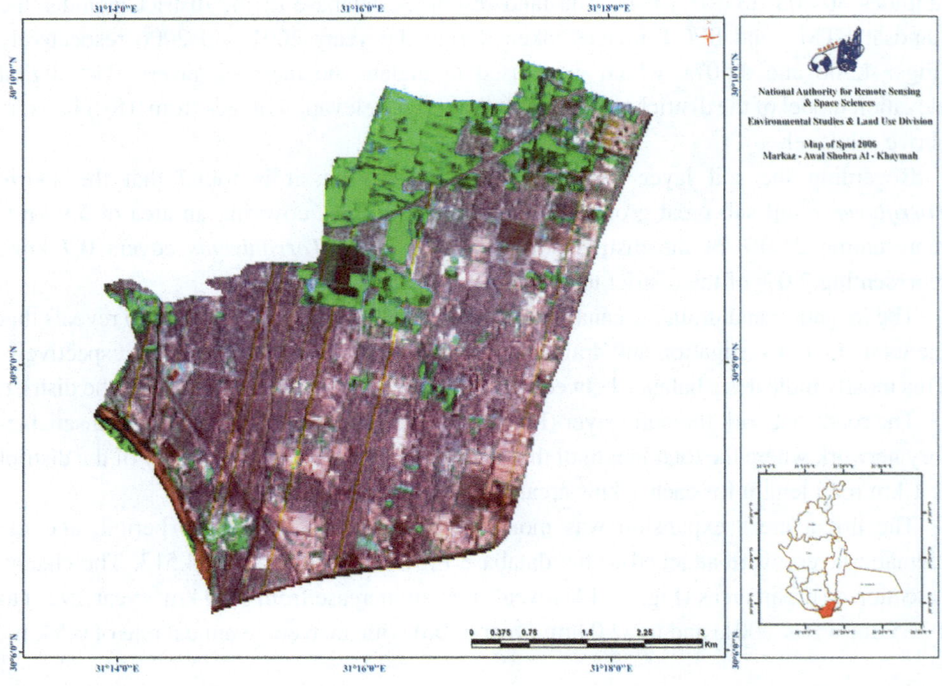

Fig. 4.507 Spot image of Awal Shobra Al-Khaymah district (2006)

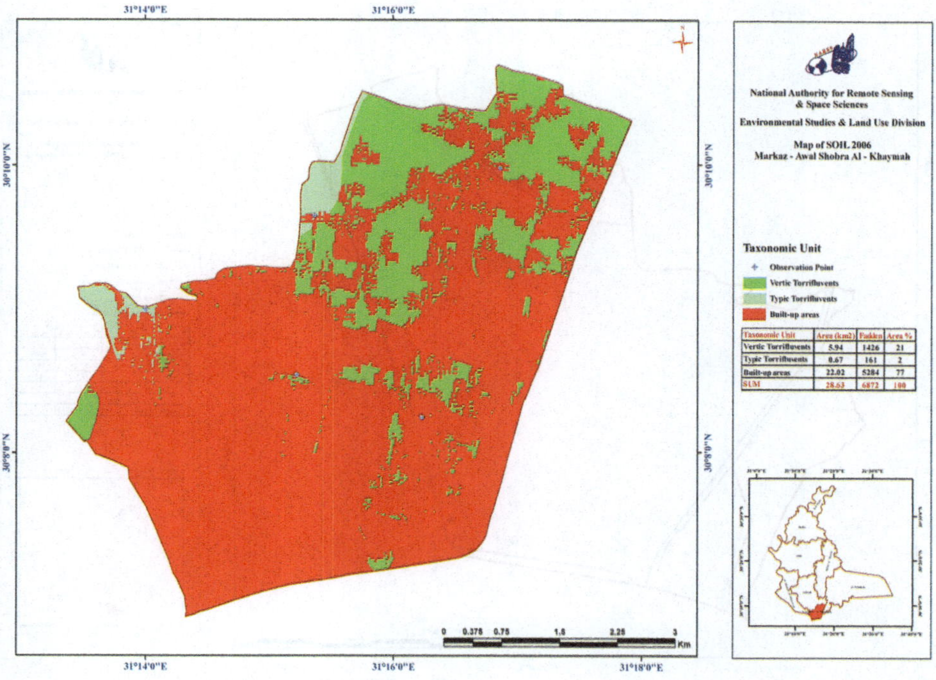

Fig. 4.508 Soil layer of Awal Shobra Al-Khaymah district

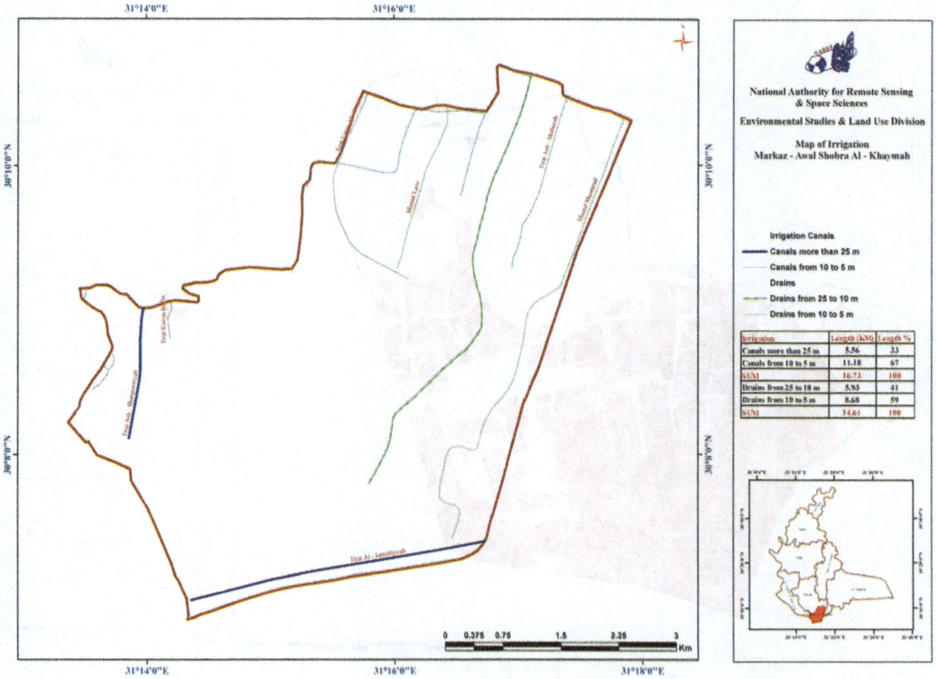

Fig. 4.509 Irrigation and drainage canals network of Awal Shobra Al-Khaymah district

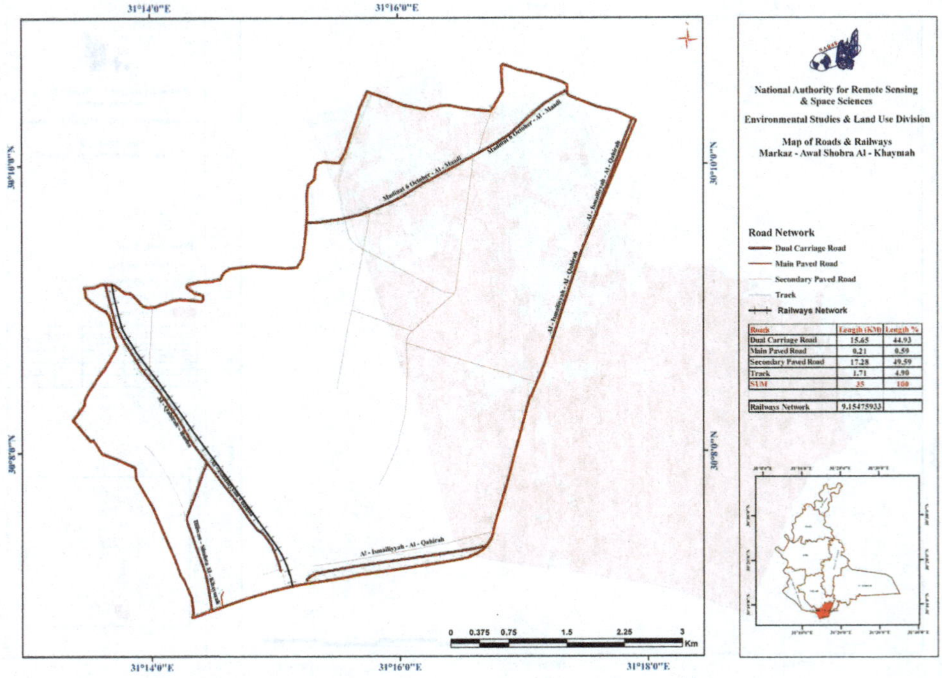

Fig. 4.510 Roads network of Awal Shobra Al-Khaymah district

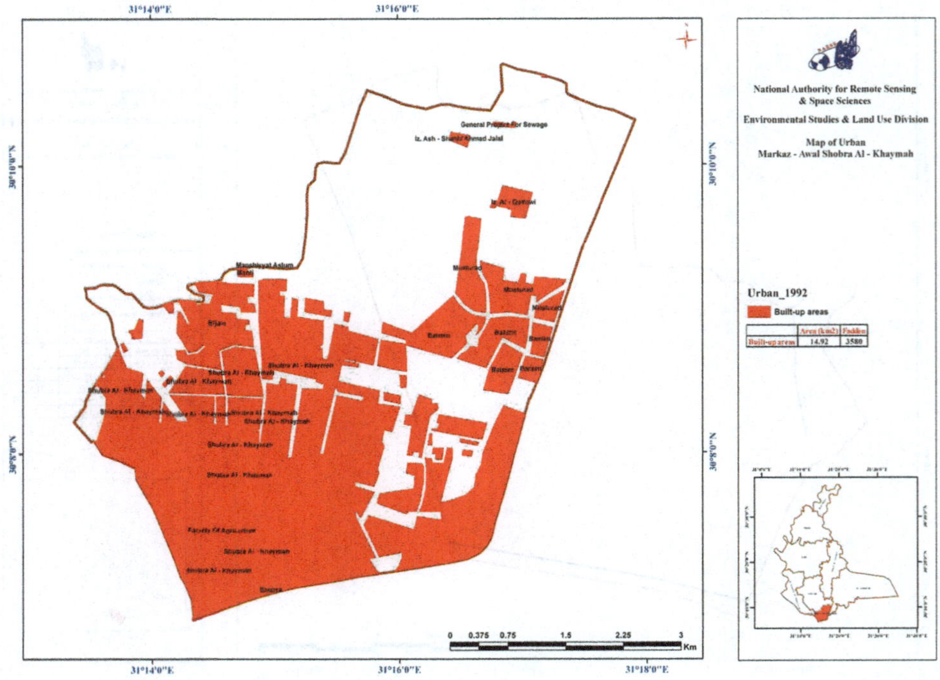

Fig. 4.511 Urban areas of Awal Shobra Al-Khaymah district (1992)

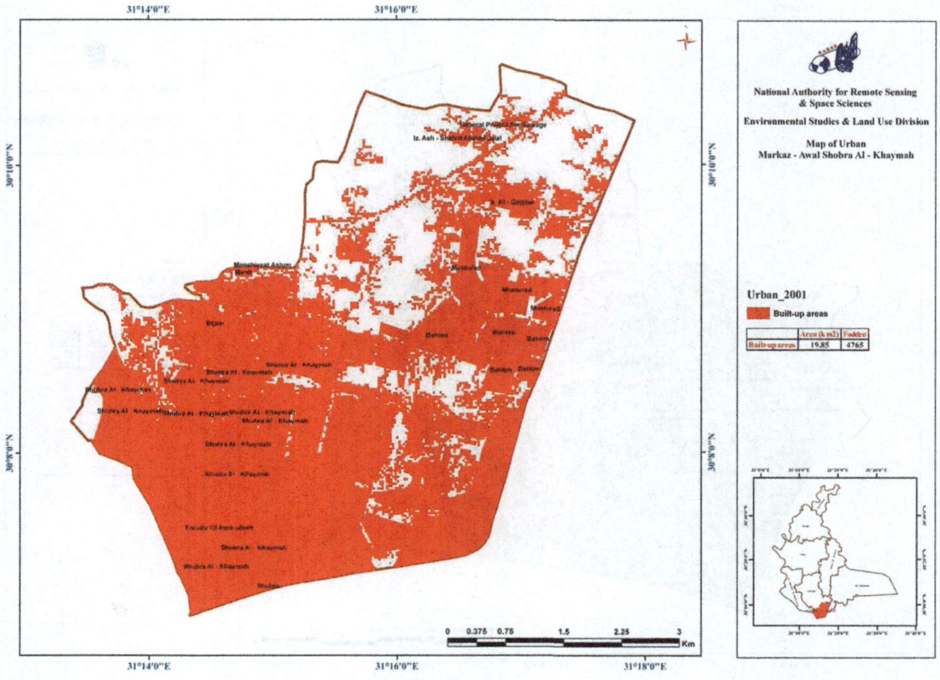

Fig. 4.512 Urban areas of Awal Shobra Al-Khaymah district (2001)

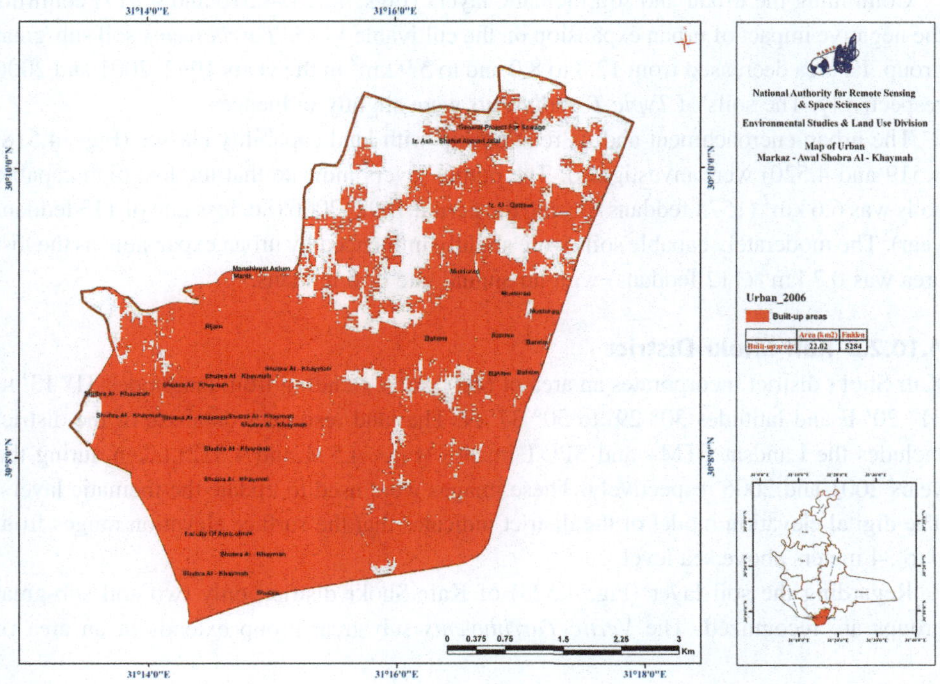

Fig. 4.513 Urban areas of Awal Shobra Al-Khaymah district (2006)

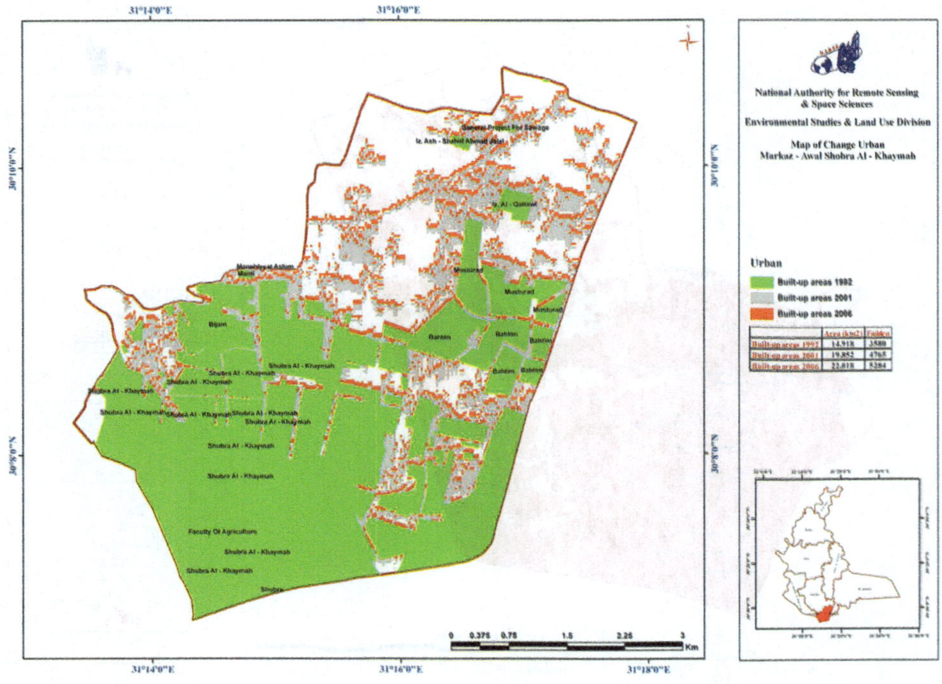

Fig. 4.514 Urban areas of Awal Shobra Al-Khaymah district (1992, 2001, 2006)

Combining the urban and soil thematic layers (Figs. 4.515, 4.516 and 4.517) confirms the negative impact of urban expansion on the cultivable *Vertic Torrifluvents* soil sub-great group. Its area decreased from 12.8 to 8.0 and to 5.9 km^2 in the years 1992, 2001 and 2006 respectively. The soils of *Typic Torrifluvents* were slightly influenced.

The urban encroachment and its relationship with land capability classes (Figs. 4.518, 4.519 and 4.520) were investigated. The output layers indicate that the lost high capable soils was 6.6 km^2 (1578 feddans) during the period 1992–2006 (i.e. loss rate of 113 feddan/year). The moderately capable soils were slightly influenced by urban expansion, as the lost area was 0.7 km^2 (112 feddans) with an annual rate of 8 feddans.

4.10.2.5 Kafr Shokr District

Kafr Shokr district incorporates an area of 60.6 km^2, extending from longitudes 31° 13′ to 31° 20′ E and latitudes 30° 29′ to 30° 37′ N. The land resources database of the district includes the Landsat ETM+ and SPOT images (Figs. 4.521 and 4.522) taken during the years 2001 and 2006 respectively. These images were used to update the thematic layers. The digital elevation model of the district indicates that the surface elevation ranges from 0 to 34 meters above sea level.

Regarding the soil layer (Fig. 4.523) of Kafr Shokr district, only two soil sub-great groups are recognized. The *Vertic Torrifluvents* sub great group extends in an area of

Fig. 4.515 Urban areas draped over the soil layer Awal Shobra Al-Khaymah district (1992)

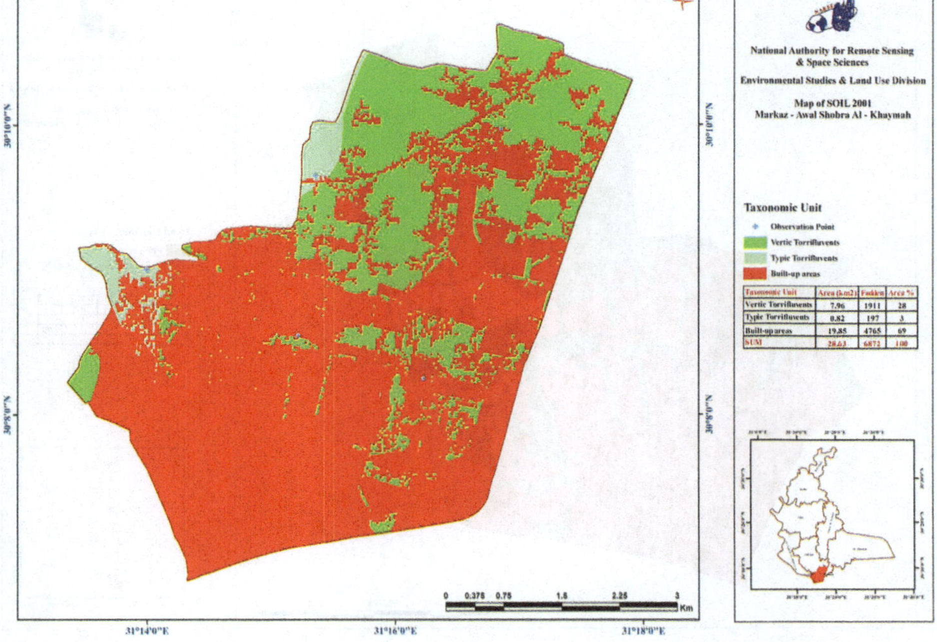

Fig. 4.516 Urban areas draped over the soil layer of Awal Shobra Al-Khaymah district (2001)

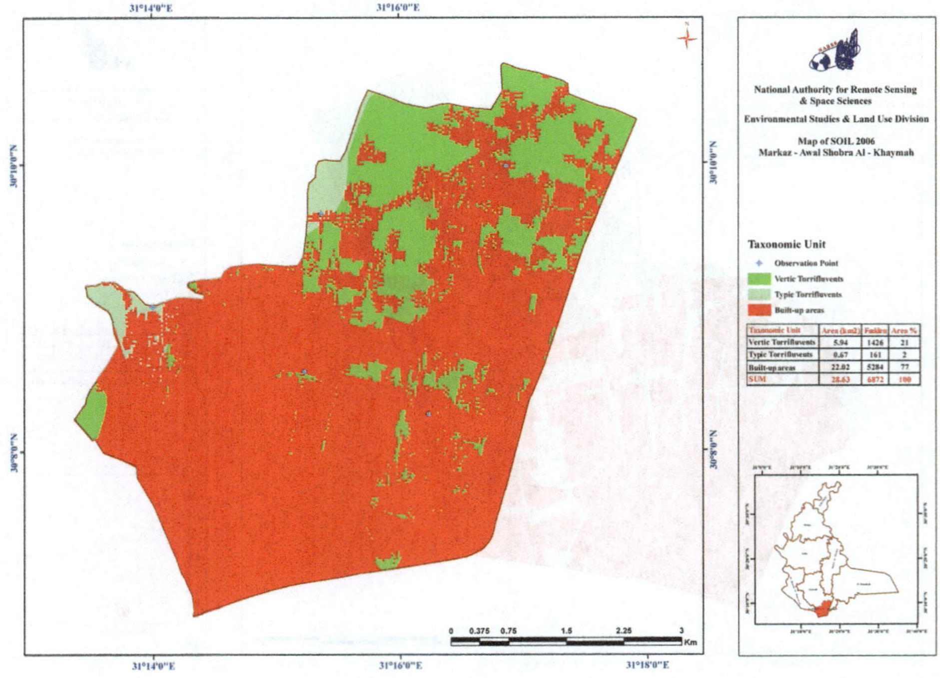

Fig. 4.517 Urban areas draped over the soil layer of Awal Shobra Al-Khaymah district (2006)

Fig. 4.518 Urban areas draped over the soil capability of Awal Shobra Al-Khaymah district (1992)

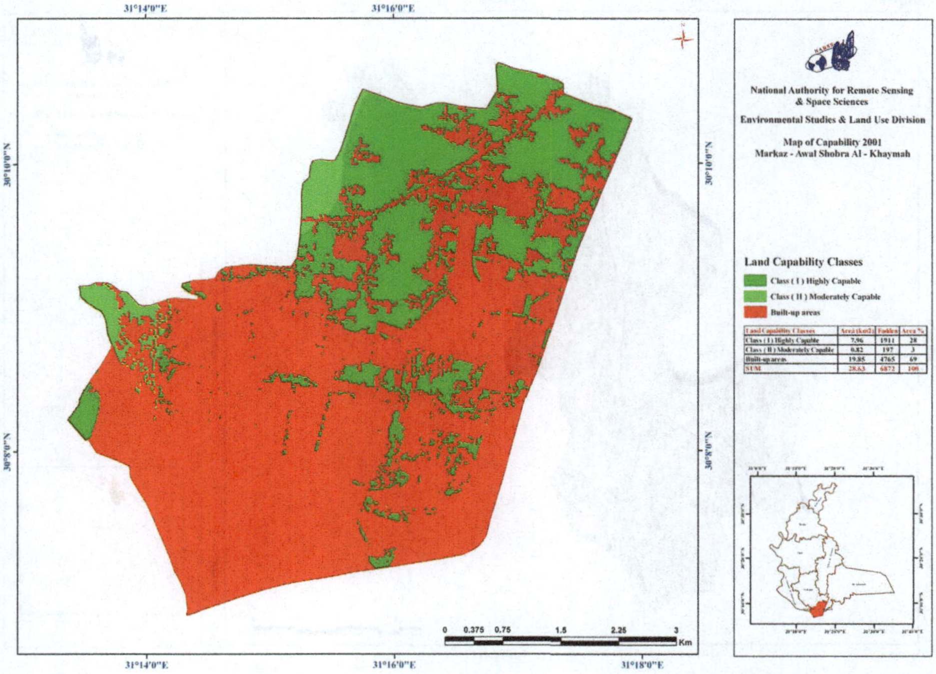

Fig. 4.519 Urban areas draped over the soil capability of Awal Shobra Al-Khaymah district (2001)

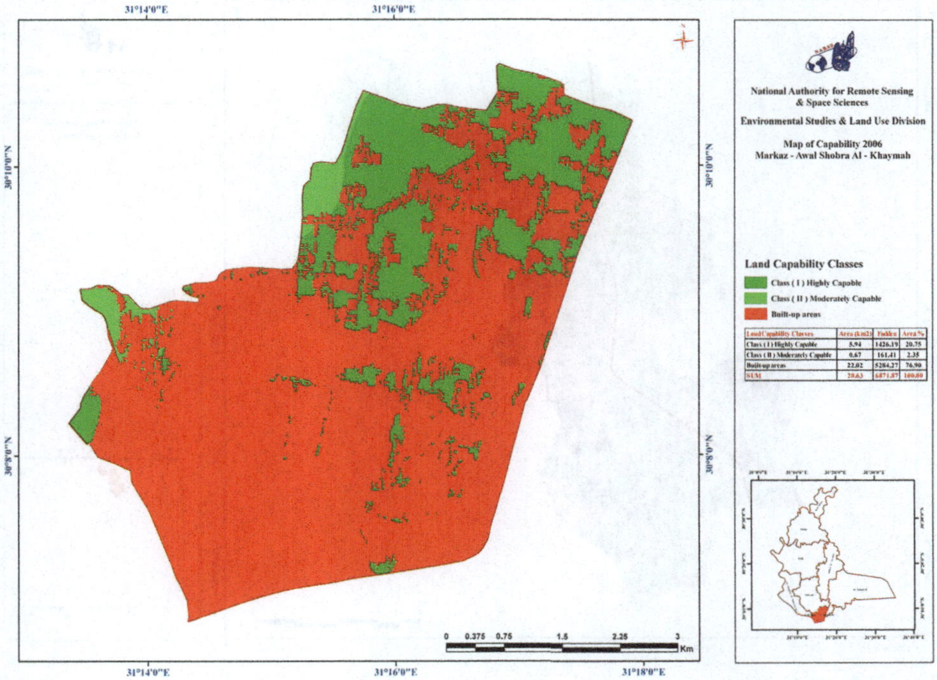

Fig. 4.520 Urban areas draped over the soil capability of Awal Shobra Al-Khaymah district (2006)

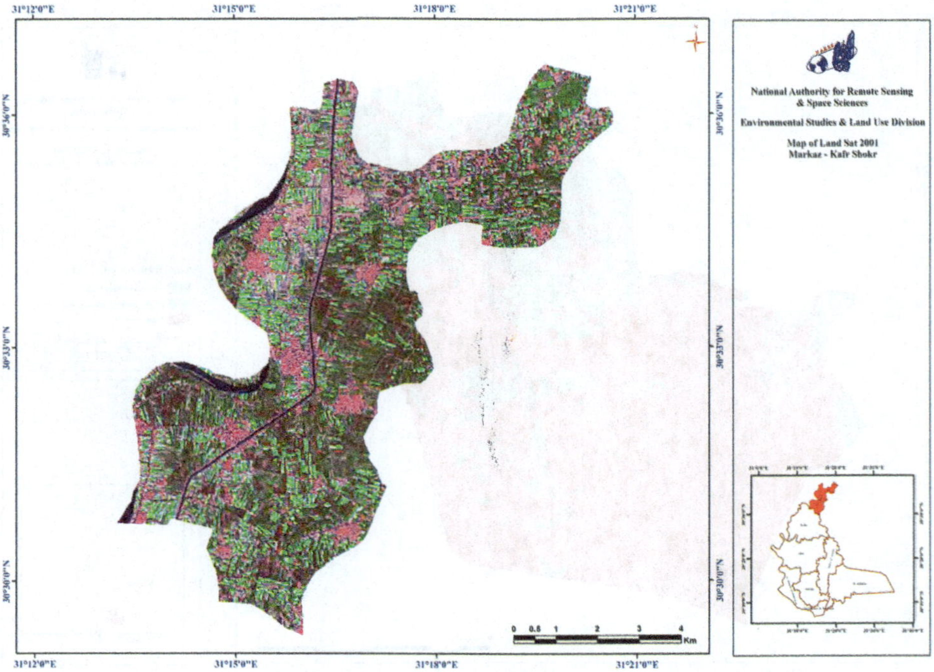

Fig. 4.521 Landsat ETM+ image of Kafr Shokr district (2001)

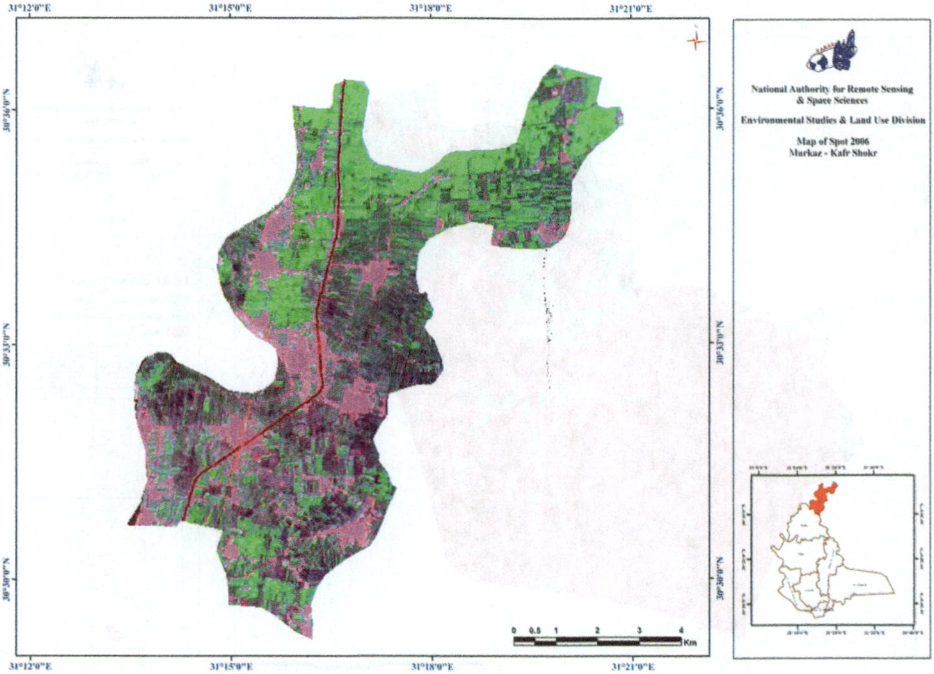

Fig. 4.522 SPOT image of Kafr Shokr district (2006)

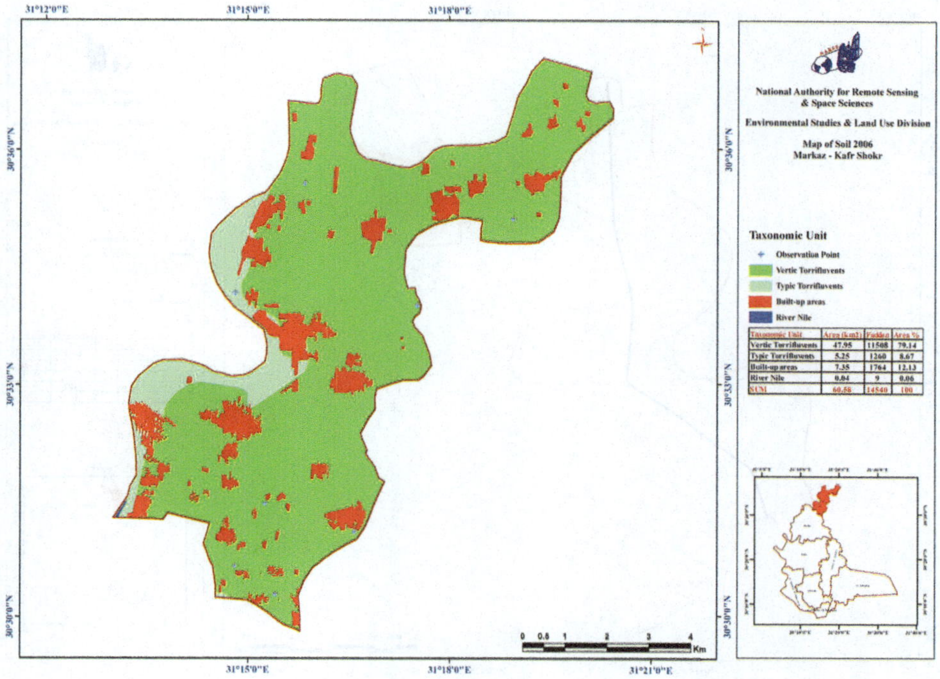

Fig. 4.523 Soil layer of Kafr Shokr district

48.0 km^2 representing 79.1% of the total district area. The *Typic Torrifluvents* also occupy an area of 5.3 km^2, representing 8.7% of the total district area.

The irrigation and drainage canals thematic layer of the district (Fig. 4.524) reveals that the total length of irrigation canals is 61.9 km. The total length of drainage network is 18.3 km, which may indicate a wide gap between the drainage and irrigation network lengths.

The roads network thematic layer (Fig. 4.525) indicates that the district have a good and satisfactory network. A total length of roads network is found to be 106.0 km, which refers to a road length density of 1.8 km for 1 km^2.

The urban multi-temporal expansion progress was monitored during the 1992–2006 period and the thematic layers (Figs. 4.526, 4.527 and 4.528) were attached to the database. Detecting the changes (Fig. 4.529) reveals that the urban area increased from 4.8 km^2 (year 1992) to 6.0 km^2 (year 2001) and to 7.4 km^2 (year 2006) with very small average annual rate of 0.18 km^2 per year (43.5 feddan/year).

Attaching the soil and land capability layers (Figs. 4.530, 4.531 and 4.532), in one hand, and the multi-temporal urban areas, in another hand, was performed. The area of the cultivable soils of *Vertic Torrifluvents* sub-great group is influenced by the urban expansion; it decreased from 50.1 to 49.1 to 48.0 km^2 in the years 1992, 2001 and 2006 respectively. The soils of *Typic Torrifluvents* were slightly influenced, since the soils

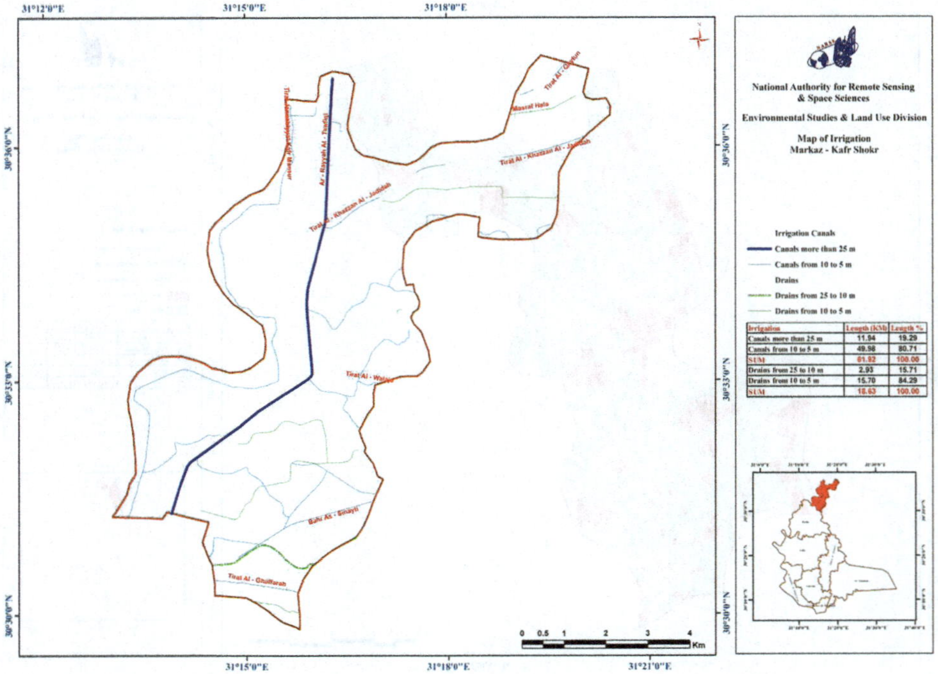

Fig. 4.524 Irrigation and drainage canals network of Kafr Shokr district

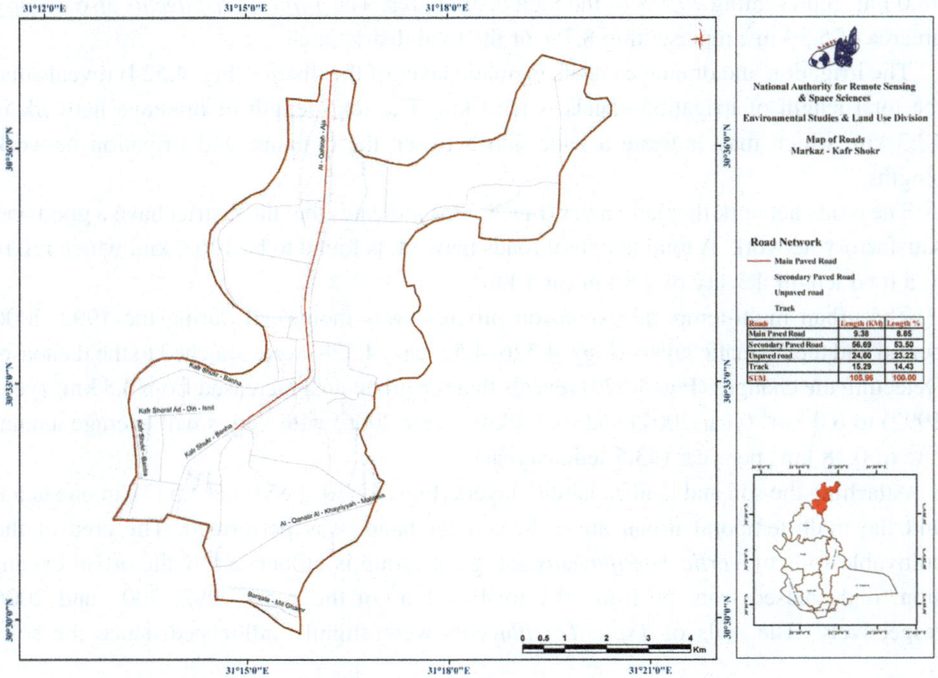

Fig. 4.525 Roads network of Kafr Shokr district

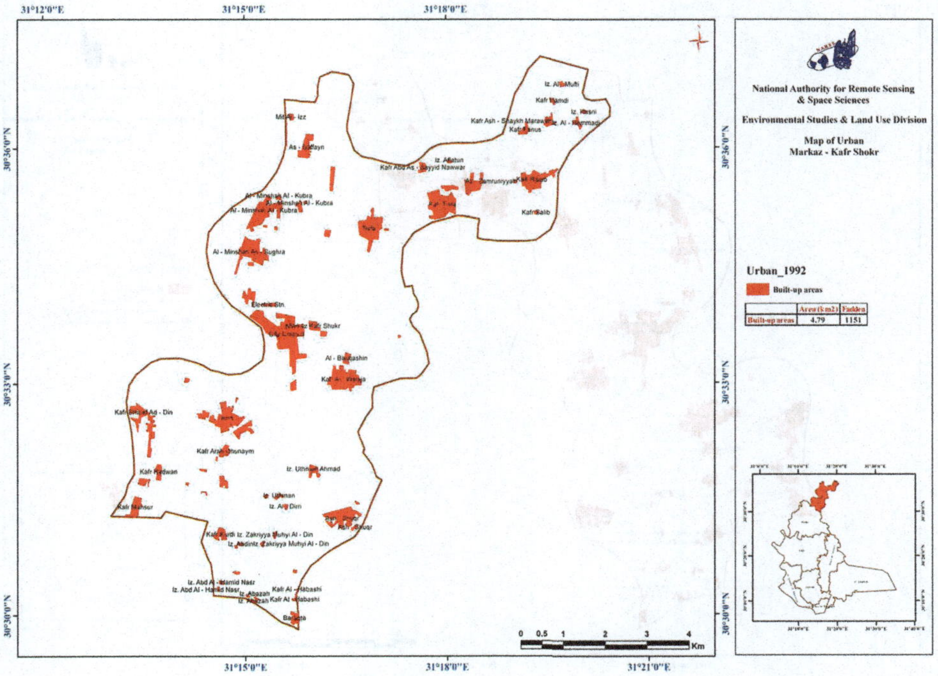

Fig. 4.526 Urban areas of Kafr Shokr district (1992)

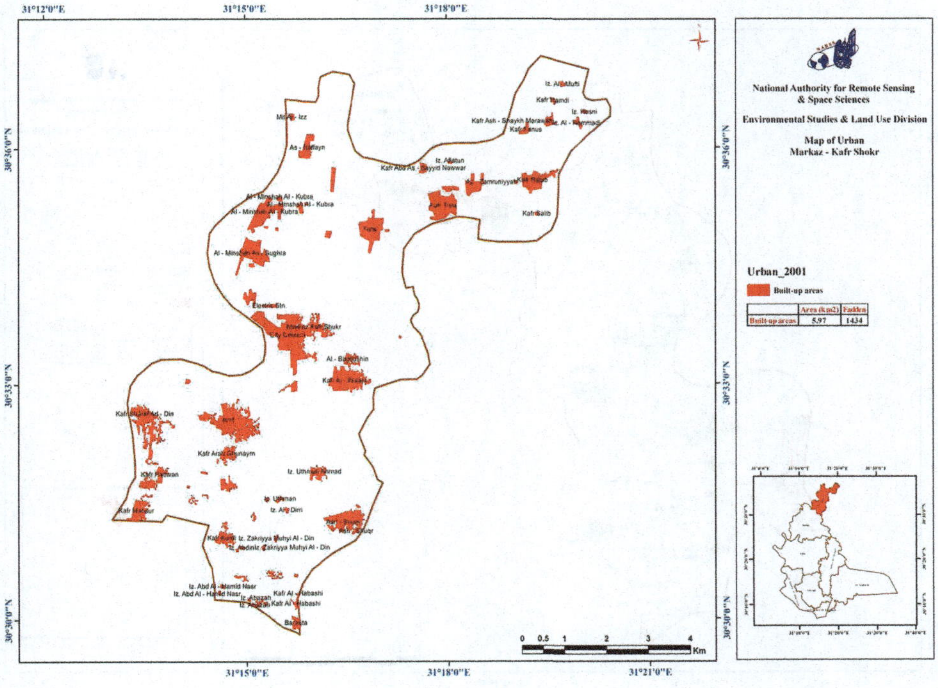

Fig. 4.527 Urban areas of Kafr Shokr district (2001)

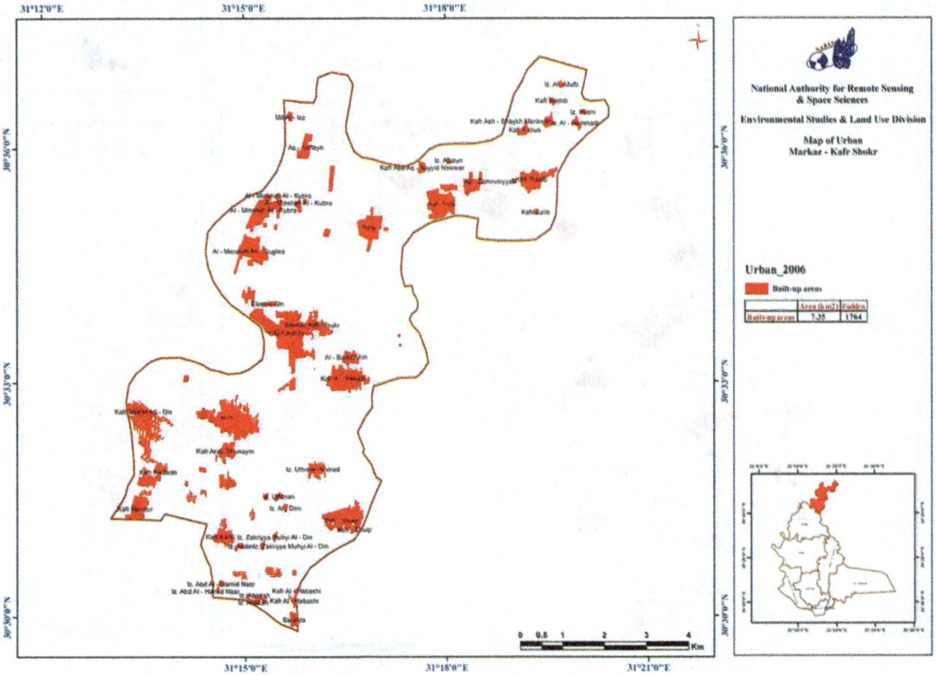

Fig. 4.528 Urban areas of Kafr Shokr district (2006)

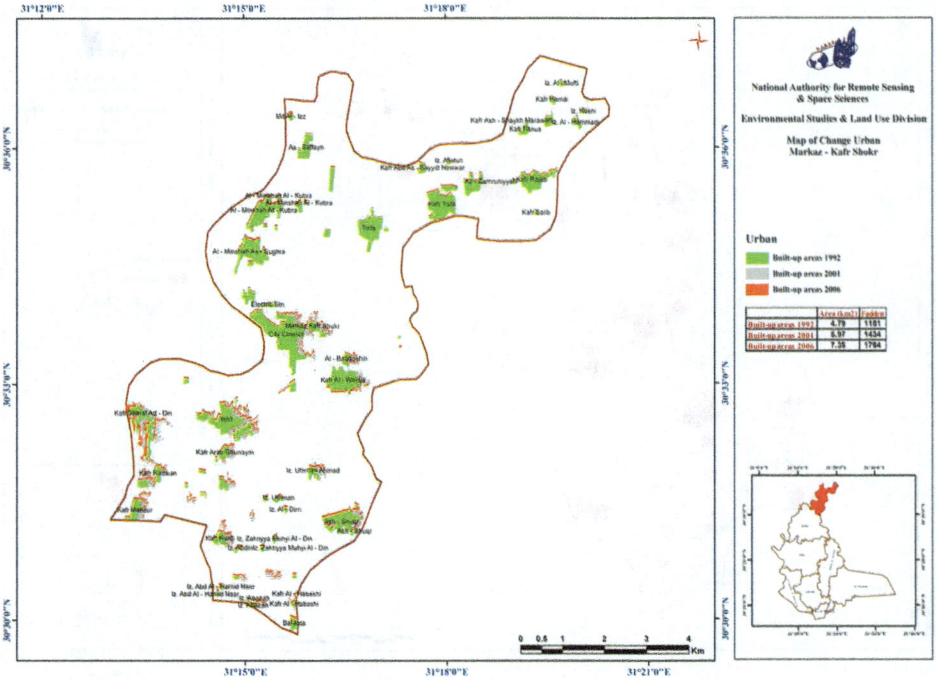

Fig. 4.529 Urban areas of Kafr Shokrdistrict (1992, 2001, 2006)

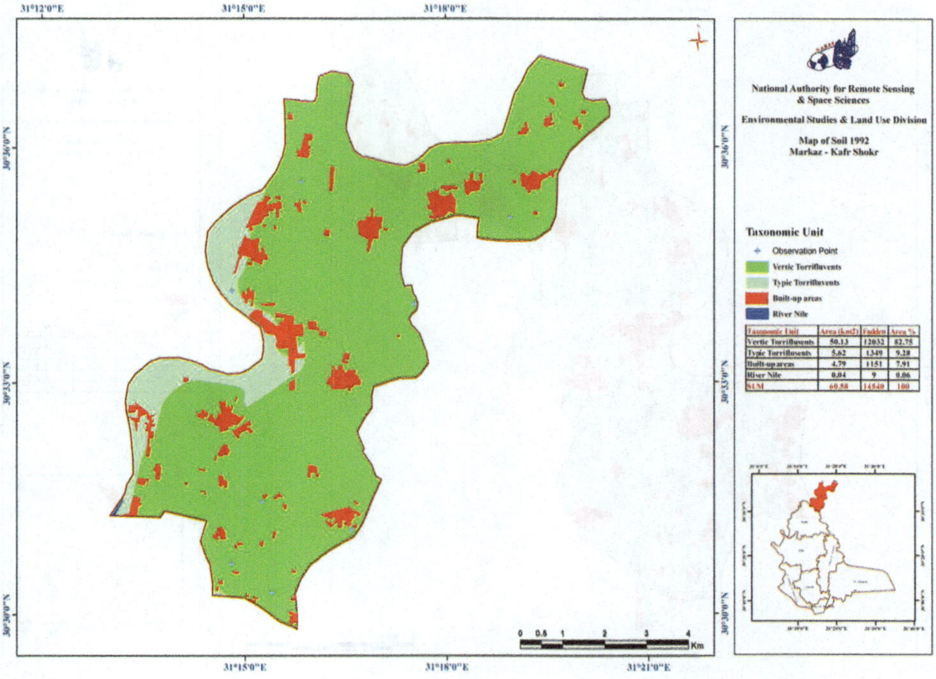

Fig. 4.530 Urban areas draped over the soil layer of Kafr Shokr district (1992)

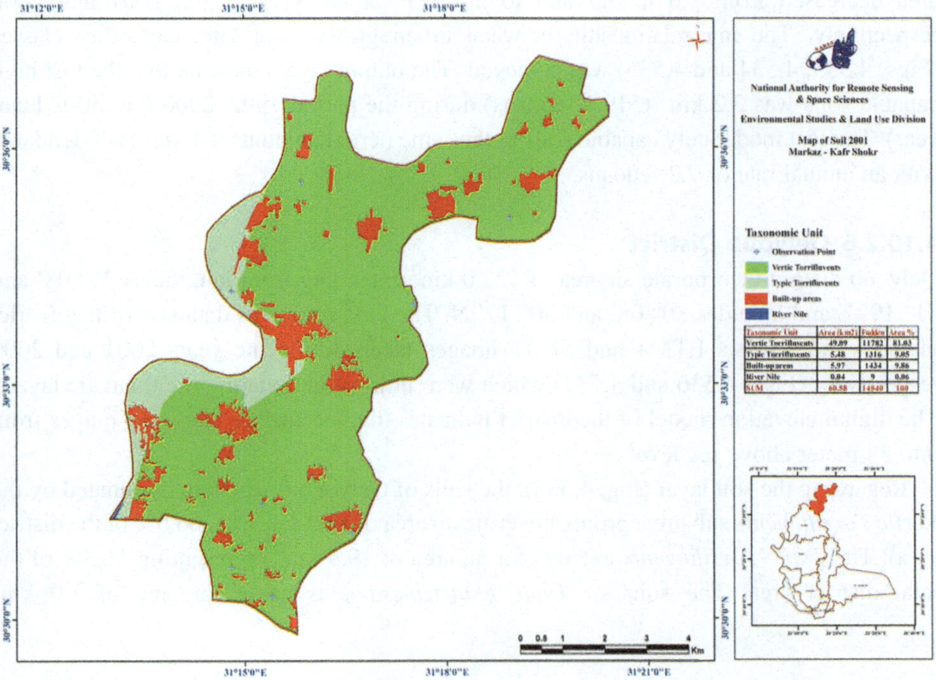

Fig. 4.531 Urban areas draped over the soil layer of Kafr Shokr district (2001)

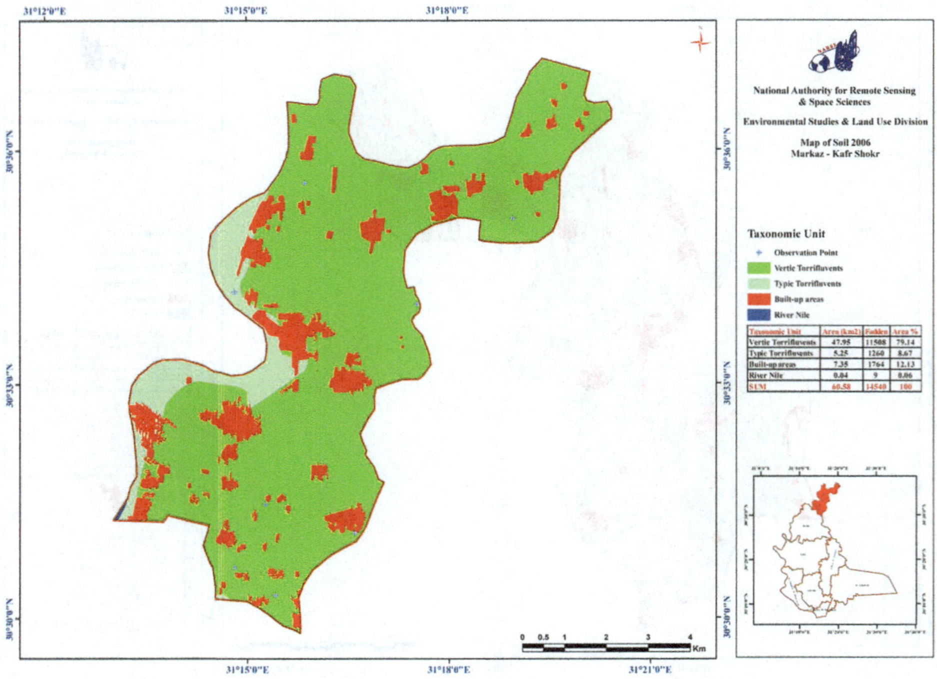

Fig. 4.532 Urban areas draped over the soil layer of Kafr Shokr district (2006)

area decreased from 5.6 to 5.5 and to 5.3 km^2 in the years 1992, 2001 and 2006 respectively. The interrelationship between urban growth and land capability classes (Figs. 4.533, 4.534 and 4.535) was achieved. The output layers indicate that the lost high capable soils was 2.2 km^2 (519.1 feddans) during the period 1992–2006 (i.e. 30 feddan/year). The lost moderately capable soils in the same period amounts 0.4 km^2 (100 feddans) with an annual rate of 7.2 feddans.

4.10.2.6 Qelyoub District

Qelyoub district incorporate an area of 122.0 km^2 extending from longitudes 31° 08′ and 31° 19′ E and latitudes 30° 08′ and 30° 17′ N. The land resources database of the district includes the Landsat ETM+ and SPOT images taken during the years 2001 and 2006 respectively (Figs. 4.536 and 4.537) which were utilized in updating the thematic layers. The digital elevation model of the district indicates that the surface elevation ranges from 0 to 34 meter above sea level.

Regarding the soil layer (Fig. 4.538), the soils of Qelyoub district are dominated by the *Vertic Torrifluvents* sub great group, covering an area of 81.1 km^2 (i.e. 66.5% of the district area). The *Typic Torrifluvents* extends for an area of 19.9 km^2 representing 16.3% of the total district area. The soils of *Typic Quartzipssaments* cover an area of 1.0 km^2

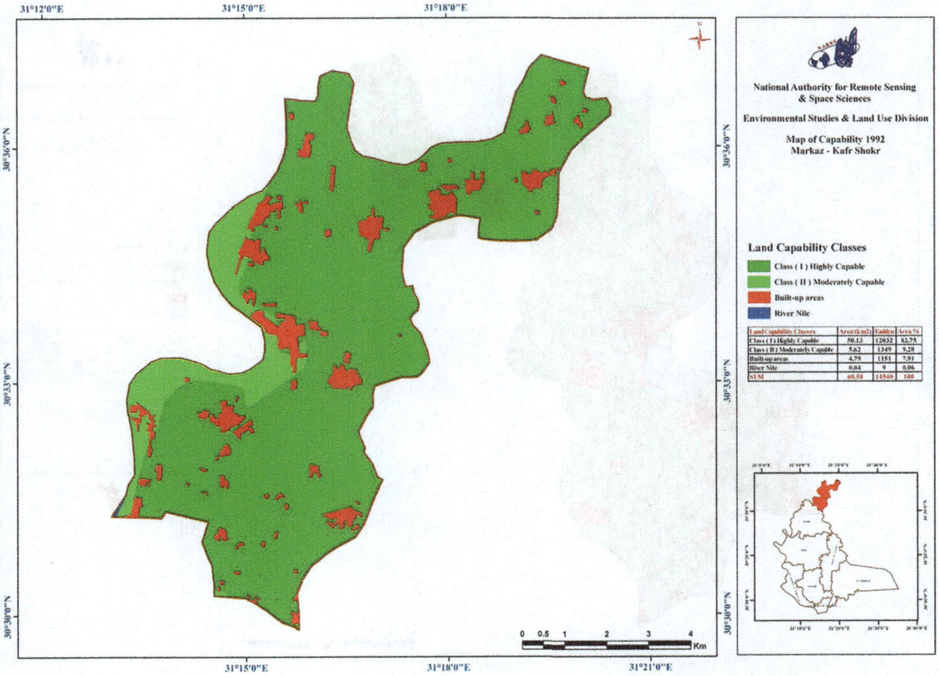

Fig. 4.533 Urban areas draped over the soil capability of Kafr Shokr district (1992)

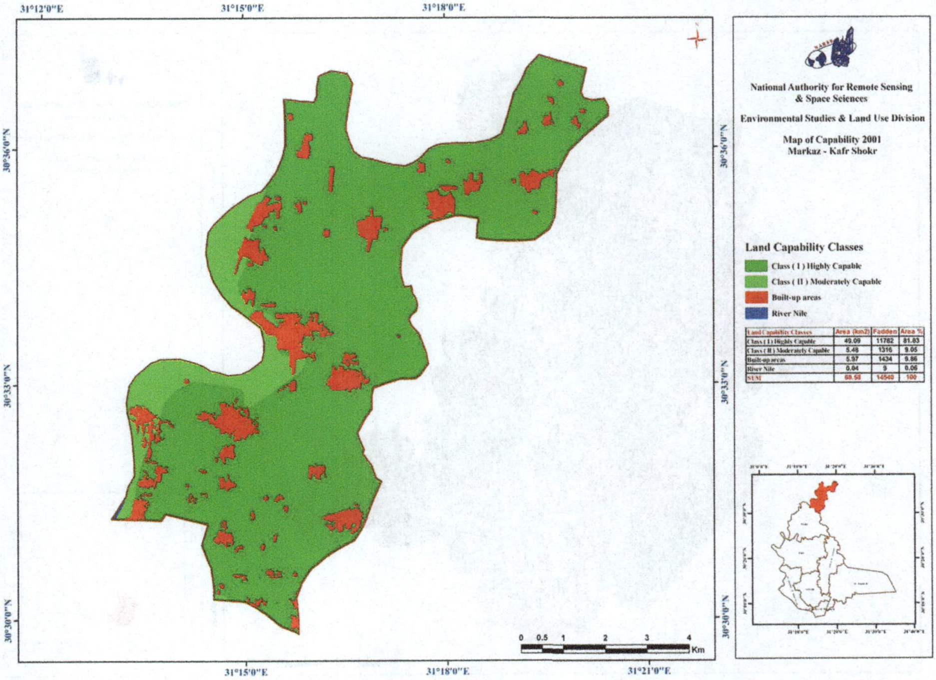

Fig. 4.534 Urban areas draped over the soil capability of Kafr Shokr district (2001)

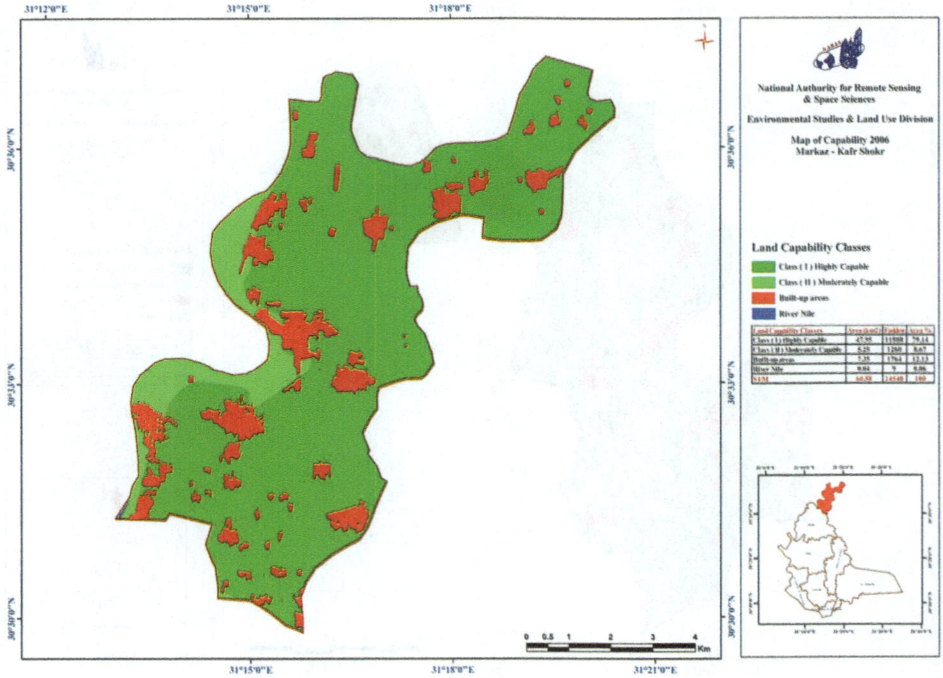

Fig. 4.535 Urban areas draped over the soil capability of Kafr Shokr district (2006)

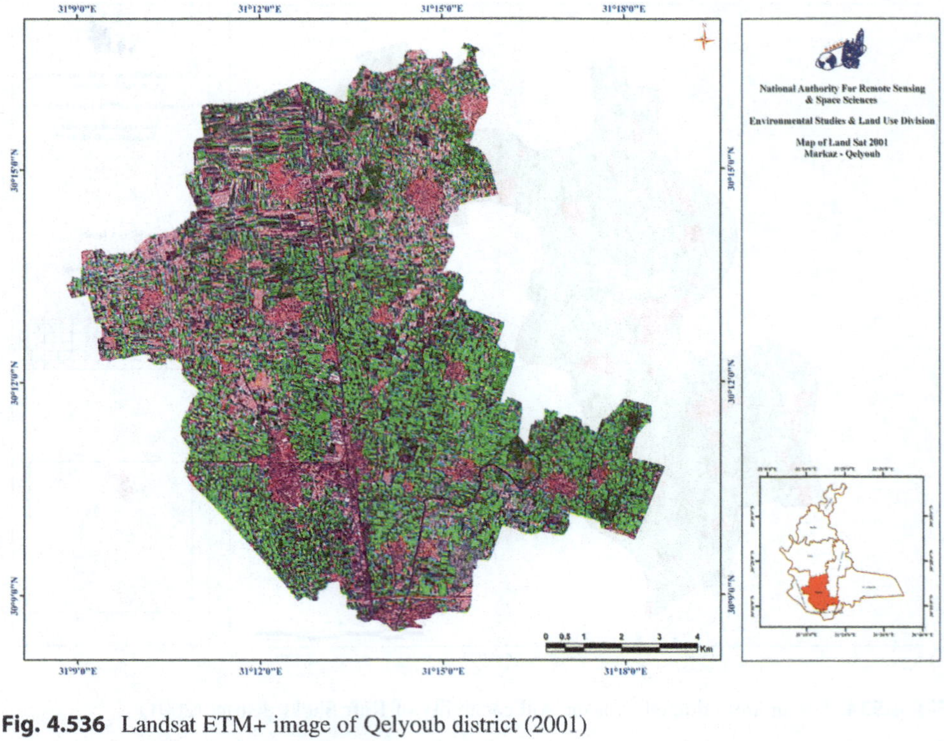

Fig. 4.536 Landsat ETM+ image of Qelyoub district (2001)

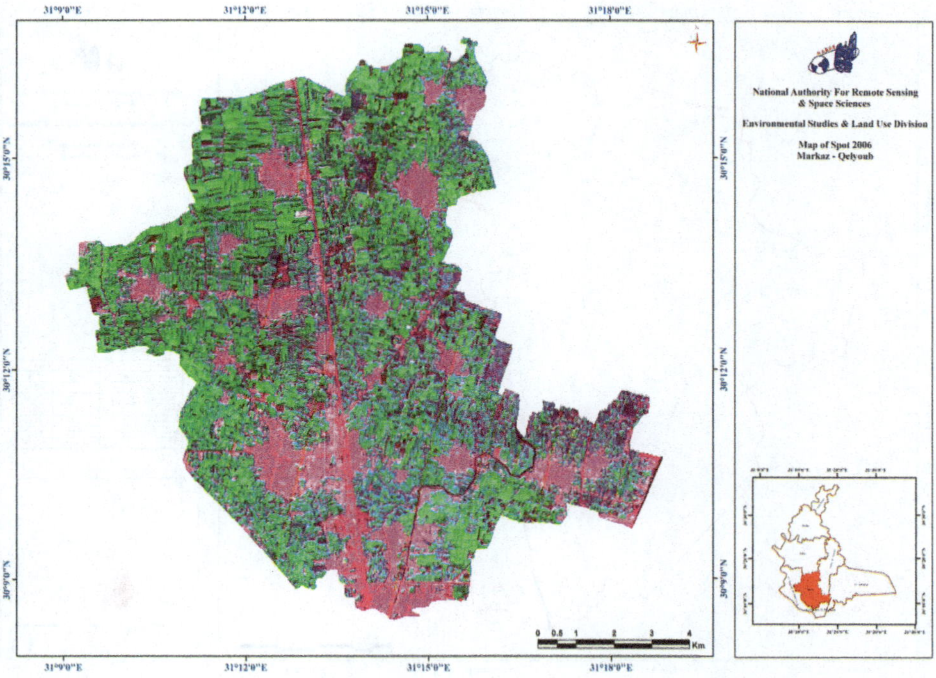

Fig. 4.537 SPOT image of Qelyoub district (2006)

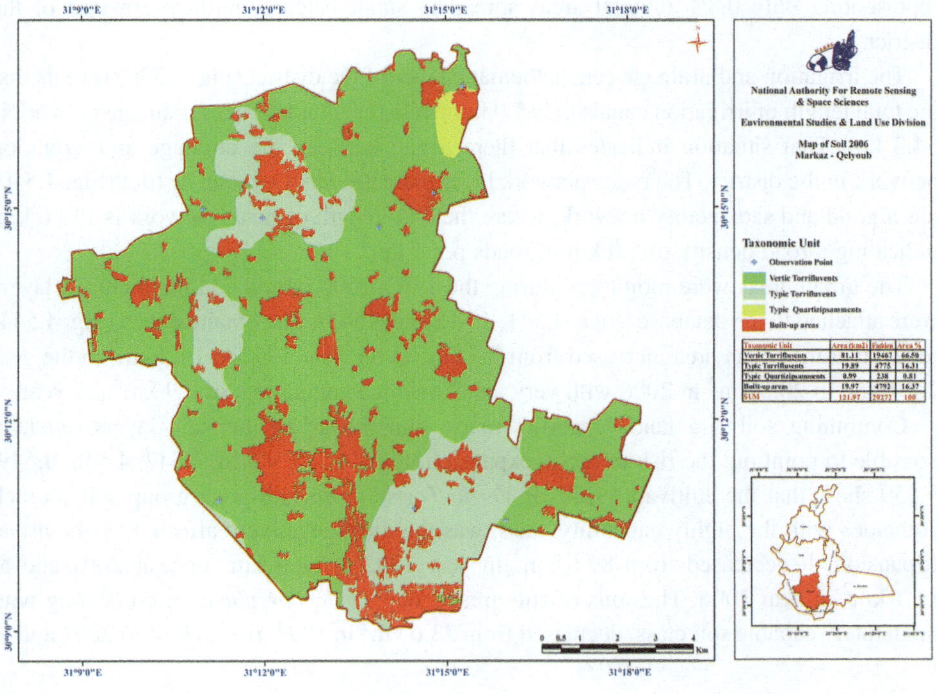

Fig. 4.538 Soil layer of Qelyoub district

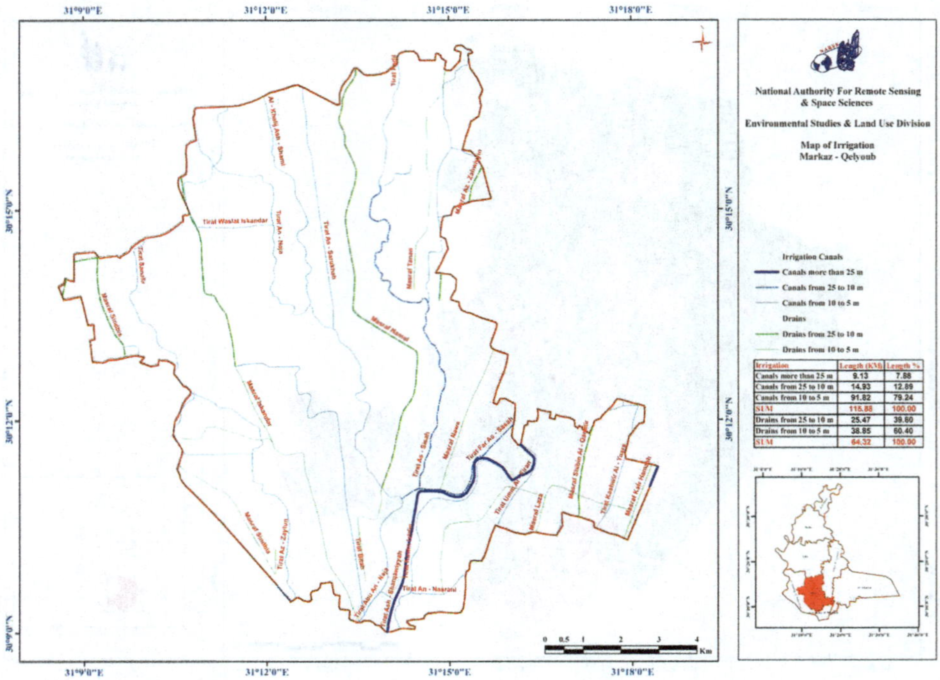

Fig. 4.539 Irrigation and drainage canals network of Qelyoub district

representing only 0.8% of total areas spread in small patches northeastern part of the district.

The irrigation and drainage canals thematic layer of the district (Fig. 4.539) reveals that the total length of irrigation canals is 115.9 km while the total length of drainage network is 64.3 km. That situation indicates that there a gap between the drainage and irrigation network in the district. The roads network layer indicates that Qelyoub district (Fig. 4.540) has a good and satisfactory network, where the total length of roads network is 118.6 km, indicating a road density of 1.0 km of roads per 1 km^2.

The urban areas were monitored during the 1992–2006 period and the thematic layers were attached to the database (Figs. 4.541, 4.542 and 4.543). The obtained data (Fig. 4.544) reveal that the urban area increased from 7.7 km^2 in the year 1992 to 14.5 km^2 in the year 2001 and to 20.0 km^2 in 2006 with very small average annual rate of 0.9 km^2 per year.

Combining soil and land capability maps, and the urban thematic layers, made it possible to point out the risk of urban expansion, Figs. 4.545, 4.546, 4.547, 4.548, 4.549, 4.550 show that the cultivable soils of *Vertic Torrifluvents* sub great group soils, which coincides with the highly capability class, was the most negatively affected by the urban expansion. It decreased from 89.6 km^2 in year 1992 to 84.9 km^2 in year 2001 and to 81.1 km^2 in year 2006. The soils of sub great group *Typic Torrifluvents,* coinciding with moderately capable soil class, decreased from 23.6 km^2 in 1992 to 21.5 km^2 in 2001 and to

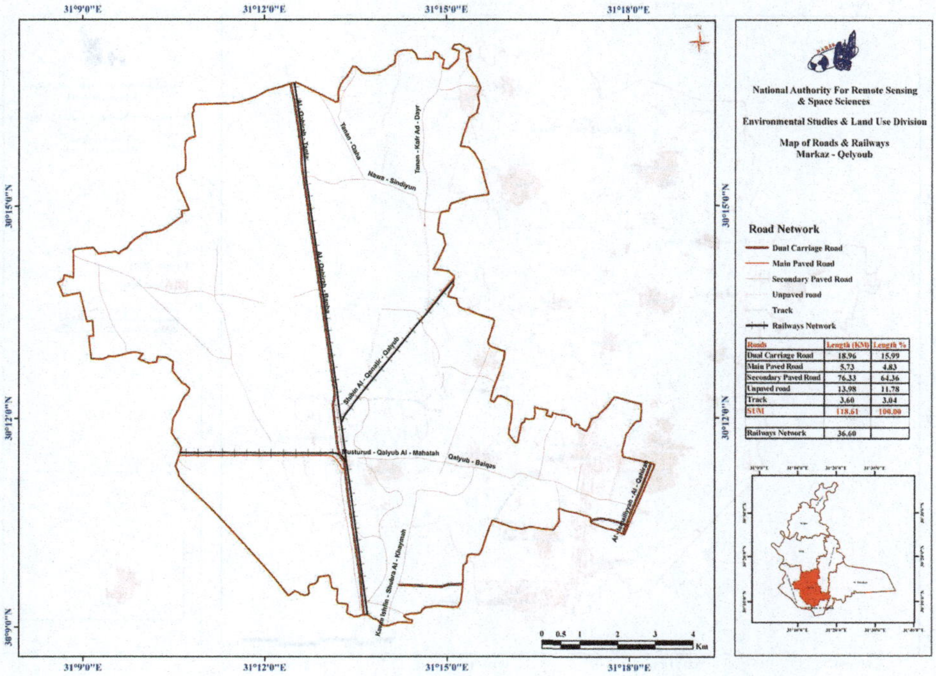

Fig. 4.540 Roads network of Qelyoub district

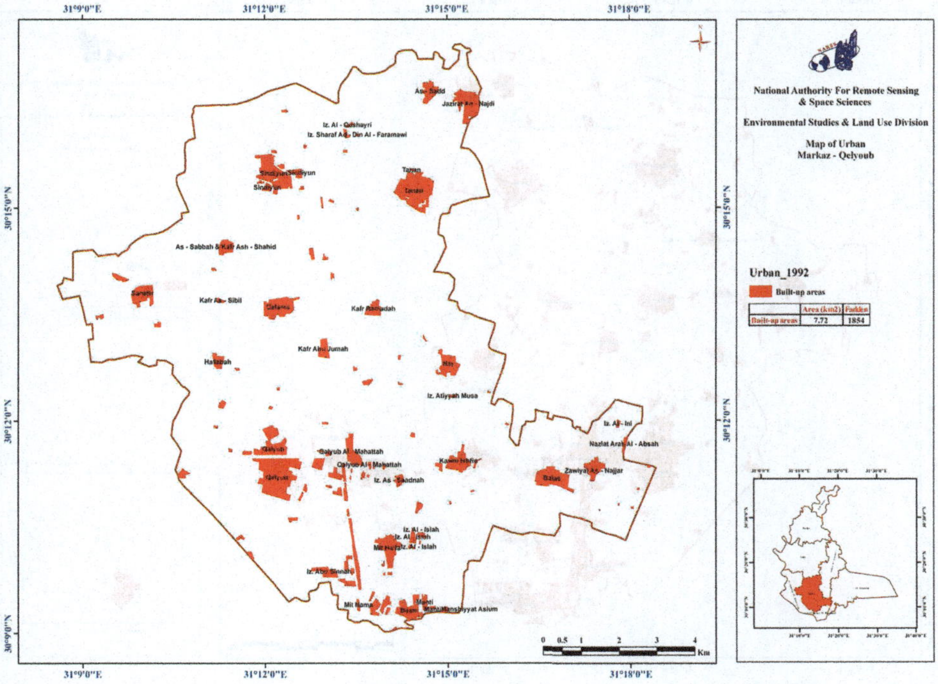

Fig. 4.541 Urban areas of Qelyoub district (1992)

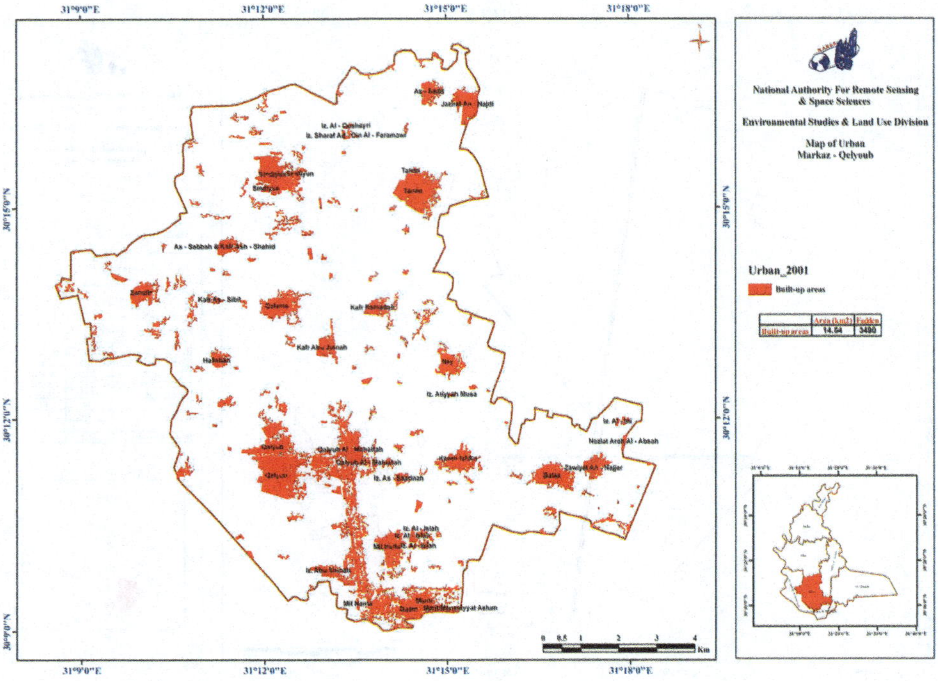

Fig. 4.542 Urban areas of Qelyoub district (2001)

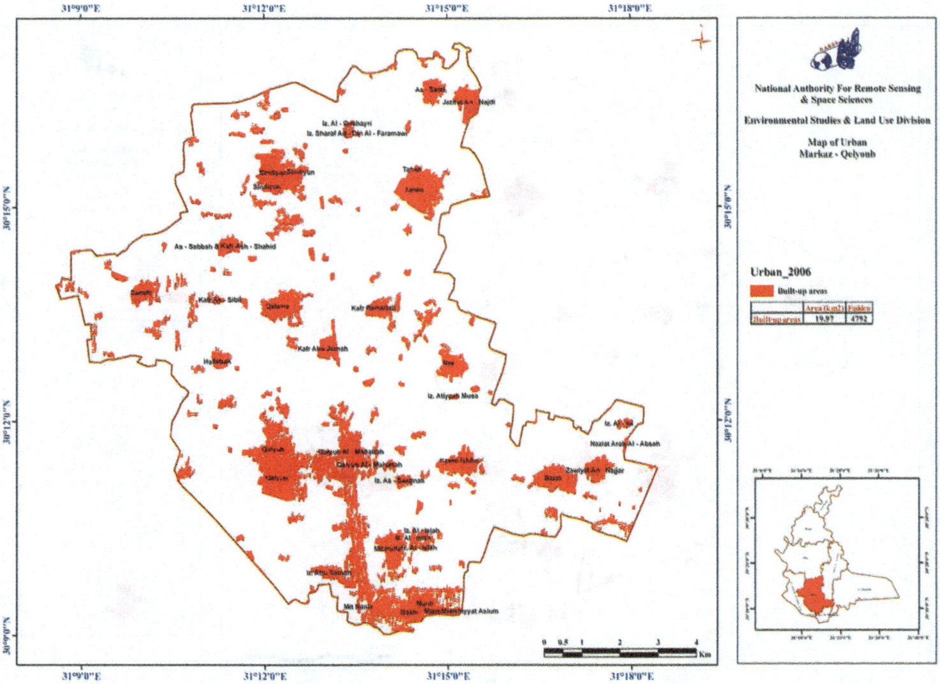

Fig. 4.543 Urban areas of Qelyoub district (2006)

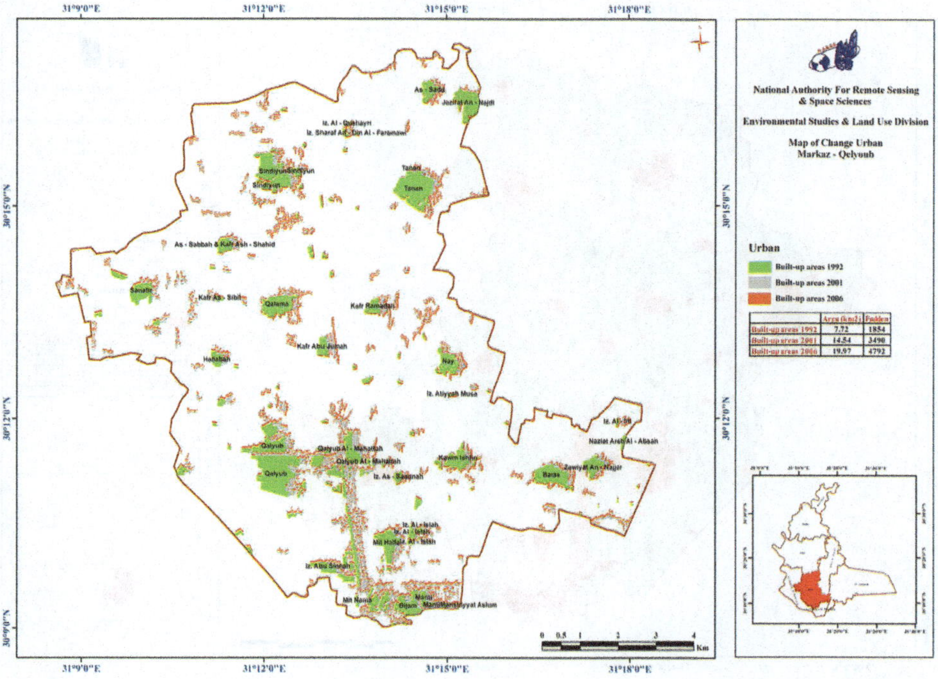

Fig. 4.544 Urban areas of Qelyoub district (1992, 2001, 2006)

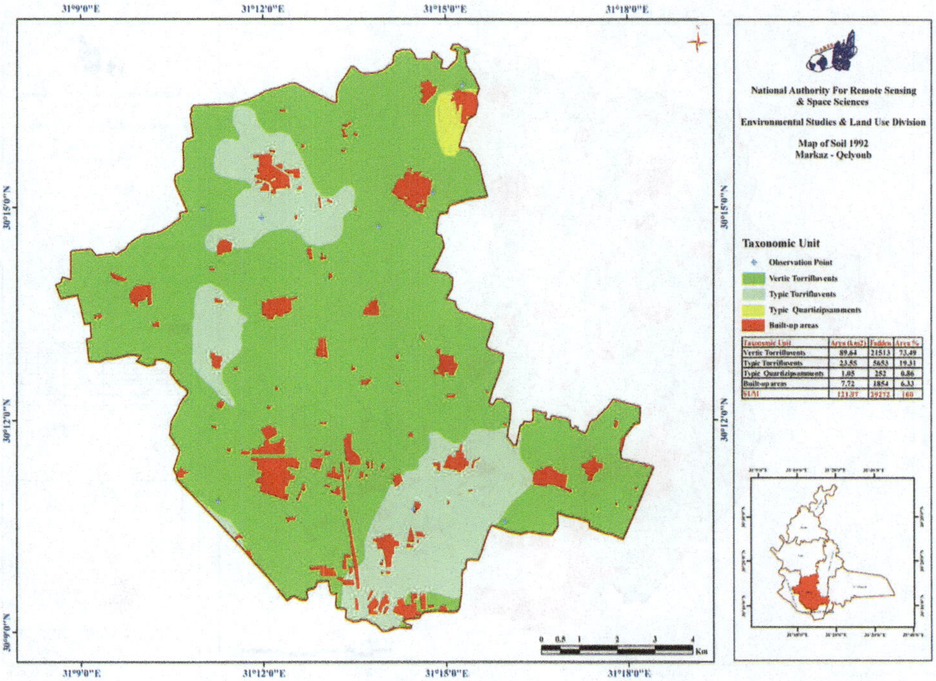

Fig. 4.545 Urban areas draped over the soil layer of Qelyoub district (1992)

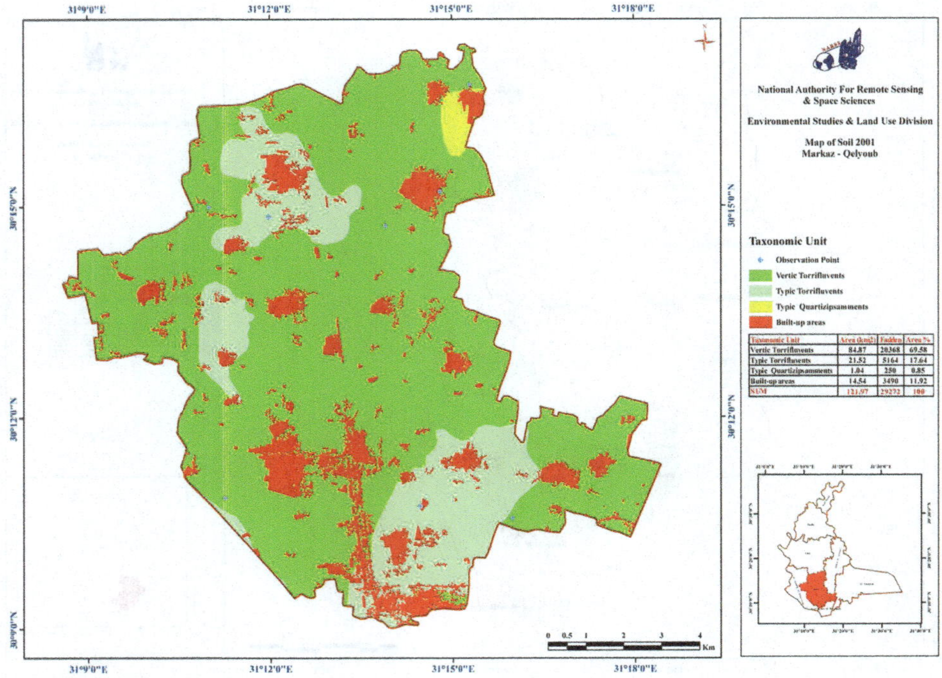

Fig. 4.546 Urban areas draped over the soil layer of Qelyoub district (2001)

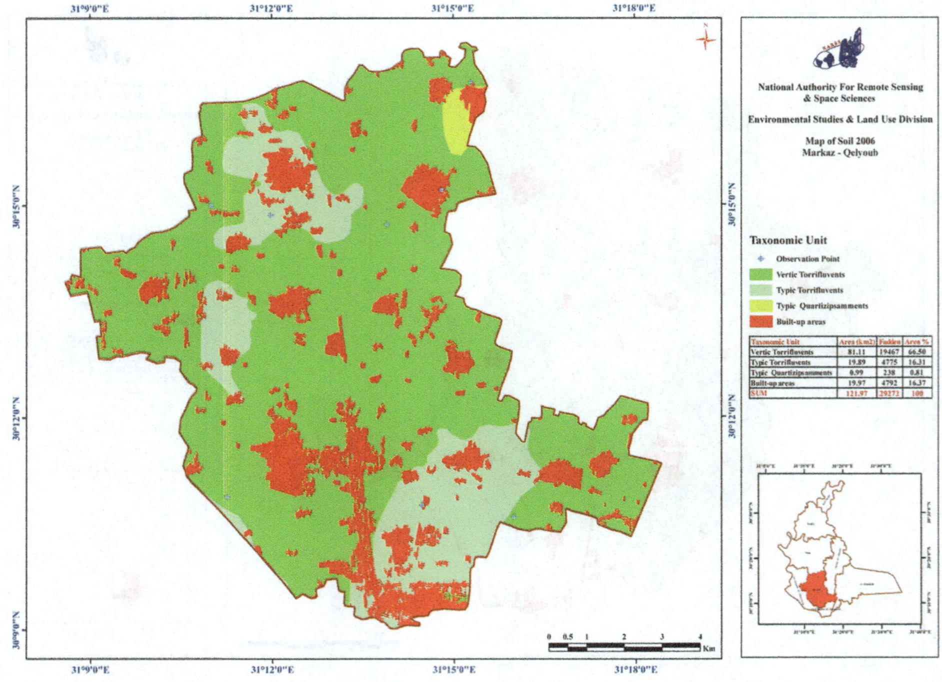

Fig. 4.547 Urban areas draped over the soil layer of Qelyoub district (2006)

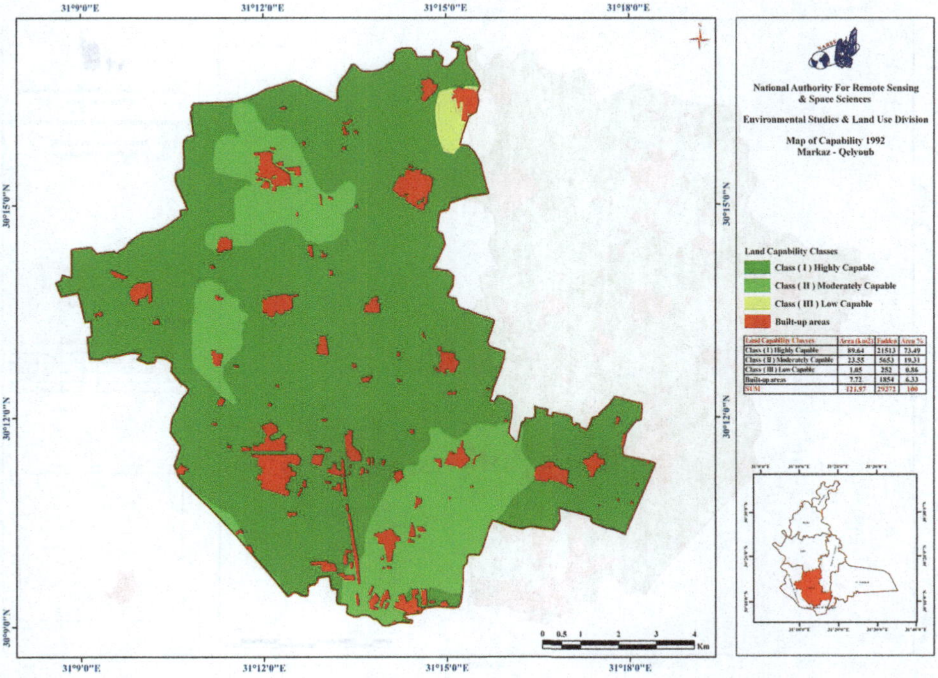

Fig. 4.548 Urban areas draped over the soil capability of Qelyoub district (1992)

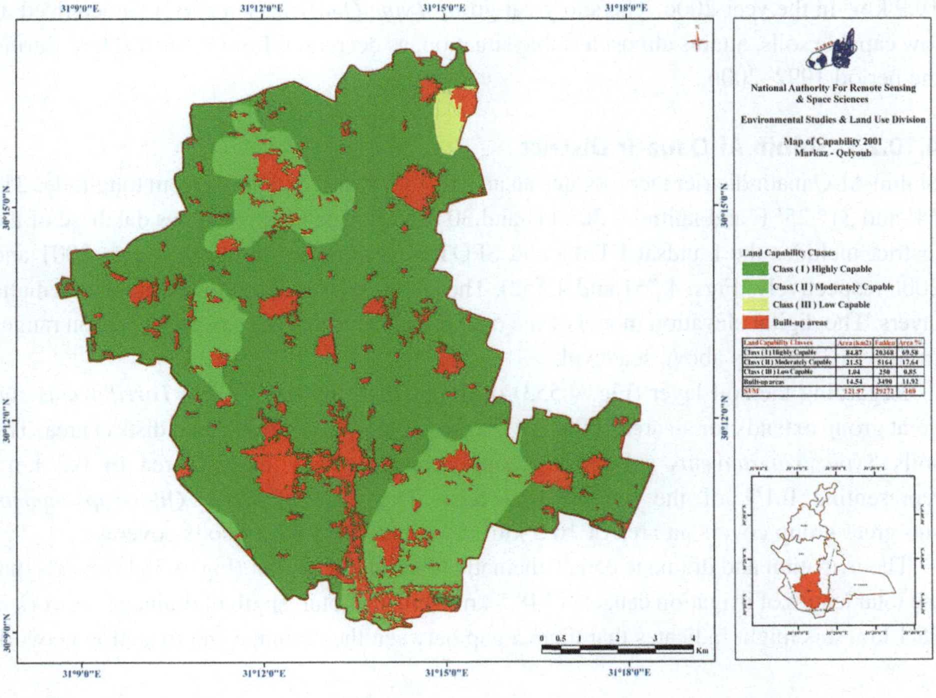

Fig. 4.549 Urban areas draped over the soil capability of Qelyoub district (2001)

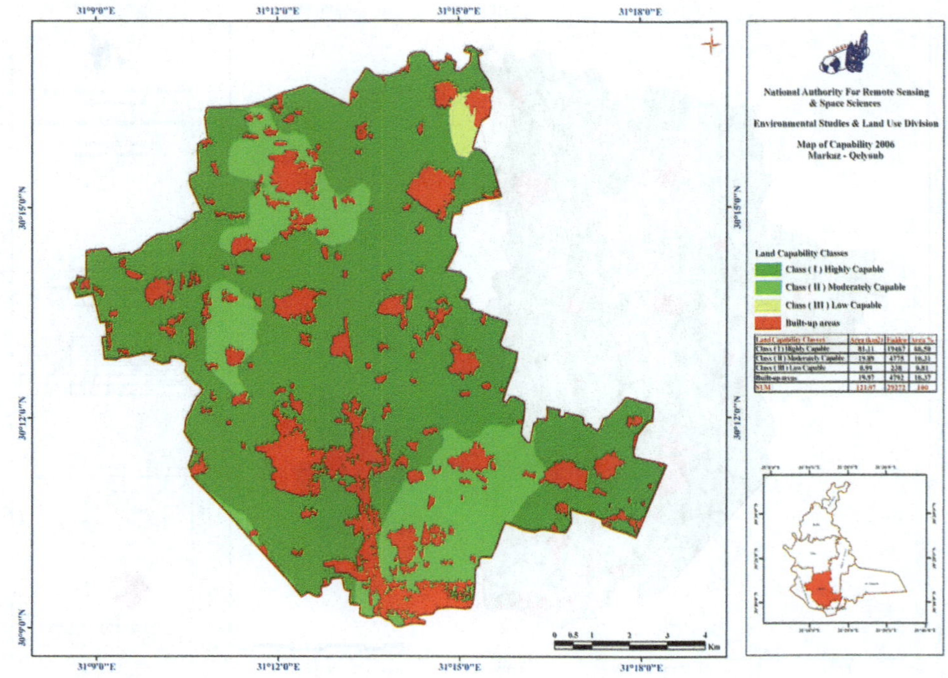

Fig. 4.550 Urban areas draped over the soil capability of Qelyoub district (2006)

19.9 km^2 in the year 2006. The sub great group *Typic Quartzipssaments,* characterized as low capable soils, attains almost a stable situation, as decreased from 1.1 to 1.0 km^2 during the period 1992–2006.

4.10.2.7 Shibin Al-Qanatir District

Shibin Al-Qanatir district incorporates an area of 138.4 km^2, extending from longitudes 31° 14′ and 31° 25′ E and latitudes 30° 11′ and 30° 25′ N. The land resources database of the district includes the Landsat ETM+ and SPOT images taken during the years 2001 and 2006 respectively (Figs. 4.551 and 4.552). These images were used to update the thematic layers. The digital elevation model of the district indicates that the surface elevation ranges from 0 to 34 meters above sea level.

Regarding the soil layer (Fig. 4.553) of the district, the soils *Vertic Torrifluvents* sub great group extends for an area 100.4 km^2, representing 72.5% of the total district area. The soils *Typic Torrifluvents* sub-great group occupies a very limited area of 0.2 km^2, representing 0.1% of the total district area. The soils of *Typic Quartizipsamment* sub-great group covers an area of 16.3 km^2 representing 11.8% of soils coverage.

The irrigation and drainage canals thematic layer of the district (Fig. 4.554) reveals that the total length of irrigation canals is 148.5 km while the total length of drainage network is 85.1 km, this might indicates that there a gap between the drainage and irrigation network

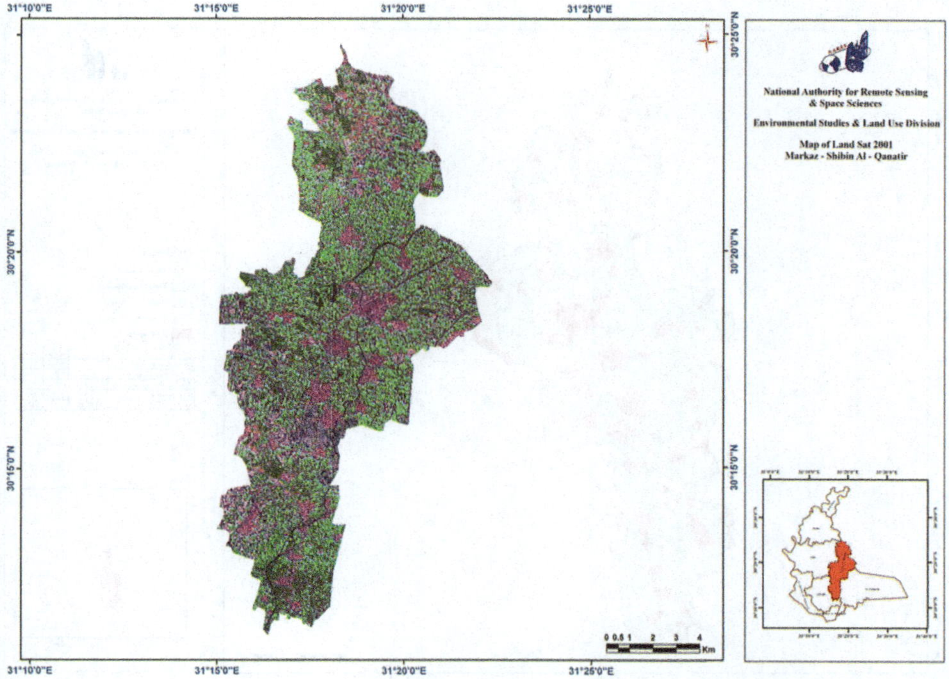

Fig. 4.551 Landsat ETM+ image of Shibin Al-Qanatir district (2001)

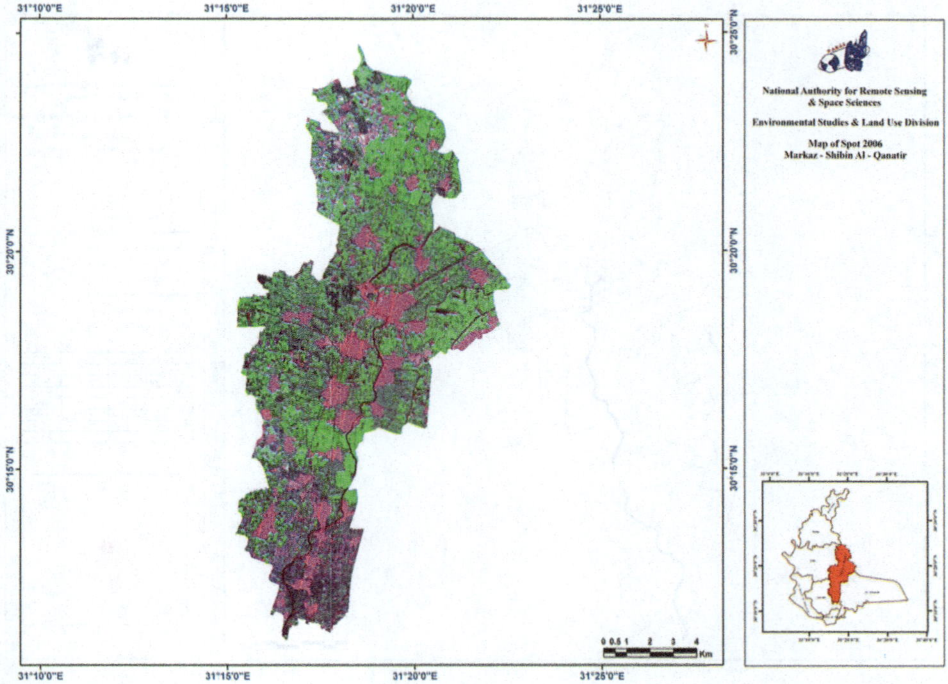

Fig. 4.552 SPOT image of Shibin Al-Qanatir district (2006)

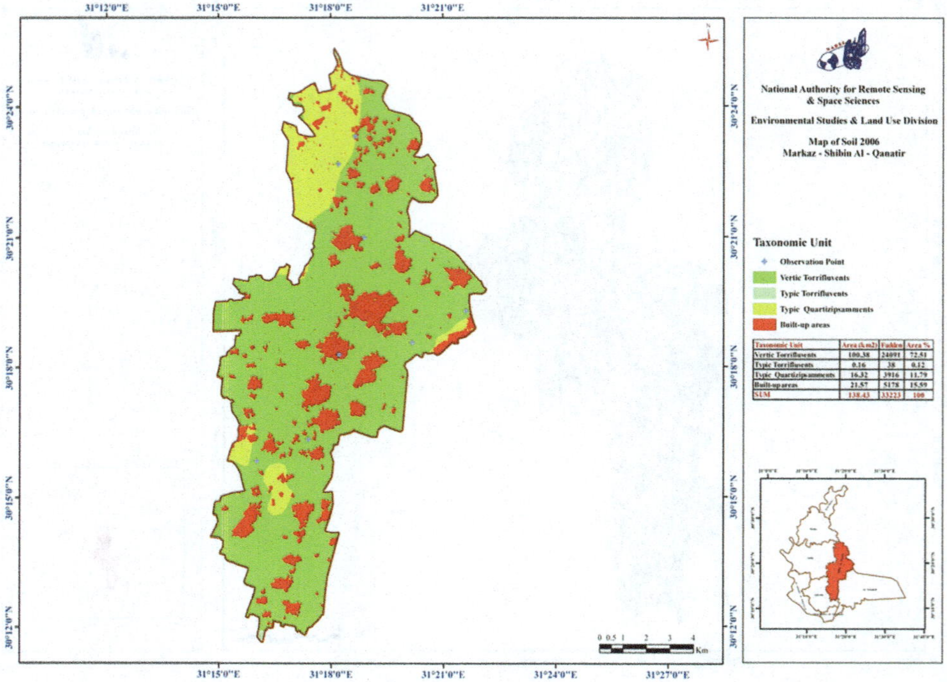

Fig. 4.553 Soil layer of Shibin Al-Qanatir district

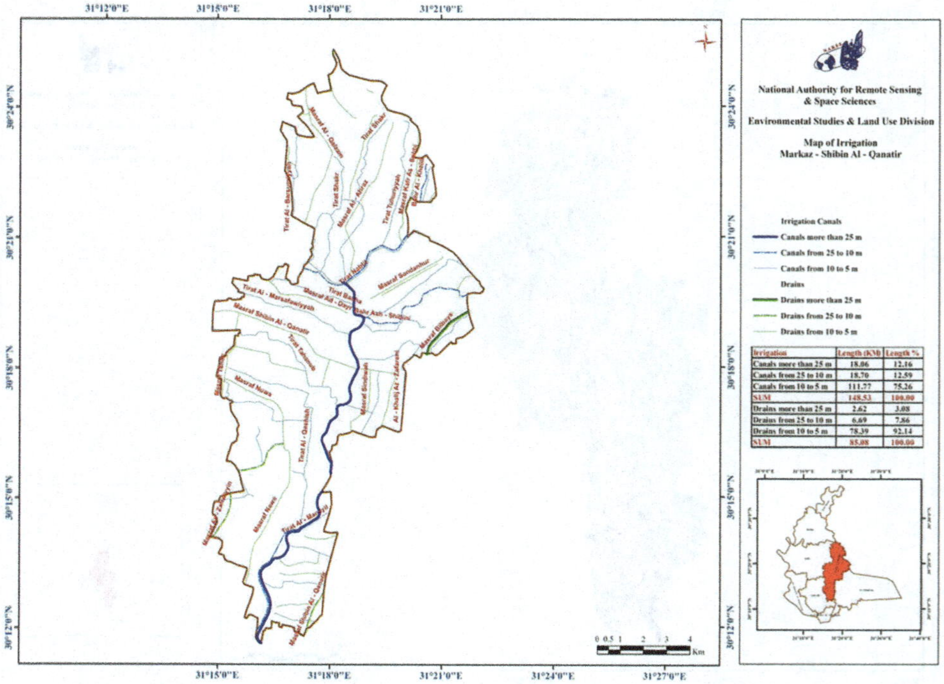

Fig. 4.554 Irrigation and drainage canals network of Shibin Al-Qanatir district

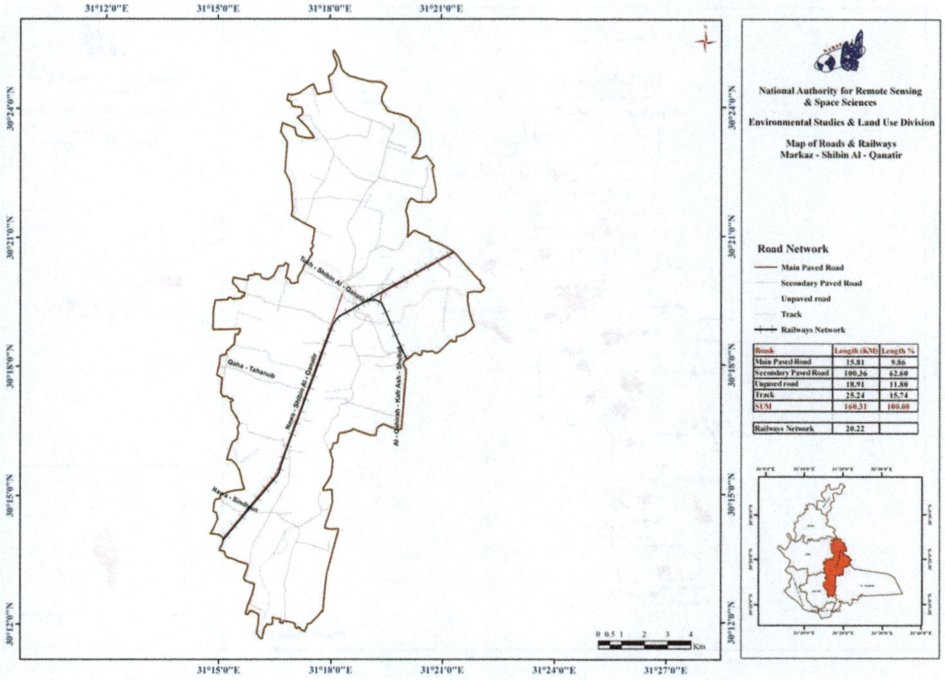

Fig. 4.555 Roads network of Shibin Al-Qanatir district

in the Shibin Al-Qanatir district. The roads network layer (Fig. 4.555) indicates that the district have a good and satisfactory network, where the total length of the roads network is 160.0 km, indicating a road network density of 1.2 km for each 1.0 km^2.

The urban areas were monitored, on basis of satellite images 1992, 2001 and 2006. The obtained thematic layers (Figs. 4.556, 4.557, 4.558) were attached to the created land resources database. Change detection was applied on urban expansion (Fig. 4.559) and resulted in pointing out an urban increase from 10.2 km^2 in the year 1992 to 15.4 km^2 in the year 2001 and to 21.6 km^2 in 2006. The urban coverage in 2006 is mostly double the area of 1992, however its average annual growth rate is 0.81 km^2 (193 feddan) per year.

Combining the soil and land capability maps, and the urban thematic layers, pointed out the risk of urban expansion. Figures 4.560, 4.561, 4.562, 4.563, 4.564, 4.565 show that the cultivable soils of *Vertic Torrifluvents* sub great group, which coincides with the highly capable class, was the most negatively affected by the urban expansion. It decreased from 110.7 km^2 in year 1992 to 105.9 km^2 in year 2001 and to 100.38 km^2 in year 2006. The soils of sub great group *Typic Torrifluvents,* coinciding with moderately capable soil class, was found stable (i.e. 0.2 km^2) at the three monitoring periods. The sub great group *Typic Quartizipssaments,* characterized as Low capable soils, attains very slight changes, as decreased from 17.4 to 16.3 km^2 during the period 1992–2006.

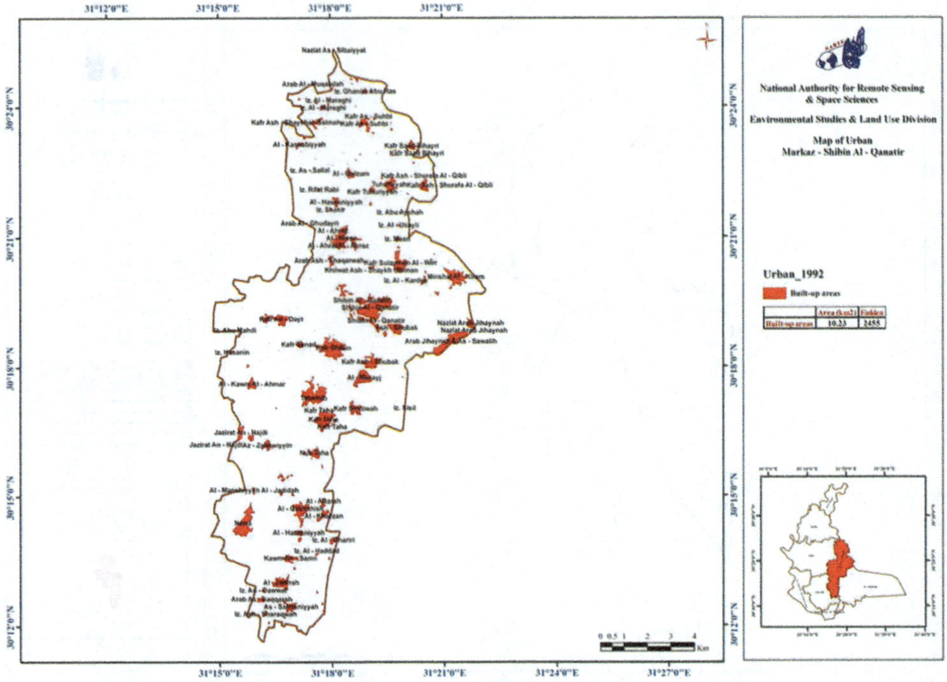

Fig. 4.556 Urban areas of Shibin Al-Qanatir district (1992)

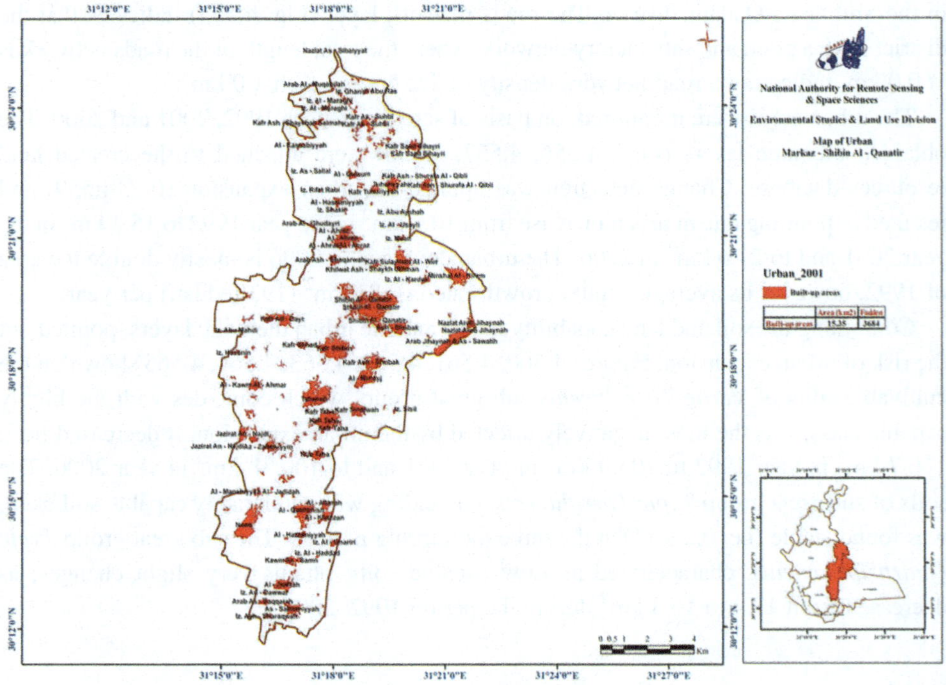

Fig. 4.557 Urban areas of Shibin Al-Qanatir district (2001)

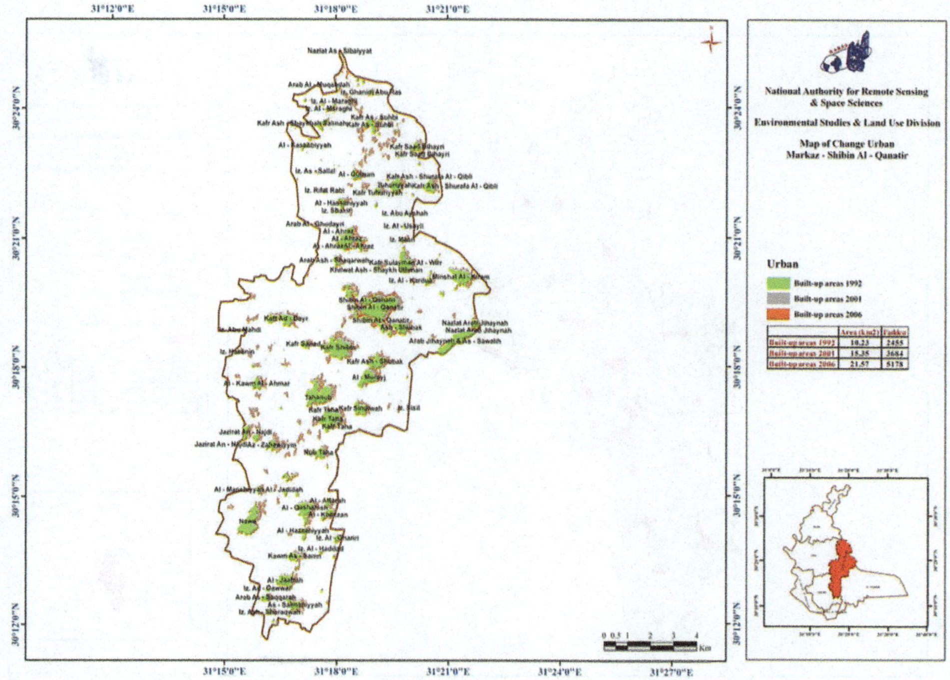

Fig. 4.559 Urban areas of Shibin Al-Qanatir district (1992, 2001, 2006)

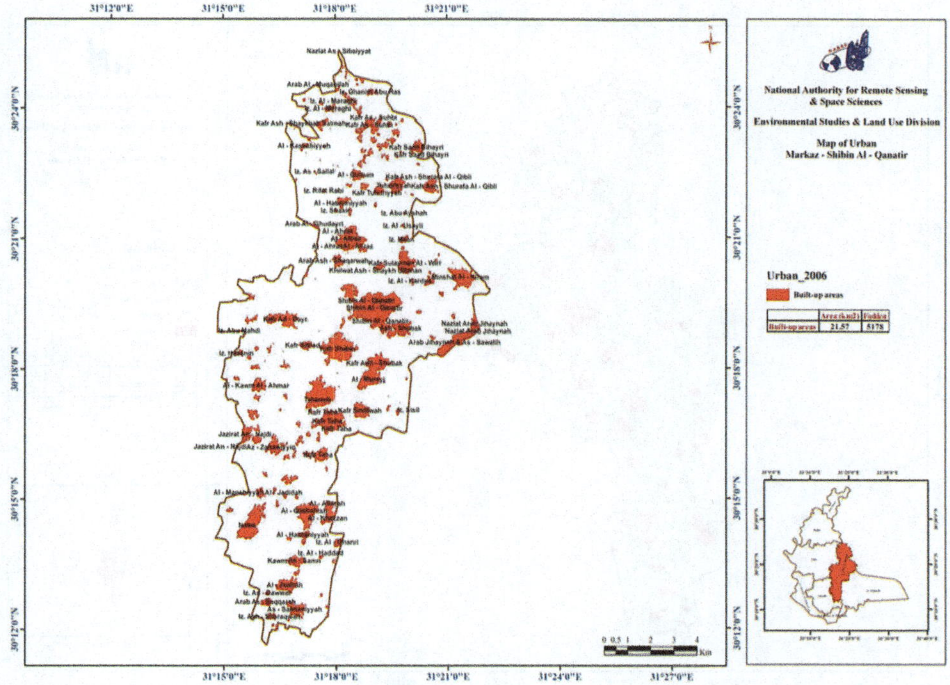

Fig. 4.558 Urban areas of Shibin Shibin Al-Qanatir district (2006)

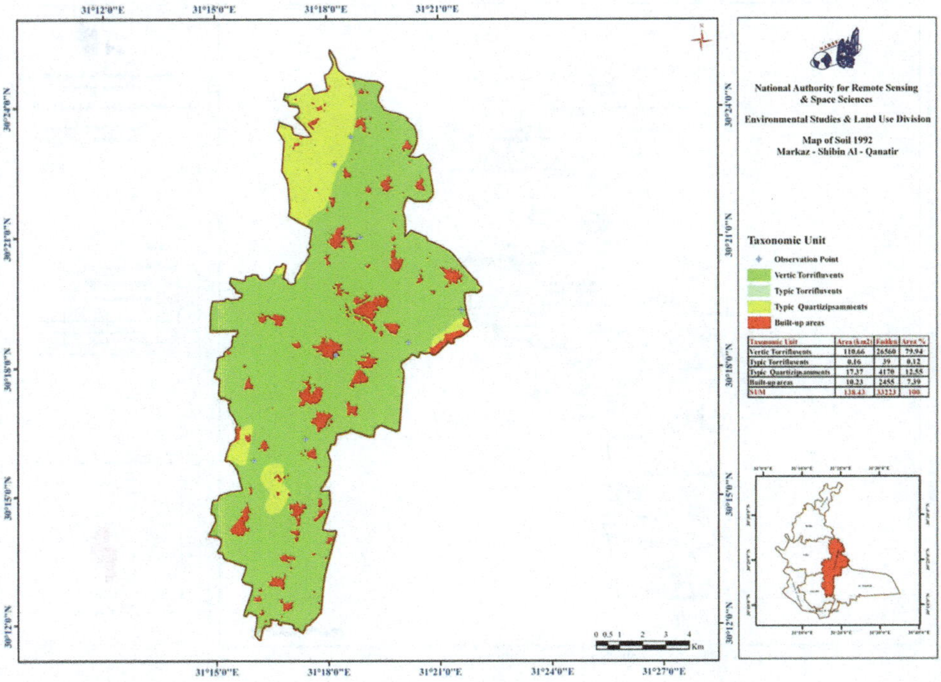

Fig. 4.560 Urban areas draped over the soil layer of Shibin Al-Qanatir district (1992)

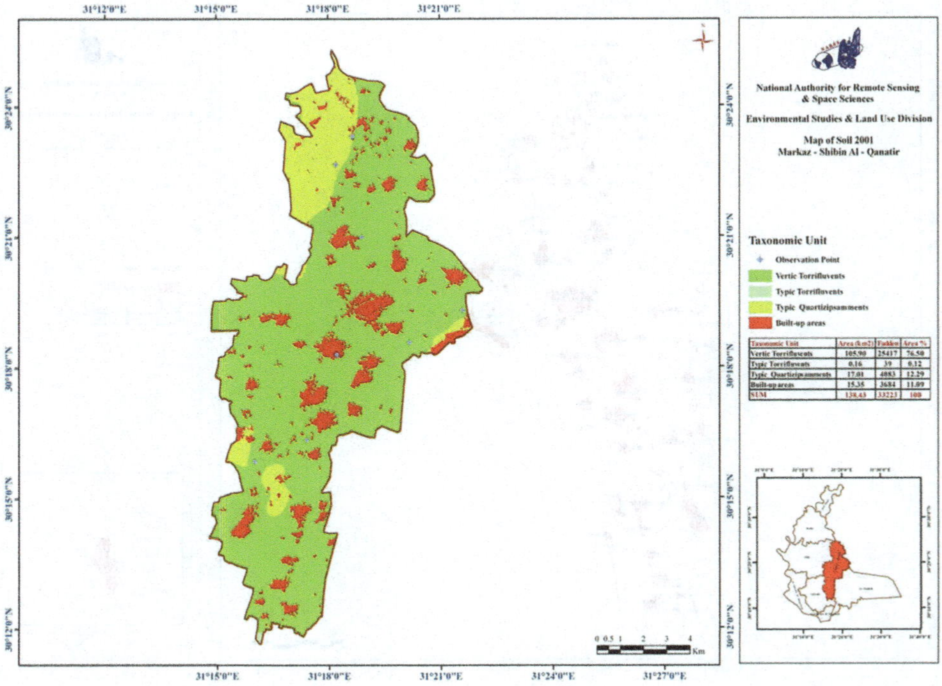

Fig. 4.561 Urban areas draped over the soil layer of Shibin Al-Qanatir district (2001)

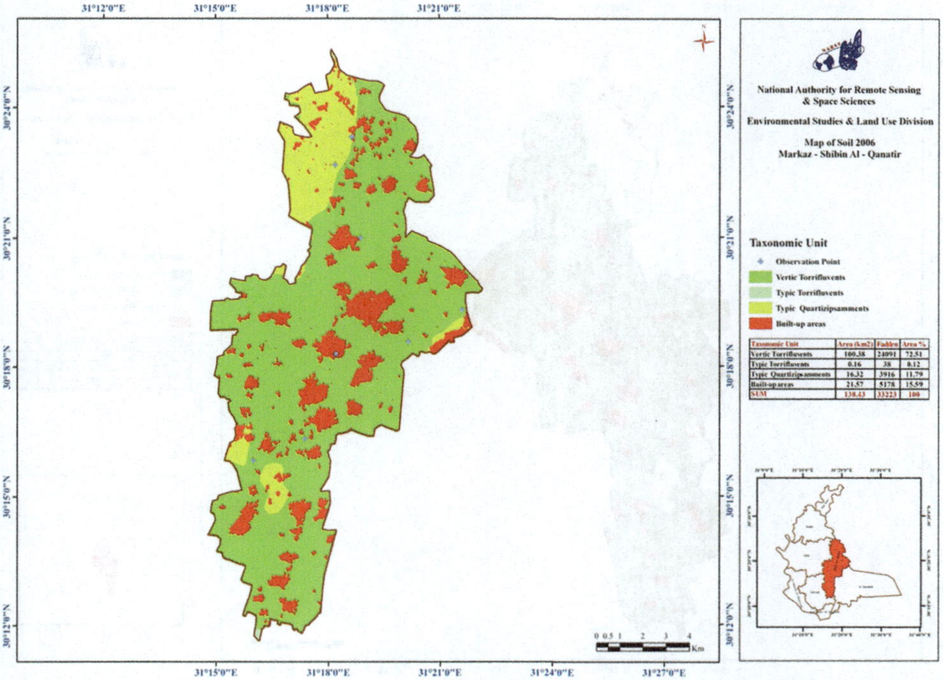

Fig. 4.562 Urban areas draped over the soil layer of Shibin Al-Qanatir district (2006)

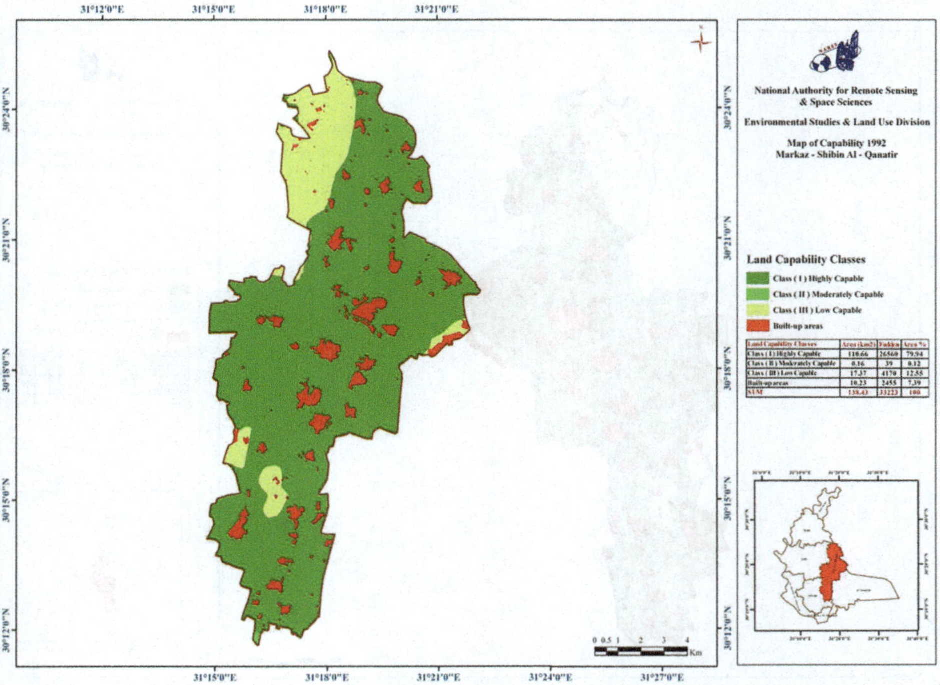

Fig. 4.563 Urban areas draped over the soil capability of Shibin Al-Qanatir district (1992)

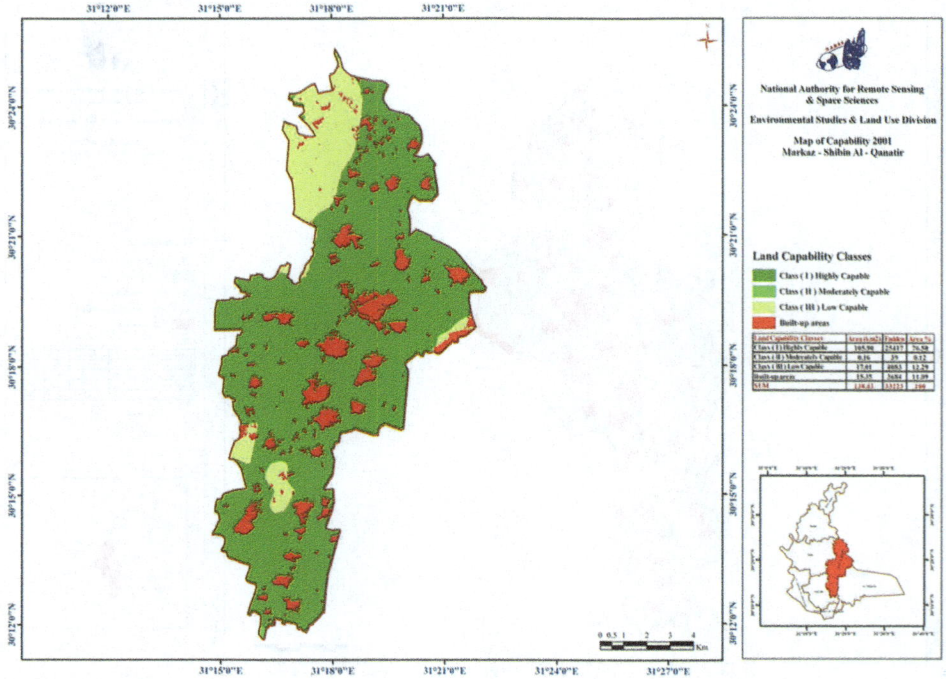

Fig. 4.564 Urban areas draped over the soil capability of Shibin Al-Qanatir district (2001)

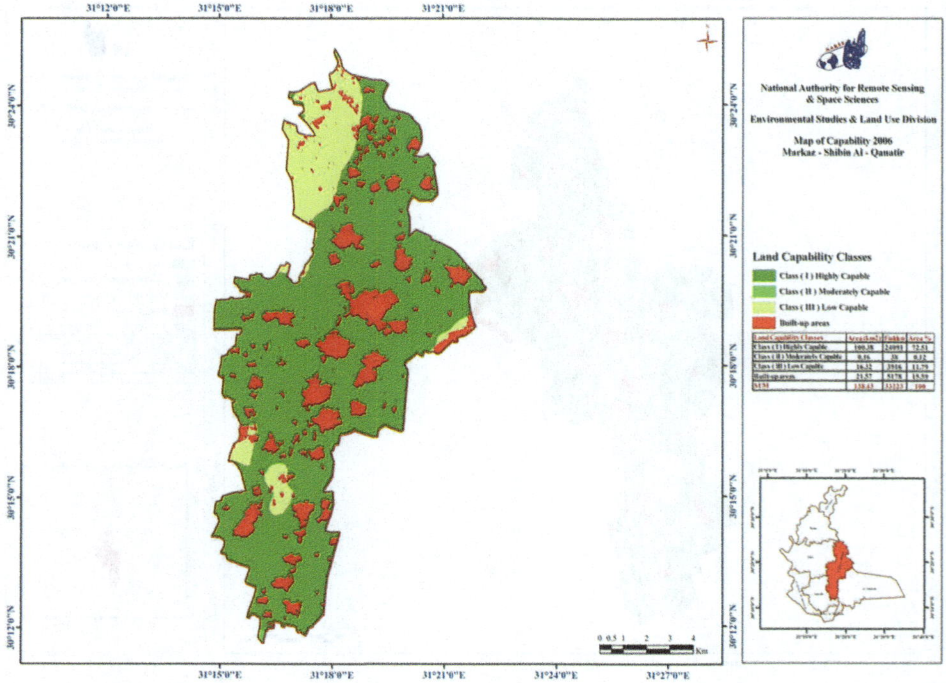

Fig. 4.565 Urban areas draped over the soil capability of Shibin Al-Qanatir district (2006)

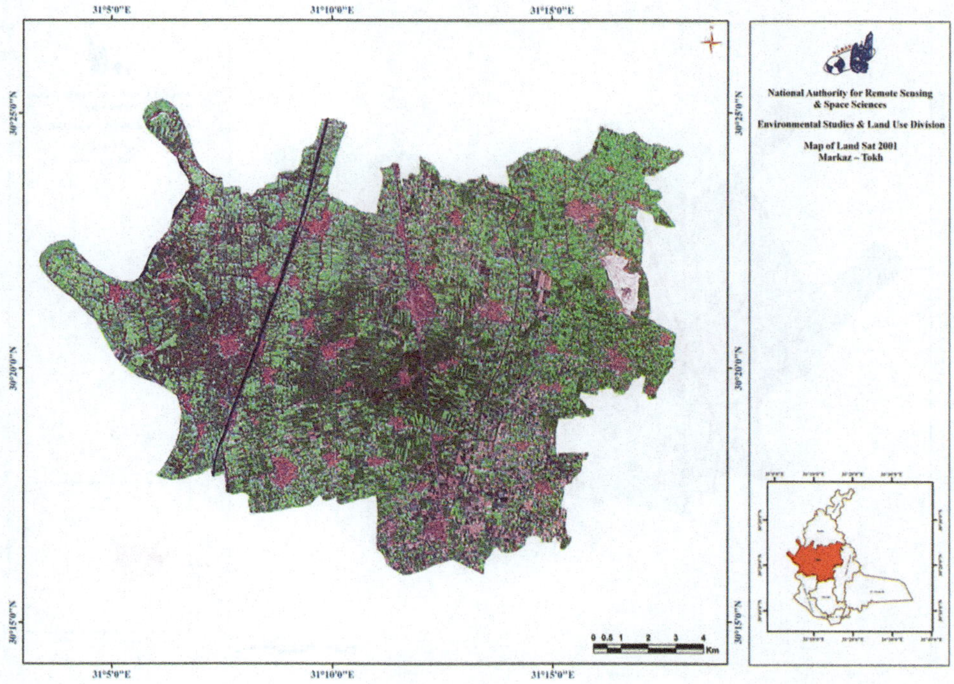

Fig. 4.566 Landsat ETM+ image of Tokh district (2001)

4.10.2.8 Tokh District

Tokh district incorporates an area of 229.7 km^2, extending from longitudes 31° 03′ to 31° 17′ E and latitudes 30° 17′ to 30° 26′ N. The land resources database of the district includes the Landsat ETM+ and SPOT images taken during the years 2001 and 2006 respectively (Figs. 4.566 and 4.567) which were used to update the thematic layers. The digital elevation model of the district indicates that the surface elevation ranges from 0 to 34 meter above sea level.

Regarding the soil layer (Fig. 4.568) of Tokh district, the soils are dominated by the *Vertic Torrifluvents* sub great group with an area of 165.5 km^2 representing 72.05% of the whole district. The *Typic Torrifluvents* soil sub-great group covers an area of 25.9 km^2 (i.e. 11.3% of total area) mostly exist in the western side of the district. The soils of *Typic Quartzipssaments* extends for an area of 13.2 km^2, representing 5.7%, located in the northeastern corner of the district.

The irrigation and drainage canals thematic layer of the district (Fig. 4.569) revels that the total length of irrigation canals is 254.6 km while the total length of drainage network is 91.8 km; this indicate that there is a large gap between the drainage and irrigation network in the district. The roads network layer (Fig. 4.570) indicates that the district have a good and satisfactory network where the total length of the roads network is 283.7 km attaining a road density of 1.2 km for each 1 km^2 area.

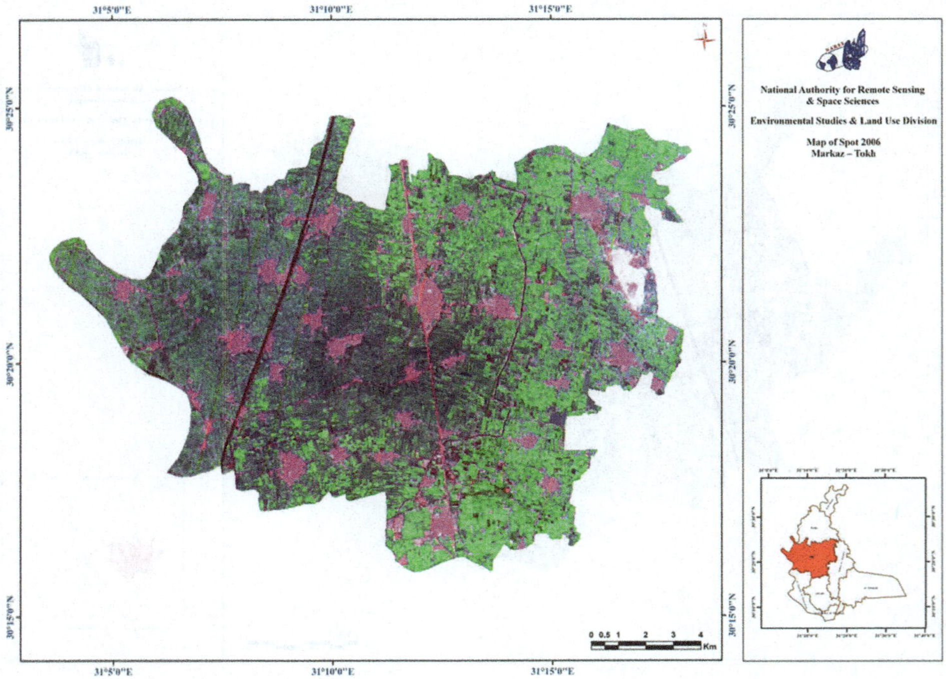

Fig. 4.567 Spot image of Tokh district (2006)

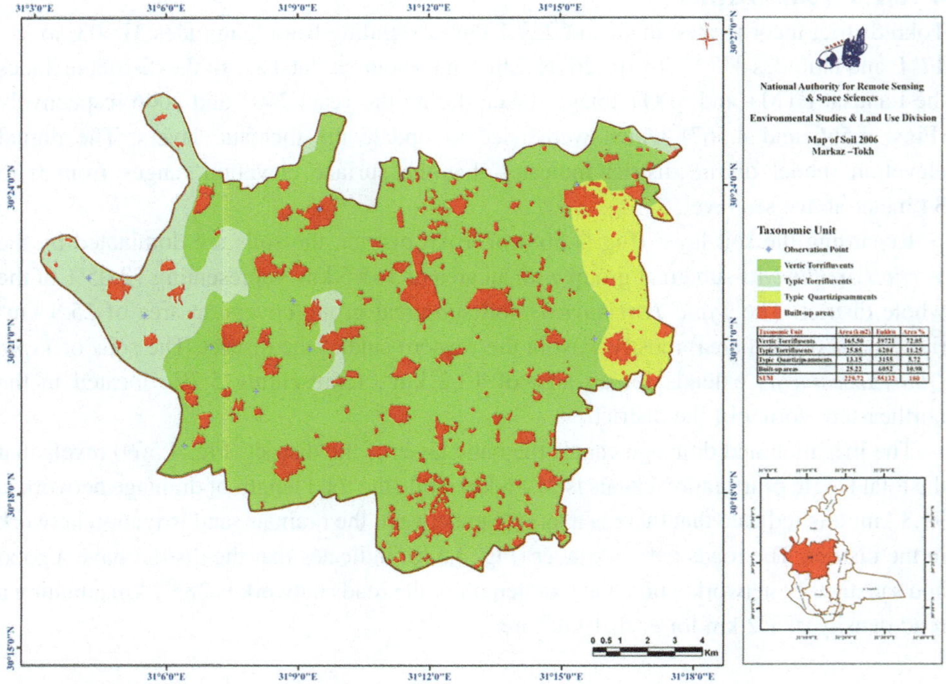

Fig. 4.568 Soil layer of Tokh district (2006)

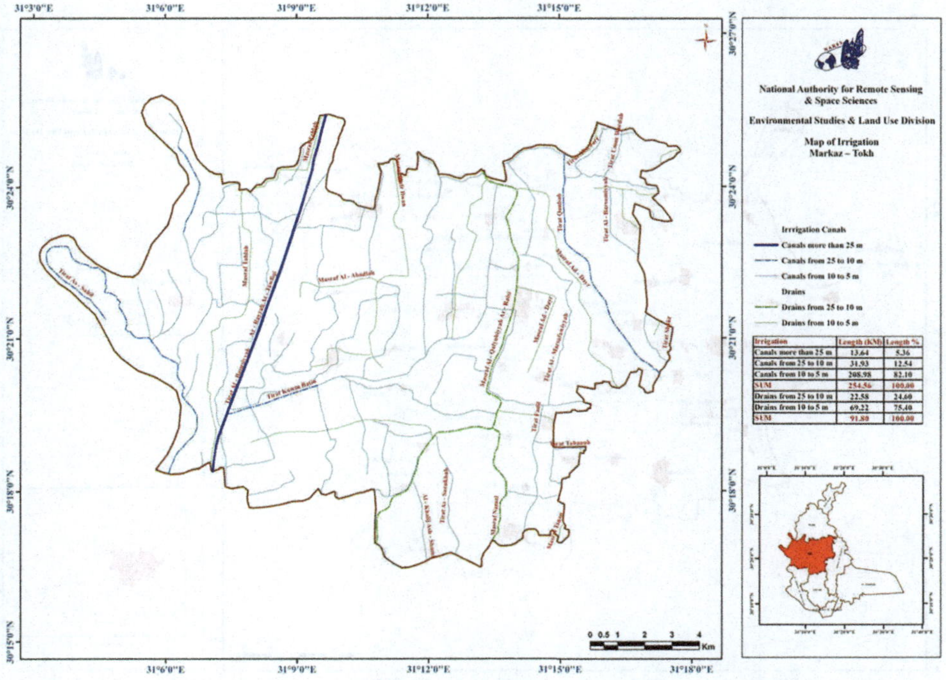

Fig. 4.569 Irrigation and drainage canals network of Tokh district

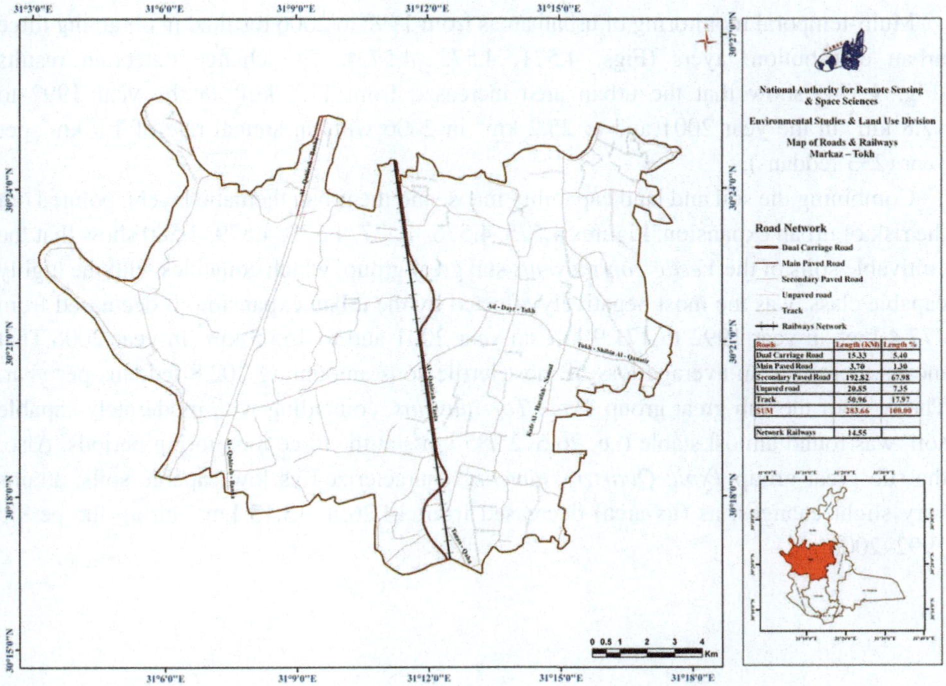

Fig. 4.570 Roads network of Tokh district

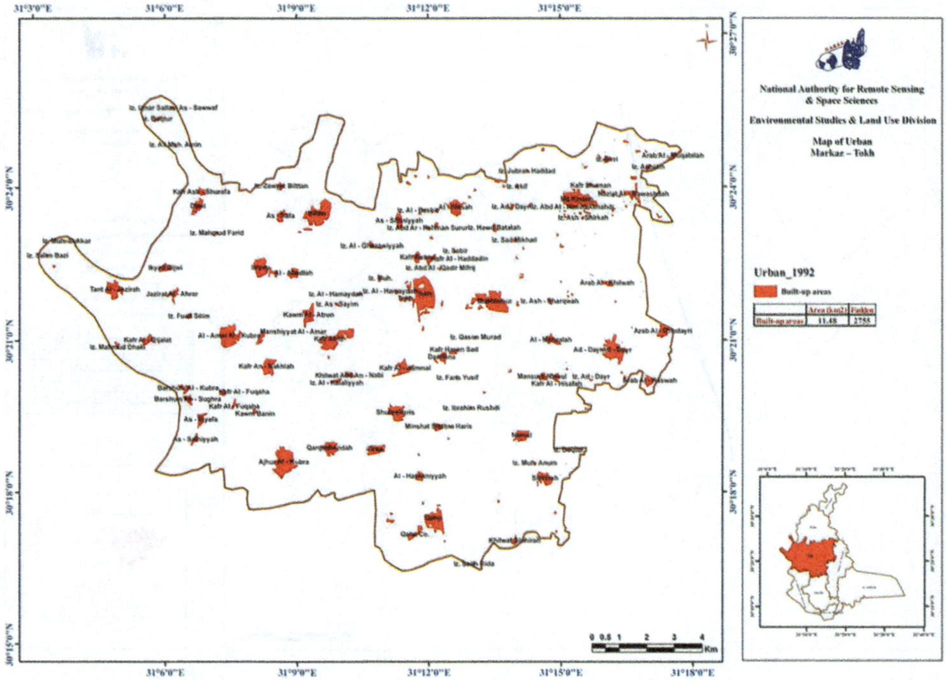

Fig. 4.571 Urban areas of Tokh district (1992)

Multi-temporal monitoring of urban areas from 1992 to 2006 resulted in obtaining three urban distribution layers (Figs. 4.571, 4.572, 4.573). The change detection results (Fig. 4.574) show that the urban area increased from 11.5 km^2 in the year 1992 to 17.8 km^2 in the year 2001 and to 25.2 km^2 in 2006 with an annual rate of 1.0 km^2 per year (233 feddans).

Combining the soil and land capability maps, and the urban thematic layers, pointed out the risk of urban expansion. Figures 4.575, 4.576, 4.577, 4.578, 4.579, 4.580 show that the cultivable soils of the *Vertic Torrifluvents* sub great group, which coincides with the highly capable class, was the most negatively affected by the urban expansion. It decreased from 177.4 km^2 in year 1992 to 171.9 km^2 in year 2001 and to 165.5 km^2 in year 2006.That means there was an average loss of most fertile soils amounting 202.8 feddans per year. The soils of the sub great group *Typic Torrifluvents,* coinciding with moderately capable soil, was found almost stable (i.e. 26.6–25.85 km^2) at the three monitoring periods. Also, the sub great group *Typic Quartzipssaments,* characterized as low capable soils, attains very slight changes, as (its area) decreased from 14.26 to 13.15 km^2 during the period 1992–2006.

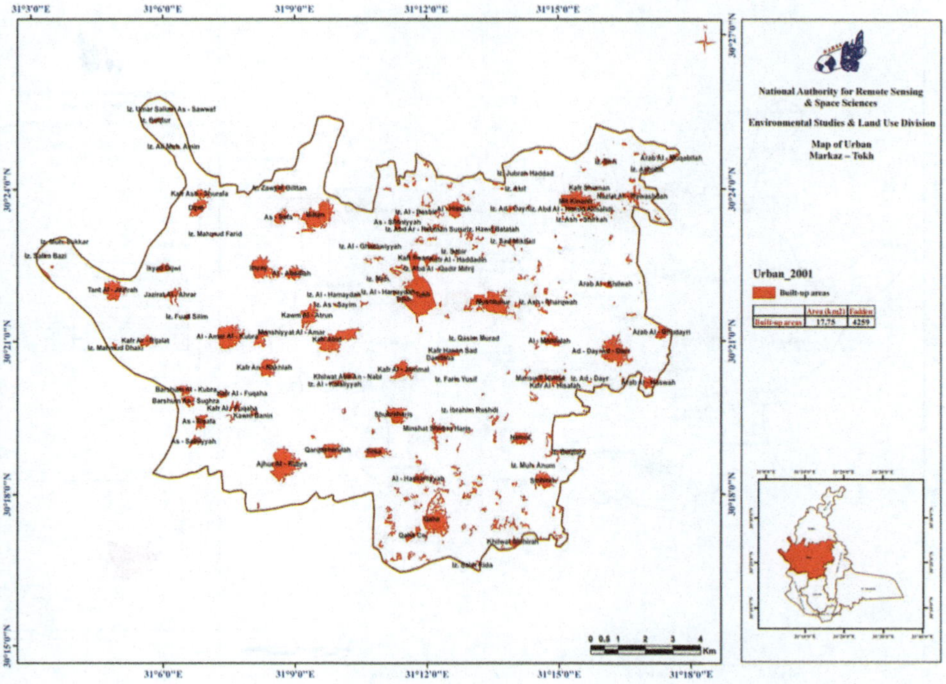

Fig. 4.572 Urban areas of Tokh district (2001)

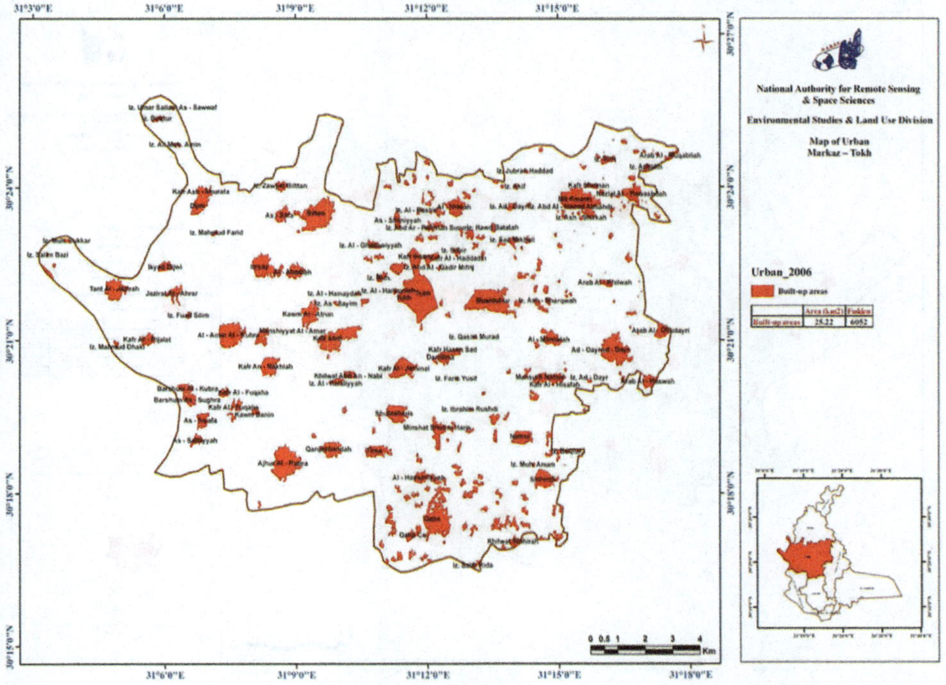

Fig. 4.573 Urban areas of Tokh district (2006)

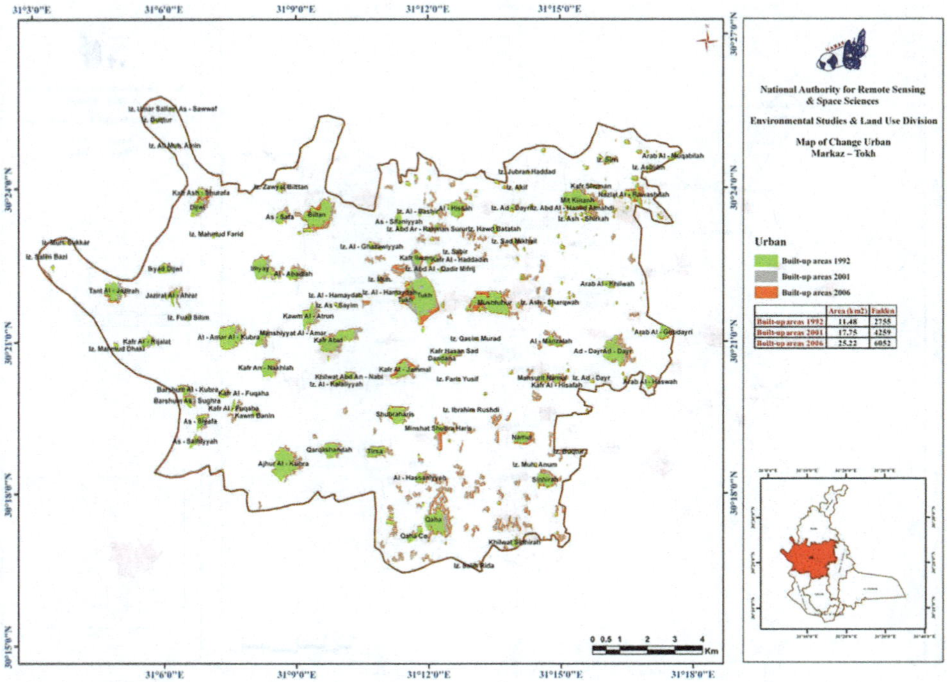

Fig. 4.574 Urban areas of Tokh district (1992, 2001, 2006)

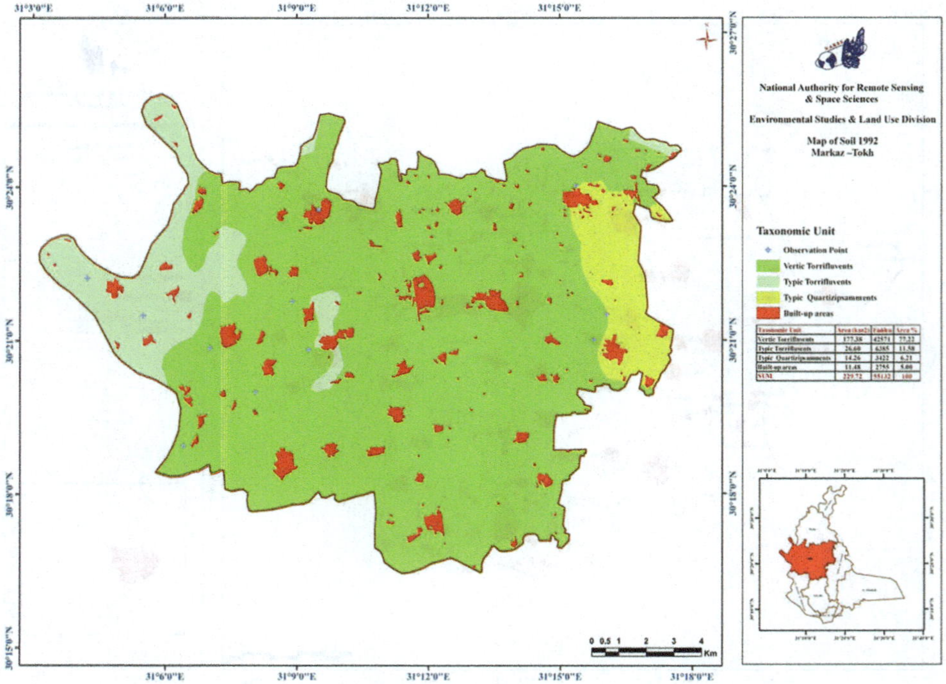

Fig. 4.575 Urban areas draped over the soil layer of Tokh district (1992)

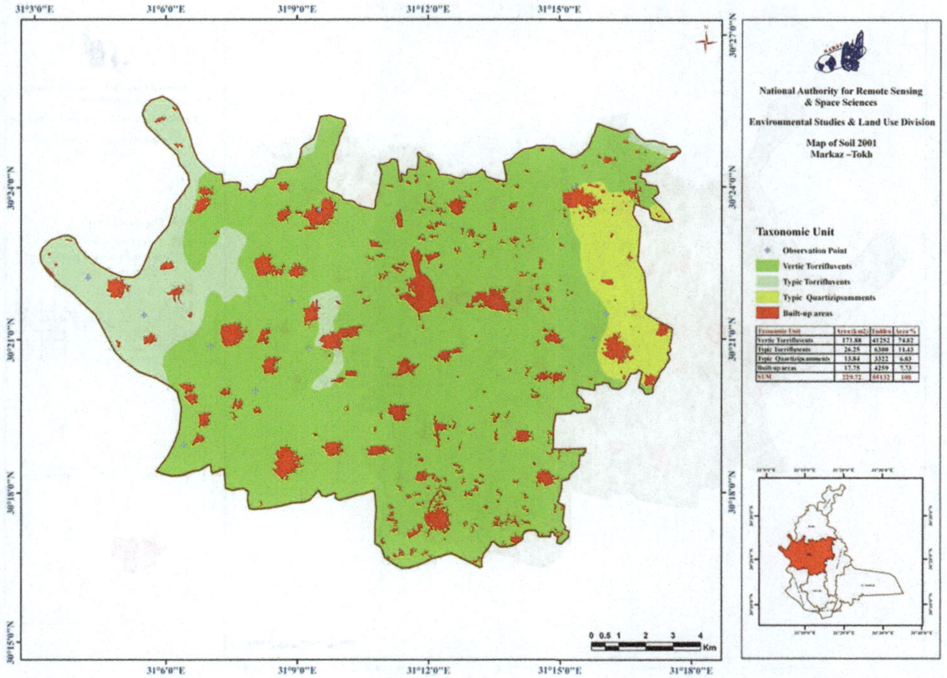

Fig. 4.576 Urban areas draped over the soil layer of Tokh district (2001)

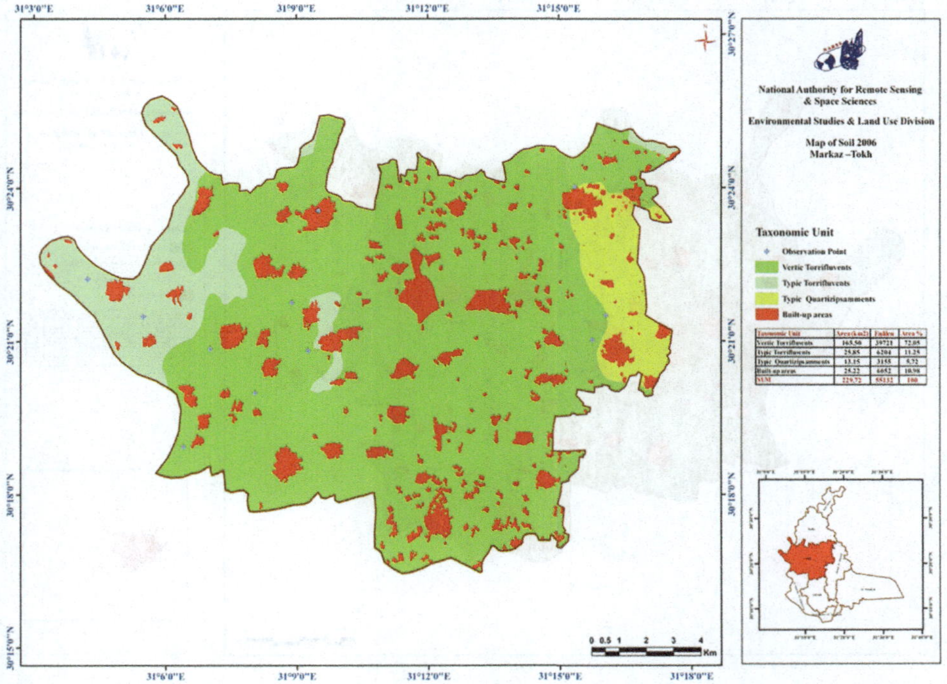

Fig. 4.577 Urban areas draped over the soil layer of Tokh district (2006)

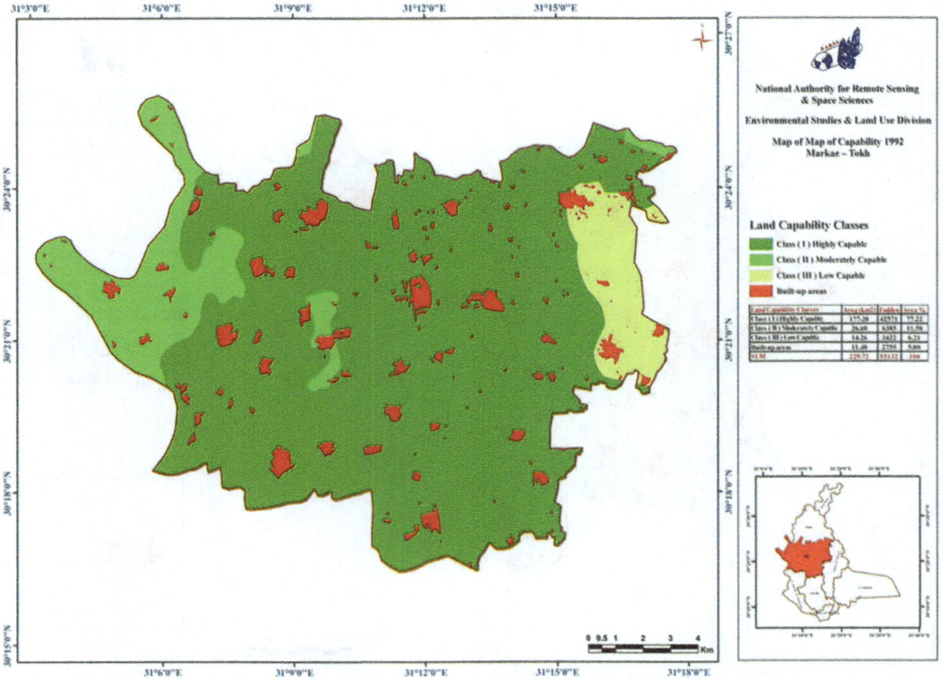

Fig. 4.578 Urban areas draped over the soil capability of Tokh district (1992)

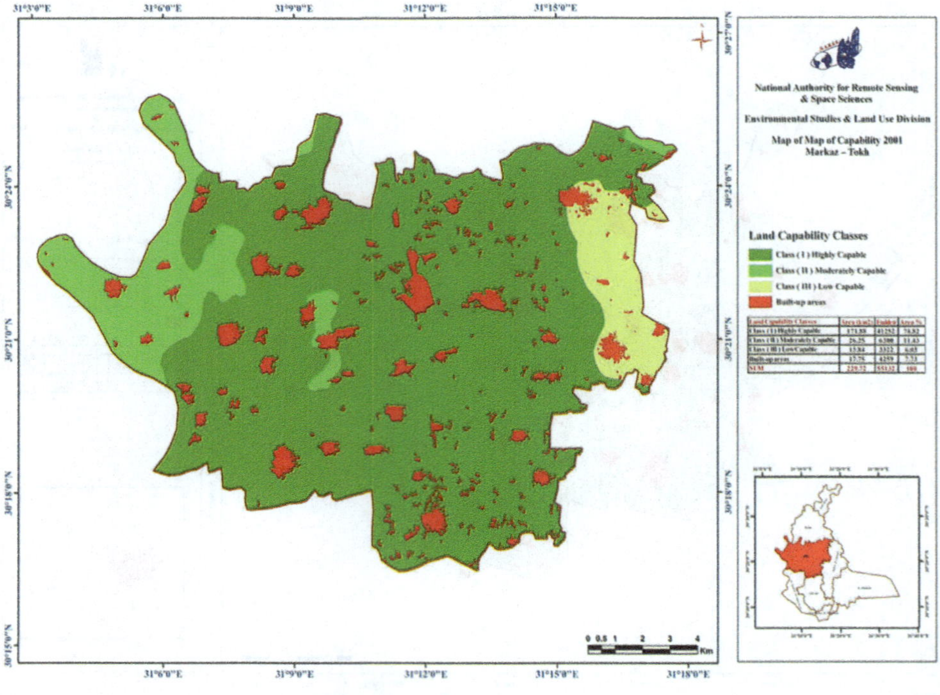

Fig. 4.579 Urban areas draped over the soil capability of Tokh district (2001)

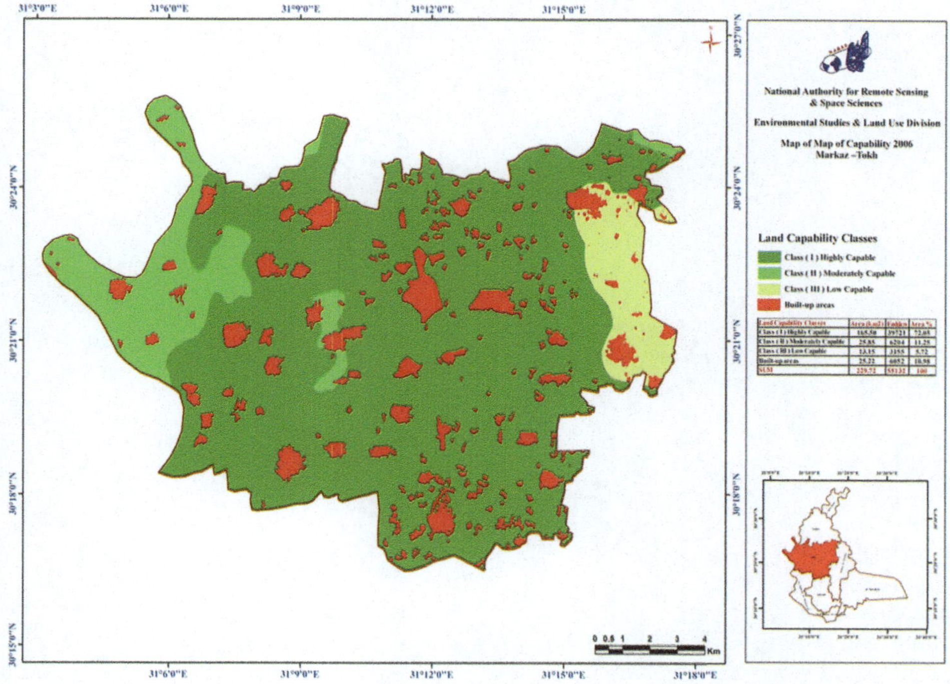

Fig. 4.580 Urban areas draped over the soil capability of Tokh district (2006)

References

Abo El-Einein SM (1999) Dune encroachment on the cultivated land in Egypt. Bull Soci'ete' de Geogaphie d'Egypte, Tome LXIV, 107–123

ASRT (1982) Soil Map of Egypt-Final Report. Academy of Scientific Research and Technology, Cairo, Egypt

ASRT (1996) Soils and land capability maps of Halaib-Shalatien region, scale 1:100,000

CLAC (2004). Central Laboratory for Agricultural Climate (CLAC) Web site: http://www.clac.edu.eg/

Hulme M, March R (1990). "Global Mean Monthly Humidity Surfaces for 1930–59, 1960–89 and Projected for 2020" UNEP/GEMS/GRID. Climatic Res. Unit, Univ. of East Anglia, Norwich, England

Said R (ed) (1990) The Geology of Egypt. A.A. Balkema, Rotterdam, 733p

USDA (2006) "Keys to Soil Taxonomy" United State Department of Agriculture, Natural Resources Conservation Service (NRCS) tenth edition, 2006

Administrations, Scientific Institutions and Research Centers

1. Academy of Scientific Research and Technology (ASRT).
2. Desert Research Center (DRS).
3. Egyptian Geological Survey and Mining Authority (EGSMA).
4. Egyptian Meteorological Authority (EMA).
5. Egyptian Survey Authority (ESA).
6. National Authority for Remote Sensing and Space Science (NARSS).
7. National Research Center (NRS).
8. National Center for Water Research.
9. Nuclear Materials Authority.
10. National Planning Institute.
11. Central Agency for Public Mobilization & Statistics.
12. Geological and Geophysical Information Center (GPC).
13. Egyptian Documentation & Information Center for Agriculture (EDICA).
14. Agricultural Research Center (ARC).
15. National Library and Archives.
16. Great Cairo Library.
17. Supreme Council of Antiquities.
18. General Organization for Physical Planning.
19. Egyptian National Science and Technology. Information Network.
20. Science and Technology Information Center.
21. Executive Office for Environmental Information.
22. National Information & Documentation Center (NIDOC).
23. Information and Decision Support Center-Ministers Cabinet.
24. National Specialized Councils.
25. Egyptian Environmental Affairs Authority (EEAA).

A.-A. Gad, *Digital Land Resources to Support Sustainable Development in Egypt*, Deltas of the World,
https://doi.org/10.1007/978-3-031-37167-7

26. African Studies and Research Institute, Cairo University.
27. Graduate Studies and Research Institute, Alexandria University.
28. Environmental Studies and Research Institute, Ain Shams University.

Arabic Publications

1. Salah Shafi'i (1959): A study of the state of rain and torrential rains in the northwest coast. Academy of Scientific Research and Technology – Cairo – Egypt – Report No. 3
2. Dawlat Muhammad Sadeq (1965): The New Valley... A Geographical Study of the Kharga Depression. The Cultural Season Volume of the Egyptian Geographical Society, pp. 105–158.
3. Ali Abdel-Wahhab Shaheen (1965): Notes on the geomorphology of the eastern region of Mariout. Journal of the Faculty of Arts, University of Alexandria, Volume 17.
4. Muhammad Safi Al-Din Abu Al-Ezz (1966): The Morphology of the Egyptian Lands, Dar Al-Nahda Al-Arabiya, Cairo.
5. Muhammad Ghanem Al-Hanafi (1971): The Impact of Agriculture on the Bedouin Settlement in the Northwest Coast Region: The Executive Authority for Desert Projects, Egypt.
6. Taha Mahmoud Gad (1974): Dakhla Depression – a geomorphological study. PhD thesis in physical geography, Faculty of Arts, Ain Shams University.
7. Hassan Ahmed Abu El-Enein (1975): Geographical features of the Western Desert in the Arab Republic of Egypt towards the reconstruction of the Western Desert. Journal of the Faculty of Arts, Alexandria University, Volume 25, pp. 183–240
8. Nabil Sayed Embabi (1978): Movement of crescent dunes and its impact on urbanization and urbanization in the Kharga Oasis Depression. Middle East Center Journal, Ain Shams University.
9. Fathi Abd al-Salam Hammad and Salah Muhammad Abd al-Mughith (1978): The importance of applications of geomorphology in land and water uses in the northwest coast. Academy of Scientific Research and Technology – Cairo – Conference on the Role of Scientific Research and Technology in the Development and Reconstruction of Western Desert, 7–9 May 1978.
10. Mamdouh Haseeb, Fayez Salib, and Muhammad Saeed Badawy (1978) (a): Technical and economical feasibility study for the optimal use of the available valleys and the maximization of production in the cultivated reeds in Siwa Oasis. Academy of Scientific Research and Technology – Cairo – Conference on the Role of Scientific Research and Technology in the Development of Western Desert, 7–9 May 1978

A.-A. Gad, *Digital Land Resources to Support Sustainable Development in Egypt*, Deltas of the World,
https://doi.org/10.1007/978-3-031-37167-7

11. Mamdouh Haseeb, Fayez Salib, and Muhammad Saeed Badawi (1978) (b): Minute report on the achievements of the reconstruction of Western Sahara and the most important problems, suggestions and recommendations. Academy of Scientific Research and Technology – Cairo – Conference on the role of scientific research and technology in the development and reconstruction of Western Sahara, 7–9 May 1978
12. Muhammad Muhammad Al-Shazly and Ibrahim Hassan Hayda. (1978): Lights on the emergence and origin of groundwater in the Western Desert as a result of attempts to determine the age of this water. Academy of Scientific Research and Technology – Cairo – Conference on the Role of Scientific Research and Technology in the Development and Reconstruction of Western Desert, 7–9 May 1978.
13. Mohamed Atef Abdel Salam and Ahmed Amin Harja (1978): Northern Coast Lands, Academy of Scientific Research and Technology – Cairo – Conference on the Role of Scientific Research and Technology in the Development and Reconstruction of Western Desert, 7–9 May 1978.
14. Muhammad Atef Abd al-Salam, Tamadir al-Hadidi, and Muhammad Abd al-Fattah al-Sibai (1978): A summary of the activity of the Desert Institute in the field of studies for the exploitation of sandy lands in the Western Desert. Academy of Scientific Research and Technology – Cairo – Conference on the Role of Scientific Research and Technology in the Development and Reconstruction of Western Desert, 7–9 May 1978.
15. Mohamed Abdel-Hadi, Ibrahim El-Qassas, and others (1978): The role of the Remote Sensing Center in the development of Western Sahara. Academy of Scientific Research and Technology – Cairo – Conference on the Role of Scientific Research and Technology in the Development and Reconstruction of Western Desert, 7–9 May 1978.
16. F. Baslereh and Abdel-Fattah Ibrahim (1978): Studies on the five-year plan for the development and reconstruction of Western Sahara 1975–2025 Academy of Scientific Research and Technology – Cairo – Conference on the role of scientific research and technology in the development and reconstruction of Western Sahara, 7–9 May 1978.
17. Salah Abdel-Moghith and Saeed Mahmoud Atwa (1978): Groundwater Quality in the Northwest Coast. Academy of Scientific Research and Technology – Cairo – Conference on the Role of Scientific Research and Technology in the Development of Western Desert, 7–9 May 1978.
18. The National Authority for Remote Sensing and Space Sciences (1997): Torrents and means of confronting their dangers to the cities of the Red Sea coast in the Egyptian lands. On a scientific report submitted to the Red Sea Governorate; August 1997; The National Authority for Remote Sensing and Space Sciences, Cairo, 294 pages
19. F. Baslereh and Abdel-Fattah Ibrahim (1978): Studies on the five-year plan for the development and reconstruction of Western Sahara 1975–2025 Academy of Scientific Research and Technology – Cairo – Conference on the role of scientific research and technology in the development and reconstruction of Western Sahara, 7–9 May 1978.

20. Salah Abdel-Moghith and Saeed Mahmoud Atwa (1978): Groundwater Quality in the Northwest Coast. Academy of Scientific Research and Technology – Cairo – Conference on the Role of Scientific Research and Technology in the Development of Western Desert, 7–9 May 1978.
21. The National Authority for Remote Sensing and Space Sciences (1997): Torrents and means of confronting their dangers to the cities of the Red Sea coast in the Egyptian lands. On a scientific report submitted to the Red Sea Governorate; August 1997; The National Authority for Remote Sensing and Space Sciences, Cairo, 294 pages.
22. Rushdi Saeed (2001): The Nile River, its origins and the use of its waters in the past and the future. Second edition, Dar Al-Hilal, Cairo, 342

Appendix I: Morphological Description of the Studied Soil Profiles

Profile No.: 1
Date of description: 9/2006
Location
Longitude: 30° 40′ E
Latitude: 31° 25′ N
Elevation: 1.5 m a.sl.
Soil taxonomy: *Typic Aquisalids*
Topography: almost flat
Land form: coastal plain
Slope: nearly level
Land use & vegetation: cultivated with sugar beet
Parent material: aeolian deposits
Drainage: moderately well drained
Ground water table: 75 cm

Depth (cm)	Description
0–40	Yellowish brown (10 YR 5/6) moist, light yellowish brown (10 YR 6/4) dry, sand, structure less, loose moderate medium and fine gravel and grits common medium to coarse discontinuous random pores, many fine roots moderate effervescence with HCl, diffuse smooth boundary.
40–75	Yellowish brown (10 YR 5/6), dry light yellowish brown (10 YR 6/4) dry, sandy, single grains, hard, cemented abundant pores, few soft lime concretions, violent effervescence. Common coarse and medium discontinuous random pores; common fine and medium roots; slight effervescence with HCl.

Profile No.: 2
Date of description: 9/2006
Location

A.-A. Gad, *Digital Land Resources to Support Sustainable Development in Egypt*,
Deltas of the World,
https://doi.org/10.1007/978-3-031-37167-7

Longitude: 30° 20′ E
Latitude: 31° 30′ N
Elevation: 1.0 m a.sl.
Soil taxonomy: *Typic Quartizipsamments*
Topography: almost flat
Land form: coastal plain
Slope: nearly level
Land use & vegetation: cultivated with sugar beet
Parent material: aeolian deposits
Drainage: well drained
Ground water table: 100 cm

Depth (cm)	Description
0–25	Yellowish brown (10 YR 5/6) moist, pale brown (10 YR 6/3) dry, sandy, single grains, loose moderate fine gravel and grits common medium to coarse discontinuous random pores, many fine roots moderate effervescence with HCl, clear smooth boundary.
25–55	Yellowish brown (10 YR 5/6) moist, light yellowish brown (10 YR 6/4) dry sand, single grains, hard, few gravels, few soft lime concretions, moderate effervescence, common fine and medium roots; slight effervescence with HCl, clear smooth boundary.
55–100	Dark yellowish brown (10 YR 4/4) moist, Yellowish brown (10 YR 5/4)dry, slightly hard medium coarse sand, single grains few lime and gypsum concretion, common coarse and medium discontinuous random pores; moderate effervescence with HCl.

Profile No.: 3
Date of description: 9/2006
Location
Longitude: 31° 05′ E
Latitude: 31° 35′ N
Elevation: 1.0 m a.sl.
Soil taxonomy: *Typic Haplosalids*
Topography: almost flat
Land form: coastal plain
Slope: nearly level
Land use & vegetation: cultivated with clover
Parent material: aeolian deposits
Drainage: well drained
Ground water table: 110 cm

Depth (cm)	Description
0–30	Yellowish brown (10 YR 5/6) moist, pale brown (10 YR 6/3) dry, sandy, single grains, loose moderate medium and fine gravel and grits common medium to coarse discontinuous random pores, many fine roots moderate effervescence with HCl, clear smooth boundary.

(continued)

Depth (cm)	Description
30–60	Yellowish brown (10 YR 5/6) moist, light yellowish brown (10 YR 6/4) dry sand, single grains, hard, few gravels, few soft white lime concretions, violent effervescence, common fine and medium roots; slight effervescence with HCl, clear smooth boundary.
60–110	Dark yellowish brown (10 YR 4/4) moist, Yellowish brown (10 YR 5/4)dry, slightly hard medium coarse sandy loam, few lime and gypsum concretion, common coarse and medium discontinuous random pores; moderate effervescence

Profile No.: 4
Date of description: 9/2006
Location:
Longitude: 30° 25′ E
Latitude: 31° 20′ N
Elevation: 1.5 m a.sl.
Soil taxonomy: *Typic Aquisalids*
Topography: almost flat
Land form: coastal plain
Slope: nearly level
Land use & vegetation: cultivated with sugar beet
Parent material: aeolian deposits
Drainage: moderately well drained
Ground water table: 70 cm

Depth (cm)	Description
0–35	Light yellowish brown (10YR 6/4) moist, very pale brown (10 Y 7/3) dry, loamy sand, weak sub angular blocky, moderate effervescence, common medium to fine discontinuous random pores, many fine roots moderate effervescence with HCl, clear smooth boundary.
35–70	Yellowish brown (10 YR 5/6) moist, light yellowish brown (10 YR 6/4) dry, sand, single grains, slightly cemented; many gravel and grits common coarse and medium discontinuous random pores; common fine and medium roots; slight effervescence with HCl.

Profile No.: 5
Date of description: 9/2006
Location
Longitude: 31° 05′ E
Latitude: 31° 35′ N
Elevation: 1.0 m a.sl.
Soil taxonomy: *Typic Quartizipsamments*
Topography: almost flat

Land form: coastal plain
Slope: nearly level
Land use & vegetation: cultivated with clover
Parent material: aeolian deposits
Drainage: well drained
Ground water table: 100 cm

Depth (cm)	Description
0–25	Yellowish brown (10 YR 5/6) moist, light yellowish brown (10 YR 6/4) dry, sandy, single grains, few fine and medium gravel, common medium to coarse discontinuous random pores, many fine roots, moderate effervescence with HCl, clear smooth boundary.
25–50	Yellowish brown (10 YR 5/6) moist, light yellowish brown (10 YR 6/4) dry, sandy, single grains, few gravels, few soft lime concretions, strong effervescence, common fine and medium roots; slight effervescence with HCl, clear smooth boundary.
50–100	Dark yellowish brown (10 YR 4/4) moist, Yellowish brown (10 YR 5/4)dry, sandy, few lime concretion, common coarse and medium discontinuous random pores; moderate effervescence with HCl.

Profile No.: 6
Date of description: 9/2006
Location:
Longitude: 30° 36′ E
Latitude: 31° 15′ N
Elevation: 2.0 m a.sl.
Soil taxonomy: Vertic Torrivluvents
Topography: flat to almost flat
Land form: fluvio-lacustrine plain/terraces
Slope: nearly level
Land use & vegetation: fallow after rice
Parent material: alluvial-lecustrine deposits
Drainage: moderately well drained
Ground water table: 110 cm

Depth (cm)	Description
0–30	Very dark brown (10YR2/2) moist, dark brown (10YR3/3) dry, clayey; string fine and medium sub-angular blocky; very sticky, very plastic, firm, hard; many fine to medium continuous and discontinuous interstitial pores; many fine and medium roots; slightly effervescence with HCl; clear gradual boundary.
30–50	Very dark grayish brown (10YR3/2) moist; dark grayish Brown (10YR4/2) dry; clayey; moderate to strong medium angular to sub angular blocky; slightly sticky slightly plastic, friable, hard; common fine and medium discontinuous random pores;

(continued)

Depth (cm)	Description
	common fine and medium roots, slightly effervescence with HCl, clear smooth boundary.
50–90	Dark brown (10YR 3/3) moist, yellowish brown (10YR 5/4) dry, clay loam weak medium angular to sub angular blocky, sticky, Plastic friable, hard; common fine and medium discontinuous random pores, few very soft lime concretion; few fine roots, slight effervescence with Hcl clear smooth boundary.
90–110	Dark brown (10YR3/3) moist; grayish Brown (10YR5/2) dry; clay loam weak fine and medium sub angular blocky; sticky, plastic, firm, hard; few fine and medium discontinuous pores; slight effervescence with HCl.

Profile No.: 7
Date of description: 9/2006
Location
Longitude: 30° 23′ E
Latitude: 31° 18′ N
Elevation: 3.0 m a.s.l.
Soil taxonomy: *Aquic Torrifluvents*
Topography: Almost flat
Land form: lacustrine plain/over flow mantle
Slope: nearly level
Land use & vegetation: fallow after rice
Parent material: lacustrine/alluvial deposits
Drainage: moderately well drained
Ground water table: 60 cm

Depth (cm)	Description
0–10	Very dark grayish brown (10YR4/2) moist, brown (10YR5/3) dry, clayey; moderate medium angular blocky; sticky, plastic; firm, hard common fine discontinuous random pores, few fresh and humified residual of organic matter, many fine and medium roots, slightly effervescence with HCl, clear smooth boundary.
10–25	Very dark grayish brown (10YR3/2) moist, dark grayish brown (10YR4/2) dry, clayey; Moderate to strong fine to medium sub-angular, sticky, plastic, firm, hard; common fine and medium continuous and discontinuous channels, few fine soft segregation of CaCO3, common fine to medium roots,. Slightly effervescence with HCl, clear smooth boundary.
25–40	Dark brown (10YR 3/3) moist, brown (10YR 4/3) dry: clayey: moderate medium angular to sub angular blocky; very sticky, very Plastic; very firm, hard; common fine and medium continuous and discontinuous random pores; few very fine lime concretion, few fine roots. Slight effervescence with HCl, clear smooth boundary.
40–60	Very dark grayish brown (10YR 3/2) moist, dark grayish brown (10YR 4/2) dry: clay loam; moderate fine to medium angular blocky; very sticky, very plastic; firm; hard; common fine random pores; common fine soft segregation of CaCO3, moderate effervescence with HCl.

Profile No.: 8
Date of description: 9/2006
Location
Longitude: 31° 04′ E
Latitude: 31° 24′ N
Elevation: 1.0 m a.s.l.
Soil taxonomy: *Typic Aquisalids*
Topography: gently undulating
Land form: lacustrine plain/basin
Slope: gently slope
Land use & vegetation: cultivated with clover after catton
Parent material: lacustrine/alluvial deposites
Drainage: imperfectly drained
Ground water table: 50 cm

Depth (cm)	Description
0–25	Dark grayish brown (10YR4/2) moist, grayish brown (10YR5/2) dry; clayey; moderate medium granular; very sticky, very plastic, friable slightly hard; common fine and medium discontinuous vughs; many fine to medium roots, slight effervescence with HCl, clear smooth boundary.
25–50	Dark brown (10YR3/3) moist, dark grayish brown (10YR4/2) dry; clayey; moderate to strong medium angular blocky; sticky, plastic, firm, hard; common fine and medium continuous and discontinuous interstitial pores; few fine residual of shell fragments, few fine roots; slight effervescence with HCl.

Profile No.: 9
Date of description: 9/2006
Location:
Longitude: 30° 52′ E
Latitude: 30° 16′ N
Elevation: 6.75 m a.s.l.
Soil taxonomy: *Typic Haplosalids*
Topography: almost flat
Land form: lacustrine plain/basin
Slope: gently slope
Land use & vegetation: cultivated with Maize
Parent material: alluvial deposits
Drainage: well drained
Ground water table: 110 cm

Depth (cm)	Description
0–20	Dark brown (10YR3/3) moist, brown (10YR4/3) dry; clay loam; moderate to strong medium granular; sticky, plastic, firm, extremely hard; common fine discontinuous random exped and interstitial pores, few humified patches of organic matter; many fine to medium roots; slight effervescence with HCl, clear smooth boundary.
20–45	Very dark grayish brown (10YR 3/2) moist, dark grayish brown (10YR 4/2) dry; clayey; moderate fine to medium sub angular blocky; very sticky, very plastic; extremely firm, extremely hard; common fine to medium continuous and discontinuous random pores; common fine to medium roots; slight effervescence with HCl, clear smooth boundary..
45–80	Very dark grayish brown (10YR 3/2) moist, dark grayish brown (10YR 4/2) dry; clayey; moderate to strong medium sub angular blocky; very sticky very plastic; very firm, very hard; common fine and medium discontinuous random pores; slight effervescence with HCl; clear smooth boundary.
80–110	Dark brown(10YR 3/3) moist, brown (10YR 4/3) dry; clayey; strong medium angular to sub angular blocky; very sticky, very plastic; very firm; very hard; common fine and medium continuous and discontinuous interstitial pores; slight effervescence with HCl.

Profile No.: 10
Date of description: 9/2006
Location:
Longitude: 31° 11′ E
Latitude: 30° 14′ N
Elevation: 9.0 m a.s.l.
Soil taxonomy: *Vertic Torrifluvents*
Topography: almost flat
Land form: flood plain/river terraces
Slope: nearly level
Land use & vegetation: cultivated with rice
Parent material: alluvial deposits
Drainage: well drained
Ground water table: 120 cm

Depth (cm)	Description
0–10	Dark yellowish brown (10YR3/3) moist, yellowish Brown (10YR5/2) dry, clayey; moderate to strong fine and medium blocky; very sticky and very plastic; very firm, very hard common fine to medium discontinuous random interstitial and few channel pores, few fresh and humified residual of organic matter, many fine to medium roots, slightly effervescence with Hal, clear smooth boundary.
10–35	Dark brown (10YR3/3) moist, Brown (10YR5/2) dry, clayey; Moderate small to medium sub angular blocky; very sticky and very plastic, very firm, very hard; common fine continuous and discontinuous random interstitial pores, thin shinny

(continued)

Depth (cm)	Description
	discontinuous organic clay films lining the voids; common fine to medium roots, slightly effervescence with HCl, gradual smooth boundary
35–70	Dark brown (10YR 3/3) moist, brown (10YR 5/3) dry: clayey: Moderate to strong medium sub angular blocky; sticky and Plastic; firm, hard; common fine to medium discontinuous random interstitial pores; few fine slight effervescence with HCl, clear smooth boundary
70–120	Dark yellowish brown (10YR 4/4) moist, yellowish brown (10YR 5/1) dry: clayey; moderate medium sub angular blocky; sticky and plastic; firms; hard; common fine to medium continuous and discontinuous interstitial pores; slightly effervescence with HCl

Profile No.: 11
Date of description: 9/2006
Location
Longitude: 31° 11′ E
Latitude: 30° 29′ N
Elevation: 13.0 m a.s.l.
Soil taxonomy: *Typic Quartizipsamments*
Topography: almost flat
Land form: flood plain/over flow basin
Slope: nearly level
Land use & vegetation: barren
Parent material: aeolian deposits
Drainage: well drained
Ground water table: 115 cm

Depth (cm)	Description
0–30	Yellowish brown (10 YR 5/6) moist, light yellowish brown (10 YR 6/4) dry, sandy, single grains, few gravels, strong effervescence, common fine and medium roots; slight effervescence with HCl, clear smooth boundary.
30–65	Yellowish brown (10 YR 5/6) moist, pale brown (10 YR 6/3) dry, sandy, single grains, loose, common medium to coarse discontinuous random pores, many fine roots moderate effervescence with HCl, clear smooth boundary.
65–115	Dark yellowish brown (10 YR 4/4) moist, Yellowish brown (10 YR 5/4)dry, slightly hard medium coarse sandy loam, common coarse and medium discontinuous random pores; moderate effervescence with HCl.

Profile No.: 12
Date of description: 9/2006
Location:
Longitude: 31° 05′ E

Latitude: 30° 35′ N
Elevation: 13.5 m a.s.l.
Soil taxonomy: *Vertic Torrivluvents*
Topography: flat to almost flat
Land form: flood plain/river terraces
Slope: gently slope
Land use & vegetation: cultivated with Cotton.
Parent material: Alluvial deposits
Drainage: well drained soil
Ground water table: 115 cm

Depth (cm)	Description
0–30	Dark grayish brown (10YR4/2) moist, grayish brown (10YR5/5) dry; clay loam; moderate small to medium granular; sticky, plastic; firm, hard; common fine to medium discontinuous random interstitial pores; many fine to medium roots; slight effervescence with HCl; clear smooth boundary.
30–55	Dark grayish brown (10YR4/2) moist, light brownish gray (10YR6/2) dry; clayey; strong small to medium sub angular blocky; sticky, plastic, firm, hard; common fine to medium discontinuous random interstitial pores; common fine to medium roots; slight effervescence with HCl, gradual smooth boundary
55–115	Very dark grayish brown (10YR 3/2) moist, dark grayish brown (10YR 4/2) dry, clayey; moderate small to medium sub angular blocky; very sticky, very Plastic, extremely firm, extremely hard; fine to medium continuous and discontinuous random interstitial pores; thin shinny discontinuous clay films lining the voids few fine roots slight effervescence with HCl.

Profile No.: 13
Date of description: 9/2006
Location:
Longitude: 30° 59′ E
Latitude: 30° 31′ N
Elevation: 14.5 m a.s.l.
Soil taxonomy: *Typic Torrifluvents*
Topography: flat to almost flat
Land form: flood plain/river terraces
Slope: nearly level
Land use & vegetation: cultivated with winter vegetables.
Parent material: alluvial deposits
Drainage: well drained soils
Ground water table: 110 cm

Depth (cm)	Description
0–20	Dark grayish brown (10YR4/2) moist, grayish Brown (10YR5/2) dry, clay loam; strong small to medium granular; moderately sticky, moderately plastic; moderately firm, moderately hard; common fine to medium continuous and discontinuous random interstitial pores; many fine to medium roots, slight effervescence with HCl.
20–45	Dark grayish brown (10YR4/2) moist, grayish brown (10YR5/2) dry, clay loam; moderately small to medium sub angular blocky; sticky, plastic, firm, hard; common fine to medium discontinuous random interstitial and few channel pores; common fine to medium roots,. slight effervescence with HCl, clear smooth boundary.
45–80	Dark grayish brown (10YR 4/2) moist, grayish brown (10YR 5/2) dry: clayey: Moderately medium sub angular blocky, sticky,Plastic; moderately firm, hard; common fine to medium discontinuous random pores; few fine roots, slight effervescence with HCl, clear smooth boundary
80–110	Very dark grayish brown (10YR 3/2) moist, dark grayish brown (10YR 5/1) dry; clayey; strong fine to medium sub angular blocky; very sticky, very plastic; very firm, very hard, very few clay skins, common fine continuous and discontinuous random pores; slight effervescence with HCl.

Profile No.: 14
Date of description: 9/2006
Location:
Longitude: 30° 51′ E
Latitude: 30° 31′ N
Elevation: 14.5 m a.s.l.
Soil taxonomy: *Vertic Torrivluvents*
Topography: gently undulating
Land form: flood plain/river terraces
Slope: gently slope
Land use & vegetation: cultivated with Maize
Parent material: alluvial deposits
Drainage: moderately well drained
Ground water table: 100 cm

Depth (cm)	Description
0–25	Dark grayish brown (10YR4/2) moist, grayish brown (10YR5/2) dry; sandy clay loam strong small to granular; sticky and plastic; firm, hard common fine to medium continuous and discontinuous and few channel pores; many fine and medium roots, slightly effervescence with HCl; gradual smooth boundary.
25–55	Dark gray (10YR4/1) moist, gray (10YR5/1) dry, clay loam; strong small to medium angular to sub angular blocky; very sticky and very plastic, extremely firm, extremely hard; common fine to medium discontinuous random interstitial pores, thin shinny continuous clay films lining the voids; few fine to medium roots, slightly effervescence with HCL, clear smooth boundary

(continued)

Depth (cm)	Description
55–100	Dark gray (10YR 4/1) moist, gray (10YR 5/1) dry: clayey: moderately to strong small to medium angular to sub angular blocky; very sticky and very Plastic; very firm, very hard; common fine discontinuous random pores; slightly effervescence with HCL.

Profile No.: 15
Date of description: 9/2006
Location:
Longitude: 31° 05′ E
Latitude: 30° 21′ N
Elevation: 14.0 m a.s.l.
Soil taxonomy: *Vertic Torrivluvents*
Topography: almost flat to gently undulating
Land form: flood plain/river terraces
Slope: gently slope
Land use & vegetation: cultivated with cotton.
Parent material: alluvial deposits
Drainage: well drained soils
Ground water table: 110 cm

Depth (cm)	Description
0–30	Dark brown (10YR3/3) moist, grayish Brown (10YR5/2) dry, sandy clay, moderate medium granular, sticky and plastic, firm, hard common fine discontinuous random interstitial and few channel pores, few fresh and humified residual of organic matter, many fine to medium roots, slight effervescence with HCl, clear smooth boundary
30–60	Dark gray (10YR 4/1) moist, gray(10YR 5/1) dry: clayey: Moderate to strong medium to coarse sub angular blocky; sticky and Plastic: very firm, very hard; common fine to medium discontinuous random interstitial pores; slight effervescence with HCL, clear smooth boundary
60–90	Dark gray (10YR 4/1) moist, gray(10YR 5/1) dry: clayey Moderate to strong medium to coarse sub angular blocky; sticky and Plastic: very firm, very hard; common fine to medium discontinuous random interstitial pores; slight effervescence with HCL, clear smooth boundary
90–110	Dark gray (10YR 4/1) moist, gray(10YR 5/1) dry: clay loam: moderate fine to medium angular to sub angular blocky, very sticky and very plastic; very firms; very hard; common fine to medium discontinuous random interstitial pores; slightly effervescence with HCL

Profile No.: 16
Date of description: 9/2006
Location:
Longitude: 30° 56′ E

Latitude: 30° 48′ N
Elevation: 12.5 m a.s.l.
Soil taxonomy: *Typic Torrifuvents*
Topography: almost flat to gently undulating
Land form: flood plain/Basin
Slope: gently slope
Land use & vegetation: fallow after cotton
Parent material: alluvial deposits
Drainage: well drained soils
Ground water table: 100 cm

Depth (cm)	Description
0–30	Very dark grayish brown (10YR3/2) moist, grayish Brown (10YR4/3) dry, silty clay; moderate fine to medium sub angular; sticky, plastic; firm, hard; common fine discontinuous random interstitial pores; common fine roots, slight effervescence with HCl, gradual smooth boundary.
30–55	Very dark grayish brown (10YR3/2) moist, dark grayish brown (10YR4/3) dry, silty clay; moderate fine to medium sub angular blocky; sticky, plastic, firm, hard; common fine continuous and discontinuous random interstitial pores; common fine to medium roots, Slight effervescence with HCl; clear smooth boundary
55–75	Dark grayish brown (10YR 4/2) moist, grayish brown (10YR 5/2) dry; silty clay; Moderate to strong medium sub angular blocky, sticky, Plastic; firm, hard; common fine continuous and discontinuous random pores; few fine roots; clear smooth boundary.
75–100	Dark grayish brown (10YR 4/2) moist, grayish brown (10YR 5/1) dry; clayey; moderate medium sub angular blocky; very sticky, very plastic; extremely firm, extremely hard; slight effervescence with HCl.

Profile No.: 17
Date of description: 9/2006
Location:
Longitude: 30° 51′ E
Latitude: 30° 16′ N
Elevation: 15.0 m a.s.l.
Soil taxonomy: *Vertic Torrivluvents*
Topography: almost flat
Land form: flood plain/river levees
Slope: gently slope
Land use & vegetation: cultivated with orchards (orange-mango).
Parent material: alluvial deposits
Drainage: well drained
Ground water table: 110 cm

Depth (cm)	Description
0–25	Dark gray (10YR4/1) moist, gray (10YR5/1) dry; clayey; moderate medium sub angular blocky; sticky, plastic; moderately firm, moderately hard; common fine to medium discontinuous random exped interstitial pores; few fresh and humified residual of organic matter; many fine to coarse roots; slight effervescence with Hcl; clear smooth boundary.
25–65	Very dark grayish brown (10YR3/2) moist, dark grayish brown (10YR4/3) dry; clayey; moderate fine to medium sub angular blocky; sticky, plastic; moderately firm, moderately hard; common fine to medium discontinuous random exped interstitial pores; common fine to coarse roots; slight effervescence with Hcl, clear smooth boundary.
65–110	Very dark grayish brown (10YR 3/2) moist, dark grayish brown (10YR 4/3) dry, clayey; moderate fine to medium sub angular blocky; sticky, plastic; firm, hard; common fine and medium discontinuous random interstitial pores; slight effervescence with Hcl.

Profile No.: 18
Date of description: 9/2006
Location:
Longitude: 31° 12′ E
Latitude: 30° 11′ N
Elevation: 7.0 m a.s.l.
Soil taxonomy: *Typic Torrivluvents*
Topography: almost flat
Land form: flood plain/river terraces
Slope: gently slope
Land use & vegetation: fallow after Maize.
Parent material: alluvial deposits
Drainage: moderately well drained
Ground water table: 90 cm

Depth (cm)	Description
0–15	Dark grayish brown (10YR4/2) moist, grayish brown (10YR5/2) dry; clay loam strong medium angular to sub angular blocky; moderate sticky, moderate plastic, firm, hard; common fine to medium discontinuous random exped medium interstitial pores; few humified residual of organic matter; common fine to medium roots; slight effervescence with HCl; gradual smooth boundary.
15–35	Dark grayish brown (10YR4/2) moist, grayish brown (10YR5/2) dry; clay loam; moderate medium sub angular blocky; sticky, plastic, firm, hard; common fine to medium discontinuous random interstitial pores; common fine to medium roots; slight effervescence with HCl, clear smooth boundary
35–90	Dark grayish brown (10YR 4/2) moist, grayish brown (10YR 5/2) dry; silty clay; moderate to strong medium sub angular blocky; sticky, Plastic, very firm, very hard; common fine and medium discontinuous random interstitial pores; slight effervescence with HCl.

Profile No.: 19
Date of description: 9/2006
Location
Longitude: 31° 18′ E
Latitude: 30° 24′ N
Elevation: 23.0 m a.s.l.
Soil taxonomy: *Typic Quartizipsamments*
Topography: almost flat
Land form: flood plain
Slope: nearly level
Land use & vegetation: barren
Parent material: aeolian deposits
Drainage: moderately well drained
Ground water table: 85 cm

Depth (cm)	Description
0–20	Light yellowish brown (10YR 6/4) moist, very pale brown (10 YR 7/3) dry, sandy, single grains, few gravels, strong effervescence, common fine and medium roots; slight effervescence with HCl, diffuse smooth boundary.
20–45	Yellowish brown (10 YR 5/6) moist, pale brown (10 YR 6/3) dry, sandy loam, single grains, common gravels, loose, common fine to medium discontinuous random pores, many fine roots moderate effervescence with HCl, diffuse smooth boundary.
45–85	Dark yellowish brown (10 YR 4/4) moist, Yellowish brown (10 YR 5/4)dry, slightly hard medium coarse loamy sand, common coarse and medium discontinuous random pores; moderate effervescence with HCl.

Profile No.: 20
Date of description: 9/2006
Location:
Longitude: 31° 30′ E
Latitude: 30° 31′ N
Elevation: 20.0 m a.s.l.
Soil taxonomy: *Vertic Torrivluvents*
Topography: almost flat
Land form: flood plain/overflow mantle
Slope: nearly level to gently slope
Land use & vegetation: fallow after maize
Parent material: alluvial deposits
Drainage: poor drained soils
Ground water table: 90 cm

Depth (cm)	Description
0–25	Dark brown (10YR2/3) moist, grayish Brown (10YR5/2) dry, clayey; moderate fine to medium granular; sticky, plastic; firm, hard; common fine to medium discontinuous random exped and interstitial pores; many fine to medium roots, slight effervescence with Hcl, clear smooth boundary.
25–60	Dark grayish brown (10YR4/2) moist, grayish brown (10YR5/4) dry, clayey; moderate fine to medium sub angular blocky; very sticky, very plastic, very firm, extremely hard; common fine to medium discontinuous random exped and interstitial pores; few humified residual of organic matter; common fine to medium roots,. Slight effervescence with Hcl; clear smooth boundary
60–90	Dark brown (10YR 3/3) moist, grayish brown (10YR 5/3) dry; clayey; strong medium sub angular blocky; very sticky, very plastic; extremely firm, extremely hard; common fine discontinuous random interstitial pores; slight effervescence with Hcl.

Profile No.: 21
Date of description: 9/2006
Location:
Longitude: 31° 41′ E
Latitude: 30° 40′ N
Elevation: 28.0 m a.sl.
Soil taxonomy: *Typic Torriorthents*
Topography: undulating
Land form: aeolian plain
Slope: gently slope
Land use & vegetation: barren
Parent material: aeolian deposits
Drainage: poor drained soils
Ground water table: 90 cm

Depth (cm)	Description
0–25	Yellowish brown (10 YR 5/6) moist, pale brown (10 YR 6/3) dry, sandy, single grains, moderate fine gravel and grits common medium to coarse discontinuous random pores, common lime concretions, common fine to medium roots strong effervescence with HCl, diffuse smooth boundary.
25–60	Yellowish brown (10 YR 5/6) moist, light yellowish brown (10 YR 6/4) dry, sandy, single grains, hard, common coarse gravels, common lime concretions, moderate effervescence, common fine and medium roots; strong effervescence with HCl, clear smooth boundary.
60–90	Dark yellowish brown (10 YR 4/4) moist, Yellowish brown (10 YR 5/4)dry, sandy, few lime and gypsum concretion, common coarse and medium discontinuous random pores; moderate effervescence with HCl.

Profile No.: 22
Date of description: 9/2006
Location:
Longitude: 31° 56′ E
Latitude: 30° 42′ N
Elevation: 31.0 m a.sl.
Soil taxonomy: *Typic Quartizipsamments*
Topography: gently undulating
Land form: sand sheet
Slope: gently slope
Land use & vegetation: cultivated with sugar beet
Parent material: aeolian deposits
Drainage: well drained soils
Ground water table: 110 cm

Depth (cm)	Description
0–30	Light yellowish brown (10YR 6/4) moist, very pale brown (10 Y 7/3) dry, sandy, single grains, moderate effervescence, common medium to fine discontinuous random pores, many fine roots moderate effervescence with HCl, clear smooth boundary
30–60	Yellowish brown (10 YR 5/6) moist, light yellowish brown (10 YR 6/4) dry sand, single grains, hard, few gravels, few soft lime concretions, moderate effervescence, common fine and medium roots; slight effervescence with HCl, clear smooth boundary.
60–110	Dark yellowish brown (10 YR 4/4) moist, Yellowish brown (10 YR 5/4)dry, sandy, single grain, few lime concretion, common coarse and medium discontinuous random pores; moderate effervescence with HCl.

Profile No.: 23
Date of description: 9/2006
Location:
Longitude: 32° 08′ E
Latitude: 30° 14′ N
Elevation: 24.5 m a.sl.
Soil taxonomy: *Typic Torripsamments*
Topography: almost flat
Land form: sandy plain
Slope: nearly level
Land use & vegetation: cultivated with sugar beet
Parent material: aeolian deposits
Drainage: moderately well drained
Ground water table: not hit

Depth (cm)	Description
0–35	Yellowish brown (10 YR 5/6) moist, light yellowish brown (10 YR 6/4) dry, sandy, structure less, moderate medium and fine gravel, common medium to coarse discontinuous random pores, few lime concretion, common fine roots moderate effervescence with HCl, diffuse smooth boundary.
35–70	Yellowish brown (10 YR 5/6), dry light yellowish brown (10 YR 6/4) dry, sandy, single grains, hard, cemented abundant pores, common lime concretions, strong effervescence with HCl. common coarse and medium discontinuous random pores; slight effervescence with HCl.

Profile No.: 24
Date of description: 9/2006
Location:
Longitude: 32° 18′ E
Latitude: 31° 16′ N
Elevation: 1.5 m a.sl.
Soil taxonomy: *Typic Petrogypsids*
Topography: almost flat
Land form: sandy plain
Slope: nearly level
Land use & vegetation: barren
Parent material: aeolian deposits
Drainage: poor drained soils
Ground water table: not hit

Depth (cm)	Description
0–20	Light yellowish brown (10YR 6/4) moist, very pale brown (10 Y 7/3) dry, clay loam, weak sub angular blocky, moderate medium and fine gravel, common medium to coarse discontinuous random pores, few lime and common gypsum concretion, common fine roots moderate effervescence with HCl, diffuse smooth boundary.
20–75	Dark yellowish brown (10 YR 4/4) moist, Yellowish brown (10 YR 5/4)dry, clay loam, moderate fine sub angular blocky, hard, cemented abundant pores, common lime and gypsum concretions, strong effervescence with HCl. common coarse and medium discontinuous random pores; slight effervescence with HCl.

Profile No.: 25
Date of description: 9/2006
Location:
Longitude: 32° 14′ E
Latitude: 30° 50′ N
Elevation: 5.0 m a.s.l.
Soil taxonomy: *Typic Haplogypsids*

Topography: almost flat to gently undulating
Land form: lacustrine deposits
Slope: gently slope
Land use & vegetation: barren, common natural vegetation.
Parent material: alluvial deposits
Drainage: moderately well drained soils
Ground water table: 85 cm

Depth (cm)	Description
0–15	Dark gray (10YR4/1) moist, gray (10YR5/1) dry; sandy loam; moderate fine to medium angular to sub angular blocky; sticky, plastic; moderately firm, moderately hard; common fine to medium continuous and discontinuous random exped medium interstitial pores; few fresh and humified residual of organic matter; common fine to coarse roots; slight effervescence with Hcl; clear smooth boundary.
15–50	Dark gray (10YR4/1) moist, gray (10YR5/1) dry; sandy clay loam; moderate medium sub angular blocky; sticky, plastic; moderately firm, moderately hard; common fine to medium discontinuous random interstitial pores; thin shinny continuous clay films lining the voids; many medium to coarse roots; slight effervescence with Hcl, clear smooth boundary .
50–85	Very dark grayish brown (10YR 3/2) moist, dark grayish brown (10YR 4/3) dry, sandy clay; moderate to strong fine to medium sub angular blocky; sticky, plastic; firm, very hard; common fine to medium discontinuous random interstitial pores; slight effervescence with Hcl.

Profile No.: 26
Date of description: 9/2006
Location:
Longitude: 31° 57′ E
Latitude: 31° 02′ N
Elevation: 6.0 m a.s.l.
Soil taxonomy: *Typic Aquisalids*
Topography: Almost flat to gently undulating
Land form: alluvial/lacustrine plain/basin
Slope: Nearly level to gently slope
Land use & vegetation: Fallow after maize.
Parent material: alluvial/lacustrine deposits
Drainage: moderately well drained
Ground water table: 75 cm

Depth (cm)	Description
0–15	Dark brown (10YR2/3) moist, grayish brown (10YR5/2) dry; sandy clay loam; moderate medium angular to sub angular blocky; very sticky, very plastic; very firm, extremely hard; common fine to medium discontinuous random interstitial pores; many fine to medium roots; slight effervescence with Hcl; clear smooth boundary.
15–35	Dark brown (10YR2/3) moist, grayish brown (10YR5/2) dry; sandy clay loam; moderate medium angular blocky; sticky, plastic; very firm, very hard; common fine to medium discontinuous random interstitial pores; common fine to medium roots; slight effervescence with Hcl, gradual smooth boundary.
35–75	Dark brown (10YR 2/3) moist, grayish brown (10YR 5/2) dry, sandy clay loam; moderate medium to fine angular blocky; sticky, plastic; very firm, very hard; common fine to medium discontinuous random interstitial pores; slight effervescence with Hcl.

Profile No.: 27
Date of description: 9/2006
Location:
Longitude: 31° 42′ E
Latitude: 31° 40′ N
Elevation: 6.5 m a.s.l.
Soil taxonomy: *Typic Torrivluvents*
Topography: almost flat
Land form: flood plain/river terraces
Slope: nearly level
Land use & vegetation: cultivated with clover
Parent material: alluvial deposits
Drainage: moderately well drained soils
Ground water table: 70 cm

Depth (cm)	Description
0–20	Dark brown (10YR2/3) moist, grayish brown (10YR5/2) dry; clay loam; moderately fine to medium angular to sub angular blocky; very sticky, very plastic, very firm, extremely hard; common fine discontinuous random exped and interstitial pores, few humified patches of organic matter; many fine to coarse roots; slight effervescence with HCl, clear smooth boundary.
20–50	Dark grayish brown (10YR 4/2) moist, grayish brown (10YR 5/4) dry; clayey; strong fine to medium sub angular blocky; sticky, plastic; firm, hard; common fine to medium continuous and discontinuous random interstitial pores; few humified patches of organic matter; many fine to coarse roots; gradual smooth boundary.
50–70	Dark grayish brown (10YR 4/2) moist, grayish brown (10YR 5/4) dry; clayey; moderate fine to medium sub angular blocky; very sticky very plastic: very firm, very hard; common fine and medium discontinuous random exped and interstitial pores; slight smooth boundary.

Profile No.: 28
Date of description: 9/2006
Location:
Longitude: 31° 03′ E
Latitude: 30° 55′ N
Elevation: 6.25 m a.s.l.
Soil taxonomy: *Vertic Torrivluvents*
Topography: almost flat to gently undulating
Land form: flood plain/river terraces
Slope: nearly level to gently slope
Land use & vegetation: fallow after Rice
Parent material: alluvial deposits
Drainage: moderately well drained
Ground water table: 90 cm

Depth (cm)	Description
0–30	Very dark grayish brown (10YR3/2) moist, dark grayish brown (10YR4/3) dry; sandy clay loam; moderate to strong medium sub angular blocky; very sticky, very plastic, extremely firm, extremely hard; common fine discontinuous random exped and medium interstitial pores; many fine to medium roots; moderately effervescence with HCl; clear smooth boundary.
30–60	Dark brown (10YR2/3) moist, grayish brown (10YR5/2) dry; clay loam; strong medium sub angular blocky; very sticky, very plastic, extremely firm, extremely hard; common fine to medium discontinuous random interstitial pores; thin shinny discontinuous clay films lining the voids; few fine to medium roots; slight effervescence with HCl, clear smooth boundary .
60–90	Dark brown (10YR 2/3) moist, grayish brown (10YR 5/2) dry; clayey; strong medium sub angular blocky; very sticky, very Plastic, extremely firm, extremely hard; common fine and medium discontinuous random interstitial pores and exped; slight effervescence with HCl.

Profile No.: 29
Date of description: 9/2006
Location
Longitude: 30° 27′ E
Latitude: 31° 48′ N
Elevation: 2.5 m a.s.l.
Soil taxonomy: *Typic Torrivluvents*
Topography: gently undulating
Land form: flood plain/basin
Slope: gently slope
Land use & vegetation: fallow after rice
Parent material: alluvial deposits

Drainage: imperfectly drained
Ground water table: 110 cm

Depth (cm)	Description
0–20	Very dark grayish brown (10YR3/2) moist, dark grayish brown (10YR4/3) dry, sandy clay; moderate medium subangular blocky; sticky and plastic; firm, hard common fine discontinuous random interstitial pores, few fresh and humified residual of organic matter, many fine to medium roots, slightly effervescence with HCl, clear smooth boundary.
20–35	Dark brown (10YR4/3) moist, Brown (10YR5/3) dry, clayey; Moderate medium angular to subangular blocky; very sticky, very plastic, firm, hard; common fine to medium discontinuous random pores, thin shinny discontinuous organic clay films lining the voids; common fine to medium roots, slightly effervescence with HCl, clear smooth boundary.
35–65	Dark brown (10YR 3/3) moist, dark grayish brown (10YR 4/3) dry: clayey: Moderate to strong medium angular to subangular blocky; very sticky, very Plastic; very firm, very hard; common fine to medium continuous and discontinuous random pores; few fine roots slight effervescence with HCl, clear smooth boundary.
65–110	Dark yellowish brown (10YR 3/3) moist, dark grayish brown (10YR 4/3) dry: clayey; moderate to strong medium subangular blocky; very sticky and very plastic; very firm; very hard; common fine random pores; common fine soft segregation of $CaCO_3$, moderate effervescence with HCl.

Profile No.: 30
Date of description: 9/2006
Location
Longitude: 30° 51′ E
Latitude: 31° 15′ N
Elevation: 41.5 m a.s.l.
Soil taxonomy: *Typic Torriorthents*
Topography: undulating
Land form: old deltaic plain
Slope: gently slope
Land use & vegetation: barren
Parent material: old deltaic deposits
Drainage: moderately well drained
Ground water table: not hit

Depth (cm)	Description
0–15	Yellowish brown (10 YR 5/6) moist, light yellowish brown (10 YR 6/4) dry, loamy, massive, common coarse gravels, common lime concretions, moderate effervescence, common fine and medium roots; strong effervescence with HCl, clear smooth boundary.

(continued)

Depth (cm)	Description
15–35	Yellowish brown (10 YR 5/6) moist, pale brown (10 YR 6/3) dry, loamy, massive, common coarse gravels and grits common medium to coarse discontinuous random pores, common lime concretions, common fine to medium roots strong effervescence with HCl, diffuse smooth boundary.
35–75	Yellowish brown (10 YR 5/6) moist, light yellowish brown (10 YR 6/4) dry, clay loam, weak sub angular blocky, hard, common coarse gravels, common lime concretions, strong effervescence, common fine and medium roots; strong effervescence with HCl, clear smooth boundary.
75–130	Dark yellowish brown (10 YR 4/4) moist, Yellowish brown (10 YR 5/4)dry, clay loam, weak sub angular blocky, few lime concretion, common coarse and medium discontinuous random pores; moderate effervescence with HCl.

Profile No.: 31
Date of description: 9/2006
Location:
Longitude: 30° 36′ E
Latitude: 30° 34′ N
Elevation: 45.5 m a.sl.
Soil taxonomy: *Typic Haplocalcids*
Topography: undulating
Land form: sandy plain
Slope: nearly level
Land use & vegetation: barren
Parent material: old deltaic deposits
Drainage: poor drained soils
Ground water table: not hit

Depth (cm)	Description
0–25	Yellowish brown (10 YR 5/6) moist, light yellowish brown (10 YR 6/4) dry, sandy loam, structure less, moderate medium and fine gravel, common medium to coarse discontinuous random pores, common fine roots moderate effervescence with HCl, diffuse smooth boundary.
25–70	Yellowish brown (10 YR 5/6), dry light yellowish brown (10 YR 6/4) dry, loamy sand, single grains, hard, cemented abundant pores, common lime concretions, strong effervescence with HCl. common coarse and medium discontinuous random pores; slight effervescence with HCl.

Profile No.: 32
Date of description: 9/2006
Location:
Longitude: 32° 10′ E

Latitude: 30° 36′ N
Elevation: 50.0 m a.sl.
Soil taxonomy: *Typic Haplogypsids*
Topography: undulating
Land form: sandy plain
Slope: gently slope
Land use & vegetation: barren
Parent material: aeolian deposits
Drainage: poor drained soils
Ground water table: not hit

Depth (cm)	Description
0–15	Light yellowish brown (10YR 6/4) moist, very pale brown (10 Y 7/3) dry, sandy, massive, moderate medium and fine gravel, common medium to coarse discontinuous random pores, few lime and common gypsum concretion, common fine roots moderate effervescence with HCl, diffuse smooth boundary.
15–50	Dark yellowish brown (10 YR 4/4) moist, Yellowish brown (10 YR 5/4)dry, sandy, massive, hard, cemented abundant pores, common lime and gypsum concretions, strong effervescence with HCl. common coarse and medium discontinuous random pores; slight effervescence with HCl.
50–120	Yellowish brown (10 YR 5/6), dry light yellowish brown (10 YR 6/4) dry, sandy, massive, hard, cemented abundant pores, common lime concretions, strong effervescence with HCl. common coarse and medium discontinuous random pores; slight effervescence with HCl.

Profile No.: 33
Date of description: 9/2006
Location:
Longitude: 30° 33′ E
Latitude: 30° 07′ N
Elevation: 54.0 m a.sl.
Soil taxonomy: *Typic Torripsamments*
Topography: gently undulating
Land form: sand sheet
Slope: gently slope
Land use & vegetation: cultivated with sugar beet
Parent material: aeolian deposits
Drainage: well drained soils
Ground water table: not hit

Depth (cm)	Description
0–15	Light yellowish brown (10YR 6/4) moist, very pale brown (10 Y 7/3) dry, loamy sand, single grains, moderate effervescence, common medium to fine discontinuous random pores, many fine roots slight effervescence with HCl, clear smooth boundary
15–50	Yellowish brown (10 YR 5/6) moist, light yellowish brown (10 YR 6/4) dry, sandy, single grains, moderate medium and fine gravel, common medium to coarse discontinuous random pores, few lime concretion, common fine roots moderate effervescence with HCl, diffuse smooth boundary..
50–120	Dark yellowish brown (10 YR 4/4) moist, Yellowish brown (10 YR 5/4)dry, sandy, single grain, few lime concretion, common coarse and medium discontinuous random pores; moderate effervescence with HCl.

Profile No.: 34
Date of description: 9/2006
Location:
Longitude: 30° 40′ E
Latitude: 30° 01′ N
Elevation: 55.0 m a.sl.
Soil taxonomy: *Typic Haplocalcids*
Topography: undulating
Land form: sandy plain
Slope: gently slope
Land use & vegetation: barren
Parent material: aeolian deposits
Drainage: poor drained soils
Ground water table: not hit

Depth (cm)	Description
0–20	Yellowish brown (10 YR 5/6) moist, light yellowish brown (10 YR 6/4) dry, sandy, structure less, common medium to coarse discontinuous random pores, common fine roots moderate effervescence with HCl, diffuse smooth boundary.
20–110	Yellowish brown (10 YR 5/6), dry light yellowish brown (10 YR 6/4) dry, sand, single grains, hard, common coarse pores, common lime concretions, strong effervescence with HCl. common coarse and medium discontinuous random pores; slight effervescence with HCl.

Profile No.: 35
Date of description: 9/2006
Location:
Longitude: 30° 42′ E
Latitude: 29 15′ N
Elevation: 24.0 m a.sl.

Soil taxonomy: *Typic Calcigypsids*
Topography: undulating
Land form: lacustrine plain
Slope: gently slope
Land use & vegetation: barren
Parent material: lacustrine deposits
Drainage: poor drained soils
Ground water table: not hit

Depth (cm)	Description
0–15	Dark yellowish brown (10 YR 4/4) moist, Yellowish brown (10 YR 5/4)dry, clay, moderate medium sub angular blocky, hard, cemented abundant pores, common gypsum concretions, strong effervescence with HCl. common coarse and medium discontinuous random pores; slight effervescence with HCl.
15–40	Dark yellowish brown (10 YR 4/4) moist, Yellowish brown (10 YR 5/4)dry, clay loam, moderate fine to medium sub angular blocky, hard, cemented abundant pores, common gypsum concretions, strong effervescence with HCl. common coarse and medium discontinuous random pores; slight effervescence with HCl.
40–60	Yellowish brown (10 YR 5/6), dry light yellowish brown (10 YR 6/4) dry, sandy clay loam, moderate medium sub angular blocky, hard, cemented abundant pores, common lime concretions, strong effervescence with HCl. common coarse and medium discontinuous random pores; slight effervescence with HCl, clear smooth boundary.
60–90	Yellowish brown (10 YR 5/6), dry light yellowish brown (10 YR 6/4) dry, loamy sand, weak sub angular blocky, hard, common coarse pores, common lime concretions, strong effervescence with HCl. common coarse and medium discontinuous random pores; slight effervescence with HCl.

Profile No.: 36
Date of description: 9/2006
Location:
Longitude: 31° 02′ E
Latitude: 29° 18′ N
Elevation: 50.0 m a.sl.
Soil taxonomy: *not soil*
Topography: undulating
Land form: rock outcrops
Slope: gently slope
Land use & vegetation: barren
Parent material: limestone
Drainage: poor drained soils.
Ground water table: not hit
Remarks: *Site description to check the rock land units*

Depth (cm)	Description
Rock land	

Profile No.: 37
Date of description: 9/2006
Location
Longitude: 30° 59′ E
Latitude: 29° 22′ N
Elevation: 1.5 m. A.S.L
Soil taxonomy: *Typic Haplosalids*
Topography: gently undulating
Land form: fluvio-lacustrine plain/basin
Slope: gently slope
Land use & vegetation: fallow after rice
Parent material: alluvial-lacustrine deposites
Drainage: imperfectly drained
Ground water table: 110 cm

Depth (cm)	Description
0–30	Dark grayish brown (10YR4/2) moist, grayish brown (10YR5/3) dry; sandy loam; moderate to strong medium subanglar blocky, sticky, plasic, firm, hard; common fine discontinuous random interstitial pores; common fine to medium roots, slight effervescence with HCl, gradual smooth boundary.
30–50	Dark grayish brown (10YR4/2) moist, grayish brown (10YR5/2) dry, sandy loam; moderate medium sub angular blocky; slightly sticky, slightly plastic friable, soft, common medium to coarse discontinuous random interstitial pores, few fine gypsum concretion, common fine to medium roots; slight effervescence with HCl; gradual smooth boundary.
50–80	Very dark grayish brown (10YR3/2) moist, dark grayish brown (10YR4/2) dry; sandy clay loam; moderate medium to coarse sub angular blocky; sticky, plastic, firm, hard; common fine and medium continuous and discontinuous random pores; few fresh and humified residuals of organic matter; few fine soft modules of gypsum, many fine and medium roots; slight effervescence with HCl; clear smooth boundary.
80–110	Dark grayish brown (10YR4/2) moist, grayish brown (10YR5/2) dry, clay loam; moderate fine to medium sub angular blocky; slightly sticky, slightly plastic slightly firm, soft; common fine to coarse discontinuous random interstitial pores; few fine to medium gypsum concretion, slight effervescence with HCl.

Profile No.: 38
Date of description: 9/2006
Location
Longitude: 30° 55′ E
Latitude: 31° 04′ N

Elevation: 25.0 m a.s.l.
Soil taxonomy: *Typic Torrifluvents*
Topography: gently undulating
Land form: flood plain/basins
Slope: gently slope
Land use & vegetation: fallow after maize
Parent material: alluvial deposits
Drainage: moderately well drained
Ground water table: 100 cm

Depth (cm)	Description
0–25	Dark brown (10YR3/3) moist, brown (10YR4/3). dry; clay; moderate medium granular, sticky, plastic, fine, hard; common fine and; medium discontinuous random interstitial pores; many fine and, medium roots; slight effervescence with HCl; clear smooth boundary.
25–55	Very dark grayish brown (10YR3/2) moist; dark grayish brown (10YR4/2) dry; silty clay; blocky; sticky, firm, hard; common fine and medium continuous and discontinuous random interstitial pores, few fine to medium roots, slightly effervescence with HCl, gradual smooth boundary.
55–100	Dark gray (10YR 4/1) moist, gray (10YR 5/1) dry, clay; very sticky, very Plastic, firm, hard; common fine to medium continuous and discontinuous vughs, few very fine lime concretion, moderate effervescence with HCl.

Profile No.: 39
Date of description: 9/2006
Location
Longitude: 30° 52′ E
Latitude: 30° 12′ N
Elevation: 18.0 m a.s.l.
Soil taxonomy: *Vertic Torrifluvents*
Topography: gently undulating
Land form: flood plain/basin
Slope: gently slope
Land use & vegetation: fallow after maize
Parent material: alluvial deposits
Drainage: moderately well drained
Ground water table: not hit
Remarks: surface cracks 5 cm width & 25 cm depth

Depth (cm)	Description
0–20	Dark grayish brown (10YR4/2) moist, grayish brown (10YR5/3) dry, clayey; moderate to strong medium granular, very sticky, very plastic, very firm,; very hard; common fine discontinuous random interstitial pores; few fresh and humified residuals of organic matter, many fine to medium roots; slight effervescence with HCl; clear smooth boundary.
20–45	Very dark brown (10YR2/2) moist; Brown (10YR3/3) dry; clay loam; moderate to strong medium angular to sub angular blocky; very sticky, very plastic, very firm, extremely hard; common fine and medium continuous and discontinuous random interstitial pores, few fine roots, slightly effervescence with HCl, gradual smooth boundary.
45–80	Very dark grayish brown (10YR3/2) moist; dark grayish brown (10YR4/2) dry; clay loam; blocky; sticky, firm, hard; common fine and medium continuous and discontinuous random interstitial pores, few fine to medium roots, slightly effervescence with HCl, gradual smooth boundary.
80–110	Very dark grayish brown (10YR 3/2) moist, dark grayish brown (10YR 4/2) dry, clay loam; Moderate to strong medium angular to sub angular blocky, very sticky, very Plastic, firm, hard; common to fine medium continuous and discontinuous random pores; few fine roots, few very fine lime concretion, slight effervescence with HCl.

Profile No.: 40
Date of description: 9/2006
Location:
Longitude: 30° 25′ E
Latitude: 30° 45′ N
Elevation: 35.0 m a.sl.
Soil taxonomy: *Typic Torripsamments*
Topography: gently undulating
Land form: old deltaic plain
Slope: gently slope
Land use & vegetation: orchards
Parent material: old deltaic deposits
Drainage: well drained soils
Ground water table: not hit

Depth (cm)	Description
0–30	Light yellowish brown (10YR 6/4) moist, very pale brown (10 Y 7/3) dry, sandy, massive, moderate effervescence with HCl, common medium to fine discontinuous random pores, many fine roots moderate effervescence with HCl, clear smooth boundary
30–60	Yellowish brown (10 YR 5/6) moist, light yellowish brown (10 YR 6/4) dry sand, single grains, hard, many gravels, few soft lime concretions, moderate effervescence, common fine and medium roots; slight effervescence with HCl, clear smooth boundary.

(continued)

Depth (cm)	Description
60–100	Dark yellowish brown (10 YR 4/4) moist, Yellowish brown (10 YR 5/4)dry, sandy, single grain, common lime concretion, common coarse and medium discontinuous random pores; moderate effervescence with HCl.

Profile No.: 41
Date of description: 9/2006
Location:
Longitude: 30° 29′ E
Latitude: 30° 38′ N
Elevation: 31.0 m a.sl.
Soil taxonomy: *Typic Torripsamments*
Topography: gently undulating
Land form: old deltaic plain
Slope: gently slope
Land use & vegetation: orchards
Parent material: old deltaic deposits
Drainage: well drained soils
Ground water table: not hit

Depth (cm)	Description
0–15	Yellowish brown (10 YR 5/6) moist, light yellowish brown (10 YR 6/4) dry, sandy, massive, common medium and fine gravel, common medium to coarse discontinuous random pores, few lime concretion, common fine roots, strong effervescence with HCl, diffuse smooth boundary
15–40	Yellowish brown (10 YR 5/6) moist, light yellowish brown (10 YR 6/4) dry sand, massive, hard, many coarse gravels, common lime concretions, strong effervescence, common fine and medium roots; slight effervescence with HCl, clear smooth boundary.
40–130	Dark yellowish brown (10 YR 4/4) moist, Yellowish brown (10 YR 5/4) dry, sandy, massive, few lime concretion, few gravels common coarse and medium discontinuous random pores; moderate effervescence with HCl.

Profile No.: 42
Date of description: 9/2006
Location:
Longitude: 31° 56′ E
Latitude: 30° 42′ N
Elevation: 31.0 m a.sl.
Soil taxonomy: *Typic Quartizipsamments*
Topography: gently undulating
Land form: sand sheet

Slope: gently slope
Land use & vegetation: orchards
Parent material: aeolian deposits
Drainage: well drained soils
Ground water table: not hit

Depth (cm)	Description
0–25	Light yellowish brown (10YR 6/4) moist, very pale brown (10 Y 7/3) dry, sandy, single grains, slight effervescence with HCl, common medium to fine discontinuous random pores, many fine roots moderate effervescence with HCl, clear smooth boundary
25–100	Light yellowish brown (10YR 6/4) moist, light yellowish brown (10 YR 6/4) dry sand, single grains, few gravels, slight effervescence with HCl; slight effervescence with HCl, clear smooth boundary.

Soil Profiles Description

(Nile Valley)

Profile No.: 43
Date of description: 10/2006
Location:
Longitude: 31° 41′ E
Latitude: 30° 40′ N
Elevation: 28.0 m a.sl.
Soil taxonomy: *Typic Torriorthents*
Topography: undulating
Land form: aeolian plain
Slope: gently slope
Land use & vegetation: barren
Parent material: fluvial/aeolian deposits
Drainage: poor drained soils
Ground water table: not hit

Depth (cm)	Description
0–20	Yellowish brown (10 YR 5/6) moist, pale brown (10 YR 6/3) dry, sandy, single grains, moderate fine gravel and grits common medium to coarse discontinuous random pores, common lime concretions, common fine to medium roots strong effervescence with HCl, diffuse smooth boundary.

(continued)

Depth (cm)	Description
20–40	Yellowish brown (10 YR 5/6) moist, light yellowish brown (10 YR 6/4) dry, sandy, single grains, hard, common coarse gravels, common lime concretions, moderate effervescence, common fine and medium roots; strong effervescence with HCl, clear smooth boundary.
40–85	Dark yellowish brown (10 YR 4/4) moist, Yellowish brown (10 YR 5/4)dry, sandy, few lime and gypsum concretion, common coarse and medium discontinuous random pores; moderate effervescence with HCl.

Profile No.: 44
Date of description: 10/2006
Location:
Longitude: 31° 03′ E
Latitude: 30° 55′ N
Elevation: 14.0 m a.s.l.
Soil taxonomy: *Vertic Torrifluvents*
Topography: almost flat to gently undulating
Land form: flood plain/river terraces
Slope: nearly level to gently slope
Land use & vegetation: fallow after rice
Parent material: alluvial deposits
Drainage: moderately well drained
Ground water table: +120 cm

Depth (cm)	Description
0–25	Very dark grayish brown (10YR3/2) moist, dark grayish brown (10YR4/3) dry; sandy clay loam; moderate to strong medium sub angular blocky; very sticky, very plastic, extremely firm, extremely hard; common fine discontinuous random exped and medium interstitial pores; many fine to medium roots; moderately effervescence with HCl; clear smooth boundary.
25–65	Dark brown (10YR2/3) moist, grayish brown (10YR5/2) dry; clay loam; strong medium sub angular blocky; very sticky, very plastic, extremely firm, extremely hard; common fine to medium discontinuous random interstitial pores; thin shinny discontinuous clay films lining the voids; few fine to medium roots; slight effervescence with HCl, clear smooth boundary .
65–120	Dark brown (10YR 2/3) moist, grayish brown (10YR 5/2) dry; clayey; strong medium sub angular blocky; very sticky, very Plastic, extremely firm, extremely hard; common fine and medium discontinuous random interstitial pores and exped; slight effervescence with HCl.

Profile No.: 45
Date of description: 10/2006
Location:

Longitude: 31° 42′ E
Latitude: 31° 40′ N
Elevation: 12.0 m a.s.l.
Soil taxonomy: *Typic Torrifluvents*
Topography: almost flat
Land form: flood plain/river terraces
Slope: nearly level
Land use & vegetation: cultivated with clover
Parent material: alluvial deposits
Drainage: moderately well drained soils
Ground water table: +100 cm

Depth (cm)	Description
0–20	Dark brown (10YR2/3) moist, grayish brown (10YR5/2) dry; clay loam; moderately fine to medium angular to sub angular blocky; very sticky, very plastic, very firm, extremely hard; common fine discontinuous random exped and interstitial pores, few humified patches of organic matter; many fine to coarse roots; slight effervescence with HCl, clear smooth boundary.
20–45	Dark grayish brown (10YR 4/2) moist, grayish brown (10YR 5/4) dry; clayey; strong fine to medium sub angular blocky; sticky, plastic; firm, hard; common fine to medium continuous and discontinuous random interstitial pores; few humified patches of organic matter; many fine to coarse roots; gradual smooth boundary.
45–100	Dark grayish brown (10YR 4/2) moist, grayish brown (10YR 5/4) dry; clayey; moderate fine to medium sub angular blocky; very sticky very plastic: very firm, very hard; common fine and medium discontinuous random exped and interstitial pores; slight smooth boundary.

Profile No.: 46
Date of description: 10/2006
Location:
Longitude: 31° 56′ E
Latitude: 30° 42′ N
Elevation: 31.0 m a.sl.
Soil taxonomy: *Typic Quartizipsamments*
Topography: gently undulating
Land form: sand sheet
Slope: gently slope
Land use & vegetation: barren
Parent material: aeolian deposits
Drainage: well drained soils
Ground water table: not hit

Depth (cm)	Description
0–25	Light yellowish brown (10YR 6/4) moist, very pale brown (10 Y 7/3) dry, sandy, single grains, slight effervescence with HCl, common medium to fine discontinuous random pores, many fine roots moderate effervescence with HCl, clear smooth boundary
25–60	Yellowish brown (10 YR 5/6) moist, pale brown (10 YR 6/3) dry, loamy, massive, common coarse gravels and grits common medium to coarse discontinuous random pores, common lime concretions, common fine to medium roots strong effervescence with HCl, diffuse smooth boundary.
60–100	Light yellowish brown (10YR 6/4) moist, light yellowish brown (10 YR 6/4) dry sand, single grains, few gravels, slight effervescence with HCl; slight effervescence with HCl, clear smooth boundary.

Profile No.: 47
Date of description: 10/2006
Location:
Longitude: 32° 10′ E
Latitude: 30° 36′ N
Elevation: 50.0 m a.sl.
Soil taxonomy: *Typic Haplogypsids*
Topography: undulating
Land form: sandy plain
Slope: gently slope
Land use & vegetation: barren
Parent material: aeolian deposits
Drainage: poor drained soils
Ground water table: not hit

Depth (cm)	Description
0–15	Light yellowish brown (10YR 6/4) moist, very pale brown (10 Y 7/3) dry, sandy, massive, moderate medium and fine gravel, common medium to coarse discontinuous random pores, few lime and common gypsum concretion, common fine roots moderate effervescence with HCl, diffuse smooth boundary.
15–60	Yellowish brown (10 YR 5/6), dry light yellowish brown (10 YR 6/4) dry, sandy, massive, hard, cemented abundant pores, common lime concretions, strong effervescence with HCl. common coarse and medium discontinuous random pores; slight effervescence with HCl. clear smooth boundary.
60–120	Dark yellowish brown (10 YR 4/4) moist, Yellowish brown (10 YR 5/4) dry, sandy, massive, hard, cemented abundant pores, common lime and gypsum concretions, strong effervescence with HCl. common coarse and medium discontinuous random pores; slight effervescence with HCl.

Profile No.: 48
Date of description: 10/2006
Location:
Longitude: 32° 14′ E
Latitude: 30° 50′ N
Elevation: 32.0 m a.s.l.
Soil taxonomy: *Typic Petrogypsids*
Topography: almost flat to gently undulating
Land form: lacustrine deposits
Slope: gently slope
Land use & vegetation: barren.
Parent material: alluvial deposits
Drainage: moderately well drained soils
Ground water table: not hit cm

Depth (cm)	Description
0–20	Dark gray (10YR4/1) moist, gray (10YR5/1) dry; sandy clay loam; moderate medium sub angular blocky; sticky, plastic; moderately firm, moderately hard; common fine to medium discontinuous random interstitial pores; thin shinny continuous clay films lining the voids; many medium to coarse roots; slight effervescence with Hcl. clear smooth boundary
20–60	Dark yellowish brown (10 YR 4/4) moist, Yellowish brown (10 YR 5/4)dry, clay loam, moderate fine sub angular blocky, hard, cemented abundant pores, common lime and gypsum concretions, strong effervescence with HCl. common coarse and medium discontinuous random pores; slight effervescence with HCl
+60	Hardpan

Profile No.: 49
Date of description: 10/2006
Location:
Longitude: 30° 51′ E
Latitude: 30° 16′ N
Elevation: 22.0 m a.s.l.
Soil taxonomy: *Vertic Torrifluvents*
Topography: almost flat
Land form: flood plain/river levees
Slope: gently slope
Land use & vegetation: cultivated with orchards (orange-mango).
Parent material: alluvial deposits
Drainage: well drained
Ground water table: 120 cm

Depth (cm)	Description
0–25	Very dark grayish brown (10YR3/2) moist, dark grayish brown (10YR4/3) dry; clayey; moderate fine to medium sub angular blocky; sticky, plastic; moderately firm, moderately hard; common fine to medium discontinuous random exped interstitial pores; common fine to coarse roots; slight effervescence with Hcl, clear smooth boundary.
25–60	Dark gray (10YR4/1) moist, gray (10YR5/1) dry; clayey; moderate medium sub angular blocky; sticky, plastic; moderately firm, moderately hard; common fine to medium discontinuous random exped interstitial pores; few fresh and humified residual of organic matter; many fine to coarse roots; slight effervescence with Hcl; clear smooth boundary.
60–120	Very dark grayish brown (10YR 3/2) moist, dark grayish brown (10YR 4/3) dry, clayey; moderate fine to medium sub angular blocky; sticky, plastic; firm, hard; common fine and medium discontinuous random interstitial pores; slight effervescence with Hcl.

Profile No.: 50
Date of description: 10/2006
Location:
Longitude: 31° 12′ E
Latitude: 30° 11′ N
Elevation: 24.0 m a.s.l.
Soil taxonomy: *Typic Torrifluvents*
Topography: almost flat
Land form: flood plain/river terraces
Slope: gently slope
Land use & vegetation: fallow after Maize
Parent material: alluvial deposits
Drainage: moderately well drained
Ground water table: 90 cm

Depth (cm)	Description
0–20	Dark grayish brown (10YR4/2) moist, grayish brown (10YR5/2) dry; clay loam; moderate medium sub angular blocky; sticky, plastic, firm, hard; common fine to medium discontinuous random interstitial pores; common fine to medium roots; slight effervescence with HCl, clear smooth boundary
20–40	Dark grayish brown (10YR4/2) moist, grayish brown (10YR5/2) dry; clay loam strong medium angular to sub angular blocky; moderate sticky, moderate plastic, firm, hard; common fine to medium discontinuous random exped medium interstitial pores; few humified residual of organic matter; common fine to medium roots; slight effervescence with HCl; gradual smooth boundary.
40–90	Dark grayish brown (10YR 4/2) moist, grayish brown (10YR 5/2) dry; silty clay; moderate to strong medium sub angular blocky; sticky, Plastic, very firm, very hard; common fine and medium discontinuous random interstitial pores; slight effervescence with HCl.

Profile No.: 51
Date of description: 10/2006
Location
Longitude: 31° 18′ E
Latitude: 30° 24′ N
Elevation: 50.0 m a.s.l.
Soil taxonomy: *Typic Quartizipsamments*
Topography: almost flat
Land form: flood plain
Slope: nearly level
Land use & vegetation: barren
Parent material: aeolian deposits
Drainage: moderately well drained
Ground water table: not hit

Depth (cm)	Description
0–30	Light yellowish brown (10YR 6/4) moist, very pale brown (10 YR 7/3) dry, sandy, single grains, few gravels, strong effervescence, common fine and medium roots; slight effervescence with HCl, diffuse smooth boundary.
30–70	Yellowish brown (10 YR 5/6) moist, pale brown (10 YR 6/3) dry, sandy loam, single grains, common gravels, loose, common fine to medium discontinuous random pores, many fine roots moderate effervescence with HCl, diffuse smooth boundary.
70–120	Dark yellowish brown (10 YR 4/4) moist, Yellowish brown (10 YR 5/4)dry, slightly hard medium coarse loamy sand, common coarse and medium discontinuous random pores; moderate effervescence with HCl.

Profile No.: 52
Date of description: 10/2006
Location:
Longitude: 30° 40′ E
Latitude: 30° 01′ N
Elevation: 4.0 m a.sl.
Soil taxonomy: *Typic Haplocalcids*
Topography: undulating
Land form: sandy plain
Slope: gently slope
Land use & vegetation: barren
Parent material: aeolian deposits
Drainage: poor drained soils
Ground water table: not hit

Depth (cm)	Description
0–35	Yellowish brown (10 YR 5/6), dry light yellowish brown (10 YR 6/4) dry, sand, single grains, hard, common coarse pores, common lime concretions, strong effervescence with HCl, common coarse and medium discontinuous random pores; slight effervescence with HCl, diffuse smooth boundary.
35–100	Yellowish brown (10 YR 5/6) moist, light yellowish brown (10 YR 6/4) dry, sandy, structure less, common medium to coarse discontinuous random pores, common fine roots moderate effervescence with HCl.

Profile No.: 53
Date of description: 10/2006
Location:
Longitude: 30° 25′ E
Latitude: 30° 45′ N
Elevation: 35.0 m a.sl.
Soil taxonomy: *Typic Torripsamments*
Topography: gently undulating
Land form: old deltaic plain
Slope: gently slope
Land use & vegetation: barren
Parent material: old deltaic deposits
Drainage: well drained soils
Ground water table: not hit

Depth (cm)	Description
0–25	Light yellowish brown (10YR 6/4) moist, very pale brown (10 Y 7/3) dry, sandy, massive, moderate effervescence with HCl, common medium to fine discontinuous random pores, many fine roots moderate effervescence with HCl, clear smooth boundary
25–60	Yellowish brown (10 YR 5/6) moist, light yellowish brown (10 YR 6/4) dry sand, single grains, hard, many gravels, few soft lime concretions, moderate effervescence, common fine and medium roots; slight effervescence with HCl, clear smooth boundary.
60–100	Dark yellowish brown (10 YR 4/4) moist, Yellowish brown (10 YR 5/4)dry, sandy, single grain, common lime concretion, common coarse and medium discontinuous random pores; moderate effervescence with HCl.

Profile No.: 54
Date of description: 10/2006
Location
Longitude: 30° 51′ E
Latitude: 31° 15′ N

Elevation: 41.5 m a.s.l.
Soil taxonomy: *Typic Torriorthents*
Topography: undulating
Land form: old deltaic plain
Slope: gently slope
Land use & vegetation: barren
Parent material: fluvial deposits
Drainage: moderately well drained
Ground water table: not hit

Depth (cm)	Description
0–25	Yellowish brown (10 YR 5/6) moist, light yellowish brown (10 YR 6/4) dry, loamy, massive, common coarse gravels, common lime concretions, moderate effervescence, common fine and medium roots; strong effervescence with HCl, clear smooth boundary.
25–60	Yellowish brown (10 YR 5/6) moist, pale brown (10 YR 6/3) dry, loamy, massive, common coarse gravels and grits common medium to coarse discontinuous random pores, common lime concretions, common fine to medium roots strong effervescence with HCl, diffuse smooth boundary.
60–130	Yellowish brown (10 YR 5/6) moist, light yellowish brown (10 YR 6/4) dry, clay loam, weak sub angular blocky, hard, common coarse gravels, common lime concretions, strong effervescence, common fine and medium roots; strong effervescence with HCl, clear smooth boundary.

Profile No.: 55
Date of description: 10/2006
Location:
Longitude: 31° 56′ E
Latitude: 30° 42′ N
Elevation: 46.0 m a.sl.
Soil taxonomy: *Typic Quartizipsamments*
Topography: gently undulating
Land form: sand sheet
Slope: gently slope
Land use & vegetation: barren
Parent material: aeolian deposits
Drainage: well drained soils
Ground water table: +130 cm

Depth (cm)	Description
0–30	Light yellowish brown (10YR 6/4) moist, very pale brown (10 Y 7/3) dry, sandy, single grains, moderate effervescence, common medium to fine discontinuous random pores, many fine roots moderate effervescence with HCl, clear smooth boundary

(continued)

Depth (cm)	Description
30–70	Yellowish brown (10 YR 5/6) moist, light yellowish brown (10 YR 6/4) dry sand, single grains, hard, few gravels, few soft lime concretions, moderate effervescence, common fine and medium roots; slight effervescence with HCl, clear smooth boundary.
70–130	Dark yellowish brown (10 YR 4/4) moist, Yellowish brown (10 YR 5/4)dry, sandy, single grain, few lime concretion, common coarse and medium discontinuous random pores; moderate effervescence with HCl.

Profile No.: 56
Date of description: 10/2006
Location:
Longitude: 31° 03′ E
Latitude: 30° 55′ N
Elevation: 28.0 m a.s.l.
Soil taxonomy: *Vertic Torrifluvents*
Topography: almost flat to gently undulating
Land form: flood plain/river terraces
Slope: nearly level to gently slope
Land use & vegetation: fallow after rice
Parent material: alluvial deposits
Drainage: moderately well drained
Ground water table: 110 cm

Depth (cm)	Description
0–30	Very dark grayish brown (10YR3/2) moist, dark grayish brown (10YR4/3) dry; sandy clay loam; moderate to strong medium sub angular blocky; very sticky, very plastic, extremely firm, extremely hard; common fine discontinuous random exped and medium interstitial pores; many fine to medium roots; moderately effervescence with HCl; clear smooth boundary.
30–65	Dark brown (10YR2/3) moist, grayish brown (10YR5/2) dry; clay loam; strong medium sub angular blocky; very sticky, very plastic, extremely firm, extremely hard; common fine to medium discontinuous random interstitial pores; thin shinny discontinuous clay films lining the voids; few fine to medium roots; slight effervescence with HCl, clear smooth boundary .
65–110	Dark brown (10YR 2/3) moist, grayish brown (10YR 5/2) dry; clayey; strong medium sub angular blocky; very sticky, very Plastic, extremely firm, extremely hard; common fine and medium discontinuous random interstitial pores; slight effervescence with HCl.

Profile No.: 57
Date of description: 10/2006
Location

Longitude: 30° 51′ E
Latitude: 31° 15′ N
Elevation: 62.0 m a.s.l.
Soil taxonomy: *Typic Torriorthents*
Topography: undulating
Land form: old deltaic plain
Slope: gently slope
Land use & vegetation: barren
Parent material: fluvial deposits
Drainage: moderately well drained
Ground water table: not hit

Depth (cm)	Description
0–15	Yellowish brown (10 YR 5/6) moist, light yellowish brown (10 YR 6/4) dry, loamy sand, massive, many coarse gravels, common lime concretions, moderate effervescence, common fine and medium roots; strong effervescence with HCl, clear smooth boundary.
15–40	Yellowish brown (10 YR 5/6) moist, pale brown (10 YR 6/3) dry, loamy, massive, common coarse gravels and grits common medium to coarse discontinuous random pores, common lime concretions, common fine to coarse roots strong effervescence with HCl, clear smooth boundary.
40–65	Yellowish brown (10 YR 5/6) moist, light yellowish brown (10 YR 6/4) dry, clay loam, weak sub angular blocky, moderate, common coarse gravels, common lime concretions, moderate effervescence, common fine and medium roots; strong effervescence with HCl, clear smooth boundary.
65–110	Dark yellowish brown (10 YR 4/4) moist, Yellowish brown (10 YR 5/4)dry, clay loam, weak sub angular blocky, few lime concretion, common coarse and medium discontinuous random pores; moderate effervescence with HCl.

Profile No.: 58
Date of description: 10/2006
Location
Longitude: 30° 51′ E
Latitude: 31° 15′ N
Elevation: 68.0 m a.s.l.
Soil taxonomy: *Typic Torriorthents*
Topography: undulating
Land form: old deltaic plain
Slope: gently slope
Land use & vegetation: barren
Parent material: fluvial deposits
Drainage: moderately well drained
Ground water table: not hit

Depth (cm)	Description
0–40	Yellowish brown (10 YR 5/6) moist, light yellowish brown (10 YR 6/4) dry, gravelly sand, massive, slightly cemented common fine pores, many gravel and grits, strong effervescence, clear smooth boundary.
40–120	Yellowish brown (10 YR 5/6) moist, pale brown (10 YR 6/3) dry, loamy, massive, many coarse gravels and grits common, medium to coarse discontinuous random pores, common lime concretions, common fine to medium roots, strong effervescence with HCl.

Profile No.: 59
Date of description: 10/2006
Location:
Longitude: 30° 33′ E
Latitude: 30° 07′ N
Elevation: 54.0 m a.sl.
Soil taxonomy: *Typic Quartizipsamments*
Topography: gently undulating
Land form: sand sheet
Slope: gently slope
Land use & vegetation: barren
Parent material: aeolian deposits
Drainage: well drained
Ground water table: not hit

Depth (cm)	Description
0–20	Light yellowish brown (10YR 6/4) moist, very pale brown (10 Y 7/3) dry, loamy sand, single grains, moderate effervescence, common medium to fine discontinuous random pores, many fine roots slight effervescence with HCl, clear smooth boundary
20–60	Yellowish brown (10 YR 5/6) moist, light yellowish brown (10 YR 6/4) dry, sandy, single grains, moderate medium and fine gravel, common medium to coarse discontinuous random pores, few lime concretion, common fine roots moderate effervescence with HCl, diffuse smooth boundary..
60–130	Dark yellowish brown (10 YR 4/4) moist, Yellowish brown (10 YR 5/4)dry, sandy, single grain, few lime concretion, common coarse and medium discontinuous random pores; moderate effervescence with HCl.

Profile No.: 60
Date of description: 10/2006
Location
Longitude: 30° 55′ E
Latitude: 31° 04′ N
Elevation: 32.0 m a.s.l.

Soil taxonomy: *Vertic Torrifluvents*
Topography: almost flat
Land form: flood plain/basins
Slope: gently slope
Land use & vegetation: fallow after maize
Parent material: alluvial deposits
Drainage: moderately well drained
Ground water table: 100 cm
Remarks: surface cracks 5 cm width & 30 cm depth

Depth (cm)	Description
0–25	Very dark grayish brown (10YR3/2) moist; dark grayish brown (10YR4/2) dry; silty clay; blocky; sticky, firm, hard; common fine and medium continuous and discontinuous random interstitial pores, few fine to medium roots, slightly effervescence with HCl, gradual smooth boundary.
25–50	Dark brown (10YR3/3) moist, brown (10YR4/3). dry; clay; moderate medium granular, sticky, plastic, fine, hard; common fine and; medium discontinuous random interstitial pores; many fine and, medium roots; slight effervescence with HCl; clear smooth boundary.
50–100	Dark gray (10YR 4/1) moist, gray (10YR 5/1) dry, clay; very sticky, very Plastic, firm, hard; common fine to medium continuous and discontinuous vughs, few very fine lime concretion, moderate effervescence with HCl.

Profile No.: 61
Date of description: 10/2006
Location
Longitude: 30° 52′ E
Latitude: 30° 12′ N
Elevation: 28.0 m a.s.l.
Soil taxonomy: *Vertic Torrifluvents*
Topography: almost flat
Land form: flood plain/terraces
Slope: gently slope
Land use & vegetation: fallow after maize
Parent material: alluvial deposits
Drainage: moderately well drained
Ground water table: not hit
Remarks: surface cracks 6 cm width & 30 cm depth

Depth (cm)	Description
0–15	Dark grayish brown (10YR4/2) moist, grayish brown (10YR5/3) dry, clayey; moderate to strong medium granular, very sticky, very plastic, very firm,; very hard;

(continued)

Depth (cm)	Description
	common fine discontinuous random interstitial pores; few fresh and humified residuals of organic matter, many fine to medium roots; slight effervescence with HCl; clear smooth boundary.
15–45	Very dark brown (10YR2/2) moist; Brown (10YR3/3) dry; clay loam; moderate to strong medium angular to sub angular blocky; very sticky, very plastic, very firm, extremely hard; common fine and medium continuous and discontinuous random interstitial pores, few fine roots, slightly effervescence with HCl, gradual smooth boundary.
45–85	Very dark grayish brown (10YR3/2) moist; dark grayish brown (10YR4/2) dry; clay loam; blocky; sticky, firm, hard; common fine and medium continuous and discontinuous random interstitial pores, few fine to medium roots, slightly effervescence with HCl, gradual smooth boundary.
85–120	Very dark grayish brown (10YR 3/2) moist, dark grayish brown (10YR 4/2) dry, clay loam; Moderate medium angular to sub angular blocky, very sticky, very Plastic, firm, hard; common to fine medium continuous and discontinuous random pores; few fine roots, few fine lime concretion, slight effervescence with HCl.

Profile No.: 62
Date of description: 10/2006
Location:
Longitude: 31° 42′ E
Latitude: 31° 40′ N
Elevation: 6.5 m a.s.l.
Soil taxonomy: *Typic Torrifluvents*
Topography: almost flat
Land form: flood plain/river terraces
Slope: nearly level
Land use & vegetation: cultivated with clover
Parent material: alluvial deposits
Drainage: moderately well drained soils
Ground water table: 90 cm

Depth (cm)	Description
0–20	Dark brown (10YR2/3) moist, grayish brown (10YR5/2) dry; clay loam; moderately fine to medium sub angular blocky; very sticky, very plastic, very firm, hard; common fine discontinuous random exped and interstitial pores, few humified patches of organic matter; many fine to coarse roots; slight effervescence with HCl, clear smooth boundary.
20–50	Dark grayish brown (10YR 4/2) moist, grayish brown (10YR 5/4) dry; clayey; strong fine to medium sub angular blocky; sticky, plastic; firm, hard; common fine to medium continuous and discontinuous random interstitial pores; few humified patches of organic matter; many fine to coarse roots; gradual smooth boundary.

(continued)

Depth (cm)	Description
50–90	Dark grayish brown (10YR 4/2) moist, grayish brown (10YR 5/4) dry; clayey; moderate fine to medium sub angular blocky; very sticky very plastic: very firm, very hard; common fine and medium discontinuous random exped and interstitial pores.

Profile No.: 63
Date of description: 10/2006
Location:
Longitude: 31° 30′ E
Latitude: 30° 31′ N
Elevation: 22.0 m a.s.l.
Soil taxonomy: *Vertic Torrifluvents*
Topography: almost flat
Land form: flood plain/overflow mantle
Slope: nearly level to gently slope
Land use & vegetation: fallow after maize
Parent material: alluvial deposits
Drainage: well drained
Ground water table: 120 cm

Depth (cm)	Description
0–25	Dark brown (10YR2/3) moist, grayish Brown (10YR5/2) dry, clayey; moderate fine to medium granular; sticky, plastic; firm, hard; common fine to medium discontinuous random interstitial pores; many fine to medium roots, slight effervescence with Hcl, clear smooth boundary.
25–50	Dark grayish brown (10YR4/2) moist, grayish brown (10YR5/2) dry; clay loam; moderate medium sub angular blocky; sticky, plastic, firm, hard; common fine to medium discontinuous random interstitial pores; common fine to medium roots; slight effervescence with HCl, clear smooth boundary.
50–80	Dark grayish brown (10YR4/2) moist, grayish brown (10YR5/4) dry, clayey; moderate fine to medium sub angular blocky; very sticky, very plastic, very firm, extremely hard; common fine to medium discontinuous random exped and interstitial pores; few humified residual of organic matter; common fine to medium roots,. Slight effervescence with Hcl; clear smooth boundary
80–120	Dark brown (10YR 3/3) moist, grayish brown (10YR 5/3) dry; clayey; strong medium sub angular blocky; very sticky, very plastic; extremely firm, extremely hard; common fine discontinuous random interstitial pores; slight effervescence with Hcl.

Profile No.: 64
Date of description: 10/2006
Location:
Longitude: 30° 56′ E

Latitude: 30° 48′ N
Elevation: 20.5 m a.s.l.
Soil taxonomy: *Typic Torrifuvents*
Topography: almost flat to gently undulating
Land form: flood plain/Basin
Slope: gently slope
Land use & vegetation: fallow after cotton
Parent material: alluvial deposits
Drainage: well drained soils
Ground water table: 110 cm

Depth (cm)	Description
0–25	Very dark grayish brown (10YR3/2) moist, dark grayish brown (10YR4/3) dry, clay loam; moderate fine to medium sub angular blocky; sticky, plastic, firm, hard; common fine continuous and discontinuous random interstitial pores; common fine to medium roots, Slight effervescence with HCl; clear smooth boundary.
25–55	Very dark grayish brown (10YR3/2) moist, grayish Brown (10YR4/3) dry, silty clay; moderate fine to medium sub angular; sticky, plastic; firm, hard; common fine discontinuous random interstitial pores; common fine roots, slight effervescence with HCl, gradual smooth boundary.
55–80	Dark grayish brown (10YR 4/2) moist, grayish brown (10YR 5/2) dry; silty clay; Moderate to strong medium sub angular blocky, sticky, Plastic; firm, hard; common fine continuous and discontinuous random pores; few fine roots; clear smooth boundary.
80–110	Dark grayish brown (10YR 4/2) moist, grayish brown (10YR 5/1) dry; clayey; moderate medium sub angular blocky; very sticky, very plastic; extremely firm, extremely hard; slight effervescence with HCl.

Profile No.: 65
Date of description: 10/2006
Location:
Longitude: 30° 51′ E
Latitude: 30° 16′ N
Elevation: 56.0 m a.s.l.
Soil taxonomy: *Vertic Torriorthents*
Topography: almost flat
Land form: flood plain/river levees
Slope: gently slope
Land use & vegetation: barren
Parent material: alluvial deposits
Drainage: moderately well drained
Ground water table: 110 cm

Depth (cm)	Description
0–30	Dark yellowish brown (10 YR 4/4) moist, Yellowish brown (10 YR 5/6) dry; gravelly sand, structure less, loose moderate medium and fine gravel and grits common fine to medium discontinuous random exped interstitial pores; many fine roots moderate effervescence with Hcl, diffuse smooth boundary.
30–75	Yellowish brown (10 YR 5/6) moist, Yellowish brown (10 YR 5/6) dry, gravelly sand; massive; coarse sand, massive, loose, few fine gravel moderate effervescence, strong effervescence with Hcl, diffuse smooth boundary.
75–110	Yellowish brown (10 YR 5/6) moist, Yellowish brown (10 YR 5/6) dry, dark grayish brown sand, massive, hard, cemented abundant pores, many gravel, few soft white lime concretions, violent effervescence with Hcl.

Profile No.: 66
Date of description: 10/2006
Location:
Longitude: 32° 08′ E
Latitude: 30° 14′ N
Elevation: 64.5 m a.sl.
Soil taxonomy: *Typic Torriorthents*
Topography: almost flat
Land form: sandy plain
Slope: nearly level
Land use & vegetation: barren
Parent material: fluvial/aeolian deposits
Drainage: moderately well drained
Ground water table: not hit

Depth (cm)	Description
0–35	Yellowish brown (10 YR 5/6) moist, light yellowish brown (10 YR 6/4) dry, sandy, gravelly sand; massive; coarse sand, loose, few fine gravel moderate effervescence, strong effervescence with Hcl, diffuse smooth boundary.
35–120	Yellowish brown (10 YR 5/6), dry light yellowish brown (10 YR 6/4) dry, sandy, single grains, hard, cemented abundant pores, common lime concretions, common coarse and medium discontinuous random pores; moderate effervescence with HCl.

Profile No.: 67
Date of description: 10/2006
Location
Longitude: 30° 27′ E
Latitude: 31° 48′ N
Elevation: 25.5 m a.s.l.
Soil taxonomy: *Vertic Torrifluvents*

Topography: gently undulating
Land form: flood plain/basin
Slope: gently slope
Land use & vegetation: cultivated with alfalfa
Parent material: alluvial deposits
Drainage: moderately well drained
Ground water table: 120 cm
Remarks: surface cracks 5 cm width & 30 cm depth

Depth (cm)	Description
0–15	Very dark grayish brown (10YR3/2) moist, dark grayish brown (10YR4/3) dry, sandy clay; moderate medium sub angular blocky; sticky and plastic; firm, hard common fine discontinuous random interstitial pores, few fresh and humified residual of organic matter, many fine to medium roots, slightly effervescence with HCl, clear smooth boundary.
15–40	Dark brown (10YR4/3) moist, Brown (10YR5/3) dry, clayey; Moderate medium angular to sub angular blocky; very sticky, very plastic, firm, hard; common fine to medium discontinuous random pores, thin shinny discontinuous organic clay films lining the voids; common fine to medium roots, slightly effervescence with HCl, clear smooth boundary.
40–75	Dark brown (10YR 3/3) moist, dark grayish brown (10YR 4/3) dry: clayey: Moderate to strong medium angular to sub angular blocky; very sticky, very Plastic; very firm, very hard; common fine to medium continuous and discontinuous random pores; few fine roots slight effervescence with HCl, clear smooth boundary.
75–120	Dark yellowish brown (10YR 3/3) moist, dark grayish brown (10YR 4/3) dry: clayey; strong medium sub angular blocky; very sticky and very plastic; firm; hard; common fine random pores; moderate effervescence with HCl.

Profile No.: 68
Date of description: 10/2006
Location:
Longitude: 30° 36′ E
Latitude: 31° 15′ N
Elevation: 20.0 m a.sl.
Soil taxonomy: *Typic Torrifluvents*
Topography: flat to almost flat
Land form: fluvio-lacustrine plain/terraces
Slope: nearly level
Land use & vegetation: fallow after rice
Parent material: alluvial-lecustrine deposits
Drainage: moderately well drained
Ground water table: 110 cm

Depth (cm)	Description
0–20	Very dark brown (10YR2/2) moist, dark brown (10YR3/3) dry, clayey; string fine and medium sub-angular blocky; very sticky, very plastic, firm, hard; many fine to medium continuous and discontinuous interstitial pores; many fine and medium roots; slightly effervescence with HCl; clear gradual boundary.
20–55	Very dark grayish brown (10YR3/2) moist; dark grayish Brown (10YR4/2) dry; clayey; moderate to strong medium sub angular blocky; slightly sticky slightly plastic, friable, hard; common fine and medium discontinuous random pores; common fine and medium roots, slightly effervescence with HCl, clear smooth boundary.
55–110	Dark brown (10YR3/3) moist; grayish Brown (10YR5/2) dry; clay loam weak fine and medium sub angular blocky; sticky, plastic, firm, hard; few fine and medium discontinuous pores; slight effervescence with HCl,

Profile No.: 69
Date of description: 10/2006
Location:
Longitude: 32° 08′ E
Latitude: 30° 14′ N
Elevation: 58.5 m a.sl.
Soil taxonomy: *Typic Torriorthents*
Topography: almost flat
Land form: sandy plain
Slope: gently slope
Land use & vegetation: barren
Parent material: fluvial/aeolian deposits
Drainage: moderately well drained
Ground water table: not hit
Remarks: common gravels (about 30%) cover the land surface

Depth (cm)	Description
0–40	Yellowish brown (10 YR 5/6) moist, light yellowish brown (10 YR 6/4) dry, sandy, gravelly sand; massive; common gravels, loose, few fine gravel moderate effervescence, moderate effervescence with Hcl, diffuse smooth boundary.
40–120	Yellowish brown (10 YR 5/6), dry light yellowish brown (10 YR 6/4) dry, sandy, single grains, hard, common lime concretions, common coarse and medium discontinuous random pores; strong effervescence with HCl.

Profile No.: 70
Date of description: 10/2006
Location:
Longitude: 31° 41′ E
Latitude: 30° 40′ N

Elevation: 54.0 m a.sl.
Soil taxonomy: *Typic Torriorthents*
Topography: undulating
Land form: aeolian plain
Slope: gently slope
Land use & vegetation: barren
Parent material: fluvial/aeolian deposits
Drainage: poor drained soils
Ground water table: not hit

Depth (cm)	Description
0–20	Dark yellowish brown (10 YR 4/4) moist, Yellowish brown (10 YR 5/4) dry, gravelly sand, massive, few lime concretion, common gravel, common coarse and medium discontinuous random pores; moderate effervescence with HCl.
20–50	Yellowish brown (10 YR 5/6) moist, light yellowish brown (10 YR 6/4) dry, sandy, massive, hard, common coarse gravels, common lime concretions, moderate effervescence, common fine and medium roots; strong effervescence with HCl, clear smooth boundary.
50–100	Yellowish brown (10 YR 5/6) moist, pale brown (10 YR 6/3) dry, sandy, massive, moderate fine gravel and grits common medium to coarse discontinuous pores, common lime concretions, strong effervescence with HCl, diffuse smooth boundary.

Profile No.: 71
Date of description: 10/2006
Location
Longitude: 30° 23′ E
Latitude: 31° 18′ N
Elevation: 3.0 m a.s.l.
Soil taxonomy: *Vertic Torrifluvents*
Topography: Almost flat
Land form: alluvial plain/terrace
Slope: nearly level
Land use & vegetation: cultivated with alfalfa
Parent material: lacustrine/alluvial deposits
Drainage: moderately well drained
Ground water table: 60 cm
Remarks: surface cracks 5 cm width & 30 cm depth

Depth (cm)	Description
0–20	Very dark grayish brown (10YR4/2) moist, brown (10YR5/3) dry, clayey; moderate medium angular blocky; sticky, plastic; firm, hard common fine discontinuous random pores, few fresh and humified residual of organic matter, many fine and medium roots, slightly effervescence with HCl, clear smooth boundary.

(continued)

Depth (cm)	Description
20–60	Very dark grayish brown (10YR3/2) moist, dark grayish brown (10YR4/2) dry, clayey; Moderate to strong fine to medium sub-angular, sticky, plastic, firm, hard; common fine and medium continuous and discontinuous channels, few fine soft segregation of CaCO3, common fine to medium roots,. Slightly effervescence with HCl, clear smooth boundary.
60–100	Very dark grayish brown (10YR 3/2) moist, dark grayish brown (10YR 4/2) dry: clay loam; moderate fine to medium angular blocky; very sticky, very plastic; firm; hard; common fine random pores; common fine soft lime segregation, moderate effervescence with HCl.

Profile No.: 72
Date of description: 9/2006
Location:
Longitude: 30° 25′ E
Latitude: 30° 45′ N
Elevation: 35.0 m a.sl.
Soil taxonomy: *Typic Torrifluvents*
Topography: gently undulating
Land form: flood plain
Slope: gently slope
Land use & vegetation: orchards
Parent material: alluvial deposits
Drainage: well drained soils
Ground water table: not hit

Depth (cm)	Description
0–15	Light yellowish brown (10YR 6/4) moist, very pale brown (10 Y 7/3) dry, sandy clay, moderate medium sub angular blocky, moderate effervescence with HCl, common medium to fine discontinuous random pores, many fine roots moderate effervescence with HCl, clear smooth boundary
15–50	Dark yellowish brown (10 YR 4/4) moist, Yellowish brown (10 YR 5/4) dry, clay loam, moderate medium sub angular blocky, few lime concretion, common coarse and medium discontinuous random pores; moderate effervescence with HCl., clear smooth boundary.
50–100	Yellowish brown (10 YR 5/6) moist, light yellowish brown (10 YR 6/4) dry clay loam, single grains, hard, many gravels, few soft lime concretions, slight effervescence, common fine and medium roots; slight effervescence with HCl.

Soil Profiles Description

(Northwestern Coast)

Profile No.: 73
Date of description: 10/2006
Location:
Longitude: 29° 12′ 14″ E
Latitude: 30° 46′ 11″ N
Elevation: 2.0 m a.sl.
Soil taxonomy: *Aquic Haplosalids*
Topography: gently undulating
Land form: coastal plain
Slope: gently slope
Land use & vegetation: orchards
Parent material: sand deposits
Drainage: poorly drained soils
Ground water table: 45 cm

Depth (cm)	Description
0–15	Yellow)IO YR 7/6, dry); fine sand; loose, non-sticky, non-plastic; moderate efferv. With Hcl; Common fine and medium roots smooth boundary.
15–30	Yellowish brown (10YR 5/8, dry); fine to medium sand; loose; non-sticky, non-plastic; moderate efferv. With Hcl; few fine and medium roots; diffuse boundary
30–45	Yellowish brown (10YR 5/8, dry); coarse sand; loose; non-sticky, non-plastic; moderate efferv. With Hcl.

Profile No.: 74
Date of description: 10/2006
Location:
Longitude: 29° 07′ 56″ E
Latitude: 30° 47′ 21″ N
Elevation: 2.0 m a.sl.
Soil taxonomy: *Aquic Haplosalids*
Topography: gently undulating
Land form: coastal plain
Slope: gently slope
Land use & vegetation: barren
Parent material: sand deposits
Drainage: poorly drained soils
Ground water table: 50 cm

(continued)

Depth (cm)	Description
Depth (cm)	Description
0–20	Yellowish brown (10YR 5/8, wet); sandy clay loam; moderate medium sub-angular blocky Str.; slightly sticky, slightly plastic; moderate efferv. With Hcl; diffuse snooth boundary.
20–50	Yellowish brown (10YR 5/8, wet); sandy clay loam; moderate medium sub-angular blocky stru.; slightly sticky, slightly plastic; moderate efferv. With Hcl.;

Profile No.: 75
Date of description: 10/2006
Location:
Longitude: 29° 17′ 51″ E
Latitude: 30° 51′ 13″ N
Elevation: 32.0 m a.sl.
Soil taxonomy: *Typic Haplocalcids*
Topography: gently undulating
Land form: alluvial fans
Slope: gently slope
Land use & vegetation: orchards
Parent material: limestone
Drainage: well drained soils
Ground water table: not hit

Depth (cm)	Description
0–15	Yellowish brown (10 YR 5/6, dry); Sandy; strong coarse sub-angular blocky str., non sticky, non plastic; strong efferv. With Hcl; white spot of $caco_3$ concretions; few fine and medium roots; diffuse smooth boundary.
15–50	Brownish yellow (10 YR 6/6 dry); sandy to sandy loam; moderate medium sub-angular blocky str.; non sticky non plastic; strong efferv. With Hcl; white spot of $caco_3$ concretions; few small crystals of gypsum; diffuse smooth boundary.
50–120	Yellowish brown (10 YR 5/8, wet); loamy; moderate medium sub-angular blocky str.; slightly sticky, slightly plastic; some small crystals of gypsum; strong efferv. With Hcl.

Profile No.: 76
Date of description: 10/2006
Location:
Longitude: 29° 12′ 14″ E
Latitude: 30° 46′ 11″ N
Elevation: 30.0 m a.sl.
Soil taxonomy: *Typic Haplocalcids*

Topography: gently undulating
Land form: alluvial fans
Slope: gently slope
Land use & vegetation: barren
Parent material: limestone
Drainage: moderately well drained soils
Ground water table: not hit

Depth (cm)	Description
0–25	Brownish Yellow (10 YR 6/6; dry); sandy; single grains; non-sticky, non-plastic; medium to strong efferv. With Hcl; clear smooth boundary.
25–60	Light Yellowish brown (10 YR 6/4, dry); sandy, weak; fine sub-angular blocky str.; non-sticky, non-plastic; strong efferv. With Hcl; some small broken shells; hard pan; many medium and large gravels; diffuse smooth boundary.
60–120	Light Yellowish brown (10 YR 6/4, dry); sandy; weak fine sub-angular blocky str.; non-sticky, non plastic; hard pan; some to many medium grarvels; strong efferv. With Hcl.

Profile No.: 77
Date of description: 10/2006
Location:
Longitude: 29° 15′ 22″ E
Latitude: 30° 47′ 30″ N
Elevation: 30.0 m a.sl.
Soil taxonomy: *Typic Petrogypsids*
Topography: gently undulating
Land form: Coastal plain
Slope: slopping
Land use & vegetation: barren
Parent material: limestone
Drainage: poor drained soils
Ground water table: not hit

Depth (cm)	Description
0–15	Brownish Yellow (10 YR 6/6, dry); sandy loam; moderate medium sub-angular blocky str.; slightly sticky, slightly plastic; strong efferv, with Hcl., some lime concretions; few fine roots; diffuse smooth boundary.
15–60	Yellowish brown (10 YR 5/8 dry). Sandy loam; week fine to medium sub-angular blocky str.; slightly sticky, slightly plastic; strong efferv. With Hcl.; few fine roots.

Profile No.: 78
Date of description: 10/2006

Location:
Longitude: 29° 15′ 27″ E
Latitude: 30° 47′ 59″ N
Elevation: 30.0 m a.sl.
Soil taxonomy: *Typic Petrogypsids*
Topography: gently undulating
Land form: Coastal plain
Slope: gently slope
Land use & vegetation: barren
Parent material: limestone
Drainage: poor drained soils
Ground water table: not hit

Depth (cm)	Description
0–25	Brownish Yellow (10 YR 6/6; dry); sandy; moderate medium sub-angular blocky str.; non-sticky, non plastic; strong efferv. With Hcl.; hard, common fine and medium roots; diffuse smooth boundary.
25–80	Dark yellowish brown (10 YR 4/5, dry); sandy; moderate medium sub-angular blocky str.; non sticky; non plastic; many broken white lime stone.

Profile No.: 79
Date of description: 10/2006
Location:
Longitude: 31° 41′ E
Latitude: 30° 40′ N
Elevation: 54.0 m a.sl.
Soil taxonomy: *Typic Torriorthents*
Topography: undulating
Land form: aeolian plain
Slope: gently slope
Land use & vegetation: barren
Parent material: fluvial/aeolian deposits
Drainage: poor drained soils
Ground water table: not hit

Depth (cm)	Description
0–20	Dark yellowish brown (10 YR 4/4) moist, Yellowish brown (10 YR 5/4) dry, gravelly sand, massive, few lime concretion, common gravel, common coarse and medium discontinuous random pores; moderate effervescence with HCl.
20–60	Yellowish brown (10 YR 5/6) moist, light yellowish brown (10 YR 6/4) dry, sandy, massive, hard, common coarse gravels, common lime concretions, strong

(continued)

Depth (cm)	Description
	effervescence, common fine and medium roots; strong effervescence with HCl, clear smooth boundary.
60–100	Yellowish brown (10 YR 5/6) moist, pale brown (10 YR 6/3) dry, sandy, massive, moderate fine gravel and grits common medium to coarse discontinuous pores, common lime concretions, strong effervescence with HCl, diffuse smooth boundary.

Profile No.: 80
Date of description: 10/2006
Location:
Longitude: 31° 41′ E
Latitude: 30° 40′ N
Elevation: 54.0 m a.sl.
Soil taxonomy: *Typic Torriorthents*
Topography: undulating
Land form: aeolian plain
Slope: gently slope
Land use & vegetation: barren
Parent material: fluvial/aeolian deposits
Drainage: poor drained soils
Ground water table: not hit.

Depth (cm)	Description
0–20	Dark yellowish brown (10 YR 4/4) moist, Yellowish brown (10 YR 5/4) dry, gravelly sand, massive, few lime concretion, common gravel, common coarse and medium discontinuous random pores; moderate effervescence with HCl.
20–60	Yellowish brown (10 YR 5/6) moist, light yellowish brown (10 YR 6/4) dry, sandy, massive, hard, common coarse gravels, common lime concretions, moderate effervescence, common fine and medium roots; strong effervescence with HCl, clear smooth boundary.
60–120	Light Yellowish brown (10 YR 6/4, dry); sandy; moderate medium sub-angular blocky str.; non-sticky, non-plastic, hard; moderate efferv. With Hcl.; few broken shells.

Profile No.: 81
Date of description: 10/2006
Location:
Longitude: 29° 15′ 20″ E
Latitude: 30° 44′ 49″ N
Elevation: 54.0 m a.sl.
Soil taxonomy: *Typic Torripsamments*
Topography: undulating

Land form: aeolian plain
Slope: gently slope
Land use & vegetation: barren
Parent material: aeolian deposits
Drainage: well drained soils
Ground water table: not hit

Depth (cm)	Description
0–20	Yellowish brown (10 YR 5/6, dry). Sandy; single grain str.; non sticky, non plastic; sligh efferv. With Hcl.; hard; few to some medium gravels, diffuse smooth boundary.
20–40	Yellowish brown (10 YR 5/6, dry). Sandy, single grain; non-sticky; non-plastic; moderate efferv. With Hcl.; some broken rocks; diffuse smooth boundary.
40–120	Light Yellowish brown (10 YR 6/4, dry); sandy; single grain str.; non-sticky, non-plastic, hard; moderate efferv. With Hcl.; few broken shells.

Profile No.: 82
Date of description: 10/2006
Location:
Longitude: 29° 16′ 51″ E
Latitude: 30° 43′ 49″ N
Elevation: 54.0 m a.sl.
Soil taxonomy: *Typic Torripsamments*
Topography: almost flat
Land form: aeolian plain
Slope: gently slope
Land use & vegetation: barren
Parent material: aeolian deposits
Drainage: well drained soils
Ground water table: not hit

Depth (cm)	Description
0–30	Dark yellowish brown (10 YR 4/4) moist, Yellowish brown (10 YR 5/4) dry, sandy, single grain, few lime concretion, common gravel, common coarse and medium discontinuous random pores; moderate effervescence with HCl.
30–60	Yellowish brown (10 YR 5/6) moist, light yellowish brown (10 YR 6/4) dry, sandy, single grain, hard, common coarse gravels, common lime concretions, sligh effervescence, common fine and medium roots; strong effervescence with HCl, clear smooth boundary.
60–130	Light Yellowish brown (10 YR 6/4, dry); sandy; single grain.; non-sticky, non-plastic, hard; moderate efferv. With Hcl.; few broken shells.

Profile No.: 83
Date of description: 10/2006
Location:
Longitude: 29° 01′ 06″ E
Latitude: 30° 46′ 53″ N
Elevation: 30.0 m a.sl.
Soil taxonomy: Lithic *Torripsamments*
Topography: gently undulating
Land form: Coastal plain
Slope: gently slope
Land use & vegetation: barren
Parent material: limestone
Drainage: poor drained soils
Ground water table: not hit

Depth (cm)	Description
0–40	Very pale Brown (10 YR 7/4, dry); sandy; single grains; non sticky, non-plastic; moderate efferv. With Hcl; very hard and firm; compact layer. Ped rock (lime stone).

Profile No.: 84
Date of description: 10/2006
Location:
Longitude: 29° 04′ 30″ E
Latitude: 30° 46′ 05″ N
Elevation: 30.0 m a.sl.
Soil taxonomy: Lithic *Torripsamments.*
Topography: gently undulating
Land form: coastal plain
Slope: gently slope
Land use & vegetation: barren
Parent material: limestone
Drainage: poor drained soils
Ground water table: not hit

Depth (cm)	Description
0–30	Very pale Brown (10 YR 7/4, dry); sandy; single grains; non sticky, non-plastic; slight efferv. With Hcl; very hard and firm; clear smooth boundary.
30–45 +45	Very pale Brown (10 YR 7/4, dry); sandy; single grains; non sticky, non-plastic; strong efferv. With Hcl; very hard and firm; compact leyer Ped rock (lime stone).

Soil Profiles Description

(Western Desert Oases)

Bahariya Oases

Profile No.: 85
Date of description: 9/2006
Location:
Longitude: 28° 53′ 18″ E
Latitude: 28° 20′ 20″ N
Elevation: 7.0 m a.sl.
Soil taxonomy: *Typic Torrifluvents*
Topography: gently undulating
Land form: alluvial basin
Slope: gently slope
Land use & vegetation: olive
Parent material: alluvial deposits
Drainage: well drained soils
Ground water table: not hit

Depth (cm)	Description
0–15	Dark brown (10 YR 3/3) clay loam, weak sub-angular blocky, massive, friable, slightly, sticky, few fine pores, few fine and medium roots, strong effervescence, clear smooth boundary
15–50	Dark brown (10 YR 4/3) moist, loam, weak sub-angular blocky, friable, few fine pores, few fine roots, effervescence, diffuse smooth boundary.
50–110	Dark yellowish brown (10 YR 4/4) moist, sandy loam, moderate sub-angular blocky, friable few fine pores, strong effervescence.

Profile No.: 86
Date of description: 10/2006
Location:
Longitude: 29° 07′ 22″ E
Latitude: 28° 18′ 10″ N
Elevation: 11.0 m a.sl.
Soil taxonomy: *Typic Quartizipsamments*
Topography: gently undulating
Land form: sand sheet
Slope: gently slope
Land use & vegetation: barren
Parent material: aeolian deposits
Drainage: well drained

Ground water table: not hit

Depth (cm)	Description
0–30	Yellowish brown (10 YR 5/1) moist, sand, single grains,, common coarse and medium discontinuous random pores, many fine roots, weak effervescence, diffuse boundary.
30–70	Light yellowish brown (10 YR 6 M) moist, sand, single grains, common coarse and medium discontinuous random pores, loose, weak effervescence.
70–130	Dark yellowish brown (10 YR 4/4) moist, Yellowish brown (10 YR 5/4)dry, sandy, single grain, few lime concretion, common coarse and medium discontinuous random pores; moderate effervescence with HCl.

Profile No.: 87
Date of description: 10/2006
Location:
Longitude: 29° 00′ 10″ E
Latitude: 28° 05′ 10″ N
Elevation: 54.0 m a.sl.
Soil taxonomy: *Typic Torripsamments*
Topography: almost flat
Land form: aeolian plain
Slope: gently slope
Land use & vegetation: barren
Parent material: aeolian deposits
Drainage: well drained soils
Ground water table: not hit

Depth (cm)	Description
0–30	Dark yellowish brown (10 YR 4/4) moist, Yellowish brown (10 YR 5/4) dry, sandy, single grain, common gravel, common coarse and medium discontinuous random pores; moderate effervescence with HCl.
30–65	Yellowish brown (10 YR 5/6) moist, light yellowish brown (10 YR 6/4) dry, sandy, single grain, hard, common coarse gravels, slight effervescence, common fine and medium roots; strong effervescence with HCl, clear smooth boundary.
65–150	Light Yellowish brown (10 YR 6/4, dry); sandy; single grain; non-sticky, non-plastic, hard; few lime concretion moderate efferv. With Hcl.; few broken shells.

Profile No.: 88
Date of description: 10/2006
Location:
Longitude: 28° 55′ 00″ E
Latitude: 28° 25′ 10″ N
Elevation: 30.0 m a.sl.

Soil taxonomy: Typic *Torripsamments*
Topography: gently undulating
Land form: sand sheet
Slope: gently slope
Land use & vegetation: barren
Parent material: limestone
Drainage: poor drained soils
Ground water table: not hit

Depth (cm)	Description
0–25	Very pale Brown (10 YR 7/4, dry); sandy; single grains; non sticky, non-plastic; slight efferv. With Hcl; very hard and firm; few lime concretion, clear smooth boundary.
25–50 +50	Very pale Brown (10 YR 7/4, dry); sandy; single grains; non sticky, non-plastic; strong efferv. With Hcl; very hard and firm; common lime concretion, compact leyer Ped-rock (lime stone).

Profile No.: 89
Date of description: 10/2006
Location:
Longitude: 28° 50′ 05″ E
Latitude: 28° 22′ 10″ N
Elevation: 54.0 m a.sl.
Soil taxonomy: *Typic Torriorthents*
Topography: undulating
Land form: depression
Slope: gently slope
Land use & vegetation: barren
Parent material: aeolian deposits
Drainage: well drained soils
Ground water table: not hit

Depth (cm)	Description
0–15	Yellowish brown (10 YR 5/6) moist, light yellowish brown (10 YR 6/4) dry, gravelly sand, massive, hard, common coarse gravels, common lime concretions, moderate effervescence, common fine and medium roots; strong effervescence with HCl, clear smooth boundary.
15–35	Dark yellowish brown (10 YR 4/4) moist, Yellowish brown (10 YR 5/4) dry, sandy, massive, few lime concretion, common gravel, common coarse and medium discontinuous random pores; moderate effervescence with HCl.
35–110	Light Yellowish brown (10 YR 6/4, dry); sandy; moderate medium sub-angular blocky str.; non-sticky, non-plastic, hard; moderate efferv. With Hcl.; few broken shells.

Profile No.: 90
Date of description: 9/2006
Location:
Longitude: 28° 53′ 18″ E
Latitude: 28° 20′ 20″ N
Elevation: 7.0 m a.sl.
Soil taxonomy: *Typic Haplosalids*
Topography: gently undulating
Land form: alluvial basin
Slope: gently slope
Land use & vegetation: olive
Parent material: alluvial deposits
Drainage: well drained soils
Ground water table: 100 cm

Depth (cm)	Description
0–30	Dark brown (10 YR 3/3) clay loam, moderate medium sub-angular blocky, massive, friable, slightly, sticky, few fine pores, few fine and medium roots, strong effervescence, clear smooth boundary
30–45	Dark brown (10 YR 4/3) moist, loam, weak medium sub-angular blocky, friable, few fine pores, few fine roots, effervescence, diffuse smooth boundary.
45–100	Dark yellowish brown (10 YR 4/4) moist, sandy loam, weak fine sub-angular blocky, friable few fine pores, strong effervescence.

Soil Profiles Description

(Western Desert Oases)

Siwa Oases

Profile No.: 91
Date of description: 10/2006
Location:
Longitude: 25° 28′ 05″ E
Latitude: 29° 18′30″ N
Elevation: 1.0 m a.sl.
Soil taxonomy: *Typic Torriorthents*
Topography: undulating
Land form: depression
Slope: gently slope

Land use & vegetation: barren
Parent material: alluvium/aeolian deposits
Drainage: well drained soils
Ground water table: not hit

Depth (cm)	Description
0–15	Yellowish brown (10 YR 5/6) moist, light yellowish brown (10 YR 6/4) dry, gravelly sand, massive, hard, common medium gravels, common lime concretions, moderate effervescence, few fine and medium roots; slight effervescence with HCl, clear smooth boundary.
15–40	Dark yellowish brown (10 YR 4/4) moist, Yellowish brown (10 YR 5/4) dry, sandy, massive, few lime concretion, common medium gravel, common coarse and medium discontinuous random pores; moderate effervescence with HCl.
40–110	Light Yellowish brown (10 YR 6/4, dry); sandy; moderate medium sub-angular blocky str.; non-sticky, non-plastic, hard; moderate efferv. With Hcl.

Profile No.: 92
Date of description: 10/2006
Location:
Longitude: 28° 30′ 40″ E
Latitude: 29° 14′ 10″ N
Elevation: 2.0 m a.sl.
Soil taxonomy: *Typic Torripsamments*
Topography: almost flat
Land form: aeolian plain
Slope: gently slope
Land use & vegetation: barren
Parent material: aeolian deposits
Drainage: well drained soils
Ground water table: not hit

Depth (cm)	Description
0–30	Light Yellowish brown (10 YR 6/4, dry); light yellowish brown (10 YR 6/4) dry, sandy, single grain, common gravel, few fine roots, common coarse and medium discontinuous random pores; moderate effervescence with HCl.
30–65	Yellowish brown (10 YR 5/6) moist, light yellowish brown (10 YR 6/4) dry, sandy, single grain, hard, common coarse gravels, slight effervescence, few fine and medium roots; strong effervescence with HCl, clear smooth boundary.
65–140	Dark yellowish brown (10 YR 4/4) moist, Yellowish brown (10 YR 5/4) dry, sandy; single grain; non-sticky, non-plastic, hard; few lime concretion moderate efferv. With Hcl.; few broken shells.

Profile No.: 93
Date of description: 10/2006
Location:
Longitude: 29° 18′ 08″ E
Latitude: 25° 20′ 05″ N
Elevation: 2.0 m a.sl.
Soil taxonomy: *Typic Haplaquents*
Topography: almost flat
Land form: sand sheet
Slope: gently slope
Land use & vegetation: barren
Parent material: limestone
Drainage: poor drained soils
Ground water table: not hit

Depth (cm)	Description
0–20	Dark yellowish brown (10 YR 4/4) moist, Yellowish brown (10 YR 5/4) dry, sandy; single grains; non sticky, non-plastic; slight efferv. With Hcl; hard and firm; few fine roots, common lime concretion, clear smooth boundary.
20–50 +50	Light Yellowish brown (10 YR 6/4, dry); light yellowish brown (10 YR 6/4) dry, sandy; single grains; non sticky, non-plastic; strong efferv. With Hcl; very hard and firm; few fine and medium roots, common lime concretion, compact leyer Water table.

Profile No.: 94
Date of description: 10/2006
Location:
Longitude: 29° 15′ 08″ E
Latitude: 25° 30′ 07″ N
Elevation: 1.0 m a.sl.
Soil taxonomy: *Typic Haplosalids*
Topography: almost flat
Land form: sand sheet
Slope: gently slope
Land use & vegetation: barren
Parent material: lacustrine deposits
Drainage: poor drained soils
Ground water table: not hit

Depth (cm)	Description
0–25	Dark yellowish brown (10 YR 4/4) moist, Yellowish brown (10 YR 5/4) dry, sandy; single grains; non sticky, non-plastic; slight efferv. With Hcl; few fine roots, few lime concretion, clear smooth boundary.
25–60 +60	Light Yellowish brown (10 YR 6/4, dry); light yellowish brown (10 YR 6/4) dry, sandy; single grains; non sticky, non-plastic; strong efferv. With Hcl; common lime concretion, compact leyer Water table.

Profile No.: 95
Date of description: 10/2006
Location:
Longitude: 25° 32′ 11″ E
Latitude: 29° 18′ 09″ N
Elevation: 2.0 m a.sl.
Soil taxonomy: *Typic Haplocalcids*
Topography: gently undulating
Land form: aeolian plain
Slope: gently slope
Land use & vegetation: barren
Parent material: aeolian deposits
Drainage: well drained soils
Ground water table: not hit

Depth (cm)	Description
0–30	Light Yellowish brown (10 YR 6/4, dry); light yellowish brown (10 YR 6/4) dry, sandy, single grain, common gravel, few fine roots, common coarse and medium discontinuous random pores; strong effervescence with HCl.
30–60	Yellowish brown (10 YR 5/6) moist, light yellowish brown (10 YR 6/4) dry, sandy, single grain, hard, common coarse gravels, common lime concretion strong effervescence, few fine and medium roots; strong effervescence with HCl, clear smooth boundary.
60–150	Dark yellowish brown (10 YR 4/4) moist, Yellowish brown (10 YR 5/4) dry, sandy; single grain; non-sticky, non-plastic, hard; common lime concretion strong efferv. With Hcl.; few broken shells.

Profile No.: 96
Date of description: 10/2006
Location:
Longitude: 25° 44′ 31″ E
Latitude: 29° 15′ 05″ N
Elevation: 2.0 m a.sl.

Soil taxonomy: *Calcic Haplosalids*
Topography: gently undulating
Land form: aeolian plain
Slope: gently slope
Land use & vegetation: barren
Parent material: aeolian/lacustrine deposits
Drainage: well drained soils
Ground water table: not hit

Depth (cm)	Description
0–30	Light Yellowish brown (10 YR 6/4, dry); light yellowish brown (10 YR 6/4) dry, sandy, single grain, common gravel, common coarse and medium discontinuous random pores; strong effervescence with HCl.
30–65 +65	Yellowish brown (10 YR 5/6) moist, light yellowish brown (10 YR 6/4) dry, sandy, single grain, hard, common coarse gravels, common lime concretion strong effervescence, strong effervescence with HCl, clear smooth boundary. Water table

Soil Profiles Description

(Western Desert Oases)

Kharga Oases

Profile No.: 97
Date of description: 10/2006
Location:
Longitude: 30° 30′ 10″ E
Latitude: 25° 30′ 20″ N
Elevation: 4.5 m a.s.l.
Soil taxonomy: *Typic Torrifluvents*
Topography: flat to almost flat
Land form: depression
Slope: nearly level
Land use & vegetation: cultivated vegetables
Parent material: alluvial deposits
Drainage: well drained soils
Ground water table: 110 cm

Depth (cm)	Description
0–15	Dark grayish brown (10YR4/2) moist, grayish Brown (10YR5/2) dry, sandy clay loam; strong small to medium granular; moderately sticky, moderately plastic; moderately firm, moderately hard; common fine to medium continuous and discontinuous random interstitial pores; many fine to medium roots, slight effervescence with HCl,
15–30	Dark grayish brown (10YR4/2) moist, grayish brown (10YR5/2) dry, clay loam; moderately small to medium sub angular blocky; sticky, plastic, firm, hard; common fine to medium discontinuous random interstitial and few channel pores; common fine to medium roots,. slight effervescence with HCl, clear smooth boundary
30–75	Dark grayish brown (10YR 4/2) moist, grayish brown (10YR 5/2) dry: clayey: Moderately medium sub angular blocky, sticky,Plastic; moderately firm, hard; common fine to medium discontinuous random pores; few fine roots, slight effervescence with HCl, clear smooth boundary
75–120	Very dark grayish brown (10YR 3/2) moist, dark grayish brown (10YR 5/1) dry; clayey; strong fine to medium sub angular blocky; very sticky, very plastic; very firm, very hard, very few clay skins, common fine continuous and discontinuous random pores; slight effervescence with HCl.

Profile No.: 98
Date of description: 10/2006
Location
Longitude: 30° 35′ 08″ E
Latitude: 25° 50′ 03″ N
Elevation: 1.0 m a.sl.
Soil taxonomy: *Typic Torripsamments*
Topography: almost flat
Land form: aeolian plain
Slope: nearly level
Land use & vegetation: barren
Parent material: aeolian deposits
Drainage: well drained
Ground water table: 150 cm

Depth (cm)	Description
0–25	Yellowish brown (10 YR 5/6) moist, pale brown (10 YR 6/3) dry, sandy, single grains, loose, moderate fine gravel, common medium to coarse discontinuous random pores, many fine roots moderate effervescence with HCl, clear smooth boundary.
25–55	Yellowish brown (10 YR 5/6) moist, light yellowish brown (10 YR 6/4) dry, sand, single grains, hard, few gravels, few soft lime concretions, moderate effervescence, common fine and medium roots; slight effervescence with HCl, clear smooth boundary.

(continued)

Depth (cm)	Description
55–130	Dark yellowish brown (10 YR 4/4) moist, Yellowish brown (10 YR 5/4)dry, slightly hard medium coarse sand, single grains few lime and gypsum concretion, common coarse and medium discontinuous random pores; moderate effervescence with HCl.

Profile No.: 99
Date of description: 10/2006
Location
Longitude: 30° 35′ 15″ E
Latitude: 25° 45′.04″ N
Elevation: 1.0 m a.sl.
Soil taxonomy: Typic *Torripsamments*
Topography: almost flat
Land form: aeolian plain
Slope: nearly level
Land use & vegetation: barren
Parent material: aeolian deposits
Drainage: moderately well drained
Ground water table: not hit

Depth (cm)	Description
0–15	Yellowish brown (10 YR 5/6) moist, light yellowish brown (10 YR 6/4) dry, sandy, single grains, few fine and medium gravel, common medium to coarse discontinuous random pores, many fine roots, strong effervescence with HCl, clear smooth boundary.
15–30	Yellowish brown (10 YR 5/6) moist, light yellowish brown (10 YR 6/4) dry, sandy, single grains, few gravels, few soft lime concretions, strong effervescence, common fine and medium roots; strong effervescence with HCl, clear smooth boundary.
30–50	Dark yellowish brown (10 YR 4/4) moist, Yellowish brown (10 YR 5/4)dry, sandy, few lime concretion, common coarse and medium discontinuous random pores; strong effervescence with HCl.

Profile No.: 100
Date of description: 10/2006
Location
Longitude: 30° 30′ 00″ E
Latitude: 25° 28′ 04″ N
Elevation: 1.0 m a.sl.
Soil taxonomy: *Typic Haplotorrerts*
Topography: almost flat
Land form: basin
Slope: nearly level

Land use & vegetation: cultivated with barly
Parent material: alluvial deposits
Drainage: moderately well drained
Ground water table: not hit

Depth n (cm)	Description
0–15	Dark brown (10 YR 3/3) moist, Dark grayish brown (10 YR 4/2) dry, clay, weak sub-angular blocky, firm, sticky, plastic, many fine and medium pores, few hard lime concretions, many fine roots, moderate effervescence, diffuse boundary.
15–30	Dark greyish brown (10 YR 4/3) moist, Dark grey (10 YR 4/1) dry, clay, moderate angular blocky, firm, sticky, plastic, many fine and medium pores, many soft lime concretions, weak effervescence, diffuse boundary.
30–100	Very dark greyish brown (10 YR 3/3) moist, Dark grey (10 YR 4/1) dry, clay, strong fine angular blocky, firm, sticky, plastic, slickensides, many fine and medium pores, many soft lime concretions, woak effervescence.

Profile No.: 101
Date of description: 10/2006
Location:
Longitude: 30° 20′ 00″E
Latitude: 25° 25′10″ N
Elevation: 6.0 m a.sl.
Soil taxonomy: *Chromic Haplotorrrrts*
Topography: flat to almost flat
Land form: basin
Slope: nearly level
Land use & vegetation: fallow after rice
Parent material: alluvial deposits
Drainage: moderately well drained
Ground water table: not hit

Depth (cm)	description
0–25	Dark reddish brown (5 YR 5/3) moist, dark reddish brown (5 YR 5/3) dry, clayey; string fine and medium sub-angular blocky; very sticky, very plastic, firm, hard; many fine to medium continuous and discontinuous interstitial pores; many fine and medium roots; slightly effervescence with HCl; clear gradual boundary.
25–50	Dark reddish brown (5 YR 5/3) moist, dark reddish brown (5 YR 5/3) dry, clayey; moderate to strong medium angular to sub angular blocky; slightly sticky slightly plastic, friable, hard; common fine and medium discontinuous random pores; common fine and medium roots, slightly effervescence with HCl, clear smooth boundary.
50–80	Dark reddish brown (5 YR 5/3) moist moist, Reddish brown (7.4 YR 6/8) dry, clay loam weak medium angular to sub angular blocky, sticky, Plastic friable, hard; common fine and medium discontinuous random pores, few very soft lime concretion; few fine roots, slight effervescence with Hcl clear smooth boundary.

(continued)

Depth (cm)	description
80–140	Dark reddish brown (5 YR 5/3) moist moist, Reddish brown (7.4 YR 6/8) dry; clay loam weak fine and medium sub angular blocky; sticky, plastic, firm, hard; few fine and medium discontinuous pores; slight effervescence with HCl,

Soil Profiles Description

(Western Desert Oases)

Dakhla Oases

Profile No.: 102
Date of description: 10/2006
Location:
Longitude: 28° 28′ 10″ E
Latitude: 25° 18′ 20″ N
Elevation: 8.5 m a.s.l.
Soil taxonomy: *Typic Torrifluvents*
Topography: flat to almost flat
Land form: depression
Slope: nearly level
Land use & vegetation: cultivated vegetables
Parent material: alluvial deposits
Drainage: well drained soils
Ground water table: not hit

Depth (cm)	Description
0–20	Dark grayish brown (10YR4/2) moist, grayish Brown (10YR5/2) dry, sandy clay loam; strong medium granular; moderately sticky, moderately plastic; moderately firm, moderately hard; common fine continuous and discontinuous random interstitial pores; many fine to medium roots, slight effervescence with HCl,
20–40	Dark grayish brown (10YR4/2) moist, grayish brown (10YR5/2) dry, clay loam; moderately medium sub angular blocky; sticky, plastic, firm, hard; common fine to medium discontinuous random interstitial and few channel pores; common fine to medium roots,. slight effervescence with HCl, clear smooth boundary
40–100	Dark grayish brown (10YR 4/2) moist, grayish brown (10YR 5/2) dry: clayey: Moderately medium sub angular blocky, sticky,Plastic; moderately firm, hard; common fine to medium discontinuous random pores; few fine roots, slight effervescence with HCl.

Profile No.: 103
Date of description: 10/2006
Location
Longitude: 28° 30′ 03″ E
Latitude: 25° 19′ 13″ N
Elevation: 1.0 m a.sl.
Soil taxonomy: *Typic Torripsamments*
Topography: almost flat
Land form: aeolian plain
Slope: nearly level
Land use & vegetation: barren
Parent material: aeolian deposits
Drainage: well drained
Ground water table: not hit

Depth (cm)	Description
0–30	Yellowish brown (10 YR 5/6) moist, pale brown (10 YR 6/3) dry, sandy, single grains, loose, few fine gravel, common medium to coarse discontinuous random pores, many fine roots moderate effervescence with HCl, clear smooth boundary.
30–70	Yellowish brown (10 YR 5/6) moist, light yellowish brown (10 YR 6/4) dry, sand, single grains, hard, few medium gravels, few soft lime concretions, moderate effervescence, common fine and medium roots; slight effervescence with HCl, clear smooth boundary.
70–120	Dark yellowish brown (10 YR 4/4) moist, Yellowish brown (10 YR 5/4)dry, sand, single grains few lime and gypsum concretion, common coarse and medium discontinuous random pores; moderate effervescence with HCl.

Profile No.: 104
Date of description: 10/2006
Location
Longitude: 28° 34′ 15″ E
Latitude: 25° 28′ 00″ N
Elevation: 1.0 m a.sl.
Soil taxonomy: Typic *Torripsamments*
Topography: almost flat
Land form: aeolian plain
Slope: sloppig
Land use & vegetation: barren
Parent material: aeolian deposits
Drainage: poor drained
Ground water table: not hit

Depth (cm)	Description
0–30	Yellowish brown (10 YR 5/6) moist, light yellowish brown (10 YR 6/4) dry, sandy, single grains, few fine and medium gravel, common medium to coarse discontinuous random pores, many fine roots, strong effervescence with HCl, clear smooth boundary.
30–45 +50	Yellowish brown (10 YR 5/6) moist, light yellowish brown (10 YR 6/4) dry, sandy, single grains, few gravels, few soft lime concretions, strong effervescence, common fine and medium roots; strong effervescence with HCl, clear smooth boundary. Ped rock (limestone)

Profile No.: 105
Date of description: 10/2006
Location
Longitude: 28° 45′ 01″ E
Latitude: 25° 31′ 35 N
Elevation: 1.0 m a.sl.
Soil taxonomy: *Typic Haplotorrerts*
Topography: almost flat
Land form: basin
Slope: nearly level
Land use & vegetation: cultivated with barly
Parent material: alluvial deposits
Drainage: moderately well drained
Ground water table: not hit

Depth (cm)	Description
0–25	Dark brown (10 YR 3/3) moist, Dark grayish brown (10 YR 4/2) dry, clay, moderate fine sub-angular blocky, firm, sticky, plastic, many fine and medium pores, few hard lime concretions, many fine roots, moderate effervescence, diffuse boundary.
25–70	Dark greyish brown (10 YR 4/3) moist, Dark grey (10 YR 4/1) dry, clay, moderate medium sub-angular blocky, sticky, plastic, many fine and medium pores, many soft lime concretions, weak effervescence, diffuse boundary.
70–120	Very dark greyish brown (10 YR 3/3) moist, Dark grey (10 YR 4/1) dry, clay, strong medium angular blocky, firm, sticky, plastic, slickensides, many fine and medium pores, many soft lime concretions, woak effervescence.

Profile No.: 106
Date of description: 10/2006
Location:
Longitude: 28° 30′ 30″ E
Latitude: 25° 48′40″ N
Elevation: 10.0 m a.sl.
Soil taxonomy: *Chromic Haplotorrrrts*

Topography: flat to almost flat
Land form: basin
Slope: nearly level
Land use & vegetation: fallow after maize
Parent material: alluvial deposits
Drainage: moderately well drained
Ground water table: not hit

Depth (cm)	Description
0–30	Dark reddish brown (5 YR 5/3) moist, dark reddish brown (5 YR 5/3) dry, clayey; string fine and medium sub-angular blocky; very sticky, very plastic, firm, hard; many fine to medium continuous and discontinuous interstitial pores; many fine and medium roots; slightly effervescence with HCl; clear gradual boundary.
30–65	Dark reddish brown (5 YR 5/3) moist, dark reddish brown (5 YR 5/3) dry, clayey; moderate to strong medium angular to sub angular blocky; slightly sticky slightly plastic, friable, hard; common fine and medium discontinuous random pores; common fine and medium roots, slightly effervescence with HCl, clear smooth boundary.
65–120	Dark reddish brown (5 YR 5/3) moist moist, Reddish brown (7.4 YR 6/8) dry; clay loam weak fine and medium sub angular blocky; sticky, plastic, firm, hard; few fine and medium discontinuous pores; slight effervescence with HCl

Annex I: Morphological Description of the Studied Soil Profiles

Al Fayoum Governorate

Profile No.: 107
Date of description: 10/2006
Location:
Longitude: 30° 58′ E
Latitude: 29° 42′ N
Elevation: 31.0 m a.sl.
Soil taxonomy: *Typic Quartizipsamments*
Topography: gently undulating
Land form: Lacustrine plain
Slope: gently slope
Land use & vegetation: barren
Parent material: aeolian deposits
Drainage: well drained soils
Ground water table: not hit

Depth (cm)	Description
0–25	Light yellowish brown (10YR 6/4) moist, very pale brown (10 Y 7/3) dry, sandy, single grains, slight effervescence with HCl, common medium to fine discontinuous random pores, many fine roots moderate effervescence with HCl, clear smooth boundary
25–60	Yellowish brown (10 YR 5/6) moist, pale brown (10 YR 6/3) dry, sandy, massive, common coarse gravels and grits common medium to coarse discontinuous random pores, common lime concretions, common fine to medium roots strong effervescence with HCl, diffuse smooth boundary.

(continued)

A.-A. Gad, *Digital Land Resources to Support Sustainable Development in Egypt*,
Deltas of the World,
https://doi.org/10.1007/978-3-031-37167-7

Depth (cm)	Description
60–90	Light yellowish brown (10YR 6/4) moist, light yellowish brown (10 YR 6/4) dry sandy, single grains, few gravels, slight effervescence with HCl; slight effervescence with HCl, clear smooth boundary.

Profile No.: 108
Date of description: 10/2006
Location
Longitude: 30° 53′ E
Latitude: 29° 45′ N
Elevation: 41.5 m a.s.l.
Soil taxonomy: *Typic Torriorthents*
Topography: undulating
Land form: Lacustrine plain
Slope: gently slope
Land use & vegetation: barren
Parent material: fluvial deposits
Drainage: moderately well drained
Ground water table: not hit

Depth (cm)	Description
0–25	Yellowish brown (10 YR 5/6) moist, light yellowish brown (10 YR 6/4) dry, sandy loam, massive, common coarse gravels, common lime concretions, moderate effervescence, common fine and medium roots; strong effervescence with HCl, clear smooth boundary.
25–60	Yellowish brown (10 YR 5/6) moist, pale brown (10 YR 6/3) dry, sandy, massive, common coarse gravels and grits common medium to coarse discontinuous random pores, common lime concretions, common fine to medium roots strong effervescence with HCl, diffuse smooth boundary.
60–120	Yellowish brown (10 YR 5/6) moist, light yellowish brown (10 YR 6/4) dry, sandy loam, weak sub angular blocky, hard, common coarse gravels, common lime concretions, strong effervescence, common fine and medium roots; strong effervescence with HCl, clear smooth boundary.

Profile No.: 109
Date of description: 10/2006
Location:
Longitude: 30° 48′ 20″ E
Latitude: 29° 42′ 19″ N
Elevation: 54.0 m a.sl.
Soil taxonomy: *Typic Torripsamments*
Topography: undulating
Land form: Lacustrine plain
Slope: gently slope

Land use & vegetation: barren
Parent material: aeolian deposits
Drainage: well drained soils
Ground water table: not hit

Depth (cm)	Description
0–20	Yellowish brown (10 YR 5/6, dry). Sandy; single grain str.; non sticky, non plastic; sligh efferv. With Hcl.; hard; few to some medium gravels, diffuse smooth boundary.
20–40	Yellowish brown (10 YR 5/6, dry). Sandy, single grain; non-sticky; non-plastic; moderate efferv. With Hcl.; some broken rocks; diffuse smooth boundary.
40–120	Light Yellowish brown (10 YR 6/4, dry); sandy; single grain str.; non-sticky, non-plastic, hard; moderate efferv. With Hcl.; few broken shells.

Profile No.: 110
Date of description: 10/2006
Location:
Longitude: 30° 35′ 11″ E
Latitude: 29° 32′ 49″ N
Elevation: 92.0 m a.sl.
Soil taxonomy: *Typic Haplocalcids*
Topography: gently undulating
Land form: aeolian plain
Slope: gently slope
Land use & vegetation: barren
Parent material: aeolian deposits
Drainage: well drained soils
Ground water table: not hit

Depth (cm)	Description
0–30	Light Yellowish brown (10 YR 6/4, dry); light yellowish brown (10 YR 6/4) dry, sandy, single grain, common gravel, few fine roots, common coarse and medium discontinuous random pores; strong effervescence with HCl.
30–60	Yellowish brown (10 YR 5/6) moist, light yellowish brown (10 YR 6/4) dry, sandy, single grain, hard, common coarse gravels, common lime concretion strong effervescence, few fine and medium roots; strong effervescence with HCl, clear smooth boundary.
60–100	Dark yellowish brown (10 YR 4/4) moist, Yellowish brown (10 YR 5/4) dry, sandy; single grain; non-sticky, non-plastic, hard; common lime concretion strong efferv. With Hcl.; few broken shells.

Profile No.: 111
Date of description: 10/2006
Location:

Longitude: 30° 30′ 06″ E
Latitude: 29° 33′ 53″ N
Elevation: 130.0 m a.sl.
Soil taxonomy: *Lithic Calciorthids*
Topography: gently undulating
Land form: Lacustrine plain
Slope: gently slope
Land use & vegetation: barren
Parent material: limestone
Drainage: poor drained soils
Ground water table: not hit

Depth (cm)	Description
0–40 40–55 +55	Very pale Brown (10 YR 7/4, dry); sandy; single grains; non sticky, non-plastic; strong efferv. With Hcl; very hard and firm; compact layer. Very pale Brown (10 YR 7/4, dry); sandy; single grains; non sticky, non-plastic; strong efferv. With Hcl; very hard and firm; compact leyer Ped rock (lime stone).

Profile No.: 112
Date of description: 10/2006
Location:
Longitude: 30° 16′ 22″ E
Latitude: 29° 18′ 10″ N
Elevation: 111.0 m a.sl.
Soil taxonomy: *Typic Quartizipsamments*
Topography: gently undulating
Land form: sand sheet
Slope: gently slope
Land use & vegetation: barren
Parent material: aeolian deposits
Drainage: well drained
Ground water table: not hit

Depth (cm)	Description
0–30	Yellowish brown (10 YR 5/1) moist, sand, single grains,, common coarse and medium discontinuous random pores, many fine roots, weak effervescence, diffuse boundary.
30–70	Light yellowish brown (10 YR 6 M) moist, sand, single grains, common coarse and medium discontinuous random pores, loose, weak effervescence.
70–90	Dark yellowish brown (10 YR 4/4) moist, Yellowish brown (10 YR 5/4)dry, sandy, single grain, few lime concretion, common coarse and medium discontinuous random pores; moderate effervescence with HCl.

Profile No.: 113
Date of description: 10/2006
Location:
Longitude: 30° 23′ E
Latitude: 29° 17′ N
Elevation: 54.0 m a.sl.
Soil taxonomy: *Typic Quartizipsamments*
Topography: gently undulating
Land form: sand sheet
Slope: gently slope
Land use & vegetation: barren
Parent material: aeolian deposits
Drainage: well drained
Ground water table: not hit

Depth (cm)	Description
0–20	Light yellowish brown (10YR 6/4) moist, very pale brown (10 Y 7/3) dry, loamy sand, single grains, moderate effervescence, common medium to fine discontinuous random pores, many fine roots slight effervescence with HCl, clear smooth boundary
20–60	Yellowish brown (10 YR 5/6) moist, light yellowish brown (10 YR 6/4) dry, sandy, single grains, moderate medium and fine gravel, common medium to coarse discontinuous random pores, few lime concretion, common fine roots moderate effervescence with HCl, diffuse smooth boundary..
60–110	Dark yellowish brown (10 YR 4/4) moist, Yellowish brown (10 YR 5/4)dry, sandy, single grain, few lime concretion, common coarse and medium discontinuous random pores; moderate effervescence with HCl.

Profile No.: 114
Date of description: 10/2006
Location:
Longitude: 30° 31′ E
Latitude: 29° 08′ N
Elevation: 25.0 m a.sl.
Soil taxonomy: *Typic Torripsamments*
Topography: gently undulating
Land form: sand sheet
Slope: gently slope
Land use & vegetation: barren
Parent material: aeolian deposits
Drainage: well drained soils
Ground water table: not hit

Depth (cm)	Description
0–25	Light yellowish brown (10YR 6/4) moist, very pale brown (10 Y 7/3) dry, sandy, massive, moderate effervescence with HCl, common medium to fine discontinuous random pores, many fine roots moderate effervescence with HCl, clear smooth boundary
25–50	Yellowish brown (10 YR 5/6) moist, light yellowish brown (10 YR 6/4) dry sand, single grains, hard, many gravels, few soft lime concretions, moderate effervescence, common fine and medium roots; slight effervescence with HCl, clear smooth boundary.
50–100	Dark yellowish brown (10 YR 4/4) moist, Yellowish brown (10 YR 5/4)dry, sandy, single grain, common lime concretion, common coarse and medium discontinuous random pores; moderate effervescence with HCl.

Profile No.: 115
Date of description: 10/2006
Location:
Longitude: 30° 17′ 11″ E
Latitude: 29° 08′ 10″ N
Elevation: 92.0 m a.sl.
Soil taxonomy: *Typic Haplocalcids*
Topography: gently undulating
Land form: plateau
Slope: gently slope
Land use & vegetation: orchards
Parent material: limestone
Drainage: well drained soils
Ground water table: not hit

Depth (cm)	Description
0–25	Yellowish brown (10 YR 5/6, dry); Sandy; strong coarse sub-angular blocky str., non sticky, non plastic; strong efferv. With Hcl; white spot of $caco_3$ concretions; few fine and medium roots; diffuse smooth boundary.
25–50	Brownish yellow (10 YR 6/6 dry); sandy to sandy loam; moderate medium sub-angular blocky str.; non sticky non plastic; strong efferv. With Hcl; white spot of $caco_3$ concretions; few small crystals of gypsum; diffuse smooth boundary.
50–110	Yellowish brown (10 YR 5/8, wet); loamy; moderate medium sub-angular blocky str.; slightly sticky, slightly plastic; some small crystals of gypsum; strong efferv. With Hcl.

Profile No.: 116
Date of description: 10/2006
Location:

Longitude: 30° 15′ 52″ E
Latitude: 29° 24′ 10″ N
Elevation: 130.0 m a.sl.
Soil taxonomy: *Typic Petrogypsids*
Topography: gently undulating
Land form: plateau
Slope: slopping
Land use & vegetation: barren
Parent material: limestone
Drainage: poor drained soils
Ground water table: not hit

Depth (cm)	Description
0–15	Brownish Yellow (10 YR 6/6, dry); sandy loam; moderate medium sub-angular blocky str.; slightly sticky, slightly plastic; strong efferv, with Hcl., some lime concretions; few fine roots; diffuse smooth boundary.
15–60	Yellowish brown (10 YR 5/8 dry). Sandy loam; week fine to medium sub-angular blocky str.; slightly sticky, slightly plastic; strong efferv. With Hcl.; few fine roots.

Profile No.: 117
Date of description: 10/2006
Location:
Longitude: 30° 19′ E
Latitude: 29° 14′ N
Elevation: 74.0 m a.sl.
Soil taxonomy: *Typic Torriorthents*
Topography: undulating
Land form: aeolian plain
Slope: gently slope
Land use & vegetation: barren
Parent material: fluvial/aeolian deposits
Drainage: poor drained soils
Ground water table: not hit

Depth (cm)	Description
0–20	Dark yellowish brown (10 YR 4/4) moist, Yellowish brown (10 YR 5/4) dry, gravelly sand, massive, few lime concretion, common gravel, common coarse and medium discontinuous random pores; moderate effervescence with HCl.
20–60	Yellowish brown (10 YR 5/6) moist, light yellowish brown (10 YR 6/4) dry, sandy, massive, hard, common coarse gravels, common lime concretions, strong effervescence, common fine and medium roots; strong effervescence with HCl, clear smooth boundary.

(continued)

Depth (cm)	Description
60–90	Yellowish brown (10 YR 5/6) moist, pale brown (10 YR 6/3) dry, sandy, massive, moderate fine gravel and grits common medium to coarse discontinuous pores, common lime concretions, strong effervescence with HCl, diffuse smooth boundary.

Profile No.: 118
Date of description: 10/2006
Location:
Longitude: 30° 29′ E
Latitude: 29° 10′ N
Elevation: 66.0 m a.sl.
Soil taxonomy: *Typic Quartizipsamments*
Topography: gently undulating
Land form: sand sheet
Slope: gently slope
Land use & vegetation: barren
Parent material: aeolian deposits
Drainage: well drained soils
Ground water table: +130 cm

Depth (cm)	Description
0–30	Light yellowish brown (10YR 6/4) moist, very pale brown (10 Y 7/3) dry, sandy, single grains, moderate effervescence, common medium to fine discontinuous random pores, many fine roots moderate effervescence with HCl, clear smooth boundary
30–70	Yellowish brown (10 YR 5/6) moist, light yellowish brown (10 YR 6/4) dry sandy, single grains, hard, few gravels, few soft lime concretions, moderate effervescence, common fine and medium roots; slight effervescence with HCl, clear smooth boundary.
70–130	Dark yellowish brown (10 YR 4/4) moist, Yellowish brown (10 YR 5/4)dry, sandy, single grain, few lime concretion, common coarse and medium discontinuous random pores; moderate effervescence with HCl.

Profile No.: 119
Date of description: 10/2007
Location:
Longitude: 30° 24′ E
Latitude: 30° 31′ N
Elevation: 88.0 m a.sl.
Soil taxonomy: *Typic Torriorthents*
Topography: undulating
Land form: aeolian plain

Slope: gently slope
Land use & vegetation: barren
Parent material: fluvial/aeolian deposits
Drainage: poor drained soils
Ground water table: not hit

Depth (cm)	Description
0–20	Yellowish brown (10 YR 5/6) moist, pale brown (10 YR 6/3) dry, sandy, single grains, moderate fine gravel and grits common medium to coarse discontinuous random pores, common lime concretions, common fine to medium roots strong effervescence with HCl, diffuse smooth boundary.
20–40	Yellowish brown (10 YR 5/6) moist, light yellowish brown (10 YR 6/4) dry, sandy, single grains, hard, common coarse gravels, common lime concretions, moderate effervescence, common fine and medium roots; strong effervescence with HCl, clear smooth boundary.
40–85	Dark yellowish brown (10 YR 4/4) moist, Yellowish brown (10 YR 5/4)dry, sandy, few lime and gypsum concretion, common coarse and medium discontinuous random pores; moderate effervescence with HCl.

Profile No.: 120
Date of description: 9/2006
Location:
Longitude: 30° 49′ 10″ E
Latitude: 29° 30′ 04″ N
Elevation: 1.0 m a.sl.
Soil taxonomy: *Typic Haplosalids*
Topography: gently undulating
Land form: alluvial basin
Slope: gently slope
Land use & vegetation: fallow field
Parent material: lacustrin deposits
Drainage: poor drained soils
Ground water table: 100 cm

Depth (cm)	Description
0–30	Dark brown (10 YR 3/3) clay loam, moderate medium sub-angular blocky, massive, friable, slightly, sticky, few fine pores, few fine and medium roots, strong effervescence, clear smooth boundary
30–45	Dark brown (10 YR 4/3) moist, loam, weak medium sub-angular blocky, friable, few fine pores, few fine roots, effervescence, diffuse smooth boundary.
45–80	Dark yellowish brown (10 YR 4/4) moist, sandy loam, weak fine sub-angular blocky, friable few fine pores, strong effervescence.

Profile No.: 121
Date of description: 10/2006
Location:
Longitude: 30° 41′ 08″ E
Latitude: 29° 27′ 07″ N
Elevation: 1.0 m a.sl.
Soil taxonomy: *Typic Haplosalids*
Topography: almost flat
Land form: sand sheet
Slope: gently slope
Land use & vegetation: barren
Parent material: lacustrine deposits
Drainage: poor drained soils
Ground water table: not hit

Depth (cm)	Description
0–25	Dark yellowish brown (10 YR 4/4) moist, Yellowish brown (10 YR 5/4) dry, sandy; single grains; non sticky, non-plastic; slight efferv. With Hcl; few fine roots, few lime concretion, clear smooth boundary.
25–60 +60	Light Yellowish brown (10 YR 6/4, dry); light yellowish brown (10 YR 6/4) dry, sandy; single grains; non sticky, non-plastic; strong efferv. With Hcl; common lime concretion, compact leyer Water table.

Profile No.: 122
Date of description: 9/2006
Location:
Longitude: 30° 29′ 40″ E
Latitude: 29° 24′ 34″ N
Elevation: 0.0 m a.sl.
Soil taxonomy: *Typic Haplosalids*
Topography: gently undulating
Land form: alluvial basin
Slope: gently slope
Land use & vegetation: fallow field
Parent material: lacustrin deposits
Drainage: poor drained soils
Ground water table: 100 cm

Depth (cm)	Description
0–25	Dark brown (10 YR 3/3) sandy clay loam, moderate medium sub-angular blocky, massive, friable, slightly, sticky, few fine pores, many fine and medium roots, strong effervescence, clear smooth boundary
25–40	Dark brown (10 YR 4/3) moist, loam, weak medium sub-angular blocky, friable, few fine pores, few fine roots, effervescence, diffuse smooth boundary.
40–80	Dark yellowish brown (10 YR 4/4) moist, sandy loam, weak fine sub-angular blocky, friable few fine pores, strong effervescence.

Profile No.: 123
Date of description: 10/2006
Location:
Longitude: 31° 01′ E
Latitude: 29° 32′ N
Elevation: 36.0 m a.s.l.
Soil taxonomy: *Typic Gypsiorthids*
Topography: almost flat to gently undulating
Land form: lacustrine deposits
Slope: gently slope
Land use & vegetation: barren
Parent material: lacustrine deposits
Drainage: moderately well drained soils
Ground water table: not hit cm

Depth (cm)	Description
0–30	Dark gray (10YR4/1) moist, gray (10YR5/1) dry; sandy clay loam; moderate medium sub angular blocky; sticky, plastic; moderately firm, moderately hard; common fine to medium discontinuous random interstitial pores; thin shinny continuous clay films lining the voids; many medium to coarse roots; slight effervescence with Hcl. clear smooth boundary
30–65	Dark yellowish brown (10 YR 4/4) moist, Yellowish brown (10 YR 5/4)dry, clay loam, moderate fine sub angular blocky, hard, cemented abundant pores, common lime and gypsum concretions, strong effervescence with HCl. common coarse and medium discontinuous random pores; slight effervescence with HCl
+65	Hardpan

Profile No.: 124
Date of description: 10/2006
Location:
Longitude: 31° 00′ E
Latitude: 29° 28′ N
Elevation: 44.0 m a.sl.

Soil taxonomy: *Typic Haplocalcids*
Topography: undulating
Land form: sandy plain
Slope: gently slope
Land use & vegetation: barren
Parent material: limestone
Drainage: poor drained soils
Ground water table: not hit

Depth (cm)	Description
0–40	Yellowish brown (10 YR 5/6), dry light yellowish brown (10 YR 6/4) dry, sand, single grains, hard, common coarse pores, common lime concretions, strong effervescence with HCl, common coarse and medium discontinuous random pores; slight effervescence with HCl, diffuse smooth boundary.
40–100	Yellowish brown (10 YR 5/6) moist, light yellowish brown (10 YR 6/4) dry, sandy, structure less, common medium to coarse discontinuous random pores, common fine roots moderate effervescence with HCl.

Profile No.: 125
Date of description: 10/2006
Location:
Longitude: 30° 32′ 14″ E
Latitude: 29° 22′ 11″ N
Elevation: 37.0 m a.sl.
Soil taxonomy: *Typic Haplocalcids*
Topography: gently undulating
Land form: alluvial fans
Slope: gently slope
Land use & vegetation: barren
Parent material: limestone
Drainage: moderately well drained soils
Ground water table: not hit

Depth (cm)	Description
0–20	Brownish Yellow (10 YR 6/6; dry); sandy; single grains; non-sticky, non-plastic; medium to strong efferv. With Hcl; clear smooth boundary.
20–60	Light Yellowish brown (10 YR 6/4, dry); sandy, weak; fine sub-angular blocky str.; non-sticky, non-plastic; strong efferv. With Hcl; some small broken shells; hard pan; many medium and large gravels; diffuse smooth boundary.
60–90	Light Yellowish brown (10 YR 6/4, dry); sandy; weak fine sub-angular blocky str.; non-sticky, non plastic; hard pan; some to many medium grarvels; strong efferv. With Hcl.

Profile No.: 126
Date of description: 10/2006
Location:
Longitude: 30° 39′ E
Latitude: 29° 21′ N
Elevation: 10.0 m a.s.l.
Soil taxonomy: *Vertic Torrifluvents*
Topography: almost flat
Land form: flood plain/river levees
Slope: gently slope
Land use & vegetation: cultivated with field crops
Parent material: alluvial deposits
Drainage: well drained
Ground water table: 120 cm

Depth (cm)	Description
0–30	Very dark grayish brown (10YR3/2) moist, dark grayish brown (10YR4/3) dry; clayey; moderate fine to medium sub angular blocky; sticky, plastic; moderately firm, moderately hard; common fine to medium discontinuous random exped interstitial pores; common fine to coarse roots; slight effervescence with Hcl, clear smooth boundary.
30–60	Dark gray (10YR4/1) moist, gray (10YR5/1) dry; clayey; moderate medium sub angular blocky; sticky, plastic; moderately firm, moderately hard; common fine to medium discontinuous random exped interstitial pores; few fresh and humified residual of organic matter; many fine to coarse roots; slight effervescence with Hcl; clear smooth boundary.
60–110	Very dark grayish brown (10YR 3/2) moist, dark grayish brown (10YR 4/3) dry, clayey; moderate fine to medium sub angular blocky; sticky, plastic; firm, hard; common fine and medium discontinuous random interstitial pores; slight effervescence with Hcl.

Profile No.: 127
Date of description: 10/2006
Location:
Longitude: 30° 42′ E
Latitude: 29° 15′ N
Elevation: 8.0 m a.s.l.
Soil taxonomy: *Vertic Torrifluvents*
Topography: almost flat to gently undulating
Land form: flood plain/river terraces
Slope: nearly level to gently slope
Land use & vegetation: fallow after rice

Parent material: alluvial deposits
Drainage: moderately well drained
Ground water table: 110 cm

Depth (cm)	Description
0–35	Very dark grayish brown (10YR3/2) moist, dark grayish brown (10YR4/3) dry; sandy clay loam; moderate to strong medium sub angular blocky; very sticky, very plastic, extremely firm, extremely hard; common fine discontinuous random exped and medium interstitial pores; many fine to medium roots; moderately effervescence with HCl; clear smooth boundary.
35–65	Dark brown (10YR2/3) moist, grayish brown (10YR5/2) dry; clay loam; strong medium sub angular blocky; very sticky, very plastic, extremely firm, extremely hard; common fine to medium discontinuous random interstitial pores; thin shinny discontinuous clay films lining the voids; few fine to medium roots; slight effervescence with HCl, clear smooth boundary .
65–110	Dark brown (10YR 2/3) moist, grayish brown (10YR 5/2) dry; clayey; strong medium sub angular blocky; very sticky, very Plastic, extremely firm, extremely hard; common fine and medium discontinuous random interstitial pores; slight effervescence with HCl.

Profile No.: 128
Date of description: 10/2006
Location
Longitude: 30° 50′ E
Latitude: 29° 15′ N
Elevation: 2.0 m a.s.l.
Soil taxonomy: *Vertic Torrifluvents*
Topography: almost flat
Land form: flood plain/basins
Slope: gently slope
Land use & vegetation: fallow after maize
Parent material: alluvial deposits
Drainage: moderately well drained
Ground water table: 100 cm
Remarks: surface cracks 5 cm width & 30 cm depth

Depth (cm)	Description
0–25	Very dark grayish brown (10YR3/2) moist; dark grayish brown (10YR4/2) dry; silty clay; blocky; sticky, firm, hard; common fine and medium continuous and discontinuous random interstitial pores, few fine to medium roots, slightly effervescence with HCl, gradual smooth boundary.
25–45	Dark brown (10YR3/3) moist, brown (10YR4/3). dry; clay; moderate medium granular, sticky, plastic, fine, hard; common fine and; medium discontinuous random interstitial pores; many fine and, medium roots; slight effervescence with HCl; clear smooth boundary.

(continued)

Depth (cm)	Description
45–100	Dark gray (10YR 4/1) moist, gray (10YR 5/1) dry, clay; very sticky, very Plastic, firm, hard; common fine to medium continuous and discontinuous vughs, few very fine lime concretion, moderate effervescence with HCl.

Profile No.: 129
Date of description: 10/2006
Location:
Longitude: 30° 52′ E
Latitude: 29° 28′ N
Elevation: 4.0 m a.s.l.
Soil taxonomy: *Vertic Torrifluvents*
Topography: almost flat
Land form: flood plain/overflow mantle
Slope: nearly level to gently slope
Land use & vegetation: fallow after maize
Parent material: alluvial deposits
Drainage: moderately well drained
Ground water table: 90 cm

Depth (cm)	Description
0–25	Dark brown (10YR2/3) moist, grayish Brown (10YR5/2) dry, clayey; moderate fine to medium granular; sticky, plastic; firm, hard; common fine to medium discontinuous random interstitial pores; many fine to medium roots, slight effervescence with Hcl, clear smooth boundary.
25–50	Dark grayish brown (10YR4/2) moist, grayish brown (10YR5/2) dry; clay loam; moderate medium sub angular blocky; sticky, plastic, firm, hard; common fine to medium discontinuous random interstitial pores; common fine to medium roots; slight effervescence with HCl, clear smooth boundary.
50–90	Dark grayish brown (10YR4/2) moist, grayish brown (10YR5/4) dry, clayey; moderate fine to medium sub angular blocky; very sticky, very plastic, very firm, extremely hard; common fine to medium discontinuous random exped and interstitial pores; few humified residual of organic matter; common fine to medium roots,. Slight effervescence with Hcl; .

Profile No.: 130
Date of description: 10/2006
Location:
Longitude: 30° 47′ E
Latitude: 29° 25′ N
Elevation: 8.5 m a.s.l.
Soil taxonomy: *Typic Torrifuvents*

Topography: almost flat
Land form: flood plain/Basin
Slope: gently slope
Land use & vegetation: fallow after cotton
Parent material: alluvial deposits
Drainage: well drained soils
Ground water table: 100 cm

Depth (cm)	Description
0–25	Very dark grayish brown (10YR3/2) moist, dark grayish brown (10YR4/3) dry, clay loam; moderate fine to medium sub angular blocky; sticky, plastic, firm, hard; common fine continuous and discontinuous random interstitial pores; common fine to medium roots, Slight effervescence with HCl; clear smooth boundary.
25–55	Very dark grayish brown (10YR3/2) moist, grayish Brown (10YR4/3) dry, silty clay; moderate fine to medium sub angular; sticky, plastic; firm, hard; common fine discontinuous random interstitial pores; common fine roots, slight effervescence with HCl, gradual smooth boundary.
55–100	Dark grayish brown (10YR 4/2) moist, grayish brown (10YR 5/2) dry; silty clay; Moderate to strong medium sub angular blocky, sticky, Plastic; firm, hard; common fine continuous and discontinuous random pores; few fine roots; clear smooth boundary.

Profile No.: 131
Date of description: 10/2006
Location:
Longitude: 30° 40′ E
Latitude: 29° 21′ N
Elevation: 9.0 m a.sl.
Soil taxonomy: *Typic Torrifluvents*
Topography: flat to almost flat
Land form: fluvio-lacustrine plain/terraces
Slope: nearly level
Land use & vegetation: fallow after rice
Parent material: alluvial-lecustrine deposits
Drainage: moderately well drained
Ground water table: 110 cm

Depth (cm)	Description
0–20	Very dark brown (10YR2/2) moist, dark brown (10YR3/3) dry, clayey; string fine and medium sub-angular blocky; very sticky, very plastic, firm, hard; many fine to medium continuous and discontinuous interstitial pores; many fine and medium roots; slightly effervescence with HCl; clear gradual boundary.

(continued)

Depth (cm)	Description
20–45	Very dark grayish brown (10YR3/2) moist; dark grayish Brown (10YR4/2) dry; clayey; moderate to strong medium sub angular blocky; slightly sticky slightly plastic, friable, hard; common fine and medium discontinuous random pores; common fine and medium roots, slightly effervescence with HCl, clear smooth boundary.
45–110	Dark brown (10YR3/3) moist; grayish Brown (10YR5/2) dry; clay loam weak fine and medium sub angular blocky; sticky, plastic, firm, hard; few fine and medium discontinuous pores; slight effervescence with HCl,

Profile No.: 132
Date of description: 10/2006
Location:
Longitude: 30° 39′ 20″ E
Latitude: 29° 20′ 10″ N
Elevation: 22.0 m a.sl.
Soil taxonomy: *Typic Haplocalcids*
Topography: gently undulating
Land form: alluvial fans
Slope: gently slope
Land use & vegetation: orchards
Parent material: limestone
Drainage: well drained soils
Ground water table: not hit

Depth (cm)	Description
0–15	Yellowish brown (10 YR 5/6, dry); Sandy; strong coarse sub-angular blocky str., non sticky, non plastic; strong efferv. With Hcl; white spot of $caco_3$ concretions; few fine and medium roots; diffuse smooth boundary.
15–40	Brownish yellow (10 YR 6/6 dry); sandy to sandy loam; moderate medium sub-angular blocky str.; non sticky non plastic; strong efferv. With Hcl; white spot of $caco_3$ concretions; few small crystals of gypsum; diffuse smooth boundary.
40–90	Yellowish brown (10 YR 5/8, wet); loamy; moderate medium sub-angular blocky str.; slightly sticky, slightly plastic; some small crystals of gypsum; strong efferv. With Hcl.

Profile No.: 133
Date of description: 10/2006
Location
Longitude: 30° 48′ E
Latitude: 29° 11′ N
Elevation: 4.0 m a.s.l.
Soil taxonomy: *Vertic Torrifluvents*

Topography: Almost flat
Land form: alluvial plain/terrace
Slope: nearly level
Land use & vegetation: cultivated with alfalfa
Parent material: lacustrine/alluvial deposits
Drainage: moderately well drained
Ground water table: 90 cm
Remarks: surface cracks 8 cm width & 25 cm depth

Depth (cm)	Description
0–30	Very dark grayish brown (10YR4/2) moist, brown (10YR5/3) dry, clayey; moderate medium angular blocky; sticky, plastic; firm, hard common fine discontinuous random pores, few fresh and humified residual of organic matter, many fine and medium roots, slightly effervescence with HCl, clear smooth boundary.
30–45	Very dark grayish brown (10YR3/2) moist, dark grayish brown (10YR4/2) dry, clayey; Moderate to strong fine to medium sub-angular, sticky, plastic, firm, hard; common fine and medium continuous and discontinuous channels, few fine soft segregation of CaCO3, common fine to medium roots,. Slightly effervescence with HCl, clear smooth boundary.
45–100	Very dark grayish brown (10YR 3/2) moist, dark grayish brown (10YR 4/2) dry: clay loam; moderate fine to medium angular blocky; very sticky, very plastic; firm; hard; common fine random pores; common fine soft lime segregation, moderate effervescence with HCl.

Profile No.: 134
Date of description: 10/2006
Location
Longitude: 30° 52′ E
Latitude: 29° 13′ N
Elevation: 7.5 m a.s.l.
Soil taxonomy: *Vertic Torrifluvents*
Topography: gently undulating
Land form: flood plain/basin
Slope: gently slope
Land use & vegetation: cultivated with alfalfa
Parent material: alluvial deposits
Drainage: moderately well drained
Ground water table: 120 cm
Remarks: surface cracks 8 cm width & 30 cm depth

Depth (cm)	Description
0–20	Very dark grayish brown (10YR3/2) moist, dark grayish brown (10YR4/3) dry, sandy clay; moderate medium sub angular blocky; sticky and plastic; firm, hard common fine discontinuous random interstitial pores, few fresh and humified residual of organic matter, many fine to medium roots, slightly effervescence with HCl, clear smooth boundary.
20–45	Dark brown (10YR4/3) moist, Brown (10YR5/3) dry, clayey; Moderate medium angular to sub angular blocky; very sticky, very plastic, firm, hard; common fine to medium discontinuous random pores, thin shinny discontinuous organic clay films lining the voids; common fine to medium roots, slightly effervescence with HCl, clear smooth boundary.
45–80	Dark brown (10YR 3/3) moist, dark grayish brown (10YR 4/3) dry: clayey: Moderate to strong medium angular to sub angular blocky; very sticky, very Plastic; very firm, very hard; common fine to medium continuous and discontinuous random pores; few fine roots slight effervescence with HCl, clear smooth boundary.

Profile No.: 135
Date of description: 10/2006
Location:
Longitude: 30° 50′ 24″ E
Latitude: 29° 09′ 19″ N
Elevation: 60.0 m a.sl.
Soil taxonomy: *Petrogypsic Gypsiorthids*
Topography: gently undulating
Land form: sand sheets
Slope: gently slope
Land use & vegetation: barren
Parent material: lacustraine deposits
Drainage: poor drained soils
Ground water table: not hit

Depth (cm)	Description
0–35	Brownish Yellow (10 YR 6/6; dry); sandy; moderate medium sub-angular blocky str.; non-sticky, non plastic; strong efferv. With Hcl.; hard, common fine and medium roots; diffuse smooth boundary.
38–70	*Dark yellowish brown (10 YR 4/5, dry); sandy; moderate medium sub-angular blocky str.; non sticky; non plastic; many broken white lime stone.*

Profile No.: 136
Date of description: 10/2006
Location:
Longitude: 30° 44′ 56″ E

Latitude: 29° 28′ 42″ N
Elevation: 2.0 m a.sl.
Soil taxonomy: *Typic Haplosalids*
Topography: gently undulating
Land form: laustraine plain
Slope: gently slope
Land use & vegetation: barren
Parent material: lacustraine deposits
Drainage: poorly drained soils
Ground water table: 60 cm

Depth (cm)	Description
0–30	Yellowish brown (10YR 5/8, wet); sandy clay loam; moderate medium sub-angular blocky Str.; slightly sticky, slightly plastic; moderate efferv. With Hcl; diffuse snooth boundary.
30–60	Yellowish brown (10YR 5/8, wet); sandy clay loam; moderate medium sub-angular blocky stru.; slightly sticky, slightly plastic; moderate efferv. With Hcl.

Annex II: Morphological Description of the Studied Soil Profiles

AL-Qalubiyea Governorate

Profile No.: 137
Date of description: 9/2006
Location:
Longitude: 30° 17′ E
Latitude: 30° 10′ N
Elevation: 6.0 m a.sl.
Soil taxonomy: *Vertic Torrivluvents*
Topography: flat to almost flat
Land form: fluvio-lacustrine plain/terraces
Slope: nearly level
Land use & vegetation: fallow fields
Parent material: alluvial-lecustrine deposits
Drainage: moderately well drained
Ground water table: 100 cm

Depth (cm)	Description
0–25	Very dark brown (10YR2/2) moist, dark brown (10YR3/3) dry, clayey; string fine and medium sub-angular blocky; very sticky, very plastic, firm, hard; many fine to medium continuous and discontinuous interstitial pores; many fine and medium roots; slightly effervescence with HCl; clear gradual boundary.
25–50	Very dark grayish brown (10YR3/2) moist; dark grayish Brown (10YR4/2) dry; clayey; moderate to strong medium angular to sub angular blocky; slightly sticky slightly plastic, friable, hard; common fine and medium discontinuous random pores; common fine and medium roots, slightly effervescence with HCl, clear smooth boundary.

(continued)

A.-A. Gad, *Digital Land Resources to Support Sustainable Development in Egypt*,
Deltas of the World,
https://doi.org/10.1007/978-3-031-37167-7

Depth (cm)	Description
50–100	Dark brown (10YR 3/3) moist, yellowish brown (10YR 5/4) dry, clay loam weak medium angular to sub angular blocky, sticky, Plastic friable, hard; common fine and medium discontinuous random pores, few very soft lime concretion; few fine roots, slight effervescence with Hcl.

Profile No.: 138
Date of description: 9/2006
Location:
Longitude: 31° 15′ E
Latitude: 30° 11′ N
Elevation: 11.0 m a.s.l.
Soil taxonomy: *Typic Torrifluvents*
Topography: flat to almost flat
Land form: flood plain/river terraces
Slope: nearly level
Land use & vegetation: cultivated with winter vegetables
Parent material: alluvial deposits
Drainage: moderately well drained soils
Ground water table: 100 cm

Depth (cm)	Description
0–25	Dark grayish brown (10YR4/2) moist, grayish Brown (10YR5/2) dry, clay loam; strong small to medium granular; moderately sticky, moderately plastic; moderately firm, moderately hard; common fine to medium continuous and discontinuous random interstitial pores; many fine to medium roots, slight effervescence with HCl,
25–45	Dark grayish brown (10YR4/2) moist, grayish brown (10YR5/2) dry, clay loam; moderately small to medium sub angular blocky; sticky, plastic, firm, hard; common fine to medium discontinuous random interstitial and few channel pores; common fine to medium roots,. slight effervescence with HCl, clear smooth boundary
45–100	Dark grayish brown (10YR 4/2) moist, grayish brown (10YR 5/2) dry: clayey: Moderately medium sub angular blocky, sticky,Plastic; moderately firm, hard; common fine to medium discontinuous random pores; few fine roots, slight effervescence with HCl.

Profile No.: 139
Date of description: 9/2006
Location:
Longitude: 31° 18′ E
Latitude: 30° 10′ N
Elevation: 9.0 m a.s.l.
Soil taxonomy: *Vertic Torrifluvents*

Topography: almost flat
Land form: flood plain/river terraces
Slope: nearly level
Land use & vegetation: cultivated with rice
Parent material: alluvial deposits
Drainage: well drained
Ground water table: 100 cm

Depth (cm)	Description
0–20	Dark yellowish brown (10YR3/3) moist, yellowish Brown (10YR5/2) dry, clayey; moderate to strong fine and medium blocky; very sticky and very plastic; very firm, very hard common fine to medium discontinuous random interstitial and few channel pores, few fresh and humified residual of organic matter, many fine to medium roots, slightly effervescence with Hal, clear smooth boundary.
20–55	Dark brown (10YR3/3) moist, Brown (10YR5/2) dry, clayey; Moderate small to medium sub angular blocky; very sticky and very plastic, very firm, very hard; common fine continuous and discontinuous random interstitial pores, thin shinny discontinuous organic clay films lining the voids; common fine to medium roots, slightly effervescence with HCl, gradual smooth boundary
55–100	Dark brown (10YR 3/3) moist, brown (10YR 5/3) dry: clayey: Moderate to strong medium sub angular blocky; sticky and Plastic; firm, hard; common fine to medium discontinuous random interstitial pores; few fine slight effervescence with HCl.

Profile No.: 140
Date of description: 9/2006
Location:
Longitude: 31° 12′ E
Latitude: 30° 12′ N
Elevation: 12 m a.s.l.
Soil taxonomy: *Vertic Torrivluvents*
Topography: flat to almost flat
Land form: flood plain/river terraces
Slope: gently slope
Land use & vegetation: cultivated with Cotton
Parent material: Alluvial deposits
Drainage: well drained soil
Ground water table: 120 cm

Depth (cm)	Description
0–25	Dark grayish brown (10YR4/2) moist, grayish brown (10YR5/5) dry; clay loam; moderate small to medium granular; sticky, plastic; firm, hard; common fine to medium discontinuous random interstitial pores; many fine to medium roots; slight effervescence with HCl; clear smooth boundary.
25–55	Dark grayish brown (10YR4/2) moist, light brownish gray (10YR6/2) dry; clayey; strong small to medium sub angular blocky; sticky, plastic, firm, hard; common fine to medium discontinuous random interstitial pores; common fine to medium roots; slight effervescence with HCl, gradual smooth boundary
55–120	Very dark grayish brown (10YR 3/2) moist, dark grayish brown (10YR 4/2) dry, clayey; moderate small to medium sub angular blocky; very sticky, very Plastic, extremely firm, extremely hard; fine to medium continuous and discontinuous random interstitial pores; thin shinny discontinuous clay films lining the voids few fine roots slight effervescence with HCl.

Profile No.: 141
Date of description: 9/2006
Location:
Longitude: 30° 11′ E
Latitude: 30° 14′ N
Elevation: 12.0 m a.s.l.
Soil taxonomy: *Typic Torrifuvents*
Topography: almost flat to gently undulating
Land form: flood plain/Basin
Slope: gently slope
Land use & vegetation: fallow after cotton
Parent material: alluvial deposits
Drainage: well drained soils
Ground water table: 120 cm

Depth (cm)	Description
0–30	Very dark grayish brown (10YR3/2) moist, grayish Brown (10YR4/3) dry, silty clay; moderate fine to medium sub angular; sticky, plastic; firm, hard; common fine discontinuous random interstitial pores; common fine roots, slight effervescence with HCl, gradual smooth boundary.
30–60	Very dark grayish brown (10YR3/2) moist, dark grayish brown (10YR4/3) dry, silty clay; moderate fine to medium sub angular blocky; sticky, plastic, firm, hard; common fine continuous and discontinuous random interstitial pores; common fine to medium roots, Slight effervescence with HCl; clear smooth boundary
60–120	Dark grayish brown (10YR 4/2) moist, grayish brown (10YR 5/2) dry; silty clay; Moderate to strong medium sub angular blocky, sticky, Plastic; firm, hard; common fine continuous and discontinuous random pores; few fine roots.

Profile No.: 142
Date of description: 9/2006
Location:
Longitude: 31° 09′ E
Latitude: 30° 11′ N
Elevation: 7.0 m a.s.l.
Soil taxonomy: *Typic Torrivluvents*
Topography: almost flat
Land form: flood plain/river terraces
Slope: gently slope
Land use & vegetation: fallow after Maize
Parent material: alluvial deposits
Drainage: moderately well drained
Ground water table: 110 cm

Depth (cm)	Description
0–25	Dark grayish brown (10YR4/2) moist, grayish brown (10YR5/2) dry; clay loam strong medium angular to sub angular blocky; moderate sticky, moderate plastic, firm, hard; common fine to medium discontinuous random exped medium interstitial pores; few humified residual of organic matter; common fine to medium roots; slight effervescence with HCl; gradual smooth boundary.
25–50	Dark grayish brown (10YR4/2) moist, grayish brown (10YR5/2) dry; clay loam; moderate medium sub angular blocky; sticky, plastic, firm, hard; common fine to medium discontinuous random interstitial pores; common fine to medium roots; slight effervescence with HCl, clear smooth boundary
50–110	Dark grayish brown (10YR 4/2) moist, grayish brown (10YR 5/2) dry; silty clay; moderate to strong medium sub angular blocky; sticky, Plastic, very firm, very hard; common fine and medium discontinuous random interstitial pores; slight effervescence with HCl.

Profile No.: 143
Date of description: 9/2006
Location:
Longitude: 31° 16′ E
Latitude: 31° 15′ N
Elevation: 7.0 m a.s.l.
Soil taxonomy: *Typic Torrivluvents*
Topography: almost flat
Land form: flood plain/river terraces
Slope: nearly level
Land use & vegetation: cultivated with clover
Parent material: alluvial deposits
Drainage: moderately well drained soils
Ground water table: 70 cm

Depth (cm)	Description
0–20	Dark brown (10YR2/3) moist, grayish brown (10YR5/2) dry; clay loam; moderately fine to medium angular to sub angular blocky; very sticky, very plastic, very firm, extremely hard; common fine discontinuous random exped and interstitial pores, few humified patches of organic matter; many fine to coarse roots; slight effervescence with HCl, clear smooth boundary.
20–45	Dark grayish brown (10YR 4/2) moist, grayish brown (10YR 5/4) dry; clayey; strong fine to medium sub angular blocky; sticky, plastic; firm, hard; common fine to medium continuous and discontinuous random interstitial pores; few humified patches of organic matter; many fine to coarse roots; gradual smooth boundary.
45–80	Dark grayish brown (10YR 4/2) moist, grayish brown (10YR 5/4) dry; clayey; moderate fine to medium sub angular blocky; very sticky very plastic: very firm, very hard; common fine and medium discontinuous random exped and interstitial pores; slight smooth boundary.

Profile No.: 144
Date of description: 9/2006
Location:
Longitude: 31° 19′ E
Latitude: 30° 12′ N
Elevation: 31.0 m a.sl.
Soil taxonomy: *Typic Quartizipsamments*
Topography: gently undulating
Land form: sand sheet
Slope: gently slope
Land use & vegetation: barren
Parent material: aeolian deposits
Drainage: well drained soils
Ground water table: not hit

Depth (cm)	Description
0–30	Light yellowish brown (10YR 6/4) moist, very pale brown (10 Y 7/3) dry, sandy, single grains, moderate effervescence, common medium to fine discontinuous random pores, many fine roots moderate effervescence with HCl, clear smooth boundary
30–60	Yellowish brown (10 YR 5/6) moist, light yellowish brown (10 YR 6/4) dry sand, single grains, hard, few gravels, few soft lime concretions, moderate effervescence, common fine and medium roots; slight effervescence with HCl, clear smooth boundary.
60–150	Dark yellowish brown (10 YR 4/4) moist, Yellowish brown (10 YR 5/4)dry, sandy, single grain, few lime concretion, common coarse and medium discontinuous random pores; moderate effervescence with HCl.

Profile No.: 145
Date of description: 9/2006
Location
Longitude: 31° 28′ E
Latitude: 30° 14′ N
Elevation: 33.0 m a.s.l.
Soil taxonomy: *Typic Quartizipsamments*
Topography: almost flat
Land form: flood plain
Slope: nearly level
Land use & vegetation: barren
Parent material: aeolian deposits
Drainage: moderately well drained
Ground water table: not hit

Depth (cm)	Description
0–20	Light yellowish brown (10YR 6/4) moist, very pale brown (10 YR 7/3) dry, sandy, single grains, few gravels, strong effervescence, common fine and medium roots; slight effervescence with HCl, diffuse smooth boundary.
20–45	Yellowish brown (10 YR 5/6) moist, pale brown (10 YR 6/3) dry, sandy loam, single grains, common gravels, loose, common fine to medium discontinuous random pores, many fine roots moderate effervescence with HCl, diffuse smooth boundary.
45–110	Dark yellowish brown (10 YR 4/4) moist, Yellowish brown (10 YR 5/4)dry, slightly hard medium coarse loamy sand, common coarse and medium discontinuous random pores; moderate effervescence with HCl.

Profile No.: 146
Date of description: 9/2006
Location:
Longitude: 31° 30′ E
Latitude: 30° 13′ N
Elevation: 45.0 m a.s.l.
Soil taxonomy: *Typic Torriorthents*
Topography: undulating
Land form: aeolian plain
Slope: gently slope
Land use & vegetation: barren
Parent material: aeolian deposits
Drainage: poor drained soils
Ground water table: not hit

Depth (cm)	Description
0–30	Yellowish brown (10 YR 5/6) moist, pale brown (10 YR 6/3) dry, sandy, single grains, moderate fine gravel and grits common medium to coarse discontinuous random pores, common lime concretions, common fine to medium roots strong effervescence with HCl, diffuse smooth boundary.
30–60	Yellowish brown (10 YR 5/6) moist, light yellowish brown (10 YR 6/4) dry, sandy, single grains, hard, common coarse gravels, common lime concretions, moderate effervescence, common fine and medium roots; strong effervescence with HCl, clear smooth boundary.
60–100	Dark yellowish brown (10 YR 4/4) moist, Yellowish brown (10 YR 5/4)dry, sandy, few lime and gypsum concretion, common coarse and medium discontinuous random pores; moderate effervescence with HCl.

Profile No.: 147
Date of description: 9/2006
Location:
Longitude: 31° 30′ E
Latitude: 30° 15′ N
Elevation: 31.0 m a.sl.
Soil taxonomy: *Typic Qartizipsamments*
Topography: gently undulating
Land form: old deltaic plain
Slope: gently slope
Land use & vegetation: barren
Parent material: old deltaic deposits
Drainage: well drained soils
Ground water table: not hit

Depth (cm)	Description
0–30	Yellowish brown (10 YR 5/6) moist, light yellowish brown (10 YR 6/4) dry, sandy, massive, common medium and fine gravel, common medium to coarse discontinuous random pores, few lime concretion, common fine roots, strong effervescence with HCl, diffuse smooth boundary
30–65	Yellowish brown (10 YR 5/6) moist, light yellowish brown (10 YR 6/4) dry sand, massive, hard, many coarse gravels, common lime concretions, strong effervescence, common fine and medium roots; slight effervescence with HCl, clear smooth boundary.
65–120	Dark yellowish brown (10 YR 4/4) moist, Yellowish brown (10 YR 5/4) dry, sandy, massive, few lime concretion, few gravels common coarse and medium discontinuous random pores; moderate effervescence with HCl.

Profile No.: 148
Date of description: 9/2006
Location:
Longitude: 31° 56′ E
Latitude: 30° 42′ N
Elevation: 31.0 m a.sl.
Soil taxonomy: *Typic Quartizipsamments*
Topography: gently undulating
Land form: sand sheet
Slope: gently slope
Land use & vegetation: barren
Parent material: aeolian deposits
Drainage: well drained soils
Ground water table: not hit

Depth (cm)	Description
0–30	Light yellowish brown (10YR 6/4) moist, very pale brown (10 Y 7/3) dry, sandy, single grains, slight effervescence with HCl, common medium to fine discontinuous random pores, many fine roots moderate effervescence with HCl, clear smooth boundary
30–90	Light yellowish brown (10YR 6/4) moist, light yellowish brown (10 YR 6/4) dry sand, single grains, few gravels, slight effervescence with HCl; slight effervescence with HCl, clear smooth boundary.

Profile No.: 149
Date of description: 9/2006
Location
Longitude: 31° 12′ E
Latitude: 30° 15′ N
Elevation: 5.0 m a.s.l.
Soil taxonomy: *Typic Torrifluvents*
Topography: gently undulating
Land form: flood plain/basins
Slope: gently slope
Land use & vegetation: fallow after maize
Parent material: alluvial deposits
Drainage: moderately well drained
Ground water table: 110 cm

Depth (cm)	Description
0–25	Dark brown (10YR3/3) moist, brown (10YR4/3). dry; clay; moderate medium granular, sticky, plastic, fine, hard; common fine and; medium discontinuous random interstitial pores; many fine and, medium roots; slight effervescence with HCl; clear smooth boundary.
25–45	Very dark grayish brown (10YR3/2) moist; dark grayish brown (10YR4/2) dry; silty clay; blocky; sticky, firm, hard; common fine and medium continuous and discontinuous random interstitial pores, few fine to medium roots, slightly effervescence with HCl, gradual smooth boundary.
45–110	Dark gray (10YR 4/1) moist, gray (10YR 5/1) dry, clay; very sticky, very Plastic, firm, hard; common fine to medium continuous and discontinuous vughs, few very fine lime concretion, moderate effervescence with HCl.

Profile No.: 150
Date of description: 9/2006
Location
Longitude: 31° 13′ E
Latitude: 30° 16′ N
Elevation: 9.0 m a.s.l.
Soil taxonomy: *Vertic Torrifluvents*
Topography: gently undulating
Land form: flood plain/basin
Slope: gently slope
Land use & vegetation: fallow after maize
Parent material: alluvial deposits
Drainage: moderately well drained
Ground water table: not hit
Remarks: surface cracks 5 cm width & 30 cm depth

Depth (cm)	Description
0–20	Dark grayish brown (10YR4/2) moist, grayish brown (10YR5/3) dry, clayey; moderate to strong medium granular, very sticky, very plastic, very firm,; very hard; common fine discontinuous random interstitial pores; few fresh and humified residuals of organic matter, many fine to medium roots; slight effervescence with HCl; clear smooth boundary.
20–40	Very dark brown (10YR2/2) moist; Brown (10YR3/3) dry; clay loam; moderate to strong medium angular to sub angular blocky; very sticky, very plastic, very firm, extremely hard; common fine and medium continuous and discontinuous random interstitial pores, few fine roots, slightly effervescence with HCl, gradual smooth boundary.
40–100	Very dark grayish brown (10YR3/2) moist; dark grayish brown (10YR4/2) dry; clay loam; blocky; sticky, firm, hard; common fine and medium continuous and

(continued)

Depth (cm)	Description
	discontinuous random interstitial pores, few fine to medium roots, slightly effervescence with HCl, gradual smooth boundary.

Profile No.: 151
Date of description: 9/2006
Location
Longitude: 31° 08′ E
Latitude: 30° 21′ N
Elevation: 2.5 m a.s.l.
Soil taxonomy: *Typic Torrivluvents*
Topography: gently undulating
Land form: flood plain/basin
Slope: gently slope
Land use & vegetation: fallow after rice
Parent material: alluvial deposits
Drainage: well drained
Ground water table: 120 cm

Depth (cm)	Description
0–30	Very dark grayish brown (10YR3/2) moist, dark grayish brown (10YR4/3) dry, sandy clay; moderate medium subangular blocky; sticky and plastic; firm, hard common fine discontinuous random interstitial pores, few fresh and humified residual of organic matter, many fine to medium roots, slightly effervescence with HCl, clear smooth boundary.
30–60	Dark brown (10YR4/3) moist, Brown (10YR5/3) dry, clayey; Moderate medium angular to subangular blocky; very sticky, very plastic, firm, hard; common fine to medium discontinuous random pores, thin shinny discontinuous organic clay films lining the voids; common fine to medium roots, slightly effervescence with HCl, clear smooth boundary.
60–120	Dark brown (10YR 3/3) moist, dark grayish brown (10YR 4/3) dry: clayey: Moderate to strong medium angular to subangular blocky; very sticky, very Plastic; very firm, very hard; common fine to medium continuous and discontinuous random pores; few fine roots slight effervescence with HCl, clear smooth boundary.

Profile No.: 152
Date of description: 9/2006
Location:
Longitude: 31° 17′ E
Latitude: 30° 22′ N
Elevation: 14.5 m a.s.l.
Soil taxonomy: *Typic Quartizipsamments*

Topography: almost flat
Land form: sandy plain
Slope: nearly level
Land use & vegetation: cultivated with sugar beet
Parent material: aeolian deposits
Drainage: moderately well drained
Ground water table: not hit

Depth (cm)	Description
0–40	Yellowish brown (10 YR 5/6) moist, light yellowish brown (10 YR 6/4) dry, sandy, structure less, moderate medium and fine gravel, common medium to coarse discontinuous random pores, few lime concretion, common fine roots moderate effervescence with HCl, diffuse smooth boundary.
40–130	Yellowish brown (10 YR 5/6), dry light yellowish brown (10 YR 6/4) dry, sandy, single grains, hard, cemented abundant pores, common lime concretions, strong effervescence with HCl. common coarse and medium discontinuous random pores; slight effervescence with HCl.

Profile No.: 153
Date of description: 9/2006
Location:
Longitude: 31° 20′ E
Latitude: 30° 20′ N
Elevation: 6.25 m a.s.l.
Soil taxonomy: *Vertic Torrivluvents*
Topography: almost flat to gently undulating
Land form: flood plain/river terraces
Slope: nearly level to gently slope
Land use & vegetation: fallow after rice
Parent material: alluvial deposits
Drainage: moderately well drained
Ground water table: 90 cm

Depth (cm)	Description
0–35	Very dark grayish brown (10YR3/2) moist, dark grayish brown (10YR4/3) dry; sandy clay loam; moderate to strong medium sub angular blocky; very sticky, very plastic, extremely firm, extremely hard; common fine discontinuous random exped and medium interstitial pores; many fine to medium roots; moderately effervescence with HCl; clear smooth boundary.
35–70	Dark brown (10YR2/3) moist, grayish brown (10YR5/2) dry; clay loam; strong medium sub angular blocky; very sticky, very plastic, extremely firm, extremely hard; common fine to medium discontinuous random interstitial pores; thin shinny discontinuous clay films lining the voids; few fine to medium roots; slight effervescence with HCl, clear smooth boundary .

(continued)

Depth (cm)	Description
70–100	Dark brown (10YR 2/3) moist, grayish brown (10YR 5/2) dry; clayey; strong medium sub angular blocky; very sticky, very Plastic, extremely firm, extremely hard; common fine and medium discontinuous random interstitial pores and exped; slight effervescence with HCl.

Profile No.: 154
Date of description: 9/2006
Location:
Longitude: 31° 17′ E
Latitude: 30° 18′ N
Elevation: 12.0 m a.s.l.
Soil taxonomy: *Vertic Torrivluvents*
Topography: almost flat to gently undulating
Land form: flood plain/river terraces
Slope: gently slope
Land use & vegetation: cultivated with cotton
Parent material: alluvial deposits
Drainage: well drained soils
Ground water table: 100 cm

Depth (cm)	Description
0–25	Dark brown (10YR3/3) moist, grayish Brown (10YR5/2) dry, sandy clay, moderate medium granular, sticky and plastic, firm, hard common fine discontinuous random interstitial and few channel pores, few fresh and humified residual of organic matter, many fine to medium roots, slight effervescence with HCl, clear smooth boundary
25–60	Dark gray (10YR 4/1) moist, gray(10YR 5/1) dry: clayey: Moderate to strong medium to coarse sub angular blocky; sticky and Plastic: very firm, very hard; common fine to medium discontinuous random interstitial pores; slight effervescence with HCL, clear smooth boundary
60–100	Dark gray (10YR 4/1) moist, gray(10YR 5/1) dry: clayey Moderate to strong medium to coarse sub angular blocky; sticky and Plastic: very firm, very hard; common fine to medium discontinuous random interstitial pores; slight effervescence with HCL, clear smooth boundary

Profile No.: 155
Date of description: 9/2006
Location:
Longitude: 30° 09′ E
Latitude: 30° 20′ N
Elevation: 15.0 m a.s.l.
Soil taxonomy: *Vertic Torrivluvents*

Topography: almost flat
Land form: flood plain/river levees
Slope: gently slope
Land use & vegetation: cultivated with orchards (orange-mango).
Parent material: alluvial deposits
Drainage: well drained
Ground water table: 110 cm

Depth (cm)	Description
0–30	Dark gray (10YR4/1) moist, gray (10YR5/1) dry; clayey; moderate medium sub angular blocky; sticky, plastic; moderately firm, moderately hard; common fine to medium discontinuous random exped interstitial pores; few fresh and humified residual of organic matter; many fine to coarse roots; slight effervescence with Hcl; clear smooth boundary.
30–65	Very dark grayish brown (10YR3/2) moist, dark grayish brown (10YR4/3) dry; clayey; moderate fine to medium sub angular blocky; sticky, plastic; moderately firm, moderately hard; common fine to medium discontinuous random exped interstitial pores; common fine to coarse roots; slight effervescence with Hcl, clear smooth boundary.
65–110	Very dark grayish brown (10YR 3/2) moist, dark grayish brown (10YR 4/3) dry, clayey; moderate fine to medium sub angular blocky; sticky, plastic; firm, hard; common fine and medium discontinuous random interstitial pores; slight effervescence with Hcl.

Profile No.: 156
Date of description: 9/2006
Location:
Longitude: 31° 18′ E
Latitude: 30° 21′ N
Elevation: 4.0 m a.sl.
Soil taxonomy: *Typic Quartizipsamments*
Topography: gently undulating
Land form: sand sheet
Slope: gently slope
Land use & vegetation: cultivated with sugar beet
Parent material: aeolian deposits
Drainage: well drained soils
Ground water table: not hit

Depth (cm)	Description
0–30	Light yellowish brown (10YR 6/4) moist, very pale brown (10 Y 7/3) dry, loamy sand, single grains, moderate effervescence, common medium to fine discontinuous random pores, many fine roots slight effervescence with HCl, clear smooth boundary
30–50	Yellowish brown (10 YR 5/6) moist, light yellowish brown (10 YR 6/4) dry, sandy, single grains, moderate medium and fine gravel, common medium to coarse discontinuous random pores, few lime concretion, common fine roots moderate effervescence with HCl, diffuse smooth boundary..
50–130	Dark yellowish brown (10 YR 4/4) moist, Yellowish brown (10 YR 5/4)dry, sandy, single grain, few lime concretion, common coarse and medium discontinuous random pores; moderate effervescence with HCl.

Profile No.: 157
Date of description: 10/2006
Location:
Longitude: 31° 19′ E
Latitude: 30° 16′ N
Elevation: 14.0 m a.s.l.
Soil taxonomy: *Vertic Torrifluvents*
Topography: almost flat to gently undulating
Land form: flood plain/river terraces
Slope: nearly level to gently slope
Land use & vegetation: fallow after rice
Parent material: alluvial deposits
Drainage: moderately well drained
Ground water table: +110 cm

Depth (cm)	Description
0–25	Very dark grayish brown (10YR3/2) moist, dark grayish brown (10YR4/3) dry; sandy clay loam; moderate to strong medium sub angular blocky; very sticky, very plastic, extremely firm, extremely hard; common fine discontinuous random exped and medium interstitial pores; many fine to medium roots; moderately effervescence with HCl; clear smooth boundary.
25–70	Dark brown (10YR2/3) moist, grayish brown (10YR5/2) dry; clay loam; strong medium sub angular blocky; very sticky, very plastic, extremely firm, extremely hard; common fine to medium discontinuous random interstitial pores; thin shinny discontinuous clay films lining the voids; few fine to medium roots; slight effervescence with HCl, clear smooth boundary .
70–110	Dark brown (10YR 2/3) moist, grayish brown (10YR 5/2) dry; clayey; strong medium sub angular blocky; very sticky, very Plastic, extremely firm, extremely hard; common fine and medium discontinuous random interstitial pores and exped; slight effervescence with HCl.

Profile No.: 158
Date of description: 10/2006
Location
Longitude: 31° 13′ E
Latitude: 30° 18′ N
Elevation: 8.0 m a.s.l.
Soil taxonomy: *Vertic Torrifluvents*
Topography: almost flat
Land form: flood plain/terraces
Slope: gently slope
Land use & vegetation: fallow after maize
Parent material: alluvial deposits
Drainage: moderately well drained
Ground water table: not hit
Remarks: surface cracks 6 cm width & 30 cm depth

Depth (cm)	Description
0–15	Dark grayish brown (10YR4/2) moist, grayish brown (10YR5/3) dry, clayey; moderate to strong medium granular, very sticky, very plastic, very firm,; very hard; common fine discontinuous random interstitial pores; few fresh and humified residuals of organic matter, many fine to medium roots; slight effervescence with HCl; clear smooth boundary.
15–45	Very dark brown (10YR2/2) moist; Brown (10YR3/3) dry; clay loam; moderate to strong medium angular to sub angular blocky; very sticky, very plastic, very firm, extremely hard; common fine and medium continuous and discontinuous random interstitial pores, few fine roots, slightly effervescence with HCl, gradual smooth boundary.
45–90	Very dark grayish brown (10YR3/2) moist; dark grayish brown (10YR4/2) dry; clay loam; blocky; sticky, firm, hard; common fine and medium continuous and discontinuous random interstitial pores, few fine to medium roots, slightly effervescence with HCl, gradual smooth boundary.

Profile No.: 159
Date of description: 10/2006
Location
Longitude: 31° 14′ E
Latitude: 31° 21′ N
Elevation: 15.0 m a.s.l.
Soil taxonomy: *Vertic Torrifluvents*
Topography: gently undulating
Land form: flood plain/basin
Slope: gently slope
Land use & vegetation: cultivated with alfalfa

Parent material: alluvial deposits
Drainage: moderately well drained
Ground water table: +90 cm
Remarks: surface cracks 5 cm width & 30 cm depth

Depth (cm)	Description
0–15	Very dark grayish brown (10YR3/2) moist, dark grayish brown (10YR4/3) dry, sandy clay; moderate medium sub angular blocky; sticky and plastic; firm, hard common fine discontinuous random interstitial pores, few fresh and humified residual of organic matter, many fine to medium roots, slightly effervescence with HCl, clear smooth boundary.
15–40	Dark brown (10YR4/3) moist, Brown (10YR5/3) dry, clayey; Moderate medium angular to sub angular blocky; very sticky, very plastic, firm, hard; common fine to medium discontinuous random pores, thin shinny discontinuous organic clay films lining the voids; common fine to medium roots, slightly effervescence with HCl, clear smooth boundary.
40–90	Dark brown (10YR 3/3) moist, dark grayish brown (10YR 4/3) dry: clayey: Moderate to strong medium angular to sub angular blocky; very sticky, very Plastic; very firm, very hard; common fine to medium continuous and discontinuous random pores; few fine roots slight effervescence with HCl, clear smooth boundary.

Profile No.: 160
Date of description: 10/2006
Location
Longitude: 31° 08′ E
Latitude: 30° 21′ N
Elevation: 3.0 m a.s.l.
Soil taxonomy: *Vertic Torrifluvents*
Topography: Almost flat
Land form: alluvial plain/terrace
Slope: nearly level
Land use & vegetation: cultivated with alfalfa
Parent material: lacustrine/alluvial deposits
Drainage: moderately well drained
Ground water table: 60 cm
Remarks: surface cracks 5 cm width & 30 cm depth

Depth (cm)	Description
0–20	Very dark grayish brown (10YR4/2) moist, brown (10YR5/3) dry, clayey; moderate medium angular blocky; sticky, plastic; firm, hard common fine discontinuous random pores, few fresh and humified residual of organic matter, many fine and medium roots, slightly effervescence with HCl, clear smooth boundary.

(continued)

Depth (cm)	Description
20–50	Very dark grayish brown (10YR3/2) moist, dark grayish brown (10YR4/2) dry, clayey; Moderate to strong fine to medium sub-angular, sticky, plastic, firm, hard; common fine and medium continuous and discontinuous channels, few fine soft segregation of CaCO3, common fine to medium roots,. Slightly effervescence with HCl, clear smooth boundary.
50–100	Very dark grayish brown (10YR 3/2) moist, dark grayish brown (10YR 4/2) dry: clay loam; moderate fine to medium angular blocky; very sticky, very plastic; firm; hard; common fine random pores; common fine soft lime segregation, moderate effervescence with HCl.

Profile No.: 161
Date of description: 9/2006
Location:
Longitude: 31° 18′ E
Latitude: 30° 26′ N
Elevation: 10.0 m a.sl.
Soil taxonomy: *Typic Torrifluvents*
Topography: gently undulating
Land form: flood plain
Slope: gently slope
Land use & vegetation: orchards
Parent matrial: alluvial deposits
Drainage: well drained soils
Ground water table: not hit

Depth (cm)	Description
0–20	Light yellowish brown (10YR 6/4) moist, very pale brown (10 Y 7/3) dry, sandy clay, moderate medium sub angular blocky, moderate effervescence with HCl, common medium to fine discontinuous random pores, many fine roots moderate effervescence with HCl, clear smooth boundary
20–50	Dark yellowish brown (10 YR 4/4) moist, Yellowish brown (10 YR 5/4) dry, clay loam, moderate medium sub angular blocky, few lime concretion, common coarse and medium discontinuous random pores; moderate effervescence with HCl., clear smooth boundary.
50–120	Yellowish brown (10 YR 5/6) moist, light yellowish brown (10 YR 6/4) dry clay loam, single grains, hard, many gravels, few soft lime concretions, slight effervescence, common fine and medium roots; slight effervescence with HCl.

Profile No.: 162
Date of description: 10/2006
Location:

Longitude: 31° 11′ E
Latitude: 30° 27′ N
Elevation: 10.0 m a.sl.
Soil taxonomy: *Vertic Torrifluvents*
Topography: flat to almost flat
Land form: fluvio-lacustrine plain/terraces
Slope: nearly level
Land use & vegetation: fallow after rice
Parent material: alluvial-lecustrine deposits
Drainage: moderately well drained
Ground water table: 110 cm
Remarks: surface cracks 8 cm width & 30 cm depth

Depth (cm)	Description
0–20	Very dark brown (10YR2/2) moist, dark brown (10YR3/3) dry, clayey; string fine and medium sub-angular blocky; very sticky, very plastic, firm, hard; many fine to medium continuous and discontinuous interstitial pores; many fine and medium roots; slightly effervescence with HCl; clear gradual boundary.
20–60	Very dark grayish brown (10YR3/2) moist; dark grayish Brown (10YR4/2) dry; clayey; moderate to strong medium sub angular blocky; slightly sticky slightly plastic, friable, hard; common fine and medium discontinuous random pores; common fine and medium roots, slightly effervescence with HCl, clear smooth boundary.
60–120	Dark brown (10YR3/3) moist; grayish Brown (10YR5/2) dry; clay loam weak fine and medium sub angular blocky; sticky, plastic, firm, hard; few fine and medium discontinuous pores; slight effervescence with HCl,

Profile No.: 163
Date of description: 10/2006
Location:
Longitude: 31° 11′ E
Latitude: 30° 29′ N
Elevation: 6.5 m a.s.l.
Soil taxonomy: *Typic Torrifluvents*
Topography: almost flat
Land form: flood plain/river terraces
Slope: nearly level
Land use & vegetation: cultivated with clover
Parent material: alluvial deposits
Drainage: moderately well drained soils
Ground water table: 90 cm

Depth (cm)	Description
0–20	Dark brown (10YR2/3) moist, grayish brown (10YR5/2) dry; clay loam; moderately fine to medium sub angular blocky; very sticky, very plastic, very firm, hard; common fine discontinuous random exped and interstitial pores, few humified patches of organic matter; many fine to coarse roots; slight effervescence with HCl, clear smooth boundary.
20–45	Dark grayish brown (10YR 4/2) moist, grayish brown (10YR 5/4) dry; clayey; strong fine to medium sub angular blocky; sticky, plastic; firm, hard; common fine to medium continuous and discontinuous random interstitial pores; few humified patches of organic matter; many fine to coarse roots; gradual smooth boundary.
45–90	Dark grayish brown (10YR 4/2) moist, grayish brown (10YR 5/4) dry; clayey; moderate fine to medium sub angular blocky; very sticky very plastic: very firm, very hard; common fine and medium discontinuous random exped and interstitial pores.

Profile No.: 164
Date of description: 10/2006
Location:
Longitude: 31° 18′ E
Latitude: 30° 32′ N
Elevation: 8.0 m a.s.l.
Soil taxonomy: *Vertic Torrifluvents*
Topography: almost flat
Land form: flood plain/river terraces
Slope: gently slope
Land use & vegetation: fallow after Maize
Parent material: alluvial deposits
Drainage: moderately well drained
Ground water table: 90 cm

Depth (cm)	Description
0–20	Dark grayish brown (10YR4/2) moist, grayish brown (10YR5/2) dry; clay loam; moderate medium sub angular blocky; sticky, plastic, firm, hard; common fine to medium discontinuous random interstitial pores; common fine to medium roots; slight effervescence with HCl, clear smooth boundary
20–50	Dark grayish brown (10YR4/2) moist, grayish brown (10YR5/2) dry; clay loam strong medium angular to sub angular blocky; moderate sticky, moderate plastic, firm, hard; common fine to medium discontinuous random exped medium interstitial pores; few humified residual of organic matter; common fine to medium roots; slight effervescence with HCl; gradual smooth boundary.
50–90	Dark grayish brown (10YR 4/2) moist, grayish brown (10YR 5/2) dry; silty clay; moderate to strong medium sub angular blocky; sticky, Plastic, very firm, very hard; common fine and medium discontinuous random interstitial pores; slight effervescence with HCl.

Profile No.: 165
Date of description: 10/2006
Location:
Longitude: 31° 18′ E
Latitude: 31° 35′ N
Elevation: 12.0 m a.s.l.
Soil taxonomy: *Vertic Torrifluvents*
Topography: almost flat
Land form: flood plain/river terraces
Slope: nearly level
Land use & vegetation: cultivated with clover
Parent material: alluvial deposits
Drainage: moderately well drained soils
Ground water table: +100 cm

Depth (cm)	Description
0–20	Dark brown (10YR2/3) moist, grayish brown (10YR5/2) dry; clay loam; moderately fine to medium angular to sub angular blocky; very sticky, very plastic, very firm, extremely hard; common fine discontinuous random exped and interstitial pores, few humified patches of organic matter; many fine to coarse roots; slight effervescence with HCl, clear smooth boundary.
20–35	Dark grayish brown (10YR 4/2) moist, grayish brown (10YR 5/4) dry; clayey; strong fine to medium sub angular blocky; sticky, plastic; firm, hard; common fine to medium continuous and discontinuous random interstitial pores; few humified patches of organic matter; many fine to coarse roots; gradual smooth boundary.
35–100	Dark grayish brown (10YR 4/2) moist, grayish brown (10YR 5/4) dry; clayey; moderate fine to medium sub angular blocky; very sticky very plastic: very firm, very hard; common fine and medium discontinuous random exped and interstitial pores; slight smooth boundary.

Profile No.: 166
Date of description: 10/2006
Location
Longitude: 31° 11′ E
Latitude: 30° 31′ N
Elevation: 11.0 m a.s.l.
Soil taxonomy: *Typic Quartizipsamments*
Topography: almost flat
Land form: flood plain
Slope: nearly level
Land use & vegetation: barren
Parent material: aeolian deposits

Drainage: moderately well drained
Ground water table: not hit

Depth (cm)	Description
0–30	Light yellowish brown (10YR 6/4) moist, very pale brown (10 YR 7/3) dry, sandy, single grains, few gravels, strong effervescence, common fine and medium roots; slight effervescence with HCl, diffuse smooth boundary.
30–75	Yellowish brown (10 YR 5/6) moist, pale brown (10 YR 6/3) dry, sandy loam, single grains, common gravels, loose, common fine to medium discontinuous random pores, many fine roots moderate effervescence with HCl, diffuse smooth boundary.
75–120	Dark yellowish brown (10 YR 4/4) moist, Yellowish brown (10 YR 5/4)dry, slightly hard medium coarse loamy sand, common coarse and medium discontinuous random pores; moderate effervescence with HCl.

Annex III: Laboratory Analyses of the Studied Soil Profiles

Particle size distribution of the representative Nile Delta soil profiles

A.-A. Gad, *Digital Land Resources to Support Sustainable Development in Egypt*,
Deltas of the World,
https://doi.org/10.1007/978-3-031-37167-7

Taxonomic unit	Representative profile	Depth in cm	Particle size %					Texture class
			Gravel	Coarse sand	Fine sand	Silt	Clay	
Typic Aquisalids	1	0–40	0.00	65.1	33.22	0.5	1.0	S
		40–75	0.00	69.4	30.30	0.2	0.8	S
Typic Quartzipsamments	2	0–25	0.00	67.8	27.5	1.3	3.2	S
		25–55	0.00	70.4	22.6	4.5	2.5	S
		55–100	0.00	61.6	30.2	4.5	3.5	S
Typic Haplosalids	3	0–30	0.00	71.8	25.1	1.2	1.8	S
		30–60	0.00	72.5	23.2	1.5	2.5	S
		60–110	0.00	72.9	21.4	2.6	2.4	S
Typic Aquisalids	4	0–35	0.00	52.1	34.7	1.5	9.5	LS
		35–70	0.00	55.1	35.7	3.5	4.5	S
Typic Quartzipsamments	5	0–25	0.00	63.5	36.2	0.8	0.5	S
		25–50	0.00	69.3	24.5	0.5	2.5	S
		50–100	0.00	73.4	18.8	3.2	2.5	S
Vertic Torrivluvents	6	0–30	0.00	0.95	34.5	16.6	45.1	C
		30–50	0.00	0.39	28.4	27.5	45.9	C
		50–90	0.00	0.2	25.1	18.5	44.5	C
		90–110	0.00	0.3	28.6	24.5	45.5	C
Typic Haplosalids	9	0–20	0.00	0.2	8.1	15.5	75.0	C
		20–45	0.00	0.2	6.2	4.5	76.1	C
		45–80	0.00	0.9	3.3	18.0	78.8	C
		80–110	0.00	0.2	2.0	15.5	79.6	C
Typic Quartzipsamments	11	0–30	0.00	72.2	22.4	1.7	3.3	S
		30–65	0.00	70.7	21.5	4.1	3.4	S
		65–115	0.00	59.1	33.3	2.4	5.1	S

Vertic Torrivluvents	12	0–30	0.00	0.7	29.8	30.1	35.4	CL
		30–55	0.00	0.3	26.7	26.5	45.5	C
		55–115	0.00	0.4	27.8	28.3	42.2	C
Vertic Torrivluvents	14	0–25	0.00	1.8	46.8	19.5	30.0	SCL
		25–55	0.00	0.9	40.6	21.5	35.1	CL
		55–100	0.00	1.5	36.4	20.2	39.7	C
Vertic Torrivluvents	15	0–30	0.00	2.2	49.9	9.5	36.2	SC
		30–60	0.00	2.5	40.6	11.5	42.5	C
		60–90	0.00	0.4	29.7	16.2	51.3	C
		90–110	0.00	0.4	20.1	27.4	50.9	C
Typic Torrivluvents	16	0–30	0.00	0.6	14.1	26.5	56.5	C
		30–55	0.00	0.4	10.9	29.5	57.1	C
		55–75	0.00	0.3	9.1	25.5	63.5	C
		75–100	0.00	1.8	45.5	19.5	30.0	CL
Vertic Torrivluvents	17	0–25	0.00	1.6	21.5	27.2	47.8	C
		25–65	0.00	2.7	25.7	17.3	51.5	C
		65–110	0.00	0.4	46.1	12.8	38.5	SC
Typic Torrivluvents	18	0–15	0.00	2.8	25.1	50.3	48.3	C
		15–35	0.00	0.1	50.7	17.2	28.3	SCL
		35–90	0.00	0.2	41.5	22.9	21.6	CL
Typic Quartizipsamments	19	0–20	0.00	21.7	72.2	1.5	4.1	S
		20–45	0.00	3.4	74.4	5.2	15.8	SL
		45–85	0.00	7.5	74.9	3.4	9.6	LS
Vertic Torrivluvents	20	0–25	0.00	3.3	28.8	12.6	55.5	C
		25–60	0.00	2.8	29.4	16.9	50.7	C
		60–90	0.00	1.9	34.6	12.8	50.4	C
Typic Torriorthents	21	0–25	10.1	65.2	30.22	3.5	1.0	S
		25–60	9.2	70.5	29.8	0.5	0.5	S

(continued)

Taxonomic unit	Representative profile	Depth in cm	Particle size %					Texture class
			Gravel	Coarse sand	Fine sand	Silt	Clay	
		60–90	5.1	61.9	33.1	0.8	3.2	S
Typic Quartzipsamments	22	0–30	5.5	67.3	27.5	1.6	3.4	S
		30–60	6.2	71.2	25.7	1.6	1.5	S
		60–110	4.1	63.6	32.2	0.5	3.5	S
Typic Torripsamments	23	0–35	2.0	57.1	34.7	1.6	4.4	S
		35–70	6.8	57.1	34.7	2.5	4.5	S
Typic Aquisalids	26	0–15	3.5	0.2	10.75	26.5	20.5	SCL
		15–35	5.6	0.3	38.2	25.5	32.5	SCL
		35–75	8.9	5.1	49.2	20.5	23.5	SCL
Typic Torrivluvents	27	0–20	2.3	0.8	29.7	30.2	35.3	CL
		20–50	3.5	0.3	26.7	26.5	45.5	C
		50–70	4.2	0.4	27.8	28.3	42.2	C
Vertic Torrivluvents	28	0–30	1.2	1.8	50.8	15.5	30.1	SCL
		30–60	2.3	0.9	40.6	21.5	35.2	CL
		60–90	1.6	1.7	36.7	20.5	40.5	C
Typic Torrivluvents	29	0–20	1.3	2.2	49.9	9.5	36.1	SC
		20–35	0.9	2.7	40.4	11.2	42.8	C
		35–65	0.5	0.4	29.7	16.5	51.2	C
		65–110	0.6	1.8	36.1	19.5	40.5	C
Typic Haplogypsids	32	0–30	11.5	36.5	52.5	2.1	8.9	S
		30–60	12.6	32.5	56.4	1.8	9.3	S
		60–120	10.4	32.6	54.5	1.7	11.2	S

Rock land	36	Limestone						
Typic Haplosalids	37	0–25	2.5	1.1	43.9	27.3	26.2	SCL
		25–60	6.3	1.1	35.2	31.5	30.1	CL
		60–90	8.8	0.4	57.1	19.4	21.7	
Typic Torrivluvents	38	0 = 25	1.2	1.5	56.3	9.1	30.4	SCL
		25–55	2.1	0.4	63.5	8.5	25.5	SCL
		55–100	1.3	0.2	57.2	9.2	31.3	SCL
Vertic Torrivluvents	39	0–20	2.1	0.6	23.5	23.1	50.4	C
		20–45	1.1	0.2	41.5	22.5	34.0	CL
		45–80	0.9	0.2	37.3	25.5	35.4	CL
		80–110	1.2	0.9	40.6	21.2	35.3	CL
Typic Torripsamments	40	0–30	6.2	30.8	58.0	1.5	9.7	S
		30–60	5.8	32.4	57.6	1.5	8.5	S
		60–100	7.2	15.9	51.3	9.5	18.5	S

Typic Torriorthents	43	0–20	9.2	70.5	29.8	0.5	0.5	S
		20–40	10.1	65.2	30.22	3.5	1.0	S
		40–85	5.1	61.9	33.1	0.8	3.2	S
Vertic Torrifluvents	44	0–25	0.00	2.7	25.7	17.3	51.5	C
		25–65	0.00	2.5	40.6	11.5	42.5	C
		65–120	0.00	0.4	29.7	16.2	51.3	C
Typic Torrifluvents	45	0–20	0.00	0.3	26.7	26.5	45.5	C
		20–45	0.00	0.9	40.6	21.5	35.1	CL
		45–100	0.00	1.5	36.4	20.2	39.7	C
Typic Quartizipsamments	46	0–25	12.10	24.30	51.50	7.70	4.40	S
		25–60	7.40	29.60	51.40	7.20	4.80	S
		60–100	7.70	26.30	55.80	6.10	4.10	S
Typic Haplogypsids	47	0–15	6.3	8.4	40.1	30.1	15.1	SL
		15–60	2.04	10.8	37.04	22.5	27.62	SCL
		60–120	7.5	30.5	22.88	10.02	29.1	SCL
Typic Petrogypsids	48	0–20	9.10	0.70	32.6	32.5	25.1	L
		20–60	8.30	0.70	18.5	30.1	42.4	C
Vertic Torrifluvents	49	0–25	0.00	0.4	10.9	29.5	57.1	C
		25–60	0.00	2.7	25.7	17.3	51.5	C
		60–120	0.00	0.4	46.1	12.8	38.5	SC
Typic Torrifluvents	50	0–20	0.00	0.2	41.5	22.9	21.6	CL
		20–40	0.00	2.7	25.7	17.3	51.5	C
		40–90	0.00	0.4	46.1	12.8	38.5	SC
Typic Quartizipsamments	51	0–30	9.2	70.5	29.8	0.5	0.5	S
		30–70	6.2	71.2	25.7	1.6	1.5	S
		70–120	4.1	63.6	32.2	0.5	3.5	S
Typic Haplocalcids	52	0–35	10.20	8.62	43.22	29.85	8.11	SL
		35–100	16.40	22.62	36.46	18.81	5.71	SL
Typic Torripsamments	53	0–25	8.2	30.1	51.5	1.5	8.7	S
		25–60	5.8	31.4	53.1	1.5	8.2	S
		60–100	6.2	25.5	58.3	2.5	7.5	S
Typic Torriorthents	54	0–25	10.1	65.2	30.22	3.5	1.0	S
		25–60	9.07	26.73	52.90	7.00	4.43	S
		60–130	9.10	0.70	32.6	32.5	25.1	L
Typic Quartizipsamments	55	0–30	5.1	61.9	33.1	0.8	3.2	S
		30–70	5.5	67.3	27.5	1.6	3.4	S
		70–130	6.2	71.2	25.7	1.6	1.5	S
Vertic Torrifluvents	56	0–30	1.2	1.8	50.8	15.5	30.1	SCL
		30–65	2.3	0.9	40.6	21.5	35.2	CL
		65–110	1.6	1.7	36.7	20.5	40.5	C

(continued)

Typic Torriorthents	57	0–15	11.5	36.5	52.5	2.1	8.9	S
		15–40	12.6	32.5	56.4	1.8	9.3	S
		40–65	10.4	32.6	54.5	1.7	11.2	S
		65–110	5.8	31.4	53.1	1.5	8.2	S
Typic Torriorthents	58	0–40	8.2	30.1	51.5	1.5	8.7	S
		40–120	5.8	31.4	53.1	1.5	8.2	S
Typic Quartizipsamments	59	0–20	6.2	25.5	58.3	2.5	7.5	S
		20–60	11.5	36.5	52.5	2.1	8.9	S
		60–130	12.6	32.5	56.4	1.8	9.3	S
Vertic Torrifluvents	60	0–25	2.1	0.6	23.5	23.1	50.4	C
		25–50	2.1	0.4	63.5	8.5	25.5	SCL
		50–100	1.3	0.2	57.2	9.2	31.3	SCL
Vertic Torrifluvents	61	0–15	0.00	0.9	40.6	21.5	35.1	CL
		15–45	0.00	1.5	36.4	20.2	39.7	C
		45–85	0.00	2.2	49.9	9.5	36.2	SC
		85–120	0.00	2.5	40.6	11.5	42.5	C
Typic Torrifluvents	62	0–20	1.2	1.2	54.1	12.1	31.4	SCL
		20–50	2.7	1.4	61.2	10.4	24.3	SCL
		50–90	2.3	0.9	56.2	8.3	32.3	SCL
Vertic Torrifluvents	63	0–25	0.00	0.5	7.5	34.0	58.0	C
		25–50	0.00	0.4	8.6	35.0	56.0	C
		50–80	0.00	0.3	5.7	44.0	50.0	SiC
		80–120	0.00	1.8	18.2	38.0	42.0	C
Typic Torrifuvents	64	0–25	0.00	0.6	6.3	41.0	52.1	SiC
		25–55	0.00	0.2	7.5	42.0	50.3	SiC
		55–80	0.00	2.5	18.3	36.9	42.3	C
		80–110	0.00	0.4	6.8	44.6	48.2	SiC
Typic Torriorthents	65	0–30	14.52	1.90	26.49	32.71	24.38	L
		30–75	12.74	2.50	29.52	35.13	20.11	L
		75–110	14.52	1.90	26.49	32.71	24.38	L
Typic Torriorthents	66	0–35	11.00	62.5	26.2	0.8	0.5	S
		35–120	10.00	63.4	18.8	3.2	2.5	S
Vertic Torrifluvents	67	0–15	0.00	0.8	6.9	33.1	59.2	C
		15–40	0.00	0.4	18.8	36.5	44.3	C
		40–75	0.00	1.4	14.9	38.5	45.2	C
		75–120	0.00	1.4	17.3	35.1	46.2	C
Typic Torrifluvents	68	0–20	0.00	2.8	20.2	45.0	32.0	CL
		20–55	0.00	0.6	6.3	41.0	52.1	SiC
		55–110	0.00	0.2	7.5	42.0	50.3	SiC
Typic Torriorthents	69	0–40	3.00	52.1	35.7	3.5	4.5	S
		40–120	7.00	56.5	36.2	0.8	0.5	S

(continued)

Typic Torriorthents	70	0–20	5.00	67.2	22.4	1.7	3.3	S
		20–50	6.00	64.7	21.5	4.1	3.4	S
		50–100	4.00	55.1	33.3	2.4	5.1	S
Vertic Torrifluvents	71	0–20	0.00	1.80	10.30	45.40	42.50	SiC
		20–60	0.00	1.30	11.50	41.50	45.70	SiC
		60–100	0.00	1.50	10.90	41.30	46.30	SiC
Typic Torrifluvents	72	0–15	2.8	5.1	38.6	28.3	25.2	SCL
		15–50	7.3	4.1	30.2	30.3	28.1	CL
		50–100	8.8	7.4	43.1	20.4	20.3	SCL
Aquic Haplosalids	73	0–15	0.00	1.10	36.90	19.80	42.20	C
		15–30	0.00	1.60	26.80	33.30	38.30	CL
		30–45	0.00	1.40	23.20	38.30	37.10	CL
Aquic Haplosalids	74	0–20	0.00	0.20	21.70	17.50	60.60	C
		20–50	0.00	0.30	20.90	19.30	59.50	C
Typic Haplocalcids	75	0–15	9.2	13.62	43.22	25.83	8.13	SL
		15–50	15.4	29.62	39.46	12.81	2.71	LS
		50–120	11.8	20.12	42.34	20.32	5.42	SL
Typic Haplocalcids	76	0–25	0.00	15.36	28.49	18.72	37.43	CL
		25–60	0.00	11.35	37.14	27.16	25.35	SCL
		60–120	0.00	25.65	55.91	7.78	10.66	LS
Typic Petrogypsids	77	0–15	4.10	7.79	34.10	29.00	26.10	CL
		15–60	3.33	5.74	18.54	30.07	42.32	CL
Typic Petrogypsids	78	0–25	6.50	28.6	45.30	10.1	8.5	S
		25–80	5.30	26.30	50.95	9.60	7.85	LS
Typic Torriorthents	79	0–20	3.00	18.69	67.19	3.36	7.76	S
		20–60	10.75	20.79	55.74	3.91	7.81	S
		60–100	11.28	26.50	59.96	1.25	1.01	S
Typic Torriorthents	80	0–20	12.25	6.30	71.75	8.20	1.50	S
		20–60	20.70	6.33	62.03	10.72	0.22	S
		60–120	24.99	8.27	55.45	10.29	1.00	S
Typic Torripsamments	81	0–20	9.33	40.51	42.71	7.14	0.31	S
		20–40	9.68	49.53	31.16	9.52	0.11	S
		40–120	11.72	26.6	51.22	10.17	0.29	S
Typic Torripsamments	82	0–30	12.57	42.01	36.14	15.31	0.12	S
		30–60	14.42	47.32	21.92	16.2	0.14	S
		60–130	16.85	33.66	39.12	10.11	0.26	S
Lithic *Torripsamments*	83	0–40	23.92	40.39	29.25	5.79	0.65	S
Lithic *Torripsamments*	84	0–30	13.23	42.4	41.33	3.12	0.12	S
		30–45	11.39	47.22	30.0	10.24	1.15	S
Typic Torrifluvents	85	0–15	14.25	37.74	21.2	6.17	21.34	SL
		15–50	1.72	26.6	31.22	20.17	20.29	L
		50–110	2.57	42.01	16.14	15.31	30.12	L

(continued)

Typic Quartizipsamments	86	0–30	5.31	38.76	45.67	9.26	1.0	S
		30–70	4.33	40.51	47.71	7.14	0.31	S
		70–130	4.68	49.53	36.16	9.52	0.11	S
Typic Torripsamments	87	0–30	2.77	46.6	41.94	7.41	1.28	S
		30–65	3.21	47.38	42.11	6.51	0.79	S
		65–150	1.36	42.92	45.0	10.2	0.01	S
Typic Torripsamments	88	0–25	7.56	35.67	52.27	3.15	1.35	S
		25–50	11.23	37.76	46.2	4.16	0.35	S
Typic Torriorthents	89	0–15	12.57	42.01	36.14	15.31	0.12	S
		15–35	4.25	41.74	46.2	7.17	1.34	S
		35–110	2.56	47.14	41.19	6.17	0.44	S
Typic Haplosalids	90	0–30	0.85	39.66	49.12	10.11	0.26	S
		30–45	0.92	43.39	49.25	5.79	0.65	S
		45–100	0.00	46.32	46.15	6.14	0.32	S
Typic Torriorthents	91	0–15	22.77	36.6	31.94	7.41	1.28	S
		15–40	23.21	27.38	42.11	6.51	0.79	S
		40–110	31.36	22.92	40	5.2	0.01	S
Typic Torripsamments	92	0–30	8.56	39.19	48.14	3.17	0.44	S
		30–65	9.31	38.76	41.17	9.26	0.5	S
		65–140	4.33	42.51	45.71	7.14	0.31	S
Typic Haplaquents	93	0–20	2.10	27.3	51.50	12.7	6.4	LS
		20–50	5.90	26.1	53.40	9.2	5.4	S
Typic Haplosalids	94	0–25	1.70	30.3	55.80	6.1	6.1	S
		25–60	4.30	27.30	55.90	5.7	6.80	LS
Typic Haplocalcids	95	0–30	5.36	43.72	46.18	4.64	0.1	S
		30–60	3.23	43.4	48.33	4.12	0.12	S
		60–150	6.23	37.76	51.25	4.16	0.30	S
Calcic Haplosalids	96	0–30	2.30	8.4	50.18	24.02	15.10	SL
		30–65	2.04	9.82	41.04	21.50	25.60	SCL
Typic Torrifluvents	97	0–15	0.00	0.2	15.3	53.0	31.5	SCL
		15–30	0.00	1.2	9.5	50.1	39.2	SiC
		30–75	0.00	0.8	10.2	45.5	43.5	SiC
		75–120	1.28	8.23	32.70	23.51	34.61	SCL
Typic Torripsamments	98	0–25	2.50	42.08	46.14	15.31	0.12	S
		25–55	4.40	47.34	31.92	16.2	0.14	S
		55–130	6.85	38.66	39.12	15.11	0.26	S
Typic Torripsamments	99	0–15	13.20	42.43	41.33	3.12	0.12	S
		15–30	14.23	34.76	46.2	4.11	0.40	S
		30–50	10.56	37.19	48.14	3.16	0.45	S
Typic Haplotorrerts	100	0–15	0.00	3.4	19.6	41.5	35.5	CL
		15–30	0.00	1.6	18.4	37.5	42.5	C
		30–100	0.00	6.6	16.4	31.6	45.4	C

(continued)

Chromic Haplotorrrrts	101	0–25	1.5	1.2	7.2	34.1	56.1	C
		25–50	1.2	0.6	6.9	32.1	59.2	C
		50–80	0.5	0.4	18.3	36.6	44.2	C
		80–140	0.5	0.2	8.3	38.5	52.5	C
Typic Torrifluvents	102	0–20	0.00	1.3	9.0	36.4	53.3	C
		20–40	0.00	0.4	6.5	43.0	50.1	SiC
		40–100	0.00	0.2	14.4	38.8	46.6	C
Typic Torripsamments	103	0–30	6.13	49.44	40.21	2.6	1.62	S
		30–70	7.25	37.74	48.4	6.15	1.34	S
		70–120	6.56	42.14	45.19	5.18	0.43	S
Typic Torripsamments	104	0–30	11.5	28.8	51.24	8.17	0.29	S
		30–45	23.90	41.39	28.27	5.79	0.65	S
Typic Haplotorrerts	105	0–25	2.5	1.2	9.8	39.0	47.5	C
		25–70	1.0	0.5	6.5	41.0	51.0	SiC
		70–120	1.5	0.7	5.8	37.5	54.5	C
Chromic Haplotorrrrts	106	0–30	2.6	1.4	15.7	34.1	46.2	C
		30–65	1.3	2.8	19.2	44.7	32.0	CL
		65–120	2.4	3.5	16.5	46.0	32.6	CL
Typic Quartizipsamments	107	0–25	5.1	61.9	33.1	0.8	3.2	S
		25–60	5.5	67.3	27.5	1.6	3.4	S
		60–90	6.2	71.2	25.7	1.6	1.5	S
Typic Torriorthents	108	0–25	10.1	65.2	30.22	3.5	1.0	S
		25–60	9.07	26.73	52.90	7.00	4.43	S
		60–120	9.10	0.70	32.6	32.5	25.1	L
Typic Torripsamments	109	0–20	8.2	30.1	51.5	1.5	8.7	S
		20–40	5.8	31.4	53.1	1.5	8.2	S
		40–120	6.2	25.5	58.3	2.5	7.5	S
Typic Haplocalcids	110	0–30	0.00	15.36	28.49	18.72	37.43	CL
		30–60	0.00	11.35	37.14	27.16	25.35	SCL
		60–100	0.00	25.65	55.91	7.78	10.66	LS
Lithic Haplocalcids	111	0–40	5.36	43.72	46.18	4.64	0.1	S
		40–55	3.23	43.4	48.33	4.12	0.12	S
Typic Quartizipsamments	112	0–30	6.2	25.5	58.3	2.5	7.5	S
		30–70	11.5	36.5	52.5	2.1	8.9	S
		70–90	12.6	32.5	56.4	1.8	9.3	S
Typic Quartizipsamments	113	0–20	12.10	24.30	51.50	7.70	4.40	S
		20–60	7.40	29.60	51.40	7.20	4.80	S
		60–110	7.70	26.30	55.80	6.10	4.10	S
Typic Torripsamments	114	0–25	9.33	40.51	42.71	7.14	0.31	S
		25–50	9.68	49.53	31.16	9.52	0.11	S
		50–100	11.72	26.6	51.22	10.17	0.29	S

(continued)

Typic Haplocalcids	115	0–25	9.2	13.62	43.22	25.83	8.13	SL
		25–50	15.4	29.62	39.46	12.81	2.71	LS
		50–110	11.8	20.12	42.34	20.32	5.42	SL
Typic Petrogypsids	116	0–25	6.50	28.6	45.30	10.1	8.5	S
		25–80	5.30	26.30	50.95	9.60	7.85	LS
Typic Torriorthents	117	0–20	3.00	18.69	67.19	3.36	7.76	S
		20–60	10.75	20.79	55.74	3.91	7.81	S
		60–90	11.28	26.50	59.96	1.25	1.01	S
Typic Quartizipsamments	118	0–30	9.2	70.5	29.8	0.5	0.5	S
		30–70	6.2	71.2	25.7	1.6	1.5	S
		70–130	4.1	63.6	32.2	0.5	3.5	S
Typic Torriorthents	119	0–20	9.2	70.5	29.8	0.5	0.5	S
		20–40	10.1	65.2	30.22	3.5	1.0	S
		40–85	5.1	61.9	33.1	0.8	3.2	S
Typic Haplosalids	120	0–30	0.85	39.66	49.12	10.11	0.26	S
		30–45	0.92	43.39	49.25	5.79	0.65	S
		45–80	0.00	46.32	46.15	6.14	0.32	S
Typic Haplosalids	121	0–25	0.00	0.20	21.70	17.50	60.60	C
		20–60	0.00	0.30	20.90	19.30	59.50	C
Typic Haplosalids	122	0–25	0.00	1.10	36.90	19.80	42.20	C
		25–40	0.00	1.60	26.80	33.30	38.30	CL
		40–80	0.00	1.40	23.20	38.30	37.10	CL
Typic Haplogypsids	123	0–30	4.10	7.79	34.10	29.00	26.10	CL
		30–65	3.33	5.74	18.54	30.07	42.32	CL
Typic Haplocalcids	124	0–30	11.5	36.5	52.5	2.1	8.9	S
		30–60	12.6	32.5	56.4	1.8	9.3	S
Typic Torriorthents	125	0–20	12.57	42.01	36.14	15.31	0.12	S
		20–60	4.25	41.74	46.2	7.17	1.34	S
		60–90	2.56	47.14	41.19	6.17	0.44	S
Vertic Torrifluvents	126	0–30	0.00	0.9	40.6	21.5	35.1	CL
		30–60	0.00	1.5	36.4	20.2	39.7	C
		60–110	0.00	2.2	49.9	9.5	36.2	SC
Vertic Torrifluvents	127	0–35	2.1	0.6	23.5	23.1	50.4	C
		35–65	2.1	0.4	63.5	8.5	25.5	SCL
		65–110	1.3	0.2	57.2	9.2	31.3	SCL
Vertic Torrifluvents	128	0–25	1.2	1.8	50.8	15.5	30.1	SCL
		25–45	2.3	0.9	40.6	21.5	35.2	CL
		45–100	1.6	1.7	36.7	20.5	40.5	C
Vertic Torrifluvents	129	0–25	0.00	2.7	25.7	17.3	51.5	C
		25–50	0.00	2.5	40.6	11.5	42.5	C
		50–90	0.00	0.4	29.7	16.2	51.3	C

(continued)

Typic Torrifluvents	130	0–25	0.00	0.2	41.5	22.9	21.6	CL
		25–55	0.00	2.7	25.7	17.3	51.5	C
		55–100	0.00	0.4	46.1	12.8	38.5	SC
Typic Torrfifuvents	131	0–20	0.00	0.6	6.3	41.0	52.1	SiC
		20–45	0.00	0.2	7.5	42.0	50.3	SiC
		45–110	0.00	2.5	18.3	36.9	42.3	C
Typic Haplocalcids	132	0–15	5.00	67.2	22.4	1.7	3.3	S
		15–40	6.00	64.7	21.5	4.1	3.4	S
		40–90	4.00	55.1	33.3	2.4	5.1	S
Vertic Torrifluvents	133	0–30	0.00	0.8	6.9	33.1	59.2	C
		30–45	0.00	0.4	18.8	36.5	44.3	C
		45–100	0.00	1.4	14.9	38.5	45.2	C
Vertic Torrifluvents	134	0–20	0.00	0.5	7.5	34.0	58.0	C
		20–45	0.00	0.4	8.6	35.0	56.0	C
		45–80	0.00	0.3	5.7	44.0	50.0	SiC

Table (3) Particle size distribution of the representative Qalubiyea Governorate soil profiles

			Particle size %					
Taxonomic unit	Representative profile	Depth in cm	Gravel	Coarse sand	Fine sand	Silt	Clay	Texture class
Petrogypsic Gypsiorthids	135	0–35	2.30	8.4	50.18	24.02	15.10	SL
		35–70	2.04	9.82	41.04	21.50	25.60	SCL
Typic Haplosalids	136	0–30	1.70	30.3	55.80	6.1	6.1	S
		30–60	4.30	27.30	55.90	5.7	6.80	LS
Vertic Torrivluvents	137	0–25	0.00	0.95	34.5	16.6	45.1	C
		25–50	0.00	0.39	28.4	27.5	45.9	C
		50–1400	0.00	0.2	25.1	18.5	44.5	C
Typic Torrivluvents	138	0–25	0.00	2.8	25.1	50.3	48.3	C
		25–45	0.00	0.1	50.7	17.2	28.3	SCL
		45–100	0.00	0.2	41.5	22.9	21.6	CL
Vertic Torrivluvents	139	0–20	0.00	0.7	29.8	30.1	35.4	CL
		20–55	0.00	0.3	26.7	26.5	45.5	C
		55–100	0.00	0.4	27.8	28.3	42.2	C
Vertic Torrivluvents	140	0–25	0.00	2.2	49.9	9.5	36.2	SC
		25–55	0.00	2.5	40.6	11.5	42.5	C
		55–120	0.00	0.4	29.7	16.2	51.3	C

(continued)

Taxonomic unit	Representative profile	Depth in cm	Particle size % Gravel	Coarse sand	Fine sand	Silt	Clay	Texture class
Typic Torrivluvents	141	0–30	0.00	0.6	14.1	26.5	56.5	C
		30–60	0.00	0.4	10.9	29.5	57.1	C
		60–120	0.00	0.3	9.1	25.5	63.5	C
Typic Torrivluvents	142	0–25	2.3	0.8	29.7	30.2	35.3	CL
		25–50	3.5	0.3	26.7	26.5	45.5	C
		50–110	4.2	0.4	27.8	28.3	42.2	C
Typic Torrivluvents	143	0–20	1.3	2.2	49.9	9.5	36.1	SC
		20–45	0.9	2.7	40.4	11.2	42.8	C
		45–80	0.5	0.4	29.7	16.5	51.2	C
Typic Quartizipsamments	144	0–30	5.5	67.3	27.5	1.6	3.4	S
		30–60	6.2	71.2	25.7	1.6	1.5	S
		60–150	4.1	63.6	32.2	0.5	3.5	S
Typic Quartizipsamments	145	0–20	0.00	21.7	72.2	1.5	4.1	S
		20–45	0.00	3.4	74.4	5.2	15.8	SL
		45–110	0.00	7.5	74.9	3.4	9.6	LS
Typic Torriorthents	146	0–30	10.1	65.2	30.22	3.5	1.0	S
		30–60	9.2	70.5	29.8	0.5	0.5	S
		60–100	5.1	61.9	33.1	0.8	3.2	S
Typic Quartizipsamments	147	0–30	0.00	72.2	22.4	1.7	3.3	S
		30–65	0.00	70.7	21.5	4.1	3.4	S
		65–120	0.00	59.1	33.3	2.4	5.1	S
Typic Quartizipsamments	148	0–30	0.00	63.5	36.2	0.8	0.5	S
		30–90	0.00	69.3	24.5	0.5	2.5	S
Typic Torrivluvents	149	0–25	1.3	2.2	49.9	9.5	36.1	SC
		25–45	0.9	2.7	40.4	11.2	42.8	C
		45–110	0.5	0.4	29.7	16.5	51.2	C
Vertic Torrivluvents	150	0–20	0.00	3.3	28.8	12.6	55.5	C
		20–40	0.00	2.8	29.4	16.9	50.7	C
		40–100	0.00	1.9	34.6	12.8	50.4	C
Typic Torrivluvents	151	0–30	1.2	1.5	56.3	9.1	30.4	SCL
		30–60	2.1	0.4	63.5	8.5	25.5	SCL
		60–120	1.3	0.2	57.2	9.2	31.3	SCL

(continued)

Taxonomic unit	Representative profile	Depth in cm	Particle size %					Texture class
			Gravel	Coarse sand	Fine sand	Silt	Clay	
Typic Quartizipsamments	152	0–40	6.2	30.8	58.0	1.5	9.7	S
		40–130	5.8	32.4	57.6	1.5	8.5	S
Vertic Torrivluvents	153	0–35	2.1	0.6	23.5	23.1	50.4	C
		35–70	1.1	0.2	41.5	22.5	34.0	CL
		70–100	0.9	0.2	37.3	25.5	35.4	CL
Vertic Torrivluvents	154	0–25	0.00	1.8	46.8	19.5	30.0	SCL
		25–60	0.00	0.9	40.6	21.5	35.1	CL
		60–100	0.00	1.5	36.4	20.2	39.7	C
Vertic Torrivluvents	155	0–30	0.00	2.2	49.9	9.5	36.2	SC
		30–65	0.00	2.5	40.6	11.5	42.5	C
		65–110	0.00	0.4	29.7	16.2	51.3	C
Typic Quartizipsamments	156	0–30	0.00	63.5	36.2	0.8	0.5	S
		30–50	0.00	69.3	24.5	0.5	2.5	S
		50–130	0.00	73.4	18.8	3.2	2.5	S
Vertic Torrivluvents	157	0–25	0.00	1.8	46.8	19.5	30.0	SCL
		25–70	0.00	0.9	40.6	21.5	35.1	CL
		70–110	0.00	1.5	36.4	20.2	39.7	C
Vertic Torrivluvents	158	0–15	0.00	2.2	49.9	9.5	36.2	SC
		15–45	0.00	2.5	40.6	11.5	42.5	C
		45–90	0.00	0.4	29.7	16.2	51.3	C
Vertic Torrivluvents	159	0–15	0.00	1.6	21.5	27.2	47.8	C
		15–40	0.00	2.7	25.7	17.3	51.5	C
		40–90	0.00	0.4	46.1	12.8	38.5	SC
Vertic Torrivluvents	160	0–20	1.2	1.8	50.8	15.5	30.1	SCL
		20–50	2.3	0.9	40.6	21.5	35.2	CL
		50–100	1.6	1.7	36.7	20.5	40.5	C
Typic Torrivluvents	161	0–20	0.00	0.6	14.1	26.5	56.5	C
		20–50	0.00	0.4	10.9	29.5	57.1	C
		50–120	0.00	0.3	9.1	25.5	63.5	C
Vertic Torrifluvents	162	0–20	0.00	1.80	10.30	45.40	42.50	SiC
		20–60	0.00	1.30	11.50	41.50	45.70	SiC
		60–120	0.00	1.50	10.90	41.30	46.30	SiC

(continued)

Taxonomic unit	Representative profile	Depth in cm	Particle size % Gravel	Coarse sand	Fine sand	Silt	Clay	Texture class
Typic Torrifluvents	163	0–20	2.8	5.1	38.6	28.3	25.2	SCL
		20–45	7.3	4.1	30.2	30.3	28.1	CL
		45–90	8.8	7.4	43.1	20.4	20.3	SCL
Vertic Torrifluvents	164	0–20	0.00	0.4	10.9	29.5	57.1	C
		20–50	0.00	2.7	25.7	17.3	51.5	C
		50–90	0.00	0.4	46.1	12.8	38.5	SC
Vertic Torrifluvents	165	0–20	0.00	3.4	19.6	41.5	35.5	CL
		20–35	0.00	1.6	18.4	37.5	42.5	C
		35–100	0.00	6.6	16.4	31.6	45.4	C
Typic Quartzipsamments	166	0–30	0.00	67.8	27.5	1.3	3.2	S
		30–75	0.00	70.4	22.6	4.5	2.5	S
		75–120	0.00	61.6	30.2	4.5	3.5	S

Annex III Some chemical analyses of the representative soil profiles

Taxonomic unit	Profile No.	Depth in cm	pH 1: 2.5	EC dS/m	$CaCO_3$%	OM %	ESP	CEC/ meq/ 100 g. soil	Soluble cations meq/L				Soluble anions meq/L			
									Na^+	K^+	Ca^{++}	Mg^{++}	CO_3^{--}	HCO_3^-	Cl^-	SO_4^{--}
Typic Aquisalids	1	0–40	7.32	26.63	1.22	1.23	11.4	11.5	290.7	5.3	104.6	12.8	0.00	1.3	272.3	139.8
		40–75	7.35	56.2	156	1.51	10.3	9.41	770.5	13.5	173.5	191.8	0.00	1.6	788.4	359.3
Typic Quartizipsamments	2	0–25	7.62	0.93	1.92	1.71	12.9	12.9	4.1	0.1	4.2	1.3	0.00	4.5	3.1	2.1
		25–55	7.73	1.72	0.82	1.43	13.8	14.3	3.6	0.1	3.6	1.4	0.00	2.9	3.2	2.6
		55–100	7.55	1.63	0.91	1.65	11.6	13.1	5.1	0.1	3.5	2.1	0.00	3.2	2.8	4.8
Typic Haplosalids	3	0–30	7.52	18.23	0.20	0.85	14.6	14.2	140.2	5.4	43.5	23.7	0.00	1.9	158.9	52.0
		30–60	7.66	17.31	0.11	0.69	16.8	12.6	144.6	3.5	19.8	24.9	0.00	1.9	154.9	36.0
		60–110	8.12	74.5	0.35	0.48	20.3	129	950.2	24.8	82.6	233.0	0.00	1.9	1143.9	144.8
Typic Aquisalids	4	0–35	7.82	40.71	1.12	0.87	22.3	15.6	460.8	15.2	37.4	86.1	0.00	2.3	514.7	82.5
		35–70	8.21	55.69	0.62	1.32	19.5	16.8	509.6	18.2	92.4	90.6	0.00	1.8	523.5	185.5
Typic Quartizipsamments	5	0–25	7.86	24.63	3.22	0.51	19.3	12.5	167.8	1.2	39.2	63.6	0.00	1.1	212.5	58.2
		25–50	7.72	25.74	166	0.33	15.9	10.7	203.4	2.4	47.3	44.2	0.00	1.8	263.2	32.3
		50–100	7.52	36.22	2.14	0.12	20.4	9.9	224.8	3.2	35.8	101.7	0.00	3.6	251.4	110.5
Vertic Torrivluvents	6	0–30	7.82	4.63	2.85	1.25	19.8	44.6	33.5	1.2	9.2	6.9	0.00	2.1	46.3	2.4
		30–50	7.77	6.22	2.64	1.31	17.6	48.3	50.2	1.2	15.5	17.9	0.00	1.6	36.2	47.0
		50–90	7.69	7.85	1.98	1.22	22.3	46.2	51.5	1.2	17.6	16.8	0.00	1.9	61.5	23.7
		90–110	7.92	8.66	2.14	0.44	24.5	48.6	54.5	1.5	26.9	24.6	0.00	1.7	37.5	68.3

Typic Haplosalids	9	0–20	7.50	49.71	1.83	1.56	22.3	64.5	760.5	13.5	174.5	190.8	0.00	1.5	787.5	414.8
		20–45	7.74	31.54	1.95	1.44	19.8	58.3	290.5	5.5	104.6	12.8	0.00	1.5	272.1	198.1
		45–80	7.55	8.56	1.22	0.81	24.3	56.7	70.5	1.7	5.7	12.9	0.00	1.2	90.5	57.0
		80–110	7.42	20.44	1.23	0.57	26.4	59.2	250.5	5.4	46.7	70.5	0.00	3.6	317.5	111.2
Typic Quartizipsamments	11	0–30	7.66	1.43	1.41	1.11	11.6	12.6	4.5	0.2	6.8	1.7	0.00	3.5	6.5	3.2
		30–65	7.94	1.33	1.96	0.62	14.3	10.8	3.9	0.1	4.7	1.8	0.00	2.5	3.5	4.5
		65–115	7.81	1.20	1.32	0.35	12.8	11.8	3.4	0.1	3.6	0.5	0.00	1.8	3.7	2.1

Vertic Torrivluvents	12	0–30	7.88	1.22	1.32	1.80	12.5	38.4	4.1	0.1	3.7	1.8	0.00	4.3	3.2	40.6
		30–55	7.85	0.95	1.45	1.12	14.3	42.6	3.6	0.1	3.6	1.4	0.00	3.5	2.5	45.3
		55–115	7.94	1.53	0.93	0.40	15.4	45.7	5.0	0.1	3.7	2.3	0.00	3.3	2.7	50.8
Vertic Torrivluvents	14	0–25	7.77	0.75	2.50	1.95	17.2	39.4	4.5	0.7	3.6	2.4	0.00	3.2	2.8	44.6
		25–55	7.95	1.34	2.41	1.87	13.5	41.5	8.8	0.2	2.6	2.4	0.00	2.5	3.0	50.0
		55–100	7.63	1.58	1.36	0.86	12.8	44.3	7.7	0.1	1.5	1.5	0.00	2.7	5.8	46.6
Vertic Torrivluvents	15	0–30	7.82	2.42	1.62	1.74	13.6	41.2	11.5	0.7	9.5	8.2	0.00	4.6	5.4	19.9
		30–60	7.75	1.62	1.23	1.52	14.5	39.5	8.8	0.3	2.7	1.9	0.00	3.7	5.8	4.2
		60–90	8.12	1.56	1.54	1.10	16.8	52.6	8.3	0.2	4.3	3.3	0.00	3.7	6.3	6.1
		90–110	8.23	1.10	0.64	0.53	22.4	51.4	8.4	0.4	4.3	3.0	0.00	2.6	7.4	6.1
Typic Torrivluvents	16	0–30	7.85	1.22	1.91	1.83	18.6	46.4	5.5	0.2	1.3	2.7	0.00	1.7	7.8	0.2
		30–55	7.65	2.86	2.31	1.66	17.2	42.3	18.7	0.4	5.9	4.3	0.00	1.9	21.4	6.0
		55–75	7.76	1.12	2.11	0.88	15.4	40.3	6.9	0.3	2.1	1.3	0.00	2.6	4.9	3.1
		75–100	7.98	1.43	1.50	0.41	16.3	38.5	6.2	0.6	2.5	1.3	0.00	1.8	4.1	4.7
Vertic Torrivluvents	17	0–25	8.10	1.15	2.25	1.81	17.2	48.6	6.3	0.1	4.3	2.8	0.00	5.4	3.6	4.5
		25–65	8.21	2.41	2.67	1.61	15.6	52.4	14.6	0.4	3.6	1.4	0.00	2.8	8.2	9.0
		65–110	8.33	1.85	1.64	0.88	14.3	40.3	12.1	0.1	5.4	1.7	0.00	3.3	5.7	10.3
Typic Torrivluvents	18	0–15	7.84	1.23	0.86	1.62	13.4	51.4	4.7	0.2	3.2	1.9	0.00	2.9	4.1	3.0
		15–35	7.74	0.92	1.11	1.20	11.2	39.2	4.5	0.2	2.2	0.9	0.00	3.5	3.5	0.8
		35–90	7.65	1.42	0.64	0.46	11.5	30.4	7.7	0.3	2.9	1.8	0.00	2.8	5.8	4.1
Typic Quartizipsamments	19	0–20	7.63	4.87	4.23	0.73	12.6	11.2	29.1	1.8	8.0	10.0	0.00	4.0	37.5	7.4
		20–45	7.64	4.31	6.84	0.26	12.8	21.6	30.4	2.2	7.0	7.0	0.00	5.0	31.5	10.1
		45–85	7.92	6.78	9.52	0.17	10.3	13.5	65.9	1.2	5.1	5.1	0.00	3.7	49.9	23.7

Vertic Torrivluvents	20	0–25	7.90	1.23	2.51	1.52	14.6	52.3	6.8	0.1	2.2	1.9	0.00	3.2	2.2	5.6
		25–60	7.91	1.42	2.44	1.00	17.8	48.6	8.6	0.1	1.6	2.5	0.00	2.6	6.1	4.1
		60–90	7.82	1.74	1.65	0.43	22.3	45.9	12.6	0.1	1.5	2.1	0.00	2.6	4.2	9.5
Typic Torriorthents	21	0–25	7.35	12.71	6.22	0.22	12.5	10.2	91.6	1.8	107.5	73.6	0.00	1.7	260.5	12.3
		25–60	7.85	4.60	5.41	0.10	11.6	11.6	26.8	0.5	27.1	19.1	0.00	2.1	47.0	24.4
		60–100	7.74	3.50	3.94	0.10	14.6	12.2	28.3	0.4	28.5	16.1	0.00	2.3	24.1	46.9
Typic Quartizipsamments	22	0–30	7.95	1.06	4.25	0.40	12.6	10.9	16.1	0.5	9.5	7.2	0.00	4.1	10.5	18.7
		30–60	8.44	0.74	3.45	0.21	12.8	10.5	13.1	0.2	4.5	4.7	0.00	4.1	8.5	9.9
		60–110	7.67	11.51	1.93	0.10	14.3	12.3	9.13	1.4	11.5	11.8	0.00	2.7	6.5	24.6
Typic Torripsamments	23	0–30	7.41	9.57	18.3	0.31	14.3	13.4	68.7	1.7	30.5	22.5	0.00	2.1	112.5	8.8
		30–70	7.91	3.35	8.74	0.1	12.3	10.8	38.4	0.8	17.5	14.5	0.00	1.3	36.5	33.4
Typic Aquisalids	26	0–15	7.62	22.50	0.84	0.74	16.5	25.3	154.5	0.7	77.7	53.6	0.00	0.8	226.4	59.3
		15–35	7.62	20.62	1.26	0.85	14.7	28.7	150.2	0.4	63.6	49.0	0.00	1.5	215.5	46.2
		35–75	7.81	31.5	1.40	1.10	12.5	28.6	194.3	0.9	87.2	70.4	0.00	1.5	285.5	65.8
Typic Torrivluvents	27	0–20	7.52	7.22	3.66	1.81	22.6	29.5	39.5	1.3	26.7	23.9	0.00	1.8	22.2	67.4
		20–50	7.60	8.51	4.52	1.22	24.8	31.5	63.5	1.2	16.2	15.7	0.00	1.1	19.8	75.7
		50–70	8.10	9.53	6.12	0.33	27.4	30.2	87.3	1.2	18.4	18.8	0.00	1.2	18.7	105.8
Vertic Torrivluvents	28	0–30	8.12	0.52	2.14	2.32	19.4	41.3	3.8	0.1	2.5	2.2	0.00	1.3	2.2	5.1
		30–60	8.20	0.81	2.31	1.3	14.2	42.1	5.5	0.3	1.1	16.1	0.00	2.1	1.4	19.5
		60–90	8.20	0.66	1.55	0.41	14.6	39.5	6.8	0.2	1.2	1.2	0.00	3.2	4.1	2.1
Typic Torrivluvents	29	0–20	7.88	11.52	2.64	1.80	14.3	44.6	75.5	0.7	30.5	36.2	0.00	1.7	106.3	34.9
		20–35	7.74	6.22	3.51	1.25	17.8	55.2	53.5	0.2	14.5	12.1	0.00	1.2	39.8	39.3
		35–65	8.10	7.10	2.41	0.81	19.3	47.3	59.4	0.8	17.5	17.1	0.00	1.9	45.1	47.8
		65–110	8.10	8.32	2.10	0.3	22.5	41.3	65.5	0.9	26.5	15.5	0.00	1.6	39.4	67.4

Typic Haplogypsids	32	0–30	7.72	5.22	12.81	1.23	17.6	18.5	6.8	1.7	27.2	5.2.	0.00	4.2	6.4	35.9
		30–60	7.72	5.41	16.84	0.42	19.2	12.3	6.4	2.2	34.8	2.6	0.00	2.9	5.6	37.8
		60–120	7.80	6.23	22.44	0.33	23.5	12.4	15.4	2.8	26.1	5.2	0.00	2.6	7.8	38.6
	36	Limestone														
Typic Haplosalids	37	0–25	7.8	31.94	3.36	0.62	15.9	33.5	304.2	2.2	72.0	55.9	0.00	4.3	380.5	49.5
		25–60	8.46	29.88	3.68	0.33	22.6	31.6	201.1	1.8	120.0	102.0	0.00	3.0	331.7	90.2
		60–90	8.52	28.50	2.34	0.29	19.7	27.9	165.3	2.3	95.9	87.5	0.00	4.0	290.4	56.6
Typic Torrivluvents	38	0–25	8.20	0.84	7.28	1.55	14.6	32.4	3.1	0.2	2.2	1.2	0.00	2.3	2.1	2.3
		25–55	8.20	0.65	5.63	1.32	15.8	284	3.2	0.2	2.3	2.1	0.00	2.1	2.4	3.3
		55–100	7.80	0.80	2.44	0.45	18.7	24.6	2.8	0.1	2.7	1.3	0.00	1.8	2.9	2.2
Vertic Torrivluvents	39	0–20	7.90	0.56	0.80	1.84	20.5	31.5	3.4	0.2	2.7	1.4	0.00	2.3	1.8	3.6
		20–45	7.90	0.42	0.60	1.40	21.2	36.4	3.8	0.3	3.3	1.8	0.00	2.2	1.9	5.1
		45–80	7.63	0.60	0.65	0.42	24.5	34.6	4.2	0.2	2.2	1.7	0.00	2.6	1.9	3.8
		80–110	7.64	0.82	0.59	0.21	31.2	31.6	3.2	0.1	1.7	1.4	0.00	2.0	2.9	1.5
Typic Torripsamments	40	0–30	8.08	4.87	1.68	0.62	11.5	10.1	29.1	1.8	8.0	10.0	0.00	4.0	37.5	7.4
		30–60	8.29	4.31	2.1	0.53	13.5	9.5	30.4	2.1	7.0	7.0	0.00	5.0	31.5	10.1
		60–100	7.76	6.78	1.26	0.43	29.4	11.2	65.9	1.2	5.1	5.1	0.00	3.7	49.9	22.8

Some chemical analyses of the representative soil profiles

Taxonomic unit	profile	Depth in cm	pH 1:2.5	EC dS/m	$CaCO_3$ %	OM %	ESP	CEC meq/100 g. soil	Soluble cations meq/L				Soluble anions meq/L			
									Na^+	K^+	Ca^{++}	Mg^{++}	CO_3^{--}	HCO_3^-	Cl^-	SO_4^{--}
Typic Torriorthents	43	0–20	8.20	3.10	4.10	1.80	22.30	8.50	13.10	1.40	9.00	7.50	0.00	1.90	18.00	10.90
		20–40	8.10	2.00	3.10	1.80	21.50	19.50	8.00	1.50	6.00	4.50	0.00	2.40	12.30	5.30
		40–85	8.10	1.60	1.70	0.58	18.00	7.20	6.60	1.40	4.70	3.30	0.00	2.30	8.00	5.70
Vertic Torrifluvents	44	0–25	8.00	4.50	2.53	1.12	11.70	33.50	21.00	1.40	17.60	6.90	0.00	4.50	23.00	18.90
		25–65	8.10	5.80	1.82	0.64	9.80	39.70	29.00	2.50	23.30	8.70	0.00	4.50	30.00	29.50
		65–120	8.20	7.10	1.54	0.36	15.60	28.70	38.00	2.70	26.80	10.90	0.00	3.50	34.50	31.10
Typic Torrifluvents	45	0–20	8.10	15.50	4.58	1.31	23.00	39.20	124.00	1.90	15.80	54.30	0.00	6.00	144.00	46.30
		20–45	8.10	7.80	3.22	0.41	21.80	37.00	45.00	2.70	20.20	14.00	0.00	6.00	59.00	17.30
		45–100	7.9	11.40	3.11	0.22	29.00	39.30	68.00	3.20	30.90	28.10	0.00	5.50	84.00	41.30
Typic Quartzipsamments	46	0–25	7.62	1.26	4.4	0.42	14.1	8.9	4.1	0.1	4.2	1.5	0.00	4.7	3.1	2.1
		25–60	7.45	2.35	3.6	0.61	15.8	8.5	3.5	0.2	3.7	1.3	0.00	2.9	3.2	2.6
		60–100	7.47	3.55	1.2	0.12	20.3	6.3	5.1	0.1	3.4	2.1	0.00	3.2	3.7	3.8
Typic Haplogypsids	47	0–15	7.52	11.52	10.3	0.45	14.7	16.5	57.35	1.8	55.4	80.1	0.00	3.5	100.5	90.65
		15–60	7.45	6.22	8.2	1.81	11.5	15.3	172.5	2.35	120.4	106.5	0.00	4.74	362.5	34.51
		60–120	7.77	7.10	8.7	1.43	12.6	15.4	154.5	2.25	100.3	62.1	0.00	3.82	267	48.33
Typic Petrogypsids	48	0–20	7.73	5.62	8.6	0.84	12.5	14.5	10.4	2.5	34.5	2.6	0.00	2.9	9.6	37.8
		20–60	7.65	8.24	12.5	0.91	11.6	11.3	15.2	2.6	26.1	5.6	0.00	2.6	9.8	36.6
Vertic Torrifluvents	49	0–25	8.06	6.50	3.2	2.01	19.50	46.25	218.50	18.00	31.00	80.33	0.00	6.80	137.00	104.03
		25–60	8.09	4.30	3.51	1.32	20.50	49.90	198.00	27.00	31.00	63.20	0.00	5.20	256.00	58.03
		60–120	7.86	5.60	4.12	0.89	21.00	53.50	195.50	19.00	26.00	290.23	0.00	7.00	221.00	41.73
Typic Torrifluvents	50	0–20	7.92	3.77	1.55	2.20	28.22	34.40	21.00	1.70	8.00	9.81	0.00	7.40	20.00	13.81
		20–40	7.85	3.19	2.40	1.80	10.24	32.50	23.00	12.20	5.00	4.62	0.00	5.20	23.00	7.62
		40–90	7.93	4.17	1.90	0.90	20.22	40.90	28.00	4.00	7.00	8.46	0.00	3.80	30.00	11.66
Typic Quartzipsamments	51	0–30	8.60	0.43	0.80	0.60	15.50	4.50	1.70	4.00	1.20	1.00	0.00	0.70	2.50	1.10
		30–70	8.50	3.55	1.20	0.35	22.20	5.40	17.30	2.00	9.20	7.00	0.00	0.40	20.20	14.90
		70–120	7.90	24.00	2.50	2.61	25.40	32.20	17.20	8.50	68.30	42.20	0.00	7.40	171.80	120.80
Typic Haplocalcids	52	0–35	7.54	6.31	12.8	1.12	14.2	12.4	61.4	0.8	17.8	17.1	0.00	1.9	46.3	48.8
		35–100	7.67	7.22	22.5	0.59	12.6	10.2	64.5	0.9	26.7	15.5	0.00	1.8	39.4	66.4

Typic Torripsamments	53	0–25	7.81	1.83	6.4	0.86	20.5	6.1	145.87	3.45	20.6	26.5	0.00	6.4	136.4	53.62
		25–60	7.38	1.71	9.2	1.10	16.6	4.3	137.8	2.25	22.9	17.9	0.00	5.8	135.2	39.85
		60–100	7.36	1.15	11.3	1.81	18.8	5.7	171.5	2.4	35.7	31.5	0.00	5.1	194.9	41.1
Typic Torriorthents	54	0–25	7.6	3.65	2.53	1.38	9.24	18.5	31.34	0.49	3.44	7.56	0.00	8.15	18.17	16.51
		25–60	7.8	5.41	4.66	1.12	11.52	15.77	47.38	0.66	4.05	6.30	0.00	7.42	32.06	18.91
		60–130	7.8	9.77	3.15	0.38	13.64	12.36	85.22	1.12	7.71	15.43	0.00	6.82	61.12	41.54
Typic Quartzipsamments	55	0–30	7.8	8.61	5.12	1.56	12.2	8.57	69.40	0.81	8.24	19.18	0.00	3.76	59.73	34.18
		30–70	7.8	7.32	3.21	0.74	19.8	6.41	54.94	0.59	9.77	21.93	0.00	4.08	55.66	27.49
		70–130	7.8	13.51	2.53	0.57	14.9	3.38	109.77	2.19	10.35	20.26	0.00	6.22	95.72	40.63
Vertic Torrifluvents	56	0–30	8.60	2.51	2.93	2.11	13.91	53.31	14.99	0.31	5.16	6.47	0.00	4.65	12.27	10.01
		30–65	7.8	4.33	2.17	0.83	11.17	51.96	26.98	1.8	4.28	13.47	0.00	4.45	19.60	22.48
		65–110	7.8	6.66	2.51	0.68	12.93	45.31	55.27	0.48	4.98	10.14	0.00	5.02	37.07	28.78
Typic Torriorthents	57	0–15	8.20	3.10	4.10	1.80	22.30	8.50	13.10	1.40	9.00	7.50	0.00	1.90	18.00	10.90
		15–40	8.10	2.00	3.10	1.10	21.50	19.50	8.00	1.50	6.00	4.50	0.00	2.40	12.30	5.30
		40–65	8.10	1.60	1.70	0.32	18.00	7.20	6.60	1.40	4.70	3.30	0.00	2.30	8.00	5.70
		65–110	8.20	0.85	1.70	0.12	17.40	6.30	3.30	0.80	2.50	2.20	0.00	1.50	5.10	1.90
Typic Torriorthents	58	0–40	7.44	15.90	1.84	2.10	28.22	34.40	148.00	12.00	6.00	40.49	0.00	4.00	139.00	73.49
		40–120	7.52	8.56	2.36	1.21	10.24	32.50	39.00	21.00	26.00	26.99	0.00	5.80	47.00	60.69
Typic Quartzipsamments	59	0–20	8.20	3.10	4.10	1.10	22.30	8.50	13.10	1.40	9.00	7.50	0.00	1.90	18.00	10.90
		20–60	8.10	2.00	3.10	0.51	21.50	9.50	8.00	1.50	6.00	4.50	0.00	2.40	12.30	5.30
		60–130	8.10	1.60	1.70	0.31	18.00	7.20	6.60	1.40	4.70	3.30	0.00	2.30	8.00	5.70
Vertic Torrifluvents	60	0–25	7.9	2.1	2.8	2.11	17.33	40.11	11.63	0.28	7.34	2.25	0.00	2.73	13.61	5.16
		25–50	8.3	3.5	3.2	1.74	16.22	40.13	10.36	0.24	4.51	9.52	0.00	3.08	10.11	11.44
		50–100	8.3	4.9	2.1	1.32	15.88	42.32	11.64	0.23	7.53	11.31	0.00	3.30	8.25	19.14
Vertic Torrifluvents	61	0–15	7.8	2.47	2.85	1.16	7.77	49.31	18.25	0.17	7.68	3.35	0.00	1.40	16.74	11.31
		15–45	8.1	1.88	2.63	0.81	14.83	31.57	11.34	0.15	5.33	3.72	0.00	1.62	14.38	4.54
		45–85	2.8	3.31	2.93	0.61	18.92	33.71	20.18	0.15	9.44	4.59	0.00	1.83	19.72	13.81
		85–120	8.1	5.68	4.97	5.51	12.65	38.13	41.58	0.31	12.51	5.53	0.00	2.12	35.16	22.65

Typic Torrifluvents	62	0–20	7.74	1.54	2.1	0.43	14.2	30.5	5.5	0.2	1.3	2.7	0.00	1.7	7.8	0.2
		20–50	7.75	2.33	3.4	0.22	14.6	31.5	18.7	0.4	5.9	4.3	0.00	1.9	21.4	6.0
		50–90	7.62	3.62	1.3	0.10	14.3	35.2	6.9	0.3	2.1	1.3	0.00	2.6	4.9	3.1
Vertic Torrifluvents	63	0–25	7.7	3.73	3.11	1.71	9.18	48.35	28.92	1.35	7.51	6.28	0.00	5.05	33.25	7.76
		25–50	7.8	1.98	2.36	0.93	9.35	50.81	16.57	0.58	3.15	2.79	0.00	2.79	13.34	6.96
		50–80	8.1	4.21	1.78	0.72	14.72	52.13	32.06	0.69	7.21	5.14	0.00	5.93	30.75	8.42
		80–120	8.1	6.81	2.68	0.39	16.31	40.30	63.22	1.04	12.31	9.38	0.00	7.87	64.23	13.85
Typic Torrifuvents	64	0–25	8.1	4.62	2.51	0.99	19.31	46.32	32.73	0.58	7.02	8.30	0.00	4.79	24.88	18.96
		25–55	8.0	8.12	2.63	0.91	14.25	44.88	66.88	0.46	5.77	10.69	0.00	6.84	45.27	3169
		55–80	8.1	12.37	2.41	0.71	14.50	41.31	113.54	1.81	6.19	17.29	0.00	8.77	88.90	41.16
		80–110	8.3	17.69	3.11	0.43	28.81	37.51	22.38	2.76	15.79	53.09	0.00	10.58	181.53	101.91
Typic Torriorthents	65	0–30	8.10	8.51	13.4	0.40	12.1	18.5	89.6	1.8	107.5	73.6	0.00	1.6	258.5	12.4
		30–75	7.80	9.53	14.2	0.31	14.3	12.3	27.6	0.5	27.1	19.1	0.00	2.1	48.0	24.2
		75–110	7.6	3.65	2.53	1.38	9.24	18.5	31.34	0.49	3.44	7.56	0.00	8.15	18.17	16.51
Typic Torriorthents	66	0–35	7.8	5.41	4.66	1.12	11.52	15.77	47.38	0.66	4.05	6.30	0.00	7.42	32.06	18.91
		35–120	7.8	9.77	3.15	0.38	13.64	12.36	85.22	1.12	7.71	15.43	0.00	6.82	61.12	41.54
Vertic Torrifluvents	67	0–15	8.1	4.62	2.51	0.99	19.31	46.32	32.73	0.58	7.02	8.30	0.00	4.79	24.88	18.96
		15–40	8.0	8.12	2.63	0.91	14.25	44.88	66.88	0.46	5.77	10.69	0.00	6.84	45.27	3169
		40–75	8.1	12.37	2.41	0.71	14.50	41.31	113.54	1.81	6.19	17.29	0.00	8.77	88.90	41.16
		75–120	8.3	17.69	3.11	0.43	28.81	37.51	22.38	2.76	15.79	53.09	0.00	10.58	181.53	101.91
Typic Torrifluvents	68	0–20	7.7	11.7	2.2	1.23	8.31	48.11	50.33	0.81	26.68	16.16	0.00	4.68	57.66	30.95
		20–55	7.7	8.7	1.1	1.67	9.22	48.83	40.99	0.73	23.12	21.20	0.00	3.18	51.16	31.70
		55–110	7.9	5.1	0.7	0.94	11.17	52.13	30.66	0.51	14.64	12.01	0.00	3.64	39.61	14.57
Typic Torriorthents	69	0–40	8.21	7.51	9.12	0.51	16.84	28.3	27.17	0.64	13.75	8.25	0.00	9.03	25.81	14.97
		40–120	8.05	8.11	11.23	0.12	21.61	46.1	26.74	1.67	13.30	13.90	0.00	8.52	28.40	18.69
Typic Torriorthents	70	0–20	8.32	1.55	1.44	0.30	11.9	9.6	27.39	0.30	9.45	8.51	0.00	0.80	16.52	28.32
		20–50	7.88	28.90	1.18	0.45	14.8	12.6	244.12	1.20	45.55	16.31	0.00	1.10	176.61	30.08
		50–100	8.10	15.23	1.31	0.38	13.3	11.1	135.76	0.75	27.50	12.41	0.00	0.95	96.57	29.20

Vertic Torrifluvents	71	0–20	8.24	3.52	3.1	1.87	9.3	42.3	5.1	0.2	2.3	1.2	0.00	2.4	4.1	2.3
		20–60	7.81	2.81	2.5	1.55	8.3	40.1	4.2	0.2	2.4	2.1	0.00	2.1	3.4	3.4
		60–100	7.82	1.66	2.3	1.32	9.5	37.5	4.8	0.2	2.8	1.4	0.00	1.8	4.9	2.5
Typic Torrifluvents	72	0–15	7.64	42.5	4.1	1.53	23.8	28.4	750.5	13.5	174.5	190.8	0.00	1.3	777.7	414.8
		15–50	7.91	30.3	5.3	1.00	27.4	35.4	285.5	5.5	104.6	12.8	0.00	1.5	272.1	193.1
		50–100	7.54	15.5	7.4	0.2	19.4	33.2	68.5	1.7	5.7	12.9	0.00	1.2	88.5	57.0
Aquic Haplosalids	73	0–15	7.87	26.34	2.4	1.12	17.8	22.6	6.2	0.6	2.5	1.3	0.00	1.8	4.1	4.7
		15–30	7.94	56.52	5.3	0.74	19.3	28.3	6.3	0.1	4.3	2.8	0.00	5.4	3.6	4.5
		30–45	7.60	45.40	8.18	1.05	44.60	51.80	429.00	1.90	47.10	159.8	0.00	4.50	595.00	38.50
Aquic Haplosalids	74	0–20	7.90	31.50	7.52	1.03	36.10	39.90	318.00	8.10	33.50	140.30	0.00	6.00	351.00	142.80
		20–50	7.90	38.90	9.71	0.51	41.70	38.40	308.00	10.40	45.30	55.10	0.00	3.50	323.00	92.70
Typic Haplocalcids	75	0–15	7.7	11.7	12.2	1.23	8.31	48.11	50.33	0.81	26.68	16.16	0.00	4.68	57.66	30.95
		15–50	7.7	8.7	21.1	1.67	9.22	48.83	40.99	0.73	23.12	21.20	0.00	3.18	51.16	31.70
		50–120	7.9	5.1	30.7	0.94	11.17	52.13	30.66	0.51	14.64	12.01	0.00	3.64	39.61	14.57
Typic Haplocalcids	76	0–25	8.1	5.1	21.51	1.45	16.90	50.85	40.17	0.78	18.35	16.82	0.00	2.01	36.55	25.16
		25–60	8.1	6.1	18.83	1.56	16.83	48.35	43.18	0.57	24.30	18.23	0.00	1.96	20.20	64.12
		60–120	7.9	7.7	28.77	1.03	17.91	50.36	53.19	0.52	26.30	20.11	0.00	1.36	44.47	54.29
Typic Petrogypsids	77	0–15	8.41	4.33	12.19	1.52	11.9	25.4	23.48	0.67	8.85	4.88	0.00	7.81	22.31	7.76
		15–60	8.11	5.62	16.73	0.41	12.7	45.3	25.43	0.95	3.50	2.45	0.00	0.37	24.67	7.29
Typic Petrogypsids	78	0–25	7.93	3.49	11.95	0.21	12.6	9.5	38.78	0.24	12.81	10.50	0.00	0.70	43.02	18.61
		25–80	8.00	3.51	22.03	0.02	13.5	9.3	32.61	0.19	5.65	16.13	0.00	0.60	8.01	45.97
Typic Torriorthents	79	0–20	7.6	3.65	2.53	1.38	9.24	58.5	31.34	0.49	3.44	7.56	0.00	8.15	18.17	16.51
		20–60	7.8	5.41	4.66	1.12	11.52	55.77	47.38	0.66	4.05	6.30	0.00	7.42	32.06	18.91
		60–100	7.8	9.77	3.15	0.38	13.64	42.36	85.22	1.12	7.71	15.43	0.00	6.82	61.12	41.54
Typic Torriorthents	80	0–20	7.8	8.61	5.12	1.56	12.2	48.57	69.40	0.81	8.24	19.18	0.00	3.76	59.73	34.18
		20–60	7.8	7.32	3.21	0.74	19.8	46.41	54.94	0.59	9.77	21.93	0.00	4.08	55.66	27.49
		60–120	7.8	13.51	2.53	0.57	14.9	43.38	109.77	2.19	10.35	20.26	0.00	6.22	95.72	40.63

Typic Torripsamments	81	0–20	7.72	5.03	7.74	0.36	12.5	12.3	32.81	0.75	25.65	16.03	0.00	1.70	45.51	28.06
		20–40	7.73	7.08	9.15	0.32	14.3	10.6	38.77	0.41	44.45	17.51	0.00	1.30	64.50	35.32
		40–120	7.61	5.92	3.65	0.05	9.8	11.4	29.82	0.29	39.35	16.05	0.00	0.90	54.51	29.66
Typic Torripsamments	82	0–30	7.60	7.62	1.51	0.41	14.8	8.9	53.69	0.54	28.61	12.20	0.00	0.60	86.01	8.43
		30–60	7.93	3.49	1.95	0.21	12.6	9.5	38.78	0.24	12.81	10.50	0.00	0.70	43.02	18.61
		60–130	8.00	3.51	2.03	0.02	13.5	9.3	32.61	0.19	5.65	16.13	0.00	0.60	8.01	45.97
Lithic *Torripsamments*	83	0–40	8.34	0.485	9.28	0.31	10.75	7.75	6.95	0.34	4.08	1.89	0.00	1.85	3.76	7.65
Lithic *Torripsamments*	84	0–30	7.61	5.92	3.65	0.05	9.8	11.4	29.82	0.29	39.35	16.05	0.00	0.90	54.51	29.66
		30–45	7.69	6.01	6.85	0.24	12.2	11.4	33.80	0.48	36.48	16.53	0.00	1.30	54.84	31.01
Typic Torrifluvents	85	0–15	8.2	5.38	1.97	1.63	14.9	33.31	37.74	0.41	3.47	15.31	0.00	6.11	32.41	18.41
		15–50	8.2	7.26	1.53	0.64	15.23	37.72	53.58	0.37	9.14	10.87	0.00	8.18	38.39	27.39
		50–110	8.4	12.26	2.31	0.41	27.31	36.92	94.70	0.78	7.37	16.27	0.00	7.22	68.42	43.48
Typic Quartizipsamments	86	0–30	8.41	4.33	5.19	1.52	11.9	7.4	23.48	0.67	8.85	4.88	0.00	7.81	22.31	7.76
		30–70	8.11	5.62	6.73	0.41	12.7	8.3	25.43	0.95	3.50	2.45	0.00	0.37	24.67	7.29
		70–130	8.41	11.65	8.65	1.11	24.2	9.5	10.22	1.18	9.75	7.31	0.00	3.32	14.46	10.86
Typic Torripsamments	87	0–30	7.96	14.71	7.33	0.42	26.4	8.6	11.26	1.85	24.00	16.80	0.00	0.16	10.70	43.05
		30–65	8.23	3.15	4.86	1.13	13.7	11.6	12.83	0.18	14.75	4.50	0.00	10.00	22.18	0.08
		65–150	8.41	4.33	5.19	1.52	11.9	12.4	23.48	0.67	8.85	4.88	0.00	7.81	22.31	7.76
Typic *Torripsamments*	88	0–25	7.74	6.10	8.61	0.57	14.7	14.4	26.61	5.15	16.90	13.52	0.00	5.81	25.28	31.09
		25–50	7.70	5.80	6.92	0.45	14.9	10.8	26.00	4.74	16.25	12.18	0.00	5.62	24.44	29.11
Typic Torriorthents	89	0–15	7.86	2.52	9.52	0.23	15.7	14.8	10.22	2.51	7.00	5.21	0.00	2.11	9.19	13.64
		15–35	8.11	5.62	6.73	0.41	12.7	15.3	25.43	0.95	3.50	2.45	0.00	0.37	24.67	7.29
		35–110	8.05	7.11	7.12	0.10	18.6	19.5	23.48	1.72	4.55	3.64	0.00	3.78	16.44	13.17
Typic Haplosalids	90	0–30	7.87	18.90	4.41	0.64	14.8	17.4	48.91	0.64	13.75	10.31	0.00	0.75	49.86	23.00
		30–45	8.05	30.60	7.22	0.41	15.6	12.4	63.04	3.33	15.00	11.25	0.00	0.94	62.37	29.31
		45–100	7.64	16.20	9.71	0.42	13.4	18.2	54.35	3.85	37.5	28.8	0.00	0.77	51.63	72.10
Typic Torriorthents	91	0–15	7.92	3.65	13.52	0.12	25.6	12.8	24.57	0.69	12.65	8.85	0.00	0.32	21.13	25.31
		15–40	8.11	5.62	6.73	0.41	12.7	20.3	25.43	0.95	3.50	2.45	0.00	0.37	24.67	7.29
		40–110	8.05	7.11	7.12	0.10	18.6	19.5	23.48	1.72	4.55	3.64	0.00	3.78	16.44	13.17

Typic Torripsamments	92	0–30	7.95	0.93	1.92	1.71	12.9	12.9	4.1	0.1	4.2	1.3	0.00	4.5	3.1	2.1
		30–65	8.44	1.72	0.82	1.43	13.8	11.3	3.6	0.1	3.6	1.4	0.00	2.9	3.2	2.6
		65–140	7.67	1.63	0.91	1.65	11.6	13.1	5.1	0.1	3.5	2.1	0.00	3.2	2.8	4.8
Typic Haplaquents	93	0–20	7.41	6.50	3.2	1.32	19.50	46.25	218.50	18.00	31.00	80.33	0.00	6.80	137.00	104.03
		20–50	7.91	4.30	3.51	0.89	20.50	49.90	198.00	27.00	31.00	63.20	0.00	5.20	256.00	58.03
Typic Haplosalids	94	0–25	7.62	20.62	1.26	0.85	14.7	24.5	37.74	0.41	3.47	15.31	0.00	6.11	32.41	18.41
		25–60	7.62	31.5	1.40	1.10	12.5	18.3	53.58	0.37	9.14	10.87	0.00	8.18	38.39	27.39
Typic Haplocalcids	95	0–30	7.81	5.22	12.81	1.23	17.6	13.5	6.8	1.7	27.2	5.2	0.00	4.2	6.4	35.9
		30–60	7.52	5.41	16.84	0.42	19.2	12.3	6.4	2.2	34.8	2.6	0.00	2.9	5.6	37.8
		60–150	7.60	6.23	22.44	0.33	23.5	11. 4	15.5	2.8	26.1	5.1	0.00	2.6	7.8	38.6
Calcic Haplosalids	96	0–30	8.10	22.94	10.81	0.33	14.6	27.9	21.00	1.70	8.00	9.81	0.00	7.40	20.00	13.81
		30–65	8.12	29.88	17.80	0.29	15.8	32.4	23.00	12.20	5.00	4.62	0.00	5.20	23.00	7.62
Typic Torrifluvents	97	0–15	8.20	4.62	2.51	0.99	19.31	46.32	32.73	0.58	7.02	8.30	0.00	4.79	24.88	18.96
		15–30	8.21	8.12	2.63	0.91	14.25	44.88	66.88	0.46	5.77	10.69	0.00	6.84	45.27	3169
		30–75	7.90	12.37	2.41	0.71	14.50	41.31	113.54	1.81	6.19	17.29	0.00	8.77	88.90	41.16
		75–120	7.91	17.69	3.11	0.43	28.81	37.51	22.38	2.76	15.79	53.09	0.00	10.58	181.53	101.91
Typic Torripsamments	98	0–25	8.20	3.10	4.10	0.58	21.50	19.50	8.00	1.50	6.00	4.50	0.00	2.40	12.30	5.30
		25–55	8.10	2.00	3.10	0.59	18.00	7.20	6.60	1.40	4.70	3.30	0.00	2.30	8.00	5.70
		55–130	8.10	1.60	1.70	0.34	17.40	6.30	3.30	0.80	2.50	2.20	0.00	1.50	5.10	1.90
Typic Torripsamments	99	0–15	8.60	0.43	0.80	0.60	15.50	4.50	1.70	4.00	1.20	1.00	0.00	0.70	2.50	1.10
		15–30	8.50	3.55	1.20	0.35	22.20	5.40	17.30	2.00	9.20	7.00	0.00	0.40	20.20	14.90
		30–50	7.90	8.00	2.50	0.11	25.40	9.20	17.20	8.50	68.30	42.20	0.00	7.40	171.80	120.80
Typic Haplotorrerts	100	0–15	7.67	11.7	2.2	1.23	8.31	48.11	50.33	0.81	26.68	16.16	0.00	4.68	57.66	30.95
		15–30	7.90	8.7	1.1	1.67	9.22	48.83	40.99	0.73	23.12	21.20	0.00	3.18	51.16	31.70
		30–100	7.91	5.1	0.7	0.94	11.17	52.13	30.66	0.51	14.64	12.01	0.00	3.64	39.61	14.57
Chromic Haplotorrrrts	101	0–25	7.82	11.8	1.66	1.59	9.37	47.11	83.12	1.30	30.22	23.91	0.00	1.66	104.5	32.39
		25–50	7.35	13.6	1.75	1.48	10.77	44.66	86.33	0.66	18.23	25.95	0.00	2.83	90.18	38.16
		50–80	7.85	8.1	1.53	0.98	11.33	50.37	63.51	1.62	29.21	21.38	0.00	2.48	56.88	30.33
		80–140	7.74	9.7	1.50	0.83	12.01	52.41	52.84	0.63	15.47	29.25	0.00	4.88	57.08	36.23

Typic Torrifluvents	102	0–20	8.60	2.51	2.93	2.11	13.91	53.31	14.99	0.31	5.16	6.47	0.00	4.65	12.27	10.01
		20–40	7.8	4.33	2.17	0.83	11.17	51.96	26.98	1.8	4.28	13.47	0.00	4.45	19.60	22.48
		40–100	7.8	6.66	2.51	0.68	12.93	45.31	55.27	0.48	4.98	10.14	0.00	5.02	37.07	28.78
Typic Torripsamments	103	0–30	8.08	4.87	1.68	0.62	11.5	10.1	11.81	0.71	4.37	2.83	0.00	2.5	7.32	9.90
		30–70	8.29	4.31	2.1	0.53	13.5	9.5	41.32	1.12	14.6	10.91	0.00	2.22	35.18	30.55
		70–120	7.76	6.78	1.26	0.43	29.4	11.2	36.89	1.71	12.27	7.81	0.00	2.81	30.51	25.36
Typic Torripsamments	104	0–30	7.35	4.60	6.41	0.15	11.6	10.6	24.81	1.23	9.66	10.83	0.00	3.25	15.55	27.73
		30–45	7.85	3.50	3.94	0.10	14.6	9.2	40.12	1.22	15.61	16.38	0.00	2.81	23.77	46.75
Typic Haplotorrerts	105	0–25	7.90	34.40	8.22	1.03	16.10	39.90	318.00	8.10	33.50	140.30	0.00	6.00	351.00	142.80
		25–70	7.90	31.50	7.52	0.51	21.70	38.40	308.00	10.40	45.30	55.10	0.00	3.50	323.00	92.70
		70–120	7.90	38.90	9.71	0.32	23.70	35.9	338.00	10.90	40.40	150.30	0.00	4.20	401.00	134.40
Chromic Haplotorrrrts	106	0–30	8.4	1.4	2.01	1.92	26.34	52.51	8.21	0.40	3.99	2.89	0.00	2.88	5.66	6.95
		30–65	8.3	2.8	3.39	1.51	28.77	50.11	18.23	0.61	5.83	4.37	0.00	3.02	15.31	10.71
		65–120	8.1	6.9	5.62	1.02	28.93	50.32	23.16	0.51	11.42	24.39	0.00	7.18	40.98	11.32

Appendix II: Some Laboratory Analyses of the Studied Soil Profiles

Al Fayoum Governorate

Some chemical analyses of the representative Al Fayoum Governorate soil profiles

A.-A. Gad, *Digital Land Resources to Support Sustainable Development in Egypt*,
Deltas of the World,
https://doi.org/10.1007/978-3-031-37167-7

Taxonomic unit	profile	Depth in cm	pH 1: 2.5	EC dS/m	$CaCO_3$%	OM %	ESP	CEC meq/100 g. soil	Soluble cations meq/L				Soluble anions meq/L			
									Na^+	K^+	Ca^{++}	Mg^{++}	CO_3^{--}	HCO_3^-	Cl^-	SO_4^{--}
Typic Quartizipsamments	107	0–25	7.62	1.26	4.4	0.42	14.1	8.9	4.1	0.1	4.2	1.5	0.00	4.7	3.1	2.1
		25–60	7.45	2.35	3.6	0.61	15.8	8.5	3.5	0.2	3.7	1.3	0.00	2.9	3.2	2.6
		60–90	7.47	3.55	1.2	0.12	20.3	6.3	5.1	0.1	3.4	2.1	0.00	3.2	3.7	3.8
Typic Torriorthents	108	0–25	7.6	3.65	2.53	1.38	9.24	18.5	31.34	0.49	3.44	7.56	0.00	8.15	18.17	16.51
		25–60	7.8	5.41	4.66	1.12	11.52	15.77	47.38	0.66	4.05	6.30	0.00	7.42	32.06	18.91
		60–120	7.8	9.77	3.15	0.38	13.64	12.36	85.22	1.12	7.71	15.43	0.00	6.82	61.12	41.54
Typic Torripsamments	109	0–20	7.72	5.03	7.74	0.36	12.5	12.3	32.81	0.75	25.65	16.03	0.00	1.70	45.51	28.06
		20–40	7.73	7.08	9.15	0.32	14.3	10.6	38.77	0.41	44.45	17.51	0.00	1.30	64.50	35.32
		40–120	7.61	5.92	3.65	0.05	9.8	11.4	29.82	0.29	39.35	16.05	0.00	0.90	54.51	29.66
Typic Haplocalcids	110	0–30	7.81	5.22	12.81	1.23	17.6	13.5	6.8	1.7	27.2	5.2	0.00	4.2	6.4	35.9
		30–60	7.52	5.41	16.84	0.42	19.2	12.3	6.4	2.2	34.8	2.6	0.00	2.9	5.6	37.8
		60–100	7.60	6.23	22.44	0.33	23.5	11. 4	15.5	2.8	26.1	5.1	0.00	2.6	7.8	38.6
Lithic Haplocalcids	111	0–15	8.60	0.43	0.80	0.60	15.50	4.50	1.70	4.00	1.20	1.00	0.00	0.70	2.50	1.10
		15–30	8.50	3.55	1.20	0.35	22.20	5.40	17.30	2.00	9.20	7.00	0.00	0.40	20.20	14.90
Typic Quartizipsamments	112	0–30	8.41	4.33	5.19	1.52	11.9	7.4	23.48	0.67	8.85	4.88	0.00	7.81	22.31	7.76
		30–70	8.11	5.62	6.73	0.41	12.7	8.3	25.43	0.95	3.50	2.45	0.00	0.37	24.67	7.29
		70–90	8.41	11.65	8.65	1.11	24.2	9.5	10.22	1.18	9.75	7.31	0.00	3.32	14.46	10.86
Typic Quartizipsamments	113	0–20	8.20	3.10	4.10	1.10	22.30	8.50	13.10	1.40	9.00	7.50	0.00	1.90	18.00	10.90
		20–60	8.10	2.00	3.10	0.51	21.50	9.50	8.00	1.50	6.00	4.50	0.00	2.40	12.30	5.30
		60–110	8.10	1.60	1.70	0.31	18.00	7.20	6.60	1.40	4.70	3.30	0.00	2.30	8.00	5.70
Typic Torripsamments	114	0–25	7.60	7.62	1.51	0.41	14.8	8.9	53.69	0.54	28.61	12.20	0.00	0.60	86.01	8.43
		25–50	7.93	3.49	1.95	0.21	12.6	9.5	38.78	0.24	12.81	10.50	0.00	0.70	43.02	18.61
		50–100	8.00	3.51	2.03	0.02	13.5	9.3	32.61	0.19	5.65	16.13	0.00	0.60	8.01	45.97
Typic Haplocalcids	115	0–25	7.6	6.4	4.2	0.00	5.2.	27.2	1.7	6.8	18.5	17.6	0.00	12.81	5.22	35.9
		25–50	7.8	5.6	2.9	0.00	2.6	34.8	2.2	6.4	12.3	19.2	0.00	16.84	5.41	37.8
		50–110	7.6	7.8	2.6	0.00	5.2	26.1	2.8	15.4	12.4	23.5	0.00	22.44	6.23	38.6
Typic Petrogypsids	116	0–25	8.10	22.94	10.81	0.33	14.6	27.9	21.00	1.70	8.00	9.81	0.00	7.40	20.00	13.81
		25–80	8.12	29.88	17.80	0.29	15.8	32.4	23.00	12.20	5.00	4.62	0.00	5.20	23.00	7.62

Typic Torriorthents	117	0–20	7.8	260.5	1.7	0.00	73.6	107.5	1.8	91.6	10.2	12.5	0.00	6.22	12.71	12.3
		20–60	8.2	47.0	2.1	0.00	19.1	27.1	0.5	26.8	11.6	11.6	0.00	5.41	4.60	24.4
		60–90	7.9	24.1	2.3	0.00	16.1	28.5	0.4	28.3	12.2	14.6	0.00	3.94	3.50	46.9
Typic Quartizipsamments	118	0–30	7.7	10.5	4.1	0.00	7.2	9.5	0.5	16.1	10.9	12.6	0.00	4.25	1.06	18.7
		30–70	7.5	8.5	4.1	0.00	4.7	4.5	0.2	13.1	10.5	12.8	0.00	3.45	0.74	9.9
		70–130	7.6	6.5	2.7	0.00	11.8	11.5	1.4	9.13	12.3	14.3	0.00	1.93	11.51	24.6
Typic Torriorthents	119	0–20	7.86	2.52	9.52	0.23	15.7	14.8	10.22	2.51	7.00	5.21	0.00	2.11	9.19	13.64
		20–40	8.11	5.62	6.73	0.41	12.7	15.3	25.43	0.95	3.50	2.45	0.00	0.37	24.67	7.29
		40–85	8.05	7.11	7.12	0.10	18.6	19.5	23.48	1.72	4.55	3.64	0.00	3.78	16.44	13.17
Typic Haplosalids	120	0–30	7.87	18.90	4.41	0.64	14.8	17.4	48.91	0.64	13.75	10.31	0.00	0.75	49.86	23.00
		30–45	8.05	30.60	7.22	0.41	15.6	12.4	63.04	3.33	15.00	11.25	0.00	0.94	62.37	29.31
		45–80	7.64	16.20	9.71	0.42	13.4	18.2	54.35	3.85	37.5	28.8	0.00	0.77	51.63	72.10
Typic Haplosalids	121	0–25	7.87	26.34	2.4	1.12	17.8	22.6	6.2	0.6	2.5	1.3	0.00	1.8	4.1	4.7
		20–60	7.94	56.52	5.3	0.74	19.3	28.3	6.3	0.1	4.3	2.8	0.00	5.4	3.6	4.5
Typic Haplosalids	122	0–25	7.60	45.40	8.18	1.05	44.60	51.80	429.00	1.90	47.10	159.8	0.00	4.50	595.00	38.50
		25–40	7.90	31.50	7.52	1.03	36.10	39.90	318.00	8.10	33.50	140.30	0.00	6.00	351.00	142.80
		40–80	7.90	38.90	9.71	0.51	41.70	38.40	308.00	10.40	45.30	55.10	0.00	3.50	323.00	92.70
Typic Haplogypsids	123	0–30	8.41	4.33	12.19	1.52	11.9	25.4	21.48	0.67	8.85	4.88	0.00	7.81	22.31	9.76
		30–65	8.11	5.62	16.73	0.41	12.7	45.3	22.43	0.95	3.50	2.45	0.00	0.37	24.67	10.29
Typic Haplocalcids	124	0–30	8.1	5.1	21.51	1.45	16.90	50.85	40.17	0.78	18.35	16.82	0.00	2.01	36.55	25.16
		30–60	8.1	6.1	18.83	1.56	16.83	48.35	43.18	0.57	24.30	18.23	0.00	1.96	20.20	64.12
Typic Torriorthents	125	0–20	7.6	3.65	2.53	1.38	9.24	58.5	31.34	0.49	3.44	7.56	0.00	8.15	18.17	16.51
		20–60	7.8	5.41	4.66	1.12	11.52	55.77	47.38	0.66	4.05	6.30	0.00	7.42	32.06	18.91
		60–90	7.8	9.77	3.15	0.38	13.64	42.36	85.22	1.12	7.71	15.43	0.00	6.82	61.12	41.54

Vertic Torrifluvents	126	0–30	7.67	11.7	2.2	1.23	8.31	48.11	50.33	0.81	26.68	16.16	0.00	4.68	57.66	30.95
		30–60	7.90	8.7	1.1	1.67	9.22	48.83	40.99	0.73	23.12	21.20	0.00	3.18	51.16	31.70
		60–110	7.91	5.1	0.7	0.94	11.17	52.13	30.66	0.51	14.64	12.01	0.00	3.64	39.61	14.57
Vertic Torrifluvents	127	0–35	7.82	11.8	1.66	1.59	9.37	47.11	83.12	1.30	30.22	23.91	0.00	1.66	104.5	32.39
		35–65	7.35	13.6	1.75	1.48	10.77	44.66	86.33	0.66	18.23	25.95	0.00	2.83	90.18	38.16
		65–110	7.85	8.1	1.53	0.98	11.33	50.37	63.51	1.62	29.21	21.38	0.00	2.48	56.88	30.33
Vertic Torrifluvents	128	0–25	7.67	11.7	2.2	1.23	8.31	48.11	50.33	0.81	26.68	16.16	0.00	4.68	57.66	30.95
		25–45	7.90	8.7	1.1	1.67	9.22	48.83	40.99	0.73	23.12	21.20	0.00	3.18	51.16	31.70
		45–100	7.91	5.1	0.7	0.94	11.17	52.13	30.66	0.51	14.64	12.01	0.00	3.64	39.61	14.57
Vertic Torrifluvents	129	0–25	8.4	1.4	2.01	1.92	26.34	52.51	8.21	0.40	3.99	2.89	0.00	2.88	5.66	6.95
		25–50	8.3	2.8	3.39	1.51	28.77	50.11	18.23	0.61	5.83	4.37	0.00	3.02	15.31	10.71
		50–90	8.1	6.9	5.62	1.02	28.93	50.32	23.16	0.51	11.42	24.39	0.00	7.18	40.98	11.32
Typic Torrifluvents	130	0–25	7.64	42.5	4.1	1.53	23.8	28.4	750.5	13.5	174.5	190.8	0.00	1.3	777.7	414.8
		25–55	7.91	30.3	5.3	1.00	27.4	35.4	285.5	5.5	104.6	12.8	0.00	1.5	272.1	193.1
		55–100	7.54	15.5	7.4	0.2	19.4	33.2	68.5	1.7	5.7	12.9	0.00	1.2	88.5	57.0
Typic Torrfifuvents	131	0–20	8.60	2.51	2.93	2.11	13.91	53.31	14.99	0.31	5.16	6.47	0.00	4.65	12.27	10.01
		20–45	7.8	4.33	2.17	0.83	11.17	51.96	26.98	1.8	4.28	13.47	0.00	4.45	19.60	22.48
		45–110	7.8	6.66	2.51	0.68	12.93	45.31	55.27	0.48	4.98	10.14	0.00	5.02	37.07	28.78
Typic Haplocalcids	132	0–15	7.96	14.71	7.33	0.42	26.4	8.6	11.26	1.85	24.00	16.80	0.00	0.16	10.70	43.05
		15–40	8.23	3.15	18.86	1.13	13.7	11.6	12.83	0.18	14.75	4.50	0.00	10.00	22.18	0.08
		40–90	8.41	4.33	25.19	1.52	11.9	12.4	23.48	0.67	8.85	4.88	0.00	7.81	22.31	7.76
Vertic Torrifluvents	133	0–30	8.24	3.52	3.1	1.87	9.3	42.3	5.1	0.2	2.3	1.2	0.00	2.4	4.1	2.3
		30–45	7.81	2.81	2.5	1.55	8.3	40.1	4.2	0.2	2.4	2.1	0.00	2.1	3.4	3.4
		45–100	7.82	1.66	2.3	1.32	9.5	37.5	4.8	0.2	2.8	1.4	0.00	1.8	4.9	2.5
Vertic Torrifluvents	134	0–20	8.2	5.38	1.97	1.63	14.9	33.31	37.74	0.41	3.47	15.31	0.00	6.11	32.41	18.41
		20–45	8.2	7.26	1.53	0.64	15.23	37.72	53.58	0.37	9.14	10.87	0.00	8.18	38.39	27.39
		45–80	8.4	12.26	2.31	0.41	27.31	36.92	94.70	0.78	7.37	16.27	0.00	7.22	68.42	43.48

Petrogypsic Gypsiorthids	135	0–35	7.93	3.49	11.95	0.21	12.6	9.5	38.78	0.24	12.81	10.50	0.00	0.70	43.02	18.61
		35–70	8.00	3.51	22.03	0.02	13.5	9.3	32.61	0.19	5.65	16.13	0.00	0.60	8.01	45.97
Typic Haplosalids	136	0–30	7.90	31.50	7.52	1.03	36.10	39.90	318.00	8.10	33.50	140.30	0.00	6.00	351.00	142.80
		30–60	7.90	38.90	9.71	0.51	41.70	38.40	308.00	10.40	45.30	55.10	0.00	3.50	323.00	92.70

AL-Qalubiyea Governorate

Some chemical analyses of the representative Qalubiyea Governorate soil profiles

Taxonomic unit	Profile	Depth in cm	pH 1: 2.5	EC dS/m	$CaCO_3$%	OM %	ESP	CEC meq/100 g. soil	Soluble cations meq/L				Soluble anions meq/L			
									Na^+	K^+	Ca^{++}	Mg^{++}	CO_3^{--}	HCO_3^-	Cl^-	SO_4^{--}
Vertic Torrivluvents	137	0–25	8.00	4.50	2.53	1.12	11.70	33.50	21.00	1.40	17.60	6.90	0.00	4.50	23.00	18.90
		25–50	8.10	5.80	1.82	0.64	9.80	39.70	29.00	2.50	23.30	8.70	0.00	4.50	30.00	29.50
		50–100	8.20	7.10	1.54	0.36	15.60	28.70	38.00	2.70	26.80	10.90	0.00	3.50	34.50	31.10
Typic Torrivluvents	138	0–25	8.10	15.50	4.58	1.31	23.00	39.20	124.00	1.90	15.80	54.30	0.00	6.00	144.00	46.30
		25–45	8.10	7.80	3.22	0.41	21.80	37.00	45.00	2.70	20.20	14.00	0.00	6.00	59.00	17.30
		45–100	7.9	11.40	3.11	0.22	29.00	39.30	68.00	3.20	30.90	28.10	0.00	5.50	84.00	41.30
Vertic Torrivluvents	139	0–20	8.06	6.50	3.2	2.01	19.50	46.25	218.50	18.00	31.00	80.33	0.00	6.80	137.00	104.03
		20–55	8.09	4.30	3.51	1.32	20.50	49.90	198.00	27.00	31.00	63.20	0.00	5.20	256.00	58.03
		55–100	7.86	5.60	4.12	0.89	21.00	53.50	195.50	19.00	26.00	290.23	0.00	7.00	221.00	41.73
Vertic Torrivluvents	140	0–25	8.60	2.51	2.93	2.11	13.91	53.31	14.99	0.31	5.16	6.47	0.00	4.65	12.27	10.01
		25–55	7.8	4.33	2.17	0.83	11.17	51.96	26.98	1.8	4.28	13.47	0.00	4.45	19.60	22.48
		55–120	7.8	6.66	2.51	0.68	12.93	45.31	55.27	0.48	4.98	10.14	0.00	5.02	37.07	28.78
Typic Torrivluvents	141	0–30	7.92	3.77	1.55	2.20	28.22	34.40	21.00	1.70	8.00	9.81	0.00	7.40	20.00	13.81
		30–60	7.85	3.19	2.40	1.80	10.24	32.50	23.00	12.20	5.00	4.62	0.00	5.20	23.00	7.62
		60–120	7.93	4.17	1.90	0.90	20.22	40.90	28.00	4.00	7.00	8.46	0.00	3.80	30.00	11.66
Typic Torrivluvents	142	0–25	7.74	1.54	2.1	0.43	14.2	30.5	5.5	0.2	1.3	2.7	0.00	1.7	7.8	0.2
		25–50	7.75	2.33	3.4	0.22	14.6	31.5	18.7	0.4	5.9	4.3	0.00	1.9	21.4	6.0
		50–110	7.62	3.62	1.3	0.10	14.3	35.2	6.9	0.3	2.1	1.3	0.00	2.6	4.9	3.1
Typic Torrivluvents	143	0–20	8.1	4.62	2.51	0.99	19.31	46.32	32.73	0.58	7.02	8.30	0.00	4.79	24.88	18.96
		20–45	8.0	8.12	2.63	0.91	14.25	44.88	66.88	0.46	5.77	10.69	0.00	6.84	45.27	3169
		45–80	8.1	12.37	2.41	0.71	14.50	41.31	113.54	1.81	6.19	17.29	0.00	8.77	88.90	41.16
Typic Quartizipsamments	144	0–30	7.8	8.61	5.12	1.56	12.2	8.57	69.40	0.81	8.24	19.18	0.00	3.76	59.73	34.18
		30–60	7.8	7.32	3.21	0.74	19.8	6.41	54.94	0.59	9.77	21.93	0.00	4.08	55.66	27.49
		60–150	7.8	13.51	2.53	0.57	14.9	3.38	109.77	2.19	10.35	20.26	0.00	6.22	95.72	40.63
Typic Quartizipsamments	145	0–20	7.81	1.83	6.4	0.86	20.5	6.1	145.87	3.45	20.6	26.5	0.00	6.4	136.4	53.62
		20–45	7.38	1.71	9.2	1.10	16.6	4.3	137.8	2.25	22.9	17.9	0.00	5.8	135.2	39.85
		45–110	7.36	1.15	11.3	1.81	18.8	5.7	171.5	2.4	35.7	31.5	0.00	5.1	194.9	41.1

Typic Torriorthents	146	0–30	8.20	3.10	4.10	1.80	22.30	8.50	13.10	1.40	9.00	7.50	0.00	1.90	18.00	10.90
		30–60	8.10	2.00	3.10	1.10	21.50	19.50	8.00	1.50	6.00	4.50	0.00	2.40	12.30	5.30
		60–100	8.10	1.60	1.70	0.32	18.00	7.20	6.60	1.40	4.70	3.30	0.00	2.30	8.00	5.70
Typic Quartizipsamments	147	0–30	7.72	5.03	7.74	0.36	12.5	12.3	32.81	0.75	25.65	16.03	0.00	1.70	45.51	28.06
		30–65	7.73	7.08	9.15	0.32	14.3	10.6	38.77	0.41	44.45	17.51	0.00	1.30	64.50	35.32
		65–120	7.61	5.92	3.65	0.05	9.8	11.4	29.82	0.29	39.35	16.05	0.00	0.90	54.51	29.66
Typic Quartizipsamments	148	0–30	7.95	0.93	1.92	1.71	12.9	12.9	4.1	0.1	4.2	1.3	0.00	4.5	3.1	2.1
		30–90	8.44	1.72	0.82	1.43	13.8	11.3	3.6	0.1	3.6	1.4	0.00	2.9	3.2	2.6
Typic Torrivluvents	149	0–25	7.7	11.7	2.2	1.23	8.31	48.11	50.33	0.81	26.68	16.16	0.00	4.68	57.66	30.95
		25–45	7.7	8.7	1.1	1.67	9.22	48.83	40.99	0.73	23.12	21.20	0.00	3.18	51.16	31.70
		45–110	7.9	5.1	0.7	0.94	11.17	52.13	30.66	0.51	14.64	12.01	0.00	3.64	39.61	14.57
Vertic Torrivluvents	150	0–20	8.1	4.62	2.51	0.99	19.31	46.32	32.73	0.58	7.02	8.30	0.00	4.79	24.88	18.96
		20–40	8.0	8.12	2.63	0.91	14.25	44.88	66.88	0.46	5.77	10.69	0.00	6.84	45.27	3169
		40–100	8.1	12.37	2.41	0.71	14.50	41.31	113.54	1.81	6.19	17.29	0.00	8.77	88.90	41.16
Typic Torrivluvents	151	0–30	7.9	2.1	2.8	2.11	17.33	40.11	11.63	0.28	7.34	2.25	0.00	2.73	13.61	5.16
		30–60	8.3	3.5	3.2	1.74	16.22	40.13	10.36	0.24	4.51	9.52	0.00	3.08	10.11	11.44
		60–120	8.3	4.9	2.1	1.32	15.88	42.32	11.64	0.23	7.53	11.31	0.00	3.30	8.25	19.14
Typic Quartizipsamments	152	0–40	7.81	1.83	6.4	0.86	20.5	6.1	145.87	3.45	20.6	26.5	0.00	6.4	136.4	53.62
		40–130	7.38	1.71	9.2	1.10	16.6	4.3	137.8	2.25	22.9	17.9	0.00	5.8	135.2	39.85
Vertic Torrivluvents	153	0–35	7.8	2.47	2.85	1.16	7.77	49.31	18.25	0.17	7.68	3.35	0.00	1.40	16.74	11.31
		35–70	8.1	1.88	2.63	0.81	14.83	31.57	11.34	0.15	5.33	3.72	0.00	1.62	14.38	4.54
		70–100	2.8	3.31	2.93	0.61	18.92	33.71	20.18	0.15	9.44	4.59	0.00	1.83	19.72	13.81
Vertic Torrivluvents	154	0–25	7.7	3.73	3.11	1.71	9.18	48.35	28.92	1.35	7.51	6.28	0.00	5.05	33.25	7.76
		25–60	7.8	1.98	2.36	0.93	9.35	50.81	16.57	0.58	3.15	2.79	0.00	2.79	13.34	6.96
		60–100	8.1	4.21	1.78	0.72	14.72	52.13	32.06	0.69	7.21	5.14	0.00	5.93	30.75	8.42

Vertic Torrivluvents	155	0–30	8.24	3.52	3.1	1.87	9.3	42.3	5.1	0.2	2.3	1.2	0.00	2.4	4.1	2.3
		30–65	7.81	2.81	2.5	1.55	8.3	40.1	4.2	0.2	2.4	2.1	0.00	2.1	3.4	3.4
		65–110	7.82	1.66	2.3	1.32	9.5	37.5	4.8	0.2	2.8	1.4	0.00	1.8	4.9	2.5
Typic Quartizipsamments	156	0–30	7.72	5.03	7.74	0.36	12.5	12.3	32.81	0.75	25.65	16.03	0.00	1.70	45.51	28.06
		30–50	7.73	7.08	9.15	0.32	14.3	10.6	38.77	0.41	44.45	17.51	0.00	1.30	64.50	35.32
		50–130	7.61	5.92	3.65	0.05	9.8	11.4	29.82	0.29	39.35	16.05	0.00	0.90	54.51	29.66
Vertic Torrivluvents	157	0–25	7.90	34.40	8.22	1.03	16.10	39.90	318.00	8.10	33.50	140.30	0.00	6.00	351.00	142.80
		25–70	7.90	31.50	7.52	0.51	21.70	38.40	308.00	10.40	45.30	55.10	0.00	3.50	323.00	92.70
		70–110	7.90	38.90	9.71	0.32	23.70	35.9	338.00	10.90	40.40	150.30	0.00	4.20	401.00	134.40
Vertic Torrivluvents	158	0–15	7.90	8.7	1.1	1.67	9.22	48.83	40.99	0.73	23.12	21.20	0.00	3.18	51.16	31.70
		15–45	7.91	5.1	0.7	0.94	11.17	52.13	30.66	0.51	14.64	12.01	0.00	3.64	39.61	14.57
		45–90	7.82	11.8	1.66	1.59	9.37	47.11	83.12	1.30	30.22	23.91	0.00	1.66	104.5	32.39
Vertic Torrivluvents	159	0–15	7.60	45.40	8.18	1.05	44.60	51.80	429.00	1.90	47.10	159.8	0.00	4.50	595.00	38.50
		15–40	7.81	2.81	2.5	1.55	8.3	40.1	4.2	0.2	2.4	2.1	0.00	2.1	3.4	3.4
		40–90	7.82	1.66	2.3	1.32	9.5	37.5	4.8	0.2	2.8	1.4	0.00	1.8	4.9	2.5
Vertic Torrivluvents	160	0–20	8.1	6.81	2.68	0.39	16.31	40.30	63.22	1.04	12.31	9.38	0.00	7.87	64.23	13.85
		20–50	8.1	4.62	2.51	0.99	19.31	46.32	32.73	0.58	7.02	8.30	0.00	4.79	24.88	18.96
		50–100	8.0	8.12	2.63	0.91	14.25	44.88	66.88	0.46	5.77	10.69	0.00	6.84	45.27	3169
Typic Torrivluvents	161	0–20	7.62	3.62	1.3	0.10	14.3	35.2	6.9	0.3	2.1	1.3	0.00	2.6	4.9	3.1
		20–50	7.7	3.73	3.11	1.71	9.18	48.35	28.92	1.35	7.51	6.28	0.00	5.05	33.25	7.76
		50–120	7.8	1.98	2.36	0.93	9.35	50.81	16.57	0.58	3.15	2.79	0.00	2.79	13.34	6.96
Vertic Torrifluvents	162	0–20	8.3	4.9	2.1	1.32	15.88	42.32	11.64	0.23	7.53	11.31	0.00	3.30	8.25	19.14
		20–60	7.8	2.47	2.85	1.16	7.77	49.31	18.25	0.17	7.68	3.35	0.00	1.40	16.74	11.31
		60–120	8.1	1.88	2.63	0.81	14.83	31.57	11.34	0.15	5.33	3.72	0.00	1.62	14.38	4.54
Typic Torrifluvents	163	0–20	8.1	1.88	2.63	0.81	14.83	31.57	11.34	0.15	5.33	3.72	0.00	1.62	14.38	4.54
		20–45	2.8	3.31	2.93	0.61	18.92	33.71	20.18	0.15	9.44	4.59	0.00	1.83	19.72	13.81
		45–90	8.1	5.68	4.97	5.51	12.65	38.13	41.58	0.31	12.51	5.53	0.00	2.12	35.16	22.65

Vertic Torrifluvents	164	0–20	7.86	5.60	4.12	0.89	21.00	53.50	195.50	19.00	26.00	290.23	0.00	7.00	221.00	41.73
		20–50	7.92	3.77	1.55	2.20	28.22	34.40	21.00	1.70	8.00	9.81	0.00	7.40	20.00	13.81
		50–90	7.85	3.19	2.40	1.80	10.24	32.50	23.00	12.20	5.00	4.62	0.00	5.20	23.00	7.62
Vertic Torrifluvents	165	0–20	8.20	7.10	1.54	0.36	15.60	28.70	38.00	2.70	26.80	10.90	0.00	3.50	34.50	31.10
		20–35	8.10	15.50	4.58	1.31	23.00	39.20	124.00	1.90	15.80	54.30	0.00	6.00	144.00	46.30
		35–100	8.10	7.80	3.22	0.41	21.80	37.00	45.00	2.70	20.20	14.00	0.00	6.00	59.00	17.30
Typic Quartizipsamments	166	0–30	7.81	1.83	6.4	0.86	20.5	6.1	145.87	3.45	20.6	26.5	0.00	6.4	136.4	53.62
		30–75	7.38	1.71	9.2	1.10	16.6	4.3	137.8	2.25	22.9	17.9	0.00	5.8	135.2	39.85
		75–120	7.36	1.15	11.3	1.81	18.8	5.7	171.5	2.4	35.7	31.5	0.00	5.1	194.9	41.1

Bibliography

Abdel Rahman SL, El-Taweel MI (1995) Assessment and evaluation of salt-affected soils in El-Tina plain, NW Sinai Penisula, Egypt. International symposium on salt affected Lagoon ecosystems, Valencia, Spain, 18–25 September 1995, Abstracts Volume, pp. 155

Abdel-Hady MA, El Shazly EM, Moustafa HH, El Said MM, El Kassas I.A (1980) Atlas of Sinai Pennisula from Landsat images (with annotations). Remote Sensing Center, Academy of Scientific Research and Technology, Cairo, Egypt, 87 p. (in Arabic and English)

Abdel-Hady MA, Younes HA, Gad AG; Abdel Rahman MA, El Taweel MI; Abdel Rahman SI, Ghabour TK (1994)

Abdel-Hady MA, Abdel Samei, Younes HA, El-Rakaiby ML, Gad A (1992) Desertification of Arable Lands in Egypt. Final Report-NARSS, Cairo, 295 p

Abdel-Hady MA, Yehia MA, Ayoub A (1987) Remote sensing and satellite data for desert development. Proceedings of the second international conference on desert development, Cairo, Egypt

Abdel-Kader F, Ayyad MA, Le-Floch E (1981) Semi-detailed soil survey, El-Omayed pilot area. REMDENE Project Progress Report No. 2, Vol. III, pp. 8–33

Abdel-Rahman M et al (1977) Some geomorphologic aspects of siwa region, monograph on the geology and prehistory of Siwa Oasis Washington state University, Pullman

Abdel-Salam MA, Elwan AA (1970) Evaluation of the physiolographic features of Dhabaa Area, Mediterranean coastal zone of U.A.R. from aerial photographs and their bearing on land use; Bull Desert Inst Cairo, Egypt 20(1):77

Abdel-Salam MA, Hammad MA (1970) Geomorphology, geology, hydrology and soils of Wadi El-Natrun-Maryut Agric. Project: Publ. Instit. Desert, October, 1970, (App)

Abel-Rahman SI, El-Nahry AH, Yehia MA (2000) lLand use-Land cover classification of Siwa Oasis in the Western Desert of Egypt. Proceedings of the 2nd International Conference on EOEI, 11—14 November 2000, Cairo, Egypt

Abu Al-Izz MS (1971) Landforms of Egypt. The American University in Cairo Press, Cairo, p 281p

Ball J (1939) Contribution to the geography of Egypt. Cairo

Blamfield RM (1921) Desert flora of northern Egypt: Hort Rev, Cairo 7, 4–6

Bonham-Carter GF (1994) Geographic information systems for geoscientists: modelling with GIS. Geological Survey of Canada, Love Printing Service Ltd, Ontario, Canada

Butzer KW, Hasen CL (1968) Desert and River in Nubia, geomorphology and pre-history environments at Aswan reservoir. Wisconsin University Press, Madison, Wiscousin, 562 p

A.-A. Gad, *Digital Land Resources to Support Sustainable Development in Egypt*,
Deltas of the World,
https://doi.org/10.1007/978-3-031-37167-7

Chumakov IS (1967) Pliocene and pleistocene deposits of the Nile Valley in Nubia and Upper Egypt. Transactions of Geological Institute, Academy of Sciences of the IUSSR, Vol. 170, 115 p. (Russian)

Egyptian Meteorological Authority (1975) Climatic Atlas of Egypt. Egyptian Meteorological Authority, Cairo, Egypt, 157p

Egyptian Military Survey (1981) Topographic Maps Scales 1:250,000. Egyptian Military Survey, Cairo, Egypt

Egyptian Survey Authority (1983) Topographic Maps, Scales 1:100,000 and 1:250,000. Egyptian Survey Authority, Cairo

El-Fishawi NM (1977) Sedimentological Studies of the Present Nile Delta Sediments on Some Accretional and Erosional Areas between Burullus and Gamassa. M.Sc. Thesis, Faculty of Science, Alexandria University, 143 p

El Gammal EA (1999) Geomorphology of Southern Sinai, using landsat images. Egypt J Remote Sens Space Sci 2:35–56

El Shazly EM, Abdel Hady MA, El-Kassas IA (1977a) Delineation of land features in Egypt by land satellite images In: Remote sensing of earth resources, F. Shaharoki, 6, 227–294, Oklahoma

El Shazly EM (1964) Geology, pedology and hydrology of Mersa Matruh, Western Mediteranean Littoral, Egypt. Ph.D. Thesis, Fac. Sciences, Cairo Univ., Egypt

El Shazly EM, Abdel Hady MA, El-Ghawaby MA Khawasik SM, El-Shazly MM (1977b) Application of Landsat satellite imagery in assessing the regional, geological, structural, environment and groundwater conditions in the Qattara depression area, Egypt: Remote Sensing Center, International Astronautical Federation XXVIII Congress. Praha, Czechoslovakia

El Shazly EM, Abdel Hady MA, El-Ghawaby MA Khawasik SM, El-Shazly MM (1976) Geologic Interpretation of Landsat satellite images for Qattara depression area, Egypt: Remote Sensing Center, Academy of Scientific Research and Technology, Cairo

El-Baz F (1978) Egypt as seen by Landsat: Dar El-Maaref, Cairo, Egypt. (Maps)

El-Bouseily AM, Frihy OE (1984) Textural and mineralogical evidence denoting the position of the mouth of the Old Canonic Nile branch on the mediterranean coast, Egypt. J Afr Earth Sci 2:103–107

El-Demerdash S (1970) Studies on the nature of interference between the Alluvial and desert soils in the Western border of Nile Delta: Ph.D. Thesis, Fac. Agriculture (Soils), Cairo Univ., Egypt

El-Etr HA, Embabi N, Yousif MS (1980) Geologic, Geomorphologic studies in the Egyptian Deserts: Ain Shams University, Cairo, Egypt

El-Fisahwi NM, El-Askary MA (1981) Characteristic features of coastal sand dunes along Burullus-Gamassa stretch, Egypt. Acta Min Pet Szeded 25(1):63–76

El-Fishawi NM, Siestini G, Fahmy M, Shawki A (1976) Grain Size of the Nile Delta Beach Sands. Proceedings of the Seminar on Nile Delta Sediments, Alexandria, October 1975, pp. 79–94

El-Shazly EM, Abdel-Hady M, El-Ghawaby M, Salman A, El-Assy I, El-Rakaiby M (1980) Sinai Penninsula: landsat imagery interpretation maps, 1:250,000. Remote Sensing Center, ASRT, Cairo, Egypt

El-Shazly EM, Abdel-Hady MA, El-Ghawaby MA, El-Kassa IA, Khawasik SM,El-Shazly MM, Samad S (1975) Geologic interpretation of landsat satellite images for West Nile Delta Area. Remote Sensing Research Project, Academy of Scientific Research and Technology, Cairo, Egypt

Ezzat MA (1973) Groundwater in the Arab Republic of Egypt (in Arabic), Ground water resources in the Arab Countries: ALECSO, Cairo, Egypt, p. 99–142

Ezzat MA (1976) Groundwater resources in the North Western coastal zone, Part I: Ministry of Irrigation, Cairo, Egypt

FAO (1968) Definitions of soils units for the soil map of the world. World soil resources. Rep. 33, FAO, United Nations, Rome

FAO (2006) "Guidelines for soil description" 4th edn, FAO, Rome, ISBN 92–5-1055211

Fathi A, Abdel Salam MA, Khalil KE, Tawakol MS, Awadalla SY (1973) Some physical properties affecting soil moisture characteristics of Ras El-Hikma soils, A.R. Egypt. Bull. Desert Institute, Cairo, Egypt, 23(2):407–417

Folk L (1974) Petrology of Sedimentary Rocks. Hemplill, Austin, Texas, 182 p

Frihy OE, Lotfy MF (1994) Mineralogy and texture of beach in relation to erosion and accretion along the rosetta promontory of the Nile Delta, Egypt. J Coast Res 10:588–599

Frihy OE, El-Fishawi M, El-Askary MA (1988) Geomorphological features of the Nile delta coastal plain: a review. Acta Adriat 29(1/2):51–65

Gifford AW, Warner DM, El-Baz F (1979) Orbital observations of sand distribution in the Western Desert of Egypt. Apollo-Soyuz Test Project Summary Science Report, Vol. II, Earth Observations and Photography, NASA. Sp-41z, pp. 219–236

Gindy AR, El-Askary MA (1969) Stratigraphy, structure and origin of Siwa Depression, Western Desert, Egypt. Am Assoc Petrol Geol Bull 53(3):265–306

Hamdi H (1961a) Mineralogy of the soils of Egypt. Bull Central Glass and Ceramic. Research Institute, India, 8, 87

Hamdi H (1961b) The alluvial soils of Egypt. Bull Fac Agriculture, Cairo Univ., Cairo

Hamdi H (1967) The Minerolgoy of the fine fraction of the alluvial soils of Egypt. J Soil Sci Egypt 7(1):15–21

Hamdi H, Abdel Salam NA, Mady F, Abdel Salalam A (1969) Soil morphology and classification of the Dhabaa area. Bull Desert Institute, Egypt 19(1):119–145

Hammad FA (1972) Geology of the Soils and water resources in the area between Ras El-Hekma and Ras Alam El-Rum. Ph.D. Thesis, Fac. Sciences, Cairo Univ., Egypt

Hammad MA (1964) A study of soils of Mersa Matruh. M.Sc. Thesis, Fac Agriculture, Ain Shams Univ. Cairo, Egypt

Hammad MA (1968) Gensis of the soils of the Western Mediterranean Coast of Egypt. Ph.D. Thesis, Fac. Agriculture, Ain Shams Univ. Egypt

Harga AA (1967) Morphological and physio-chemical studies of Burg El-Arab, El Hammam area. M. Sc. Thesis, Fac. Agriculture, Ain Shams Unv., Cairo, Egypt

Harga AA (1972) Salinity Appraisal under the highly calcareous soils condition of the Western coastal littoral of Egypt: Bull. Desert Institute, Cairo, Egypt 22(2):369–378

Harga AA (1974) The classification of the soils of the Mediterranean coastal littoral into the 7th approximatgion: Egypt J Soil Science, Cairo, Egypt 14(2): 177–186

Harga AA (1976) The Soils of the cross section at Mersa Matruh to Siwa depression: Bull Desert Institute. Cairo, Egypt 26(1):93–114

Harga AA, Abdel Salam MA, Mady F, Hamdy H (1968) Study of the Burg El-Arab–El Hamman area. The Western Mediterranean coastal zone of U.A.R. Bull. Desert Institute, Cairo, Egypt, 18(1):45–68

Hasaneen MN (1983) Lithological and Geomorphological Studies on the Coastal Area between Ras El-Bar and Kitschner Drain. Msc, Thesis, Faculty of Science, Manoura University, Mansoura, 129 p

Hilmym ME (1951) Beach sands of mediterranean coast of Egypt. I. Sed. Petrol 21:109–120

Ismail H, Ayyad MA, Le-Floch E (1983) Pedological Study of El-Omayed Toposeqeunce. ORSROM Bull. Pedologie

Kaddah MT (1948) The Soils of Egypt. Soil Sci 65(5):357–365. (Pa)

Kaddah MT (1954) The Soils and water supply of Burg El-Arab: Bull. Desert Institute, Cairo, Egypt 4(1)

Kandil MF (1961) Studies on structure and aggregate analysis of Egyptian Alluvial soils. M.sc. Thesis, Fac. Agriculture, Cairo Univ. Egypt

Kassas M (1953a) Habitat and plant communities in the Egyptian desert 2. The features of the desert community. J Ecol 41:248–256

Kassas M (1953b) Landform and Plant cover in the Egyptian desert. Bull Soc Geog Egypt 26:193–205

Mostafa MAM (1974) A study of the soil of Mersa Matrouh area. M.Sc. Thesis, Fac. Agriculture, Ain Shams University, Cairo, Egypt

Murray GW (1952) The Artesian water of Egypt. Egypt. Survey Dept., Cairo, Paper No. 52

Murray GW (1953) The water beneath the Egyptian Western Desert. Geogr J 118, pt. 4, London, p 443–452

Murray GW (1951) Egyptian climate and historical outline. Geogr J

Nasr Y (1998) Soil qualities with special reference to mineralogy and fertility status as related to soil environments of some areas in Sinai, Egypt, Ph.D. Thesis, Faculty of Science, Cairo University, Cairo, Egypt

Rashed MA (1978) Sedimentological and Mineralogical Studies of the Coastal Samples of Abu Quir Bay, Alexandria. M.Sc. Thesis, Faculty of Science, Alexandria University, 157 p

Research Institute for Groundwater, RIGW (1988) Hydrogeological Map of Egypt, scale 1: 2,000,000. Research Institute for Groundwater, Cairo, Egypt

Research Institute for Groundwater, RIGW (1998) Hydrogeological Map of Egypt, Scale 1:500,000, El Sad El Ali sheet. First Edition. Research Institute for Groundwater, Cairo, Egypt

Said R (1958) Remarks on geomorphology of deltaic coast between Rosetta and Port Said. Bull Soc Geog Egypt 31:115–125

Said R (1962) The geology of Egypt. EL Sevier, Amsterdam, p 377p

Said R (1981) The geological evolution of the River Nile. Springer, New York, p 151

Said R (1993) The River Nile. Pergamon Press, London, p 320

Soil Map of Sinai Pennisula using Remote Sensing Techniques. Technical Report, NARSS, Cairo, December, 1994

Suliman I, Raga M (1991) An Economic Study of Cropland Market in the Egyptian village The Egyptian Society of Agricultural Economists 1(1), March

Suliman MK (1991) Universities and Development of the Desert land in the ARE". The second annual University Conference, 2–5 November 1991, Cairo

The manufacturer's authorised representative in the EU is Springer Nature Customer Service Centre GmbH, Europaplatz 3, 69115 Heidelberg, Germany. If you have any concerns regarding our products, please contact ProductSafety@springernature.com

Printed and bound by CPI Group (UK) Ltd, Croydon, CR0 4YY
30/09/2026
02225509-0020